AF540473

NANOWIRES AND NANOBELTS

NANOWIRES AND NANOBELTS

A.S. BHATIA

Foreword by

DR. A. CHAWLA
Professor
Department of Mechanical Engineering
Indian Institute of Technology, Delhi
Hauz Khas, New Delhi

from

DEEP & DEEP PUBLICATIONS PVT. LTD.
F-159, Rajouri Garden, New Delhi - 110 027

NANOWIRES AND NANOBELTS

ISBN 978-81-8450-329-6

Typeset by THE COMPOSERS
260 C.A. Apt., Paschim Vihar, New Delhi - 110 063

Printed in India at MAYUR ENTERPRISES
WZ Plot No. 3, Gujjar Market, Tihar Village, New Delhi - 110 018

NuTech Books are published by DEEP & DEEP PUBLICATIONS PVT. LTD.
F-159, Rajouri Garden, New Delhi - 110 027 • Phone : 25435369, 25440916
E-mail : ddpubs@gmail.com • ddpbooks@yahoo.co.in
Showroom :
2/13, Ansari Road, Daryaganj, New Delhi - 110 002 • Telefax : 23245122

Contents

Foreword

The ability to study and manipulate matter at the nanoscale is the defining feature of 21st century science. Nanotechnology is no longer a subdiscipline of chemistry, physics, engineering or any other field and therefore demands a new paradigm for teaching. It gives me great pleasure to write a foreword for this book "*Nanowires and Nanobelts*" by Mr A.S. Bhatia. The objective of this book is to cover the synthesis, properties and device applications of nanowires and nanobelts based on metals and semiconductor materials. This book will be useful to leading scientists, technologists, educationists, researchers from all areas of nano-science and technology.

DR. A. CHAWLA
Professor
Department of Mechanical Engineering
Indian Institute of Technology, Delhi
Hauz Khas, New Delhi

Preface

Nanotechnology has exploded in the last few years, due in small parts to its rapidly increasing impact in the area of the development of nanowires. Nanowires, nanobelts, nanoribbons are a class of 1-dimensional materials. These materials exhibit superior electrical, mechanical, optical and thermal properties and are used in making chemical and biological nanosensors, FETs, logic circuits, nanogenerators etc.

This book on nanowires and nanobelts has twenty chapters which are self contained on various aspects of their synthesis, characterization and modifications. It includes scientific and technological details along with detailed figures and graphs. The book will be useful to researchers educationists, active in nanoscience and technology, in industry and academia, medical professionals, students. It opens with a general introduction (Chapter 1),followed by their Fabrcation and Synthesis (Chapter 2) and Growoth (Chapter 3). It also covers the growth of ZnO nanowires (Chapter 4).

The properties of NWs and NBs are described in Chapter 7. The elastic properties and field emission properties are taken up in Chapter 8 and 9 respectively. The book also focusses on the synthesis, properties and applications of NWs and NBs based on functional materials like functional oxides such as ZnO (Chapter 4), SnO_2 (Chapter 12) and composite materials such a Sic-SiO_2 (Chapter 11). Single element semiconductor, silicon NWs are covered in Chapter 10 whereas other semiconductor NWs and NBs are in Chapter 17 and –Fe O are described in Chapter 13.

The phenomenal growth and staggering variety of applications of NWs and NBs prevent any reference from providing a complete picture of the field. The book contains the most imporetant chapter 19, the most promising technologies and the fast growing development of current interest particula, nanogenerateor, as in particular it gives the fundamental theory.

Finally ZnS nanowires (Chapter 14), core shell naonwires (Chapter 15), semiconductor NWs and nanoribbons (Chapter 16), use of NWs in electronics (Chapter 18) and nanopieoelectronics are also covered.

New Delhi

A.S. BHATIA

Introduction

Nanostructures-structures that are defined as having at least one dimension between 1 and 100nm - have received steadily growing interests as a result of their peculiar and fascinating properties, and applications superior to their' bulk counterparts. The ability to generate such minuscule structures is essential to much of modern science and technology. There are large number of opportunities that may be realized by making new types of nanostructures, or simply by down-sizing existing microstructures into the 1-100nm regime. The most successful example is provided by microelectronics, where "smaller" means greater performance ever since the invention of integrated circuits: more components per chip, faster operation, lower cost, and less power consumption. Miniaturization also represents the trend in a range of other technologies. In information storage, for example, there are many active efforts to develop magnetic and optical storage components with critical dimensions as small as tens of nanometers. A wealth of interesting and new phenomena are associated with nanometer-sized structures, examples include the size-dependent excitation or emission, quantized (or ballistic) conductance, Coulomb blockade (or single-electron tunneling, SET), and metal insulator transition etc. The quantum confinement of electrons by the potential wells of nanometer-sized structures provides, a most powerful (and yet versatile) means to control the electrical optical, magnetic, and thermoelectric properties of a solid-state functional material.

One dimensional (1D) nanostructures such as wires, rods, belts, and tubes have become the focus of intensive research owing to their applications in mesoscopic physics and fabrication of nanoscale devices.

There are a few names being used for describing 1D nanostructures: nanorods, nanowires naonofibers, nanobelts and nanoribbons etc. Nanowires usually means a linear structure that has specific growth direction, but its side surfaces and shape of cross-section may not be well defined or uniform (Figure 1.1(a)). Less restriction about the shape of the cross-section and uniformity broadens the family of nanowires to include more nanostructures. But for achieving property control, it is important to early define the side surfaces of a nanowire. A nanorod is a nanowire with a shorter length (Figure 1.1(b)). A nanotube is a 1D nanostructure with a hollow

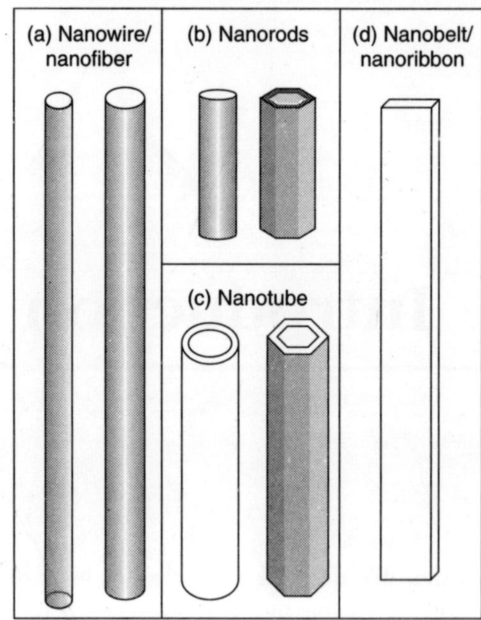

Figure 1.1 *Typical morphologies of one-dimensional nanostructures: nanowire, nanorods, nanotubes and nanobelts.*

interior channel (Figure 1.1(c)). Nanobelts/nanoribbons are 1D nanostructure with well-defined side facets (Figure 1.1(d)), and they have more restrictive shape and uniformity than the nanowires.

One-dimensional (1D) and quasi-one dimensional nanostructures are attracting a great deal of attention due to their unique properties and novel applications. Besides carbon tubular structures, a variety of nanostructures have been synthesized, including cages cylindrical wires, rods, nails, coaxial and bi-axial cables, ribbons or belts, sheets, diskettes, springs, rings, propellers and more. These novel structures have semiconductive, ferroelectric, magnetic and/or piezoelectric properties and are of great importance to nanotechnology.

One-dimensional (1D) nanostructures have unique physical properties and potential to revolutionize broad areas of nanotechnology. There are three main type of 1D nanomaterials carbon nanotubes, semiconductor nanowires, and semiconducting oxide nanobelts. It is well-known that a single-walled carbon nanotube can be metallic or semiconductive, depending on the helical angle at which the graphitic sheet is rolled up. It is unfortunate that synthesis of nanotubes with control helical angles is not feasible. They are the fundamental building blocks for constructing nanodevices and nanosystem that exhibit superior performances. They have intrinsic nanoscale geometry and rich, sometimes unique, electrical, optical, and mechanical properties due to their small size and the enhanced charge carrier mobility.

For 1D nanostructures, when their diameter reaches a few nanometers, their properties are strongly affected by the structure of the side surfaces. For applications, it is important to determine the surface structure of 1D nanostructures. ID nanomaterials that have not only controlled shapes and crystal structure, are not only grown but also designed for electrical and optical properties in applications as sensors, field-emitters, p-n diodes, and the diluted magnetic semiconductors (DMS) for spintronics. A key requirement, for many of these applications is the doping of ZnO with various elements for enhancing and controlling its electrical and optical performance.

As size-dependent effects can be found when the volume of a nanostructure is so small that surface effects start to be relevant. This usually happens for dimensions smaller than tens of nanometers, while for larger sizes, the Young's modulus approaches its bulk value. For example, the Young's modulus of carbon nanotubes increase significantly with decreasing size for diameters smaller than 4 nm Conversely, the modulus of GaN nanowires increases with increasing diameter, reaching the bulk value at 84 nm. The origin of the elastic modulus size dependence is related to different effects such as the presence of defects and the balance between surface and bulk properties as the surface-to-volume ratio varies.

Transmission electron microscopy (TEM), as one of the most powerful tools in nanotechnology, has played an important role in characterizing 1D nanostructures, not only in determining crystal and surface structure, but also chemical structure.

Semiconducting nanowires (NWs) represent an important and broad class of nanometer scale wire structure, which can be rationally and predictably synthesized in single crystal form with all key parameters controlled during growth: chemical composition, diameter, length, and doping. Semiconductor NWs thus represent one best-defined and controlled classes of nanoscale building blocks, which correspondingly have enabled a wide range of devices and integration strategies to be pursued in a rational manner. For example, semiconductor NWs have been assembled into nanometer scale field effect transistors (FETs), p-n diodes, light emitting diodes (LEDs) bipolar junction transistors, complementary inverters, complex logic gates and even computational circuits that have been used to carry out basic digital calculations. In contrast to NTs, devices can also be assembled in a rational and predictable manner because the size, interfacial properties, and electronic properties of the NWs can be precisely controlled during synthesis, and moreover, reliable methods exist for their parallel assembly In addition, it should be recognized that it is possible to combine distinct NW building blocks in ways not possible in conventional electronics and to leverage the knowledge base that exists for the chemical modification of inorganic surfaces to produce semiconductor NW devices that achieve new function and correspondingly could lead to unexpected device concepts.

Semiconducting oxide nanobelts (NBs) are another unique group of quasi-one-dimensional nanomaterials, which have been systematically studied for a wide range of materials with distinct chemical composition and crystallographic structures. Nanobelts are nanowires that have a well defined geometrical shape and side surfaces.

The oxides have two unique structural features: missed cation valences and an adjustable oxygen deficiency, which are the bases for creating and tuning many novel materials properties, from electric, chemical, and optical to magnetic. The synthesis of nanostructures of functional oxides, with a controlled structure and morphology, is critical for scientific and technological applications.

The nanobelts have a ribbon shape, and nanobelt structures of a group of semiconducting oxides have been synthesized, which are structurally perfect and geometrically uniform. The nanobelts have a uniform width of 30–300 nm, along its entire length, thicknesses of 10–30 nm, and lengths of up to a few millimeters. A nanobelt grows along $[01\bar{1}0]$, with top and bottom flat surfaces $\pm(2\bar{1}\bar{1}0)$and side surfaces $\pm(0001)$. The cross section of the nanobelt is rectangular.

Utilizing the high surface area of nanobelt structures, nanoscale devices, such as field effect transistors and gas sensors, have been fabricated, and they have been found to exhibit a superior performance.

Belt-like, quasi-one-dimensional nanostructures (so called nanobelts or nanoribbons) are synthesized for semiconducting oxides of zinc, tin, indium, cadmium and gallium, by simply evaporating the desired commercial metal oxide powders at high temperatures.

The as-synthesized oxide nanobelts are pure, structurally uniform, single crystalline and most of them free from dislocations; they have a rectangular cross-section with a specific growth direction, side surfaces, and top/bottom surfaces. Their geometrical shape is very uniform with atomically flat surfaces. The most typical examples are the transparent conducting oxides such as ZnO, SnO_2, In_2O_3 and CdO.

The belt like morphology appears to be a unique and common structural characteristic for the family of semiconducting oxides with cations of different valence states and materials of distinct crystallographic structures. Field effect transistors and ultra-sensitive nano-size gas sensors, nanoresonators and nanocantilevers have also been fabricated based on individual nanobelts. Thermal transport along the nanobelt has also been measured. Very recently, nanobelts, nanosprings and nanorings that exhibit piezoelectric properties have been synthesized, which are candidates for nano scale traducers, actuators and sensors.

Mechanical behavior of 1D nanomaterials is one of the most important properties that dictate their applications in nanotechnology. Various methods are developed for quantifying the mechanical property of ID nanomaterials and they are classified into three categories. The first approach is based on dynamic resonance of a ID nanostructure that is affixed at one end and free at the other: the mechanical resonance is excited by an externally applied oscillating electrical field, and the observation is made through electron microscopy. The second approach is quantifying the static axial tensile stretching of a ID nanostructure using an atomic force microscope (AFM) tip, which is installed inside a scanning electron microscope (SEM). The third approach is based on AFM and nanoindenter. One of the most important and common strategies is deforming a ID nanostructure that is supported at the two ends using an AFM tip which pushes the ID nanostructure at its middle point. Quantifying the Middle poing force displacement curve gives the elastic modulus. The accuracy of this measurement is, however, limited by the size of the tip and accuracy of positioning the AFM tip right at the middle of the ID nanostructure due to the unavoidable hysteresis of the piezoceramic actuator of the AFM cantilever.

The bottom-up approach of nanotechnology has yielded many high-quality, single –crystal, and defect-free structures such as nanobelts.

A Wide range of techniques have been developed for the synthesis of ID nanostructures, including solid-vapor deposition, lithography, laser ablation, sol-gel and template assisted methods; and nanowires. The nanobelt of a monocrystal structure is free from dislocations but planer defects such as twins and stacking faults are observed.

In general it is believed that nanowires less than 10nm have novel and unique physical and chemical properties due to quantum confinement. Nanowires of a few nanometers in size have been synthesized for InP and Si.

ZINC OXIDE (ZnO) NANOWIRE AND NANOBELTS

Nanowires and Nanobelts of ZnO Why?

ZnO is the main material used for one dimensional nanostructures, such as nanowires and nanobelts. Wurtzite-structured materials such as CdS-based NGs and GaN also show similar effects. The nanowire-based energy (harvesting technology) offers a few advantages:

(i) The NW/NB can be subjected to extremely large elastic deformation (90° bending) without plastic deformation or fracture.
(ii) Due to their small diameter, NWs/NBs are most likely free of dislocations, and thus, expected to have a high resistance to fatigue, possibly extending the lifetime of the device.
(iii) NWs/NBs can be bent under an extremely small applied force. This is unique for harvesting energy created by weak mechanical disturbance.
(iv) The large surface area offered by NWs/NBs provides a unique opportunity for surface functionalization to improve physical and chemical properties.
(v) ZnO NW arrays can be easily grown via chemical synthesis at 350K on any shaped substrate made of any material (crystalline or amorphous, hard or soft), it can be easily integrated with technologically important materials, such as silicon or polymers, at low cost.
(vi) ZnO is a biocompatible, degradable, and nontoxic material with a wide range of applications in medicine and cosmetics; thus, it has the most profound potential for implantable and flexible power sources.
(vii) ZnO is an environmentally “green” material.

ZnO nanowires (NWs) and Nanobelts (NB) are equally important as carbon nanotubes and silicon NWs are for electronics, optoelectronics, sensors and energy science, because of their diverse and unique semiconductive, optic, piezoelectric, and pyroelectric properties.

ZnO a piezoelectric material and with a wide bandgap semiconductor (3.37eV), a large excitation energy (60 meV) at room temperature, is an important and versatile functional semiconductor material. It has been extensively used in several industrial products such as ceramics, rubber additives, pigments, and medicines. New phenomena and novel properties are expected at the nanoscale due to the size confinement effect. As a result Since the first discovery of ZnO nanobelts research in one dimensional (1D) nanostructures of functional oxide materials has attracted great attention. It is suitable for short wavelength optoelectronic applications. The wide band gap and the high exciton binding energy in ZnO crystal ensures efficient excitonic emission at room temperature as ultraviolet (UV) luminescence is observed. A considerable effort was devoted to the synthesis and characterization of ZnO nanostructures.

The semiconductor zinc oxide (ZnO) has drawn considerable interest due to its piezoelectric properties and band gap in the ultraviolet in a wide range of applications. Various ZnO nanostructures, such as nanocombs, nanorings nanosprings, nanobelts, nanowires and nanocages nano-propeller, nano-sisal and alignéd nanowire arrays among others, have been synthesized.

For single crystal structured ZnO, devices such as field emission transistors (FET), nanolasers and cantilevers have also been made. In addition, the coupling between semiconduction and piezoelectricity enable ZnO nanowire to work as nanogenerator for converting mechanical energy into electric energy, which has great potential for harvesting energy from the environment and self-powering nanodevices. These nanomaterials exhibit structure-dependent physical and chemical properties, which are different from those of bulk materials and make ideal building blocks for future electronic components gas sensors field emission displays and nanoelectromechanical systems (NEMs). All of these applications require the knowledge and the ability to control the mechanical behavior of ZnO nanostructures. In particular, it is of crucial importance to understand the size dependence of the elastic properties.

Zinc oxide (ZnO) is one of the most important nanomaterials for nano-optoelectronics sensors, transistors and nanopiezotronies Because of the unique piezoelectric and semiconducting dual properties, ZnO nanowire (NWS) and nanobelts (NBs) are the fundamental material for nanogenerators, which convert mechanical energy into electricity. ZnO has huge promise for optical applications, such as UV detection, however, because of the presence of point defects and confined dimensionality the UV sensitivity of ZnO NWs and NBs is limited.

ZnO is also a material that is biocompatible and biosafe applications as implantable biosensors.

ZnO, which is widely used as a transparent conducting oxide material optical device and gas sensors, is a good candidate for one-dimensional nanomaterials. After fabricating ZnO nanotube nanowire, and nanobelts; metal-semiconductor core-shell Zn-ZnO nanobelts and nanowires have been synthesized. From simple structure analysis it is found that the Zn core and the sheathing ZnO shell have an epitaxial relationship. Considering the large difference between their lattice parameters, the misfit relaxation in the core-shell interface is of interest. It is known from two dimensional materials, such as semiconductor ferroelectrics thin films, the interface plays a key role in their properties. In order to use metal semiconductor Zn-ZnO core-shell nanobelts as devices the detailed structure, especially how to relax the large misfit, must first be understood .

Zinc oxide (ZnO) has key applications in catalysts, sensors, piezoelectric transducers transparent conductors and surface acoustic wave devices. The structure of ZnO, for example is described as a number of alternating planes composed of tetrahedrally coordinated O^{2-} and Zn^{2+} ions, stacked alternately along the *c-axis* (Fig. 1.2a). The oppositely charged ions produce positively charged Zn (0001) and negatively charged O $(000\bar{1})$ polar surfaces, resulting in a normal dipole moment and spontaneous polarization along the *c-axis* (Fig. 1.2b). If the elastic deformation energy is largely suppressed by reducing the thickness of a nanobelt, the polar nanobelt can self-assemble into different shapes as driven by minimizing the electrostatic energy coming from the ionic charges on the polar surfaces (Fig. 1.2(c) and (d)), analogous to' the charge configurations of a RNA molecule.

As integrating nanophotonics with nanoelectronics is essential for developing the technologies for the next generation communication, computing, and biomedical imaging, lowdimensional nanostructures, such as nanodots, nanotubes, and nanowires (NWs), were therefore extensively

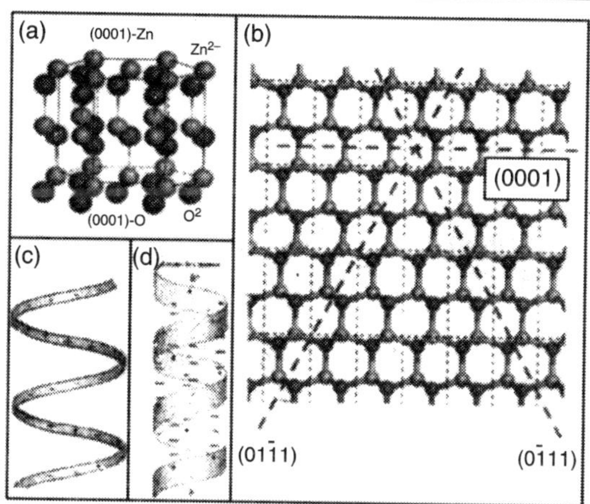

Figure 1.2 *(a) Structural model of ZnO. (b) The projected structure of ZnO along a-axis, showing the {0001}and {01$\bar{1}$ 1}polar-surfaces. (c)Spiral model of ZnO polar nanobelt. (d) Charge model of an RNA molecule.*

studied for high sensitive optical detection, among which ZnO is a typical example. It exhibits the most diverse and abundant configurations of nanostructures such as hierarchical nanostructures, NWs, nanobelts (NBs) nanosprings, nanorings, nanobows, and nanohelices. Due to semiconducting and piezoelectric properties of ZnO, it has also potential applications in electronics and optoelectronics, such as optically pumped nanolaser nanolaser acoustic resonator, piezoelectric gated diode,' field emitter, photodiode, and photoconductor.

Since the large surface-to-volume ratio of one dimensional ZnO nanostructures and the presence of deep level surface trap state in NWs/NBs, ZnO exhibits long life time photocarriers, and it usually exhibits lower photosensitivity than photodiodes.

Semiconducting nanowires (NWs) and nanobelts (NBs) as the fundamental building blocks for fabricating unique nanoscale devices, show that the electrical measurement of a field-effect transistor, is largely affected by the type of contact (Schottky or ohmic) and the contact resistance. Achieving a good ohmic contact is essential for exploring the electrical and optoelectronic properties of NWs and NBs.

Single ZnO NW has been manipulated between two electrodes for fabricating diode, field effect transistor (FET), gas sensor vacuum gauge, pressure sensor, and nanofloating gate memory. The first prototype of nanogenerator has been demonstrated using single ZnO NWs for converting mechanical energy into electricity. For practical applications, growth of aligned, patterned, and controlled NW arrays is vitally important for applications such as DC nanogenerator, solar cell, nanolasers, and field emitters. For the nanogenerator the growth of orientation, position, and size control of the NWs is the key for raising the output voltage and power. The most popular method used for growth of ZnO NWs is to use Au as catalyst and a single crystal substrate, such as GaN (0001) or a-plane alumina. The close lattice and symmetry match between the substrate surface and the ZnO results in epitaxial vertical growth of NWs. Alternatively, with the use of ZnO textured thin films as seeds, aligned ZnO NWs have been grown following the pattern generated by e-beam lithography.

ZnO nanorod and nanowire (NW) arrays have also been used in fabricating electronic optoelectronic, electrochemical, and electromechanical devices, such as solar cells, ultraviolet (UV) lasers, piezonanogenerators, light-emitting diodes, and field emission devices.

ZnO is an important material as resonators for microelectromechanical systems, optical material for fabricating blue light emitting diodes (LEDs), and piezoelectric and semiconductor materials for actuators. Field effect, transistor, gas sensor, and pH meter using ZnO nanowire/ nanobelts have been reported, the performance of which relies on the oxygen' deficiency and distribution in ZnO nanostructures.

Structurally, the wurtzite-structured ZnO is non-centrosymmetric, and its property is anisotropic. The electronic structure of ZnO nanostructures is investigated for different growth orientation. However, the diversity and lower purity of the samples prohibit the application of the well-established techniques such as XPS, AES and UPS. Electron energy-loss spectroscopy (EELS) has a high spatial resolution and gives the possibility of investigating the electronic structure of a nanobelt and a nanotube. By combining the crystallographic and chemical information provided by transmission electron microscopy, EELS may have certain advantage for the investigation of nano scale electronic structures, although its, energy resolution is limited. Wurtzite-structured ZnO has versatile properties that are important for applications in electronics, optoelectronics, photovoltaics and sensors.

Wurtzite ZnO has two important structural characteristics: noncentral symmetry and polar surfaces. Zn^{2+} and O^{2-} are tetragonally coordinated so that the center of the positive charges overlaps with that of the negative charges. If subjected to an external force, the distortion of the tetrahedron results in a dipole moment, leading to the piezoelectric effect. From the structural point of view, Zn and oxygen atoms are arranged alternatively layer-by-layer along the c axis so that the top surface is Zn^{2+}-terminated (0001) and the bottom surface is O^{2-} -terminated $(000\bar{1})$. The opposite ionic charges on the surfaces result in a spontaneous polarization. The polar surfaces of ZnO are very stable and have not been reconstructed, and they have been used to induce the formation of some novel nanostructures such as nanohelicals (nanosprings), nanorings. and nanobows. The Zn-terminated (0001) surface is also effective for growing nanostructures, but the oxygen terminated (0001) surface is chemically inert thus resulting in the growth of a "comb"-like structure in the vapor-solid process without using a catalyst. The is by a self-catalysis process due to Zn-termination at the (0001) surface.

Several techniques are employed for nanowire synthesis. One technique, vapor-liquid-solid growth (VLS) is used to form single crystal wire structures. VLS uses a liquid metal cluster or catalyst that acts as an energetically favored site for absorption of gas phase reactants. A one dimensional structure grows from the cluster as the reactants supersaturate. The diameter of the structure is limited by the diameter of the liquid metal cluster that can be grown under equilibrium conditions. Another VLS method uses Au to confine silicon nanowire growth. The diameters of these nanowires are relatively large and have a large size distribution-20nm to 100nm. Sorting nanowires with a large size range and analyzing their respective electronic properties is difficult to accomplish.

Laser-assisted catalytic growth (LCG) is used to grow a range of multicomponent semiconducting nanowires. LCG exploits laser ablation to generate nanometer-size clusters that define the size and direct the growth of crystalline nanowires in combination with a vapor-liquid-solid (VLS) process. This growth mechanism uses binary and pseudobinary phase diagrams to choose a specific catalyst for nanowire material composition and growth temperature such that a liquid alloy and solid nanowire material coexists. The cluster has a preferential site where the reactant absorbs, and when supersaturated, the nucleation for crystallization and ID growth occurs. LCG is used to grow elemental Si and Ge nanowires. Semiconducting GaAs and GaN nanowires (Figure 1.3 and 1.4) are grown as well, although predicting the catalyst nanowire material is more difficult due to the complexity or unavailability of information in the ternary phase diagrams.

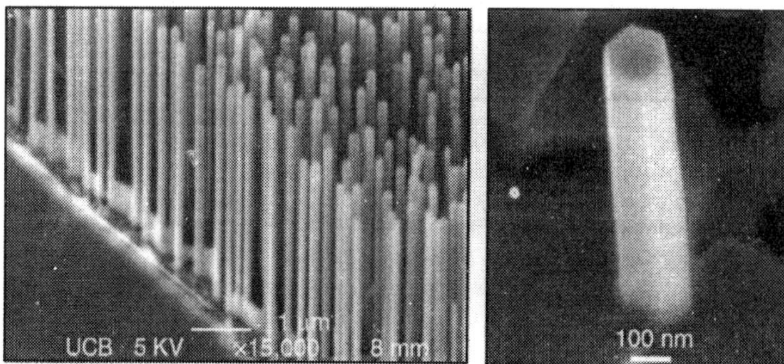

Figure 1.3 *Scanning electron microscopy images of (a)ZnO nanowire arrays on a sapphire wafer substrate and (b) aGaN Nanowire with a hexagonal cross section.*

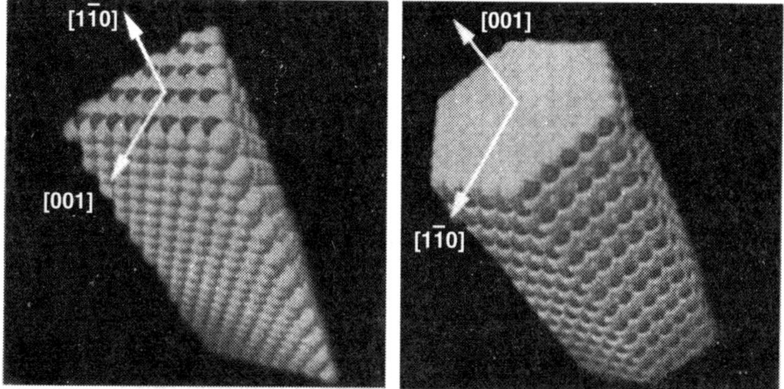

Figure 1.4 *Structural models of nanowires with(a) triangular and (b) hexagonal cross sections*

Templates are also utilized to direct nanowire growth. Nanometer-sized pores membranes and zeolites) and used to confine the growth of wires and carbon nanotubes are converted to carbide and nitride nanowires. Although template-based methods are straightforward to implement, template-based growth produces polycrystalline materials with diameter large enough to observe the effects of quantum confinement.

Molecular beam epitaxy is used to grow and align GaAs nanowires on nanochannel alumina (NCA) templates. Gold, is deposited onto the substrate using e-beam evaporation. GaAs are then grown using MBE using the deposited gold as a catalyst. Wires are grown perpendicular to the substrate, but no electronic characterization are performed.

In nanowire fabrication techniques, measuring their electron transport properties is problematic because the wires need to be connected to electrodes. Nanowires are usually fabricated in large, tangled clumps, and finding a technique to straighten and position each wire between two electrodes for electrical analysis has not been developed. Even wires that are relatively straight stll need to be removed from the substrate in order to be connected to electrodes. Aligning nanowires is not straightforward although several techniques exist which address the problem.

Electron transport properties are usually measured by randomly dispersing nanowires on a TEM grid and fabricating electrodes around isolated wires. Since nanowires are often looped or somewhat straight, but with no directional control this process is cumbersome in that the grid needs to be scanned in order to find a relatively straight wire that is somewhat isolated in order to fabricate electrodes. Nanowires have also been prepared for analysis by using ultrasonic dispersion to distribute nanowires across a TEM grid for analysis.

Aligning nanowire is not only critical for measuring transport properties, but also for creating networks and arrays for nanoelectronics applications. One technique combines fluidic alignment with surface patterning techniques A surface is lithographically defined into channels and chemically modified. The nanowires are aligned in a microfluidics chamber. Although wires are aligned into networks, the spacing between each wire is not consistently uniform, and using microfluidics to array nanowires is laborious compared to directly depositing wires into an arrayed format.

Nanowires deposited and aligned using electric field-assisted assembly. Nanowires are positioned between two lithographically defined electrodes on a SiO_2 substrate. The nanowires are assembled due to polarization from applying an alternating electric field. Although this technique assembles nanowires for electrical characterization, the nanowires used are quite large (70 nm to 100 nm), which do not exhibit any of the quantum effects that can be used for nanoscale electronics. Although electric fields can control assembly, this method can be limited in the packing density it can achieve because of the electrostatic interference between electrodes.

Crossbar arrays are fabricated by creating a "stamp" for nanowire patterning. Thin film metallic nanowires are grown by electron beam metal evaporation and then physically transferred onto a silicon wafer. Transport properties are measured by sectioning the wires by ion etching and fabricating electrodes by e-beam lithography. Although this technique the most promising in that it produces wires that are straight for long distances and can be arrayed into interesting crossbar array patterns, this technique is limited in that it can not produce single crystal wires or semiconducting wires. In addition, since this technique places the substrate at an angle to the evaporants during evaporation, there a shadowing effect onto the nanowires, resulting non-uniform thickness across the diameter of the wire.

Oxide Nanobelts

The transmission electromicroscopy (TEM) images in Figure 1.5 show the typical shapes of oxide nanobelts of several materials made by vapor phase evaporation. Each nanobelt has a uniform width along its entire length, and the typical width of the nanobelts are in the range of

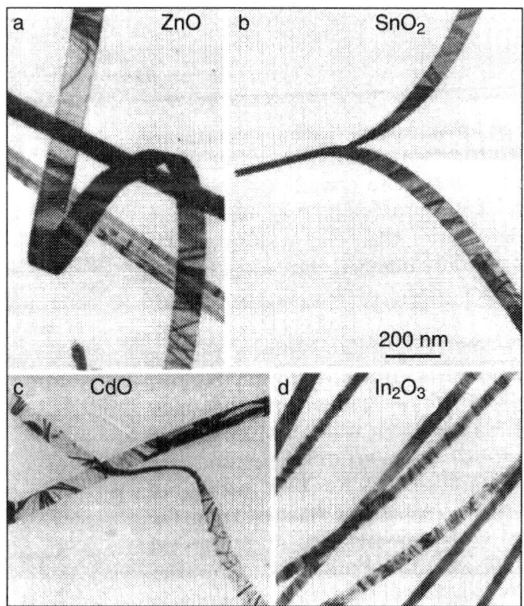

Figrue 1.5 *TEM images of nanobelts of four different oxide materials synthesized by the vapor-solid process, showing their ribbon shape. The raw material used for the synthesis of the nanobelts was the corresponding chemical powder.*

50 to 300 nm. No particle is observed at the ends of the nanobelts, thus the vapor-liquid-solid (VLS) process is (see next chapter), not a dominant growth mechanism. The ripple-like contrasts that appears in the TEM image is due to strain that resulted from the belt bending. Cross-sectional TEM image show that the nanobelts have a rectangular-like cross section, with typical thickness, in the range of 10 to 50 nm and width-to-thickness ratios of 3-10. The nanobelts grow along specific directions, and their surfaces are defined by specific crystallographic planes. This means that the nanobelts are structurally and morphologically controlled, and hence have controlled properties. Table 1 summarizes the structural characteristics of the nanobelts grown in laboratory. It seems that the nanobelt morphology is a unique and common structural characteristic for the family of semiconducting oxides.

TABLE 1 Semiconductive oxide nanobelts and their growth directions and surface planes.

Nanobelts	*Crystal structure*	*Growth direction/plane*	*Top surfaces*	*Side surfaces*
ZnO	Wurtzite	[0001]	$\pm(2\bar{1}\bar{1}0)$	$\pm(0\bar{1}\bar{1}0)$
ZnO	Wurtzite	$[01\bar{1}0]$	$\pm(2\bar{1}\bar{1}0)$	±(0001)
Ga_2O_3	Monoclinic	(010)	±(100):	$\pm(10\bar{1})$
Ga_2O_3	Monoclinic	(001)	±(100)	±1:(010)
SnO_2	Rutile	[101]	$\pm(10\bar{1})$	±(010)
In_2O_3	C-Rare earth	[001]	±(100)	±(010)
CdO	NaCI	[001]	±(100)	±(010)
PbO_2	Rutile	[010]	(201)	$\pm(10\bar{1})$

The growth of a nanobelt is related to two factors. The surface energy which determines the preferential surfaces that will grow, whereas the growth kinetics determine the final structure. CdO, for example, has a cubic structure that has equivalent growth rates along [100], because the {100} surfaces have the lowest surface energy. But the formation of a nanobelt structure with fast growth along [100] is a surprise. Figure 1.6(a) shows the nanosheet structure of CdO, which is formed by fast growth along [100] and [010], whereas growth along [001] is suppressed, resulting in the formation of a thin sheet type of structure. A similar structure is observed for Ga_2O_3 (Fig. 1.6(b)) The formation of these structures is apparently associated with the growth kinetics.

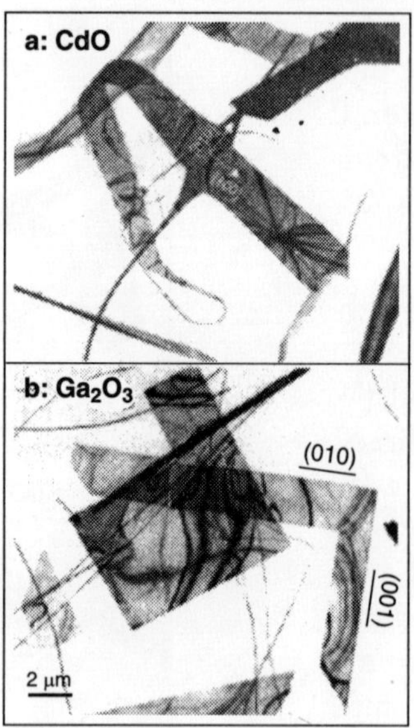

Figure 1.6 *TEM images showing the nanosheet structures of CdO (a) and GazO,(b) synthesized by the vapor-solid process. The CdO and GazO, nanosheets were made hy evaporating CdO and GazO, powder. respectively, using the experimental apparatus employed for nanobelt synthesis.*

Now, "how do one-dimensional nanostructures form?", or "what is the growth mechanism of the one-dimensional oxide nanostructures?". Depending on the presence or absence of metal catalysts in the synthesis processes, *two growth mechanisms*, i.e., VLS and vapor-solid (VS) mechanisms are adopted to account for the growth of the one-dimensional oxide nanostructures. In the VLS process, as stated above a liquid alloy droplet composed of a metal catalyst component (such as gold, iron, etc.) and a nanowire component is first formed under the reaction conditions. The liquid droplet serves as a prefrential site for the adsorption of gas-phase reactant and, when supersaturated, as the nucleation site for crystallization. Nanowire growth begins after the liquid becomes supersaturated in reactant materials and continues for as long as the catalyst alloy remains in a liquid state and the reactant is available. During growth, the catalyst droplet directs the nanowire's growth direction and defines the diameter of the nanowire.

For the VS process, for example, how atoms or other building blocks are assembled into one dimensional nanostructure with wire-like or belt-like morphologies, is unknown This process dominate the direct vaporization of the solid at higher temperature zone, with deposition occurring at a lower temperature region. This process is used for the growth of oxide nanobelts (see page 19 chapter 2).

The formation of long, uniform nanobelts are described as follows. The solid raw materials are sublimed at high temperature, and the molecular species generated have the basic structural configuration, is preserved to balance the charge for anionically bonded structures. Thus, the crystalline structure is formed. The surface that have lower energy tend to grow larger, and they will remain flat even without atom steps at the growth temperature, typically 1270K surfaces determine the enclosure surfaces of the noanobelt. As a result, no incoming molecules remain on the flat, low energy surfaces and they tend to diffuse towards the rougher surfaces, which are the growth front, leading to fast growth along this direction and forming the nanobelts.

Nanowire Growth Control

Applying the epitaxial crystal growth technique to VLS process it is possible to achieve precise orientation control during the nanowire growth. The technique, known as vapor-liquid solid epitaxial (VLSE) controls synthesis of nanowire arrays., For example, ZnO grows along the [001] direction and forms a highly oriented array when grown epitaxially on sapphire (110) substrate (Figure 1.5(a)) A similar level of growth control is achieved for GaN (Figure 1.5(a)) and Si/Ge systems. It is possible to use the VLSE technique to grow nanowire arrays with tight control over size (Diameter,<20nm) and diameter uniformity (variation of less than ~10). This size and monodispersity control is crucial for many applications for these nanowire arrays light emission and field effect transistors.

It is further possible to control the nanowire density in the range of 10^6-10^{12} cm^{-2} by adjusting the initial nanocluster density on the substrates. In addition, various ways are explored to control the nanowire growth, example; testing by e-beam lithography and self-organization of the metal nanocrystals.

Controlling the growth orientating is important for many of the proposed applications of nanowires, including vertical field-effect transistors, for example a simple approach to selectively growing semiconductor nanowires is made along chosen crystallographic directions (Figure 1.6). The use the metalorganic chemical vapor deposition (MOCVD) and appropriate substrate selection is done to control the crystallographic growth directions of high density arrays of galium nitride nanowires with distinct geometric and physical properties (for example, the ability to emit light) Epitasial growth of wurtzite gallium nitride on γ-$LiAlO_2$ (100) and MgO (111) single crystal substrates result in the selective growth of nanowires in the orthogonal $[1\bar{1}0]$ and [001] directions, thereby exhibiting triangular and hexagonal cross sections and different optical emission involving an energy shift of 100meV. The MOCVD process is compatible with GaN thin film technology, which leads to scale up and device integration. Control over the direction and of nanowire growth is required in anisotropic parameters such as thermal and electrical conductivety refractive index, piezoelectric conductivity refractive index, piezoelectric polarization and bandgap to be used to tune the physical properties of nanowires make from a given material.

Fabrication and Synthesis

A variety of nanofabrication methods, including focused ion beam patterning, electron-beam lithography, dip-pen nano-lithography, and scanning tunneling microscope lithography, are used to fabricate arrays of zero-dimensional (0D) structures with sizes from sub-100 nm to micrometer length scale. Atomic force microscopy (AFM) is used as a powerful technique for nanoindentation.

Nanowires and nanobelts of the most common semiconductors, such as ZnO, ZnS, CdSe, and CdS, can be synthesized by various techniques. These materials can be cubic zinc blende (ZB) and hexagonal wurtzite (WZ), in which each atom is tetrahedrally coordinated by atoms of the opposite species. The close-packed planes stack in ABCABC sequence in the ZB structure and ABAB in WZ structure. The stability of WZ as compared to ZB is closely related to the deviation of the c/a lattice-parameter ratio from the ideal value of 1.633 for close-packed hexagonal. Theoretical calculations show that the Madelung constant for ideal WZ configuration is about 0.2% larger than that of ZB, so that the strong ionic compounds favor WZ. It is a well known fact that ZnO crystallizes in the WZ structure under normal pressure, and the metastable ZB structured ZnO can be formed in epitaxial thin films, but freestanding ZB nano-structures of ZnO, such as nanowires and nanobelts, have never been found.

The ZB phase acts as the nucleus in the initiation of some nanostructures of ZnS, CdSe, CdS, and MnS, but it is rather unstable and quickly transforms into the WZ phase once the crystal becomes bigger. The synthesis of ZnO tetrapods has been achieved. Three different growth mechanisms about the formation or the ZnO tetrapods are there. One is based on the assumption of the existence of a ZB structured core at the center; the second is built upon the octahedral multiple twin structure; the third is that the ZB-type nucleus only exists in the high-temperature tetrapods and degenerates to multiple twins after cooling down to room temperature.

The methods for fabricating arrays of aligned NWs, include vapor phase transport, metal-organic chemical vapor deposition (MOCVD), and hydrothermal synthesis. However, vapor-phase transport and MOCVD require single-crystal substrates and high operation temperature, which are not compatible with organic substrates for applications in flexible and wearable electronics. Hydrothermal synthesis usually employs ZnO seeds in the forms of thin films or nanoparticles or requires an external electrical field to promote the growth of ZnO nanostructures. In such a case, nanowhiskers grow densely on the entire substrate with large variation in orientation, which do not meet the needs for nano-generators (see chapter 19) because it requires well-aligned and well-separated NWs. Density-controlled synthesis of oriented arrays of one-dimensional ZnO NWs or nanorods are made by nanosphere lithography, photolithography and electron-beam lithography, which are, time consuming, and usually used for small-scale experiments.

The vapor—liquid—solid (VLS) process is used in growth of quasi-one-dimensional (1D) nanowires and nanotubes (referred to as 1D nanostructures). In this process, a metal catalyst is chosen from the phase diagram by identifying a metal that is in liquid state at the growth temperature and serves as the site for adsorbing the incoming molecules, but the metal does not form a solid solution with the nanowire; thus, it is phase-separated at the growth front and leads the growth. The metal liquid droplet serves as a preferential site for absorption of gas-phase reactant. Nanowire growth begins after the liquid becomes supersaturated in reactant materials and continues as long as the catalyst alloy remains in a liquid state and the reactant is available. During the growth, the catalyst droplet directs the nanowires growth direction and defines the diameter of the nanowire. Ultimately, the growth terminates when the temperature is below the eutectic temperature of the catalyst alloy or the reactant is no longer available. As a result, the nanowires obtained from the VLS process typically have a solid catalyst nanoparticle at the ends with sizes comparable to diameters of the connected nanowires. Metal particles such as Au and Fe are effective for growing nanowires of Si, III-V compound, II-VI compound, and oxide-such as Au/ZnO, Fe/SiO_2, Co/SiO_2, Ni/Ga_2O_3, and Ga/SiO_2.

It is generally believed that the metal particle is a liquid droplet during growth, and its crystal structure in solid has no influence on the structure of the nanowires and nanobelts grown.

The vapor—liquid—solid (VLS) process is the most widely used technique for growing nanowires because of its relatively low cost and simple procedure. Metal-organic chemical vapor deposition (MOCVD) has also, been proven as an alternative method for aligned nanowires but with a much higher cost (see below). A wet chemistry process has been demonstrated as a powerful technique for growing aligned nanowires over a very large surface area. However, the density of nanowires grown on the surface cannot be controlled unless a catalyst pattern created by a mask or lithography is applied.

There is considerable interest in the methods for the fabrication of one-dimensional (1D) ordered nanostructures, by the catalytically activated vapor-liquid-solid (VLS) growth process, in which the catalyst initiates and guides, the growth and the epitaxial orientation relationship, between 1D nano-structures and the substrate leads to the alignment.

The nanowires are generally synthesized by a catalyst-mediated process involving vaporization of ZnO at high temperatures (around 1220 K). In most of the cases the synthesis is carried out first, and the structure analysis is then performed. As a result, detailed structural monitoring is missed, leaving the question of growth kinetics unknown. Although there are some transmission electron microscopy (TEM) observations of the growth of nanowires, they are limited to a small number of nanowires.

In vapor—liquid—solid (VLS) growth, it is believed that it is the catalyst particle at the tip that determines the size and growth rate of the nanowire, and the surface termination of the substrate has a little effect on the growth. Controlling the size of the catalyst particles effectively controls the size of semiconductor nanowires.

Thermal evaporation is also a highly effective technique for fabricating 1D ZnO nanostructures both in large quantities and at low cost. Gold, copper, and tin are used as catalyst to direct the growth of 1D ZnO nanostructures by a vapor-liquid-solid (VLS) process. Due to the fact that solid-vapor phase synthesis typically occurs at high temperature 970–1570 K, the synthesis of small-size (<10 nm) ZnO nanobelts and nanobelts with large yield is cumbersome. Fabrication of nanoscale biosensors based on nanowires (NWs), nanotubes (NTs), and other nanomaterials has attracted enormous attention. In comparison to nanoparticles, 1D NWs and NTs have higher sensitivity because of depletion or accumulation of charge carriers at the surface that is caused by binding of charged biological macro-molecules at the surface, and affects the entire cross-sectional conduction pathway. Among all 1D nanomaterials, Si NWs and carbon NTs are the materials as biosensors. Functionalized Si NWs and carbon NTs are used for detecting proteins, DNA and DNA sequence variations, and cancer markers. However, the biocompatibility and biodegradability of these nanostructures remain to be known. For example, carbon NTs injected into human blood vessels might accumulate and occlude capillaries in the human brain, which an cause serious damage or be fatal.

A general approach is a direct deposition of a single NW/NB from solution onto prefabricated electrodes, which usually leads to a high contact resistance and even nonsymmetric contacts at the two ends. A better electrical contact can be achieved by depositing metal on the top of the NW/NB using focused ion beam (FIB) microscopy. FIB-based metal deposition is a slow process of depositon from one contact to the next. There are also other approaches to make contact with NWs, such as electron beam lithography, focused electron beam, and shadow masking. It is highly desirable to have a parallel process for economically fabricating the NW devices over a large array of contacts.

SYNTHESIS OF OXIDE NANOWIRES AND NANOBELTS

1. Vapour-Liquid-Solid (VLS)

(a) Nanowires Using Catalyst Particles

A general strategy that has received increasing focus involves exploiting a 'catalyst' to confine growth in 1D. Depending on the phases involved in the reaction, this approach is typically defined as vapor-liquid-solid (VLS), solution-liquid-solid (SLS), or vapor-solid (VS) growth.

VLS is one of the most popular and powerful growth of NWs (Fig. 2.1), in which the catalyst particle is used to direct 1D growth of single crystal materials. Here the catalyst is envisioned as a nanocluster or nanodroplet that defines in most of cases the diameter of and serves as the site that directs preferentially the addition of reactant to the end of a growing NW much like a living polymerization catalyst directs the addition of monomers to a growing polymer chain. In principle, the vapor-solid technique is a simple process in which condensed or powder source

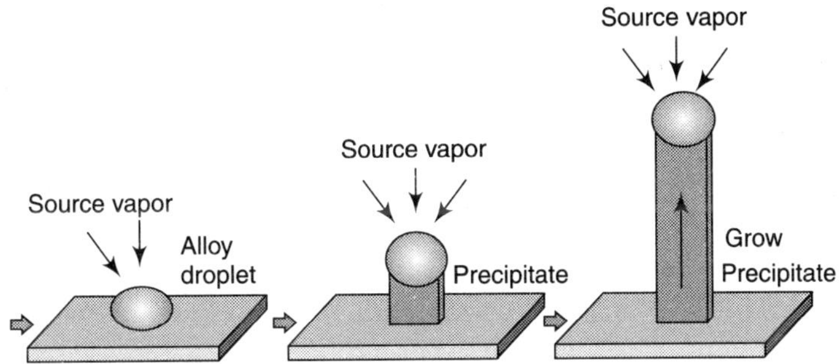

Figure 2.1 *The growth of aligned nanowire arrays via the vapor-liquid-solid process.*

material(s) is vaporized at elevating temperature and then the resultant vapor phase(s) condenses at certain conditions (temperature, pressure, atmosphere, substrate, etc.) to form the desired product(s). The processes are conducted in a tube furnace, as shown in Figure 2.2. It consists of a horizontal tube furnace, a rotary pump system and a gas supply and control system. A view window is set up at the left end of the alumina tube, which is used to monitor the growth process. The right end of the alumina tube is connected to the rotary pump. Both ends are sealed by rubber O-rings. The ultimate vacuum for this configuration is ~2×10^{-3} mm Hg.

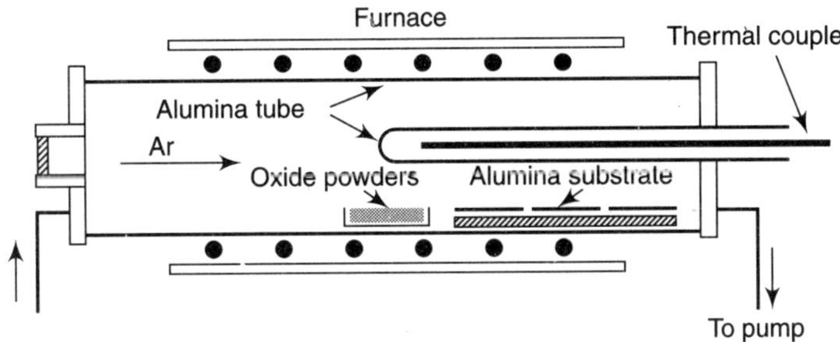

Figure 2.2 *An apparatus for growth of oxides nanostructures.*

The carrying gas comes in from the left end of the alumina tube and is pumped out at the right end. The source material is loaded on an alumina boat and positioned at the center of the alumina tube. Several alumina strip plates (60 × 10 mm) are placed downstream one-by-one inside the alumina tube, which acts as substrates for collecting growth products.

There are several processing parameters such as temperature, pressure, carrier gas (including gas species and its flow rate), substrate and evaporation time period, which are controlled and selected properly before and/or during the thermal evaporation. The source temperature selection depends on volatility of the source material(s). The pressure is determined according to evaporation rate or vapor pressure of source material(s). The substrate temperature drops as the distance of its location is away from the position of source material(s). The far the distance is, the lower the substrate temperature is. The temperature field in the furnace tube can be changed and controlled by introducing cold finger/plate or by using a multiple-zone furnace. Selecting a proper evaporation time is also important because it influences on not only the amount but also the size and the morphology of the product(s). Also, the thermal evaporation process is very sensitive to the concentration of oxygen in the growth system. Oxygen influences not only on volatility of the source material(s) (the stoichiometry of the vapor phase), but also on formation of product(s). So, after evacuating the alumina tube to ~2×10^{-3} mm Hg, thermal evaporation is conducted at a certain temperature for 2 hours under the conditions of a pressure of 200–600 mm Hg and an Ar carrier gas of 50 sccm (standard cubic centimeters per minute).

So, in the Growth of 1D nanostructure (usually follows the vapor-liquid-solid (VLS) approach), a liquid alloy droplet is composed of a metal catalyst component (such as Au, Fe) and a nanowke component (such as Si, III-V compound, II-V compound, oxide). It is first formed under the reaction conditions. The metal catalyst is chosen from the phase diagram by identifying metals in which the nanowire component elements are soluble in the liquid phase but do not form solid compounds more stable than the desired nanowire phase, example; for the 1D ZnO nanowires grown via a VLS process, the commonly used catalyst for ZnO is Au. As a result, a nanowire obtained from the VLS process typically has a solid catalyst nanoparticle at its tip with a diameter comparable to that of the connected nanowires.

The synthesis process for patterned and aligned ZnO 1-D, nanostructures involves three main steps. The hexagonally patterned ZnO nanorod arrays are grown on a single crystal Al_2O_3 substrate, on which patterned Au catalyst particles are dispersed. First, a two-dimensional, large-area, self-assembled and ordered monolayer of sub-micron spheres is formed on a single crystal $A1_2O_3$ substrate (Fig. 2.3(a). Second, a thin layer of gold particles is deposited onto the

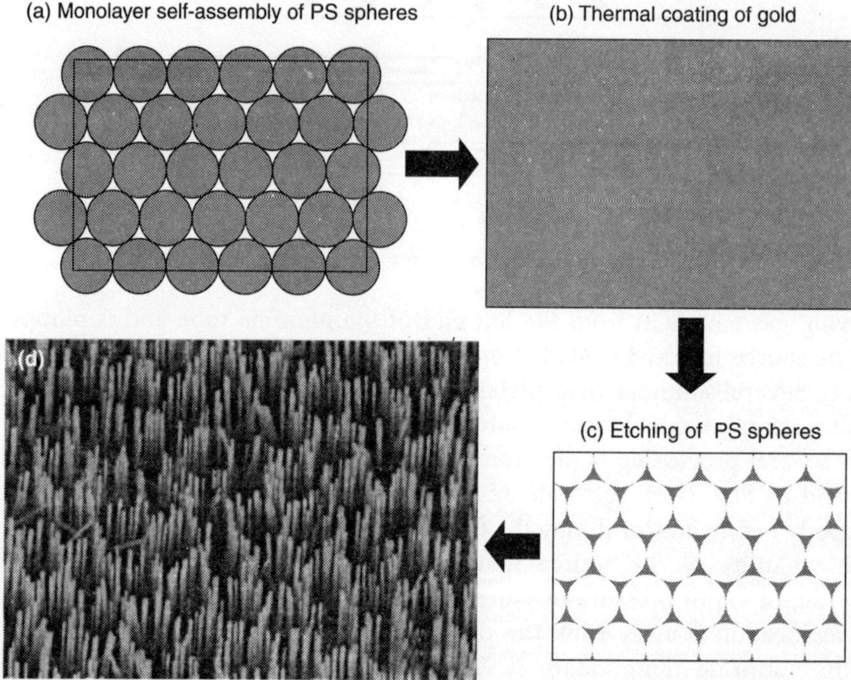

Figure 2.3 *Experimental procedure for growth of aligned nanowires. (a) Self-assembled monolayer of polystyrene spheres forming a mask. (b) Deposition of gold catalyst by thermal evaporation. (c) Removal of polystyrene spheres forming a hexagonal patterned catalyst. (d) Aligned ZnO nanowire grown on a single crystal alumina substrate with honeycomb pattern defined by the catalyst mask.*

self-assembled monolayer (Fig. 2.3(b); and then the spheres are etched away leaving a patterned gold catalyst array (Fig. 2.3(c)). Finally, NWs are grown on the substrate using the VLS process (Fig. 2.3(d)). The spatial distribution of the catalyst particles determines the pattern of the nanowires to be grown. By choosing the optimum match between the substrate lattice and the NW to be grown, the epitaxial orientation relationship between the nanowire and the substrate results in the aligned growth of nanowires normal to the substrate. The [0001] nanowires of ZnO presented in Figure 2.3(d) are grown on the $(2\bar{1}\bar{1}0)$ surface of single-crystal sapphire substrate. The distribution of the catalyst particles defines the location of the NWs, and the epitaxial growth on the substrate results in the vertical alignment.

(b) Nanobelts and Nanorings

By using ZnO powder as a source material without the presence of a catalyst, nanobelts with lengths of several tens to several hundreds of micrometers can be achieved. This growth process is dominated by a vapor-solid process, where the vapor deposits directly onto the substrate, which is held at a lower temperature.

By a method nanoribbons are formed that then transform into nanohelices. The horizontal tube furnace with ZnO powder as the source material is used. The solid-vapor deposition is carried out at 1670 K for 2 hours under a pressure of 2×10^4 Pa with Ar as the carrier gas. Nanohelices are deposited onto a polycrystalline Al_2O_3 substrate at a local temperature of 970–1070 K. 10% of the synthesized material is composed of freestanding nanohelices, with left- and right-handed chiralities in a 1:1 ratio. The formation of nanorings and nanohelices can be explained using dipole effects. The Zn-terminated <0001> surface is positively charged and the O-terminated $<000\bar{1}>$ surface negatively charged. Thus, there is a spontaneous dipole moment along the *c*-axis. For small thicknesses, as found in nanobelts (~10 nm), the spontaneous polarization produces the structures observed, such as nanorings, nanobows, nanohelices, and nanospirals. This is due to the minimization of the energy resulting from polar charges, surface area, and elastic deformation. Nanorings and nanobows are produced by a similar method. ZnO powder is heated to 1620 K for 2 hours in a horizontal tube furnace maintained at a pressure of 3×10^4 Pa with an Ar flux of 50 sccm, and the nanorings/bows are collected on an Al_2O_3 substrate at 770 K downstream of the source material. The yield in nanorings is around 5% of the total deposited product. Nanobows and nanobelts bent into semi-rings are also observed. A similar method is used to observe self-coiling of nanobelts into nanorings.

2. Vapor-Solid (VS) Process Synthesis of Nanobelts

Vapor-solid growth process is a simple and effective method for the growth of oxide nanostructures without using catalyst. There are two approach for vaporizing the source material, thermal vaporization and laser ablation. The thermal vaporization technique is a simple process, in which powder source material(s) is vaporized at elevating temperature and then the resultant vapor phase(s) condenses under certain conditions (temperature, pressure, atmosphere, substrate, etc.) to form the desired product(s). The morphology and phase structure of the synthesized product(s) depend on the source materials, growth temperature gradient, substrate, gas flow rate and pressure. But it is found that the most sensitive and important parameters for controlling the purity of the growth product are growth temperature, kinetics and source materials.

Controlling experimental conditions are the key for controlling the synthesized product. The selection of the source temperature mainly depends on volatility of the source material(s). It is usually set to be 573 K; lower than the bulk melting point of the source material. The pressure is determined according to evaporation rate or vapor pressure of source material(s). The substrate temperature drops as it does in VLS above.

Although the exact mechanisms responsible for vapor-solid growth are not completely known, many materials with interesting morphologies are made using these methods. Most significantly, nanoribbon materials (of ZnO, SnO_2, In_2O_3, and CdO) are made which have rectangular cross sections, by simply evaporating metal oxide powders at elevated temperatures. These nanoribbons are structurally uniform, with typical thicknesses from 30 to 300 nm, width-to-thickness ratios of 5–10, and lengths up to several millimeters. Finally, vapor-solid methods are utilized to form a variety of more complex morphologies. For instance, this method is used to create ZnO tetrapods and comb-like morphologies.

The ZnO nanobelts are grown by a vapor–solid process. ZnO powder and graphite powder are mixed at equal molar amounts as the source materials, and they are placed at the center of a pyramid-like quartz tube. The quartz tube is then placed in a tube furnace, which is later heated to 1370 K within 90 min. The as-grown produces is collected at the wider side of the pyramid-like quartz tube. The morphology and high-resolution images of the sample are observed with scanning electron microscopy (SEM) and high-resolution transmitted electron microscopy (HRTEM). The powder is dispersed into ethanol for optical analysis. A UV–vis absorption spectrometer and a fluorescence spectrometer are used to obtain its absorption and emission spectra, respectively. A He-Cd laser (50 mW, wavelength at 325 nm), is used to measure the photoluminescence of the sample and its stimulated emission.

(a) (b)

Figure 2.4 *SEM images of the as-synthesized zinc oxide nanowires and nanobelts.*

The morphology of zinc nanowires is shown in Figure 2.4(a) and (b). Many prismatic rods whose diameters are about 50–150 nm arc randomly distributed on the substrate. The sample is pure and clean.

Table 2.1 *Some common oxide nanobelts and their growth directions and surface planes.*

Nanobelts	*Crystal structure*	*Growth direction/plane*	*Top surfaces*	*Side surfaces*
ZnO	Wurtzite	[0001]	$\pm(2\bar{1}\bar{1}0)$	$\pm(01\bar{1}0)$
ZnO	Wurtzite	$[01\bar{1}0]$	$\pm(2\bar{1}\bar{1}0)$	$\pm(0001)$
ZnO	Wurtzite	[0001]	$\pm(2\bar{1}\bar{1}0)$	$\pm(01\bar{1}0)$
Ga_2O_3	Monoclinic	(010)	$\pm(100)$	$\pm(10\bar{1}\bar{1})$
Ga_2O_3	Monoclinic	(001)	$\pm(100)$	:$\pm(010)$
SnO_2	Rutile	[101]	$\pm(10\bar{1})$	$\pm(010)$
In_2O_3	C-Rare earth	[001]	$\pm(100)$	$\pm(010)$
CdO	NaCl	[001]	$\pm(100)$	$\pm(010)$
PbO_2	Rutile	[010]	(201)	$\pm(10\bar{1}\bar{1})$
ZnS	Wilrtzite	$[01\bar{1}0]$	$\pm(2\bar{1}\bar{1}0)$	$\pm(0001)$
CdSe	Wurtzite	$[01\bar{1}0]$	$\pm(2\bar{1}\bar{1}0)$	$\pm(0001)$

In the family of nanobelts nanoribbons (see Table 2.1), ZnO is the most extensively studied structures. Thermal evaporation of ZnO powders (purity: 99.9% melting point 2248 K; at 1673 K result in ultra-long ZnO nanobelts (Fig. 2.5(a)). The as-synthesized oxide nanobelts are pure, structurally uniform, single crystalline and most of them are free from dislocations. A ripple-like contrast, in the TEM image is due to strain resulting from the bending of the belt. Occasionally, some planar defects, such as twins and stacking faults are found, but there is no line defect. Point defects, such as oxygen vacancies, if present greatly affect the transport properties of the NBs. The NBs have a rectangular-like cross-section with typical widths of 30–300 nm, width-to-thickness ratios of 5–10 and lengths of up to a few millimeters. EDS and XRD measurements show that Wurtzite (hexagonal) structured ZnO having lattice constant of $a = 3.249$ Å and $c = 5.206$ Å, is consistent with the standard values for bulk ZnO. No particle is observed at the ends of the nanobelts. High-resolution, TEM (HRTEM) and electron diffraction show that the ZnO nanobelts are structurally uniform and single crystalline (Fig. 5(b)).

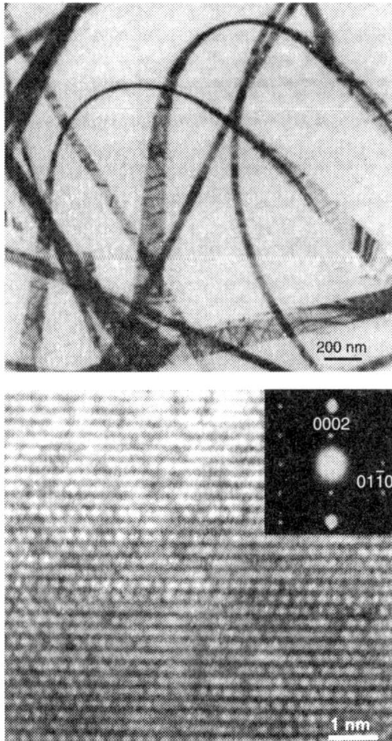

Figure 2.5 *TEM image of the as-synthesized ZnO nanobelts and a high-resolution TEM Image recorded with the incident electron perpendicular to the top surface of the nanobelt.*

The belt-like morphology is unique and common structural characteristic are observed for the family of semiconducting oxides with cations of different valence states and materials of distinct crystallographic structures. Nanobelts are synthesized from a series of oxides with different valence states and different crystal structures, including ZnO, SnO_2, In_2O_3, CdO, Ga_2O_3 and PbO_2. The nanobelts are synthesized by vaporizing the corresponding oxide powder without using catalyst. The surface morphologies of these nanostructures can be apparently identified from the SEM images.

3. MOCVD Technique

(a) MOCVD technique is employed for growing GaN, AlN, and AlGaN epilayers, which serve as the substrate for the subsequent growth of ZnO nanorod and GaAs for growing ZnO nanowires. Undoped *c*-plane-oriented GaN and AlN thin films are grown on one-side-polished *a*-plane sapphire substrates to thicknesses of 2 μm and 500 nm, respectively. $Al_{0.5}Ga_{0.5}N$ epitaxial layers with a thickness ~205 nm are grown on a 500 nm thick AlN buffer layer grown on a one-side-polished *c*-plane-oriented sapphire substrate. A 7–8 nm thick ~5×5 mm^2 gold catalyst layer is then deposited by plasma sputtering onto the epitaxial nitride substrates. The ZnO nanorods are grown through a vapor-liquid-solid process using a mixture of equal amounts (by weight) of ZnO and graphite powders (0.6 g each) that are loaded in an alumina boat located at the center of an alumina tube, which is placed in a single-zone tube furnace. Argon is used as carrier gas at a flow rate 49 sccm with additional 2% (1 sccm) oxygen to facilitate the reaction. The substrates are placed 10 cm downstream from the source materials. The source materials are heated to 1220 K at a rate of 50 K min^{-1} and the temperature is held at the peak temperature for 30 min under a pressure of 3×10^3 Pa, while the, local temperature of the substrates is 1150 K. Then the furnace is turned off, and the tube is cooled in air to room temperature within ~2 h under an argon flow.

(b) Also, Low-temperature growth for ZnO nanorods has been achieved. A two-zone furnace is used with Zn acetylacetonate hydrate placed as a precursor material into the low-temperature zone of the furnace (400–410 K). The vapor is then carried by 500 sccm N_2/O_2 flow into the higher temperature zone of the furnace with substrates kept at 770 K at 200 mm Hg. The ZnO nanorods grow directly on fused silica or Si substrates. Another low-temperature method uses Zn nitrate hexahydrate and ammonium hydroxide as a precursor. Here, the preparation of the initial ZnO buffer layer significantly affects the morphology and quality of the ZnO nanorods. Air-annealing is done to enhance the crystallinity and optical properties of the nanowires.

4. Chemical Vapor Deposition

(a) Nanowires and Nanorods

In CVD, a vapor reacts on the substrate to produce the desired product. In the case of nanowires, the vapor-liquid-solid (VLS) method is applied where the vapor is exposed to a catalyst such as Au particles. This seed layer is enriched by the vapor source material until it is saturated and the desired material starts to solidify and grow outward from the catalyst. Large-scale arrays with vertically aligned nanowires are produced by this method. The partial oxygen pressure and chamber pressure are important parameters influencing the growth mechanism that governs the final structure of the ZnO. By increasing the oxygen content under otherwise identical conditions, nanowires, dendritic side-branched/comb-like structures, and nanosheets are synthesized.

The catalyst seed layer has an immense influence *on* nanowires. Control over nanowire growth is gained by varying the Au layer thickness. Catalytic Au patterns formed by lithography are techniques are used for directed ZnO nanowire growth.

Epitaxial growth of highly aligned and nearly defect-free nanorods is observed using CVD (1:1 ZnO:graphite) on an initial ZnO film produced by pulsed laser deposition (PLD). ZnO nanorods grow on (100) sapphire at a 30° angle to the substrate normal, indicating the epitaxial crystalline growth. The growth of nanocolumns on alumina substrates via oxidation of ZnS in a tube furnace at 1220 K is achieved. The substrates are kept apart from the source at a lower temperature. The synthesized nanocolumns show a layered structure and become gradually thinner. This is assumed to be a result of a gradually decreasing supply of ZnO vapor.

(b) Other nanostructures

Nanocombs are synthesized by thermal evaporation of ZnO powder in a tube furnace (Fig. 6). The comb-like structures grow on an alumina substrate placed downstream, and growth is only observed in the region where, the temperature ranges from 1420–1620 K regardless of the

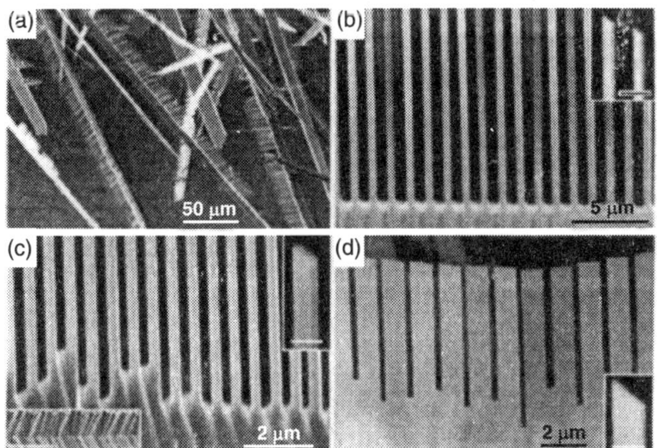

Figure 2.6 *Scanning electron microscopy (SEM) images of ZnO combs formed by evaporating ZnO powder at 1670 K for 2 hours. a) Low-magnification SEM image of ZnO combs. (b) High-magnification SEM image of a comb made of an array of rectangular ZnO nanobelts ~400 nm wide at a spacing of ~700 nm. Inset shows the growth fronts of two nanobelts with rectangular cross sections. Scale bar = 1 mm. (c) Array of nanobelts ~280 nm wide at a spacing of ~250 nm. Upper right inset shows the growth front of one rectangular nanobelt. Scale bar; 500 nm. Lower left inset is an SEM image of the stem of a comb. Scale bar; 10 mm. (d) Aligned nanobelts ~500 nm wide at a spacing of ~300 nm. inset is the growth front of a nanobelt. Scale bar = 500 nm.*

evaporation temperature. Increasing growth time influences the dimensions of the combs and even the shape of the associated nanowires. These nanocombs with periodical structures could find application as gratings in miniaturized integrated optics.

Another catalyst-free synthesis route for large-scale ZnO nanostructures involves heating a Zn foil to 970 K in a furnace in air without any additional carrier gas. By controlling the heating rate different morphologies have been achieved: porous membranes, nanowires, nanorods. rounoneedles, and nanotetrapods.

5. Synthesis and Characterization of ZnO Nanowires (CVTC Process)

An illustration of the chemical vapor transport and condensation (CVTC) system is shown in Figure 2.7 Single-crystalline silicon and sapphire of different orientations are used as substrates for the ZnO nanowire growth. The substrates are coated with a layer of Au thin film using a thermal evaporator with a quartz crystal thickness monitor to ensure an exact Au film thickness. Equal amounts of ZnO powder and graphite powder are ground together and transferred to an alumina boat. The Au-coated substrates and the alumina boat are placed into a small quartz tube. The substrates are placed ~5–10 cm from the center of the boat. This quartz tube is then placed inside a furnace quartz tube, with the center of the alumina boat positioned at the center of the furnace and the substrates placed downstream of an argon flow. The temperature of the furnace is ramped to 1070–1270 K at a rate of ~50–100 K min^{-1} and typically kept at that temperature for ~5–30 min under a constant flow of argon (~20–25 sccm). After the furnace is cooled to room temperature, light or dark gray material is found on the surface of the substrates. For the growth of ZnO nanowires using Au nanoclusters, spin-coating is used to disperse uniform Au colloids on the substrates.

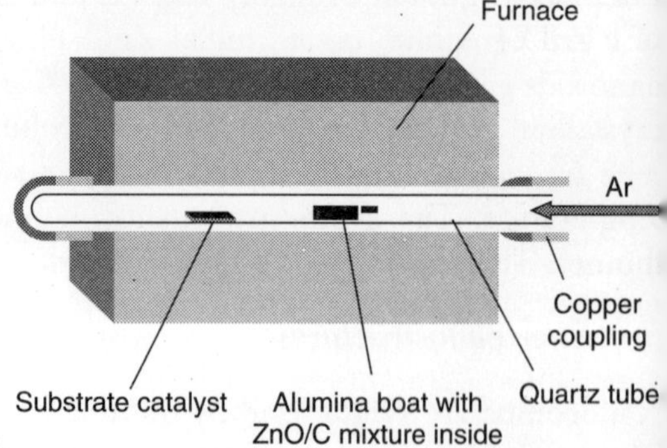

Figure 2.7 *Chemical vapor transport and condensation experimental for ZnO nanowise growth*

Although the VLS crystal growth mechanism is used for semiconductor nanowire growth, oxide nanowire growth through the VLS mechanism gets complicated by the presence of oxygen. In this process ZnO powder is reduced by carbon to form Zn and CO/CO_2 vapor in the high-temperature zone. The Zn vapor is transported and reacts with the Au solvent on substrates located downstream at a lower temperature to form alloy droplets (Fig. 2.8). As the droplets become supersaturated, crystalline ZnO nanowires are formed, by the reaction between Zn and CO/CO_2 in the low temperature zone. The presence of a small amount of CO/CO_2 do not

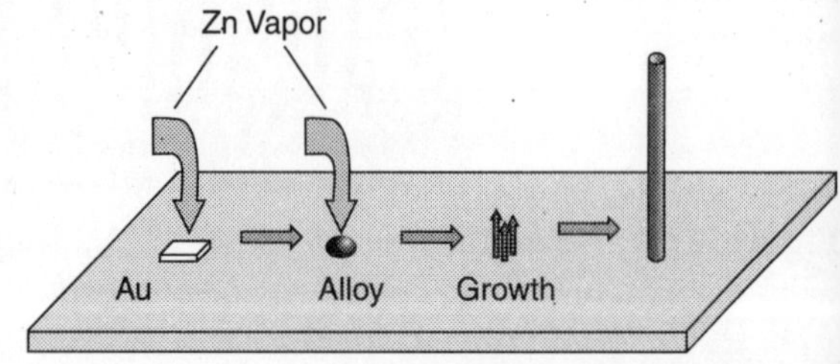

Figure 2.8 *ZnO nanowire growth mechanism.*

change the Au-Zn phase diagram but the gas mixture acts as the oxygen source during ZnO nanowire growth. Experiments without graphite addition in the starting materials, produce nothing on the Au-coated substrates, which indicate the importance of Zn vapor generated by the carbothermal reduction of ZnO.

Figure 2.9 shows a typical scanning electron microscopy (SEM) image of ZnO nanowires grown on a silicon substrate. The diameters of the nanowires normally range from 20–120 nm and their lengths are 5–20 μm. X-ray diffraction (XRD) patterns of ZnO nanowires are taken and crystal structure of the nanowires is examined. All samples give a similar XRD pattern, indicating the high crystallinity of nanowires. Figure 2.10 shows a typical XRD pattern of these ZnO nanowires. The diffraction peaks are indexed to a hexagonal structure with cell constants of $a = 3.24$ Å and $c = 5.19$ Å.

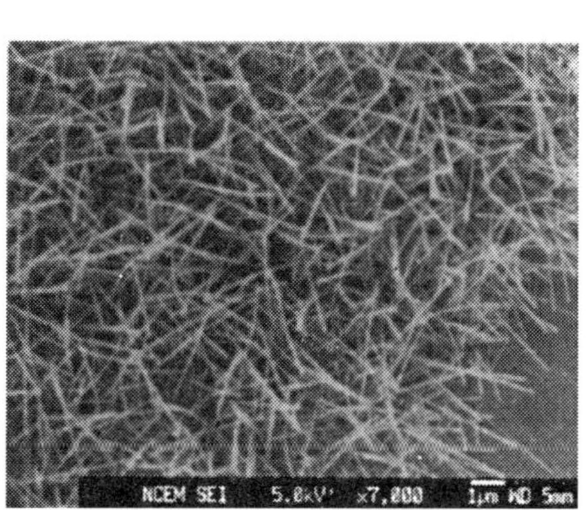

Figure 2.9 *Scanning electron microscopy images of ZnO nanowires grown on Si(100) substrate.*

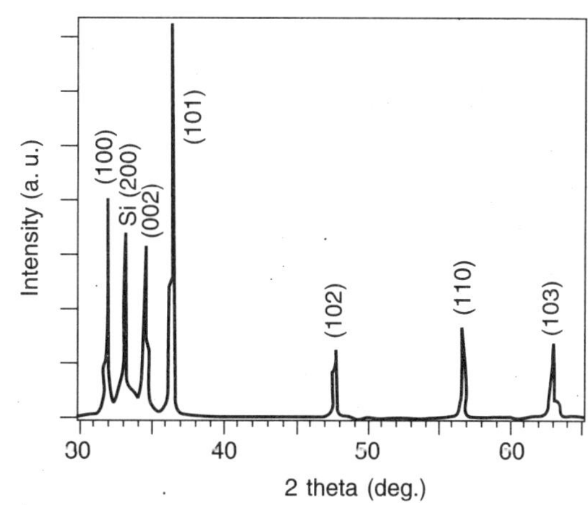

Figure 2.10 *XRD pattern of ZnO nanowires on silicon substrate.*

Additional structural characterization is carried out using transmission electron microscopy (TEM). Figure 2.11(a) is a TEM image of a thin nanowire with an alloy tip. The presence of the alloy tip is a clear indication of the VLS growth mechanism. Figure 2.11(b) shows a high-resolution TEM image of a single ZnO nanowire. The spacing of 2.56 ± 0.05 Å between

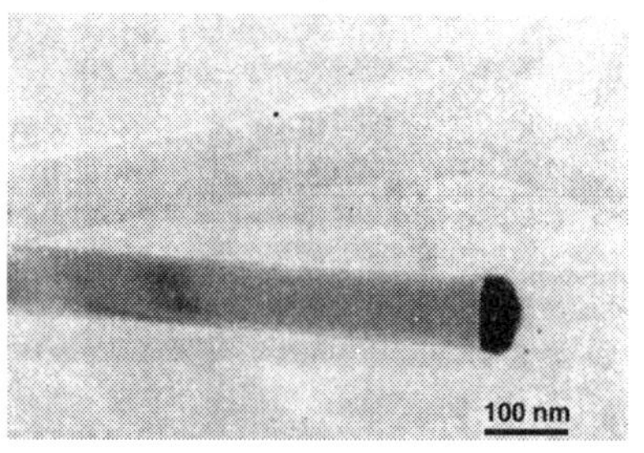

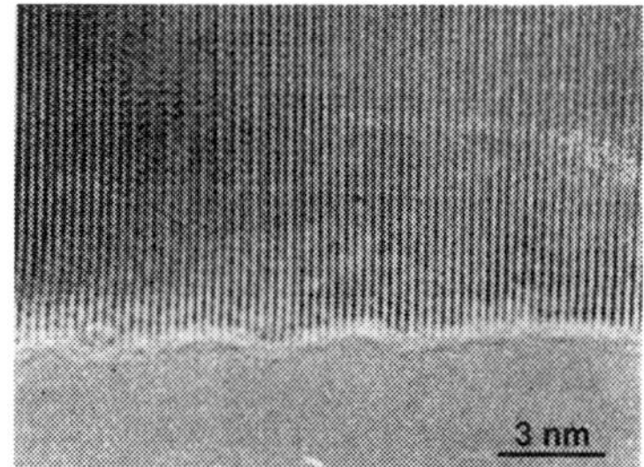

Figure 2.11 *(a) TEM imare of a ZnO nanowire with an alloy droplet on its tip. (b) High-resolution transmission electron microscopy image of an individual ZnO nanowire its <0001> growth direction. Reprinted with permission from.*

adjacent lattice planes corresponds to the distance between two (0002) crystal planes, confirming <0001> as the preferred growth direction for ZnO nanowires.

6. Fabrication of ZnO Nanowire Devices via Electrodeposition

(a) An approach to make effective contacts between zinc oxide (ZnO) NWs with microelectrodes using a selective copper-electrodeposition technique is described below. By carefully selecting the electroplating bath, the copper is deposited only on the electrodes leaving the ZnO NWs clean. Copper electrodeposition is a main interconnecting technology for ultralarge-scale integration fabrication, as well as for printed-circuit boards. Thus, this copper deposition technique allows an effective, parallel process for fabricating ZnO NW/NB devices on a large scale. This is of great importance for building functional device systems using NWs/NBs.

The ZnO NWs used here are synthesized by a solid-vapor process under controlled conditions without the presence of a catalyst, and are structurally perfect and geometrically uniform. The Au electrode patterns are defined with photolithography on a Si substrate covered with 50 nm thermally oxidized SiO_2. The microelectrodes consist of 2 μm wide fingers at a distance of 2 μm. These two fingers are connected to two 500 × 500 μm contacting pads for probe contacts as shown in Fig. 2.12. A single NW is "placed" across the electrodes using the micromanipulation

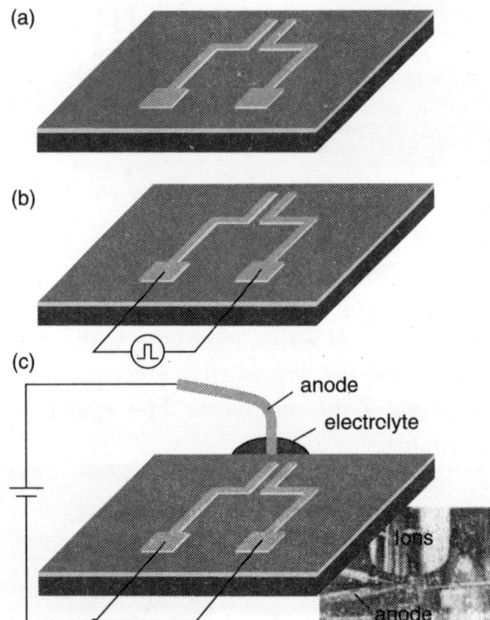

Figure 2.12 *Diagrams of the fabrication process of ZnO NW device. (a) crossing a single ZnO NW on the patterned electrodes by micromanipulating, (b) applying pulse voltage between the electrodes to achieve electroadsorption of ZnO NW on the electrodes and (c) the experimental setup of electrodeposition. The patterned electrodes are connected together as the cathode, pure copper wire is placed above the cathode to serve as the anode. A droplet of plating solution is applied between the cathode and anode to realize electrodeposition. The inset is a photograph of the experiment setup.*

technique 2.12(a). Because of the relatively weak adhesion force between the NWs and the electrodes, the NWs move or float away into the electrolyte, when it is immerged into the electrolyte at the beginning of the electrode position process, but this problem is solved by using an electroadsorption process before elcctrodeposition. After applying a pulse voltage (V_p = 5 V, f= 1000 Hz) across the two electrodes for a few minutes (Fig. 2.12b), the ZnO NWs are absorbed on the electrodes so firmly that no desorption or moving of the NW occurs during electrodeposition. A tiny movement of the NW on the electrode is observed at the beginning while applying the voltage, indicating that the NW is adhered on the electrodes by electrostatic force and tighter contacts are formed. Moreover, by applying 5 V, the current passing through the device is 1–5 μA, corresponding to a power 5–25 μW. For the small contact area (about 10^{-12} m^2) and the volume of the NW (about 10^{-13} m^3), the rise in temperature at the contacts is large. Therefore, the improvement of the adhesion force between the NW and the electrodes is caused by the thermal annealing occurring at the contact region.

The electrodeposition process, as shown in Fig. 2.12(c). is conducted on a probe station with the two finger electrodes connected together as the cathodes and a 0.2 mm diameter pure copper wire with a polished end as the anode. The anode is placed 1 mm above the cathode. After connecting the electrodes to a dc source, a droplet of the plating solution is applied between the cathode and the anode. After deposition, the wafer is rinsed with deionized water, followed by drying at 330 K for 1 h.

Two kinds of electroplating baths are used for copper deposition: pyrophosphate bath and copper nitrate bath. The copper distributions on the ZnO NW/electrode system using different plating baths are investigated by scanning electron microscope (SEM) (Fig. 2.13)). It is found that copper tends to deposit on the NW surface in pyrophosphate solution, forming a cluster of copper particles on the NW (Fig. 2.13(b)) and, consequently, deteriorating the performance of the nanodevice. However, when using copper nitrate solution, copper is only deposited on the

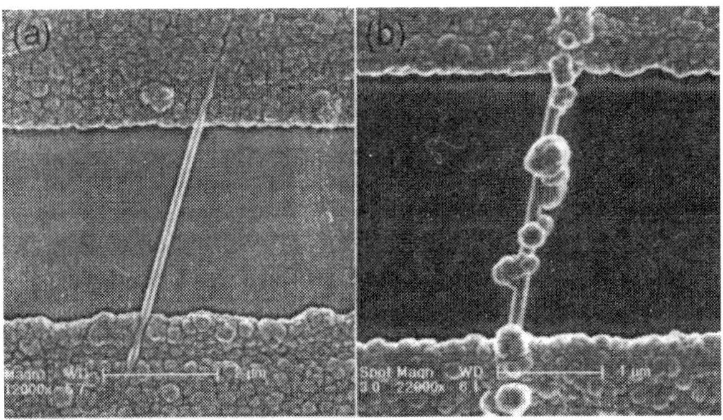

Figure 2.13 *SEM images of copper embedded ZnO NW alter electrodeposition using (a) copper nitrate bath and (b) pyrophosphate bath. The electrodeposition conditions of copper nitrate bath are $Cu(NO_3)_2$ 400 $g.L^{-1}$ Cl^- 50 ppm, PEG6000 10 ppm, current density 1 A dm^{-2} corresponding to a Cu deposition rate of 3–4 nm s^{-1}. The electrodeposition conditions of pyrophosphate bath are $Cu_2P_2O_7$ 60 g^{-1} L, $K_4P_2O_7$. $3H_2O$–280 $g.L^{-1}$ NH_3. H_2O–2 mil^{-1}, current density 0.5 A dm^{-2}, corresponding to a Cu deposition rate of 1–2 nm^{-1}.*

Au electrodes surface (Fig. 2.13(a)); therefore, the ZnO NW is enclosed in the electrodes only at the contact region. Transmission electron microscopy (TEM) is used to further examine the ZnO NWs after elcctrodeposition using copper nitrate solution (Fig. 2.14). The NW is transferred from the wafer after electrode position to a molybdenum grid for TEM observation. No copper particles are observed on the ZnO NW surface under TEM observation (Fig. 2.14(a)). Several energy-dispersive X-ray spectroscopy (EDS) spectra are taken from different positions along the ZnO NW, and no copper signature is found in the spectra (Fig. 2.14(c)). The carbon signature in the EDS spectrum is from the molybendum TEM grid coated with a carbon film. The appearance of iron and chromium signatures is caused by the impurities of the molybdenum grid. To further inspect whether a very thin film of copper is formed on the NW surface or not record a high-resolution TEM image as well as a corresponding electron diffraction pattern from the edge of the NW (Fig. 2.14(b)). It shows that a single-crystal volume with a clean surface; and no copper pattern is formed in the electron diffraction pattern.

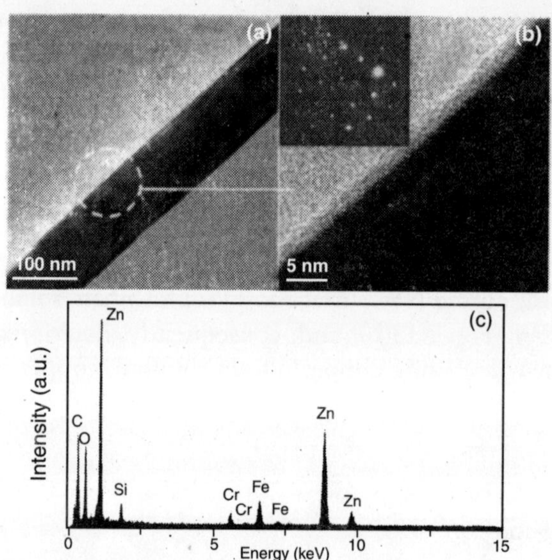

Figure 2.14 *(a) TEM image of ZnO NW after electrodeposition using copper nitrate solution. The ZnO NW is transferred onto a molybendum grid for TEM observation from the wafer after electrodeposition. (b) High-resolution TEM image and corresponding electron diffractional pattern from the edge of the NW. (c) One of the EDS spectra recorded from the ZnO NW. The results of (a)–(c) indicate that the ZnO NW surface is clean of copper.*

Generally, a metal electrodeposition process is composed of two successive steps: the reduction of metal cations to metal atoms at the cathode surface and the metal crystallization process. In the first step, the reducing reaction occurs as the cathode overpotential exceeds the reducing potential of the metal ions. In a given electrochemical system, the overpotential is determined by the current density. Due to the geometric and material inhomogeneity of the cathode surface, the distribution of the electrical field. i.e., the distribution of the current on the cathode surface is inhomogeneous, and even quite different in some positions. Plating baths with high covering power have the ability to deposit metal at regions of low current density; therefore, relatively uniform metal coatings are obtained all over the cathode surface. In low-covering-power bath, the nonuniform cathode current distribution is caused by the inhomogeneity of the cathode surface. It results in undesired nonuniform or even disconnected coatings. This phenomenon, however, is also utilized to achieve selective electrodeposition on the electrodes instead of the ZnO NW.

Because the resistivity of ZnO NWs (≈5 Ω cm) is much higher than that of Au (2.4×10^{-6} Ω cm), the electrical field, and then the cathode overpotential on the NW surface, is lower than that on the Au electrodes. Therefore, the reduction of copper ions is more difficult on the NW than on the electrodes. Figure 2.15 indicates the embedding process of the ZnO NW during electrodeposition in different plating baths. When using copper nitrate solution, copper deposition can only occur on the Au electrode surface due to the low covering power of the solution. With the growth of the coating, the ZnO NW is gradually embedded in copper at the contact sides, as shown in Fig. 2.15(a) and (c). However, with much higher covering power than the copper nitrate bath, the pyrophosphate bath neglects the difference of the cathode overpotential between the NW and the Au electrodes. Therefore, copper deposits on both the metal electrodes and the NW surface simultaneously, in the pyrophosphate bath, as shown in Fig. 2.15b and d.

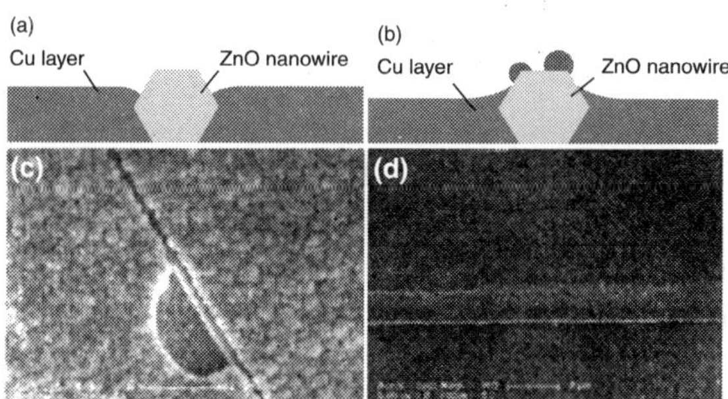

Figure 2.15 *(a) and (b) embedding process of ZnO NW using copper nitrate bath and pyrophosphate bath. (c) SEM image of embedded ZnO NW on the electrode using copper nitrate bath, showing that the NW is embedded in copper coating gradually. Region A is insulated by celluloid to prevent copper deposition for observation. (d) SEM image of embedded ZnO NW using pyrophosphate bath, shows that copper is deposited on the electrodes and NW surface simultaneously.*

The overall electrical resistance of the devices is significantly reduced by the selective electrodeposition process, as shown in Fig. 2.16, which presents the measurements on the same device before and after selective electrodeposition using copper nitrate solution. Without the treatment, the as-fabricated device exhibits a rectifying current-voltage (*I-V*) characteristic. However, the device shows a linear *I-V* curve after electrodeposition and the current flowing through the NW is increased greatly, 3–10 times. The inset in Fig. 2.16 shows the measurements of the same device at different time intervals after selective electrodeposition. The measurement performed 7 days later is consistent with the data obtained 2 h after the sample is made, indicating the stability of the device made by selective electrodeposition.

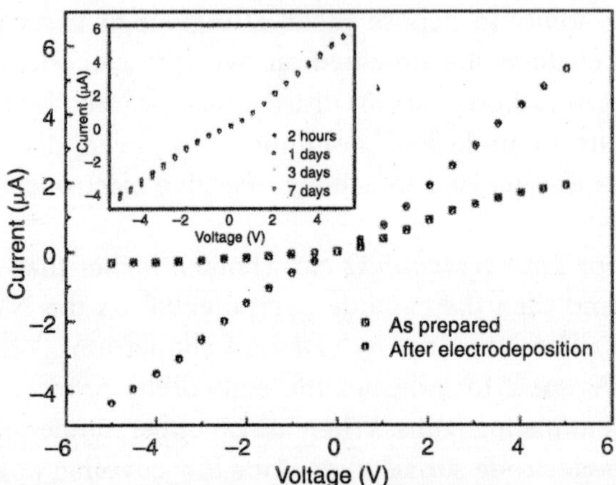

Figure 2.16 *I-V characteristics of the ZnO NW device before and after electrodeposition using copper nitrate solution, showing the contact improvement of the device by selective electrodeposition. After electrodeposition, the device is rinsed with deionized water, followed by drying at 330 K for 1 h. The inset is the I-V characteristics of a ZnO NW device at different times after electrodeposition using copper nitrate solution, showing the stability of the device. Measurements are done at room temperature.*

Columns on F-doped SnO_2 (FTO) substrate are obtained between 100 nm and 300 nm in diameter and 400 nm to 900 nm in height by varying the electrode position parameters, such as current density, deposition time, and bath temperature. Also electrodeposition of thin films is achieved. It is found that more positive electrodeposition potentials favor high-quality film growth and additional annealing results in significant enhancement and sharpening of the excitonic emission bands. The bandgap energy E_g depends on the electrodeposition potential; with increasing potential the bandgap E_g increases. Nevertheless, the bandgap E_g depends much more strongly on film thickness, which is due to the differences in grain size, lattice strain and defect states. The strong influence the nature of the substrate surface on the nucleation of nanorods is also observed. Electropolishing of Zn foil prior to electrode position hinders the nucleation of nanorods. Nonelectropolished films show much narrower and denser growth of ZnO nanorods. It is also found that a native surface oxide enhances nucleation.

Ordered porous ZnO films are fabricated by electrodeposition on Sn-doped In-oxide (ITO) glass substrates covered with a polystyrene (PS) array template. This template is created by self-assembly of an ordered PS array from a suspension of PS spheres. Subsequently, the ZnO is electrodeposited into the pores of this array. A zinc nitrate aqueous solution is used as the electrolyte. After dissolution of the PS matrix, a stable porous ZnO film is fabricated. A method to obtain ordered nanowires uses anodic alumina templates. Zn nanowires are electrodeposited with diameters of 15-90 nm into the pores of the alumina template and oxidized to ZnO in air at 570 K.

7. Aqueous Solution Growth

The growth of highly oriented ZnO nanowires and other nanostructures from aqueous solution is achieved by the aqueous solution growth method. In this method nanorods are grown conducting glass and Si substrates. For this type of growth, a ZnO seed layer is needed to initialize the uniform growth of oriented nanowires. Often, a solution of $ZnO(NO_3)_2$ and hexamethyltetramine (HMT) is used:

$$(CH_2)_6N_4 + 6H_2O \leftrightarrow 6HCHO + 4NH_3 \quad (1)$$

$$NH_3 + H_2O \leftrightarrow NH_4^+ + OH^- \quad (2)$$

$$2OH^- + Zn^{2+} \rightarrow ZnO(s) + H_2O$$

Hydroxide ions are formed by the decomposition of HMT and they react with the Zn^{2+} to form ZnO.

Figure 2.17 shows some typical nanowire arrays grown by using this method. This method operates at low temperature and a homogenous coverage of nanowires is achieved over large areas. Table 2.2 summarizes some of the growth solutions and the resulting structures.

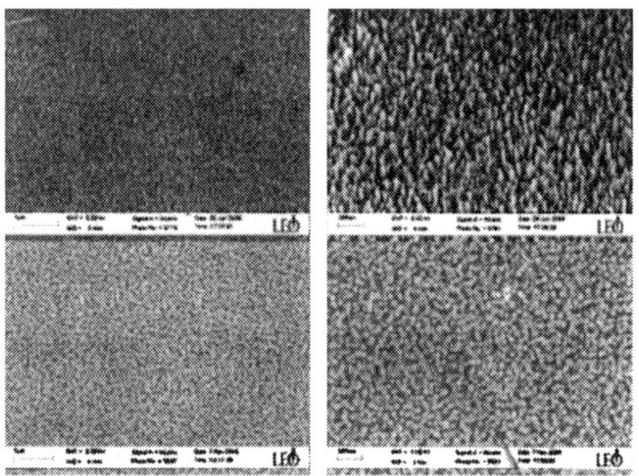

Figure 2.17 *SEM images of ZnO nanowire arrays grown over large surface areas on Si substrates with a ZnO-seed layer. Growth is performed in an aqueous solution of Zn nitrate hydrate and HMT at different concentrations (top: 0.025 M and bottom: 0.1 M). Substrate is tilted 45° in the top left image*

Table 2.2 *Summary of different results and methods for aqueous solution growth.*

Growth solution	*Resulting morphology*	*Focus of investigation*
Zinc nitrate, HMT	Nanorods, microtubes	On Si and conducting glass substrates
Zinc nitrate, HMT	Nanorods, nanotubes	Influence of substrate and seed layer.
Zinc nitrate, HMT	Aligned nanowire arrays	Influence of seed layer.
Zinc-nitrate, HMT, citriate	Oriented nanocolumns, nanoplates	Control of aspect ratio: addition of citrate anions decreases aspect ratio.
Zinc nitrate, zinc acetate, HMT	Highly aligned nanorods	Influence of substrate and seed layer
Zinc nitrate, triethanolamine, HCI (pH 5)	Ordered nanorods	Influence of substrate and counter ions in growth solution

(*Contd.*)

(*Contd.*)

Zinc nitrate, thiourea, ammonium chloride, ammonia	Nanowires, tower-like, flower-like, tube-like	Influence of reactants, substrate pretreatment, and growth time and temperature.
Zinc acetate, sodium hydroxide, citric acid	Disk-like, flower-like, nanorods	Influence of pH on growth solution
Comparison of different growth solutions	Star-like, nanorods	Influence of reaction conditions: ligand, counter-ions, pH, ionic strength, and deposition time
		Influence of substrate/seed layer
Zn foil, zinc sulfate, ammonium ions, sodium hydroxide	Nanobelt arrays, ordered nanowires	Influence of temperature and concentration of solutions

8. Hydrothermal Process

Among a rich family of solution-based chemical processes, the hydrothermal process is used for the synthesis of a vast range of solid-state compounds such as oxides, sulfides, halides, zeolites, and other microporous phases. This is approach is explored for this synthesis shape and size, and even new molecules.

Conceptually an hydrothermal route is defined as the use of water as reaction medium in a sealed reaction container when the temperature is raised above 370 K. Autogeneous pressure which is self-developed and not externally applied is generated under these conditions. The pressure within the sealed reaction container increases dramatically with temperature and also depends on other experimental factors, such as the percentage fill of the vessel and any dissolved salts as illustrated in Figure 2.18. A high percentage fill allows access to pressures of hundreds of atmospheres when the temperature is below the critical point of water, and even below 470 K.

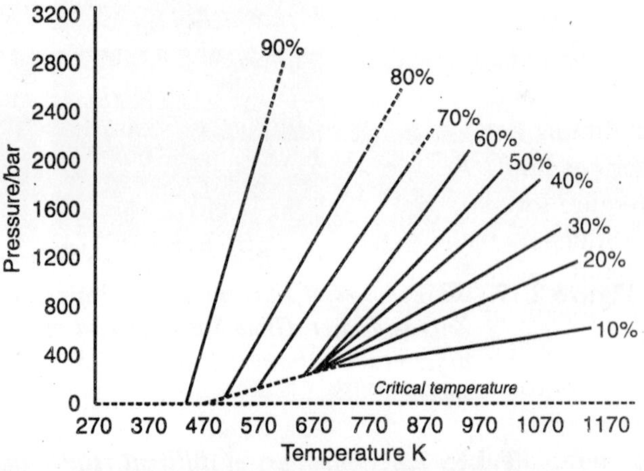

Figure 2.18 *Pressure as a function of temperature and percentage fill of water in a sealed vessel*

An advantage of the use of hydrothermal conditions is that it has significant effects on the reactivity of inorganic solids and the solubility of diverse compounds under conditions of elevated pressure and temperatures. Secondly, the chemical reactivity usually insoluble reagents are enhanced and a lot of solid state reactions are initiated under hydrothermal conditions. Hydrothermal synthesis is often applied higher temperatures above at 570 K and take place in the supercritical regime; however, the slow temperature route (temperature (<520 K) is usually applied for the synthesis of various well-known functional materials and even new solids.

The concepts embodied in hydrothermal process are extrapolated to non-aqueous systems, and the so-called '*solvothermal process* has emerged (Figure 2.19), in which an organic solvent is used as reaction medium instead of water at elevated temperature about its boiling point.

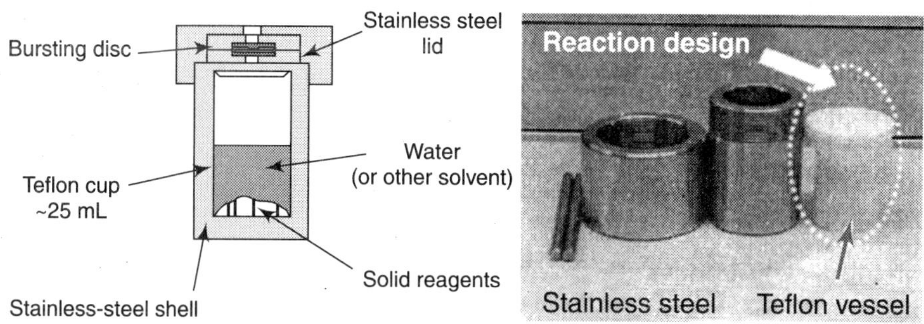

Figure 2.19 *Left: illustration of a Teflon-lined, stainless autoclave typically used in the laboratory to perform subcritical hydrothermal/solvothermal synthesis. Right: a picture of a set of the autoclave.*

9. Synthesis of Nanowires Via Porous Anodic Aluminium Oxide Template

The realization of the potential benefits of nanowires, have incidentally led to the development of a large number of fabrication techniques. Some of the important methods are electrodeposition, aquous solution growth, MOCVD technique, vapour-liquid-solid (VLS) growth hydrothermal synthesis and chemical synthesis. However, most of the above methods suffer from several serious limitations e.g. the conventional nanolithographic techniques are very cumbersome and require lot of efficiency and skill. Moreover, the techniques are also not suitable for large-scale production of nanowires with high aspect ratios. The hydrothermal and chemical synthesis techniques on the other hand are less attractive due to the difficulty in exercising control over the length and diameter of the nanowires during synthesis. For the above reason, template based synthesis of nanowires and nanorods have gained importance.

Self-ordered nanoporous membrane of molecular sieves track-etched polymer, nanochannel array glass, radiation track-etched mica and anodic aluminium oxide (AAO) are used as a mask or template for the nanowire fabrication. However, among the above mentioned templates AAO membrane is most attractive due to the regular pore distribution, high pore density and high aspect ratio of pores.

The AAO templates are fabricated easily and during fabrication precise control is exerted over the distribution, length and diameter of the pores. The AAO membranes are prepared by controlled electrochemical anodization, which involves oxidation of pure aluminium in an appropriate electrolyte. Anodization parameters such as anodization voltage, current, electrolyte bath temperature and composition are all adjusted during fabrication of the template to obtain the desired distribution, size and length of the pores. It is worthwhile to mention that by using different electrochemical deposition methods or by other processes such as evaporation, sol-gel method chemical vapour deposition melt or solution deposition and electroless deposition nanowires of different metals and semiconductor can be fabricated in the pores of this template membrane.

Considering the importance of the above aspects in the backdrop, herein the fabrication of nanowires by using the porous anodic aluminium oxide template is given below.

1. Preparation of AAO template

A thick porous aluminium oxide film is produced on the pure (99.999%) annealed aluminium by electrochemical anodization technique. Anodization is an electrolytic oxidation process in which the metal surface, when anodic is converted primarily to an adherent oxide coating having desirable thickness and structure. Oxide layers with different morphological features are formed by anodic oxidation of aluminium by varying the conditions during the electrochemical process, e.g. temperature, current density, voltage and duration of anodization.

a. *Surface preparation prior to anodization:* High purity Al foil (99.999% purity) is used for obtaining an ordered porous oxide film on aluminium by anodization, because the impurity atoms having different size and volume may induce internal stresses during of anodization, which can lead to the formation of defective structures in porous alumina template. Also Al foil is annealed for atleast 3–4 h at 770 K under nitrogen atmosphere, for the removal of crystal defects and promoting grain growth. A grain size between 100 and 200 μm is desirable for obtaining fit a uniform pore distribution during anodization.

Al foil with a thickness greater than 100 μm is cleaned in acetone prior to anodization by ultrasonic vibration. Surface impurities are removed by immersing the cleaned Al foil in a solution containing 1% HF, 10% HNO_3, 20% HCI, and 69% H_2O by volume.

The surface roughness of the annealed Al surface is reduced to nearly 3 nm by electropolishing the aluminium foil in a standard acidic electrolyte containing 165 ml of 65% $HClO_4$, 700 ml of ethanol, 100 ml of 2-butoxyethanol and 137 ml of H_2O at 50 V for 10 s.

b. *Anodization of aluminium:* The electropolished Al sample is anodized in an electrochemical bath using the sample as anode plate and Pt mesh, graphite, Al or stainless steel as the cathode. Aqueous sulfuric, oxalic or phosphoric acid is used as an electrolyte. During anodization, a thick porous alumina layer gradually grows over a very thin dense barrier oxide layer, which acts as an interface between the pure Al surface and porous alumina oxide layer. Here pure Al and few other metals like Nb, W, Zr etc which are prone to oxidation under ambient condition, develop a thin amorphous, insulating and inert oxide layer on the metal surface, (known as the barrier oxide layer). The average thickness of this barrier layer ranges between 150 and 180 Å.

But, for the preparation of ideally ordered porous structure the process parameters, such as anodization voltage, anodization current, electrochemical bath temperature, type and concentration of electrolyte are maintained. Different types of electrolytic baths used for the anodization are: anodization using 0.3M $(COOH)_2$ solution at 271 K for 2–4 h with an anodization voltage between 30 and 40 V by using 20 wt% (2 M) H_2SO_4 at 271 K with an applied voltage of 19 V for 1–2 days or 1 M H_3PO_4 under an applied voltage of 160 V for more than one day at low temperature in the range 271–280 K.

The pore diameter, cell dimension, pore distribution etc strongly depend on the applied voltage. In the anodization process the role of the anodization current is also considered to be very important. The current densities during the porous oxide film growth using 15% wt/vol.

H_2SO_4 with bath temperatures of 290, 295 and 300 K are 5, 15 and 35 mA cm^{-2}, respectively. Since the conductivity and pH values of different electrolytes are different and are influenced by the concentration, ranges of anodization voltage are also different for different electrolytes. However, in all the cases the anodization temperature is purposely kept low, preferably below the room temperature because, high bath temperature enhances the dissolution of porous oxide film in the electrolyte.

It is found that instead of one-time (single step) anodization, two-step anodization is helpful in producing the ideally ordered porous structure, for example, in the first step, Al is oxidized in 0.3 M oxalic acid at 40 V below room temperature for 15 min. The porous oxide layer so formed over the Al sample is removed by chemical etching in a mixture of 0.2 M chromic acid and 0.4 M phosphoric acid at a bath temperature of 330 K. Removal of the porous film from the Al sample leaves behind an ordered concave textured structure on the sample surface. This 'textured' Al is again anodized in the same electrolyte for a day or more. During this second step anodization process, the concave textured pattern on the aluminium sample surface acts as the sites for deep pore growth and is helpful in growing the porous oxide film uniformly.

A pre-textured Al sample can also be used to fabricate uniformly ordered alumina oxide film by anodization. In this method, a hexagonally ordered convex pattern is produced on the hard SiC using electron beam lithography technique. This patterned SiC is used as the master mould. This SiC mould is indented into the Al sample surface at high pressure (~2800 kg cm^{-2}) before anodization is carried out. This process helps to get an ideally distributed shallow order concave pattern on the Al sample surface. When the sample is anodized, the shallow ordered concave patterns act as the site for deep pore growth with uniform, similar morphological and geometrical features.

c. *Mechanism of pore growth and reactions:* During the anodization process a constant voltage is applied between the electrodes. Therefore, the applied electric field lines lie perpendicular to the sample surface (anode) towards the cathode. This applied electric field increases the dissolution of the barrier oxide layer and thus favours pore growth. Due to the competition that prevails among the growing pores, few of them stop growing while others grow. The pores grow perpendicular to the surface when the field-enhanced dissolution of the oxide at the electrolyte/oxide interface is equilibrated with the growth of oxide that occurs at the oxide/metal interface. Oxidation occurs at the metal/oxide interface by the migration of oxygen containing ions (O^{2-} or OH^-) from the electrolyte by the following reaction

$$2Al + 3O^{2-} \rightarrow Al_2O_3 + 6e^-$$

Dissolution of the oxide layer is caused mainly by the hydration reaction of the formed oxide layer as shown below

$$2Al^{3+} + 3H_2O + Al_2O_3 + 6H^+.$$

It is found that under the applied electric field the Al^{3+} ions migrate through the barrier oxide layer. This electric field also helps in the transport of the oxygen containing ions (O^{2-} or OH^-) from the electrolyte into the barrier oxide layer towards the Al substrate. The representation of the ionic migration at the pore bottom under the applied electric field is given in Figure 2.20. The electrons are ejected into the electrolyte and they combine with hydrogen ions to generate hydrogen gas at cathode.

Barrier oxide growth without pore formation occurs when the Al^{3+} ions reach the electrolyte/oxide interface and contribute to the oxide formation. Porous alumina oxide layer formed when the Al^{3+} ions drift through the oxide layer and are ejected into the solution at the oxide/electrolyte interface

$$2Al \rightarrow 2Al^{3+} + 6e^{-}.$$

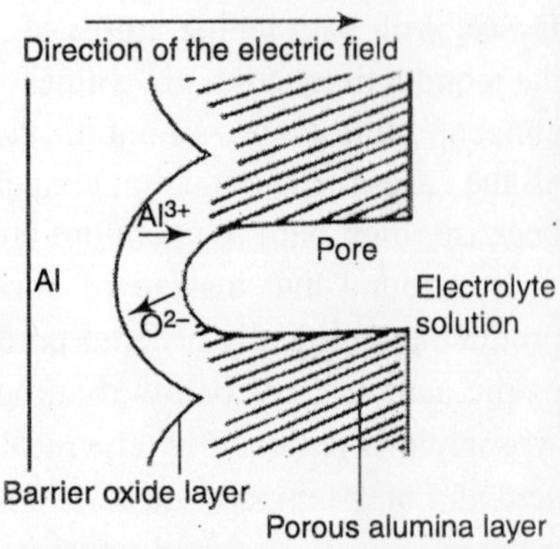

Figure 2.20 *The ionic migration under applied electric field.*

Loss of Al^{3+} ions in the electrolyte is found to be a prerequisite for porous oxide growth. Thus at steady state pore grows perpendicular to the Al surface when an equilibrium is established between the field enhanced dissolution of oxide layer at oxide/electrolyte interface and the oxide growth at metal/oxide interface.

During the anodization of aluminium, the porous layer formed is of Al_2O_3. The atomic density of aluminium in Al_2O_3 is lower than that of metallic aluminium by a factor of two. The volume expansion of Al_2O_3 is twice that of the original aluminium metal. This volume expansion takes place at the metal/oxide interface leading to a compressive stress during oxide formation. This stress is the origin of the repulsive forces that appear between the neighbouring pores and it is believed that volume expansion in the vertical direction pushes the pore wall upward.

d. *Morphology and geometrical structure of porous aluminium oxide:* Atomic force microscopic (AFM) techniques are used to show the three-dimensional surface topology of the porous film and scanning tunneling microscopic (STM) image gives an idea about the geometrical structure in real space. It is found that the pore diameter ranges from 4-200 nm, pore length from 1–50 μm and pore density in the range of 10^9–10^{11} cm^{-2}. This depends on parameters eg. Composition the electrolyte, applied voltage etc.

This porous anodic alumina produced during anodization has hexagonal closed packed nearly honeycomb like structure as shown in Figure 2.21. The pores appear at the centre of each hexagonal columnar cell. The channel like pores lie perpendicular to the aluminium sample surface. However, the geometry of AAO film obtained by anodization process is far from that of the idealized model as the arrangement of the cells is not perfectly hexagonal (see Figure 2.22).

2. *Fabrication of nanowires by pore filling through electrodeposition technique*

Methods that involve direct pore filling of the AAO template are considered to be straightforward and versatile techniques for synthesizing nanowires, are described below.

(a) *Pressure injection:* In this technique low melting point metals and semiconductors are melted and injected into the pores of AAO template by applying high pressure. The material which

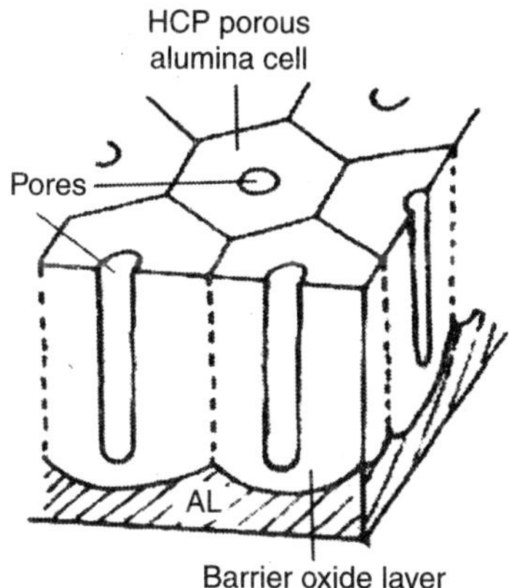

Figure 2.21 *The honeycomb like porous alumina structure.*

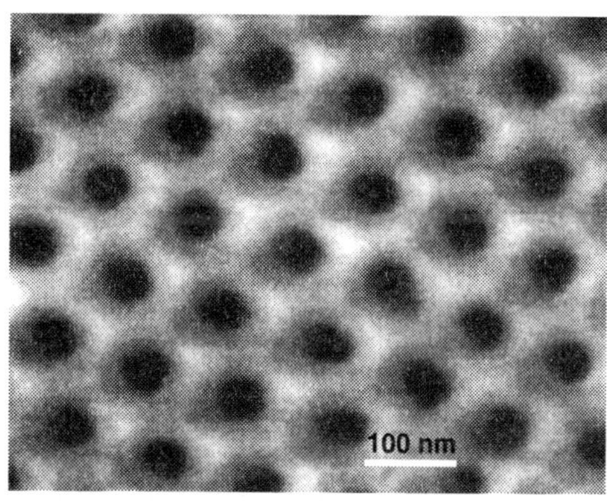

Figure 2.22 *SEM image of the top surface of AAO membrane*

infiltrates into the pores is allowed to solidify and form nanowires. The nanowires are extracted by dissolving the alumina template chemically. Nanowires of Bi, Sn, In, Al, Te, Se, GaSb and Bi_2Te_3 are fabricated by this technique.

(b) *Vapour deposition:* Physical and chemical vapour deposition techniques are often employed for pore filling and fabricating nanowires. Bi nanowires in AAO template is synthesized by introducing Bi vapour into the pores of the AAO template, which is then solidified by cooling. Compound semiconductor nanowires of GaN are also synthesized in AAO template by gas phase reaction between Ga_2O vapour and ammonia.

(c) *Vapour-liquid-solid (VLS) method:* This method involves absorption of source material from the gaseous phase into a liquid droplet that acts as a catalyst. Primarily a liquid alloy is formed which becomes supersaturated and precipitates out the source material by nucleation. The precipitate acts as a preferred site for further deposition and growth of the material at the interface of the liquid droplet, favouring growth of the precipitate into a nanowire by suppressing further nucleation on the same catalyst. It is found that the diameter of the nanowire depends on the diameter of the liquid alloy droplet, porous alumina membrane is used as an alternative to control nanowire diameter in the VLS process. Defect free Si nanowires with diameters in the range of 4–5 nm and lengths of several microns are synthesized by using a supercritical fluid solution phase approach where alka-nethiol coated Au nanocrystal (2.5 nm in diameter) are used as seeds to direct the one dimensional crystallization of Si in a solvent heated and pressurized above its critical point. The reaction pressure controls the orientation of the nanowires.

(d) *Pulsed laser deposition:* This technique is used to deposit materials on top of the porous AAO template. The material deposits inside the pores and is shaped into nanowires and the material that is deposited in the nonporous region that separate adjacent pores, assume different morphologies depending on the nature of the surface. It is found that ordered arrays of ZnO nanodots with an average diameter of 60 nm and a periodicity of 100 nm are fabricated by PLD using AAO template. Oxygen pressure used in PLD photoluminescence (PL) at 380 nm is observed in the ZnO nanodots.

(e) *Chemical conversion:* Synthesis of nanowires by chemical reaction in the pores of AAO template is another important approach. Initially nanowires of the constituent element is prepared, which is then reacted with chemicals containing the desired element to form the final product. Ag_2Se nanowires are synthesized by reacting crystalline Se nanowires with aqueous $AgNO_3$ at room temperature.

ZnO nanowires are prepared by oxidizing metallic zinc nanowires that are prepared by electrodeposition in the pores of AAO template prior to their oxidation in air at 570 K. The free standing ZnO nanowires are extracted from the template by selective dissolutions.

Hollow nanotubes of MoS_2 ~30 nm long with 50 nm external diameters are prepared by filling the pores of the AAO template with a mixture of molecular precursors, $(NH_4)_2MoS_4$ and $(NH_4)_2MO_3S_{13}$. The filled template is then heated at an elevated temperature to thermally decompose the precursor into MoS_2.

(f) *Colloidal dispersion filling:* Oxide nanowires, nanorods and nanotubes are synthesized by filling the template with colloidal dispersions. The colloidal dispersion is generally prepared through the sol-gel technique. The porous template is placed in a stable sol for appropriate length of time to drive the sol into the pores by capillary force. The filled template is then withdrawn from the sol and dried prior to firing for densification of the sol-gel derived nanowires, nanorods or nanotubes.

(g) *The Electrodeposition method:* Inspite of the fact that all the above mentioned techniques are advantageously used for the deposition of nanostructure materials (NSMs) into the pores of the nanoporous membrane, electrodeposition or electrochemical deposition is regarded as one of the most popular methods of pore filling with conducting metals to obtain continuous arrays of nanowires with large aspect ratios. Electrochemical deposition route is easy, low-cost as well as less skill dependent compared to other techniques mentioned above. Structural analysis shows that the electrodeposited nanowires tend to be densely packed, continuous and highly crystalline. Moreover, by simply monitoring the total amount of passing charge one can precisely control the aspect ratios of the metal nanowires.

(i) *Electrodeposition bath set up:* In the electrodeposition method the electrochemical bath set up is similar to that used for the electro-plating process. The porous alumina membrane with natural aluminium-substrate support is used as the electrode. Sometimes the anodic alumina oxide (AAO) membrane is separated out from the Al-substrate and a thin metal film of Ag or Au is deposited in the membrane by evaporation technique. This thin metal film supported membrane acts as the active electrode. The appropriate metal salt solution, containing the metal ions is used as the electrolyte in the electrodeposition process.

(ii) *Electrodeposition method:* The thin barrier oxide layer underneath the pores of AAO membrane has high resistivity (10^{10}–10^{12} Ωcm). A.C. electrodeposition technique is preferred because direct current cannot pass through the barrier oxide layer. Electrodeposition depends on different process parameters such as deposition current amplitude, frequency of current (voltage), electrochemical bath temperature and pH value of the electrolyte. In a.c. electrodeposition the anodic alumina template acts like a 'rectifier' when alternating current (voltage) is applied between the template and other electrodes. The ionic conduction starts when the template electrode acts as a cathode (during cathodic half cycle). Under this condition the cations from the electrolyte layer that remain attached to the template electrode diffuse to the pore bottom and are electrodeposited. On the other hand, while the template becomes anodic (during anodic half cycle), ionic conduction is cut off and the material deposited in the pore bottom of the membrane is not re-anodized. Here, note that during the cathodic half cycle the electrolyte layer, which is attached to the template electrode becomes depleted of cation, as the ion diffusion from the rest of the electrolyte to this thin attached layer is very slow. Therefore, the anodic half cycle is helpful in the sense that during this interval or period cations get the time to diffuse into the thin electrolyte layer, from rest of the electrolyte solution making electrodeposition possible during the next cathodic half cycle.

The d.c. pulsed electrodeposition is also used for pore filling, however, this process has the ability to fill-up only 10-20% of the pores of the membrane. On the other hand, a.c. electrodeposition is capable of high pore filling ratio.

(iii) *Electrodeposition of nanowires:* Metallic nanowires of Cu, Ag, Au and different transition metals such as Fe (Al and their alloys are fabricated by the electrodeposition method by filling the pores of the alumina template. Nanowires of different semiconductors, super-lattice and superconductors such as Pb is also be fabricated. SEM image of silver nanowires fabricated in alumina template by electrodeposition is shown in Figure 2.23.

Figure 2.23 *SEM image of Ag nanowires separated from the alumina template.*

It is also observed that Fe nanowires are deposited into the pores of the template by using a.c. pulse. The electrolyte solution contains 120 gL^{-1} of $FeSO_4 \cdot 7H_2O$; 45 gL^{-1} of H_3PO_3; 1 gL^{-1} of ascorbic acid and 3 ml L^{-1} of glycerin. The pH value of the solution is maintained at ≈ 3 throughout the electrodeposition process. Similarly, Co nanowires are fabricated from $CoSO_4$ salt solution containing 400 gL^{-1} of $CoSO_4 \cdot 7H_2O$ and 40 gL^{-1} of H_3BO_4 as electrolyte. In both the Fe and Co electrodeposition processes, a.c. voltage pulse of 10-20 V with frequency ranging from 50–300 Hz is used. Also by using highly concentrated electrolyte like 300 gL^{-1} $NiSO_4 \cdot 6H_2O$ or 45 gL^{-1} $NiCl_2 \cdot 6H_2O$ and 45 g of H_3BO_4 solution of pH 4.5, Ni nanowires are deposited into the pores of the alumina template under a negative current pulse of –25 mA cm^{-2} for a duration of 6 ms followed by a positive voltage pulse of +4 V for 2 ms only. Semiconductor nanowires of CdS are fabricated by electrodeposition method using the alumina oxide template. It is found that single crystal compound semiconductor nanowires of CdS, CdSe and CdTe can also be fabricated into the porous alumina template by electrodeposition.

Now, Co–Cu superlattice nanowires are synthesized by varying the cathodic potential in the electrolyte solution containing two different metal ions. These nanowires contain periodic modulation of composition along the wire axes. After pore filling with required materials, the aluminium substrate is removed by $HgCl_2$ solution. The periodic arrays of nanowires are separated out of the template by dissolving AAO membrane in hot 6 M NaOH solution.

10. TEMPLATE-DIRECTED SYNTHESIS

Template directed synthesis represents a straightforward route to 1D N.W. In this approach the templates serve as a scaffold within (or around) which a different material is generated and shaped into a nanostructure with its morphology complementary to that of the template. Examples are step edges present on the surfaces of a solid substrate; channels within a porous material: mesoscale structures self-assembled from organic surfactants or block copolymers biological macromolecules such as DNA strains or rod-shaped viruses; and existing nanostructures synthesized using other approaches. When the template is involved physically, it is necessary to remove the template using post-synthesis treatment (such as chemical etching and calcination, in order to harvest the resultant nanostructures. In a chemical process, the template is consumed as the reaction proceeds and it is possible to obtain the nanostructures as a pure product. The template-effective procedure and allows the complex topology (present on the surface of a template) to be duplicated in a single step. As a major drawback, nanostructures synthesized using template-directed methods are polycrystalline, and the quantity of structures produced in each run of synthesis is relatively limited. Given below are four templating methods, with a focus on their capability, feasibility, and potential extension.

1. Templating Against Features on Solid Substrates

Relief structures present on the surface of a solid substrate serves as a class of natural templates for generating supported 1D nanostructures. In this regard, microstructures that are conveniently patterned on the surface of a solid substrate using lithography and etching are exploited as templates to fabricate nanowires made of various materials. Decoration of these templates (usually their edges) with a different material, for example, provides a powerful route to the formation of nanowires from various metals and semiconductors. A metal nanowires as thin as 15 nm is prepared by shadow sputtering (Fig. 2.24(a)) a metal source against an array of V-grooves etched on the surface of a Si(100) wafer. Also, metal or semiconductor is applied at normal incidence using techniques based on vapor-phase deposition (e.g., MBE) or solution-phase electrochemical plating, and then allowed to reconstruct into 1D nanostructures at the bottom of each V-groove (Fig. 2.24(b)). Using this approach,

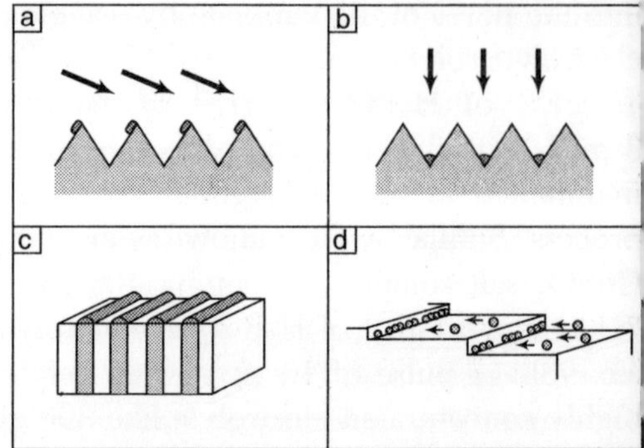

Figure 2.24 *Illustrations of procedures that generated 1D nanostructures by (a) shadow evaporation (b) reconstruction at th ebottom of V-grooves (c) cleaved-edge overgrowth on teh cross-sectopn of a multi-layer film; and (d) emplating against step edges on the surface of a solid subatrate.*

continuous thin nanowires with lengths up to hundreds of micrometers are prepared as parallel arrays on the surfaces of solid supports that is subsequently released into the free-standing form or transferred onto the surfaces of other substrates. It is possible to fabricate large parallel arrays of Ge nanowires by templating against V-grooves etched in the surfaces of Si(100) substrates. 3D arrays of iron nanowires are also fabricated by templating against relief features present on the (110) surfaces of NaCl crystals.

The cross-sections of multilayer films prepared using MBE are also exploited as templates to grow simple patterns of quantum structures from many kinds of metals and semiconductors Fig. 2.24(c). This technique is known as cleaved-edge overgrowth (CEO), and it takes advantage of the high accuracy of MBE in controlling the layer thickness of a super lattice. In this technique, a superlattice consisting of alternating layers (made of, e.g., AlGaAs and GaAs) is fabricated by MBE, and then cleaved through the thickness of the multilayer structure to produce an atomically clean surface. Next, MBE or electrochemical deposition is used to grow epitaxial layers on selected regions of the exposed surface. This approach gives intersecting quantum wells with atomic or angstrom-level control over the thickness in two directions. Devices such as quantum-wire lasers formed from the intersection of ~7 nm wide quantum wells are fabricated using this technique. For most fabrication tasks, CEO provides quantum structures with more uniform morphologies than e-beam or optical lithography, because CEO inherits the atomic uniformity and precision of MBE. This technique is, however, limited to those structures that are fabricated along the natural cleavage directions of a substrate and along lattice planes on which MBE growth occurs preferentially. In addition, .the structures must be built up from intersecting planes of a material. Although this technique is most commonly used and has the highest resolution when combined with MBE, the same basic approach is suitable for use with multilayer films grown using many other deposition techniques.

The growth of metal nanowires by templating against the steps present on a highly oriented, pyrolytic graphite using electrodeposition (Fig. 2.24(d)) is achieved. Two different types of materials are used from noble metals (e.g.. Pd, Cu, Ag, and Au) and electronically conductive metal oxides (MoO_x, MnO_2, CU_2O, and Fe_2O_3) that can be subsequently reduced to the corresponding metals (Mo, Mn, Cu, and Fe) by hydrogen gas at elevated temperatures. The nanowires nucleate and grow along the step edges present on a graphite surface into a 2D parallel array which is then transferred onto the surface of a cyanoacrylate film supported on a glass slide. For the formation of metal nanowires through an oxide precursor, the dimensional uniformity and hemicylindrical shape of the parent oxide wires retained in the H_2 reduction process, although the diameter is often reduced by as much as ~35 %. In addition to physical features present on the surface of a solid substrate, strains generated during film deposition is exploited as physical templates to direct the organization of a deposited material into arrayed 1D nanostructures.

2. *Channels in Porous Materials*

Channels in porous membranes provide another class of templates for use in the synthesis of 1D nanostructures (Fig. 2.25). Tho types of porous membranes are commonly used in such syntheses: polymer films containing track-etched channels and alumina films containing anodically etched pores. For track-etching, a polymer film (6-20 μm thick) is irradiated with heavy ions (from nuclear fission) to generate damaged spots in the surface of this film. These spots are then amplified through chemical etching to generate uniform, cylindrical pores penetrating the membrane film. The pores fabricated using this method are randomly scattered across the membrane surface; and their orientation is be tilted by as much as 34° with respect to the surface normal. Porous alumina membranes are prepared using anodization of aluminum foil in an acidic medium, and contain a hexagonally packed 2D array of cylindrical pores with a relatively uniform size. Unlike the polymer membranes fabricated by track-etching, the pores in alumina membranes have little or no tilting with respect to the surface normal and the ore density is also much higher.

A variety of materials are used with this class of templates, example metals used for semiconductors, ceramics, and organic polymers. The material is loaded into the pores using a method based on vapor-phase phase sputtering, liquid-phase injection, or solution-phase chemical or electrochemical deposition. In addition to vapor-phase evaporation and solution-phase deposition, metals (such as Bi) with relatively low melting points is directly injected as liquids into the pores of an anodic alumina membrane and subsequently solidified into highly crystalline nanowires. The use of electrophoretic deposition to fill the pores of a polymeric or alumina membrane with charged sols generated from a sol-gel precursor is made. Subsequent firing at elevated temperatures lead to the formation of uniform, ceramic nanorods with compositions such as titania and $Pb(Zr,Ti)O_3$ (PZT). In the early stage of this process, the material deposits as uniform layers on the walls of these pores to form tubular nanostructures instead of solid rods (Fig. 2.25). In both cases, the resultant nanostructures exist as well-aligned arrays within the pores, or are released from the templates and collected as an ensemble in the free-standing form.

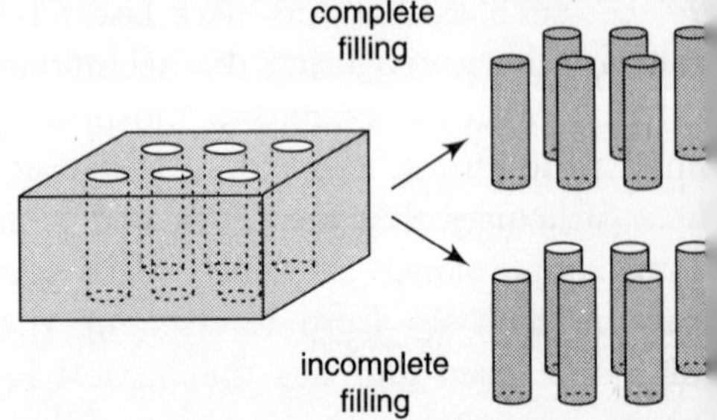

Figure 2.25 *Formation of nanowires and nanotubes by filling and partial filling the pores within a porous membrane with the desired material or a precursor to this material.*

Although the nanowires synthesized using this method are polycrystalline, single crystals are also obtained under carefully controlled conditions. For example, the use of electroless deposition in generating single-crystalline nanowires of silver in the channels of a polycarbonate membrane via a self-catalyzed process and that pulse electrodeposition can be exploited to selectively grow either single-crystalline or polycrystalline copper nanowires. It is observed that formation of single-crystalline wires of Pb require a significant departure from the equilibrium condition (e.g., with a greater overpotential) than those required for the formation of polycrystalline samples. Also observed is that titania nanowires synthesized by using the electrophoretic deposition method become single crystals once their diameters are reduced below 15 nm. The major advantage associated with membrane-based templates is that both the dimensions and compositions of nanowires are easily controlled by varying experimental conditions. This mak method is also exploited to generate nanowires containing bands of different metals with well-defined dimensions.

In addition to macroporous membranes, mesoporous materials containing much smaller, 1D channels (1.5–30 nm in diameter) are explored as physical templates to generate ultrafine nanostructures. Two types of mesoporous silica are used and examined as the templates: the MCM series (e.g., MCM-41) and the SBA family (e.g.. SBA-15). Both of them contain hexagonal arrays of mesopores. Preparation of 1D nanostructures in these mesoporous materials involve three steps: infiltration of the pores with an appropriate precursor via a vapor- or solution-based approach, conversion of this precursor to the desired material, and recovery of the 1D nanostructures by selectively removing the template. Materials incorporated into this process include noble metals (e.g., Ag, Au, Pt, and Pd), and bimetallic alloys (e.g., Au/Pt). Each individual 1D nanostructures obtained from these templates is a polycrystal. Most of them are characterized by relatively low aspect ratios, and some of them even exist as discrete nanoparticles as a result of a high volume shrinkage involved in the thermal conversion of precursor.

3. Templating Against Self-Assembled Molecular Structures

Mesophase structures self-assembled from surfactants provide another class of useful and versatile templates for generating 1D nanostructures in relatively large quantities (Fig. 2.26). It is well known that surfactant molecules spontaneously organize into rod-shaped micelles (or inverse micelles) when their concentration reaches a critical value. These anisotropic structures are used as soft templates to promote the formation of nanorods when coupled with an appropriate chemical or electrochemical reaction. The surfactant molecules are removed to collect the nanorods as a relatively pure sample. Based on this principle, the synthesis of $BaCrO_4$, $BaSO_4$, and $BaWO_4$ nanorods with monodispersed dimensions is done. The synthesis of gold nanorods by templating against rod-like micelles assembled from cetyltrimethylammonium bromide

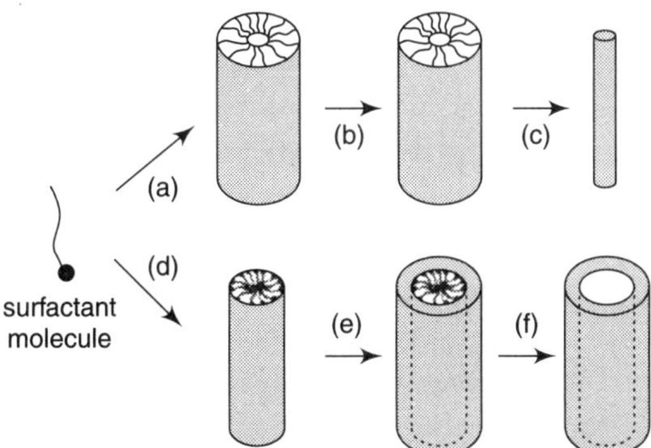

Figure 2.26 *Illustrations showing the formation of nanowires by templating against mesostructures self-assembled from surfactant molecules: (a) formation of a cylindrical micelle; (b) formation of the desired material in the aqueous phase encapsulated by the cylindrical micelle; (c) removal of the surfactant molecules with an appropriate solvent (or by calcination) to obtain an individual nanowire. (d-f). Similar to the processes illustrated in (a-c), except that the exterior surface of an inverted micelle serves as the physical template.*

(CTAB) and another hydrophobic cationic surfactant (e.g., tetraoctylammonium bromide (TOAB) can also be done. The gold is generated through electrochemical dissolution within a cell that contained a gold anode and a platinum cathode. This method is used to synthesize gold and silver nanorods with well-controlled aspect ratios and plasmon properties. Now, recently, a photochemical route is developed for the preparation of uniform gold nanorods with similar aspect ratio. Another useful route to gold nanorods is demonstrated, in which seed-mediated growth is used to generate metal nanorods with a controllable thicknesses and aspect ratios. In a typical synthesis, gold or silver nanoparticles of 3–5 nm diameter are added as seeds to a solution that contains rod-like micelles (assembled from CTAB) and a metal precursor such as $HAuCl_4$ or $AgNO_3$. When a weak reducing agent (e.g., ascorbic acid) is added, the seeds serve as nucleation sites for the growth of nanorods within the confinement of micelle structures. The lateral dimensions and aspect ratios of these nanorods are controlled by varying the ratio of seeds relative to the metal precursors. Also, arrayed crystalline nanowires of Ag with a relatively high aspect ratio are fabricated electrospinning by which is performed in a micellar phase composed of sodium bis(2-ethylhexyl)sulfosuccinate (AOT), *p*-xylene, and water. In this case, the rod-shaped micelles are assembled into a hexagonal liquid-crystalline phase (normal or reverse) by increasing the concentration of surfactants. The major advantage of this method is that metal nanowires are synthesized as regular arrays rather than randomly oriented samples.

Most recently, single-crystalline W nanowires are synthesized by templating WO_4^{2-} ions against the lamellar phase of CTAB, followed by pyrolysis in vacuo. Metal nanowires are synthesized in large quantities by using this class of templates, however the preparation and removal of the micellar phase is often difficult and tedious.

Block copolymers i.e. polymers formed by connecting two or more chemically distinct segments (or blocks) end-to-end with a covalent bond–are exploited as soft templates to generate 1D nanostructures. When the chemically distinct polymers are immiscible, a large collection of these chains separate into different phases. Under appropriate conditions (e.g., with a proper ratio between the molecular weights of different segments), regular arrays of cylinders are formed, with a structure similar to that of self-assembled surfactants. Different regions in such an arrayed structure is decorated with the precursor to a metal (or semiconductor) through physical adsorption or chemical coordination, baking block copolymers a powerful system of templates for the synthesis of 1D nanostructures. A range of different block copolymers are tested for the synthesis of silver nanowires, examples include carbosilane dendrimers and polyisocyanopeptides, and the double-hydrophilic, poly(ethylene oxide)-block-poly(methacrylic acid). Because the blocks are easily designed with functional groups to selectively interact with any specific metal ion or compound, the range of materials incorporated into this templating procedure is potentially very broad. Similar to the nanowires synthesized using other soft templates, the product obtained in this case is polycrystalline and aggregates into bundles.

4. Templating Against Existing Nanostructures

Currently existing nanowires are useful as templates (physical or chemical) to generate nanowires and other types of 1D nanostructures from various materials, some of which are difficult (or impossible) to directly synthesize as uniform samples. In an approach, the surfaces of these nanowires is directly coated with conformal sheaths made of a different material to form coaxial nanocables. Subsequent dissolution of the original nanowires leads to the formation of nanotubes. For instance, directly coated gold nanorods with polystyrene or silica (5-10 nm in thickness) are obtained which form cable-like nanostructures. Layer-by-layer (LBL) deposition of oppositely charged species on nickel nanorods is also adopted to prepare nanocables and composite nanotubes.

In addition, the sol-gel coating method is a generic route to coaxial nanocables that contain electrically conductive cores (made of metals) and insulating sheaths (in the form of amorphous silica or other dielectric materials). Figure 2.27(a) shows the TEM image of a typical sample of Ag/SiO_2 coaxial nanocables obtained by coating silver nanowires with silica derived from a sol-gel precursor. The thickness of sheath is controlled in the range from 2 to 100 nm by varying the concentration of the precursor and/or the deposition time. Selective removal of the silver cores (by etching in an ammonia solution) yield silica nanotubes with well-controlled dimensions and uniform wall structures (Fig. 2.27(b). These coaxial nanocables find use as ideal building blocks to generate 2D and 3D periodic structures through Langmuir-Blodgett self-assembly.

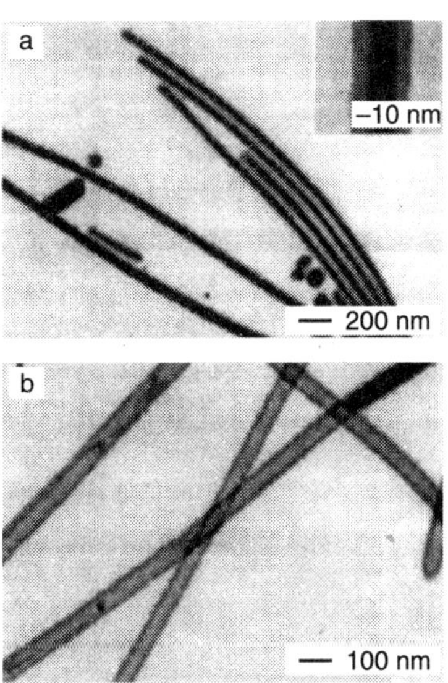

Figure 2.27 *(a) TEM images of Ag/SiO_2 coaxial nanocables that were prepared by directly coating silver nanowires with an amorphous silica sheath using the solgel method. (b) A TEM image of silica nanolubes prepared by selectively dissolving the silver cores of Ag/SiO_2 nanocables in an ammonia solution with ~pH 11.*

Single-crystalline nanowires (pre-synthesized using methods such as thermal evaporation or laser ablation) serve as substrates for the epitaxial growth of another solid to generate coaxial, bilayer nanotapes having sharp structural and compositional interfaces, example, synthesizing TiO_2/SnO_2 (Fig. 2.28(a)) and $Co_{0.05}Ti_{0.95}O_2/SnO_2$ nanotapes. The latter structures are ferromagnetic at room temperature. Semiconductor core-shell and multishell nanowire heterostructures are synthesized via epitaxial growth by modulating the composition of reactant gases in sequential steps. These demonstrate the possibility to incorporate many functions (e.g., luminescent, ferromagnetic, ferroelectric, piezoelectric, and superconducting) into an individual nanowire that find new applications in various areas.

In addition to nanowires, CNTs are also exploited as another type of physical template to generate nanorods or tubes from many materials. For example, CNTs are explored to fabricate metal nanowires through direct vapor evaporation. The pre-deposition of a thin layer of Ti is

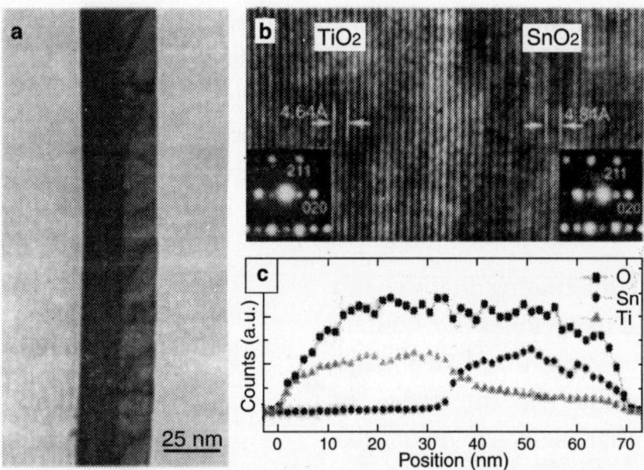

Figure 2.28 *(a) TEM image of a TiO_2/SnO_2 nanolape obtained through epitaxial growth of TiO_2 on a single-crystalline SnO_2 nanobelt. (b) A high-resolution TEM image of the atomically sharp TiO_2/SnO_2 interface. The fringe spacings of 4.64 and 4.84 Å correspond to the interplanar distances between the (010) planes of TiO_2 and SnO_2 (in the rutile structure), respectively. The insets show electron diffraction patterns taken from each side of the interface along the same zone axes of (c). Compositional line profile across the TiO_2/SnO_2 interface in the direction perpendicular to the long axis of the nanotape.*

found to be critical to the formation of continuous nanostructures from metals such as Au, Pd, Fe, Al, and Pb, because Ti improves the wettability of nanotube surfaces. Otherwise, direct deposition of these metals only lead to the formation of discrete particles as a result of dewetting. Since CNTs as long as half a centimeter have been synthesized, this approach is adopted to fabricate relatively long nanowires from a rich variety of materials. Also chain-like bio-molecules are made to guide the assembly of metal ions into linear arrays through the interaction between their functional side groups and the metal ions. These metal ions subsequently reduce to form a string of nanoparticles along the backbone of each biomolecule. If a reservoir of the metal ions are present, these nanoparticles further connect to generate continuous nanowires. For instance, electrically conductive nanowires of Ag and Pt are produced using this approach by templating against DNA strands. These metal nanowires serve as interconnects to fabricate simple electronic circuits in the prototype form. One of the major advantages with this type of template is that it generates complex patterns with arbitrary designs using many tools that are well developed in biochemistry.

Some nanostructures are converted to other materials without changing their morphology when they react with appropriate reagents under carefully controlled conditions. The concept of this method is evident from the thermal oxidation of silicon nanostructures, in which silicon is transformed into various silicon oxides. This effect provides a promising route to 1D nanostructures that, otherwise is difficult to directly synthesize or fabricate. With the concept of this method highly crystalline nanorods of metal carbides are formed by reacting CNTs with vapors of metal oxides or halides at elevated temperatures. A similar templating procedure produces crystalline nanorods of GaN, GaP, and SiC. In addition to CNTs, boron nanowires are used as templates to form highly crystalline nanowires of MgB_2, a material that exhibits interesting superconductive properties.

A number of solution-phase reactions are also demonstrated to transform existing nanowires into 1D nanostructures with other chemical compositions. For example, free-standing nanowires of noble metals (e.g., Au, Ag, Pd, and Pt) are prepared through a redox reaction that involves $LiMo_3Se_3$ molecular wires (serving as the reducing agent) and aqueous solution containing metal ions (e.g., $AuCl_4^-$, Ag^+, $PdCl_4^{2-}$, and $PtCl_4^{2-}$). Using a similar approach, highly crystalline nanotubes (Fig. 2.29) of noble metals such as Au, Pd, and Pt are synthesized galvanic displacement reactions between Ag nanowires and appropriate precursors of these metals in the aqueous medium. For example, when silver nanowires are dispersed into an aqueous $HAuCl_4$ solution, the immediately oxidize to silver ions. The resultant Au atoms are confined to the vicinity of the template surface. Once their concentration reaches a critical value, the Au atoms nucleate and grow into small clusters, and eventually evolve into a sheath-like structure around the silver template. This reaction is initiated on the facets with the highest surface energy and then proceed to those with lower energies. As a result, the thin sheath formed in the early stage is incomplete, and therefore it is possible for both reactants and products to diffuse across this layer until the silver template is completely consumed. If the reaction is continued with refluxing at an elevated temperature, the wall of each gold tube is reconstructed into a highly crystalline structure via processes such as Ostwald ripening. At the same time, the openings in the wall is also closed to form a seamless gold nanotube bounded by smooth surfaces. Based on the stoichiometry of the reaction, the thickness of the gold nanotube is one ninth of the lateral dimension of the corresponding silver template.

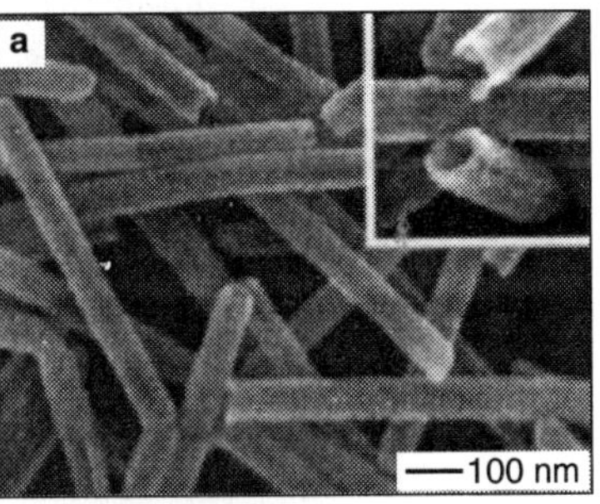

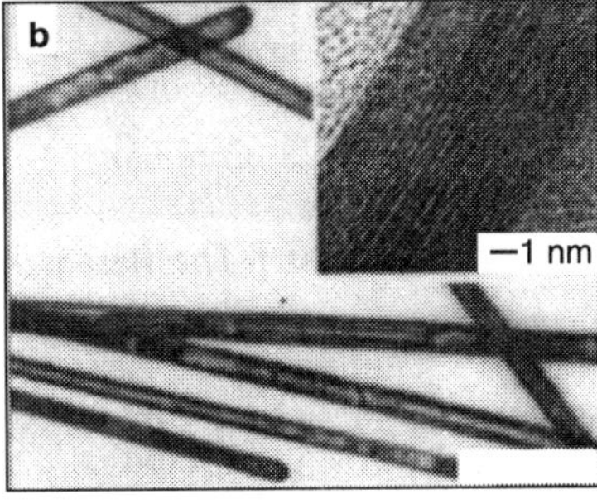

Figure 2.29 *SEM images of Pd nanotubes generated by reacting silver nnanowires with an aqueous $Pd(NO_3)_2$ solution. The nanotubes are broken via sonication for a few minutes to expose their cross-sections. (b) A TEM image of Au nanotubes prepared by reacting silver nanowires with an aqueous $HauCl_4$ solution. The inset shows a high-resolution TEM image of the edge of an individual gold nanotube, indicating its highly crystalline structure and uniformity in wall thickness.*

Single-crystalline nanowires of Ag_2Se (Fig. 2.30) synthesized through a room-temperature, topotactic reaction between single-crystalline nanowires of t-Se and aqueous $AgNO_3$ solutions. In this template-engaged process, the silver ions diffuse into the lattice of t-Se and form Ag_2Se without involving significant reorganization of the Se atoms. An interesting, diameter-dependent phase transition is observed in this new nanowire system: The Ag_2Se nanowires are found to

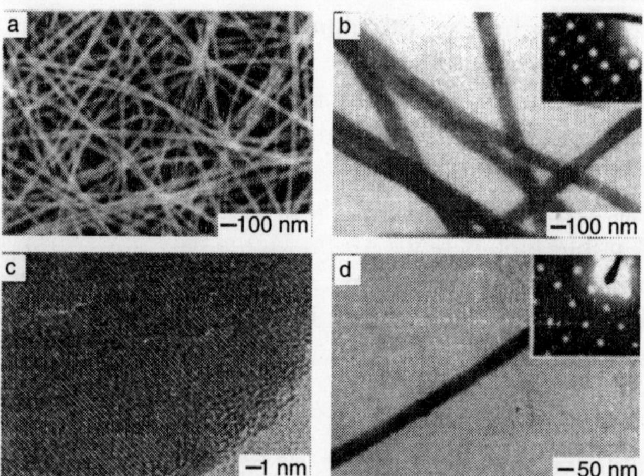

Figure 2.30 *An SEM image of a-Ag_2Se nanowires obtained by reacting t-Se nanowires with an aqueous $AgNO_3$ solution. (b) A TEM image of a-Ag_2Se nanowires and the electron diffraction pattern (inset) taken from the middle portion of an individual nanowire. The diffraction spots can be indexed to the orthorhombic structure. (c) A high-resolution TEM image obtained from the edge of an individual wire, indicating its single crystallinity. The fringe spacing of 0.35 nm corresponds to the interplanar distance of {200} planes, implying that the growth direction of this nanowire was <100>. (c) A TEM image of an Gag$_2$Se nanowire of ~50 nm in diameter. This wire was crystallized in the tetragonal structure, as revealed by its electron diffraction pattern (inset).*

crystallize in the tetragonal structure when their diameters are less than ~40 nm (Fig. 2.30(a–c). As the lateral dimensions of these nanowires increases beyond ~40 nm, the orthorhombic structure becomes the more stable one (Fig. 2.30(d). This represent the first demonstration of a template-directed synthesis that is able to generate single-crystalline nanowires in a solution phase and at room temperature. These uniform Ag_2Se nanowires are used as a superionic conductor and thermoelectric material.

Templating against nanowires pre-synthesized using other methods provides a generic and powerful approach to greatly expand the diversity of materials that is processed as uniform 1D nanostructures. One of the major problems associated with the nanowire-templated process is the difficulty in achieving a tight control over the composition and crystallinity of the final product. Nanowires synthesized using these methods are often polycrystalline in structure, and a lead to the formation of single-crystalline products. For a templating process that involves chemical reactions on the surfaces of nanowires, if the product has a larger molar volume than the initial template, the reaction automatically stops after a certain period of time when the stress accumulated around the template has reached the paramount value.

Growth

GROWTH METHODS AND GROWTH MECHANISMS

Growth mechanisms is the general phenomenon in which one-dimensional morphology is obtained. In growth method, the chemical processes incorporate the underlying mechanism to realize the synthesis of the nanostructures. A novel growth mechanism should satisfy three conditions:

(a) explain how one-dimensional growth occurs,
(b) provide a kinetic and thermodynamic rationale, and
(c) be predictable and applicable to a wide variety of systems

Growth of many one-dimensional systems are experimentally achieved without satisfactory elucidation of the underlying mechanism, example; oxide nanoribbons. Nevertheless, understanding the growth mechanism is an important aspect of developing a synthetic method for generating one-dimensional nanostructures of desired material, size, and morphology. It imparts the ability to assess which of the experimental parameters control the size, shape, and monodispersity of the nanowires for synthesis.

One-dimensional nanostructures are synthesized by the crystallization of solid-state structures along one direction. The actual mechanisms for crystal growth include:

(a) growth of an intrinsically anisotropic crystallographic structure,
(b) the use of various templates to direct the formation of one-dimensional nanostructures,
(c) the introduction of a liquid/solid interface to reduce the symmetry of a seed,
(d) use of an appropriate capping reagent to control kinetically the growth rates of various facets of a seed, and
(e) the self-assembly of OD nanostructures

Many methods utilizing these growth mechanisms, their reproducibility, product uniformity and purity, potential for scaling up, cost effectiveness, mechanism etc. are unknown. To control of size range and flexibility in materials that can be synthesized, the intrinsic limits in the growth of nanowire are described below. The quality is gauged by electron microscopy techniques and physical property measurements. Also, the ability to form heterostructures through carefully controlled doping and interfacing is responsible for the success of semiconductor integrated circuit technology; and the two-dimensional semiconductor interface is ubiquitous in optoelectronic devices such as light-emitting diodes (LEDs), laser diodes, quantum cascade lasers, and transistors. Therefore, the synthesis of one-dimensional heterostructures is important for applications including efficient light-emitting sources and thermoelectric devices. This type of one-dimensional nanoscale heterostructure can be prepared once the fundamental, one-dimensional nanostructure growth mechanism is understood. In general: two types of one-dimensional heterostructures can be formed: longitudinal heterostructures and coaxial heterostructures. Longitudinal heterostructures are nanowires composed of different stoichiometries along the length of the nanowire, and coaxial heterostructures are nanowire materials having different core and shell compositions. Nanotubes of a variety of non layered lattices are obtained by selectively etching the inner core of a coaxial heterostructure.

GROWTH OF ZnO NANOSTRUCTURES

Growth of ZnO nanostructures is achieved through various techniques, such as chemical vapour deposition (CVD), metal-organic chemical vapour deposition (MOCVD), physical vapour deposition (PVD) hydrothermal process thermal decomposition, etc. Among these, PVD produces various nanostructures with excellent crystallinity. Two basic growth modes are used for ZnO nanostructures via the PVD process. These are vapour-solid (VS) and vapour-liquid-solid (VLS). They have limited control on the location, growth direction or even size of the nanostructures. In comparison, with metal catalysts, VLS provides better control, and thus has well-defined nanowire size and growth pattern.

Aligned growth of ZnO nanorods is achieved on a solid substrate via a vapor-liquid-solid (VLS) process, by using gold and tin as catalysts, in which the catalyst initiates and guides the growth, and the epitaxial orientation relationship between the nanorods and the substrate leads to the aligned growth.

The growth of aligned ZnO nanowires on a single-crystal substrate is made by a system that is useful for vertical device fabrication. As a result, great interest in acquiring more control over the alignments, including supporting substrates, distribution of nanowires, and density of nanowires, to meet the requirements of nanodevice is needed.

Recently, growth of vertically aligned nanowire arrays have received considerable attention not only for fabricating arrays of vertical field effect transistors but also for converting mechanical energy into electrical energy in nanogenerators and nanopiezotronics and fabricating piezoelectric-semiconducting coupled devices.

The growth of ZnO nanowires on substrates such as Al_2O_3 and GaN is achieved, by systematically changing the growth conditions such as vapor pressure, oxygen partial pressure, and the thickness of catalyst layer.

Vertically aligned ZnO nanowires are grown by using Au nanoparticles as catalyst. The nanowires have a hexagonal shape with growth direction along [0001] and their side surfaces are enclosed by $\{1\bar{1}00\}$ or $\{2\bar{1}\bar{1}0\}$ facets. This is the most easy growth configuration of ZnO nanowires, However, the growth of aligned oxide NBs, is also made. Other techniques that do not use any catalyst, such as metalorganic vapor-phase epitaxial growth, template-assisted growth, and electrical field alignment, are also used for the growth of vertically aligned ZnO nanorods.

The growth of patterned and aligned 1D ZnO nanostructures shows great promise for applications in sensing, room-temperature UV lasers., optoelectronics and field emission. For many of these applications, well-aligned NW arrays are synthesized. For realizing other nanodevice applications like biotechnology and optoelectronics, it is essential that the periodicity and patterns be controlled and designed with deliberate control over interfeature distance, positions, shape, and orientation controlled ZnO nanowire/nanorod armys.

METHODS SUITABLE FOR ALL SOLID MATERIALS

The growth of nanowires from an isotropic medium is relatively simple and straightforward if the solid material has a highly anisotropic crystal structure. Uniform nanowires are easily grown from an anisotropic solid (such as trigonal phase chalcogens) with lengths up to hundreds of micrometers, no matter whether the synthesis: is carried out in a vapor or solution phase. For many solids that are characterized by isotropic crystal structures (e.g., the face-centered cubic (fcc) lattice), symmetry breaking is required in the nucleation step to induce anisotropic growth. So, a large number of approaches are made to lower the symmetry of a seed to produce nanostructures with 1 D morphologies. The vapor-liquid-solid (VLS) process represents another method, in which the symmetry is broken through the introduction of a flat solid-liquid interface. Further, the supersaturation of a system is controlled below a certain level to effectively induce arid maintain 1D growth, as is in many vapor-phase and solvothermal processes. Capping reagents are also examined to kinetically control the growth rates of various faces of a solid and thus to achieve anisotropic growth.

1. Growth of Nanowires from the Vapor Phase

Vapor-phase synthesis is mostly used for the formation of 1D nanostructures such as whiskers, nanorods, and nanowires. Usually, it is possible to process any solid material into 1D nanostructures by controlling the supersaturation at a relatively low level. As early as 1921, the formation of Hg nanofibers of 20 nm in diameter was observed (determined by measuring the Brownian motion under an optical microscope) and up to 1 mm in length when mercury vapor was condensed on a glass surface cooled below the melting point of mercury. This method was used in many other metals, and a mechanism was proposed based on axial screw dislocation to explain the 1D growth. In this mechanism, the driving force for 1D growth was determined by an axial screw dislocation, and incoming atoms could adsorb onto the entire surface of a nanowire and then migrate towards the growing tip. Although this model could explain the growth kinetics, the screw dislocation in the final product is not observed. Hence, the control of supersaturation is a prime consideration in obtaining 1D nanostructures, because there is strong evidence that the degree of supersaturation determines the prevailing growth morphology. A low supersaturation is required for whisker growth whereas a medium supersaturation supports bulk crystal growth. At high supersaturation, powders are formed by homogeneous nucleation in the vapor phase. The lateral dimensions of the whiskers can be varied by controlling a number of parameters that include the supersaturation factor, nucleation size, and growth time.

(a) Direct Vapor-Phase Methods

Although the exact mechanism responsible for 1D growth in the vapor phase is not clear, a direct-vapour-phase method is made to synthesize whiskers and derivatives from a variety of materials. Most of the products are oxides due to oxidation which produces trace amounts of O_2 in the reaction systems. The major advantage of a vapor-phase method is its simplicity and accessibility. In this process, the vapor species is first generated by evaporation, chemical reduction, and other kinds of gaseous reactions. These species are subsequently transported and condensed onto the surface of a solid substrate placed in a zone with temperature lower than that of the source material. With proper control over the supersaturation factor, 1D nanostructures are obtained in moderately large quantities. For example, Si_3N_4, SiC, Ga_2O_3, and ZnO nanowires are synthesized by heating the powders of these materials to elevated temperatures.

In addition to nanowires with a circular or square cross-section, the formation of nanobelts (or nanoribbons) with a rectangular cross-section is observed by evaporating metal oxide powders at elevated temperatures. The as-synthesized nano-belts are structurally uniform, and highly pure in elemental and phase compositions. Most of them are single crystals, and free from defects and dislocations. They have a width in the range of 30-300 nm, width-to-thickness ratios of 5–10, and lengths of up to several milli-meters. The belt-like morphology is a distinctive and common structural characteristic for the family of semiconducting oxides with cations of different valence states and materials of distinct crystallographic structures (Fig. 3.1). This type of 1D nanostructure is used for understanding dimensionally confined transport phenomena in functional oxides and building functional devices along individual nanobelts.

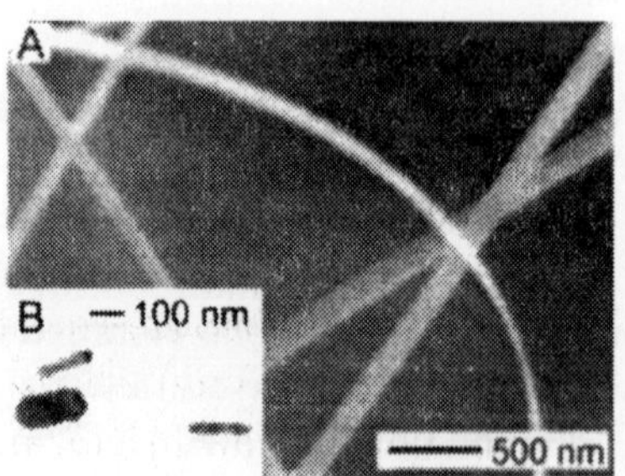

Figure 3.1 *(A) An SEM image of SnO_2 nanobelts synthesized by heating its powders at 1270 K (B) A cross-sectional transmission electron microscopy image of these nanobelts, confirming their quasi-rectangular cross-sections.*

(b) Indirect Vapor-Phase Methods

The mechanism of vapor-phase methods involve the formation of intermediates or precursors by using high temperatures. In many cases, decomposition and other types of side reactions are also considered. For example, in synthesis of MgO nanowires via a carbon thermal reduction process, (in which Mg vapor is generated through the reduction of MgO by carbon) is transported in a flow reactor to the growth zone, and then oxidized to form MgO again. Hydrogen gas and water instead of carbon are used as the reducing agent. This method is also extended to other binary oxides such as Al_2O_3, ZnO, and SnO_2. In all of these syntheses, the formation of metal oxide through a two-step process helps to keep the supersaturation of the system at, low levels.

Si nanowires can be obtained in high yields when a mixture of Si and SiO_2 powders (rather than pure Si powders) is used as the target material for laser ablation. The vapor-phase Si*x*O ($x > 1$) generated by laser ablation (or thermal evaporation) is the key intermediate in this (oxide-assisted) process. The formation of silicon occurs through the following two steps:

$$Si_xO \rightarrow Si_{x-1} + SiO \qquad (x > 1) \qquad (1)$$

$$2SiO \rightarrow Si + SiO_2 \qquad (2)$$

The TEM observations suggest that these decomposition reactions first lead to the precipitation of Si nanoparticles encapsulated in shells of silicon oxide. Some of these particles are piled up on the surface of the silicon oxide matrix, and serve as seeds for the growth of nanowires in the following step. Particles having their growth directions perpendicular to the matrix surface undergo fast growth.

Consider the following factors that determine the wire growth kinetics, example, the Si_xO ($x > 1$) layer at the tip of each nanowire seems to have a catalytic effect. This layer is in or near a molten state and is thus capable of enhancing atomic absorption, diffusion, and deposition. The SiO_2 component (from the decomposition of SiO) in the shell helps to retard the lateral growth of each nanowire. The precipitation, nucleation, and growth of Si nanowires occurred in the region close to the cold finger, which suggest that the temperature gradient provides an external driving force for the nanowire growth. This mechanism is used to synthesize GaAs nanowires via laser ablation of a mixture of GaAs and Ga_2O_3. The GaAs nanowires have a diameter in the range of 10 to 120 nm, and lengths up to tens of micrometers. Each one of them has a thin oxide layer covering a crystalline GaAs core with a [111] growth direction. The major advantage of these oxide-assisted methods is that no metal catalyst is required, and contamination of the resultant nanowires by the metal atoms is inherently eliminated.

Also synthesized are Cu_2S nanowires by oxidizing copper into Cu_2O with O_2 gas and then sulfidizing this oxide intermediate into Cu_2S with H_2S gas under ambient conditions. The synthesis of CuO nanowires by heating copper substrates (foils, grids, and wires) in air is also achieved. Figure 3.2(a) and (b) show the SEM and HRTEM images, respectively, of the surface of a millimeter-sized copper wire. Both electron diffraction and HRTEM show that each CuO nanowire is a bicrystal divided by a (111) twin plane in the middle along its longitudinal axis. Two reactions are involved in this synthesis: the oxidation of Cu to form CU_2O intermediate, and a second oxidation step to generate CuO vapor (from which uniform CuO nanowires are grown). In some cases, the involvement of an intermediate even leads to a reduction in the temperature required for the growth of nanowires. Moreover, it is also possible to greatly reduce the synthesis temperature by using an appropriate precursor to form the desired material. For example, uniform nanowires of MgO are synthesized in the temperature range of 1070 to 1171 K by generating MgO from the oxidation of MgB_2. Figure 3.2(d) and (e) show SEM and TEM images, respectively, of MgO nanowires synthesized by heating MgB_2 powders at 1170 K a mixture of H_2/Ar. In comparison, synthesis of MgO nanowires from the direct evaporation of MgO powders requires a temperature as high as 1470 K. So, one should be able to modify any chemical vapor deposition (CYD) method (e.g., by diluting the precursor gas to control supersaturation in the reaction chamber) to direct the growth of the deposited material into wire-like morphologies. This method is used for the synthesis of crystalline boron nanowires.

2. Growth of ZnO Nanowires/Nanorods by the VLS Process

In this method a sample is prepared using a two-step high-temperature vapor-solid deposition process, which is used for growing the polar-surface-dominated nanopropeller arrays of ZnO. The solid vapor deposition involves using an experimental setup consisting of a horizontal high-temperature tube furnace (~50 cm in length), an alumina tube (~75 cm in length, ~4 cm in inner diameter), a rotary pump system, and a gas controlling system. ZnO and SnO_2 powders and graphite with a molar ratio of 3:4: 1.5 are mixed, ground, and then loaded onto an alumina boat and positioned at the center of the alumina tube. The evaporation is conducted at 1370 K, for 60 min (step I) and then up to 1520 K for half an hour (step II) under a pressure of 2×10^4 Pa. The N_2 carrier gas flow rate is controlled at 20 sccm. The desired structures grow on a polycrystalline Al_2O_3 substrate in temperature zones ranging from 870 to 970 K and 1070 to 1170 K, respectively, corresponding to lower- and higher-temperature duration periods. (see Fig. 2.2 in fabrication & synthesis).

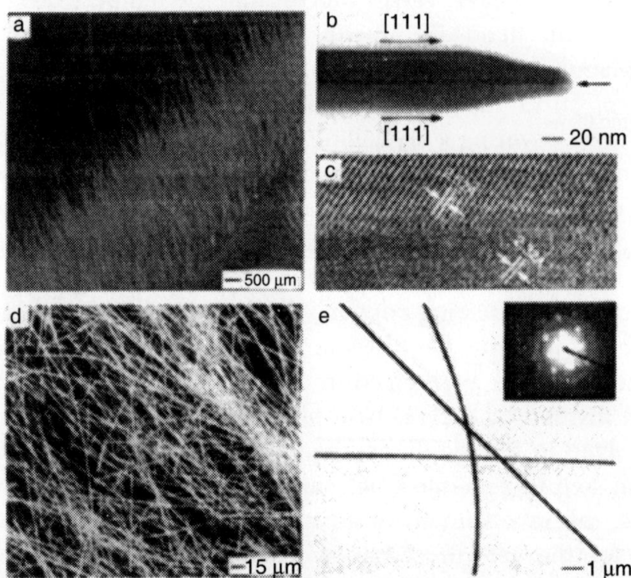

Figure 3.2 *(a) SEM and (b) TEM images of CuO nanowires synthesized by heating a copper wire (0.1 mm in diameter) in air to a temperature of 770 K for 4 h. Each CuO nanowire was a bicrystal as shown by its electron diffraction pattern and high-resolution TEM characterization (c). (d) SEM and (E) TEM images of uniform MgO nanowires (~150 nm in diameter) that are synthesized by heating MgB_2 powders in a flow of H_2/Ar mixture (1:10 by volume) to a temperature of ~1170K. The inset in (e) shows an electron diffraction pattern.*

The as-grown sample has many large crystals of ZnO formed on the alumina substrate, and these ZnO crystals serve as the substrate for growing ZnO nanostructures. Figure 3.3(a) shows a scanning electron microscopy (SEM) image of a large block of the as-grown ZnO nanostructures. It looks like a "brush" that is 40–50 μm in length, 10–15 μm in width, and several micrometers in thickness. A large ZnO crystal is the substrate with aligned ZnO nanostructures growing out of the five exposed surfaces and displaying different growth features. Energy-dispersive spectroscopy (EDS) analysis in SEM proves that the as-grown base and the as-grown nanostructures are ZnO with Sn balls at the tips. The nanostructures grown at the left-and tight-hand-side facets are distributed symmetrically, but the top and bottom surfaces have asymmetrical nanostructures of different dimensionality and morphology. On the top facet, the grown elements arc short, conelike trunks ~1 μm in height and 800 nm in width. On the bottom surfaces, uniform nanowires with a diameter of ~100 nm and a height of ~2 μm are grown. The Zn-terminated (0001) planes are active for nanowire growth, but the oxygen-terminated $(000\bar{1})$ plane is inert with Sn-guided growth: the top and bottom surfaces in Figure 3.3(a) are $(000\bar{1})$ and +(0001), respectively.

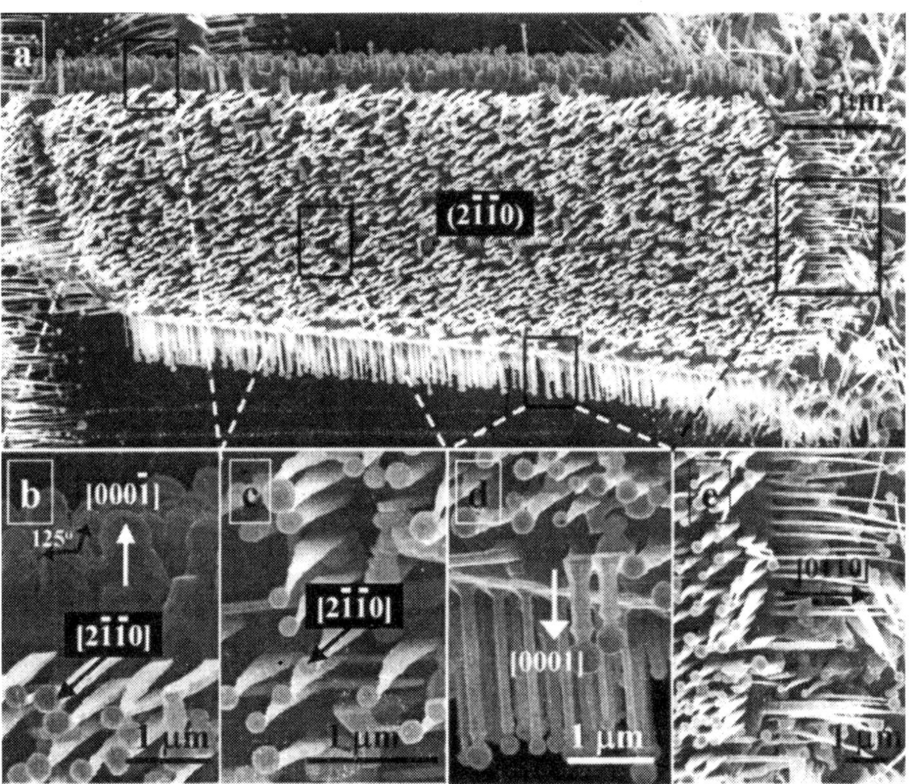

Figure 3.3 *SEM images of the as-synthesized ZnO nanostructures growing out of the five surfaces of a ZnO substrate. (b–e) Enlarged areas as marked in a. The asymmetric growth of nanostructures on the substrate surfaces of different structure characteristics is presented.*

For the nanotrunks grown on the $(000\bar{1})$ surface, the contact angle between the Sn ball and the trunk top is ~125 ± 5° (Figure 3.3b). Figure 3.3(d) is the magnified side view of the as-grown nanowires on the Zn-terminated (0001) surface. The aligned ZnO nanowires grown out of the Zn-terminated (0001) surface have a uniform diameter of ~100 nm and a length of ~2 μm and six side surfaces $\{2\bar{1}\bar{1}0\}$ (Figure 3.3d). The nanowircs are aligned, and they have an epitaxial orientation with respect to the substrate. The contact angle between the Sn particle and the nanowire growth front is ~150 ± 5°.

Figure 3.3(c) shows the well-aligned ZnO nanoribbons grown out of the $(2\bar{1}\bar{1}0)$ surface of the substrate, with epitaxial growth direction $[2\bar{1}\bar{1}0]$, top and bottom faces ±(0001), and side-stepped $(01\bar{1}0)$ surfaces. The width of the nanoribbon is not uniform along the entire length. The nanoribbon has the largest width of ~300 nm at the contact point with the substrate, the smallest width of ~ 10 nm at the contact with the Sn catalyst, and a length of ~1 μm. Figure 3.3(e) shows the normally orientated nanoribbons on the right-hand-side facet of the substrate, which grow along $[01\bar{1}0]$. It is also shown that the widths of the nanoribbons along $[2\bar{1}\bar{1}0]$ and $[0\bar{1}\bar{1}0]$ are of the same order of magnitude, tens of nanometers, and that the length of the nanoribbons along $[01\bar{1}0]$ is around 3 μm.

The nonpolar surface shows symmetric growth. Figure 3.4 is a top view along the *c* axis of the nanowires grown on a ZnO substrate. It is apparent that the triangular nanoribbons are grown symmetrically on the top and bottom nonpolar facets of $\pm(2\bar{1}\bar{1}0)$. From Figure 3.4(b), it is found that the cone trunks that formed vertically on the O-terminated $(000\bar{1})$ top surface have roughly faceted side surfaces, defined by $\{2\bar{1}\bar{1}O\}$ and displaying a hexagon-based pyramid. In Figure 3.4(c), from the broken triangular nanoribbons; the rectangular cross is section of ~300 nm in width and ~40 nm in thickness is concluded. A closer image at the top surface of the nanoribbon displays a smooth top surface (Figure 3.4d), corresponding to the chemically sluggish O-terminated $(000\bar{1})$ polar surface of ZnO.

Figure 3.5(a) shows another SEM view from the top of the brush structure. On the top and bottom $O(000\bar{1})$ and Zn-(0001) ZnO surfaces, aligned shorter and wider pyramidal ZnO trunks and longer and narrower hexagonal ZnO nanowires are grown, respectively. Different from the cases in Figures 3.3 and 3.4 the nanoribbons grown on the large $(2\bar{1}\bar{1}0)$ substrate surface have a smaller width-to-thickness ratio. The grown nanoribbons arc enclosed by relatively larger and smoother $\pm(01\bar{1}0)$ surfaces and smaller ±(0001) surfaces. Steps are introduced on the side surfaces, and the nanoribbon becomes sharper toward the tip (Figure 3.5b). The interval between the nanoribbons is quite regular, and the aligned ribbon spacing is close to periodic (Figure 3.5c).

On the basis of the SEM image in Figure 3.3 by the volumes of the as-grown ZnO nanostructures and the Sn particles is calculated covering the unit substrate surface area of the ±(0001) polar surfaces. It is assumed that the nanostructures on both polar surfaces are uniformly distributed, and the Sn particles are spherical, the ZnO nanowires are cylindrical, and the ZnO pyramidal trunks have a hexagon base. The volumes of ZnO nanostructures and Sn catalysts per unit area are 197 ± 5 (ZnO nanowires) and 600 ± 10 nm (Sn catalysts) for the $(000\bar{1})$ substrate surface, and 318 ± 9 (ZnO pyramidal trunks) and 854 ± 20 nm (catalyst Sn) for the $(000\bar{1})$ substrate surface, respectively. The volume ratio of the ZnO material per unit area on

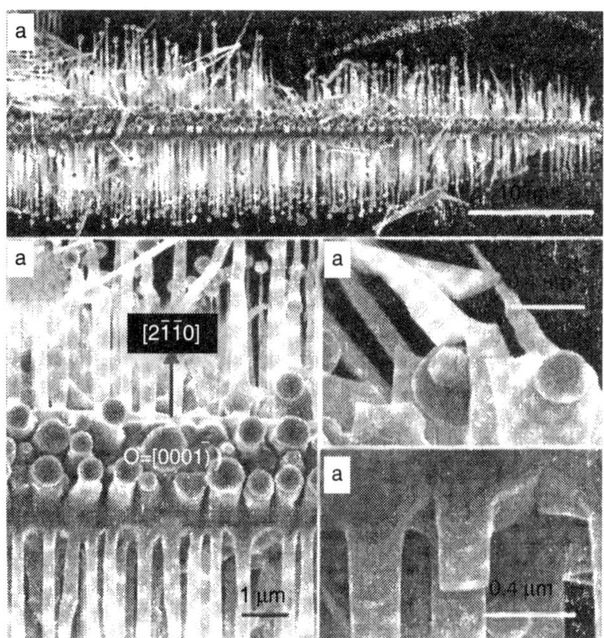

Figure 3.4 *(a) [0001] directional top view of the as-grown ZnO nanostructures, showing the symmetric growth of triangular nanoribbons along ±[2$\bar{1}\bar{1}$0]. (b) Closer top view of the nanotrunks grown out of the O-terminated (0001) ZnO surface, revealing the hexagonal base contour of nanopyramids. (c) Broken triangular nanoribbons showing rectangular cross sections. (d) Smooth top surfaces of triangular nanoribbons. corresponding to the O-terminated (000$\bar{1}$) polar surface of ZnO.*

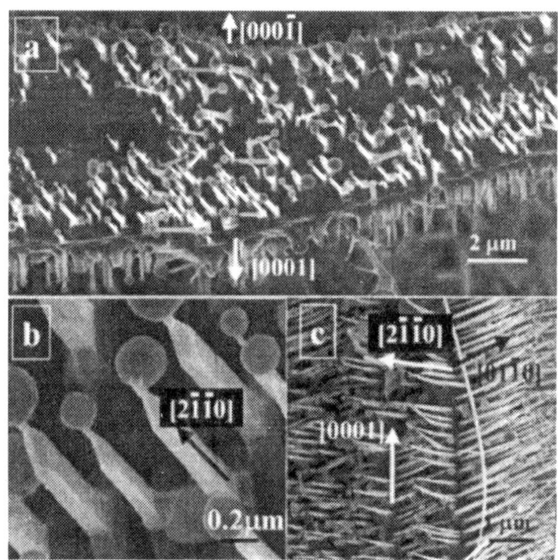

Figure 3.5 *(a) As-grown ZnO nanostructures with a (2$\bar{1}\bar{1}$0) plane that is sparsely covered by ZnO nanoribbons. (b) Closer view of the nanoribbons normal to the (2$\bar{1}\bar{1}$0) plane, indicating nanoribbons with small width/thickness ratio. (c) Periodically aligned arrays of ZnO nanoribbons with growth directions of [2$\bar{1}\bar{1}$0] and [01$\bar{1}$0] on the two surfaces, respectively.*

the $(000\bar{1})$ and $(000\bar{1})$ substrate surfaces is 1:1.6, and the volume ratio of the Sn material per unit area on the (0001) and $(000\bar{1})$ substrate surfaces is 1: 1.42. The volume ratios of ZnO/Sn on the (0001) and $(000\bar{1})$ substrate surfaces are 3 and 2.7, respectively. This indicates that the sticking rates of the ZnO vapor onto the Sn surfaces are not different on the two substrate surfaces, but the amounts of ZnO and Sn deposited on the $(000\bar{1})$ substrate surface are about 50% more than that deposited onto the (0001) substrate surface thus, indicating the high surface adsorption rate of Sn onto the (0001) substrate surface.

From Figure 3.3(b) and (d), the contacting angle between the Sn particle and the ZnO nanowire is ~150° on the Zn-terminated (0001) substrate surface and ~125° on the oxygen-terminated $(000\bar{1})$ substrate surface, indicating that the $(000\bar{1})$ surface has a higher degree of wetting contact with the Sn particle than the (0001) surface.

Consider, a growth model for the ZnO nanostructure grown on ±(0001) polar surfaces. The entire growth is dominated by VLS. The Sn catalyst is reduced from SnO_2 added to the source material. The thermal reduction of SnO_2 results in some charged species and neutral:

$$SnO_2 \leftrightarrow xSn^{2+} + (1 - x)Sn + 2xO^{2-} + (1 - x)O_2$$

where x represents the percentage of charged Sn^{2+} ions after the decomposition. The charged species may recombine and condense onto the substrate to form a charged catalyst particle $(Sn)^+$. A neutral Sn particle can also be positively charged because metal atoms tend to lose electrons. For the Zn^{2+}-terminated (0001) ZnO substrate surface with positive charges, the positively charged $(Sn)^+$ particle has a small repulsion with the substrate because of the electrostatic interaction, but it still tends to stick onto the surface because of a stronger adhesion force. The consequence of the electrostatic repulsion results in a larger contact angle between the particle and the substrate (Figure 3.6(a)). The Sn particle initiates the growth of the ZnO nanowire (Figure 3.6(b)), and the nanowire has an epitaxial orientation relationship with the substrate because of least lattice mismatch. The nanowire continues to grow following the VLS growth process (Figure 3.6(c)).

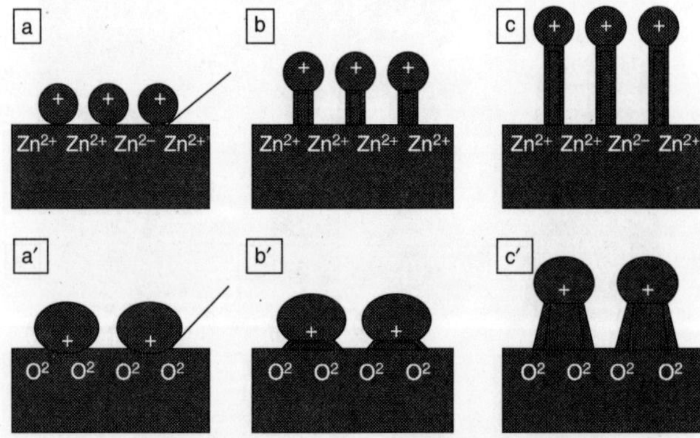

Figure 3.6 *Growth model of the ZnO nanowires on a Zn-terminated (0001) polar surface (a–c) and an O-terminated $(000\bar{1})$ polar surface (a′–c′).*

For the oxygen-terminated $(000\bar{1})$ ZnO substrate surface with negative charges, the attraction between the substrate surface and the positively charged $(Sn)^+$ particle results in a smaller contact angle and a larger contact area (Figure 3.6(a)). Naturally, there is a lower population density of Sn particles on the substrate surface, and each of them is larger. The large Sn particle initiates the growth of nanorods (Figure 3.6(b)). As the growth continus and the distance between the particle and the substrate surface is further away, the contact area between the particle and the ZnO nanorod decreases slightly because of the reduced electrostatic attraction, possibly resulting in a shrinkage in nanorod size as the growth proceeds and forming a short nanotrunk with a hexagon base (Figure 3.6(c)).

In comparison to the growth on the (0001) substrate surface, the electrostatic attraction between the $(000\bar{1})$ substrate surface and the Sn particles tends to attract a high density of Sn onto the surface in early growth stages. These Sn particles arc most effective in attracting the ZnO molecular species, resulting in a higher density of ZnO deposition onto the $(000\bar{1})$ substrate surface. But the ratio of ZnO/Sn in each case remains approximately a constant, as expected from the VLS process in which it is the catalyst surface that is responsible for adsorbing the incoming molecular species.

For the nonpolar $\pm(2\bar{1}\bar{1}0)$ and $\pm(01\bar{1}0)$ substrate surfaces, the growth on the surfaces shows symmetric features and still follows the VLS growth process. Take the $(2\bar{1}\bar{1}0)$ substrate surface as an example. Nanoribbons grow epitaxially on the substrate; they are enclosed by large $\pm(01\bar{1}0)$ and small $\pm(0001)$ surfaces, and gradual lateral growth along $\pm[0\bar{1}\bar{1}0]$ results in a triangular-like shape. (See Figure 3.3(c)).

3. Vapor-Liquid-Solid Methods

Among all vapor-based methods, the VLS process, is the most successful for generating nanowires with single-crystalline structures. The VLS process starts with the dissolution of gaseous reactants into nanosized liquid droplets of a catalyst metal, followed by nucleation and growth of single-crystalline rods and then wires. The 1D growth is induced by the liquid droplets, the sizes of which remain unchanged during the entire process of wire growth. So, each liquid droplet serves as a soft template to strictly limit the lateral growth of an individual wire. A good solvent is required for forming liquid alloy with the target material, to form eutectic compounds. All major steps involved in a VLS process are illustrated in Figure 3.7(a), with the growth of Ge nanorods as an example. Based on the Ge-Au binary phase diagram (Fig. 3.7(b)), Ge (from the decomposition of GeI_2 or other precursors) and Au form liquid alloys when the temperature is raised above the eutectic point (630 K). Once the liquid droplet is supersaturated with Ge, nanowire growth start to occur at the solid-liquid interface. The vapor pressure of Ge in the system kept sufficiently low so that secondary nucleation events are completely suppressed. Both physical methods (laser ablation, thermal evaporation, and arc discharge) and chemical methods (chemical vapor transport and deposition) are employed to generate the vapor species required for the growth of nanowires, and no difference is found in the quality of nanowires produced by these methods.

This mechanism is used to observe the growth of Ge nanorods, in the chamber of a TEM equipped with a temperature-controllable stage. GeI_2 is used as the vapor source for Ge, and

Au as the catalyst. The vapor transport is achieved by heating the system to a temperature in the range of 970 to 1170 K. Figure 3.8 shows a set of TEM images recorded during the growth of a Ge nanorod. These images show all steps illustrated in Figure 3.8: the formation of Au-Ge alloy, the nucleation of Ge nanocrystal in the droplet of Au-Ge alloy, and the growth of a Ge nanorod by pushing the liquid-solid interface forward. Based on this mechanism, the growth process from different aspects can be controlled. To the first order of approximation, the diameter of each nanowire is determined by the size of the catalyst, and smaller catalysts yield thinner nanowires. But, Si and GaP nanowires of any specific size can be obtained by controlling the diameter of monodispersed gold colloids (or clusters) serving as the catalyst.

The VLS process has now become a widely used method for generating one-dimensional nanostructures from a rich variety of pure and doped inorganic materials that include elemental semiconductors (Si, Ge), III-V semiconductors (GaN, GaAs, GaP, InP, InAs), II-VI semiconductors (ZnS, ZnSe, CdS, CdSe), oxides (indium-tin oxide, ZnO, MgO, SiO_2, CdO), carbides (SiC, B_4C), and nitrides (Si_3N_4). The nanowires produced using the VLS approach are

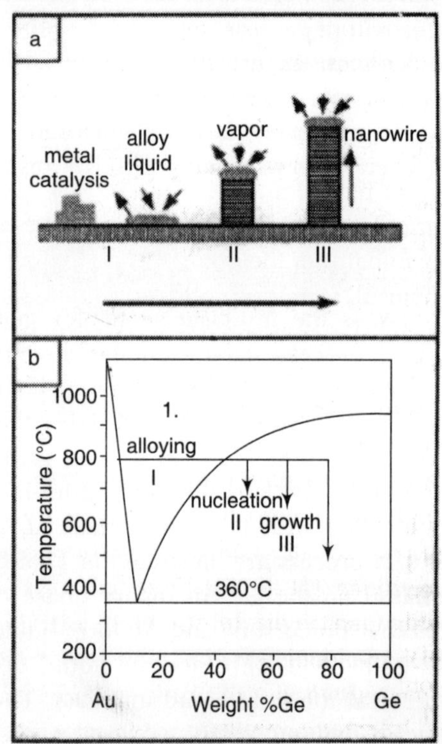

Figure 3.7 *Illustration showing the growth of a nanowire via the vapor-liquid-solid mechanism. (b) The binary phase diagram between Au and Ge, with an indication of the compositional zones responsible for alloying, nucleation, and growth.*

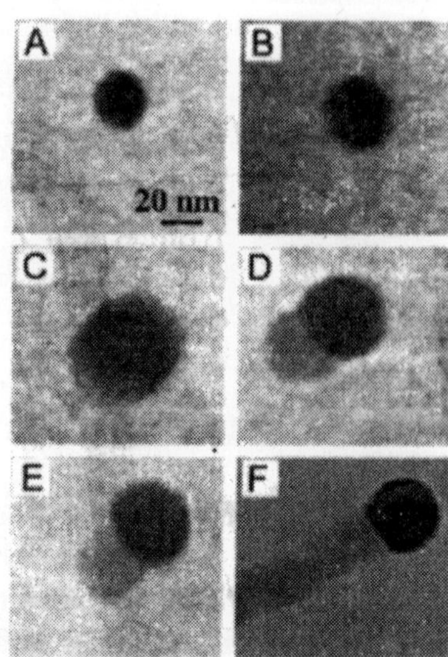

Figure 3.8 *The birth of a Ge nanowire on a Au nanucluster, as observed by TEM. The Au nanocluster start to melt after the formation of Ge-Au alloy, and this is followed by an increase in the liquid droplet size during the Ge vapor condensation process. When the droplet is supersaturated with the Ge component, a Ge nanowire grows out of this droplet of Au-Ge alloy and becomes longer as time elapses.*

remarkable for their uniformity in diameter, which is usually on the order of 10 nm over a length scale of > 1 μm. Figure 3.9 shows scanning electron microscopy (SEM), TEM, and high-resolution transmission electron microscopy (HRTEM) images of a typical sample of GaN nanowires that is prepared using a metal organic chemical vapor deposition (MOCVD) procedure. Electron diffraction and HRTEM characterization indicate that each nanowire is essentially a single crystal. The presence of a catalyst nanoparticle at one of the ends of the nanowire (Figure 3.9(b)) is clear evidence supporting the VLS mechanism. However, metal droplets may not necessarily remain on the tips of VLS-made wires because interfacial dewetting and large interfacial thermal expansion differences can dislodge catalyst tips during cooling.

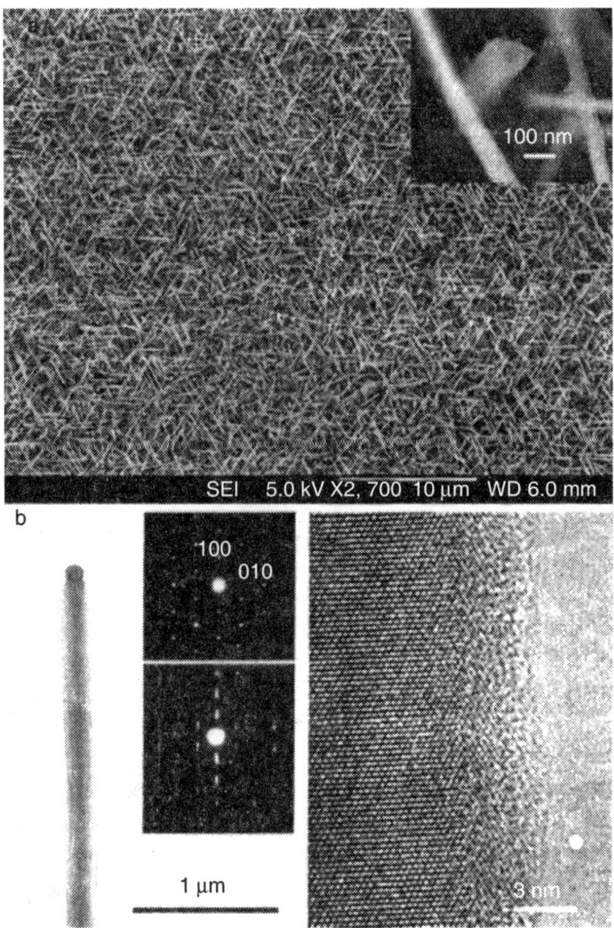

Figure 3.9 *(a) Field-effect scanning electron microscope (FESEM) image of the GaN nanowires grown on a gold-coated c-plane sapphire substrate. Inset shows a nanowire with its triangular cross section. (b) TEM image of a GaN nanowire with a gold metal alloy droplet on its tip. Insets are electron diffraction pattems taken along the [001] zone axis. The lower inset is the same electron diffraction pattern but purposely defocused to reveal the wire growth direction. (c) Lattice-resolved TEM image of the nanowire.*

Among these materials, compound semiconductors such as GaAs, GaN, ZnO, and CdSe are used since their direct bandgaps enables optical and optoelectronic applications. The nanowires produced using the VLS approach have uniform diameter, which is usually of the order of 10 nm over a length scale of > 1 μm. Figure 3.10 shows SEM, TEM, and HRTEM images of a typical sample of Ge nanowires that is prepared using the vapor-transport procedure. Characterization based on electron diffraction and HRTEM indicates that each nanowire is essentially a single crystal. Once the growth is terminated, the presence of a catalyst nanoparticle at one of the ends of the nanowire (inset of Fig. 3.10b) the VLS mechanism.

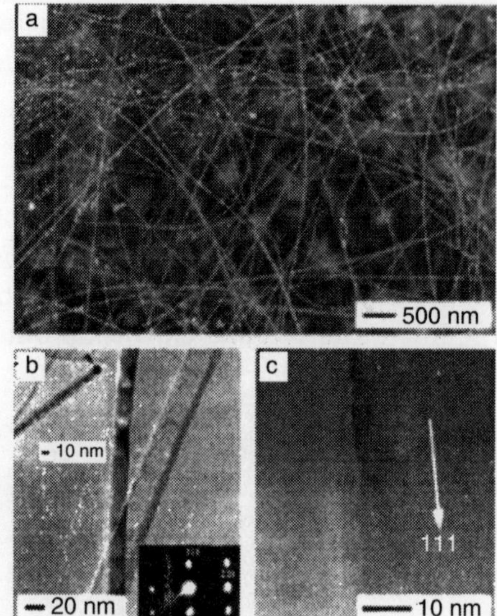

Figure 3.10 *(a) SEM and (b) TEM images of Ge nanowires prepared using the chemical vapor-transport method. Insets in (b) show the Au tip of a Ge nanowire and the electron diffraction from an individual Ge nanowires. (c) The atomically resolved TEM image of a Ge nanowire.*

One of the challenges faced by the VLS process is the selection of an appropriate catalyst that works with the solid material to be processed into 1D nanostructures. This is done by analyzing the equilibrium phase diagrams. The analysis of catalyst and growth conditions are simplified by considering the pseudo-binary phase diagram between the metal catalyst and the solid material. For instance, the pseudo-binary phase diagram of GaAs with Au exhibits a large GaAs-rich region in which liquid Au-GaAs co-exists with solid GaAs. So, single-crystalline nanowires of GaAs with diameters as small as a few nanometers are obtained by setting the Au:GaAs composition ratio and growth temperature in the proper ranges. As a limitation, it seems impossible to apply the VLS method to metals. The necessary use of a metal as the catalyst also contaminates the semiconductor nanowires and thus potentially change their properties.

4. Self-Catalytic VLS

Because nanowires of binary and more complex stoichiometries are created using the VLS mechanism, it is possible, for one of these elements, to serve as the VLS catalyst. TEM is used to observe directly self-catalytic growth of GaN nanowires by heating a GaN thin-film in a vacuum of 10^{-7} mmHg. It is known that GaN decomposes at temperatures above 1120 K high vacuum via the following process:

$$\text{GaN (s)} \rightarrow \text{Ga (1)} + 0.5\ \text{N (g)} + 0.25\ \text{N}_2\text{(g)}.$$

Also, the sublimation of GaN to the diatomic or polymeric vapor species is observed:

$$GaN\ (s) \rightarrow GaN\ (g)\ \text{or}\ [GaN]_x\ (g).$$

Initially, decomposition of the GaN film leads to the formation of isolated liquid Ga nanoparticles. The resultant vapor species, composed of the atomic nitrogen and diatomic or polymeric GaN, then redissolves into the Ga droplets and initiates VLS nanowire after supersaturating the metal and establishing a liquid-Ga/solid-GaN interface. Each step in the VLS process is observed by TEM (Figure 3.11): the alloying of the Ga droplet with the nitrogen-rich vapor species, the nucleation of the nanowire liquid-metal interface, and the subsequent axial nanowire growth.

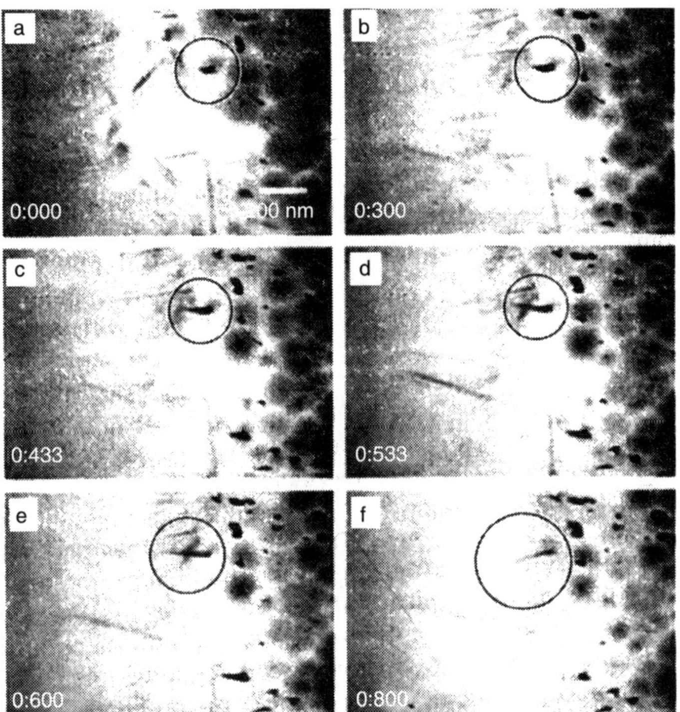

Figure 3.11 *A series of video frames grabbed from observations of GaN decomposition at ~1320 K, showing the real-time GaN nanowire growth process. The number on the bottom left comer of each frame is the time (second:millisecond).*

The major advantage of a self-catalytic process is that it avoids undesired contamination from foreign metal atoms typically used as VLS catalysts. Self-catalytic behavior takes place when the direct reaction of Ga with NH_3 or direct evaporation of GaN is used to produce GaN nanowires.

5. VLS Vapor Phase Methods

For a specific material, the method of introducing vapor species which depend on the nanowire physical properties is unknown. Certain methods of introducing vapor phase precursors allow a much greater flexibility in dopant selection, and gives greater control over the compound stoichiometry. Furthermore, integration of nanowire components into current thin-film technologies is an important consideration. Specific vapor phase methods (such as MOCVD) are more compatible with process integration than others.

Consider the case of GaN nanowires. In synthetic GaN-based devices, laser ablation, chemical vapor transport, and most recently, MOCVD are used. The highest carrier mobility values are achieved for thin films grown by MOCVD, hydride phase vapor epitaxy, or molecular beam expitaxy (MBE). Of these methods, MOCVD allows the greatest flexibility for producing nanowires with controlled dopant and other ternary nitride phase concentrations. This is partly because of the similarity of the precursor chemistries for the constituent and dopant atoms. Finally, MOCVD has similar technical platform as thin-film technologies and is therefore easily integrated into existing GaN thin-film technologies.

6. Vapor-Solid Growth Mechanisms (Synthetic Mechanism)

Consider, a experiment which produces nanostructures grown from a combination of V-S growth mechanisms (synthefic mechanism). Using thermodynamic and kinetic considerations, the formation of nanowires is achieved through (a) an anisotropic growth mechanism, (b) Frank's screw dislocation mechanism, (c) a different defect-induced growth model, or (d) self-catalytic VLS. In an anisotropic growth mechanism, one-dimensional growth is accomplished by the reactivity and binding of gas phase reactants along specific crystal facets like thermodynamic and kinetic parameters and a system to minimize surface energies. In the dislocation and defect-induced growth models, specific defects like screw dislocations are known to have larger sticking coefficients for gas phase species. This allows enhanced reactivity and deposition of gas phase reactants at these defects. Other vapor-solid growth mechanisms are the oxide-assisted growth mechanism. However, many of these vapor-solid growth mechanisms lack thermodynamic and kinetics of one-dimensional growth.

7. Nanowire Growth in Solution

One of the disadvantages of high-temperature approach to nanowire synthesis is the high cost of fabrication and scale-up and the inability to produce metallic wires. So, using solution-phase techniques results in the creation of one-dimensional nanostructures in high yields (gram scales) via selective capping mechanism. Example; silver nanowires are formed using poly(vinyl pyrroli-done) (PVP) as a capping agent. In the presence of PVP, most silver particles grow into nanowires with uniform diameter (see Fig. 3.14(b)) on page 68). Here, PVP selectively binds to the {100} facets of silver while maintaining {111} facets to allow growth. This selective passivation of Ag nanowires along the {100} faces is achieved by functionalizing nanowires, post-growth under mild conditions with a dithiol compound, and subsequently adding gold nanoparticles to the solution. The gold nanoparticles bond to the end {111} caps, so that only dithiol adhesion is there on the ends caps and not the {100} faces owing to the preferential bonding of the PVP to these faces.

In this process nanowires of silver with diameters in the range of 30–60 nm and lengths up to ~50 μm are generated. Thus, many metals are processed as nanowires through solution-phase methods by finding a chemical reagent capable of selectively interacting with various surfaces of a metal.

The growth of semiconductor nanowires is also achieved using a similar synthetic mechanism. Microrods of ZnO are produced via the hydrolysis of zinc salts in the presence of amines example: hexamethylenetetramine is used to produce dense arrays of ZnO nanowires in aqueous solution (Figure 3.12), having diameters of 30–100 nm and lengths of 2–10 μm. These nanowires can be prepared on any substrate. In this growth process a majority of the nanowires, in the array are in direct contact with the substrate and therefore, provide a continuous pathway for carrier transport, for electronic devices, based on these materials.

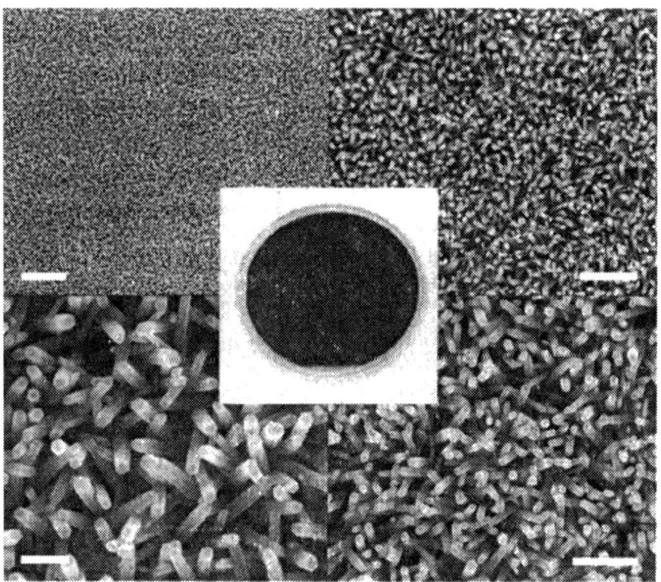

Figure 3.12 *ZnO nanowire array on a 4-inch silicon wafer. Centered is a photograph of a coated wafer, surrounded by SEM images of the army at different locations and magnifications. These images are representative of the entire surface. Scale bars, clockwise from upper left, correspond to 2 μm, 1 μm, 500 nm, and 200 nm.*

A major limitation of this growth mechanism is that most capping agents are chosen via an empirical trial-and-error approach. It is therefore advantageous to develop a library of bond strengths of various chemisorbed capping agents on specific crystal planes.

8. Solution-Liquid-Solid Methods

Based on an analogy to the VLS process, solution-liquid-solid (SLS) method is developed for the synthesis of highly crystalline nanowires of III–V semiconductors at relatively low temperatures (Fig. 3.13). Here, a metal (e.g.. In. Sn, or Bi) with a low melting point is used as the catalyst, and the desired material is generated through the decomposition of organometallic precursors. The product is a single-crystalline whiskers or filaments with lateral dimensions of 10–150 nm and lengths up to several micrometers. In principle, the operation temperature is reduced to a value below the boiling points of used aromatic solvents. For example, methanolysis of {*tert*-$Bu_2In[\mu\text{-}P(SiMe_3)_2]\}_2$ in an aromatic solvent can yield polycrystalline InP fibers 10–100 nm thick and up to 1000 nm long at temperatures in the range of 380 to 475 K. The key component of this synthesis is a molecular component whose constituent groups are eliminated to generate a non-molecular unit, from which the InP crystal lattices are assembled. The decomposition of this organometallic precursor proceeds through a sequence of isolated and fully characterized intermediates to yield complex [*tert*-$Bu_2In(\mu\text{-}PH_2)]_3$. This complex subsequently under goes an alkane elimination to generate the building blocks–$(InP)_n$ fragments. Then, the $(InP)_n$ fragments dissolve into a dispersion of droplets formed by molten indium, and recrystallize as InP fibers. The SLS process can operate at temperatures below those required by conventional VLS processes. This synthetic method is used for other highly covalent semiconductors (both binary and ternary), as well as their alloys.

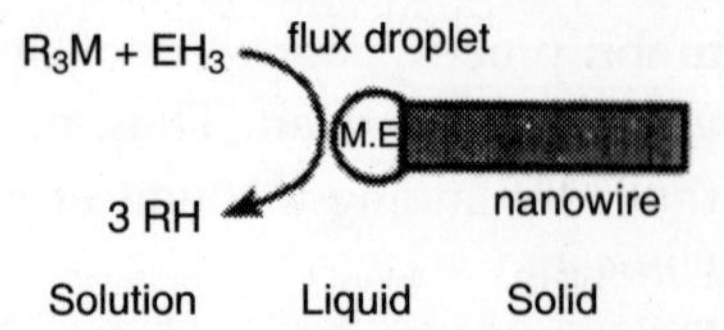

Figure 3.13 *Growth of a nanowire through the solution-liquid-solid mechanism, which shares many similarities with the vapor-liquid-solid process depicted in Figure 3.7.*

Defect-free Si nanowires with nearly uniform diameters of 4–5 nm and lengths up to several micrometers are grown by employing a supercritical fluid as the solvent for the SLS process. Monodispersed, alkanethiol-capped Au nanocrystals is used which serves as seed to direct and confine the growth of Si into nanowires having a narrow size distribution. In a procedure, the sterically stabilized Au nanocrystals and a precursor to Si, diphenylsilane, are co-dispersed in hexane that is heated and pressurized above its critical point Under these conditions, the diphenylsilane decomposes to form Si atoms. The phase diagram between Si and Au indicates that Si and Au form an alloy phase in equilibrium with pure solid Si when the Si concentration with respect to Au is greater than 18.6 mol-% and the temperature is above 635 K. As a result, the Si atoms dissolve into the sterically stabilized Au nanocrystals until a supersaturation is reached, at which point they are expelled from the Au-Si alloy particle in the form of a uniform nanowire. This growth mechanism shares some similarities with the SLS process. The supercritical fluid medium provides the extreme conditions necessary to promote the dissolution and crystallization of Si. The orientation of the Si nanowires is also controlled by varying the reaction pressure. Electron microscopic characterization indicates that the Si nanowires prepared using this method are essentially single crystals.

9. Solvothermal Methods

In solvothermal synthesis a solvent under pressures and temperatures above its critical point is used to increase the solubility of a solid and to speed up reactions between solids. It is a commonly used method for generating 1D nanostructures. Here, a precursor and possibly a reagent (such amines) capable of regulating or templating the crystal growth are added into a solvent with appropriate ratios. This mixture is then placed in an autoclave to allow, the reaction and nanowire growth, to proceed at elevated temperatures and pressures. The major advantage of this approach is that most materials can be made soluble in a proper solvent by heating and pressurizing the system close to its critical point. As a result, this approach is used with any solid material, even for generating semiconductor nanowires. Also, Ge nanowires are synthesized (at relatively low yields) by reducing $GeCl_4$ or phenyl-$GeCl_3$ with sodium in an alkane solvent heated and pressurized to 550 K and 10^7 Pa (just above the critical point). Single-crystalline nanowires of 7–30 nm in diameter and up to 10 μm in length are obtained. By this method a rich variety of materials can produce nanowires, tubes, and whiskers. Although this method is versatile in generating 1D nanostructures the products are in low yield, low purity, and poor uniformity in size or morphology.

10. Solution-Phase Methods Based on Capping Reagents

The shape of a crystal is determined by the relative specific surface energies associated with the facets of this crystal. At equilibrium, a crystal has to be bounded by facets giving a minimum total surface energy, (Wulff facets theorem). So, the shape of a single-crystalline nanostructure often reflects the intrinsic symmetry of the corresponding lattice which is a cube and not a rod. The shape of a crystal can also be considered in terms of growth kinetics, by which the fastest growing planes should disappear to leave behind the slowest growing planes as the facets of the product. So the final shape of a crystal can be controlled by introducing appropriate capping reagent(s) to change the free energies of the various crystallographic surfaces and thus to alter their growth rates. A range of compounds are used as capping reagents (or the so-called surface-modifiers) to control the shape of colloidal particles synthesized using solution-phase methods. Two of these capping reagents are described below, that are used in the synthesis of that are discussed semiconductor nanorods and metal nanowires.

(a) Semiconductor and Metal Nanorods

The use of mixed surfactants is made to control the shape of CdSe nanocrytals. It is found that CdSe quantum rods with aspect ratios as high as 10 are obtained in large quantities by adding hexylphosphonic acid (HPA) to trioctylphosphine oxide (TOPO), a stabilizer commonly used in the synthesis of CdSe quantum dots. In the initial stage of the growth process, CdSe nanocrystals rapidly grow along the *c*-axis of the wurtzite structure to form nanorods with a broad distribution in aspect ratio. As the concentration of CdSe precursor becomes depleted in the solution, the short axis grows significantly and the aspect ratio eventually decreases to nearly one. The growth conditions are optimized to obtain CdSe nanorods as monodispersed samples and with well-controlled aspect ratios. The growth of CdSe nanorods is a diffusion-controlled process, and therefore cannot be explained using the Wulff reconstruction theorem. It is found that a strong cadmium ligand is required to maintain a relatively high precursor concentration in the solution.

(b) Metal Nanowires

A polyol method is demonstrated by which silver nanowires are obtained by reducing silver nitrate with ethylene glycol in the presence of poly(vinyl pyrrolidone) (PVP). The key to the formation of 1D nanostructures is the use of PVP as a polymeric capping reagent and the introduction of a seeding step. Figure 3.14 shows a plausible mechanism for this process. When silver nitrate is reduced in the presence of seeds (Pt or Ag nanoparticles of a few nanometers across), silver nanoparticles with a bimodal size distribution are produced via heterogeneous and homogeneous nucleation processes, respectively. In the following step, silver nanoparticles with larger sizes grow at the expense of smaller ones through the Ostwald ripening process. In the presence of PVP, most silver particles are confined and directed to grow into nanowires with uniform diameters. The silver nanostructures involved in each step are confirmed by electron microscopic and spectroscopic (based on surface plasmon resonance) experiments. In this synthesis, the cubic symmetry associated with silver is reduced to a lower one when two single-crystalline seeds are fused together to form a twinned nanocrystallite.

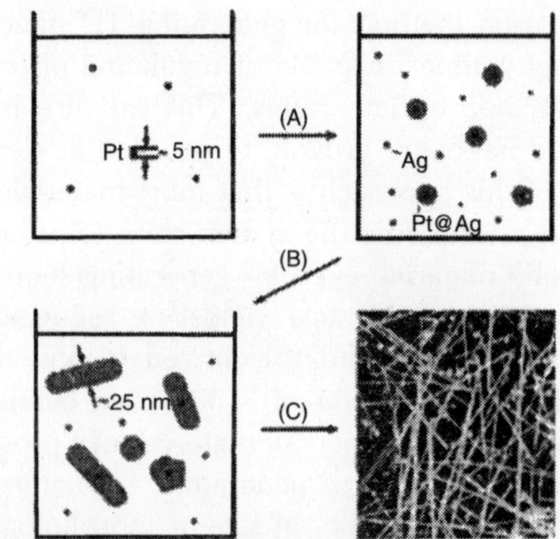

Figure 3.14 *Illustration of major experimental steps involed in the preparation of silver nanowise through a polyol process with Pt nanoparticles as the seeds: (A) Formation of bimodal silver nanoparticles through heterogeneous nucleation on Pt seeds and homogeneous nucleation: () evolution of rodshaped Ag nanostructures as directed by the cappind regent, poly (vinyl pyrolidone): and (C) growth of the Ag nanorods into wires at the expense of small Ag nanoparticles.*

The capability and feasibility of this polyol process demonstrates that bicrystalline nanowires of silver with diameters in the range of 30-60 nm and lengths up to ~50 μm are generated because in the product of synthesis, both nanowires and colloidal particles of silver are obtained. Such a mixture thus can be separated into pure components through centrifugation, with acetone added as the co-solvent. Figure 3.15(a) shows an SEM image of the purified silver nanowire thus indicating the removal of silver particles via centrifugation. These nanowires have a mean diameter of 40±5 nm. XRD indicate that these silver nanowires are crystallized purely in the fcc phase. Figure 3.15(b) is the TEM image of the nanowires, indicating the level of uniformity and perfection that is achieved using this solution-phase approach. The crystallinity and structure of these nanowires are further obtained by using HRTEM and electron diffraction (Fig. 3.15(c)). A low threshold for twinning along {111} planes of a fcc metal such as silver and gold is observed, when the nanostructures have relatively high aspect ratios. The HRTEM

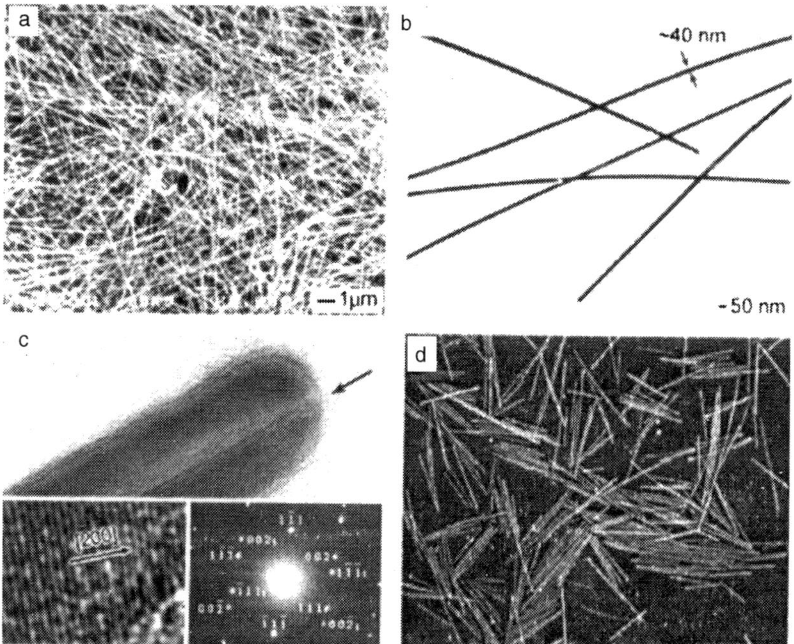

Figure 3.15 *(a) SEM and (b) TEM images of silver nanowires (after purification via centrifugation) that are synthesized at 15. (c) High-resolution TEM image of the end of an individual nanowire, with the arrow indicating a twin plane parallell to its long axis. The inset also shows the corresponding convergent-beam electron diffraction pattern, Indices without subscript refer to the left side of the nanowire shown in (c), and indices with subscript "t" refer to the right side, These two patterns have a reflection s mmetry about the {111}-type planes. (d) An SEM image of silver nanorods synthesized when the solution was heated at 455 K, with other conditions kept the same as those for (a-c).*

image and electron diffraction pattern both indicate that each individual silver nanowire contains a twin plane parallel to its longitudinal axis, with the mirror being positioned parallel to {111} planes. It is also possible to control the dimensions of silver nanowires by varying the experimental conditions. Figure 3.15(d) shows an SEM image of some silver nanowires that a synthesized at 455 K (with other conditions similar to those used for (a-c). These nanowires have a mean diameter of 39 ± 3 nm, and an average length of 1.9±0.4 μm. Note that the average length of these nanowires is greatly reduced by as much as 90%. It is found that the diameters of the silver nanowires reduces from 40 to 30 om by increasing the concentration of the seeding solution.

This synthesis can produce silver nanowires on the gram-scale. Thus, silver gold and other metals, can be processed as nanowires through solution-phase methods by finding a chemical reagent capable of selectively forming coordination bonds with various surfaces of a metal. The availability of metal nanowires in large quantities have a great impact on their use in electronic industry. For instance, the loading of metals in polymer composites can be greatly reduced by replacing those particles with nanowires with much higher aspect ratios. The reduction in metal loading can decrease the consumption of metals, as well as the weight of an electronic device. In addition, these metal nanowires are immediately useful as nanoelectrodes for electrochemical analysis and detection; they are also required (as interconnects or electrodes) in fabricating nanoelectronic devices through self-assembly approaches.

11. Other Potentially Useful Approaches

A number of radically different approaches are also used as alternative routes to the fabrication of 1D nanostructures. Two of these methods .that are based on the self-assembly of nanoparticles and size reduction, respectively are given below.

(a) Size Reduction

A number of procedures are developed to reduce the size of 1D structures that can be fabricated using conventional fabrication techniques. Here, only three of the methods are given for size reduction, that involve the use of isotropic deformation of a polycrystal-line or amorphous material, anisotropic etching of a single crystal, and near-field optical lithography with a phase-shift mask. It is possible to reduce the lateral dimensions of 1D structures from 1–10 μm to the regime ≤ 100 nm.

When a solid is uniaxially elongated, size reduction in two directions is achieved at the expense of dimensional increase in the third direction. Such a deformation can be made reversible by employing an elastomer. This reversibility adds another attractive feature to this approach in that the lateral dimension of the 1D nanostructures is continuously reduced and iteratively adjusted by controlling the extent of mechanical deformation. The deficiency of this approach is that it depends on the uniform distortion in a solid material; as the level of uniformity necessary at the < 100 nm scale is not so easy to achieve. The first successful example of this approach is the production of metal microwires. Here, a metal wire (~2 nm thick) is encapsulated within a glass capillary and reduced to many filaments of < 1 μm in diameter by drawing the system in a gas flame whose temperature is sufficiently high to melt the metal and soften the glass tube. By this process, metal filaments as thin as ~10 nm in diameter are fabricated. This method can be applied to many other solid materials, as long as they can be encapsulated by an appropriate glassy material.

In another method glass membranes are fabricated with hexagonal arrays of cylindrical holes as small as ~30 nm in diameter by repeatedly drawing a bundle of glass fibers > 1 μm in size. Each of the glass fibers consist of an etchable glass core surrounded by a thin sheath of a different, etch-resistant glass. In a third method structures (e.g., parallel arrays of metallic lines) of 100-200 nm width are obtained by mechanically manipulating the elastomeric stamp (or mold) involved in microcontact printing or replica molding. Although this method for size reduction lacks the characteristics required for registration in nanodevice fabrication, it provides easy access to 1D nanostructures that are directly useful in making sensors, arrays of nanoelectrodes, and optical diffraction gratings.

In the anisotropic etching of a single-crystalline substrate whose surface is patterned with an array of trenches, the width of the trenches decrease in a controlled fashion as etching proceeds because the etching rates along various crystallographic directions are different. For example,

V-shaped grooves with well-defined cross-sections can be generated in the surface of a Si(100) wafer by etching with an aqueous KOH solution. This technique enables the feature size to be continuously reduced from several micrometers to less than 100 nm by controlling the etching time. When combined with other tools, this technique provides a simple and convenient approach to the fabrication of supported 1D nanostructures in the form of parallel arrays. For example, tungsten nanowires as thin as 20 nm are fabricated by sputtering this metal onto patterned V-grooves from an oblique angle; metal or polymer lines of ~200 nm in width are generated by microcontact printing or photolithography with elastomer stamps cast from V-shaped grooves anisotropically etched in the surface of a Si(100) substrate.

A third method for size reduction is based on near-field optical lithography by using masks constructed from an elastomeric polymer such as poly(dimethylsiloxane) (PDMS). Such a soft mask can non-destructively come into conformal contact with a layer of photoresist over an area as large as several hundred square centimeters without the need for external force. When a light passes through this phase-shift mask, its intensity is modulated in the near-field such that an array of nulls or peaks in the intensity are formed at the edges of the relief structures patterned on the PDMS mask. As a result, nanoscale features are generated in the thin film of photoresist that is placed underneath the PDMS mask. The features patterned in the photoresist film are transferred into the underlying substrate using a reactive-ion-etching (RIE) or wet-etching process. The use of a polychromatic, incoherent light source and a conventional photoresist is made to produce parallel lines of ~90 nm in width over large areas on flat surfaces and on the surfaces of cylindrical lenses. By optimizing the design of the mask, it is also possible to generate lines as narrow as 50 nm. Since the exposure of the top surface of the resist film happens directly in the near-field the mask, it is believed that features smaller than 50 nm are be possible. The resolution is further improved by reducing the wavelength of the light source, the thickness of the resist film, the thickness of the modulating component (i.e., the surface relief of binary masks) by increasing the index of refraction of the resist, by using surface sensitive resists, or with a combination of these approaches.

With the use of simple binary PDMS masks only, it is possible to fabricate a variety of patterns that consist of lines with fixed width. The use of this technique is made to generate nanostructures of single-crystalline silicon with controllable dimensions and geometric shapes. The nanostructures are defined in a thin film of positive-tone photoresist using near-field optical lithography, and then the pattern is transferred into the underlying substrate-silicon-on-insulator (SOI) wafer–through RIE. Parallel arrays of Si nanowires ~130 nm in width and separated by ~1 μm (Fig. 3.16(a)) are obtained. The diameters of these wires can be further reduced to ~40 nm by using stress-limited oxidation in air at elevated temperatures. In addition to nanowires, the fabrication of nanorods, rings, interconnected triangles made of single-crystalline silicon is done. Figure 3.16 shows several typical SEM images of these nanostructures. This technique is used for a range of applications that require the generation of 1D nanostructures.

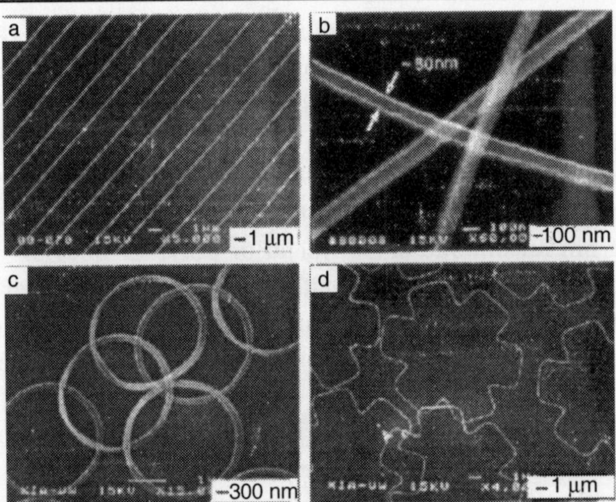

Figure 3.16 *(a) The scanning electron diffraction image of a 2D parallel array of nanobelts (130 nm wide and 100 nm thick) that are fabricated from single-crystalline silicon using near-field optical lithography. (b) An SEM image of these silicon nanobelts after they are oxidized in air at for ~1 h, and then lifted-off in HF solution. These nanowires have lateral dimensions of ~80 nm and lengths up to ~2 cm. (c) An SEM image of rings (~2 µm in diameter) that are made of single-crystalline Si wires with lateral dimensions of ~110 nm. (d) The SEM image of a more complex structure contain connecting triangles of Si wires with widths varying from 80 to 120 nm.*

(b) Self-Assembly of Nanoparticles

Self-assembly is used as a bottom-up approach for generating complex nanostructures on various scales. In particular, monodispersed colloids are used as building blocks for the formation of wire-like structures through self- or externally manipulated assembly. It is found that insulated 1D nanostructures are formed by filling the pores in an alumina or polymer membrane with gold nanoparticles or clusters. Nanowires of gold are obtained by sintering the nanoparticles confined within the pores of an alumma membrane at 570 K for several hours. In addition, it is found that ellipsoidal silver nanoparticles self-assemble into chains that are subsequently fused into bundles extending unbroken up to hundreds of nanometers in length in the solution phase. Also, gold clusters such as $Au_{55}(PPh_3)_{12}Cl_{16}$ are assembled into chains at the water-dichloromethane interface as induced by poly(*p*-phenyleneethynylene) dissolved in the liquid phase. If PVP is present in the solution phase, 2D networks of cluster-loaded polymer chains are observed at the liquid-air interface when surface compression is applied using a Langmuir-Blodgett trough. In this case, the gold clusters are mainly trapped and organized by the intersections between entangled polymer chains. The non-equilibrium self-assembly of metal nanoparticles on the surfaces of thin films are formed from di-block copolymers. The gold nanoparticles tend to aggregate into chains inside the polystyrene block (one of the blocks of the polystyrene-*block*-poly(methyl methacrylate) block copolymer) with a selectivity approaching 100 %. A Biopolymers as templates are used to guide the assembly of gold nanoparticles into chains or lattices, examples include DNA strands modified with appropriate side or end functional groups. In addition to molecule-based templates, patterned structures with mesoscale dimensions

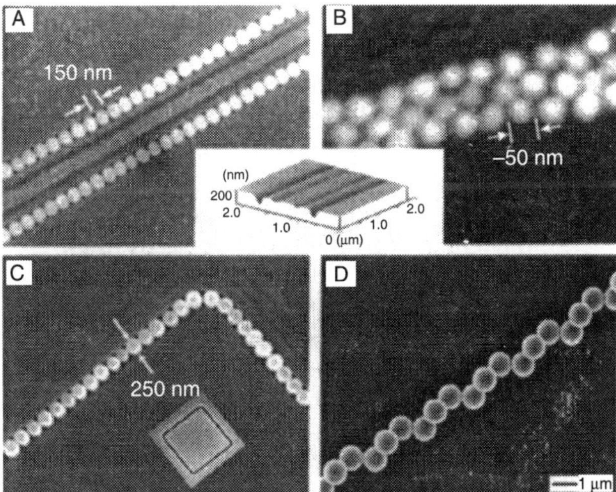

Figure 3.17 *(a-b) Structures that are assembled from 150 nm polystyrene beads (a), and 50 nm Au colloids (b), by templating against 120 nm-wide channels patterned in a thin photoresist film (see the inset). (c) An L-shaped chain of* $AuSiO_2$ *spheres assembled against a template (see the inset) patterned in a thin photoresist film (d). A spiral chain of polystyrene beads that are assembled by templaling against a V-groove etched in the surface of a Si(100) wafer.*

are explored as templates to direct the assembly of nanoparticles into structures with linear and more complex morphologies. For example, it is possible to assemble spherical colloids into a number of complex structures by templating against relief structures patterned in the surfaces of solid substrates. Figure 3.17 shows SEM images of several examples, including chains of spherical colloids that are characterized by a linear or helical conformation. Note that it is difficult to generate 1D nanostructures with the spiral morphology shown in Figure 3.17(d) using any other method. This approach is applied to building blocks with dimensions below 100 nm, as long as templates are fabricated with feature sizes on the same scale.

In addition to self-assembly, external manipulation with a field or a mechanical probe is also explored to induce or direct the organization of nanoparticles into arrayed structures. For example, metallic nanoparticles suspended in a liquid medium assemble into microwires up to 5 nm in length under dielectrophoresis. The mobility of and attractive interaction between nanoparticles are induced by an alternating electric field. Also, linear chains, circular rings, and pyramidal lattices are fabricated by manipulating the positions of individual gold nanoparticles with the probe of a scanning force microscope. Linear chains of gold nanoparticles are fabricated using a similar method and subsequently explored as templates to form continuous gold nanowires supported on the surface of a SiO_2/Si substrate via the hydroxylamine seeding process. In addition to their use as a precursor to conductive nanowires, linear chains that consist of equally spaced gold (or silver) colloids, serve as nanoscale plasmonic waveguides. Other external forces explored to direct the assembly of nanoparticles are lateral capillary forces, optical tweezers, magnetic fields, and electrostatic interactions. The use of dipole-dipole interactions is made to drive the self-assembly of monodisperse CdTe nanoparticles into highly crystalline nanowires with uniform sizes.

12. GROWTH OF ZnO NANOBELTS BY HEATING Zn IN AIR

The synthesis of ZnO nanowires from metallic zinc at moderately low temperatures (around 770 K) is achieved and attributed to a self-catalysis mechanism, in which metallic zinc selves both as source and catalyst, or to a tip-growth mechanism, in which Zn atoms diffuse through the ZnO nanostructure. Similar synthesis techniques are applied to other oxides, such as MgO and GeO_2. Some require molten zinc (melting temperature around 690 K) but ZnO nanowire synthesis is done from solid zinc at temperatures down to 620 K. The absence of an external catalyst makes these nanowires ideal for electronic applications, due to the lack of contaminating elements.

Experimental Method

For synthesis, the source material is zinc metal powder (99%), with particle sizes on the order of tens of microns. The zinc powder is placed on an alumina plate and heated isothermally in an furnace, which is left open to the air. A range of growth temperature, from 640 to 840 K is investigated. The furnace is heated up to the growth temperature at a heating rate of 25, 30 or 35°K min^{-1}. The furnace is held at the growth temperature for 80 min, then turned off and cooled in air to room temperature. The furnace is attached to a diffractometer with CuKα radiation, a Gobel minor and an X'celerator position sensitive detector. Diffraction patterns are collected at room temperature before and after heating, and every 5 min at the synthesis temperature. The scan range is chosen to be 30°–40° 2θ, to encompass the (0002) and $(10\bar{1}0)$ reflections of zinc and the $(10\bar{1}0)$, (0002) and $(10\bar{1}1)$ reflections of ZnO. The integrated area under the ZnO reflections, normalized by dividing by the integrated area under the zinc reflections in the initial room-temperature scan, is used as a measure of the volume of ZnO for kinetics analysis. After synthesis, the samples are investigated using a LEO field emission scanning electron microscope (SEM) and a high-resolution transmission electron microscope (HRTEM). The detailed results and discussion are given below:

1. Growth Morphology

The morphology of the nanostructures is investigated via scanning electron microscopy (SEM). Two classes are observed: one-dimensional and two-dimensional growth. For one-dimensional growth, several morphologies are observed i.e.. for nanowires, nanofins and branched nanostructures. The nanobelts grow out from the zinc particles, having diameters around 20–30 nm and lengths of a few microns, with no catalyst particles at the tips, as shown in the inset in Figure 3.18(a). Nanofins are observed intermingled with nanobelts. These structures are flat and are wider at the base, narrowing in steps to the tip, as in Figure 3.18(a). The nanobelt length depends on growth time. In a sample grown for 8 h, the nanobelts are around 10 μm long, as in Figure 3.18(b), and nanofins are less common.

The branching in branched nanostructures exhibits hexagonal symmetry i.e. it reflects hexagonal crystal structure of ZnO, and occurs in one plane containing the main nanobelt, (see Figure 3.18(f)). No catalyst particles are seen on the tips. In one-dimensional growth, higher growth temperatures favor straight nanobelts, and fins, while lower temperatures favor branched nanostructures. High-resolution TEM (HRTEM) analysis of nanobelts, nanofins and branched nanostructures indicate that the nanostructures are single crystalline and that the growth direction is (11$\bar{2}$0}, as shown in Figure 3.18, which is consistent with the hexagonal symmetry of the branched nanostructures. The existence of branched nanostructures and split nanofins indicate that growth occurs at the tip of the nanostructure and not at the base, thus, ruling out the self-catalysis mechanism, in which growth occurs at the zinc/ZnO interface.

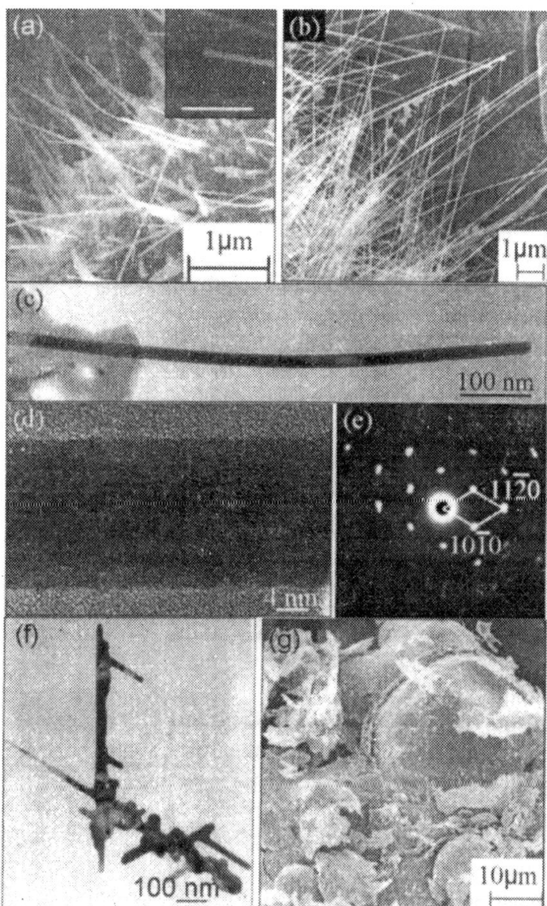

Figure 3.18 *Electron microscopy of nanobelts, including (a) SEM images of nanobelts and nanofins grown for 80 min at 840 K with a 25K min^{-1} heating rate (inset scale bar 300 nm); (b) SEM image of longer nanobelts grown for 8 h at 790 K with a 30 K min^{-1} heating rate; (c) TEM image, (d) HRTEM image and (e) electron diffraction pattern of a nanobelt grown for 80 min at 740 K with a 35 K min^{-1} heating rate; (f) branched nanostructure grown for 80 min at 740 K with a 35 K min^{-1} heating rate; (g) two-dimensional structures grown for 80 min at 790 K with a 25 K min^{-1} heating rate.*

The existence of one-dimensional or two-dimensional growth depends on both the growth temperature and the heating rate. In samples with two-dimensional growth that are heated above 670 K the two-dimensional growth layer peel away from the particles. (see Figure 3.18(g)). When cooling from the growth temperature, a mismatch in the thermal contraction causes the ZnO layers to detach from the zinc particles, as the linear thermal expansion coefficient of zinc is 'about four times that of ZnO.

2. Morphology Diagram

As mentioned above, the growth morphology is affected by the growth temperature and the heating rate to that temperature. A morphology diagram secn in Figure 3.19 is constructed to illustrate the dependence of morphology on temperature and heating rate. Faster heating rates and higher temperatures favor one-dimensional growth. (One anomaly is observed when heating up to 690 K at 25 K min^{-1}, where one sample

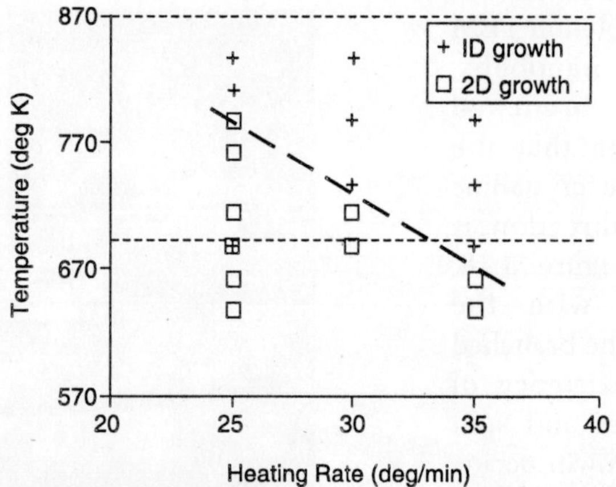

Figure 3.19 *Morphology diagram illustrating dependence of morphology on temperature and heating rate. The dashed line separates the regions of 1D and 2D growth. The dolled horizontal line indicates the melting temperature of zinc.*

exhibit one-dimensional growth, although all others exhibit two-dimensional growth.) The dependence of morphology on temperature is related to the increase in the vapor pressure of zinc with temperature. A higher partial pressure of zinc around the particles produce one-dimensional growth, by increasing the availability of Zn away from the surface.

The reason for the dependence of morphology on the heating rate is less clear. The fact that changing the heating rate only changes the beginning of a synthesis run, implies that its effect only occurs in the nucleation and perhaps initial growth stages. Another observation is that the growth mechanism does not require molten zinc, as a sample synthesized at 690 K, with a heating rate of 35 K min^{-1}, exhibit one-dimensional growth. The solid state of the zinc is evident by the presence of crystalline zinc reflections at the growth temperature. This is consistent with that found in ZnO nanobelt growth from Zinc at 620 to 670 K.

3. *Kinetics*

The kinetics of the reaction are investigated via x-ray diffraction data collected at the growth temperature. Initial scans for all samples contain zinc, with no ZnO present. Scans collected at 640 K contain no significant ZnO signal, only zinc, indicating the absence of significant growth of ZnO, as in Figure 3.20(a). Scans collected at 665 to 790 K contain both ZnO and zinc reflections. Scans collected at temperatures of 715 K and above contain only ZnO reflections, as the zinc is not crystalline. Despite the presence of molten zinc, SEM images show that the particles do not coalesce, indicating the presence of a ZnO shell that is able to contain the molten zinc. The strength of the ZnO grows with time, as seen in Figure 3.20(b). The growth is initially quick but it gradually slows down.

Figure 3.21 displays the normalized ZnO volume curves for samples with a 25 K min^{-1} heating rate at different temperatures and with different morphologies. The curves do not begin at the origin, indicating that growth initiates before reaching the growth temperature. At the

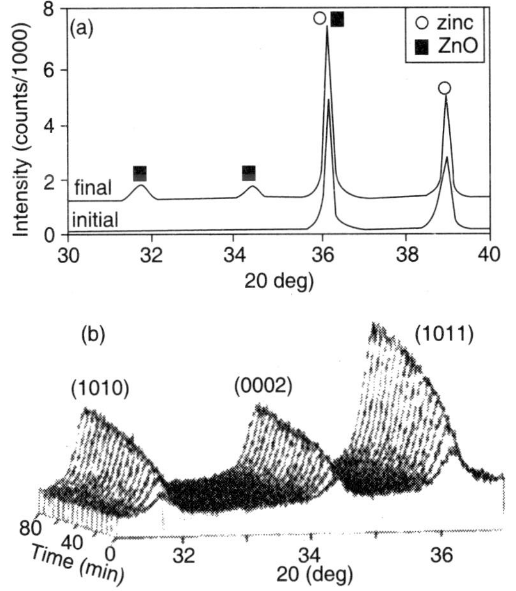

Figure 3.20 *XRD scans collected (a) al room temperature before and after growth and (b) at 770 K in 5 min intervals during growth. The heating rate is 30 K min^{-1}.*

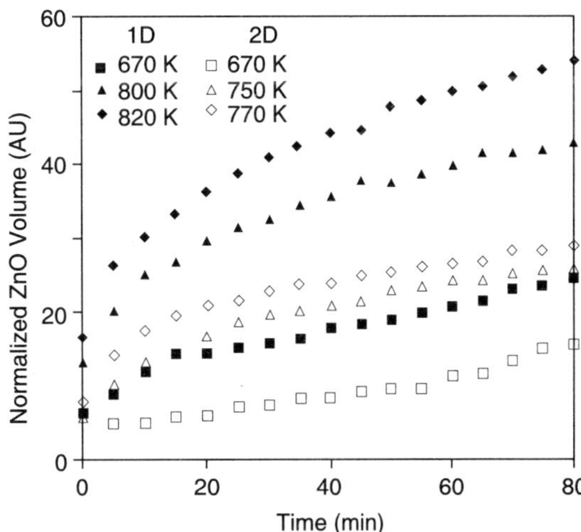

Figure 3.21 *Normalized ZnO volume versus time for samples heated at 25 K min^{-1} to different temperatures resulting in one-or two-dimensional growth.*

same heating rate and the same dimensionality of growth and at higher temperatures faster kinetics is there. This is because higher growth temperatures, lead to higher zinc vapor partial pressures and faster Zn diffusion. One-dimensional growth have faster kinetics than those with two-dimensional growth. In the case of two-dimensional growth, the formation of thicker ZnO layers on the zinc particles limit the vaporization or inhibit the diffusion of zinc, leading to slower kinetics. X-ray data is obtained for ZnO layers on the zinc particle surface in addition to ZnO nanostructures.

Finally, for one-dimensional growth, faster heating rates give slower kinetics at the same growth temperature, as illustrated in Figure 3.22, which displays the normalized ZnO volume curves for samples exhibiting one-dimensional growth at different temperatures and heating rates. The effect of the heating rate on kinetics suggests that the nucleation and initial growth stages are vital in controlling the growth of ZnO nanostructures. Similar trends occur in the growth of oxide nanostructures from other metals that have similar interactions with oxygen.

13. Growth of Self-assembled ZnO Nanowire Arrays

Experiment

The ZnO nanostructures are synthesized through a simple vapour-transport deposition process in a single-zone horizontal tube furnace equipped with an alumina tube, water-cooled end caps,

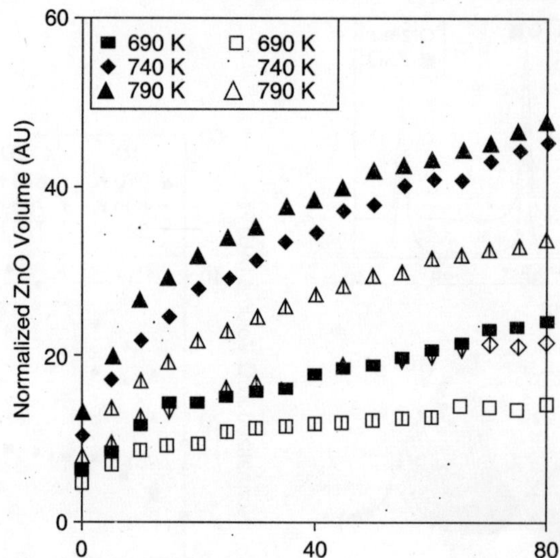

Figure 3.22 *Normalized ZnO volume versus time for samples exhibiting one-dimensional growth with different temperatures and heating rates.*

rotary pump system, a UV laser system as an assistant heating source and a gas-controlling system. During the synthesis the temperature, pressure, atmosphere and evaporation time are controlled. Serving as source materials, ZnO, SnO_2 and graphite powders (1.62; 1.51 and 0.12g, respectively) are mixed, ground and then loaded into an alumina crucible, which is placed in the middle of the alumina tube. Placed 'downstream' at a lower temperature region in the tube, a single crystalline (111)-plane silicon wafer serves as a substrate for the growth and collection of ZnO nanostructures. The alumina tube is sealed with two water-cooled end caps, to maintain a desirable temperature gradient across the tube, from the middle to the end, during the synthesis. The alumina tube is placed under vacuum (~1 Pa) for several hours before introducing Ar gas. The pressure is allowed to increase to and maintained at ~20 kPa throughout the synthesis process. The temperature in the middle of the tube is gradually elevated to 1370 to 1420 K and maintained at the peak temperature for 20–30 mm. The sublimated vapour from the source is transported by flowing Ar towards the Si substrate. Moreover, an additional target source is placed near the substrate and ablated by pulse laser to provide extra vapour. Local growth conditions near the substrate lead to the formation of various nanostructures. The resultant ZnO nanostructures are characterized by a field emission scanning electron microscopy (SEM), a SEM attached to an EDS detector and transmission electron microscopy (TEM}.

ZnO nanorod Clusters

At a peak temperature of 1370 K highly hierarchical ZnO nanowire arrays and uniaxial fuzzy nanowires are found on the Si substrate, which is in the temperature region 670–770 K. At the relatively lower temperature region of the Si substrate, numerous ZnO nanowire arrays grow from the surface, as shown in Figure 3.23. Many ZnO nanowires are roughly perpendicular to the underneath of the Si substrate. At the same time, numerous nanowires grow radially from one spot. The inset in Figure 3.23 shows that the ZnO nanowires have a tapered front tip. Although, SnO_2 is introduced in the source materials and Sn is believed to serve as a catalyst but no Sn catalyst is found on top of the ZnO nanostructures. TEM images of the nanowires are given in Figure 3.24. Figure 3.24(a) shows a typical ZnO nanowire at this temperature region, which gradually increases towards the growth front and shrinks rapidly to form a sharp tip. The high resolution TEM image in Figure 3.24(b) and the electron diffraction pattern in Figure 3.24(c) confirm that the ZnO nanowlre has a wurtzite structure and the growth direction is along [0001].

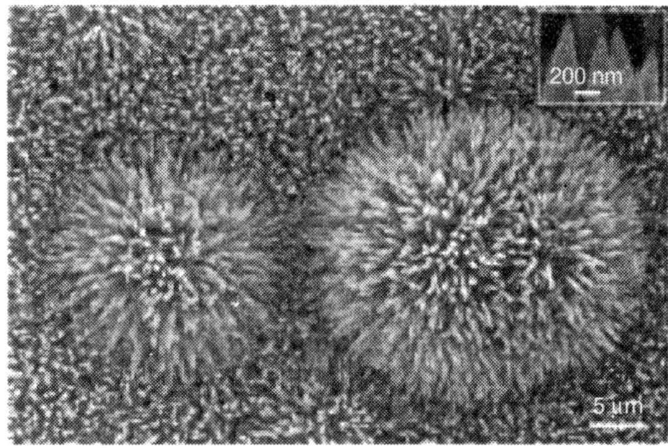

Figure 3.23 *ZnO nanowire arrays and radial nanowire clusters. showing enlarged nanowire tips in the insets.*

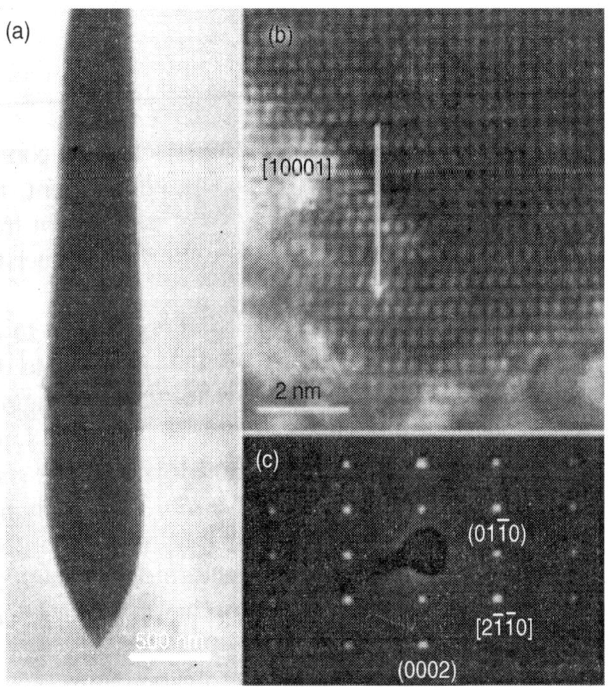

Figure 3.24 *(a) TEM image of a nanowirc taken from the ZnO nanowire array. (b) HRTEM image taken from the tip of the nanowire in (a). (c) Electron diffraction pattern of the nanowire showing a [0001] growth direction.*

Surprisingly, a more complex ZnO nanowire assembly is found in a higher temperature region on the same Si substrate. Figure 3.25(a) shows numerous ZnO nanowires growing radially from a common axis. Due to this 3-dimensional structure and ultra-dense growth, high yields of ZnO nanowires are achieved. The ZnO nanowires at the root of this cluster are slightly longer than those at the top. The tip of the nanowires also varies from the bottom to the top of the cluster. The squared areas in Figure 3.25(a), and (c) represent ZnO nanowires from the top and bottom part, respectively. The ZnO nanowires in Figure 3.25(b) have diameters of 300–400 nm, a rough surface and truncated hemispherical tips. In comparison, the ZnO nanowire in Figure 3.25(c) have diameter of 300 nm or less, a smoother surface and flat tips composed of multiple small 'fingers'. This difference is also observed in TEM characterization. Figures 3.26(a) and 3.26(b) show two types of ZnO nanowires, dissociated from the bottom and top of the cluster, respectively. The insets of Figures 3.26(a) and 3.26(b) show electron diffraction patterns from the corresponding nanowires indicating that all ZnO nanowires grow along the [0001]-direction. The HRTEM in Figure 3.26(c) from a 'finger' in Figure 3.26(b) suggests the single-crystallinity of wurtzite-structured ZnO nanowires and verifies the growth direction along [0001]. Although the ZnO nanowires have a blunt end, the component fingers have a very sharp, tapered tip, which is useful as a field emitter.

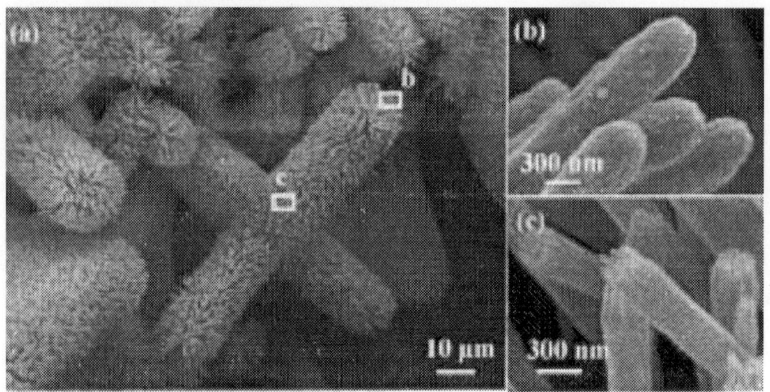

Figure 3.25 *(a) Uniaxial fuzzy nanowire clusters from a higher temperature region. (b, c) Enlarged images of nanowires from white squared area in (a).*

Assembly of ZnO Nanorods Clusters

For, the effect of temperature on ZnO nanostructures, experiments are performed with peak temperature of 1420 K at the centre and 820–920 K at the Si substrate. Self-assembled ZnO nanowire-nanobelt arrays are found on the substrates. In addition, a new ZnO nanostructure assembly, (see Figure 3.27,) is found on the same substrate in a higher temperature region (870–920 K). This nanostructure assembly remains the nanowire-nanobelt junction. Numerous ZnO radial nanowire clusters grow on the main trunk that resemble a bloom of flowers on branches. An enlarged image of one cluster in the inset of Figure 3.27 shows that all nanowires have a very uniform body and tapered tip. In addition, some ZnO nanowires have a small Sn ball at

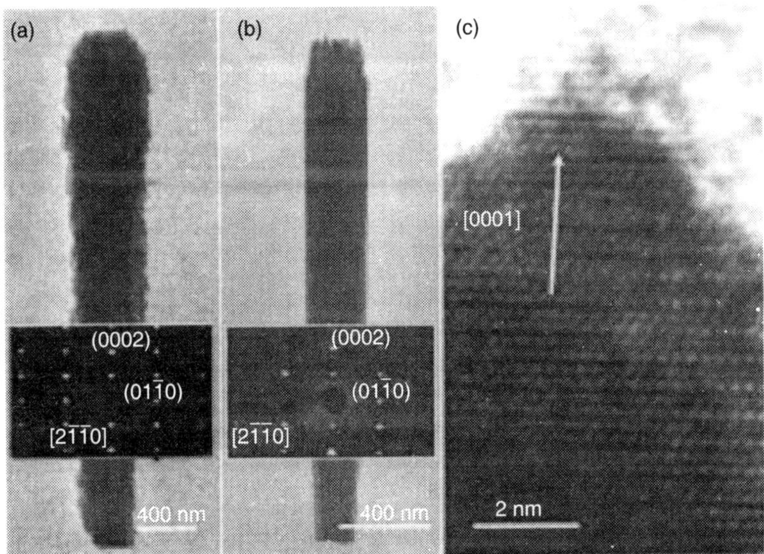

Figure 3.26 *(a. b) TEM images and corresponding electron diffraction patterns of nanowircs from uniaxial fuzzy nanowire cluster. (c) HRTEM image taken from the tip of the nanowire in (b).*

Figure 3.27 *Flower-like nanowire clusters on the nanowire-nanobelt junction with inset showing an enlarged image of a flower-like nanowire cluster.*

the end of the tip, while others have none. These ZnO nanowire clusters are mainly found on the nanowire-nanobelt junction backbones.

Figure 3.28 shows TEM of a typical nanowire-nanobelt junction structure. More interestingly, Numerous smaller secondary ZnO nanowires grow from the nanobelts, as indicated by the white arrowheads in Figures 3.28(b) and 3.28(c). The TEM image and electron diffraction pattern confirms that nanowires and nanobelts are of the same crystal unit and have a wurtzite structure. Sharing the common $\pm(2\bar{1}\bar{1}0)$ plane, ZnO nanowires and ZnO nanobelts grow along the [0001]-direction and $[01\bar{1}0]$ (and equivalent) direction, respectively. ZnO nanobelts have $\pm(0001)$ planes as side surfaces. In addition, the ZnO nanobelt in Figure 3.28(b) clearly indicates

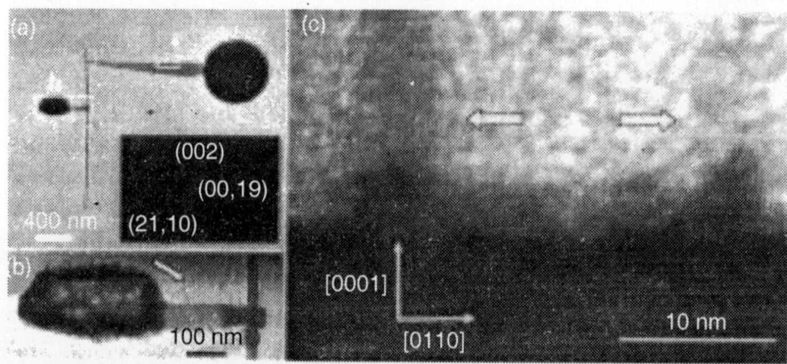

Figure 3.28 *(a) TEM image of a nanowire-nanobelt junction structure. Inset: electron diffraction pattern taken from the white circled area. (b) A nanobelt branch clearly showing asymmetrical, secondary ZnO nanowire growth from the upper surface, as indicated by white arrowhead. (c) HRTEM image taken from the white, squared area in (a), showing the single-crystal structure and growth directions of [01$\bar{1}$0] and [0001] for the nanobelt and nanowirc, respectively. White arrowheads indicate the secondary nanowires.*

that secondary ZnO nanowires grew from one side only and rarely from the opposite surface. A high-resolution TEM image of the ZnO nanobelt and secondary nanowires, from the white squared area in Figure 3.28(a), is shown in Figure 3.28(c), indicating that the ZnO nanowires grew epitaxially from the nanobelt in the [0001]-direction. All secondary ZnO nanowires display a single-crystalline structure with a diameter of about 6 nm.

Thus, ZnO nanowires are grown via a the vapour–liquid–solid (VLS) process with Sn catalyst on the tip or occasionally at the root. In the absence of other metal catalysts, ZnO nanowires are grown via a vapour–solid (VS) process and a self-catalytic effect play a role in nanostructure formation. The morphology and growth direction of nanostructures are determined by growth kinetics, which is controlled though pressure, growth temperature, etc. In the experimental setup, ZnO and SnO_2 powders give off Zn, ZnO, Sn and O vapours under the reduction of graphite and these vapours are transported to the substrate via the flow of carrier gas. Due to differences in chemical activity between Sn and Zn, SnO_2 is dissociated into Sn and O vapour before ZnO. Accordingly, at an early stage of synthesis, Sn and O vapour is the dominant source with small amounts of Zn and ZnO vapour. Sn vapour condenses on the surfaces of the Si substrate to form various liquid Sn balls. At higher temperatures, a few ZnO wires are also formed in the early stages but are quickly surrounded by a liquid Sn layer or balls as a result of the inflow of Sn-rich vapour. At lower temperatures, less ZnO wires are formed due to the low Zn and ZnO vapour concentrations. Sn vapour decreased quickly as a result of consumption of the SnO_2 source powder. As the Zn and ZnO vapour concentrations increase in the synthesis region. ZnO nanowires start growing from the Sn balls on the Si substrate and from Sn layer or balls on the few early grown ZnO nanowires. Under current growth conditions, ZnO growth along [0001] is greatly enhanced and growth along other directions is mainly suppressed, which is why only ZnO nanowires are observed along [0001] in TEM characterization. ZnO nanowires grown from tiny Sn balls on the Si substrates at the lower temperature range form nanowire arrays, as shown in Figure 3.23. Due to agglomeration, the larger Sn balls simultaneously catalyze numerous ZnO nanowires and result in radial ZnO nanowire clusters, as shown in Figure 3.23, which is similar to the growth of highly aligned silica nanowires from molten gallium balls. In comparison, Sn balls or liquid Sn layers on the ZnO wires in the higher temperature range also absorb the inflowing Zn vapours and initiate fuzzy growth of ZnO nanowires around the common axis, resulting in the highly hierarchical ZnO nanostructure assemblies, shown in Figure 3.25. A slightly better vapour source at the top of the assembly might result in a rougher surface, larger diameter and different tip morphology compared with ZnO nanowires at the lower part of this assembly.

In comparison, the nanostructures in Figures 3.27 and 3.28 follow a slightly different process. Here, the Si substrate is closer to the centre and the peak temperature is also higher. ZnO nanowires are formed before the furnace reaches a peak temperature of 1420 K Under these growth conditions, other growth directions are also possible, resulting in the ZnO nanobelt branches along $[01\bar{1}0]$. shown in Figure 3.28. with $\pm(0001)$ as side surfaces. It is well-known that Zn-terminal polar surfaces are chemically active while the opposite. O-terminal polar surface is inert during growth. The self-catalyzed growth from (0001) surfaces only results in the asymmetrical growth of secondary ZnO nanowires from one side of the ZnO nanobelt. In addition, the Sn balls at the tip of the ZnO nanobelts and other places can also initiate considerable growth of the most favourable ZnO nanowires along [0001] leading to the flower-like nanowire clusters.

14. Polar Surface Induced Growth Phenomena

Although the entire unit cell of a crystal is neutral, the distribution of the cations and anions take specific configuration as determined by crystallography. For some oxides, such as MgO, NiO and ZnO, the cations and anions in oxides produce surfaces that have net positive/ negative charges. The polar charge dominated surfaces give some unique growth phenomena. ZnO shows the characteristics of polar surfaces. It has a hexagonal structure (space group conc) with lattice parameters $a = 0.3296\%$, and $c = 0.52065$ nm. The structure of ZnO is described as a number of alternating planes composed of tetrahedrally coordinated O^{2-} and Zn^{2+} ions, stacked alternatively along the *c*-axis (Fig. 3.29(a)). The tetrahedral coordination in ZnO results

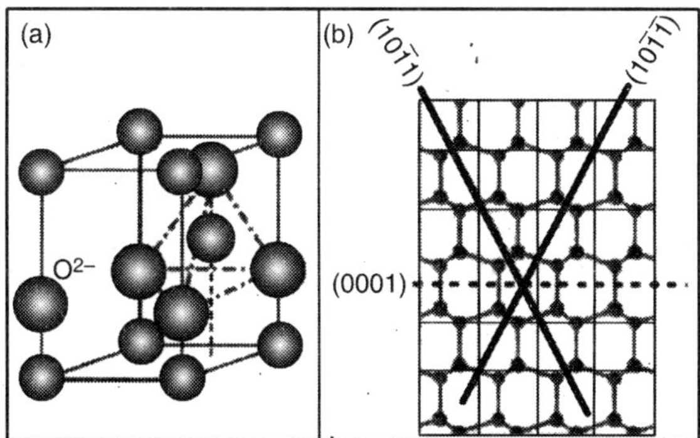

Figure 3.29 *(a) Wurtzite structure model of ZnO, which has non-central symmetry and piezoelectric effect. (b) The three types of facets of ZnO nanostructures $\pm(0001)$, $\{2\bar{1}\bar{1}0\}$ and $\{01\bar{1}0\}$.*

in non-central symmetric structure and piezo-electricity. Another important characteristic of ZnO is the polar surfaces. The most common polar surface is the basal plane. The oppositely charged ions produce positively charged Zn-(0001) and negatively charged O-(000$\bar{1}$) polar surfaces, resulting in a normal dipole moment and spontaneous polarization along the *c*-axis as well as a divergence in surface energy. The Zn-terminated (0001) surface, for example, is catalytically active, while the O-terminated (000$\bar{1}$) surface is inactive, resulting in the growth of a saw-tooth structure (Fig. 3.30). To maintain a stable structure, the polar surfaces have facets or exhibit massive surface reconstructions, but ZnO ±(0001) are exception, which are atomically flat, stable and without reconstruction.

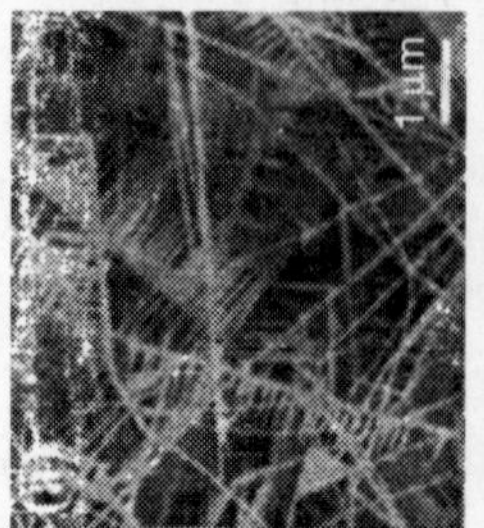

Figure 3.30

Another polar surface is the {01$\bar{1}$1}. By projecting the structure along [1$\bar{2}$10] as shown in Figure 3.39(b) beside the most typical ±(0001) polar surfaces that are terminated with Zn and oxygen, respectively, ±(10$\bar{1}$1) and ±(10$\bar{1}\bar{1}$) are also polar surfaces. The {10$\bar{1}$1} type surfaces are not common for ZnO, but they are observed in a nanohelical structure. The charges on the polar surfaces are ionic, which are non-transferable and non-flowable. Because the interaction energy among the charges depend on the distribution of the charges, the structure is arranged in such a configuration to minimize the electrostatic energy. This is the main driving force for growing the polar surface dominated nanostructures.

(a) Nanospiral and Nanosprings

Due to difference in surface energies among (0001), {01$\bar{1}$0} and {2$\bar{1}\bar{1}$0}, freestanding nanobelts and nanowires of ZnO are dominated by the lower energy, non-polar surfaces such as {01$\bar{1}$0} and {2$\bar{1}\bar{1}$0}. By introducing doping, such as In and/or Li, ZnO nanobelts dominated by the (0001) polar surfaces are grown. The nanobelt grows along [2$\bar{1}\bar{1}$0] (the *a*-axis), with its top/bottom large surface ±(0001) and the side surfaces ±(01$\bar{1}$0). Due to the small thickness of 5–20 nm and large aspect ratio of ~1:4, the flexibility and toughness of the nanobelts are extremely high. A polar surface dominated nanobelt acts as a capacitor with two parallel charged plates (Fig. 3.31)). The polar nanobelt tends to roll over into an enclosed ring to reduce the electrostatic energy (Fig. 3.31(b)). A spiral shape is also possible for reducing the electrostatic energy (Fig. 3.31(c)). The formation of the nanorings and nanohelixes is understood from the nature of the polar surfaces. If the surface charges are uncompensated during the growth, the spontaneous polarization induces electrostatic energy due to the dipole moment, but rolling up to form a circular ring minimizes or neutralizes the overall dipole moment, thereby reducing the electrostatic energy. On the other hand, bending of the nanobelt produces elastic energy. The stable shape of the nanobelt is determined by the minimization of the total energy contributed by spontaneous polarization and elasticity. If the nanobelt is rolled loop-by-loop, the repulsive forces between the charged surfaces stretch the nanospring, while the elastic deformation forces pull the loops together; the balance between the two forms the nanospring that has elasticity. The nanosprings have a uniform shape with radius of ~500–800 nm and evenly distributed pitches. Each is made of a uniformly deformed single-crystal ZnO nanobelt.

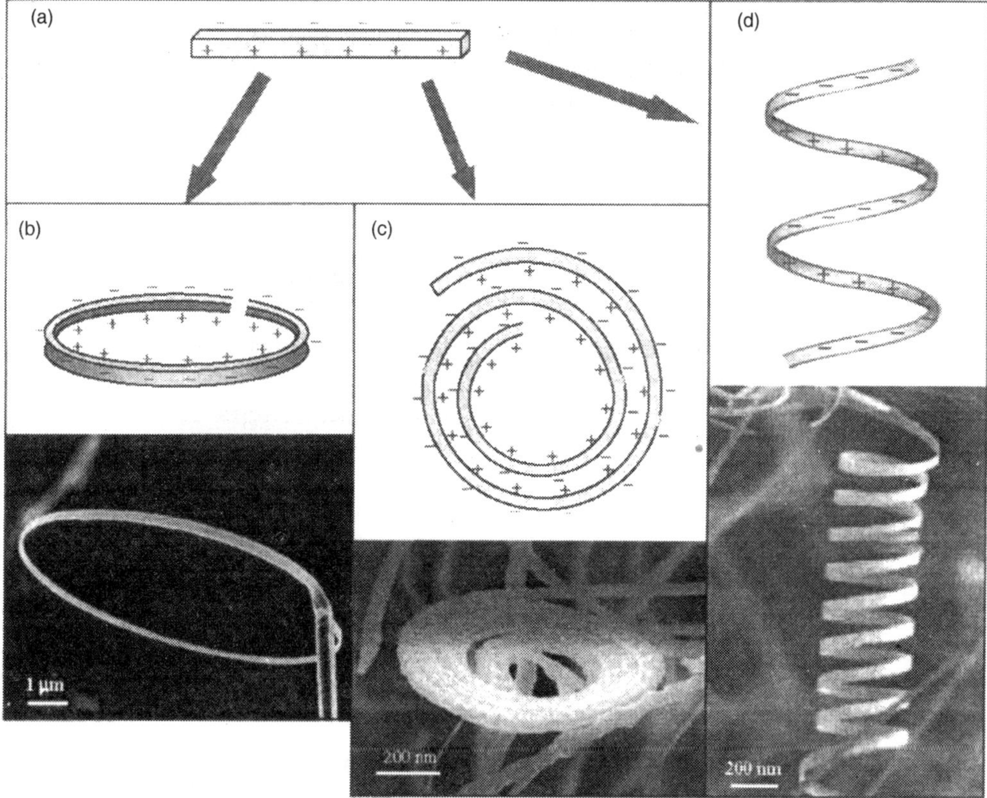

Figure 3.31 *Nanosprings of piezoelectric nanobelts. (a) A model of the polar surface dominated ZnO nanobelts. (b) Inward tending of the nanobelt results in the formation of a nanoring. (c) The formation of a nano-spiral. (d) The formation of a nanoopring.*

(b) Seamless Nanorings

By adjusting the raw materials with the introduction of impurities, such as indium, a nanoring structure is synthesized by the VS process (Fig. 3.32(a)). The as-synthesized sample is composed of many freestanding nanorings, with typical diameters ~1-4 μm, thin and wide shells of thicknesses ~10–30 nm. SEM images recorded at high-magnification present the perfect circular shape of the complete rings, with uniform shape and flat surfaces. TFM images (Fig. 3.32(b)) indicated that the nanoring is a single-crystal entity with circular shape, although there is diffraction contrast due to non-uniform deformation along the circumference. The single-crystal structure (a nanospring) is made of a single crystalline ribbon bent evenly at the curvature of the nanoring. Electron diffraction pattern of the center of the nanoring shows that the radial direction of the nanoring is $[1\bar{1}10]$, tangential direction $[10\bar{1}0]$ and nanoring axis [0001]. The nanoring is made of co-axial, uni-radius, epitaxial-coiling of a nanobelt. The trace of the coiling nanobelt is seen through the side of the nanoring. The interface between the loops is coherent, epitaxial and chemically bonded. The entire nanoring is a single crystal, although the quality of the crystallinity varies slightly across the width of the nanoring.

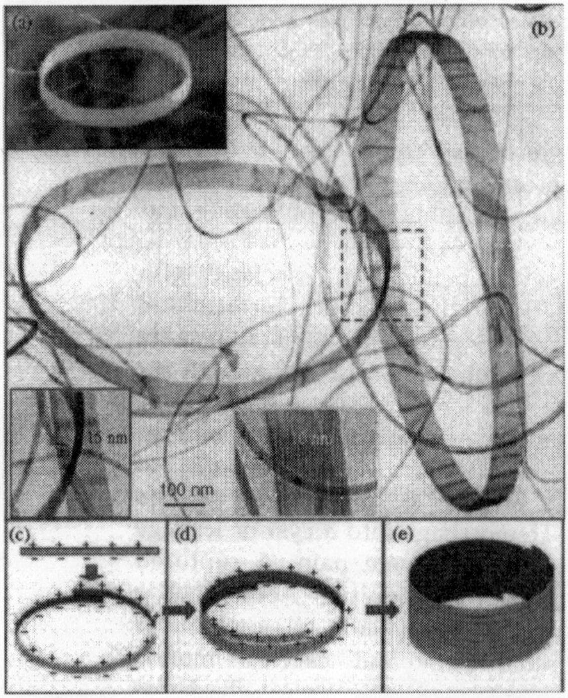

Figure 3.32 *(a) SEM image of single-crystal seamless nanoring of ZnO. (b) low magnification TEM image of the ZnO rings and ZnO nanobelts. The nanobelts have uniform shape and their widths are ~15 nm. which is about the same as the thickness of the ring shell, as measured from the tilted image inserted in the figure. The thickness of the nanobelt composing the ring is measured to be 10 nm. (c-e). The growth model shows the initiation and formation of the single-crystal nanoring via self-coiling of a polar nanobelt. The nanoring is initiated by folding a nanobelt into a loop with overlapped ends due to long range electrostatic interaction among the polar charges; the short-range chemical bonding stabilizes the coiled ring structure; and the spontaneous self-lining of the nanobelt is driven by minimizing the energy contributed by polar charges, surface area and elastic deformation.*

The growth of the nanoring structures is understood from the polar surfaces of the ZnO nanobelt. The polar-nanobelt, which is the building block of the nanoring, grows along $[10\bar{1}0]$, with side surfaces $\pm(1\bar{2}10)$ and top/bottom surfaces $\pm(0001)$, and have a width of ~15 nm and thickness ~10 nm. The nanobelt has polar charges on its top and bottom surfaces (Fig. 3.32(c)). If the surface charges are uncompensated during growth, the nanobelt folds itself as its length gets longer to minimize the area of the polar surface. One possible way is to interface the positively charged (0001)-Zn plane (top surface) with the negatively charged (0001)-O plane (bottom surface), resulting in neutralization of the local polar charges and the reduced surface area, thus, forming a loop with an overlapped end (Fig. 3.32(c)). The radius of the loop is determined by the initial folding of the nanobelt at the initial growth. The total energy involved in the process comes from polar charges, surface area and elastic deformation. The long-range electrostatic interaction produces the initial driving force for folding the nanobelt to form the first loop for the subsequent growth.

The presence of planar defect within the nanobelt leads to the fast growth of the nanobelt along $[10\bar{1}0]$, because it lowers the energy in the wurtizite-structured lattice. As the growth continues, the nanobelt is attracted onto the rim of the nanoring due to electrostatic interaction and extends parallel to the rim of the nanoring to neutralize the local polar charge and reduce the surface area, resulting in the formation of a self-coiled, co-axial, uni-radius. multi-looped nanoring structure (Fig. 3.32d)). The self-assembly is spontaneous, which means that the self-coiling along the rim proceeds as the nanobelt grows. The reduced surface area and the formation of chemical bonds (short-range force) between the loops stabilize the coiled structure. The width of the nanoring increases as more loops wind along the nanoring axis, and all of them remain in the same crystal orientation. Since the growth is carried out in a temperature region of 470–670 K "epitaxial sintering" of the adjacent loops form a single-crystal cylindrical nanoring structure, and the loops of the nanobelt are joined by chemical bonds as a single entity (Fig. 3.32(e)). A uni-radius and perfectly aligned coiling is energetically favorable because of the complete neutralization of the local polar charges inside the nanoring and the reduced surface area.

The polar-charge induced nanorings shown above are unique and distinct, and have potential applications in fundamental physical phenomena, such as the oscillations in the exciton luminescence. The unique piezoelectric and semiconducting properties of ZnO predict that the nanorings can be nano-scale sensors, transducers and resonators.

(c) Effect from Growth Kinetics

It is a common fact that the {0001} polar surfaces are the dominant facets of single-crystal nanosprings and nanoloops of ZnO. Since the polar surfaces {0001} have higher surface energy than either $\{2\bar{1}\bar{1}0\}$ or $\{10\bar{1}0\}$, finding experimental conditions that make it possible for forming the higher energy {0001} surface is important. Table 3.1 summarizes the conditions, growth morphology and dominant surfaces of ZnO nanostructures. It is shown that, despite the source peak temperature in the range of 1570–1670 K and a variation in vaporization time, both polar surfaces dominate and non-polar surface dominated nanobelts are produced. In the column of source materials used, it is seen that polar surface dominated nanobelts are synthesized with or without doping in ZnO. The pressure of the as flow is not critical to the formation of polar surfaces unless the pressure becomes very low. A common and striking fact is that a pre-evacuation of the growth chamber to $\sim 10^{-3}$ mm Hg. appears to be necessary for growing {0001} dominated nanobelts. A pre-pumping reduces the content of oxygen and other gas species in the growth chamber, which greatly reduces the possibility of molecular adsorption on the {0001} polar surfaces. This is the key for high-yield growth of polar surface dominated nanostructures. Here, the synthesis is conducted at 1660 K for 2~6 hours under a confined pressure range of 250~300 mbar after the temperature of reaction chamber is ramped to 1070 K where argon carrier gas is kept with a flow rate of 50 seem throughout the high-temperature synthesis and cooling processes. Before inletting argon carrier gas, the pre-growth pressure of 2×10^{-2} mmHg. Is maintained for 3 to 8 hours.

Table 3.1 *A summary of the ZnO belt-like nanostructures synthesized by a vapor-solid process.*

	Nanobelt growth direction/dominant flat surfaces	*Yield (%)*	*Peak Thmperature (K); time at peak temperature; source materials*	*Flow rate of Ar gas (seem)*	*Pre-growth pressure; pressure during growth (Pa)*
Nanobelts	[0001]/{10-10}; or [10-10]/{2-1-10}	100	1670 K; 120 min; ZnO 16:19 'k; 30 min; ZnO:Li2O	50	4×10^4
Nanosprings and nanosprings	[2-1-10]/{0001}	~5	1620 K; 30 min; ZnO; Li_2O	25	~10; 3.3×10^4
Nanorings	[10-10]/{0001}	~20	1670 K; 130 min; ZnO; In_2O_3; Li_2CO_3	50	~10; 6.6×10^4
Nanobows	[10-10]/{0001}	5	1620 K; 120 min; ~10 hrs; ZnO	50	$\sim 10^2$; 3×10^4
Nanosprings (current work)	[2-1-10]/{0001}	50–100	1660 K 120–360 min; ZnO	50	~10; (2.5–3) 10^4
Nanorings and nanospirals	[10-10]/{0001} (current work)	20	1660 K; 120–360 min; ZnO	50	~10; (2.5–3) 10^4

*The pressure in the growth chamber is quickly dropped to ~10 Pa towards the end of the growth.

In another experiment, it is found that a sudden drop in chamber pressure leads to formation of polar surface dominated nanobows. At a growth chamber pressure of 300 mbar, [0001] ZnO nanowires are grown (Fig. 3.33). A sudden decrease of the chamber pressure to 10 Pa results in the formation of nanorings and nanobows at the growth front, which are formed by rolling the *c*-plane dominated nanobelts. A drop in is pressure favorable of formation of polar surfaces due to the reduced probability of foreign molecule adsorption on the surface to neutralize the polar charges. The polar nanobelts, responsible for the formation of nanorings and nanobows, grow along either the $[2\bar{1}\bar{1}0]$ or $[01\bar{1}0]$ direction. A ring is formed by folding a polar nanobelt so that its *c-axis* points to the center of the ring. A change in growth direction from $[01\bar{1}0]/[2\bar{1}\bar{1}0]$ to $[10\bar{1}0]/[\bar{1}2\bar{1}0]$ could result in the formation of two crossed rings at 60° angle.

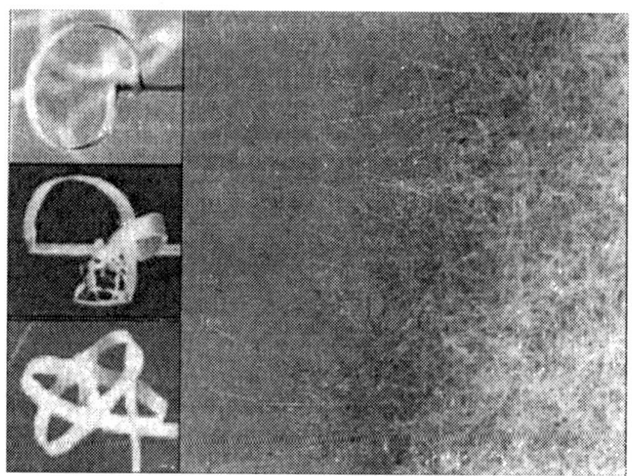

Figure 3.33 *Controlled growth of nanorings and nanobows at the tips of ZnO nanorods/nanowires by a sudden change in growth pressure towards the end of the experiment, so that the grow direction switched from [0001] 10 $[01\bar{1}0]$ or $[2\bar{1}\bar{1}0]$. The polar surface dominated nanobelts bent to form nanorings and nanobors.*

(d) Self-Catalyzed Growth Structures

Asymmetric crystal structure of a material introduces anisotropic growth structure. For ZnO, the (0001) surface is terminated with Zn, and the $(000\bar{1})$ surface is terminated with O. The former is catalytically active, while the latter is inert, and results in anisotropic growth phenomena. Here, the formations of nanocombs and tetrapoles as examples are illustrate for the self-catalysis effect.

"Comb-like" structures are observed for wurtzite structured materials, such as ZnO, ZnS, and CdSe. The comb structures (Figs. 3.34(a, b)) are ZnO and have the comb-teeth growing along [0001], the top/bottom surfaces being $\pm(01\bar{1}0)$, and side surfaces $\pm(2\bar{1}\bar{1}0)$. Using convergent beam electron diffraction (CBED), which relies on dynamic scattering effect and is an effective technique for determining the polarity of wurtzite structure, it is found that the CORD structure is an asymmetric growth along Zn-[0001]. This conclusion is received by comparison the experimentally observed CBED pattern and the theoretically calculated pattern by matching

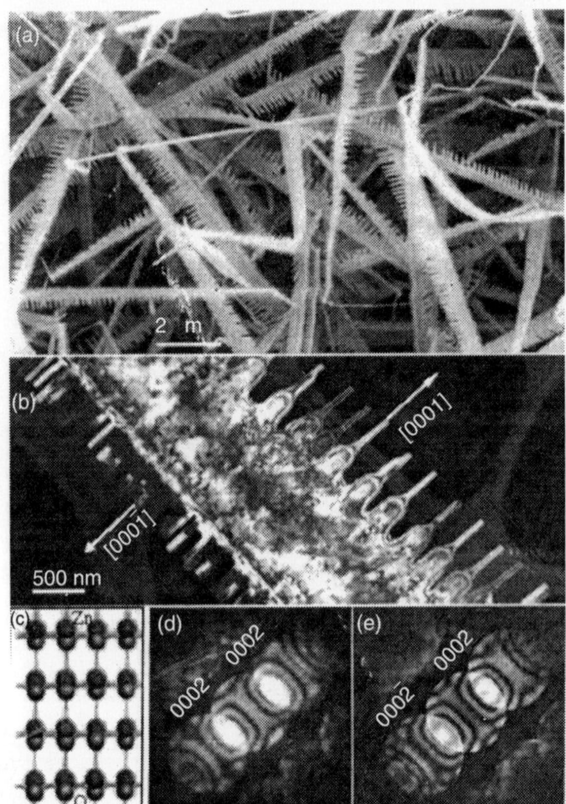

Figure 3.34 *Self-catalyzed growth of cation terminated polar surfaces. (a) SEM image of "comb-like" cantilever arrays of ZnO, which is the result of surface polarization induced growth due to the chemically active (0001)-Zn. (b) High-resolution TEM recorded from the tip of the comb-teeth, showing possible evidence of Zn regregation at the growth front, which is likely to be effective for driving self-catalyzed growth. (c) Structure model of ZnO projected along [01$\bar{1}$0], showing the termination effect of the crystal. (d, e) Experimentally observed and theoretically simulated convergent beam electron diffraction patterns for determining the polarity of the nanocombs, respectively.*

the fine detailed structure features in the (0002) and (000$\bar{1}$) diffraction disks (Figs. 3.34(d) and (e)). The positively charged Zn-(0001) surface is chemically active and the negatively charged O-(000$\bar{1}$) surface is relatively inert, resulting in a growth of long fingers along [0001]. Using HRTEM it is found that the Zn-terminated (0001) surface has tiny Zn clusters, which lead to self-catalyzed growth without the presence of foreign catalyst. The chemically inactive (000$\bar{1}$)

STRATEGIES FOR ACHIEVING 1D GROWTH

The evolution of a solid from a vapor, liquid, or solid phase involves two fundamental steps: nucleation and growth. As the concentration of the building blocks (atoms, ions, or molecules) of a solid becomes sufficiently high, they aggregate into small clusters (or nuclei) through homogeneous nucleation. With a continuous supply of the building blocks, these nuclei serve as seeds for further growth to form larger structures.

When developing a synthetic method for generating nano-structures, the most important issue is the simultaneous control over dimensions, morphology (or shape), and monodispersity (or uniformity). A variety of chemical methods have been demonstrated as the "bottom-up" approach for generating 1D nanostructures with different levels of control over these parameters. Figure 3.35 schematically illustrates some of these synthetic strategies that include (i) use of the intrinsically an isotropic crystallographic structure of a solid to accomplish 1D growth (Fig. 3.35(a)); (ii) introduction of a liquid-solid interface to reduce the symmetry of a seed (Fig. 3.35(b)); (iii) use of various templates with 1D morphologies to direct the formation of 1D nanostructures (Fig. 3.35(c)); (iv) use of supersaturation control to modify the growth habit of a seed; (v) use of appropriate capping reagent(s) to kinetically control the growth rates of various facets of a seed (Fig. 3.35(d)): (vi) self-assembly of 0D nanostructures (Fig. 3.35(e); and (vii) size eduction of 1D micro-structures (Fig. 3.35(f)). For many of these methods the characteristics (such as reproducibility, product uniformity and purity, potential for scaling up, and mechanism) are not known. Therefore their control of size range and flexibility in materials that can be synthesized as well as their intrinsic limits should be highlighted.

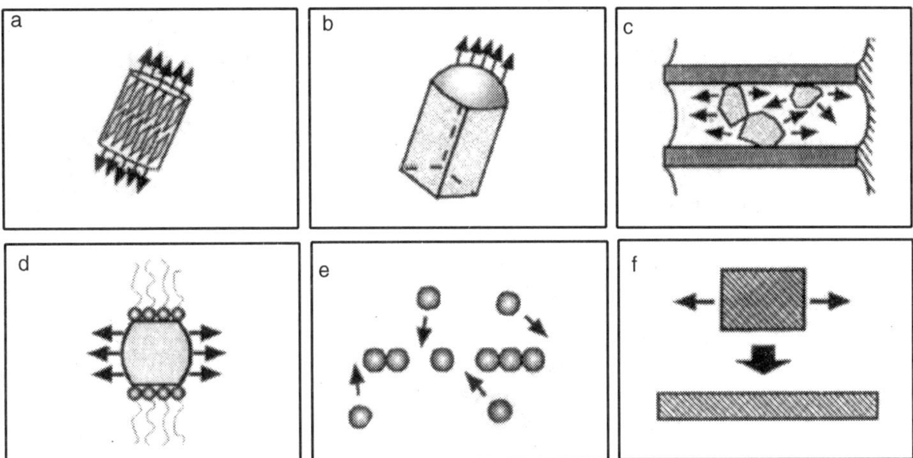

Figure 3.35 *Illustrations of six different strategies that are demonstrated for achieving 1D growth: (a) dictation by the anisotropic crystallographic structure of a solid; (b) confinement by a liquid droplet as in the vapor-liquid-solid process; (c) direction through the use of a templale; (d) kinetic conlrol provided by a capping reagent; (e) self-assembly of 0D nanoslructures; and (f) size reduction of a 1D microstructure.*

Materials with Highly Anisotropic Crystal Structures

Many solid materials naturally grow, into 1D nanostructures, and this is determined by the highly anisotropic bonding in the crystallographic structure. One of the best-known examples is poly(sulphur nitride), $(SN)_x$, an inorganic polymer that is metallic and has superconducting properties. Uniform nanowires ~20 nm in diameter and hundreds of micro-meters in length are grown from vapor phase $(SN)_x$, and some of them aggregate into bundles. The focus is on the growth of single crystals with dimensions as large as possible for conductivity measurements. Many inorganic minerals such as asbestos and chrysolite also exhibit a fibrous growth habit, which is due to their chain structures, or some other anisotropic arrangements between atoms or ion groups within their crystal lattices. Directed and confined by the anisotropic conformation of their building blocks these materials can be processed into nanowires with diameters as that of the dimensions of the seeds perpendicular to the growth direction. In addition, many polymeric and biological systems exist in the fibrous form; for example, celluloses and collagen. Two inorganic systems–molybendum chalcogenides and chalcogens are given below for their interesting electronic and optical properties.

1. Molybdenum Chalcogenide Molecular Wires

Molybdenum chalcogenides, with the general formula $M_2Mo_6X_6$ (M = Li, Na; X = Se, Te), are a family of compounds that contain hexagonal close-packed linear chains in the formula of Mo_6X_6. The Mo_6X_6 chain is considered as a prismatic column (Fig 3.36(a)) formed by staggered stacking the Mo_3X_3 triangular units, with a repeating distance of 0.45 nm. When dissolved in a highly polar solvent such as dimethylsulfoxide or *N*-methylformamide, they exist as chains ~2 nm in diameter. Some chains aggregate into bundles (or fibers) with cross-sections of ~1 μm diameter and lengths up to ~20 μm. It is possible to fabricate a polymeric matrix containing $(Mo_3Se_3^-)_n$ mono and biwires by polymerizing a dilute solution of $LiMo_3Se_3$ in vinylene carbonate. These molecular wires are 0.6–2 nm in diameter and 5–10 nm in length. The structural and electronic properties of molybendum selenide molecular wires are investigated by scanning tunneling microscopy (STM). The tunneling spectroscopic measurements indicate the existence of sharp peaks in the local density of states. No evidence for the opening of an energy gap (or the metal-to-insulator transition) is found 3.36(b)) in their conductance measurements with temperatures down to 5 K. These molecular wires are self organized into mesoscopic bundles (Fig. 3.36(b) in the presence of organic surfactants of opposite charges The transmission electron

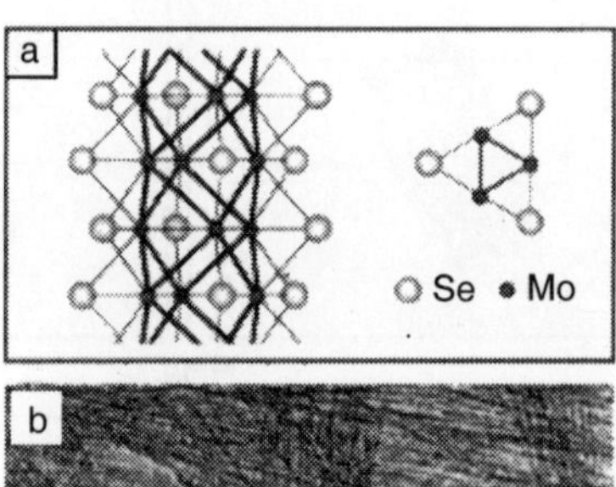

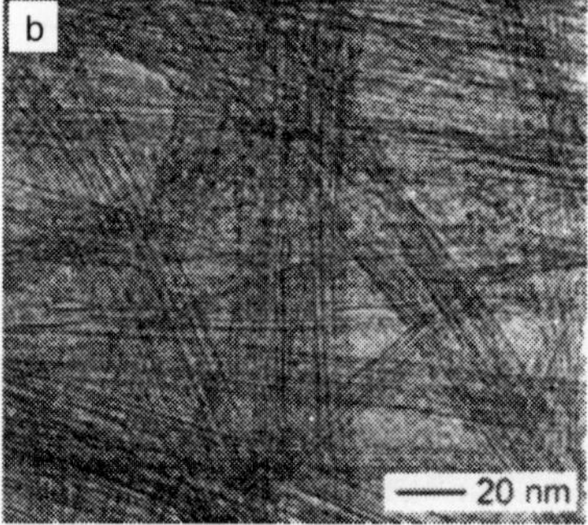

Figure 3.36 *(a) Structural model of the linear chain contained in $LiMO_6Se_6$ molecular wires that is formed by staggered stacking the triangular planar $(MO_3Se_3)^-$ units. (b) A TEM image of bundles assembled from $(MO_6Se_6)^-$ molecular wires in the presence of polymerizable cationic surfactants such as w-undecenyltrimethyl-ammonium bromide (w-UTAB).*

microscopy (TEM) and low-angle X-ray diffraction (XRD) indicates that the crystallinity along each individual molecular wire is maintained while the spacing between these inorganic wires varies in the range of 2–4 nm by changing the length of surfactant molecules. It is further found that the counter cations within the $Li_2Mo_6Se_6$ nanowires form $[Mo_3Se_3^-]$ nanowires with different counter cations, and hence electrical properties tunable from semiconducting to superconducting.

2. Selenium and Tellurium

Se and Te represent another system for generating nanostructures with 1D morphologies. The trigonal (t-) phase of these two solids is interesting because of its unique crystal structure. Unlike oxygen, Se and Te atoms tend to form polymeric helical chains through covalent bonding while oxygen exists primarily as O_2 molecules. As shown in Figure 3.37(a), the helical chains are packed into a hexagonal lattice through van der Waals interactions. Crystallization tends to occur along the *c*-axis, favoring the stronger covalent bonds over the relatively weak van der Waals forces among chains. As a result, these two solids have a strong tendency to become 1D structures even when they are crystallized from an isotropic medium. In addition to their natural anisotropy, Se and Te have a range of other intriguing properties. Not least among these is their intrinsic chirality. In principle, each individual nanostructure is composed of either *R*- or *L*-helices. These two materials have photoconductivity (~0.8×10^5 S cm^{-1} for t-Se), piezoelectricity, and high reactivity to generate a wealth of important functional materials (e.g., optoelectronic materials such as CdSe and ZnTe, and thermoelectric materials such as PbTe and Bi_2 Te_3). The above materials have been used in an array of applications that include light or temperature sensors, rectifiers, photocopying machines, inorganic pigments and piezoelectric actuators.

(a) Selenium Nanowires

A solution-phase approach to the large-scale synthesis of uniform nanowires of t-Se with lateral dimensions controllable in the range of 10 to 100 nm, and lengths up to several hundred micrometer is achieved. The first step of this approach is the formation of solid selenium in an aqueous solution through the reduction of selenious acid with excess hydrazine by refluxing this reaction mixture at an elevated temperature:

$$H_2SeO_3 + N_2H_4 \rightarrow Se(\downarrow) + N2(\uparrow) + 3H_2O \quad (1)$$

The initial product is brick-red-colored, spherical colloids of amorphous (a-) selenium with diameters of ~300 nm. When this solution is cooled to room temperature, the small amount of selenium dissolved in the hot solution precipitates out as nanocrystallites of t-Se. When this dispersion containing a mixture of a-Se colloids and t-Se nanocrystallites is aged in the dark, it undergoes the Ostwald ripening process, in which the a-Se colloids slowly dissolve in the solution as a result of their higher free energy as compared to the t-Se phase. The dissolved selenium subsequently grows as crystalline nanowires of t-Se on the seeds (Fig. 3.37(b)). In this transformation, the linear morphology of the final product is determined by the intrinsic anisotropy of the building blocks–that is, the extended, helical chains of Se atoms in the trigonal phase. Each nanowire is a single crystal (see Fig. 3.37(c)) for a high-resolution (HR)TEM

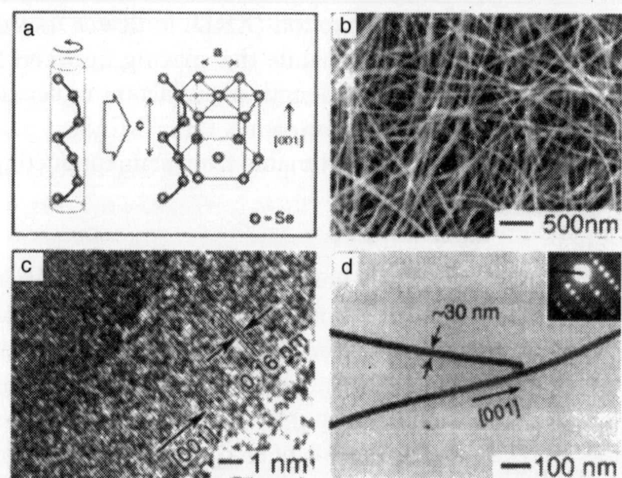

Figure 3.37 *(a) An illustration of the crystal structure of t-Se composed of hexagonally packed, helical chains of Se atoms parallel to each other along the c-axis. (b) A scanning electron microscopy (SEM) image of t-Se nanowires with a mean diameter of 32 nm. (c) A high-resolution TEM image recorded from the edge of an individual nanowire shows well-resolved interference fringe spacing of 0.16 nm that agrees well with the interplanar distance between the [003] lattice planes. (d) A TEM image of two t-Se nanowires, indicate the dimensional uniformity along wire. The inset shows an electron diffraction pattern obtained from the middle portion of an individual nanowire, confirming that the growth direction is along the [00$\bar{1}$] axis.*

image and the inset of Figure 3.37(d) for an electron diffraction pattern), characterized by a uniform diameter along its longitudinal axis (Fig. 3.37(d). The diameters of these nanowires is easily varied in a controllable fashion from 10 to 100 nm by changing the temperature at which the redox reaction is refluxed. An increase in the aging time lead to the formation of longer nanowires with essentially no variation in the wire thickness. Note that no exotic seeding materials and surfactants are used in this synthesis, and both byproducts of the reaction (nitrogen gas and water) do not cause contamination problems for the t-Se nanowires. The absence of kinks and other types of defects make these nanowires useful in fabricating nanoscale electronic and optoelectronic devices.

Formation of t-Se seeds is induced at room temperature by using sonication. Figure 3.38(a) illustrates the mechanism of such a process. Here, spherical a-Se colloids are prepared as an aqueous suspension by reducing selenious acid with an excess amount of hydrazine. This reaction is carried out at room temperature to prevent any homogeneous nucleation events as induced by a drop in temperature. The colloids are characterized by diameters in the range of 0.1 to 2 μm. They are subsequently dried and redispersed in an alcohol such as ethanol. When a short pulse of sonication is applied, these colloids undergo disruption and aggregation. In addition, small nanocrystallites of t-Se are generated on the surfaces of the colloids as a result of cavitation effects. In the following step, Se atoms are transferred from the less stable, amorphous colloids to the trigonal seeds until all a-Se colloids are consumed. This mechanism differs from the above paragraph in a number of aspects: for example, the seeds are formed primarily on

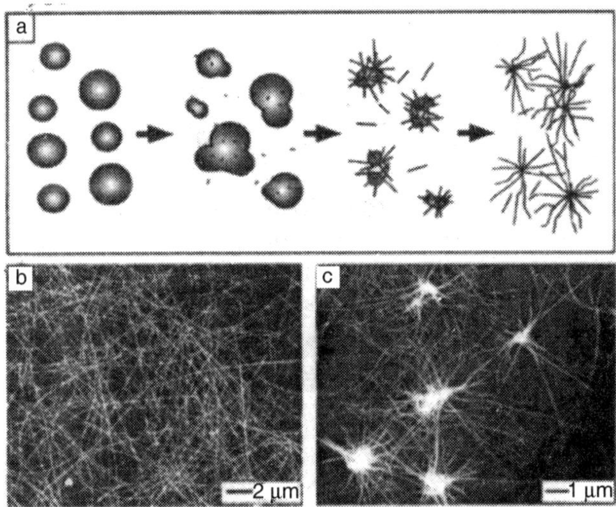

Figure 3.38 *Illustration of major steps involved in the sonochemical approach to the synthesis of t-Se nanowires: formation of t-Se seeds on the surfaces of a-Se colloids through cavitation; growth of t-Se nanowires at the expense of a-Se colloids; and continuous growth of t-Se nanowires until all a-Se colloids have been consumed. (b) An SEM image of t-Se nanowires formed in an ethanol solution after they had grown for ~5 h. (c) An SEM image of t-Se nanowires that are directly grown into an interconnected 2D network by supporting the sonicated a-Se colloids on the surface of a silicon substrate.*

the surfaces of the colloids (rather than in the dispersing medium); wire growth mainly originates from the surfaces of a-Se colloids; and the growth occurs in a mater of hours (rather than days) due to the higher solubility/mobility of selenium in an alcohol. Figure 3.38(b) shows t-Se nanowires that are synthesized using the sonochemical route with a yield approaching 100%. Note that complete transformation of a-Se colloids to t-Se wires is observed after growth is allowed to proceed for ~ 5 h. Uniform nanowires with lengths up to several hundred micrometers and controllable diameters in the range of 25 to 120 nm are achieved. Since both redox reaction and wire growth are carried out at ambient temperature and pressure, this procedure is scaled up for high-volume production.

There is another feature associated with this sonochemical approach. Once the colloidal suspension is sonicated, the disrupted a-Se colloids (whose surfaces are decorated with newly formed t-Se seeds) is deposited onto a surface and dried there. This substrate is then submersed in an alcohol (e.g., by placing a drop of ethanol on the surface in a closed, airtight container to prevent evaporation) to allow the growth of nanowires to proceed across the surface of the substrate. Figure 3.38(c) shows nanowires that are grown on a silicon substrate (using this new procedure). Here it is observed that the nanowires radiate out from a-Se colloids in a conformal 2D network. Varying the concentration of a-Se colloids in the initial dispersion controls the wire density in the network. The conformal growth of nanowires supported on solid substrates represents an important step toward self-guided growth of interconnects between nanoelectronic devices.

(b) Tellurium Nanowires, -Rods,-Tubes, and -Belts

The growth of Te whiskers in the vapor phase is achieved by growing them on solid substrates by controlling the temperature. It is demonstrated that t-Te nanowires are synthesized using a procedure similar to the one for the Se nanowires, in which the reduction of precursor acid by hydrazine generates tellurium, nitrogen gas, and water:

$$2Te(OH)_6 + 3N_2H_4 \rightarrow + 2Te\ (\downarrow) + 3N2\ (\uparrow) + 12H_2O \qquad (2)$$

The primary difference between Se and Te systems is that Te atoms form nuclei without the need for cooling of the solution. As a result, there are two types of Te products formed by this redox reaction: a-Te colloids and t-Te seeds (in the form of nanocrystallites). The transfer of material from the amorphous to the crystalline phase is essentially the same for both Se and Te. Because nucleation events occur in the reduction of orthotelluric acid, it is harder to control the monodispersity of the tellurium nanowires. Nevertheless, the t-Te nanowires prepared by this reaction are characterized by a relatively narrow distribution in size, with a typical standard deviation of < 10 % (Fig. 3.39(a)).

Similar to the vapor-phase experiments various morphologies are observed over different temperature ranges: wires formed at 290–370 K (with water as the solvent) develop a spine-like morphology having slight taper and equilateral triangular cross-sections. Wires formed at 370–470 K in ethylene glycol display, filamentary structures with pronounced taper and isosceles triangular cross-sections. Synthesis carried out at 445 K in a water/ethylene glycol mixture yielded nanorods with hexagonal cross-sections and no tapering (Fig. 3.39(b)). Here, the product is highly monodispersed in dimensions, with a mean diameter of 98±3 nm and length of

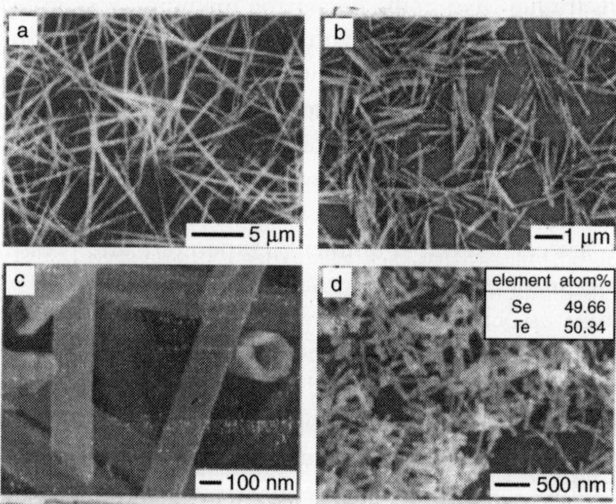

Figure 3.39 *(a, b) SEM image of t-Te nanowires and nanorods synthesized using a solution-phase method similar to that demonstrated for selenium (c) An SEM image of t-Te nanotubes that are synthesized by reducing orthotelluric acid with ethylene glycol at 470 K. (d) An SEM image of $Se_{0.5}Te_{0.5}$ nano-rods that are synthesized by reducing a mixture (1:1) of selenious acid and orthotelluric acid with hydrazine.*

1.80±0.16 μm. Electron diffraction confirms single crystallinity and growth direction along the <001> axis. These nanorods are formed because of an increased solubility and mobility for Te in ethylene glycol solution (with the addition of a small amount of water) and at high temperatures. This change ensures a higher supersaturation of Te atoms at the growing tips, and allows for the formation of defect-free nanowires with uniform hexagonal cross-sections.

A new morphology is observed when telluric acid is reduced by ethylene glycol solvent at 470 K via polyol process. The growth mechanism does not include the formation of amorphous colloids. The reaction begins with the decomposition of orthotelluric acid into tellurium dioxide, followed by the formation of t-Te hexagonally shaped seeds. The subsequent growth along the circumferential edges of these seeds lead to the evolution of a nanotube morphology. The solid walls of these nanotubes have a reasonably uniform thickness of ~30 nm, as determined by the diffusion of Te atom across the surfaces of the seeds. The lengths of these inorganic nanotubes is varied by controlling the growth time. Figure 3.39(c) shows the SEM image of a typical example of Te nanotubes crystallized in the trigonal phase. The formation of hollow structures is due to a manifestation of Te concentration profiles on the surfaces of their solid seeds. Immediately after the nucleation step, further addition of tellurium atoms to the seed surface occurs at the circumferential edges of each cylindrical seed because these sites have higher free energies than other sites on the surface. As soon as crystal growth begins mass transport to the growing regions lead to undersaturation (or complete depletion of tellurium) in the central portions of the growing faces, the {001} planes on each seed, and eventually result in the formation of nanotubes having well-defined hollow interiors. A hydrothermal route to the synthesis of single-crystalline Te nanobelts is achieved.

(c) Nanorods of Se/Te Alloys

Because trigonal-phase Se and Te solids crystallize in the same structure and the reaction conditions are the same for the generation of pure Se and Te nanowires. So by combining these two reactions 1D nanostructures can be generated from their alloys. It is possible to generate single-crystalline nanorods made of Se/Te alloys by reducing selenious and orthotelluric acids with hydrazine in one pot. These nanorods retain the elemental ratio as that of the acids used as the precursors. Each helical chain consists of domains of Se and Te atoms, as in a random organic block co-polymers. The individual Se and Te domains, do not resolved with TEM or electron diffraction, indicating that these two elements are blended on the atomic scale. Figure 3.39(d) displays an SEM image of nanorods that are obtained in a reaction that involves a 1:1 molar ratio of selenious and orthotelluric acids and excess hydrazine. Here, the mean diameter and length of these nanorods are ~50 nm and ~250 nm, respectively. Energy-dispersive X-ray (EDX) analysis of these nanorods is done, and characteristic peaks for selenium and tellurium observed and confirmed an elemental ratio of 1:1 (inset of Figure 3.39(d) for Se and Te in the nanorods. The lattice parameters calculated from the XRD diffraction pattern fall in between those values for trigonal tellurium and seleniunt, indicating the formation of a solid solution between these two elements in the alloy. By fine-tuning the elemental composition of these nanorods, it is possible to control their properties such as piezoelectricity (greater for Te) or photoconductivity (greater for Se) while still maintaining their dimensionality and single crystallinity.

3. Other Solids with Anisotropic Structures

In principle, the synthetic approach described above can be extended to a range of other solid materials whose crystallographic structures are characterized by chain-like building blocks. Many of these solids have low-dimensional semiconductors or conductors, and some of them are observed to crystallize in the form of fine needles or whiskers (the macroscopic and microscopic counterparts of nanorods). Typical examples include SbSI, a ferroelectric and optoelectronic material; $K_2[Pt(CN)_4]$ a narrow-bandgap semiconductor; and MX_3 (M = transition metal, X = S, Se, and Te), a host of semiconductors and thermoelectric materials; metallophthalocyanines, M(Pc) with M = H_2, Ni, and $[M(Pc)O]_n$ with M = Si, Ge, Sn, a group of organometallic polymers with metallic conductive and photoconductive properties. By modifying the experimental procedures that are developed for chalcogens and molybdenum chalcogenides, uniform nanowires from the reaction solutions of these solid materials can be obtained.

Growth Kinetics

Growth kinetics plays a key role in determining the morphology of nanostructures. Controlling local temperature is probably the most important parameter for controlling the grown nanostructure. Figure 3.40(a) shows an optical image of the nanostructures deposited on a substrate by vaporizing ZnO powders. The most interesting phenomenon is that the as-grown products are distributed at two distinct temperature regions, with the metallic luster is black color being the Zu nanobelts and the white color being the ZnO nanobelts. The two products are formed in the same growth chamber but they are clearly separated. Scanning electron microscopy (SEM) images show curly Zn nanobelts (Fig. 3.40(b)) and straight ZnO nanobelts (Fig. 3.40(c)). The Zn nanobelts are formed in a temperature range of 470–570 K and they are distributed across a region of >4 cm in length. The ZnO nanobelts are formed in a temperature range of 570–670 K. The transition distance between the two different products is ~0.5 cm. This shows the structural control by growth temperature and kinetics.

Figure 3.40 is obtained by using ZnO powder as the raw material in the experiment. The produced products are Zn nanowires and ZnO nanowires distributed in distinct temperature regions. It is well known that the decomposition of ZnO occurs when it is subject to high enough temperature in vacuum. The thermodynamics of these processes is given below. The solid-vapor process and the decomposition process are expressed as:

$$\mathrm{ZnO\ (s)} \leftrightarrow \mathrm{ZnO(g)} \tag{3}$$

$$\mathrm{ZnO\ (g)} \leftrightarrow \mathrm{Zn\ (g)} + \frac{1}{2}\mathrm{O_2(g)} \tag{4}$$

The equilibrium constant function K_p for processes (3) and (4) as a function of temperature is given by

$$-R \ln K_p\ (T) = \Delta \mathrm{H}^{\circ}_{298}/T + \Delta G(T) \tag{5}$$

where ΔH is the change in free energy and $\Delta G(T)$ the Gibbs energy, and $K_p = (P_{\mathrm{Zn}(g)} \cdot P^{1/2}_{\mathrm{O}_2})/P_e^{3/2}$, where P_{Zn} and P_{O_2} are the partial pressures of Zn and O_2, respectively, and P_e is the pressure in the growth chamber. Substituting the thermodynamic data into (5),

$$\ln(P_{Zn(g)} \cdot P^{1/2}_{\mathrm{O}_2}/P_e^{3/2}) = -4.855 - 2474/T \tag{6}$$

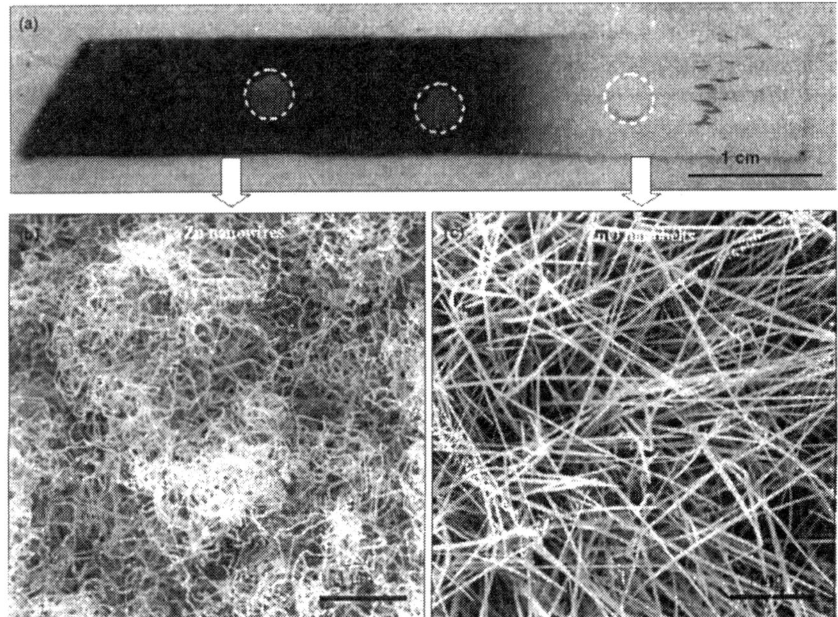

Figure 3.40 (a) *An optical micrograph of the as-synthesized sample on a silicon substrate, showing two distinct products on the surface. The three circles indicate three TEM cower grids placed on the substrate for collecting samples. (b, c) SEM images recorded from the metallic luster and white color regions, presenting the formation of pure Zn nanobelts and ZnO nanobelts.*

From equation (6), the saturation vapor pressure for Zn at 1620 K is estimated to be ~10^{-3} mm Hg, close to the pre-evacuated pressure in the growth chamber, hence, it is possible that part of the ZnO vapor decomposes into Zn vapor and O_2, Then, the ZnO vapor together with the Zn vapor are transported by Ar carrier gas to a lower temperature region of 470–670 K, which is cold enough to condense the vapor phase onto the substrate. The ZnO vapor condenses in the region of 570–670 K due to its higher sublimation temperature, resulting in the growth of ZnO nanobelts (the white region in Fig. 3.40(a)); the Zn vapor condenses in the 470–570 K region owing to lower sublimation temperature, resulting in the growth of Zn nanobelts. surface typically does not grow nanobelt structure.

4

Growth of ZnO Nanowires

CONTROLLED GROWTH

(a) Control of Orientation

Controlling the growth orientation is important for many applications of nanowires. By applying the conventional epitaxial crystal growth technique to VLS process, it is possible to achieve precise orientational control during nanowire growth. The vapor-liquid-solid epitaxy (VLSE) is particularly powerful for the controlled synthesis of nanowire arrays.

Nanowires generally have preferred growth directions. For example, Si nanowires prefer to grow along the <111> direction, while ZnO nanowires prefer to grow along the <001> direction. To grow vertically aligned nanowires properly select the substrate and control the reaction conditions, so that the nanowires grow epitaxially on the substrate. For example, sapphire is an ideal substrate with lattice constant $a = 4.75$ Å and $c = 12.94$ Å; the c surface of sapphire is composed of alternate layers of six-fold symmetric oxygen and threefold symmetric Al atoms, while in the wurtzite structure of ZnO, both O and Zn are six-fold symmetric about the ZnO *c*-axis. Since the ZnO a-axis and the sapphire *c*-axis are related by a factor of 4 (mismatch less than 0.08% at room temperature), ZnO nanowires grow epitaxially from the (110) plane of sapphire (Fig. 4.1(a)). Figure 4.1(b) shows the growth of ZnO nanowires on an *a*-plane (110) sapphire substrate. Their diameters range from 40–120 nm and lengths are between 2 and 10 μm. They are grown vertically from the substrate as expected from the epitaxial growth. The hexagonal end surface of the nanowires is there (see Figure 4.1(c)). This type of nanowire array growth is also achieved using MgO(111) as substrate. Figure 4.2 shows a typical XRD pattern recorded on these ZnO nanowire arrays. Only (001) diffraction peaks are observed, indicating excellent (001) orientation/alignment of the nanowires on a large area of the substrate.

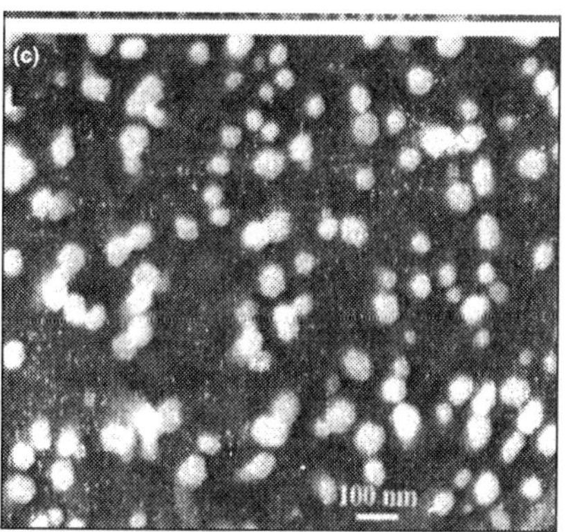

Figure 4.1 *The epitaxial growth of ZnO nanowires on a-plane (110) sapphire can be readily seen by examination of the crystal structures of ZnO and sapphire (a = 0.754 nm, c = 1,299 nm), (A) Schematic illustration of ZnO ab-plane over-lapping with the underlying (110) plane of sapphire substrate, B.C) Arrays of ZnO nanowires grown on a-plane sapphire substrate.*

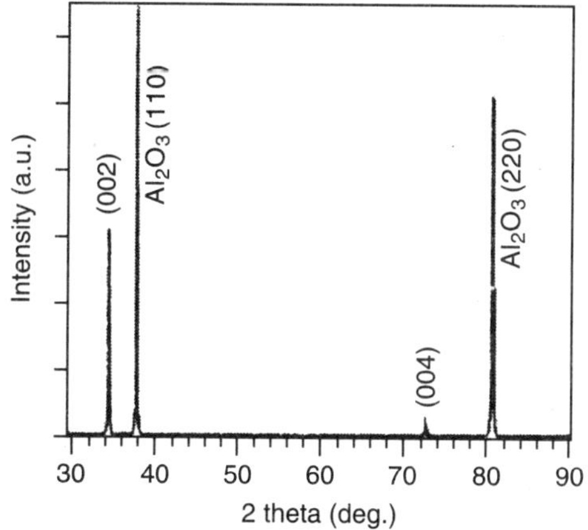

Figure 4.2 *XRD pattern of ZnO nanowires on sapphire substrate.*

(b) Control of Position

From the VLS nanowire growth mechanism, the positions of the nanowires can be controlled by the initial positions of Au clusters or thin films. Various lithographical techniques, like soft lithography, e-beam lithography, and photography, used to create patterns of Au thin films for subsequent semiconductor nanowire growth. SEM (Fig. 4.3) reveals extensive growth of fine, long, and flexible ZnO nanowires from the edges of a hexagonally patterned Au thin film. The

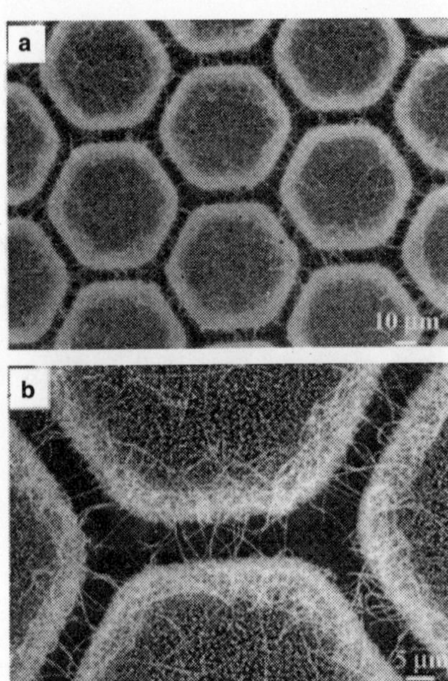

Figure 4.3 *SEM images of a patterned ZnO nanowire netrowk on Si substrate.*

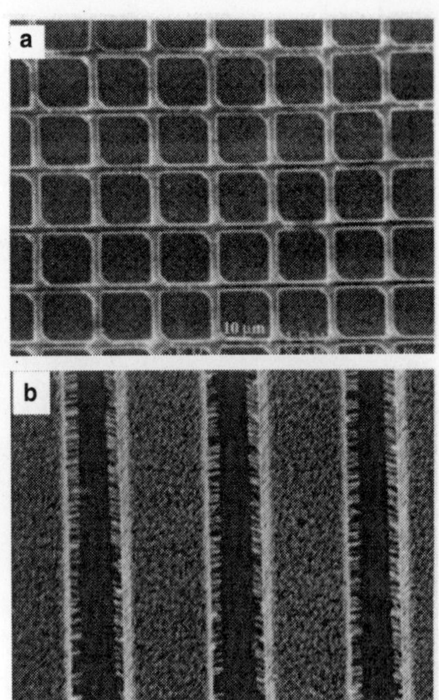

Figure 4.4 *SEM images of epitaxial growth of line- and square-patterned ZnO nanowires on a-plane sapphire substrate.*

wire growth conforms the hexagonal Au pattern with high fidelity. Most of the wires bridge the neighboring metal hexagons and form an intricate network. Figure 4.3(b) shows an SEM image at a higher magnification. The nanowires at the edge have a diameter of about 50–200 nm, and are as long as over 50 μm.

Figure 4.4 shows the SEM images of ZnO nanowires grown from the line and square Au patterns on *a*-plane sapphire substrate. Selective epitaxial nanowire growth is seen. It is clear that nanowires grow vertically from the region that is coated with Au and form designed patterns of ZnO nanowire arrays.

It is also possible to control the nanowire areal density by modifying the thin-film thickness or using solution-made Au clusters. By dispersing a different amount/density of Au clusters on the sapphire substrate, nanowire arrays with different densities are obtained (Fig. 4.5) for example, ZnO nanowire arrays have an areal density spanning 10^6–10^{10} cm^{-2}.

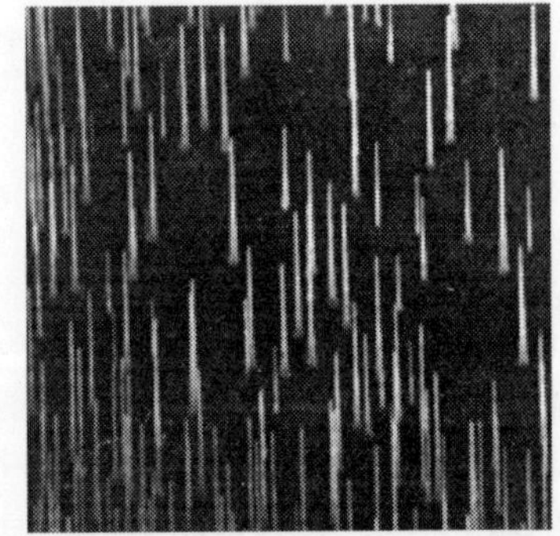

Figure 4.5 *ZnO nanowire arrays on sapphire substrate with low areal density.*

(c) Control of Diameter

To grow ZnO nanowires with controllable diameters, substrates coated with Au of different thickness are used as there is a possible direct relationship between the size of the solvent particles and the resulting diameters of the nanowires. Smaller Au droplets, (formed in the heating process during the reaction), will help in the growth of thinner wires. As the thickness of the Au thin film decreases, the width of the wires also decreases. For example, the average diameters of ZnO nanowires grown on a-plane sapphire substrates at 1170 K for 5 min are 88, 110, and 150 nm, respectively, when Au thin films of 0.5, 1, and 3 nm thick are used. The smallest diameter of ZnO nanowires achieved by using Au thin films is approximately 40 nm.

In addition, uniformly distributed Au clusters are used as solvent to control the widths of the nanowires. The Au nanoclusters are dispersed on silicon substrates or within mesoporous silica films to minimize possible particle aggregation. For example, the average diameters of ZnO nanowires are 35, 46, and 54 nm, respectively, when Au clusters of 5, 10, and 15 nm sizes are used. Au clusters produce wires with much thinner widths–as thin as 20 nm. This is due to the fact that when Au thin film melts at high temperature to form Au droplets, there is a thermodynamic limit to the minimum radius of the metal liquid clusters at high temperature, $R_{min} = 2\sigma_{LV} V_L / RT \ln s$, where σ_{LV} is the liquid-vapor surface free energy, V_L is the molar volume of liquid, and s is the vapor phase supersaturation.

(d) Control of Morphology

By using different vapor sources, it is found that certain superstructures of the nanowires readily form, for example, when Zn powder is used as vapor source, a large yield of comb-like structures made of ZnO nanowires (Fig. 4.6). These nanowires have uniform diameters and are uniformly distributed on the side of the stem. Other superstructures, such as tetra-pods and tapered nanowires, also be formed under different evaporation conditions. These different superstructures have interesting physical properties, and particularly optical properties.

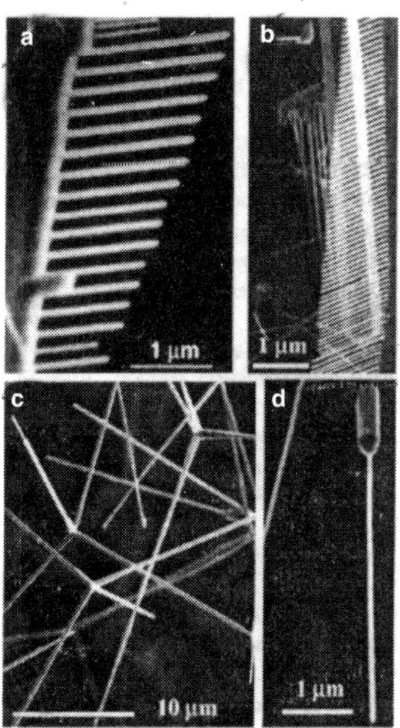

Figure 4.6 *Different morphologies and superstructures based on ZnO nanowires synthesized using Zn as the vapor source.*

DENSITY-CONTROL OF NANOWIRES

(By MOCVD Technique)

An effective, low cost, simple, and mask-free pathway for achieving density control of the grown aligned ZnO nanowires is described below. By a slight variation of the thickness of the thermally evaporated gold catalyst film, a significant change in the density of aligned ZnO nanowires is controlled. The growth processes of the nanowires, on an $Al_{0.5}Ga_{0.5}N$ substrate, based on the wetting behavior of gold catalyst with or without source vapor, and the results, classify the growth processes into three categories: separated dots initiated growth, continuous layer initiated growth, and scattered particle initiated growth.

From the growth of ZnO nanowires on nitride substrates, the effects of the thickness of the gold catalyst layer are investigated. It is found that in the VLS process, the size of the catalyst particles determine the width of nanowires. However, this applies only when the catalyst particles are small (<40 nm). Because of the wetting situation between the melted catalyst droplet and the substrate, more energy sites are created for nanowire growth with thinner catalyst layers. As a result, varying the thickness of the deposited gold film is simple than to control the density of aligned ZnO nanowire. Moreover, when the catalyst layer is thick, a continuous ZnO network is deposited at the bottom of the nanowires. This layer serve as the electrode for many electrical and device applications.

Experiment

The $Al_{0.5}Ga_{0.5}N$ substrate is fabricated by the MOCVD technique. (See fabrication & synthesis) The *c*-plane-orientated $Al_{0.5}Ga_{0.5}N$ epitaxial layer with a thickness of ~205 nm is grown on a 500 nm thick AlN buffer layer which is grown on a one-side-polished *c*-plane-oriented sapphire substrate. Eight stripes of gold layers with thicknesses from 1 to 8 nm at 1 nm gradient; as catalysts, are deposited onto a 2 × 1 cm $Al_{0.5}Ga_{0.5}N$ substrate using a gold thermal evaporator where the thickness is measured by a quartz crystal thickness monitor with a resolution of 0.1 nm. The fabrication of ZnO nanowires is carried out under experimental conditions by MOCVD

The effects of the thickness of the Au catalyst on the density of the nanowires grown is investigated. Through a thermal evaporation process, all of the nanowires are grown (under identical experimental conditions) on the above gold films with different thickness. The inset in Figure 4.7 is an optical picture of the substrate, from which the eight stripes can be clearly distinguished. From right to left, it is observed that the color of the stripes varies (from light blue to purple) which represent the aligned ZnO nanowires catalyzed by Au layers of thicknesses from 1 to 8 nm. Each stripe of as-synthesized ZnO nanowires exhibits distinct appearance even though there is only 1 nm change in the thickness of the deposited Au film. Since the blue color is from ZnO and red color is contributed by gold, the thinner layer of the gold catalyst results in more ZnO nanowires, as confirmed by scanning electronic microscopy (SEM). Each stripe of the as-synthesized ZnO nanowires exhibits differences even though there is only 1 nm change of thickness between two adjacent stripes.

Figure 4.7(a–h) represents typical SEM images recorded from the nanowires grown in the areas of the eight stripes at an identical magnification. Clearly, the 1 nm gold layer gives the highest density of aligned ZnO nanowires (Figure 4.7(a)), while the thickest gold layer results in the lowest density (Figure 4.7(h)). With the decrease of the density of the ZnO nanowires, more and more larger particles are observed on the substrate, As confirmed by energy-dispersive X-ray spectroscopy (EDS) analysis equipped on the SEM, those particles are mostly composed of gold, which makes the substrate appear more reddish in the optical image. However, despite the large difference in density, all of the ZnO nanowires exhibit very similar width and length distributions.

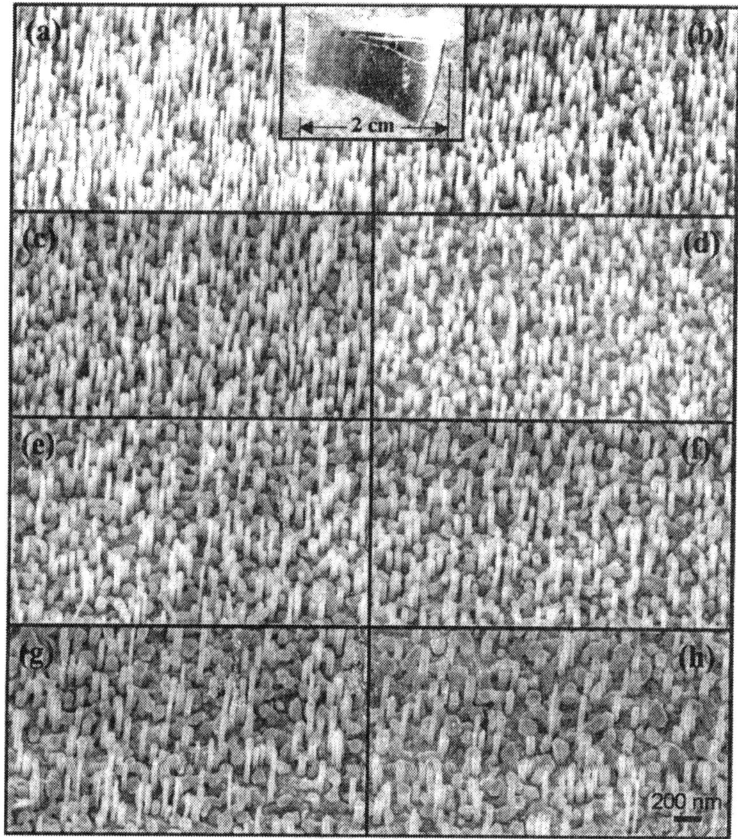

igure 4.7 *(a-h) SEM images of the aligned ZnO nanowires grown on a nitride substrate as catalyzed by a deposited gold film of thicknesses varying from 1 to 8 nm in 1 nm steps. Inset: Optic photo of the substrate showing eight stripes of the aligned ZnO nanowires from left to right corresponding to images a–h.*

Quantitative analyses are thereafter performed to reveal the density and thickness relationships f the aligned nanowires. The density is measured by top-view SEM images, as shown in the nset of Figure 4.8(a). Each bright spot corresponds to a perpendicular ZnO nanowire with a old catalyst particle on top. By counting the number of bright dots on five randomly taken EM images, the average number of nanowires per square micrometer is calculated. The anowire density at each thickness of gold catalyst layer is shown on the left-hand side axis in igure 4.8(a). An almost linear drop in density from 112 to 15 with increasing catalyst thickness s observed. The width of the nanowires is measured by transmission electronic microscopy TEM). After the aligned nanowires on each strip are transferred to different TEM grids, more han 10 TEM images are taken for each sample to calculate the average width, (See Figure .8(a)). The standard deviations of the samples are ~16% from > 100 randomly chosen nanowires. he average width remains between 30 and 40 nm despite the change in catalyst thickness. The ariation in nanowire width, TEM images nanowires catalyzed by 1 nm and 8 nm thick gold layers, are taken and are shown in Figure 4.8(b) and 4.8(c), respectively. One nanowire in Figure 4.8(c) exhibits a similar width as that in Figure 4.8(b), while the other two are about 10–15 nm wide, indicating a larger absolute width distribution for 8 nm thick gold layer catalyzed nanowires. A selective-area electron diffraction pattern is taken on the nanowire as indicated by a circle in Figure 4.8(c). As shown in the inset of Figure 4.8(c), the nanowires are grown along the [0001] direction, which is the fastest growth direction for the VLS growth of ZnO.

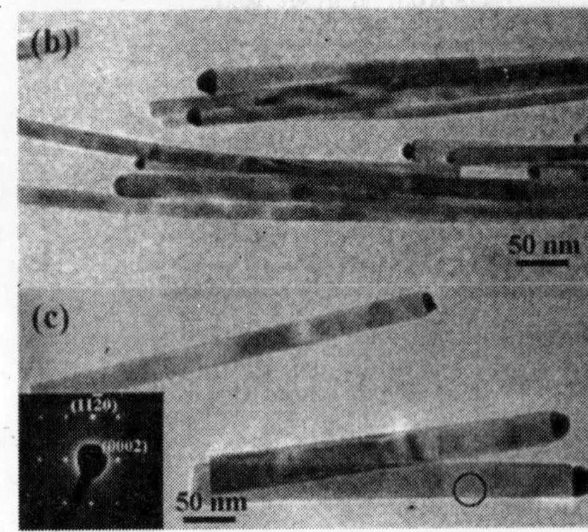

Figure 4.8 *(a) Variation of density (left-hand ver tical axis) and width (right-hand ver tical axis) of the aligned ZnO nanowire with the thickness of gold catalyst layer Inset: Top-view SEM image of th aligned ZnO nanowires used for den sity calculation, the scale bar represent 200 nm. (b, c) TEM images of ZnO nanowires catalyzed by 1 and 8 nm gold layers, respectively. Inset: Selecte area electron diffraction pattern recorde from a nanowire indicated by the circl in image c.*

Because of the linear relationship between the density of the nanowires and the thickness of the catalyst layer, thickness is control of gold catalyst is a simple and effective way to achieve density control of aligned nanowires over a large surface area. The density varies but the width remains constant is because of the wetting behavior of a gold layer on the $Al_{0.5}Ga_{0.5}N$ substrate. It is investigated by heating to the growth temperature. Two additional pieces of $Al_{0.5}Ga_{0.5}N$ substrates with the same gold catalyst distribution are prepared. Under the same condition, one piece of the substrate is heated to the growth temperature without introducing source material to reveal the formation of catalyst dots; while the other is heated to the same temperature with source material in the growth chamber for only a few minutes to reveal the initial growth stage of the nanowires. On the basis of the data received, the growth situation is divided into three categories, as shown in Figure 4.9.

The first category is the *separated dots initiated growth,* where the gold layer is only 1–2 nm in thickness. As shown in Figure 4.9(a), after it is heated to the growth condition, the 1–2 nm gold layer melts into tiny gold dots (10–20 nm) evenly distributed on the substrate surface. Once the source vapor is present, larger particles are observed because of the coarsening effect resulting from the formation of a supersaturated "alloy". (see the inset of Figure 4.9(a)). However, all of the particles remain separated from each other, thus resulting in a higher density and individual ZnO nanowire growth from the substrate surface (Figure 4.9(b)).

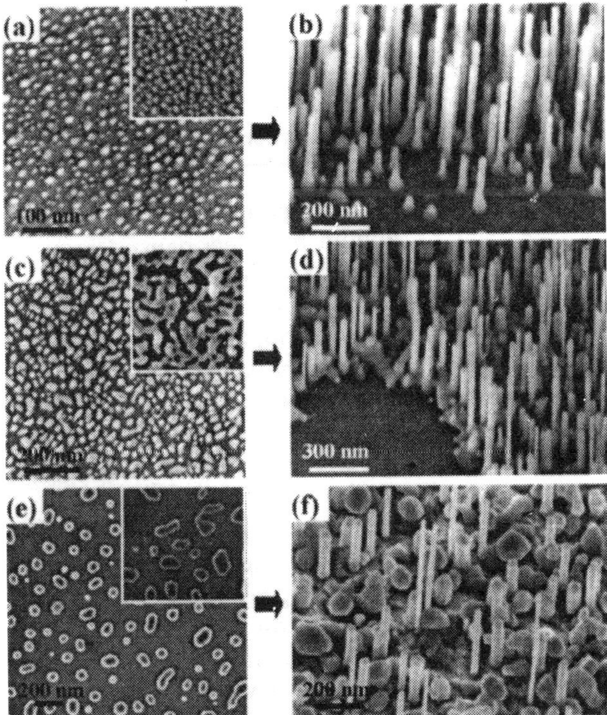

Figure 4.9 *Three categories of the growth processes based on the thickness of gold layer. (a, c, e) SEM images showing the wetting situation of gold catalyst on $Al_{0.5}Ga_{0.5}N$ substrate prior to the nucleation of the nanowires. Insets: SEM images with the same magnification showing the wetting situation when the vapor source was presented to initiate the growth. (b, d, f) SEM images of the aligned ZnO nanowires catalyzed by the corresponding gold layers.*

The second category is *continuous layer initiated growth* when the thickness of the gold deposition layer is 3–6 nm. As shown in Figure 4.9(c), the gold layer melts to separated islands with different sizes at the deposition temperature. Most of the large islands then merge into a continuous network on the substrate surface when alloying gold with the source vapor at the initial stage (inset of Figure 4.9(c), while the small islands remain isolated. As the deposition continues, the isolated small islands initiate the growth of nanowires, while the network form a complete layer covering the substrate and connect the base of all the nanowires (Figure 4.9(d). The gold in the network is eventually precipitated out forming small dots or very short rods during the cooling process. Therefore, the density of nanowires grown is relatively lower. However, every single ZnO nanowire is interconnected by the Au or ZnO layer at the root this is an alternative pathway to electrically connecting the aligned ZnO nanowires for electronic nanodevices and field emission. This is a unique advantage for solving the bottom electrical contact problem. It also has the great merit of replacing the expensive semiconducting substrates by sapphire, while preserving the electrical contact, offering great potential for industrial application.

The third category is the *scattered islands initiated growth:* This occurs when the thickness of the gold layer is greater than 7 nm. Here on melting, the catalyst layer is thick enough to form big gold particles (~100 nm). The large gold particles are widely separated and a very small number of gold dots are left (See Figure 4.9(e). Even after an alloy is formed with the source vapor, it remain unchanged (inset of Figure 4.9(e). In the deposition process, only the small dots initiate the growth of ZnO nanowires; and the big particles lead the growth of ZnO films/junks lying on the substrate, thus resulting in the lowest density of aligned ZnO nanowires.

The above process is based on moderate growth conditions, under which the evaporation of sources materials is relatively slow so that the catalyst layer has enough time to melt and form an alloy step by step, which is considered as a thermal equilibrium process. The small dots initiate growth, because the thermodynamic mechanisms are based on the pre- and postnucleation of the ZnO nanowire, as illustrated in Figure 4.10. The alloy droplet is assumed to be supersaturated at stage 1; stage 2 presents the moment when the first layer of ZnO is precipitated, during which the change in supersaturation is compensated by the diffusing of source vapor from the gas phase into the alloy. The free energies of these two stages are presented by G_1 and G_2 respectively, as

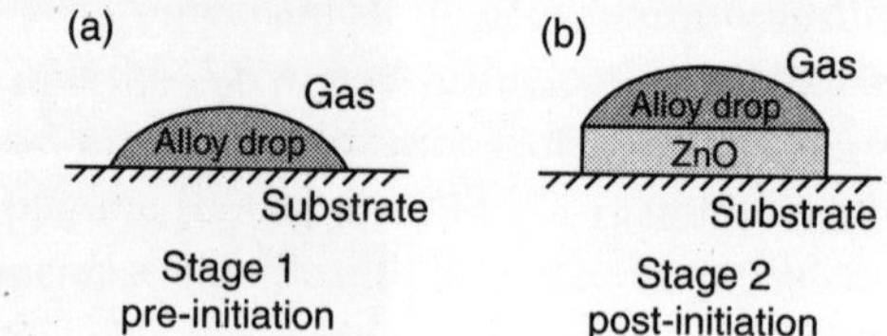

Figure 410 *Schematic of (a) pre- and (b) post-initiation of ZnO nanowire growth.*

$$G_1 = V_A G_{V1}^A + V_G G_{V1}^G + A_{AS}\gamma_{AS} + A_{AG}\gamma_{AG} + A_{GS}\gamma_{GS} \quad (1)$$

$$G_2 = V_A G_{V2}^A + V_G G_{V2}^G + A_{AZ}\gamma_{AZ} + A_{ZS}\gamma_{ZS} + A_{GZ}\gamma_{ZG} + A_{AG}\gamma_{AG} + A_{GS}\gamma_{GS} + V_Z G_V^A \quad (2)$$

where V_A, V_G, and V_Z are the volume of the alloy droplet, gas phase, and ZnO, respectively, G_{V1}^A, G_{V1}^G and G_{V2}^A, G_{V2}^G are the free energy per unit volume of alloy and gas phase at the first and second stages, respectively, G_{V2}^A is the free energy per unit volume of the precipitated ZnO crystal, and A and γ are the area and free energy of the interfaces, respectively, between gas, alloy, substrate, and ZnO surfaces as indicated by their initials in the subscriptions. The formation of ZnO results in a free energy change $\Delta G = G_2 - G_1$. By assuming the alloy composition remains the same ($G_{V1}^G = G_{V2}^G$) and the area of precipitated ZnO is the same as the contact area between the alloy and the substrate ($A_{AS} = A_{AZ} = A_{ZS}$), the ΔG can be given by

$$\Delta G = V_G(G_{V2}^G - G_{V1}^G) + A_{AS}(\gamma_{AZ} + \gamma_{ZS} - \gamma_{AS}) + A_{ZG}\gamma_{ZG} + G_V^Z \quad (3)$$

To initiate the precipitation of ZnO, ΔG has to be negative, so that,

$$V_G(G_{V1}^G - G_{V2}^G) > A_{AS}(\gamma_{AZ} + \gamma_{ZS} - \gamma_{AS}) + A_{ZG}\gamma_{ZG} + V_Z G_V^Z \quad (4)$$

The left-hand side of the inequality is the change in the free energy of the vapor phase; on the right-hand side, the first term is determined by the initial size of the catalyst, A_{AS}, and last two terms are negligible at the initiation nucleation stage since the size of the ZnO/gas interface, A_{ZG}, and the volume of ZnO, V_Z, are very small. So equation 4 can be simplified to

$$G_{V1}^G - G_{V2}^G > CA_{AS},\ C = (\gamma_{AZ} + \gamma_{ZS} - \gamma_{AS})/V_G \quad (5)$$

where C is a constant. From this equation, small A_{AS} values (e.g., small gold dots) are thermodynamically favourable sites for the initiation of aligned ZnO nanowires. Since the surface area, A_{AS}, for small size gold particles is rather small, equation 5 is usually satisfied. Therefore, growth on small size gold particles is thermodynamically favorable.

On the other hand, small gold particles are very likely to be in the liquid state at the growth temperature because of the reduced melting point for smaller particles. The melting point of gold layer below 5 nm is much lower (<1070 K) than the melting temperature of bulk gold (1335 K). These molten small gold particles result in the nucleation and growth of aligned ZnO nanowires from the VLS process, while the large gold islands are in the solid-state, thus growth by the VLS process is not possible. If the nanowires are around 30–40 nm, the deposition thickness of the gold film can be disregard.

Photoluminescence (PL) measurements are done on the sample for the light-emitting properties of the aligned ZnO nanowires with different densities. Figure 4.11 shows eight PL curves of the aligned ZnO nanowires catalyzed by gold layers with thicknesses from 1 to 8 nm. Strong luminescence peaks are observed at 377 nm from all of the eight curves, while the intensities are varied. This is consistent with the constant diameter of the nanowires for all of the samples. In general, the intensity decreases with lower density, as shown in the inset of Figure 4.11. The highest PL intensity is from the nanowires catalyzed by a 1 nm gold layer, which has the highest density. However, the lowest PL intensity is not from the sample with the lowest density. On the contrary, the sample catalyzed by an 8 nm gold layer shows an enhanced PL intensity. This is due to the increased reflectivity of the emitted light by the large gold particles lying on the substrate. The reflectance from samples 2 and 3 may also be attributed to the increased reflectivity resulting from the formation of a continuous ZnO bottom layer.

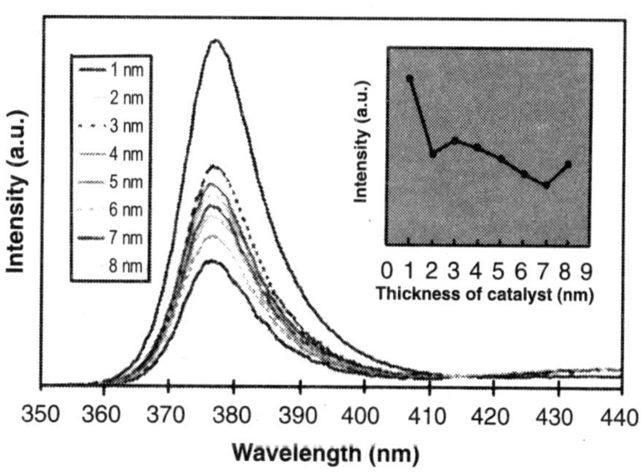

Figure 4.11 *Photoluminescence spectra of the eight stripes of the aligned ZnO nanowires catalyzed by gold layers of thicknesses from 1 to 8 nm in 1 nm steps. Inset: Variation of the intensity of PL peaks with the change of catalyst thickness.*

GROWTH OF HORIZONTAL ZNO NANOWIRE ARRAYS ON ANY SUBSTRATE

Laterally aligned ZnO NW arrays in parallel to substrate offer a benefit of fabricating integrated nanodevice arrays, but there are only a couple of reports about the growth of laterally aligned NWs. Taking advantage of lattice match between ZnO and sapphire, ZnO NWs grown from two sides of gold pads deposited on a sapphire substrate are obtained. The growth temperature is ~1170 K and thus the choice of substrate is limited, which greatly restrict the integration of NWs with silicon or polymer based devices.

A general method for the controlled growth of laterally aligned ZnO NW arrays parallel to a general substrate at low temperature (<373 K) is given below.

The substrate can be any material, inorganic, organic, crystalline, or amorphous, as long as it is flat. The growth has achieved controls over location, orientation, pattern, and uniformity of the NWs. This provides a new NW structure for fabricating device arrays on a general substrate.

Experiment

The designed growth is achieved by using different materials to activate or inhibit the growth of nanowires. Two materials are used here: ZnO seeds for the growth and a Cr layer for preventing the local growth. The first step for the growth is to fabricate a ZnO strip pattern covered with Cr at the top (Figure 4.12(a) and (b)). A (100) silicon (Si) wafer is cleaned in sequence with HF acid solution, acetone, isopropyl alcohol, and ethanol. It is then blown dry with nitrogen gas. After that, a photoresist, is spin coated on the substrate at a velocity of 4000 rpm (round per minute) to get a uniform layer. Then a pattern is produced using optical lithography. The wafer is baked at 380 K for 5 min, exposed under 405 nm light using lithography, and developed with MF-319. Magnetron sputtering is used to deposit 300 nm ZnO and 10 nm Cr on the trench patterns from prior step (deposition rate is 0.1 $Å^{-1}$, working pressure is 3 mm Hg). ZnO strips with Cr on top are achieved after lifting-off with acetone. Finally, the substrate, is put into growth solution and kept for 12 h at 350 K. The growth solution is prepared by dissolving 0.1878 g of $Zn(NO_3)_2 \cdot 6H_2O$ and 0.0881 g of hexamethylenetetramine (HMTA) in 250 mL deionized water at room temperature. The concentrations of $Zn(NO_3)_2$ and HMTA in the solution are both 0.0025 mol^{-1}. To achieve uniform control over the growth rate by the concentration of the solution, the substrate is floated upside down on the solution surface with the patterned side facing downward. NW arrays are grown from the ZnO seeds directly exposed to the solution (Figure 4.12(2)). Alternatively, a pattern Figure 4.12(a)(3): is produced by evaporating Cr or Sn onto the ZnO stripes at an angle. As a result, one side is induced (Figure 4.12(a)(4)).

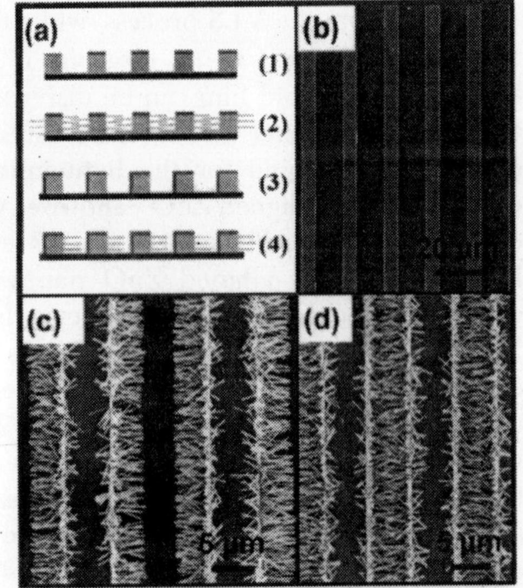

Figure 4.12 *(a) Schematic steps for growing patterned and laterally aligned ZnO NW arrays. (b) SEM image of ZnO pattern covered with 10 nm Cr shield layer prior to nanowire growth. (c and d) SEM images of ZnO NW arrays grown on Si substrates.*

Consider, Figure 4.12(c). It is the laterally grown ZnO NW arrays on Si substrate. It is seen that ZnO NW arrays grow from the lateral sides of the pattern with a good alignment. More than 70% nanowires are parallel to the substrate. Only a small fraction of disordered ZnO nanowires grow at the edge of the pattern. ZnO NWs have a diameter less than 200 nm and a length of about 4 μm. Hexagonal cross section of nanowires imply that *c* axis of ZnO NW is along its length direction. When the distance between ZnO stripes is small enough, (by controlling the growth time) the ZnO NW arrays from adjacent pattern grow toward each other, forming an interdigitated structure (Figure 114.12(d)). By increasing the growth time and renewing the growth solution, ZnO nanowires longer than 13 μm are synthesized. The diameter of ZnO NW is increases to approximately 300 nm.

Horizontal ZnO NW arrays parallel to the substrate grow from the straight patterns or from any shape patterns. Nanowires always grow vertically to the local ZnO sidewall of the pattern. Figure 4.13 shows the ZnO NW arrays grown at the ends of the patterns. ZnO nanowires grow radially and they are distributed following local curvature.

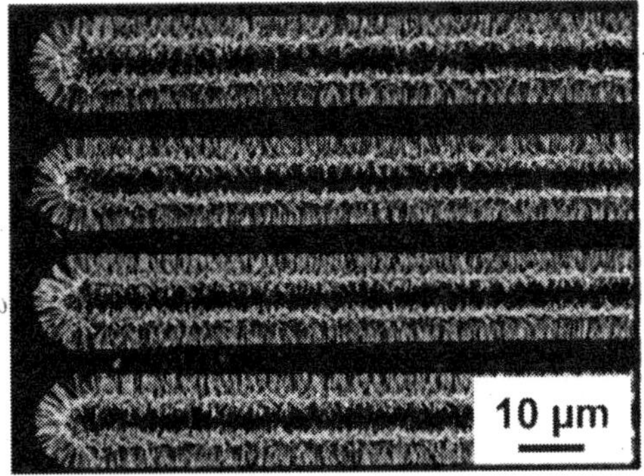

Figure 4.13 *SEM image of ZnO nanowire arrays grown laterally on an Si substrate. The orientation of the nanowires follows the surface curvature of the pattern at the end (see the left. hand side of the image).*

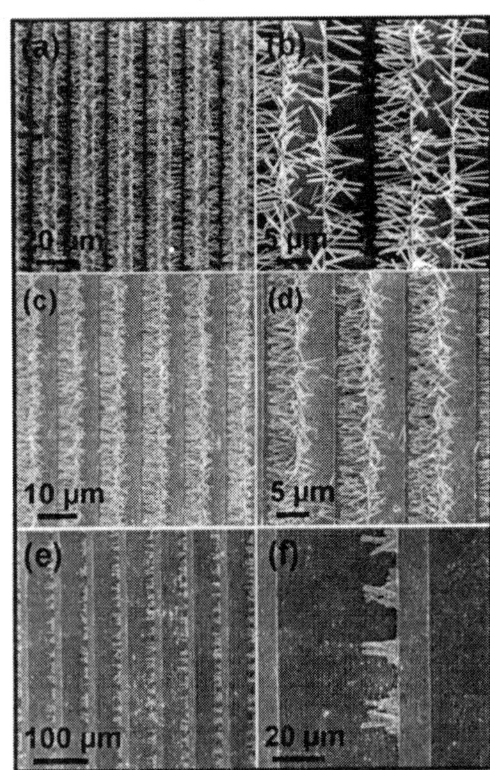

Figure 4.14 *(a and b) SEM images of ZnO NW arrays grown on patterns evaporated with 10 nrn Cr at an angle of $\theta = 70°$ in reference to the substrate nurmal. (c and d) SEM images of ZnO NW arrays grown with 50 nm Cr mask film evaporated at θ close to 90°. (e and f) SEM images of ZnO NW arrays grown with 5 nm Sn film mask evaporated at θ close to 90°.*

Some applications require control over the position and direction, as ZnO NW arrays grow only from one side of the pattern. To fulfill this Cr layer is thermally deposited at an azimuth angle θ with respect to the substrate normal along a direction perpendicular to the strips, so that one side of the strips is covered by Cr, while the other side is not. Figure 4.14 shows the influence of θ, and the type and thickness of shielding material on the growth of laterally aligned NW arrays. For the 10 nm thick Cr layer deposited at $\theta = 70°$, both sidewalls of the ZnO strips are covered but with a different surface coverage. The Cr layer is not continuous therefore, the ZnO seed layer cannot be covered completely. As a result, ZnO NWs grow from both sides of ZnO stripes and the nanowire density at one side is much larger than that at the other side (Figure 4.14(a) and (b). When the deposition angle θ approaches 90° and the Cr thickness increases to 50 nm, only one sidewall of the ZnO pattern is fully covered by the Cr layer, which leads to ZnO NW arrays growth only at one side of the strips (Figure 4.14(c) and (d).

Compared with Cr, Sn is a metal with low melting temperature and can more effectively cover the seed layer. As a 5 nm thick Sn layer is evaporated with θ close to 90°, the lateral ZnO NWs clusters grow at one sidewall of the pattern; vertical growth of nanowires is completely prevented (Figure 4.14(e) and (f).

Any substrates can be used for the lateral growth of ZnO NWs, such as inorganic, organic, single crystal, polycrystalline, or amorphous substrates. To demonstrate the freedom in choosing substrate, consideration flexible Kapton film and polyester film as substrates. The process is the same as for the growth on Si substrate. Figure 4.15(a) and (b) presents lateral ZnO NW arrays grown on the Kapton film. Horizontal ZnO NW arrays are parallel to the substrate with good alignment. For the polyester substrate, similar results are observed. (Figure 4.15(c) and (d).

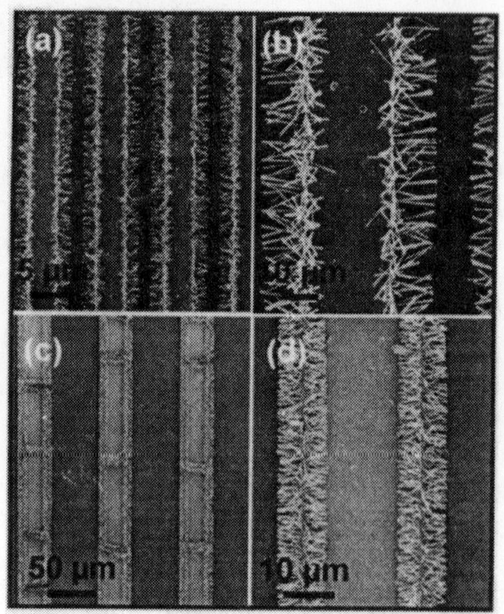

Figure 4.15 *(a and b) Low and high magnification SEM image of lateral ZnO NW arrays grown on Kapton film. (c and d) SEM images of lateral ZnO nanowire arrays grown on polyester film.*

Controlling the Growth Direction on *c*-Plane Sapphire

c-plane sapphire (0001) is the most popular substrate in the epitaxial growth of III-V and II-VI semiconductors due to its high crystalline perfection, stability at high temperature and transparency. The growth of ZnO films on this substrate is accompanied by a large number of defects at the substrate/film interface, which in part is due to the significant lattice mismatch between ZnO ($a = 3.2495$ Å, $c = 5.2069$ Å) and sapphire ($a = 4.758$ Å, $c = 12.990$ Å). In device fabrication, this mismatch is significant and results in lower charge mobility, defect related carrier concentrations and is a major obstacle to the use of ZnO films. It is found that ZnO films can be grown on *a*-plane sapphire with lower defect density relative to those on the *c*-plane. This is due to the fact that the *a* and *c* lattice parameters of ZnO and sapphire (respectively) are related by a factor of ~ four. The epitaxial fabrication of ZnO NWs on sapphire is carried out on *a*-plane sapphire. This includes standing ZnO NWs, tilted ZnO NWs, "tower-like" ZnO structures, and NWs grown on nanowalls. NWs are grown on *c*-plane sapphire substrates coated with Au films with thicknesses ranging from 1 to 10 nm. The impact of film thickness and thermal annealing of these films on NWs growth direction is given below

Standard one-side polished *c*-sapphire wafers, cut into pieces (3 mm^2) are coated with different Au thicknesses, 1 nm to 8 nm, using a thermal evaporator. Au film annealing is performed at 1170 K for 10 minutes with a rate of 97 K min^{-1}. The tube furnace is purged with ultra-pure Ar prior to annealing at room temperature and during annealing. Similar amounts of zinc oxide (99.999% purity) and graphite (99.99% purity) are ground (total mixture weight of 0.1 g) and in a Si boat transferred to the center of a small quartz tube (12 cm length, 2 cm diameter). The quartz tube is then placed inside a tube furnace (80 cm length, 5 cm diameter), such that the center of the boat is at the center of the furnace. The Au-coated sapphire substrates are placed 6 cm downstream from the center of the tube. The tube furnace is purged with Ar (0.50 L min^{-1}.) for 15 minutes at room temperature, then the temperature is increased to 1170 K (at a rate of 109 K min^{-1}) and kept constant for 20 minutes under a flow of Ar gas. The nanostructures are imaged by SEM at 1-5 kV. The crystalline quality of the ZnO NWs is assessed by X-ray diffraction (XRD) using a standard diffractometer with Cu Kα radiation in the θ-2θ configuration.

Now, consider the effects of Au thickness and Au film morphology (unannealed vs. annealed) on the growth and orientation of ZnO NWs on c-plane sapphire. Growing NWs on substrates coated with as deposited unannealed Au films with thicknesses ranging from 1 nm to 10 nm result in an interdigitated structure of NWs as shown in Figure 4.16.

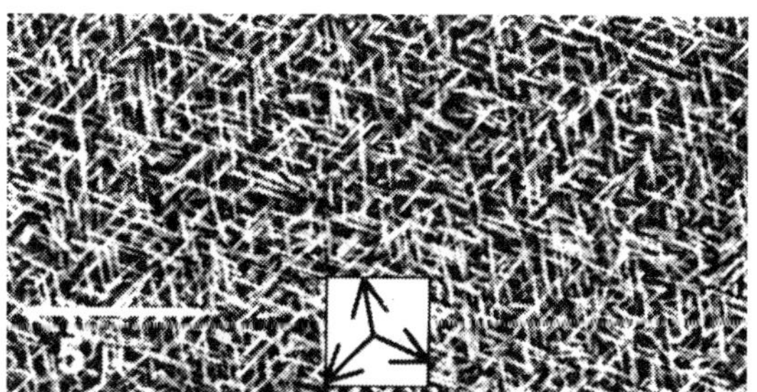

Figure 4.16 *Tilted s grown on as deposited Au film with heights ranging from 1 to 10 nm, three growth directions.*

Here NWs in addition to the normal direction, grow in six different directions with a symmetry of C_3 (inset of Fig. 4.16). Figure 4.17(a) and (b), show that at the border of the film, NWs grow normal to the surface while at the Film's center the number of tilted NWs increases. Its XRD (Fig. 4.17(c)) shows two intense ZnO reflections, (0002) (I=100%) and (101$\bar{1}$) (I=82%). The angle between the (0002) and (10$\bar{1}$1) crystallographic planes is 31.97° (in the wurtzite structure), which translates into a 58.03° tilt of [10$\bar{1}$1] direction relative to the substrate basal plane. This angle is in agreement with, the tilt angle of NWs, from SEM images which is found to be 55±3°. As a result the (0002) reflections on the XRD scan are assigned to vertically oriented NWs, while (10$\bar{1}$1) reflections are assigned to tilted NWs. In the tilted NWs, the (10$\bar{1}$1) planes are parallel to the substrate surface and the growth direction is [0001].

The XRD pattern in Figure 4.17(c) shows the presence of a weak (11$\bar{2}$0) reflection. This suggests that some of the NWs are laying flat on the substrate. As the interdigitated NWs are obtained for all unannealed Au films regardless of their film thickness. So, the effects of annealing of the Au thin films is tested for 1 to 10 i.e. its nm series its growth and orientation. It is found that annealing at 1170 K (in Argon) prior to ZnO growth promotes the vertical growth of the ZnO NWs on *c*-plane sapphire (Figure 4.18(a) and 4.18(b)).

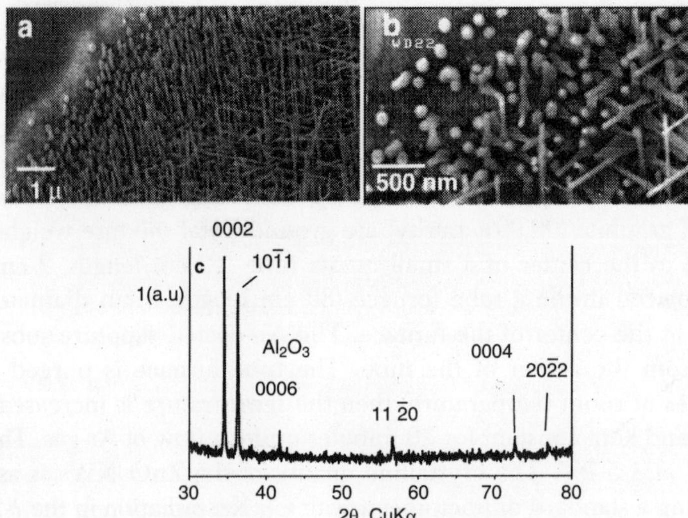

Figure 4.17 *(a) The side and (b) the top views. In parts (a and b) show that the smaller Au droplets at the outmost part of the film grow normal, while moving towards the center of the film, tilted NWs grow. (c) XRD pattern of the vertical and tilted NWs shows two intense ZnO reflections, (0002) and 10$\bar{1}$ 1. The (0002) reflection is attributed to the vertical NWs and the (10$\bar{1}$ 1) reflection to the tilted ones. Weak (11$\bar{1}$ 0) reflections indicate a presence of very small amount of ZnO NWs with c-axis parallel to the substrate surface.*

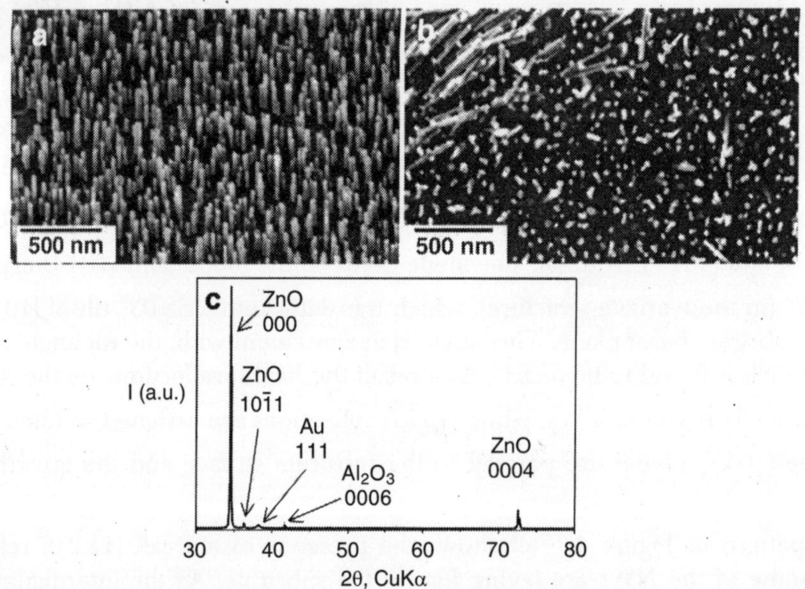

Figure 4.18 *(a) Standing ZnO NWs grown on c-plane sapphire viewed at angle and (b) from top. (c) XRD pattern of standing NWs. The strong (0002) and (0004) reflections are indicative that NWs are well oriented and grow in the c direction.*

The XRD pattern of vertical ZnO NWs is shown in Figure 4.18(c) and shows that by preannealing the Au film, the intensity of the ZnO $(10\bar{1}1)$ reflection becomes less than 2% of that of (0002). This indicates that nearly all ZnO NWs are single-crystalline and oriented in the $[0001]_{hex}$ direction normal to the substrate surface. The c-lattice parameter of ZnO is c=5.2071(5) Å, which correlates well with the c-value of 5.2069 Å for bulk ZnO. NWs thickness measurements using AFM shows that the diameter of both tilted and vertical NWs are similar and is between 30-80 nm. Annealing Au films thicker than 6 ±2 nm resulted in the growth of vertical NWs on a network of ZnO bridges (not shown here). But in the absence of Au annealin NWs grow tilted and the substrate plays a role in their growth direction

It is likely that the surface smoothness and the surface termination of the substrate are the factors that influence annealing the Au coated sapphire substrate. Annealing sapphire at high temperature (1270–1670 K) forms ultra-smooth surfaces because of surface atom reconstruction. For the Au coated sapphire, the extent of surface smoothness is not significant due to the lower annealing temperature (1170 K). The other possibility is the nature of the surface and its alteration upon annealing. If NWs are grown on *a*-plane (110) sapphire with similar experimental conditions as to that of the *c*-plane, then NWs tilt in 4 different directions, rather than the six directions observed on *c*-plane sapphire. This shows that the arrangement of atoms on the (110) plane does have impact on the growth direction. Thus change in the NW growth direction upon annealing can be a result of a change in the surface structure of the substrate relative to the unannealed surface. So, the observed growth directions, in the absence and the presence of annealing should be dictated by the surface termination of the sapphire substrate. It is assumed that in the absence of annealing the sapphire surface is mostly terminated with Al ions; (see Fig. 4.19(a)), as each Al ion on the surface is connected to three oxygen ions of the underneath layer. The symmetry of the terminal Al-O on the surface matches with that of the growth direction of the NWs (Fig. 4.19(a), (b), (c) and Fig. 4.16). The size of the facets in 4.16(a) is not realistic and to show the position of Al atoms on an Al terminated surface, further analysis is required.

In the case of the annealed Au film on sapphire, it is assumed that the surface is mostly covered with O ions. After annealing a clean sapphire surface, the surface becomes hydrophilic, which is good evidence of an increase in the surface coverage of the O ions on the surface. As shown in 4.19(d), the O ions are coplanar and have hexagonal symmetry; this is similar to the symmetry of Zn ions in ZnO along the [0001] direction. Figure 4.19(d) shows an O-terminated sapphire plane and its orientation relative to the [0001] direction. The observed normal growth direction of NWs after annealing matches with the proposed 4.19(e). Electron-back-scattered diffraction (EBSD) and high-resolution cross-section microscopy (TEM) analysis are underway which will aide in the determination of the microstructure of these NWs and their crystallographic relationship with the c-plane sapphire substrate.

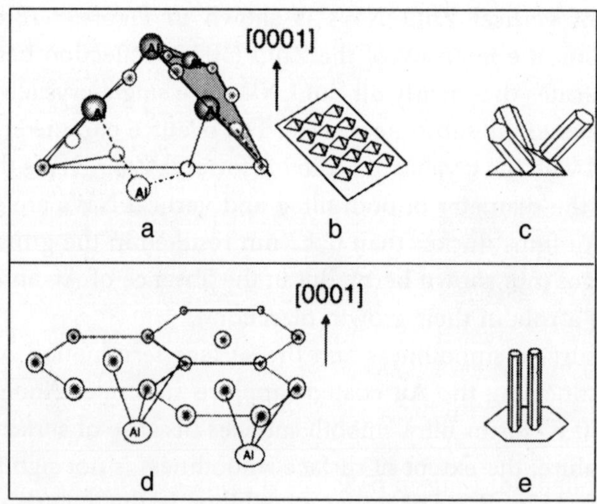

Figure 4.19 *(a, b) illustrate Al-terminated sapphire in the [0001] direction. Oxygen ions are not labeled. Al-O symmetry on the surface matches exactly with the NWs growth directions. (d) by removing Al ions from the surface in (a), an O-terminated surface is obtained. The O ions are coplanar and their hexagonal symmetry matches with that of the attached Zn ion. (e) vertical NW growth is seen in the case of surface rich oxygen substrates, which agrees with the proposed surface in d.*

Alignment of ZnO Nanowires

The large-scale perfect vertical alignment of ZnO nanowires is demonstrated on *a*-plane $(11\bar{2}0)$ crystal surface) orientated single-crystal aluminium oxide (sapphire) substrates. Here, gold nanoparticles are used as catalysts, in which the growth is initiated and guided by the Au particle and the epitaxial relationship between ZnO and Al_2O_3 leads to the alignment.

Unlike the normal vapor-liquid-solid (VLS) process, a moderate growth rate is required for the alignment since the catalyst needs to be molten, form alloys and precipitate step by step to achieve the epitaxial growth of ZnO on the sapphire surface. Therefore, a relatively low growth temperature is always applied to reduce the vapor concentration. Mixing ZnO with carbon powder, (carbon-thermal evaporation) can reduce the vaporization temperature from 1570 to 1170 K.

$$ZnO(s) + C(s) \xleftarrow{1170\ K} Zn(v) + CO(v) \tag{1}$$

The above reaction is reversible at a relatively lower temperature. So when the Zn vapor and CO are transferred, to the substrate region, they can react and reform ZnO. This is absorbed by the gold catalyst and eventually ZnO NWs are formed through the VLS process.

The typical aligned ZnO NWs on a sapphire substrate is shown in Fig. 4.20(a). It is a SEM image recorded at a 30° tilted view. All of the NWs are perpendicular to the substrate surface and the darker dot on the top of each nanowire is the gold catalyst. In this process, the growth spots are dictated by the existence of the catalyst. If the applied catalyst is a thin layer of gold, the distribution of the NWs is random (Fig. 4.20(a)). The catalyst are also prepatterned by lithography or self-assembly techniques. For example, a two-dimensional, large-area, self-assembled and ordered monolayer of submicron polystyrene spheres is formed on a sapphire substrate as a mask, through which a thin layer of gold is deposited. After etching away the spheres, a hexagonal net work of gold layer is achieved (Fig. 4.20(b)), catalyzed by which, the as-grown aligned NWs exhibit honeycomb-like distribution (Fig. 4.20(c)).

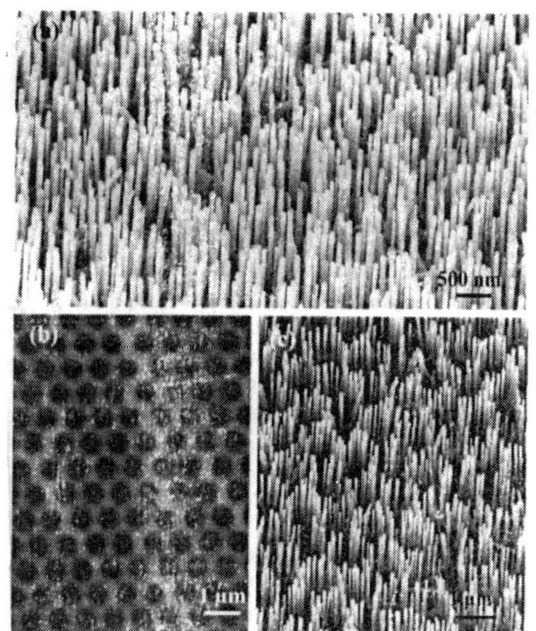

Figure 4.20 *(a) SEM image of aligned ZnO nanowires grown on a sapphire substrate using a thin layer of gold as catalyst. (b) SEM image of gold catalyst pattern using a PS sphere monolayer as a mask. (c) SEM image of aligned ZnO nanorods grown with a honeycomb pattern.*

Epitaxial relationships

Aligned ZuO NWs are grown on sapphire, GaN, AlGaN and AlN substrates through a VLS process; where the crystal structure of the substrate is crucial for the orientation of NWs. An epitaxial relationship between the substrate surface and ZnO nanowires determines whether there is an aligned growth and how good the alignment can be. NWs grown on silicon substrates are always randomly orientated because the gold catalyst tends to form an alloy with silicon at relatively low temperature and destroy the single crystalline silicon surface. The successful alignment of ZnO NWs on sapphire and nitride substrates is attributed to the very small lattice mismatches between the substrates and ZnO.

In the case of sapphire, a $(11\bar{2}0)$ plane orientated substrate is always used the smallest lattice mismatch is along the *c*-axis of Al_2O_3 and the *a*-axis of ZnO. The epitaxial relationship between a ZnO NW and an *a*-plane sapphire substrate is shown in Fig. 4.21(a), where $(0001)_{ZnO} \parallel (11\bar{2}0)_{Al_2O_3}$, $[11\bar{2}0]_{ZnO} \parallel [0001]_{Al_2O_3}$. The lattice mismatch between $4[01\bar{1}0]_{ZnO}$ ($4 \times 3.249 = 12.996$ Å) and $[0001]_{Al_2O_3}$ (12.99 Å) is almost zero, which confines the growth orientation of ZnO NWs. Nevertheless, since the $(11\bar{2}0)$ plane of Al_2O_3 is a rectangular lattice but the (0001) plane of ZnO is a hexagonal lattice, this epitaxial relationship can only hold in one direction. As illustrated in Fig. 4.21(a), along the $[1\bar{1}00]$ direction, the lattice mismatch is fairly large, which will introduce some distortion on their lattices and cause stresses around the interface. As a result, the lateral growth of ZnO side branches is always observed for *a*-plane sapphire substrates, especially on the edge of the growth area, where there is no space restriction for lateral growth (Fig. 4.21).

As for the nitride substrates, GaN, AlN and AlGaN have the same wurtzite structure as ZnO. So, the deposited ZnO NWs are confined to their six equivalent $<01\bar{1}0>$ directions and only

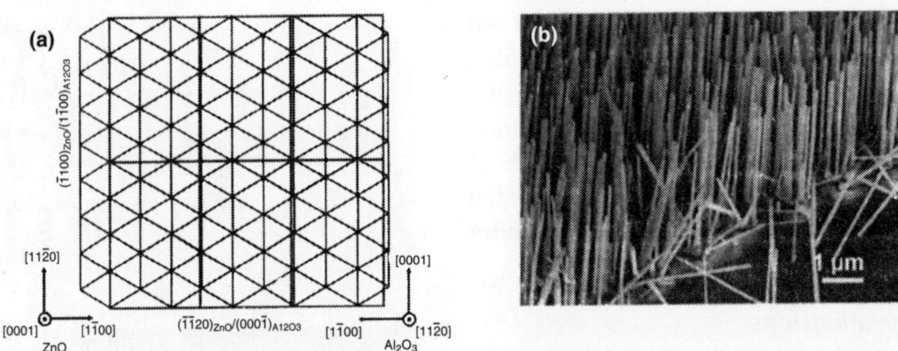

Figure 4.21 *(a) Schematic of the epitaxial relationship between the c-plane of ZnO and the a-plane of $A1_2O_3$. (b) Edge region image of the aligned ZnO nanowires grown on a sapphire substrate.*

grow along the [0001] direction, exactly following the substrate's crystal orientation, as shown in Fig. 4.22(a). In this case, the epitaxial confinement is evenly distributed along the entire 2D atomic plane. As a result, even though the lattice mismatch becomes larger and larger from GaN (1.8%,) to AIN (4.3%), the aligned growth was still kept very well and the possibility for 7 ZnO nanorods to undergo lateral growth is rare (Fig. 4.22(b)). Therefore, *c*-plane oriented $Al_xGa_{1-x}N$ substrates are ideal for the growth of aligned ZnO NWs.

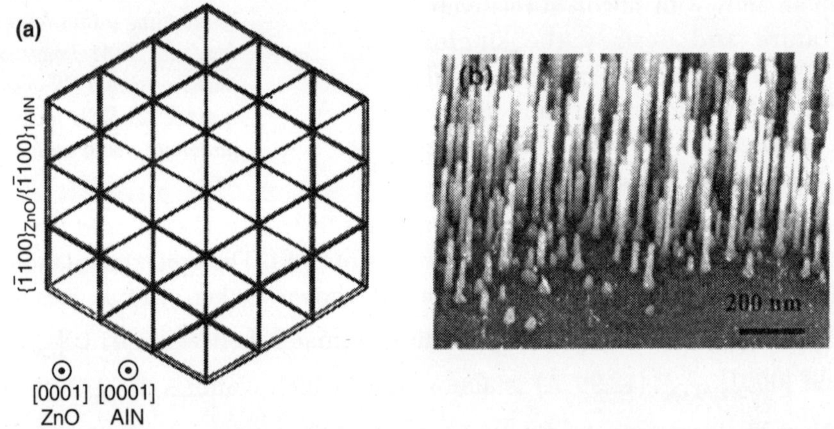

Figure 4.22 *(a) Epitaxial relationship between the c-plane of ZnO and the c-plane of AlN. (b) Edge region image of the aligned ZnO nanowires grown on an AlN substrate.*

Growth Conditions

The growth direction is controlled by the epitaxial relationship between the substrate and NWs. The aligning quality is controlled by many other factors. The growth conditions, are generalized into three basic terms: chamber pressure, oxygen partial pressure and thickness of catalyst layer.

The oxygen partial pressure and the system total pressure play key roles in the growth of ZnO nanowires. With different oxygen volume percentage and different chamber pressure, the quality and growth behavior of the ZnO nanowires are strongly affected. A number of growth experiments are performed under different growth conditions, designed to define the best combination of the O_2 partial pressure and the chamber pressure for the growth of aligned ZnO nanowires; 1150 K temperature, and is kept 0.01 m away from the source materials. The O_2 volume percentage in the chamber is varied from 1 to 4, and the system pressure is varied from 1.5×10^{-2} to 3Pa.

The results obtained are clear from in Fig. 4.23 which is a "growth diagram" for the O_2 volume percentage in the chamber and system pressure, at the optimum conditions for growing aligned ZnO nanowires. The term "growth diagram" (see Figure 4.23) or "phase diagram" (see Figure 4.30) used here is for controlled synthesis of nanowires. This phase diagram is determined for the single-zone tube furnace vapor deposition system. As shown in Fig. 4.23, the horizontal axis is the logarithm of the total chamber pressure; the vertical axis is the oxygen volume percentage in the chamber, and the quality of the grown ZnO nanowires is represented by different colors. The quality of the nanowires is characterized by their uniformity, density, length, and alignment. In the growth diagram, *dark red* represents the best growth conditions, where a perfect alignment of ZnO nanowires with a high density and uniform length and thickness are achieved. The growth is good in the red area, where the density is lower and the nanowires are shorter. In the *green and light blue* areas, the growth is poor, where only a small amount of short nanowires is found. No growth is found in the dark blue region. This growth diagram provides the road map for growing high quality aligned ZnO nanowires.

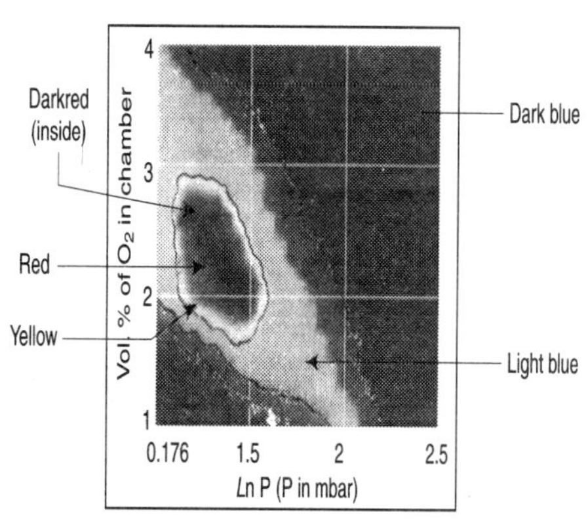

Figure 4.23 *"Growth diagram" that correlates oxygen volume percentages in the growth chamber (e.g. partial pressure) and the growth chamber pressure (plotted in logarithm and P is in mbar) for growing aligned ZnO nanowires. The point matrix is broadened to form a quasi-continuous phase diagram.*

The catalyst effect on the alignment of ZnO NWs is obtained by varying the thickness of the gold catalyst film as the density of aligned ZnO nanowires exhibit an linear relationship with the catalyst thickness. In the experiment, eight stripes of gold layer with thicknesses from 1 to 8 nm at 1 nm gradient are deposited onto a 2 cm × 1 cm $Al_{0.5}Ga_{0.5}N$ substrate through a thermal evaporation process. The fabrication of ZnO nanowires is carried out under the optimal experimental conditions. All of the nanowires are grown under identical experimental conditions. The inset in Fig. 4.24 is an optical picture of the substrate, from which the eight stripes are clearly distinguished. From left to right, the color of the stripes vary from light blue to purple, which thus, represent the aligned ZnO nanowires catalyzed by an Au layer (of thickness from 1 to 8 nm). SEM analysis show that NW densities are changed according to the variation of catalyst thickness. Quantitative analysis is performed to reveal the density and thickness relationships of the aligned nanowires. The nanowire density at each thickness of gold catalyst layer is shown on the left-hand axis in Fig. 4.24 with an average error of ~10%. An almost linear drop in density from 112 to 15 μm^{-2} with increasing catalyst thickness is observed. However, unlike the density, the average width of NWs remains between 30 and 40 nm despite the change in catalyst thickness. Owing to the linear relationship between the density of NWs and the thickness of the catalyst layer, thickness control of the gold catalyst can be very simple and effective way to achieve density control of aligned NWs over a large surface area.

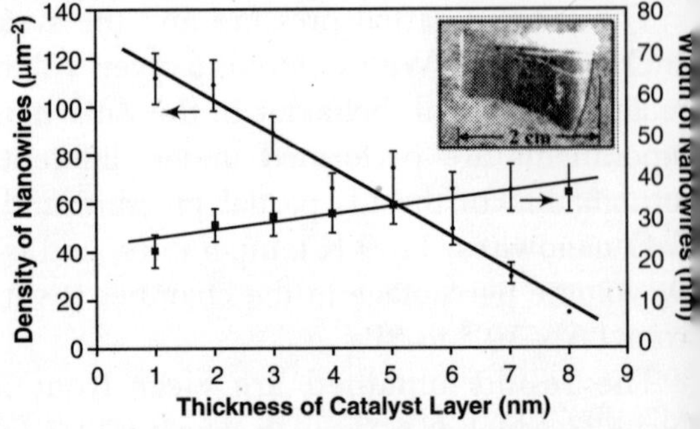

Figure 4.24 *Variation of density (left vertical axis and width (right vertical axis) of th aligned ZnO nanowires with th thickness of gold catalyst layer. Inset photo of the substrate showing eigh stripes of the aligned ZnO nanowire with different densities.*

ALIGNED GROWTH OF ZNO NANOWIRES

Aligned growth of ZnO nanowires is achieved by various methods, such as vapor-liquid-solic (VLS) process, metal-organic chemical vapor deposition (MOCVD) or sol-gel process. As th simplest and most efficient method, VLS process is employed for growing aligned ZnO nanorod on various substrates, such as silicon, sapphire, GaN, and AlGaN/AlN. However, the growtl conditions are quite diverse and without a consistent pattern. For instance, some are growr under high vacuum, whereas some are grown close to one atmosphere; some substrates are pu on the source materials, whereas some substrates are put far away from the source. As a result the size of the nanorods and the quality of the alignment varies.

(a) Aligned growth of ZNO NW with Solid Au nanoparticles as a catalyst

Synthesis of nanowires is attributed to the VLS mechanism. However, an alternate mechanisn with a solid catalyst particle during growth is called the vapor-solid-solid (VSS) mechanism The solid is catalyst particles are achieved for Au-catalysed III-V semiconductors. Ti- and Al catalysed silicon and Au-catalysed ZnSe Mostly, growth is observed at temperatures below th relevant eutectic. For Au-catalysed ZnO, nanowire growth 770 K almost 570 K below the Au rich Au-Zn eutectic temperature of 960 K is observed. The synthesis method by characterizin an array of Au-catalysed ZnO nanowires using electron microscopy and x-ray diffractio techniques, are described below. The results too provide in-depth insight into the growth o nanowires and the role played by the Au catalyst.

Experiment

The Vertically aligned ZnO nanowires are synthesized through the process explained earlier.

The aligned ZnO nanowires are characterized using scanning electron microscopy (SEM) and x-ray diffraction (XRD) methods, including texture analysis. A scanning electron microscope is used. XRD scans are collected with Cu Kα radiation. Parallel beam geometry is used to eliminate errors due to sample surface displacement. A 0.27° parallel plate collimator and 0.04 rad Soller slits controlled axial divergence collect six θ to 2θ scans for determining the ZnO lattice parameters and fourteen for the Au lattice parameter of the nanowire sample. Also collect, eight rocking curves on each of five reflections: Au{111}, ZnO{0002}, AlGaN(0002), AlN(0002) and Al_2O_3(0006), and high-temperature XRD data of the Au{111} reflection of Au-catalysed ZnO nanorods at 770 to 1270 K in 100 K increments. Data should be collected with a 0.4°ω offset in parafocusing geometry with Mo Kα radiation.

Now prepare a sample with 5 nm of Au thermally evaporated onto a GaN layer on *c*-plane oriented, single-crystal Al_2O_3 substrate, at 870 K for 30 min and then heat it upto 1120 K for 60 min. Analyse the sample using SEM and XRD, including grazing incidence (GIXRD) and θ to 2θ scans with an *w* offset to minimize the signal from the Al_2O_3 substrate.

XRD allows information to be gathered from an entire array at once. In addition, the substrate/nanostructure interface Al is preserved, which is vital, since the substrate interface plays a large role in synthesis. Texture analysis, traditionally applied to deformed metals, determines the location of poles (normals) to a specified set of planes with respect to the sample orientation. The x-ray source and detector an: fixed at a particular diffraction angle corresponding to the planes, of interest. Intensity is recorded as the sample is rotated and tilted and is then plotted in a stercographic projcction, called a pole figure. Peaks indicate crystallites oriented with the poles of the specified planes in the direction given by the rotation and tilt angles.

Texture data are collected with Cu Kα radiation, using a polycapillary x-ray lens to generate a parallel beam and placing a 0.27° parallel plate collimator and 0.04 rad Soller slits in front of the detector. The step size for both rotation and tilt is 5°, and the count time is 8 s. An Al_2O_3 $\{02\bar{2}4\}$ pole figure is collected at the beginning of each session to align the sample. Overlap from intense peaks with similar *d*-spacings causes extraneous artefacts in some pole figures. This effect is minimized by choosing strong peaks with few close reflections.

Data is analysed and orientation distribution functions (ODFs) are calculated by using computer. ODFs, combine the information in several pole figures to represent the directions of the [001] poles of a particular phase in the sample with respect to the sample geometry. The directions are represented using three Euler angles (ϕ_1, Φ and ϕ_2). The angle ϕ_1 corresponds to a rotation about the normal to the sample surface, as shown in the inset in figure 4.25. The angle Φ corresponds to a tilt of the [001] pole away from the sample and the angle ϕ_2 corresponds to a rotation about the tilted [001] pole. By examining the distribution of intensity in three-dimensional ϕ_1–Φ–ϕ_2 space, the distribution of crystal orientations is determined. ODFs are calculated from four pole figures for the AlN layer ($\{10\bar{1}0\}$, (0002), $\{10\bar{1}1\}$ and $\{10\bar{1}2\}$), three pole figures for the AlGaN layer ($\{10\bar{1}0\}$, $\{10\bar{1}2\}$ and $\{10\bar{1}3\}$), four pole figures for the ZnO nanorods ($\{10\bar{1}0\}$, $\{10\bar{1}2\}$, $\{10\bar{1}3\}$ and $\{11\bar{2}0\}$), and four pole figures for the Au particles ({111}), {200}, {220} and {311}).

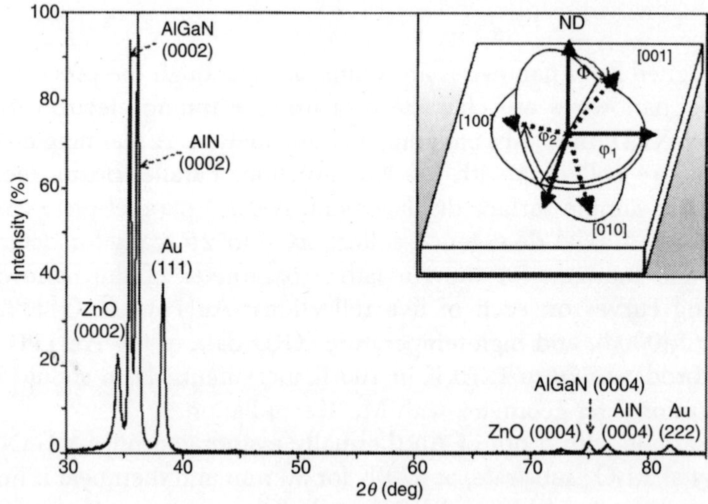

Figure 4.25 *X-ray diffraction pattern of Au-catalysed ZnO nanorods grown on a AlN/AlGaN layer on an Al_2O_3 single-crystal (0001) substrate with 1°ω offset to reduce the substrate signal. The inset are Euler angles. ϕ_1 is a rotation about the sample normal (ND). Φ is a tilt of [100] from ND. ϕ_2 is a rotation about [001].*

Results

1. Orientation Analysis

Al_2O_3, AlN, AlGaN and ZnO all have the same wurtzite crystal structure with six-fold symmetry in the basal plane. Therefore, AlN, AlGaN and ZnO take the same *c*-plane crystal orientation as the Al_2O_3 substrate. Since gold has the face-centred cubic (FCC) structure, the Au{111} planes have three-fold hexagonal-like symmetry, so a Au{111} orientation is predicted.

An XRD pattern is collected on the sample, shown in figure 4.25. The Al_2O_3 substrate reflections are minimized by using a 1°*ω* offset. For the AlN, AlGaN and ZnO, (000*l*) reflections are present. indicating that these phases are (0001)-oriented. *Au*{*hhh*} reflections are observed, indicating that the Au is {111}-oriented.

Figure 4.26 displays rocking curves collected on the Al_2O_3(0006), AlN(0002), AlGaN(0002), ZnO(0002) and Au {111} reflections. The peak widths (full width at half maximum) are 0.0243° ± 0.006, 0.5107° ± 0.0006,0.4337°± 0.002,0.8773° ± 0.1 and 1.7740° ± 0.01, respectively. This small degree of misalignment indicates that the nanostructures grow epitaxially and are well aligned.

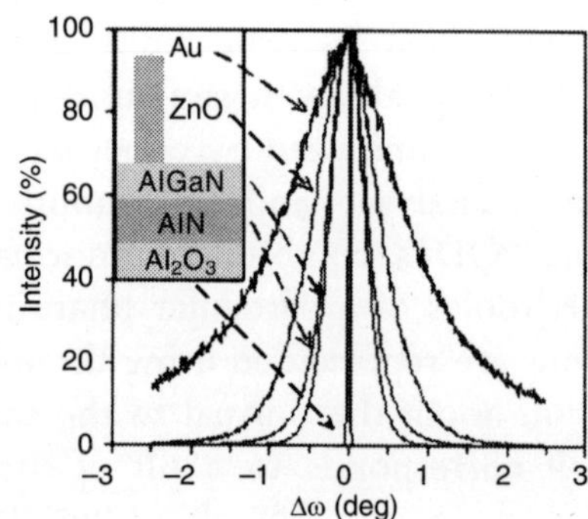

Figure 4.26 *Rocking curves collected on the Al2O3(0006), AlN(0002), AlGaN(0002), ZnO(0002) and Au {111} reflections. The inset is a schematic diagram of the ZnO nanorod sample.*

In order to further investigate the crystal orientations, texture data is collected and orientation distribution functions (figure 4.27) are calculated for the AlN, AlGaN, ZnO and Au. For AlN, only the $\Phi = 0°$ and $5°$ planes are shown, since no significant intensity is observed outside these planes it indicates that the AlN is (0001)-oriented. The spread of intensity into the $\Phi = 5°$ plane indicates a small degree of misalignment. For AlGaN, only the $\Phi = 0°$ plane is shown, since no significant intensity is observed outside that plane, it indicates that the AlGaN is (0001)-oriented.

For ZnO, only the $\Phi = 0°$ and $5°$ planes are shown, since no significant intensity is observed outside these planes it indicates that the ZnO is (0001)-oriented. The spread of intensity into the $\Phi = 5°$ plane indicates a small degree of misalignment of the nanowires. For Au only the $\Phi = 55°$ plane is shown, since no significant intensity is observed outside this plane it indicates that the Au is {111}-oriented, as seen by comparing with the calculated angle of 54.74° between the (001) and {111} poles in Au. The orientation relationships, including parallel planes and directions, are summarized in Table 1.

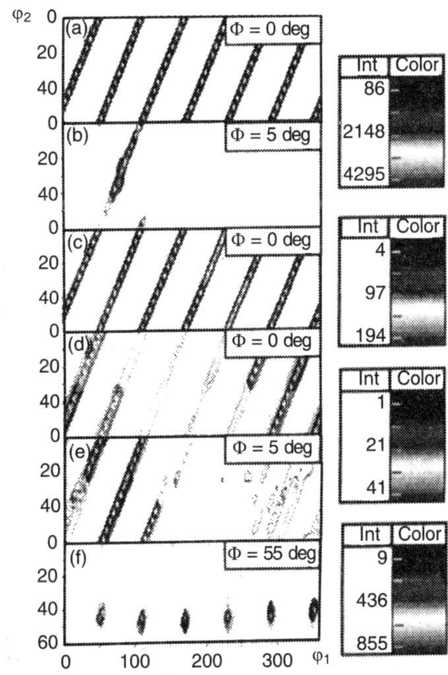

Figure 4.27 *Orientation distribution functions of (a) and (b) AIN layer. (c) AlGaN layer, (d) and (e) ZnO nanorods, and (f) Au catalyst Only Φ planes with significant intensity are displayed.*

Table 1 *Orientation relationships between the Au, ZnO, AlGaN, AlN and Al2O$_3$ phases as determined from texture analysis.*

Phase	*Parallel plane*	*Parallel direction*
Au	{111}	$\langle 11\bar{2}\rangle$
ZnO	(0001)	$\langle 01\bar{1}0\rangle$
A1GaN	(0001)	$\langle 01\bar{1}0\rangle$
AlN	(0001)	$\langle 01\bar{1}0\rangle$
Al_2O_3	(0001)	$\langle 21\bar{3}0\rangle$

An unexpected result from the texture analysis is the discovery of two orientations for the Au phase. In the Au ODF six discrete points are observed at $\phi_2 = 45°$ and $\phi_1 = 60 \times n - 10$, where n is an integer from 1 to 6. This indicates six in-plane rotations of 60° for the Au catalyst particles. However, the Au {111} planes have only three-fold symmetry. Therefore, the six rotations are due to two unique orientations related to each other by a 60° rotation about the sample normal, with the other rotations being generated by the trigonal symmetry of the Au {111} planes.

A closer examination of the texture data reveals some unexpected orientation relationships within the parallel interfacial planes. The parallel directions within the Al_2O_3 and AlN planes ($\langle 21\bar{3}0\rangle_{Al_2O_3} \parallel \langle 01\bar{1}0\rangle_{AlN}$) indicate a 30° rotation between the hexagonal symmetry of each plane. Though counterintuitive, this relationship reduces the lattice mismatch at the surface. For the observed orientation relationship, the lattice mismatch is 32% in the $\langle 21\bar{3}0\rangle \parallel \langle 01\bar{1}0\rangle_{AlN}$ directions and 13% in the $\langle 01\bar{1}0\rangle_{Al_2O_3} \parallel \langle 21\bar{3}0\rangle_{AlGaN}$ directions. Though somewhat large for epitaxial growth, this is smaller than the lattice mismatch of 53% in the $\langle 01\bar{1}0\rangle_{Al_2O_3} \parallel \langle 01\bar{1}0\rangle_{AlN}$ directions. Unlike the Al_2O_3/AlGaN interface, the hexagonal symmetries of the AlN(0001) || AlGaN(0001) || ZnO(0001) interfaces match, with $\langle 01\bar{1}0\rangle_{AlN}$ parallel to $\langle 01\bar{1}0\rangle_{AlGaN}$ parallel to $\langle 01\bar{1}0\rangle_{ZnO}$. The lattice mismatches are small, 2%.

Similar to the $A1_2O_3$/AlN interface, the hexagonal symmetry of the ZnO(0001) || Au{111} interface is rotated by 30°, with $\langle 01\bar{1}0\rangle_{ZnO}$ parallel to $\langle 11\bar{2}\rangle_{Au}$. The lattice mismatch is 30% in the $\langle 01\bar{1}0\rangle_{ZnO} \parallel \langle 11\bar{2}\rangle_{Au}$ directions and 2% in the $\langle 21\bar{3}0\rangle_{ZnO} \parallel \langle \bar{1}10\rangle_{Au}$ directions and 2% in the $\langle 21\bar{3}0\rangle_{ZnO} \parallel \langle \bar{1}10\rangle_{Au}$ directions. These mismatches are larger in one direction and smaller in the other as compared to the mismatch of 13% if the hexagonal symmetry matched, i.e. $\langle 01\bar{1}0\rangle_{ZnO} \parallel \langle \bar{1}10\rangle_{Au}$.

2. *Gold catalyst particle*

The samples are imaged with scanning electron microscopy (SEM). The nanowires have a columnar shape, with an average length of 284 ± 72 nm, and an average diameter of 35 ± 11 nm. The catalyst particle is roughly hemispherical in shape with an average diameter, 33 ± 12 nm, similar to that of the nanowire. There is a large variation in the size of the nanostructures, as evidenced by the standard deviation values above. The nanowires are predominantly aligned vertically.

The lattice parameter of the gold in the ZnO nanorod is determined from XRD scans to be 4.073 ± 0.001 Å, slightly smaller than the bulk value of 4.07894 Å. The volume ratio of ZnO to Au is 223 ± 43 to 1, as determined from the integrated intensity of each peak. Therefore, it is unlikely that any extraneous gold is present in the sample to affect the signal from the gold catalyst particles on the ZnO nanorods. The observed decrease in Au lattice parameter, as compared to the reference value, is expected due to the known lattice contraction occurring in nanoparticles.

A mode is made to quantify the change in lattice parameter, *a*, for metal nanoparticles as a function of the particle diameter (*D*) and shape, shear modulus and surface energy. For 33 nm hemispheric Au particles, the change in lattice parameter is –0.1433%, leading to an lattice parameter of 4.0731 Å, which agrees with the experimentally measured Au lattice parameter of 4.073 ± 0.001 Å.

For, Au lattice parameter, a controlled sample of Au is thermally evaporated on a GaN layer on *c*-plane oriented, single-crystal, Al_2O_3 substrate is prepared and investigated with SEM and XRD, After deposition of a 5 nm layer of Au, a small number of Au nanoparticles are observed on the sample surface, and XRD patterns show evidence of polycrystalline Au though there is also a weak Au signal. After annealing at 870 and 1120 K for 30 and 60 min, respectively, the sample is densely covered with Au nanoparticles with an average diameter of 32 ± 12 nm. An XRD pattern (with a 2° ω offset to reduce the Al_2O_3 substrate signal) indicates that the Au is primarily {111}-oriented. The Au lattice parameter of the control sample, (as determined from the XRD and 2θ offset scans) is 4.073 ± 0.002 Å, which is the same as that measured from the ZnO nanorod sample.

The finding that the gold lattice parameters measured from both the nanowire sample and the control sample, which is not exposed to ZnO, arc equal (4.073 A) establishes that there is no zinc in the catalyst particles. As is seen by using Au-rich Au/Zn, substitutional intermetallics with FCC superstructures, even 7% Zn in Au results in a significant decrease in the lattice parameter of the Au nanoparticles to 4.0562 Å. Therefore, the Au catalyst particles do not have Zn.

If Zn is not alloying into the Au catalyst particles, it cannot cause eutectic melting below the melting temperature of pure Au as in the VLS synthesis mechanism. Decreasing particle size lowers the melting temperature, but the effect is insufficient to cause melting at the synthesis temperature. Accordingly 33 nm Au particles melt at 1300 K well above the synthesis temperature of approximately 1120 K.

In order to verify the melting point of the Au catalyst particles, XRD data is collected at elevated temperatures on the Au-catalysed ZnO nanorods. Offset 2θ scans of the Au{111} reflection are collected from 770 to 1270 K in 100 K increments, seen in figure 4.28. The Au {111} reflection is present in all scans, including that at 1270 K, indicating that the Au nanoparticles do not melt. Since the synthesis temperature is around 1120 K it is likely that the Au catalyst particles remain solid during synthesis, indicating that growth is by the VSS synthesis mechanism, not the VLS synthesis mechanism.

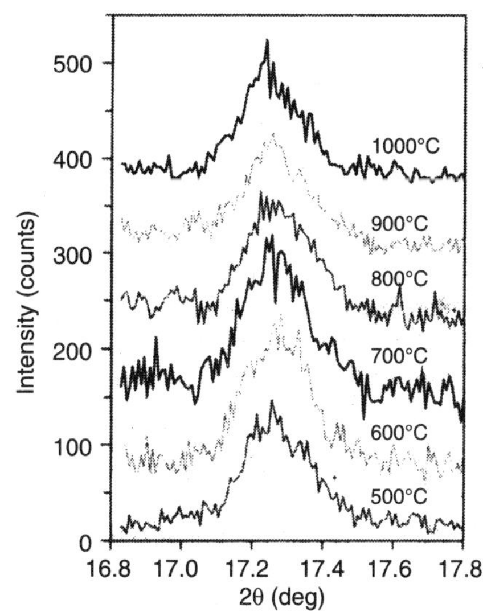

Figure 4.28 *XRD patterns of the Au{111} reflection of Au-catalysed ZnO nanorods at temperatures 770 to 1270 K.*

The absence of Zn in the Au catalyst particle also indicates that growth cannot be proceeded by a supersaturation nucleation process. A mass spectroscopy of thermally evaporated ZnO shows that the ZnO dissociates into ionic species, which do not diffuse into and alloy with metallic Au. It is more likely that growth proceeds by diffusion around the surface of the Au catalyst particle to the growing nanowire.

Surface diffusion is a significant part of the VLS mechanism because liquid surface diffusion rates are significantly higher than liquid bulk diffusion rates. This is also true for solid surface and bulk diffusion. The surface diffusion coefficient of solid Au is several orders of magnitude greater than the bulk diffusion coefficient. In addition, Au nanoparticles have a disordered, quasi-liquid layer on the surface due to surface reconstruction. In other metals, such layers have unusually high self-diffusion coefficients at temperatures approaching the melting point.

(b) Growth of Aligned ZnO Nanowires on Nitrides

By VLS, it may be assumed that nanowires would grow as long as the catalysts are supplied as well as the temperature is adequate. By applying partial pressure, oxygen plays a key role in the quality of the grown ZnO nanowires. By defining different growth results as different "phases", a "phase diagram" between the oxygen partial pressure and the growth chamber pressure for synthesizing high quality aligned ZnO nanorods on the GaN substrate using VLS process is obtained.

Here, an undoped *c*-plane oriented GaN film with a thickness of 2 μm is grown on an *a*-plane sapphire single-crystal substrate is used as the substrate. A thin layer of gold (7–8 nm) is deposited on the top or GaN via plasma sputtering, which acts as catalysts to guide the growth of ZnO nanorods. The source materials are a mixture of equal amounts (by weight) of ZnO power and graphite power, which is grounded and placed in an alumina boat. Then the boat is loaded in the center of an alumina tube (150 cm long, 4 cm inner diameter), where the substrate is placed 10 cm away from the source material at the downstream side. The tube is placed in a horizontal tube furnace with cooling water running through the outside of the tube at the ends. After pumping the system down to 2 Pa, a premixed gas (Ar and O_2) is introduced into the system with a flow rate of 50 sccm to bring the pressure back to a certain point between 1.5×10^2 Pa and 3×10^4 Pa according to the designed experiments. The furnace is then heated 1220 K at a heating rate of 50 K min^{-1} and the temperature is held at the peak temperature during the growth for 30 min. Finally, the system is slowly cooled to room temperature under flowing gas. All of the conditions are fixed during the growth except two variables: the volume ratio of O_2 to Ar gas (1% to 4%), and the chamber pressure (30 Pa to 10^4 Pa).

The as-synthesized ZnO nanorods are examined by scanning electron microscopy (SEM). Figure 4.29(a) shows a typical image of ZnO nanorods grown on GaN substrate a with a good alignment, in which almost all of the ZnO rods have the same lengths and diameters and are oriented perpendicular to the substrate. This sample is grown under the condition of 2% O_2 (by volume) and 30×10^2 Pa chamber pressure. The well-aligned structure uniformly covers all of the area of the GaN substrate. The inset in Figure 4.29(a) shows a top view of the aligned ZnO nanorods, where only the very bright gold catalyst tips arc observed. It also confirms that almost every single nanorod is perpendicular to the substrate and there is no side branch.

The crystallography of ZnO rods is characterized by transmission electron microscopy (TEM), Figure 4.29(b) shows a typical TEM image of the ZnO nanowires/nanorods. Each nanorod is a single crystal and exhibits a uniform thickness, From the ZnO nanorod, a diffraction pattern is recorded as shown in the inset of Figure 4.29(b), which reveals the growth direction is along [0001] and the six sides facets are $\{11\bar{2}0\}$.

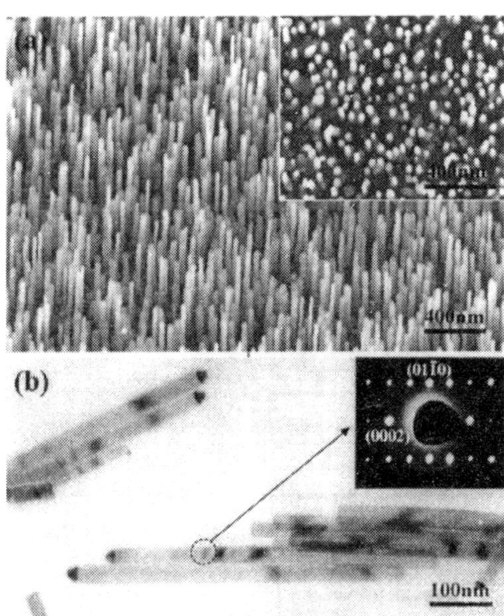

Figure 4.29 *Aligned ZnO nanorods on GaN substrate. (a) Side view SEM image, the inset if th etop view SEm image of teh samle; (b) TEM image of ZnO nanorods, the inset is an electron diffraction pattern taken from one ZnO rod.*

Now, the oxygen partial pressure and the system total pressure plays key roles in the growth of ZnO nanorods. With different oxygen volume percentage and different chamber pressure, the quality and growth behavior of the ZnO nanowires are strongly affected. Consider over 100 growth experiments under different growth conditions, which are designed to quantitatively define the best combination of the O_2 partial pressure and the chamber pressure. For consistency, collect all samples at the 1150 K which is 10 cm away zone; is from the source materials. The O_2 volume percentage in the chamber varied from 1 to 4, and the system pressure varied from 1.5×10^2 to 3×10^4 Pa as required in the growth conditions for the allignment of ZnO nanowires. Since the growth system is pumped down to 2 Pa, the system pressure is brought back to a growth pressure at a value between 1.5×10^2 to 3×10^4 Pa, the oxygen coming from the residue air only contributed 0.28–0.0014% toward the entire oxygen content, which is less than the percentage of O_2 in the flow gas. Therefore, the partial pressure of O_2 is considered to be the volume percentage of the O_2 introduced in the flow gas.

The experimental results are summarized in Figure 4.30, which is a “phase diagram” for the O_2 volume percentage in the chamber and system pressure, under Which the optimum conditions for growing aligned ZnO nanowires are presented. The term of “phase diagram” used here is for controlled synthesis of nanowires.

The effects of total chamber pressure and O_2 volume percentage to the final results, three sequences are highlighted in the phase diagram: fixed O_2 volume percentage but with variable chamber pressure (line 1); fixed system pressure but with variable O_2 volume percentage (line 2); and linear increased O_2 volume percentage with decreasing system pressure (line 3). Four

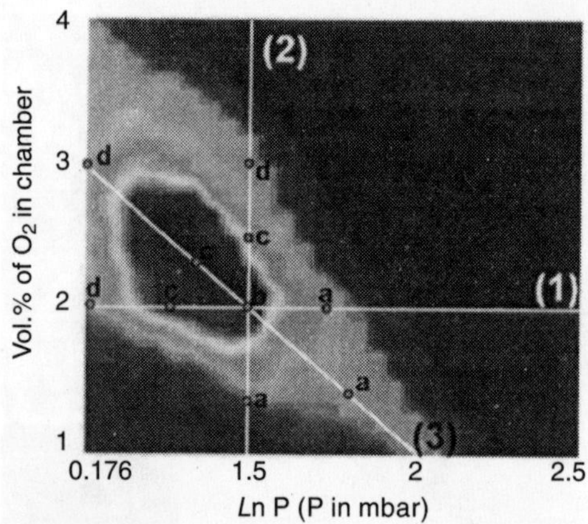

Figure 4.30 *"Phase diagram" that correlates oxygen volume percent in the growth chamber and the growth chamber pressure for growing. This phase diagram is generated from the results of a large number of experiments of different conditions. The point matrix was broadened to form a quasi-continuous phase diagram. The synthesis line (1) is the line with constant oxygen volume percentage; line (2) is the line with constant system pressure; line (3) is the line with linearly varying oxygen volume percentage and system pressure.*

representative points on each line are picked lip to present the quality of the grown nanowires under the defined growth conditions.

Figure 4.31 Presents SEM images of four samples received under the conditions of 2% fixed O_2 volume percentage (1 sccm oxygen. 49 sccm argon) but 4 different chamber pressures: 5×10^3, 3×10^3, 6×10^2 and 1.5×10^2 Pa as shown in Figure 4.31(a)–(d), respectively. In Figure 4.31(a), which represents the results with the highest system pressure among the four experiments, no nanowire is formed but only gold catalyst particles are found. With the decreasing of system pressure to 3×10^3 Pa, a perfect alignment of ZnO nanowires is achieved, as shown in Figure 4.31(b). When the system pressure drops to 10^3 Pa. ZnO nanorods become shorter and randomly orientated, as shown in Figure 4.31(c). By further dropping the system pressure to 1.5×10^2 Pa, only a few thin nanorods are deposited on a continuous ZnO film. This change in the ZnO morphology contributes to the increased Zn vapor concentration while the system pressure is decreased.

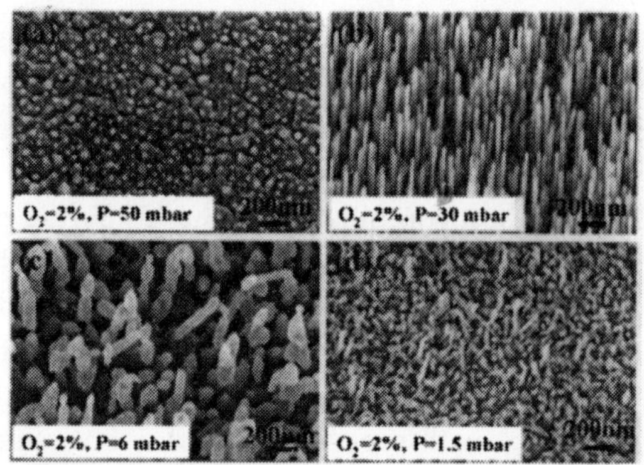

Figure 4.31 *SEM image of ZnO nanorods grown under the conditions of 4 green points along line (1) in Figure 2 under constant oxygen volume percentage (2%) but variable chamber pressure: (a) 5×10^3 Pa; (b) 3×10^3 Pa; (c) 6×10^2 Pa; (d) 1.5×10^2 Pa.*

In the carbon-thermal evaporation process, the Zn vapor source is dominated by reaction

$$ZnO(s) + C(s) \leftrightarrow Zn(v) + CO(v) \tag{1}$$

This reaction is favorable in high temperature. Once the vapors are transported to the cooler region where the substrate is located, the Zn vapor is deposited on the surface of the catalyst and reoxidized, resulting in the growth of the nanowire. Since the amount of Zn vapor is determined by the local temperature, thus, the rate at which the Zn vapor is produced can be assumed to be a constant under different chamber pressures. Consequently, when the total system pressure drops, the partial pressure of the Zn vapor increases. When the system pressure is too high, Zn vapor do not reach supersaturation, and thus, no deposition occurs on the surface of the catalyst particles, resulting in no growth at all (Figure 4.29(a)). When the system pressure is too low, Zn vapor is supersaturated, the vapor deposits not only on the surface of the Au catalyst but also the surface of the GaN substrate, resulting in formation of nanowires as well as a thin film on the substrate surface (Figure 4.32(d)). High quality aligned ZnO nanorods are achieved when a reasonable supersaturation level of Zn vapor is reached at a moderate system pressure, which is around 3×10^3 Pa under growth condition.

The effect of O_2 partial pressure is illustrated in Figure 4.32, in which SEM images of a–d represent the typical results of ZnO deposition at O_2 volume percentage of 1.3%. 2%, 2.5%, and 3%, respectively, with a fixed chamber pressure of 3×10^3 Pa. At a very low O_2 volume percentage (1.3%), only small ZnO dots are nucleated, as shown in Figure 4.32(a). By increasing the O_2 volume percentage to 2%, good alignment of ZnO nanowires is achieved (Figure 4.32(b)). When the O_2 volume percentage is increased to 2.5%, as shown in Figure 4.31(c), the density of ZnO nanorods drops, as well as their lengths. The growth of ZnO nanorods almost vanishes once the O_2 volume percentage is at 3% (Figure 4.32(d)).

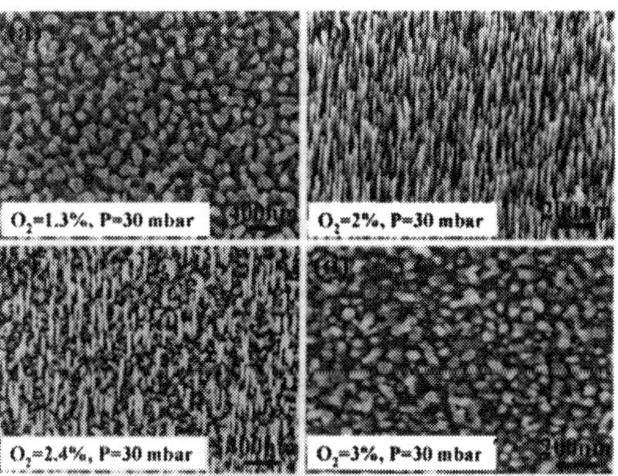

Figure 4.32 *SEM images of ZnO nanorods grown under the conditions of 4 green points along line (2) in Figure 2 under constant chamber pressure (30 mbar) but variable oxygen volume percentage: (a) 1.3%; (b) 2%; (c) 2.4%; (d) 3%.*

The three possible reactions that O_2 is involved in carbon assisted thermal evaporation process are:

$$2C + O_2 = 2CO \tag{2}$$

$$2CO + O_2 = 2CO_2 \tag{3}$$

$$2Zn + O_2 = 2ZnO \tag{4}$$

Zn vapor is provided by equation 1. The reaction 1 occurs only when the temperature is higher than 1240 K, whereas the furnace is heated to 1220 K. Thus, in the beginning when there was no other gas in the system, the main reaction involving O_2 is reaction 2, which generates heat as well as CO. The heat can be absorbed locally by the source materials (ZnO plus graphite) and facilitate reaction 1 to produce Zn vapor; whereas local concentration of Co can also increase, which hinders the Zn vapor generation. Under the influence of these two effects, reaction 1 does not accelerates, and as a result, when the O_2 partial pressure is very low, insufficient Zn vapors is are released to reach a reasonable supersaturation point for producing ZnO nanorods, as the situation shown in figure 4.32(a).

Once the partial pressure of O_2 is increases, additional O_2 reacts with CO to form CO_2 (equation 3), which lowers the concentration of CO and generates heat. Both of these effects facilitate reaction 1. Therefore, supersaturation of Zn vapor can be reached, leading to growth of high quality aligned ZnO nanorods, as shown in Figure 4.32(b). However, when the partial pressure of O_2 is further increased, additional O_2 reacts with Zn vapor generated from the source materials, to form a ZnO film/grain. By introducing more O_2 into the system it is found that a layer of large ZnO nanorods/grain is formed on the surface of the source powders. Consequently, the Zn vapor decreases again near the substrate and little growth is found (Figure 4.32d). In general, only moderate O_2 volume percentage (~2 vol %) results in good aligned ZnO nanorods.

Four typical results of ZnO deposition, under linearly increased O_2 partial pressure with decreased chamber pressure are illustrated in Figure 4.33. The experimental conditions are as follows: (a) 1.6% O_2 at 5×10^3 Pa system pressure; (b) 2% O_2 at 3×10^3 Pa system pressure; (c) 2.5% O_2 at 7×10^2 Pa system pressure; (d) 3% O_2 at 1.5×10^2 Pa system pressure. Except point (a) where both O_2 partial pressure and system pressure are intended to give a low Zn vapor concentration, the other three b, c, d points in the phase diagram exhibit a reasonable growth of ZnO nanorod This is because the decreasing of Zn vapor caused by increased O_2 compensated by reducing the chamber pressure; thus the reasonable supersaturation level of Zn vapor is maintained in a relatively large range.

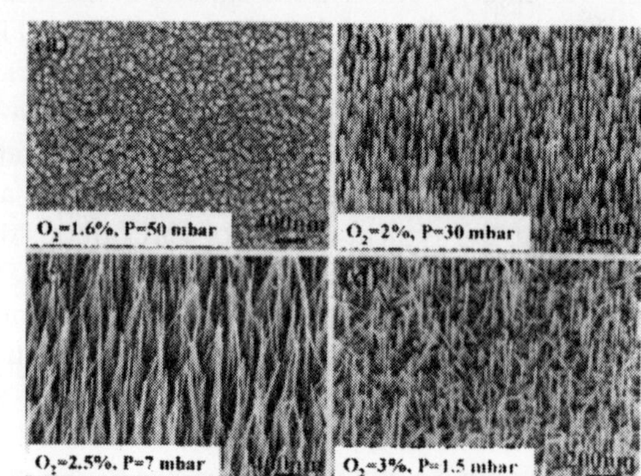

Figure 4.33 *SEM images of ZnO nanorods grown under the conditions of 4 green points along line (3) in Figure 4.25 by linearly adjusting the chamber pressure and oxygen partial pressure. (a) 1.6% oxygen volume percentage, 5×10^3 Pa chamber pressure; (b) 2% oxygen volume percentage, 3×10^3 Pa chamber pressure; (c) 2.5% oxygen volume percentage, 7×10^2 Pa chamber pressure; (d) 3% oxygen volume percentage, 1.5×10^2 Pa chamber pressure.*

GROWTH OF VERTICALLY ALIGNED ZNO

(I) Nanowire array in GaN and SiC substrates

All syntheses are performed in a high temperature tube furnace by vapor-liquid-solid process. Equal amount (by weight) of ZnO and graphite powder (0.6 g each) are loaded in an alumina boat located at the center of an alumina tube, which is then placed in a single-zone tube furnace. Argon is used as carrier gas at a flow rate of 49 sccm with additional 1 sccm oxygen to facilitate the reaction. Substrates are coated with a thin layer (7–8 nm) of Au as catalyst and placed 7–12 cm downstream from the center of the tube. The, source materials are heated to 1240 K at a rate of 50 K min^{-1}, and the temperature is held at the peak temperature for 30-90 min under a pressure of 2×10^4 Pa, while the local temperature of the substrates is between 1200–1240 K. All previous parameters are chosen after careful optimization.

The morphology of the as-grown products is examined using scanning electron microscopy (SEM). Vertically aligned nanowire arrays are found covering the entire SiC(0001) (Fig. 4.34(a)) and GaN(0001) (Fig. 4.34(b)) substrates (size is ~0.3 × 0.3 cm^2). The energy dispersive spectroscopy (EDS) experiment confirm the nanowires to be ZnO (Fig. 4.34(d) inset). However, for CdTe and Si substrates, even after a series of designed experiments, only a few of random ZnO nanowires are detected on these substrates, and the rest area is bare or covered with ZnO particles. The best nanowire growth in term of high percentage of vertically aligned nanowires and uniform length is located around 1230 K for both SiC and GaN substrates. The product is characterized using transmission electron microscopy. The growth orientation ZnO nanowires is along (0001), and a tiny Au catalyst sphere is at the growth front (Fig. 4.34(c)). Although, EDS shows the elemental composition of SiC and GaN substrates due to the short sampling depth, strong peaks of substrates are detected in X-ray diffraction (XRD) curves (Fig. 4.34(d)), indicating the high orientation of ZnO nanowires and substrates. The experiments show that perfect vertically aligned ZnO nanowire arrays (wurtzite, $a = 0.3249$ nm, $c = 0.5207$ nm) only occur on SiC(0001) (wurtzite, $a = 0.3076$ nm, $c = 0.5048$ nm) and GaN(0001) (wurtzite, $a = 0.3189$ nm, $c = 0.5185$ nm) substrates. ZnO(0001) plane has small lattice mismatch (<6%) with SiC(0001) and GaN(0001) planes, and has significant mismatch with other substrate. The small lattice mismatch between the nanowires and substrates plays a key role in heteroepitaxial growth of vertically aligned arrays. For nanowires, which have a significant surface to volume ratio, the surface energy dominates nanowires growth process, especially, at their nucleation and initial growth stage. The bigger lattice mismatch results in higher strain energy. Higher

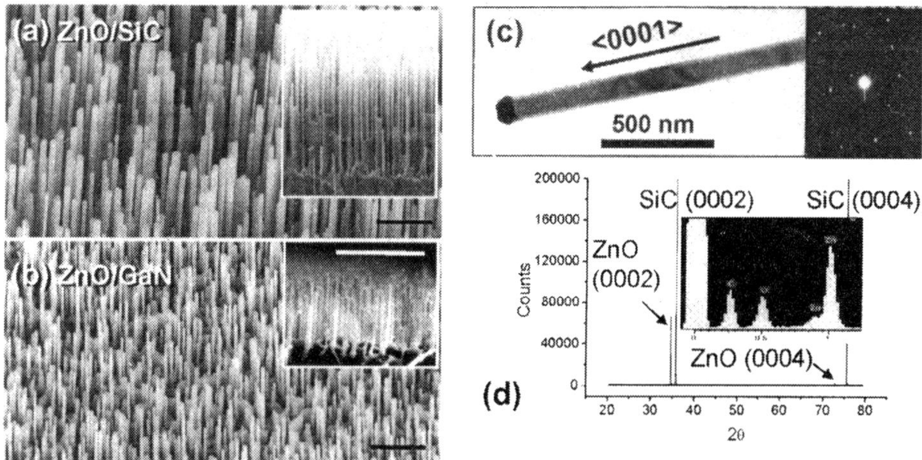

Figure 4.34 *(a) About 30° tilted and cross-sectional SEM images of vertically aligned ZnO nanowlres arrays grown on SiC(0001) substrate with 1 μm scale uar; (b) 300 tilted and cross-sectional SEM images of vertical aligned ZnO nanowire arrays grown on GaN(0001) substrate with 1 μm scale bar; (c) TEM image of a ZnO nanowire grown along (0001) direction with Au catalyst sphere, and the inset shows the corresponding TEM diffraction pattern; and (d) X-ray diffraction pattern of the same sample in (a), and the inset is EDS analysis.*

strain energy state becomes unstable and other lower strain energy states associated with other ZnO planes can occur, effectively reducing the total energy of the system. However, ZnO nanowires prefer (0001) direction as the fastest growth direction. Thus, if the (0001) plane of ZnO is not the boundary with the lowest energy, ZnO nanowires have higher possibility to lose the coherence with the substrates, and the fastest (0001) growth direction points to any random direction. Therefore, vertical aligned ZnO nanowire array is much easier to achieve when there is a small lattice mismatch between ZnO nanowires and substrates.

Samples A-F are obtained on SiC substrates at 1240, 1235, 1230, 1223, 1216 and 1207 K, temperature region, respectively, after 30 min synthesis, and their spatial distance between closest two samples is 1 cm. Five random locations (10×2 μm^2) are chosen for each sample; percentage of vertical aligned nanowires, length, diameters, and aspect ratio of nanowires are measured and calculated for all nanowires on the chosen locations. Nanowires are regarded as vertical if their axial direction is within 5° deviation of vertical line of the substrate. The length of nanowire is measured from the growth end to the nanowall that underlies of the nanowires, sinçe it is practically impossible to determine the original substrate surface, if the surface is away from the edges and has interconnecting nanowalls on it. The aspect ratio of length to diameter is also calculated. To accurately obtain these parameters, the samples are tilted to 30° or close to 90°. The results show that B has almost 100% vertically aligned nanowires, while this percentage drops sharply below 50% at 275 K or even lower temperature zone (Fig. 4.35(a)). The length of nanowires of sample B is 447 ± 10 nm, and corresponding aspect ratio of length to diameter is 8.0 ± 2.8 (Fig. 4.35(b). This means that nanowires in B are located at are the best for device application since they have relatively highest uniformity, which is required for nanodevices. The linear fitting (Fig. 4.35(a) inset) indicates that the nanowires grow at a rate of 54 nm min^{-1}. Nanowire arrays with any length flan be achieved as long as the catalyst particles is still active and the synthesis time increases. However, a poisoning of the catalyst particles limits the length of vertical nanowires.

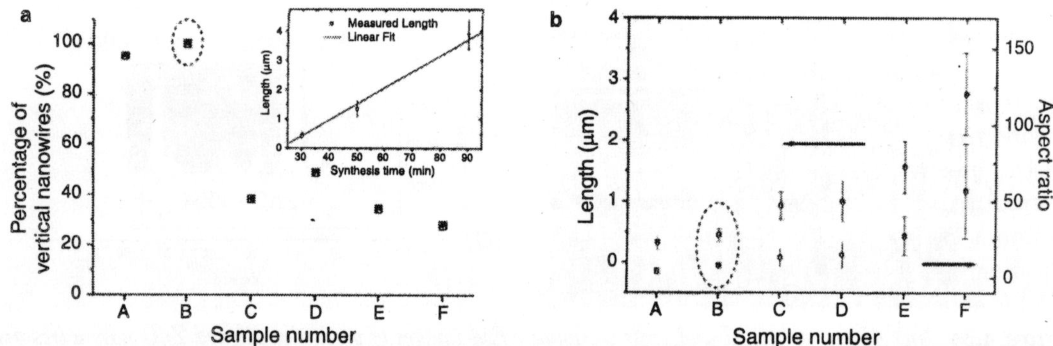

Figure 4.35 *Growth of ZnO nanowires on SiC substrates. (a) Diagram indicates the percentage of vertically aligned nanowires, and the inset shows the length dependence on synthesis time at sample B location and (b) diagram represents length distribution and aspect ratio of length to diameter distribution for samples A-F, with circles highlighting the best results.*

Generally, the vertically aligned nanowires are not separated from one another and they are connected nanowalls (Fig. 4.36(a)). Through SEM and TEM, we found that the residue Au catalyst at the bottom of aligned nanowires induce the sides (01-10) of ZnO nanowires to grow into triangular nanosheets (Fig. 4.36(b) and (c)). These triangular nanosheets meet another nanosheets or nanowires and became interconnected nanowalls. This growth mechanism is visually explained in a schematic diagram shown in Fig. 4.36(d). These conductive nanowalls are indispensable for some device applications, especially, when nanowire arrays are grown on the nonconductive substrates. The electric signals only transports to each aligned nanowire through the interconnected nanowalls instead of the underlying substrates.

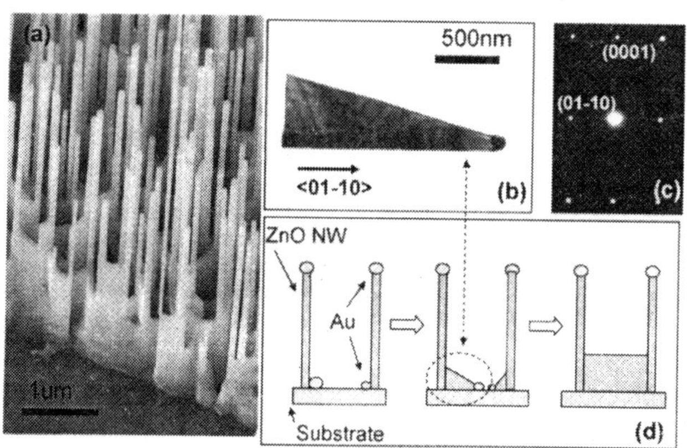

Figure 4.36 *(a) The SEM image shows nanawalls connecting individual vertically aligned nanowire; (b) the TEM image of a triangular nanosheet, which is the early stage of nanowalls; and (c) the corresponding TEM diffraction pattern of (b); Diagram demonstrating the growth mechanism of nanowalls.*

The nanowalls interconnecting ZnO nanowire arrays serve as an electrode, and the other top electrode is required for device fabrications. The top electrode can be a tiny conductive atomic force microscopy (AFM) tip, or a large piece of metal coated substrate. A new setup of device is designed to accommodate the ultra violet (UV) detector requirement. A transparent and conductive indium tin oxide (ITO) coated glass is chosen on the top of vertically aligned ZnO nanowire arrays (Fig. 4.37(a). This device is connected in a constant voltage circuit and is used as an UV etectron. The UV source used has a wavelength 365 nm and incident intensity ~20 mW cm^{-2}. As UV turns on, it penetrates through the top electrode and illuminates on the tremendous parallel ZnO nanowires. Because the photon energy of the incident UV light is higher than the band-gap of ZnO, electron-hole pairs arc generated inside ZnO nanowire arrays by light absorption. At the same time, the electron-hole pairs are separated by electric field and induce the photocurrent. The current through a typical device sharply increases three times when UV is on. The performance of device is further improved by using uniform length of nanowire array or by chemically functionalizating nanowires arrays. After layered coating a positively charged polymer poly(diallyldimethylammonium chloride) (PDADMAC) and a negatively charged polystyrene sulfate (PSS), the conductance of the device increases about eight times when UV illumination is on (Fig. 4.37). It should be noted that neither PDADMAC

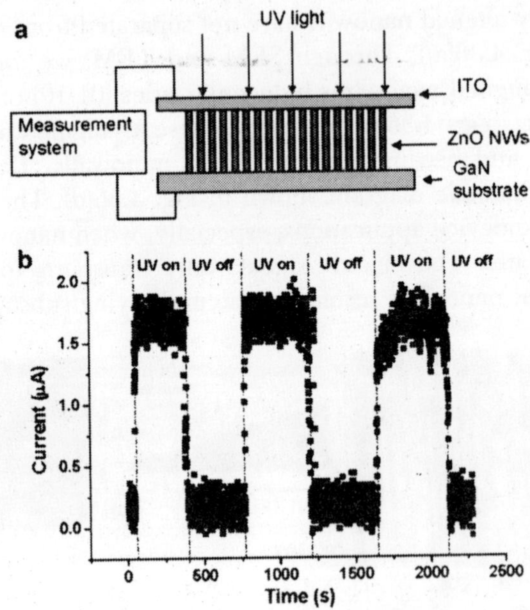

Figure 4.37 *(a) Diagram of a UV detector based on verticlaly aligned ZnO nanowires and (b) the diagram displays the current response of funcitonalized ZnO nanowire arrays to UV light.*

nor PSS is conductive itself and show no response to UV illumination. The increase in conductance for the polymer coated ZNO array must attribute to a coupling effect between the polymer and ZnO.

(II) Nanobelt Array on GaN Substrate

Experiment

The ZnO NBs are synthesized by a solid-vapor process. ZnO powder (3 g) is loaded in an alumina boat that is positioned in the middle of an alumina tube. A 2 μm thick, *c*-plane oriented GaN thin film grown on an *a*-plane sapphire wafer is used as the substrate, which is placed down stream from the source material with GaN surface facing downward in the alumipa boat. The tube is then placed in a horizontal tube furnace, in which the source material is located at the highest temperature zone. The deposition system is pumped down to ~201 m-mmHg overnight to remove the residual oxygen and water, Then the source materials are heated to 1750 K at a heating rate of 50 K min^{-1}. Argon carrier gas is introduced at a flow rate of 50 sccm (standard cubic centimeters per minute) when the temperature reaches 570 K. The source material is heated at 1750 K for 60 min. The NBs are deposited onto an alumina substrate placed in a temperature zone of 770–870 K under a pressure of 3×10^3 Pa. Then the furnace is turned off, and the tube is cooled naturally down to room temperature under an argon flow.

The as-synthesized ZnO NBs are first characterized by scanning electron microscopy (SEM). The ZnO NBs are deposited on both sides of the substrate. The top surface is unpolished sapphire, where randomly oriented and long ZnO NBs are formed. Interestingly, as shown in Figure 4.38(a), the vertically well-aligned ZnO NBs are grown at the bottom surface of the substrate, where a *c*-plane oriented GaN thin film is coated. The widths of the NBs grown on the GaN range form 40 to 200 nm and their lengths are ~5 μm. Higher magnification side view and top view SEM images are shown in Figure 4.39(b) and (c) respectively. It shows the growth direction direction change close to the tips of NBs. Also there is a particle located at the tip of each NB, which can be seen more clearly in the transmission electron microscopy (TEM) images. The presence of the Zn particle at the tip indicates a growth process by self-catalyzed vapor-liquid-solid (VLS) process, because the Zn nanoparticles are produced by the decomposition of ZnO powder.

Structural characterizations of the ZnO NBs are carried out using TEM. Figure 4.39(a) shows an enlarged TEM image of several individual ZnO NBs. Each NB is uniform in width along the growth direction, and the bending contour indicates the same thickness across the belt. Select-

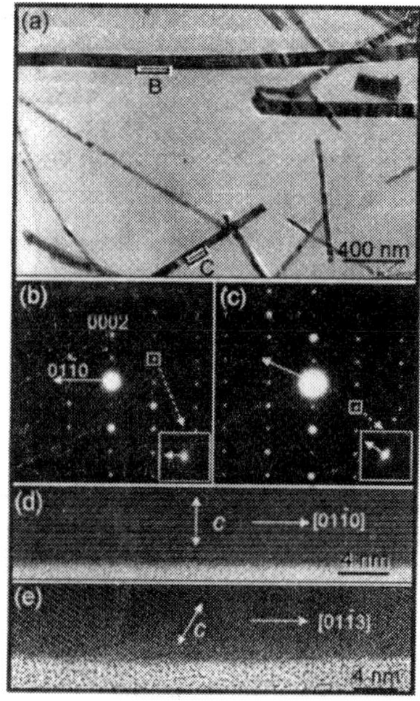

Figure 4.39 *(a) Low-magnification TEM image of ZnO nanobelts. (b and c) SAED patterns taken from rectangles B and C areas in (a), respectively. (d and e) High-resolution TEM image of nanobelts B and C. which are grow along [01$\bar{1}$0] and [01$\bar{1}$3], respectively. Note the relative rotation of the diffraction patterns in reference to the image was not adjusted for nice display purpose.*

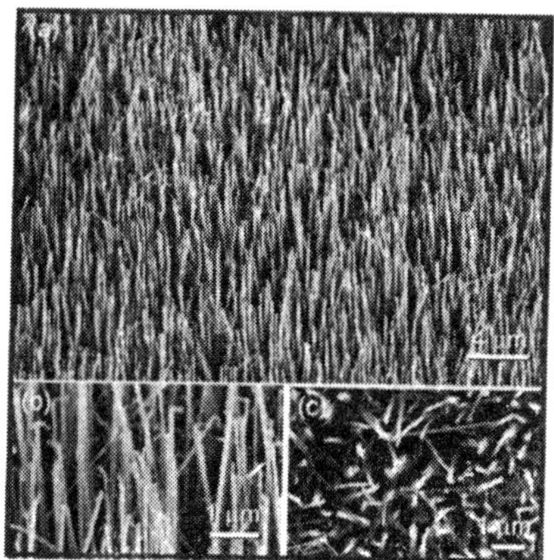

Figure 4.38 *SEM images of vertically well-aligned ZnO nanobell arrays grown on GaN substrate. (a, b) Low- and high-magnification 30° side view SEM images of the grown nanobelt array, respectively. (c) High-magnification top view SEM image of the grown nanobelt array.*

area electron diffraction (SAED) patterns from the two rectangular areas labeled B and C are shown in Figure 4.39(b) and (c) respectively. Both SAED patterns are with electron beam along $[2\bar{1}\bar{1}0)$. Due to the quasi-one-dimensional morphology, the growth directions of each NB are identified based on the shape effect in diffraction spots. The insets in Figure 4.39(b) and (c) gives the enlarged diffraction Spots, from which the streaking directions can be identified which are perpendicular to the growth direction of the NB. The growth direction of the NB at the top of Figure 4.39(a) is along $[01\bar{1}0]$, while the growth front of the bottom NB is $(01\bar{1}1)$ plane with growth direction as $[01\bar{1}3]$. The growth directions of these two NBs are further confirm their high-resolution TEM (HRTEM) images displayed in Figure 4.39(d) and (e) respectively. Among more than 20 NBs examined, 900% have either $[01\bar{1}0]$ or $[01\bar{1}3]$ as their growth directions. $[01\bar{1}0]$ is one of the most common NB growth directions, while $[01\bar{1}3]$ NB is quite rare.

Each belt is examined using TEM is only part of the whole nanostructure grown from the substrate different sections of NBs are distinguished upon their morphologies. During the vapor deposition process, a seed layer is first epitaxially grown on the substrate, then ZnO NBs grow up from such a thin film layer. So the belt in Figure 4.40(a) looks peeled off from a thin film layer, which is in fact the root of the NB binding with the GaN substrate. Figure 4.40(b) is a high-resolution TEM image of the rectangle area Figure 4.40(a). The growth direction is identified as $[01\bar{1}3]$, which is the starting growth direction. In Figure 4.40(c), an arrowhead points to a panicle at the growth front of the NB. The uniform contrast of the NB and the inserted SAED pattern indicate that it is a single crystal with two growth directions along $[01\bar{1}0]$ and $[01\bar{1}3]$, respectively. The turning points are the growth direction changed between $[01\bar{1}0]$ and $[01\bar{1}3]$ and are circled in Figure 4.40(c). One of them shows that the growth direction switched at the tip of the NB. Figure 4.40(a) confirms that the ZnO NB grow along $[01\bar{1}3]$ at the tip. From this analysis it is found that the NBs initially grew along $[01\bar{1}3]$, then switch to $[01\bar{1}0]$ during growth, and then the final part of the ZnO NBs takes the $[01\bar{1}3]$ as its growth direction.

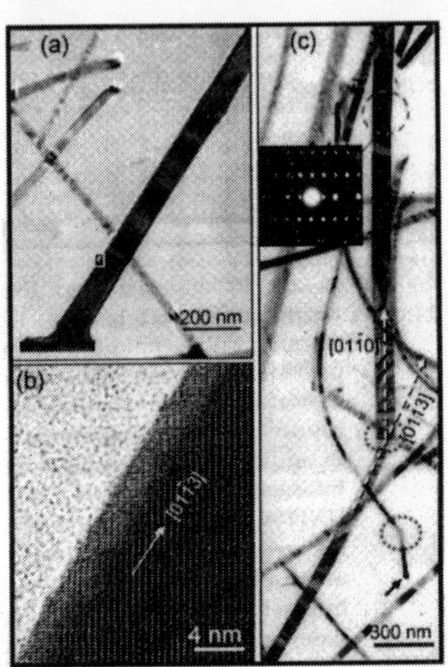

Figure 4.40 *(a) Low-magnification TEM image of the root of a ZnO nanobelt in contact with the GaN substrate. (b) High-resolution TEM image from the rectangular area in (a). The root of the nanobelt grows along $[01\bar{1}3]$ direction. (c) TEM image shows the turning point (circled area) of the growth direction from $[01\bar{1}3]$ to $[01\bar{1}1]$. Inset: the SAED pattern of the largest nanobelt shown in (c).*

The change of growth direction is mainly effected by the temperature variation during the deposition process. Since the source materials are healed, to 1750 K at a rate of 50 K min^{-1}, such a temperature is significantly higher than the temperature used for growing [0001] nanowires, which is typically ~1270 K. During the growth, take ~30 min before the furnace to reach the peak temperature. Argon is introduced at a flow rate of 50 sccm when temperature reaches 570 K. The source material, ZnO powder is decomposed before the temperature was raised to 1750 K. The NBs start growing along $[01\bar{1}3]$ at a relatively lower temperature. The growth direction switches to $[01\bar{1}0]$ when temperature is stabilized at the peak value. Once the furnace is turned off, the source material and substrate are slowly cooled down to room temperature under an argon flow. ZnO NBs takes the $[01\bar{1}3]$ as growth direction again at lower temperature.

Both SEM and TEM images show nanoparticles at the tips of the NBs. The particle is examined by HRTEM and SAED. Figure 4.41(a) shows a single crystalline Zn particle that is covered by a thin ZnO layer, which is formed due to the oxidation of Zn when exposed to air. The particle's SAED pattern is displayed in Figure (b), which is composed of two patterns; one is ZnO $[2\bar{1}\bar{1}0]$, and another one is the Zn [0001] pattern as indicated by white arrowheads. The lattice match configuration is $(2\bar{1}\bar{1}0)_{ZnO} \parallel (0001)_{Zn}$, $[0001]_{ZnO} \parallel [01\bar{1}0]_{Zn}$. The Zn particle and its lattice match to ZnO NB indicate the existence of a self-catalyzed, growth process. How only source material used is pure ZnO powder and there is no foreign catalyst deposited on the substrate.

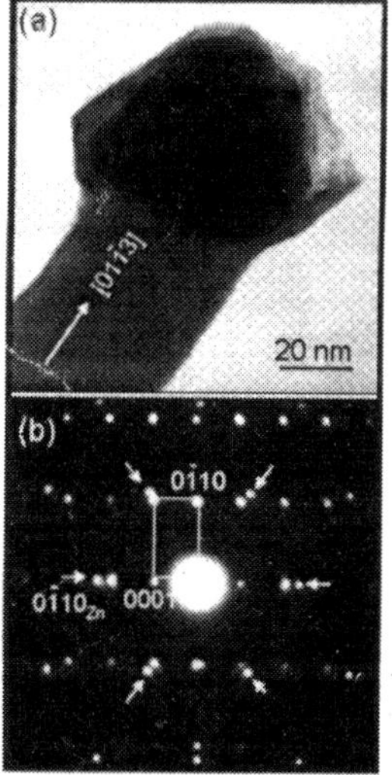

Figure 4.41 *High-resolution TEm image (a) and corresponding SAED pattern (b) of a ZnO nanobelt with a Zn particle located at ats tip. Note th relative rotation of the diffraction patterns in reference to the image was not adjusted for nice display purpose.*

Without the presence of heterocatalysis particles, the growth of a uniform belt structure usually follows vapor-solid (VS) mechanism: Here a self-catalyzed vapor-liquid-solid process explains the growth of aligned ZnO NB arrays. The Zn comes from the decomposition of ZnO. In the experiment the substrate is placed in a temperature zone of 770–870 K which is higher than the bulk melting point of Zn (690 K). The Zn is in the liquid state during growth. The liquid Zn droplet as the favorable site for the adsorption of new Zn and O_2 species) results in the growth of ZnO NB. It is possible there are a few atomic layers of locally ordered Zn at the Zn-ZnO interface due to the coherent interfacial "pinning" owing to the high melting point of ZnO (2250 K). The Zn droplet is locally crystallized or atomically ordered at the interfacial region with the ZnO to trigger the initial nucleation and subsequent growth of the NB so that the crystalline structure of the Zn determines the growth direction and subsequently the side surfaces of the NB. The Zn droplet turned to single crystal after the temperature is dropped, and its surface is oxidized to ZnO when exposed to air. There is another possibility that the growth direction is determined by substrate. The Zn particle is in the liquid state during growth, and it is crystallized after growth and matches the orientation of the NB.

(c) Growth on GaN, $Al_{0.5}$ $Ga_{0.5}$ N and AlN Substrate

Vertically aligned single-crystal ZnO nanorods are fabricated on semiconducting GaN, $Al_{0.5}Ga_{0.5}N$, and AlN substrates through a vapor-liquid-solid process. For the growth of aligned ZnO nanowires/nanorods, *a*-plane-oriented Al_2O_3 (sapphire) single crystals is used as substrates with gold particles as catalysts, in which the growth is initiated and guided by the Au particle, and the epitaxial orientation relationship between ZnO and Al_2O_3 leads to the alignment. Two intrinsic problems are associated with this technique, which limit its application to devices. Al_2O_3 is a nonconductive material, so it makes it difficult to utilize the aligned ZnO nanorods for electronic and optoelectronic devices. A lateral growth of side branches close to the substrate surface is almost inevitable during the early stages of growth. For technological applications it is highly desirable to grow ZnO nanorods on a conductive or semiconductive substrate in order to fabricate heterostructure devices and to make electronic measurements. GaN, an important optoelectronic material with the same crystal structure as ZnO. It is used as a substrate for growing aligned ZnO nanorods by the metal–organic chemical vapor deposition (MOCVD) technique and the carbon–thermal evaporation process due to their very close lattice match. MOCVD can achieve a perfect alignment of ZnO nanowires on GaN substrate but lacks the control over the nanowires' site distribution and density since no catalyst is used.

Although GaN can electrically connect to the aligned ZnO nanowires, due to the very close band gaps of GaN (3.44 eV) and ZnO (3.37 eV), the heterojunction effect is fairly weak, thus limiting their applications in semiconductor devices. As a result, it is important to achieve a reasonably large band gap difference between the substrate and the ZnO nanowires while still maintaining good electrical conductivity and aligned morphology perfectly aligned ZnO nanorods are grown on semiconducting GaN, $Al_{0.5}Ga_{0.5}N$, and AlN thin-film substrates by a(vapor-liquid-solid VLS) phase process usmg gold as a catalyst. The as-grown nanorods show no lateral growth but are vertically aligned on the substrate surface. ZnO nanorods are epitaxially grown on $Al_xGa_{1-x}N$ thin-film layers with any Al alloy composition ratio. Since the band gap of $Al_xGa_{1-x}N$ is tunable from 3.44 to 6.20 eV by changing the Al composition ham 0 to 1, a band gap tunable heterojunction array of ZnO nanorods on a thin-film substrate can be achieved, thus making them an ideal candidate structure for light-emitting diode arrays.

The as-synthesized aligned ZnO nanorod are examined under a field emission gun (FEG) scanning electron microscope (SEM) operated at 10 kV. A typical low-magnification SEM image of ZnO nanorods grown on GaN is shown in Figure 4.42(a). All of the ZnO nanorods are straight and perpendicular to the substrate with a high uniformity across the entire substrate, indicating that MOCVD technique can be scaled up for large-area production. As shown from a higher-magnification SEM image in Figure 4.42(b), the ZnO nanorods exhibit uniform diameter. Figure 4.42(c) shows a top view of the aligned ZnO nanorods, where only the very bright gold catalyst tips are observed. It also confirms that almost every single nanorod is perpendicular to the substrate and that there are no side branches, which is generally unavoidable when sapphire is used as substrate.

Highly aligned ZnO nanorods are also grown on $Al_{0.5}Ga_{0.5}N$ and AlN substrates under the same growth conditions, even though the lattice mismatch for these materials is larger than that for GaN. Figure 4.43(a) to (c) gives the X-ray diffraction (XRD) spectra of the aligned ZnO nanorod samples grown on GaN, $Al_{0.5}Ga_{0.5}N$. and AlN substrates, respectively, In all three spectra, only diffraction from (0002) and (0004) atomic planes of ZnO is observed at 34.63 and 72.76°, respectively, indicating the high degree of nanorod alignment and their epitaxial relationship with the substrate. From the lattice constants of wurtzite GaN ($a = 3.190$ Å and $c = 5.189$ Å) and wurtzitc ZnO ($a = 3.249$ Å and $c = 5.207$ Å), the lattice mismatch between the (0001) planes is 1.9%. Thus, in the XRD spectrum (Figure 4.43(a)), the peak from the GaN (0002) plane overlaps that of ZnO, while a double shoulder is observed on the (0004) peak. The

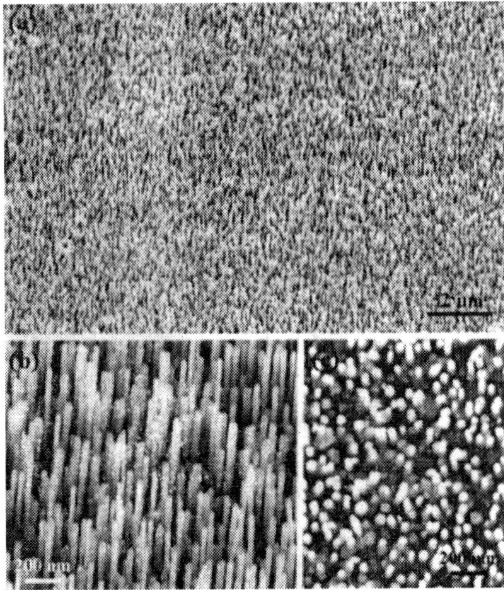

Figure 4.42 *SEM images of aligned ZnO nanorods grown on a GaN substrate. (a) Low-magnification 30° side view image. (b) High-magnification 30° side view image. (c) High-magnification top view image.*

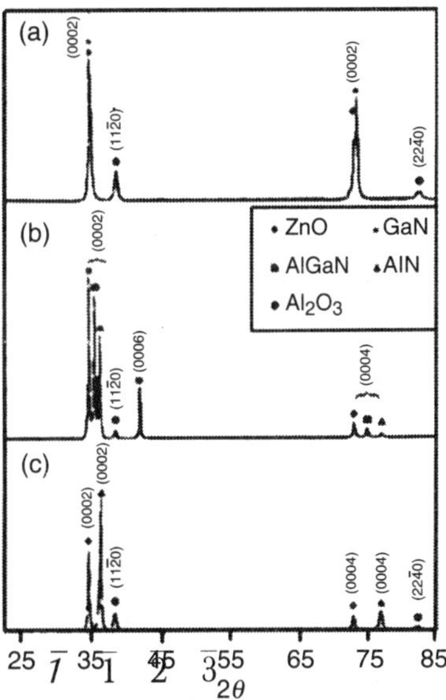

Figure 4.43 *XRD images of aligned ZnO nanorods grown on a GaN substrate. (a) Low-magnification 30° side view image. (b) High-magnification 30° side view image. (c) High-magnification top view image.*

epitaxial relationship between the ZnO nanorods and the GaN substrate layer is $(0001)_{ZnO} \parallel (0001)_{GaN}$, $[01\bar{1}0]_{ZnO} \parallel [01\bar{1}0]_{GaN}$. Although the lattice constant decreases from 3.190 (GaN) to 3.15 ($Al_{0.5}Ga_{0.5}N$) and to 3.110 (AlN) Å, and their lattice mismatch correspondingly increases from 1.8 to 3.0 to 4.3%, the epitaxial relationship is preserved as supported by the XRD spectra shown in Figure 4.43(b) and (c). This figure also shows that the diffraction peaks of $Al_{0.5}Ga_{0.5}N$ and AlN are gradually separated from ZnO peaks due to the increased lattice mismatch, whereas the ZnO (0002) and (0004) peaks remain sharp and clear.

Because both ZnO and the substrates have the same wurtzite structure, the deposited ZnO nanorods are confined in their six equivalent $<01\bar{1}0>$ directions and only grow along the [0001] direction exactly following the substrate's crystal orientation. As a result, the possibility for ZnO nanorods to undergo lateral growth is rare, even around the catalyst where conditions are favorable for lateral growth as shown in Figure 4.44(a) and 4.44(b).

The properties are investigated using energy-dispersive X-ray spectroscopy, (EDS). Panels (a) and (b) of Figure 4.44 are SEM images of ZnO nanorods grown around the edge of the gold catalyst on GaN and AlN substrates, respectively. Here, the catalyst boundary is clearly marked by the growth of aligned ZnO nanorods. EDS measurements are performed on the nanorod regions and the exposed substrate regions. Within the nanorod regions (Figure 4.44(c) and (f)), strong signals from zinc and oxygen are detected with a small gold signal originating from the Au film which acts as the catalyst and a portion of which remains at the tip of ZnO nanorods. Signals from the GaN and AlN substrates are also observed. On the exposed substrate (Figure 4.44(d) and (e)), only strong signals from the substrate materials are detected. Therefore, unlike the nanocatalyst technique, such as MOCVD, which lacks of control over the growth position, the VLS process can precisely position the aligned ZnO nanorods by patterning the gold catalyst, where the uncovered substrate area remains chemically and structurally unchanged. Moreover, an almost perfect vertical alignment of ZnO nanorods without lateral growth can be achieved using the VLS technique.

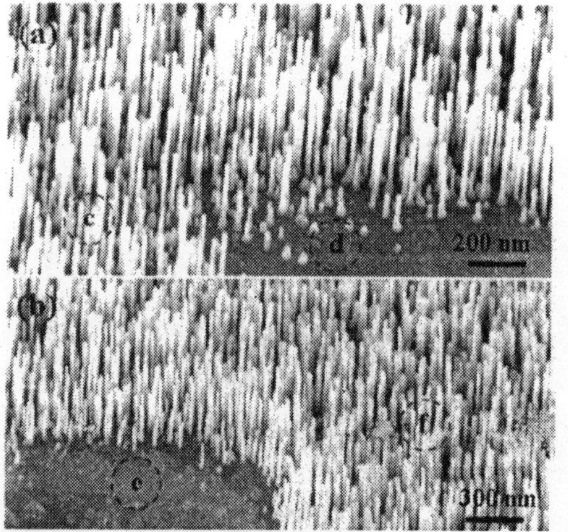

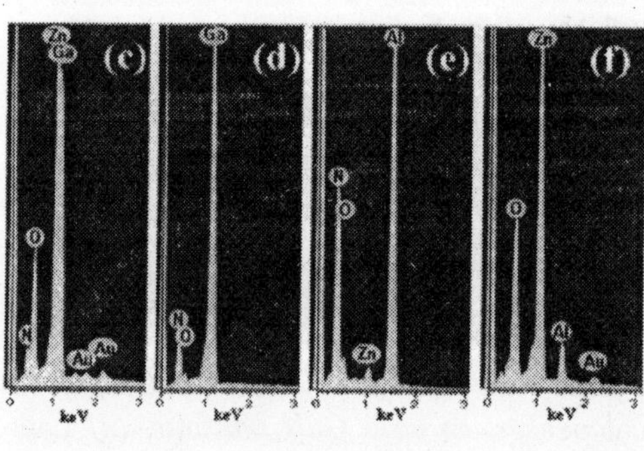

Figure 4.44 *SEM images of ZnO nanorods growing on GaN (a) and AlN (b) substrates at the edge of catalyst layer. (c-f) EDS spectra of the corresponding circled regioil in (a) and (b).*

Transmission electron microscopy (TEM) is also performed for size and crystal structure analysis at 200 kV. The lower magnification TEM images obtained for ZnO nanorods grown on GaN and AlN substrates are shown in Figure 4.45(a) and (b), respectively. All of the nanorods exhibit fairly uniform thickness along their entire length with gold catalysts at the tips. By measuring ~200 nanorods recorded on TEM images, the average length of ZnO nanorods grown on GaN is 434 nm, while the ZnO nanorods grown on AlN substrate are a little longer, ~500 nm. However, their length distributions are very close (±118 and ±120 nm: far ZnO nanorods grown on GaN and AlN substrates, respectively). The average diameter is 28 ± 6 nm for ZnO nanorods grown on GaN and 22 ± 2 nm for those deposited on AlN. The smaller size of the ZnO nanorods on AlN is due to the larger lattice mismatch between ZnO with AlN. The ZnO nanorods are single crystal (Figure 4.45(c)), the growth direction is [0001], and the six side facets are $\{11\bar{2}0\}$.

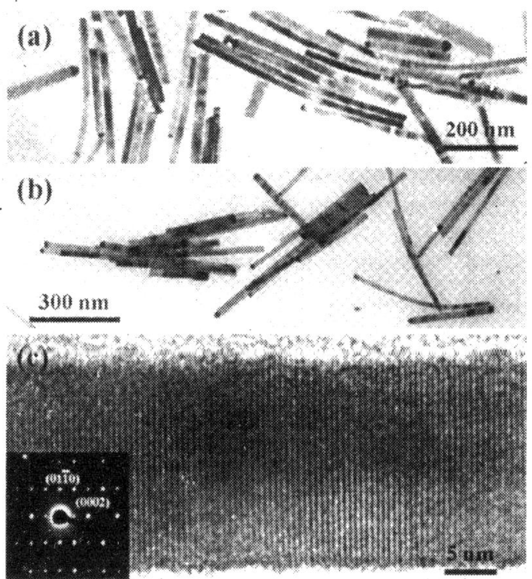

Figure 4.45 *Low-magnification TEM image of ZnO nanorods grown on GaN (a) and AlN (b) substrates. (c) High-magnification TEM image of a single ZnO nanorod; (inset) the corresponding electron diffraction pattern of the ZnO nanorod shown in (c).*

Room-temperature photoluminescence (PL) properties are measured for ZnO nanorod arrays grown under same conditions on GaN, AlGaN, and AlN substrates (Figure 4.46(a)–(c)) using a 266 nm Nd:Y AG Q-switched laser with an average power of 1.9 mW as the excitation light source. A Si photodetector is used to measure the PL. All of the samples exhibit- a strong luminescence peak at ~378 nm, corresponding to the near band gap emission of ZnO, an identical peak shape with a peak width at half intensity of ~5 nm. For the PL spectrum of ZnO nanorods 011 GaN, a small peak at 362.7 nm, is also detected, which corresponds to the band gap of GaN (3.44 eV). Due to the increased band gap when Al is introduced into the GaN lattice, the PL peaks from the substrate are blue shifted and thus no additional substrate-related peaks are observed from ZnO nanorods grown on AlGaN and AlN substrates.

Although the PL peaks have an identical shape, their peak intensity varies significantly. The highest PL signal intensity, which is more than 10 V is measured from ZnO nanorods grown on AlN (Figure 4.46(c)) and the lowest from ZnO nanorods grown on GaN, only ~4 V, as shown in Figure 4.46(a). The PL peak intensity from ZnO nanorods grown on AlGaN exhibit are intermediate peak intensity of ~7 V (Figure 4.46(b)). Since the density of nanorods grown on GaN, AlGaN, and AlN substrates is calculated from the SEM images is 146, 70, and 157 per μm^2, respectively, these relative PL intensities are not linearly related to the nanorod density. Therefore, the measured PL intensity of ZnO is affected by the absorption of GaN substrate. Due to the wide band gap of AlN (6.20 ev), the luminescence at 378 nm from ZnO is not absorbed by AlN substrate, but the band gap of GaN is very close to that of ZnO, thus the PL of ZnO is likely to be absorbed by the GaN substrate for the nanorods grown on GaN.

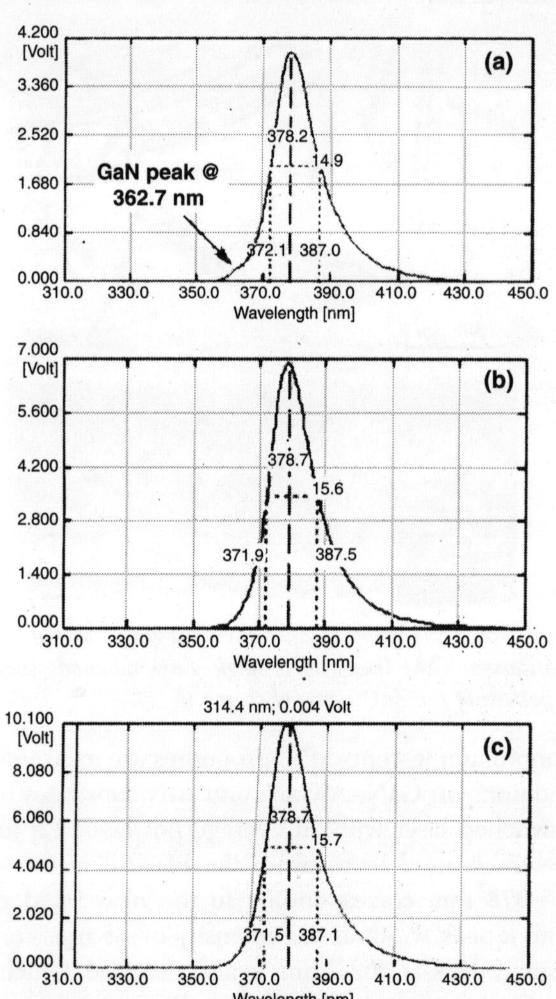

Figure 4.46 *Photoluminescence spectra (300 K) pf a;igned ZnO nanorods grown on GaN (a), Al0.5Ga0.5N (b), and AIN (c) substrates.*

ALIGNED GROWTH OF ZNO NANOWIRE ARRAYS

With the discovery of new or enhanced properties on functional NW structures, novel applications are realized in the laboratory in the fields of optics, electronics, mechanics and biomedical sciences. However, assembling nanowires into devices is challenging because of the difficulties in manipulating structures of such small sizes. New tools and approaches are needed to meet new challenges. Aligned growth of nanowires is a very efficient self assembly technique for integrating nanowires into nanodevices, such as nanowire lasers, LEDs or FETs.

Growth of aligned ZnO nanowire arrays (Chemical approach)

In the chemical approach, the density of ZnO NWs is controlled by changing the precursor concentration. Effects of both growth temperature and growth time are also investigated. By this novel synthesis technique, ZnO NW arrays grow on any substrate (polymer, glass, semiconductor, metal, and more) as long as the surface is smooth. This technique given below, is a new, low-cost, time-efficient, and scalable method for fabricating ZnO NW array for applications in field emission, vertical field effect transistor arrays, nanogenerators, and nanopiezotronics. The chemical approach is for achieving density-controlled growth of aligned ZnO NWs arrays without using ZnO seeds, (see Fig. 4.47) By adjusting the precursor concentration, the density of ZnO NW arrays is controlled within one order of magnitude (number of ZnO NWs per 100 μm^2) with one NW growing from one spot site. This novel synthesis technique does not require ZnO seeds or external electrical field, and it can be carried out at low. Temperatures and large scale on any substrate, regardless of whether it is crystalline or amorphous.

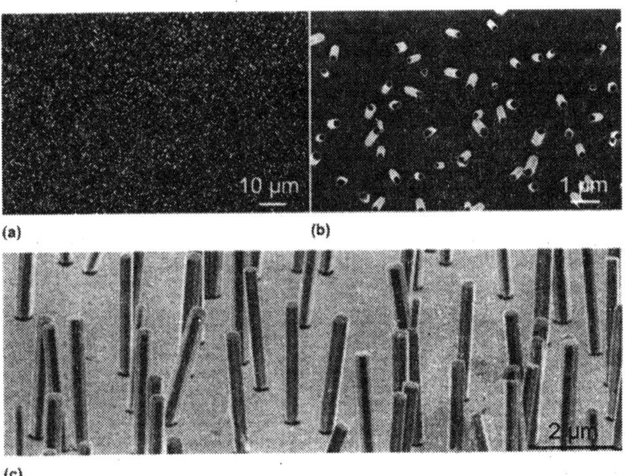

Figure 4.47 *A general view of an grown ZnO nanowire arrays at 5 mM, growing 24 at 340 K (a) top view, (b) enlarged top view, and (c) with a 60° tilt.*

Experiment

Here, Si (100) and flexible Kapton polymer is chosen as the substrates. A piece of Si (100) wafer substrate is cleaned by a standard cleaning progress. First, the wafer is ultrasonicated in acetone, ethanol, IPA (isopropyl alcohol), and de-ionized water each for 100 min; then it is blown dry with nitrogen gas and baked on a hotplate at 470 K for 5 min to eliminate any adsorbed moisture. Then a 50-nm-thick layer of Au is deposited on top of the Si wafer by magnetron plasma sputtering. This acts as an "intermediate-layer" to assist the growth. Between the Si wafer and Au layer, 20 nm Ti is deposited as an adhesion layer to buffer the large lattice mismatch between Si(100) surface with native oxide on and Au(111) surface and to improve the interface bonding. Then the substrate is annealed at 570 K for h. The next step is to prepare the nutrient solution. The nutrient solution is composed of a 1:1 ratio of zinc nitrate and hexamethylenetetramine (HMTA). The substrate is put face down at the top of the nutrient solution surface. Because of surface tension, the substrate floats at the top of the solution surface. Growth of ZnO NWs is conducted in a mechanical convection oven. The products are examined at 5 kV with scanning electron microscope (SEM).

Results

The annealing process helps the as-deposited Au layer to form a uniform crystalline thin layer on the surface of Si substrate, which is critical in the oriented growth of aligned ZnO NWs. In the growth Zinc nitrate salt provides Zn^{2+} ions required for building up the ZnO NWs. Water molecules in the solution provide O^{2-} ions. It is believed HTMA during ZnO NW growth act as a weak base, which slowly hydrolyze in water solution and gradually produce OH^-. This is critical in the synthesis process because, if the HMTA hydrolyzes too fast and produces a lot of OH^- in a short period of time, the Zn^{2+} ions in solution will precipitate out quickly due to the high pH environment. Therefore, Zn^{2+} will contribute little to the ZnO-NW-oriented growth and eventually result in fast consumption of the nutrient and prohibit further growth of ZnO NWs.

$$(CH_2)_6N_4 + 6\ H_2O \leftrightarrow 4\ NH_3 + 6\ HCHO, \quad (1)$$

(hexamethylene tetramine)

$$NH_3 + H_2O \leftrightarrow NH_3 \cdot H_2O, \quad (2)$$

$$NH_3 \cdot H_2O \leftrightarrow NH_4^+ + OH^-, \quad (3)$$

$$Zn^{2+} + 2\ OH^- \leftrightarrow Zn(OH)_2, \quad (4)$$

$$Zn(OH)_2 \rightarrow ZnO + H_2O \quad (5)$$

The growth process of ZnO NWs can be controlled through the five chemical reactions listed above. All of these reactions are in equilibrium and can be controlled by adjusting the reaction parameters, such as precursor concentration, growth temperature, and growth time, to push the reaction equilibrium forward or backward. In general, precursor concentration determines the NW density. Growth time and temperature control the ZnO NW morphology and aspect ratio. In the following parts, the growth mechanism of ZnO NW arrays is elucidated with detailed analysis.

1. Concentration

The density of ZnO NWs on the substrate are controlled by the initial concentration of the zinc salt and HMTA. For the relationship between the precursor concentration and the density of the ZnO NW arrays, a series of experiments are performed by varying the precursor concentration but keeping the ratio constant between the zinc salt and HMTA. Experimental results show that the density of the NW arrays is closely related to the precursor concentration. Detailed analysis of the measured data is shown in Fig. 4.48. (line with filled circle data points). From 0.1 to 5 mM, the ZnO NWs density, defined as number of NWs per 100 μm^2, increases as possibly zinc chemical potential inside the solution body increases with zinc concentration. To balance the increased zinc chemical potential in solution, more nucleation sites on the substrate surface are generated, and therefore, the density of ZnO NWs increases. When the zinc concentration is further increased, the density of ZnO NWs remain steady with a slight tendency to decrease. The steady/saturated density (from the nucleation and growth process) is decided by the number of nuclei formed at the beginning of the growth, which continue to grow and form nanorods (shorter NWs). The arrival of more ions on the substrate do not initiate new nuclei at a later stage because of two possible reasons. One, with consideration of the critical size required for a nucleus to grow into a crystal, no new nanorods are formed, if the sizes of the nuclei are smaller than the critical size. Second, due to the existence of the first group of grown nanorods, the newly arrived ions on the substrate have a higher probability of reaching the existing NWs than the newly formed nuclei. Thus, the size of the nuclei does not exceed the critical size, and they eventually dissolve into the solution body. In such cases, a continuous increase in solution concentration does not increase the density of the NWs when its density is larger than the saturation density. This also explains why the grown NWs have fairly uniform height. Even though the density of the ZnO NW remains steady at high precursor concentration level, the surface coverage percentage increases slightly due to the lateral growth of the NWs (Fig. 4.48, line with filled triangle data points).

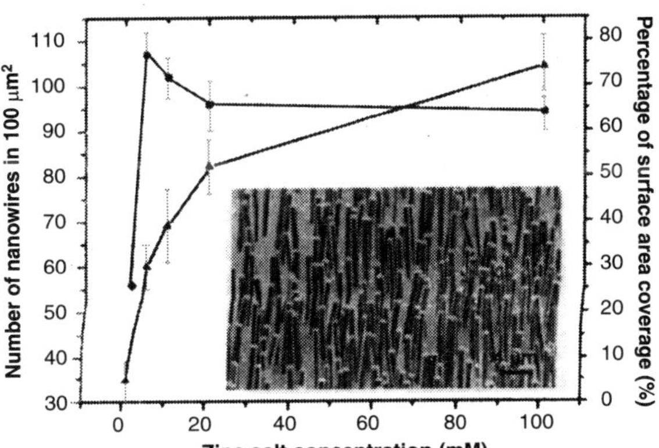

Fireug 4.48 *Density varied with concentration: Plot of ZnO nanowire density in a 100 μm^2 area (line with filled circle data points) and plot of area percentage covered by ZnO nanowires (line with tilled triangle data points). Each point is obtained from 4 different areas. Inset is a typical image of ZnO nanowires grown at 5 mM.*

2. Growth time

Wurtzite ZnO has polar surfaces, such as (0001) and (000$\bar{1}$), and non-polar surfaces such as {01$\bar{1}$0} and {2$\bar{1}\bar{1}$0). At different growth stages, including at the beginning, middle, and end, the polar surfaces axial growth is found to have a different relative growth rate from that of non-polar surfaces lateral growth. At the beginning growth period (from the start to around half an hour), lateral growth seems to be more significant than axial growth [Fig. 4.49(a)]. When the growth time exceeds half of an hour, the width of ZnO NWs is almost independent of time, indicating little lateral growth [Fig. 4.49(b)]. This indicates that, at the beginning growth stage, lateral growth is more significant than axial growth. In the middle growth stage, from half an hour to around 6 hours axial growth is dominant. At the final growth stage, from 6 to 48 hours in the experiments, lateral growth and axial growth seem to be equally significant, as can be seen from Figs. 4.49(b) and 4.49(c). In this period, ZnO NWs length is almost doubled. At the same time, ZnO NWs width is also doubled. The aspect ratio remains nearly constant.

It is also observed that the Zn-terminated (0001) surface is much more catalytically active than the O-terminated $(000\bar{1})$ surface for the growth of NWs. For the aligned array, it is observed that all of the NWs have their positive *c* axis pointing upward, which means that the (0001) surface is at the growth front. This characteristic results in the alignment of polarity and piezoelectric affect of all the NWs.

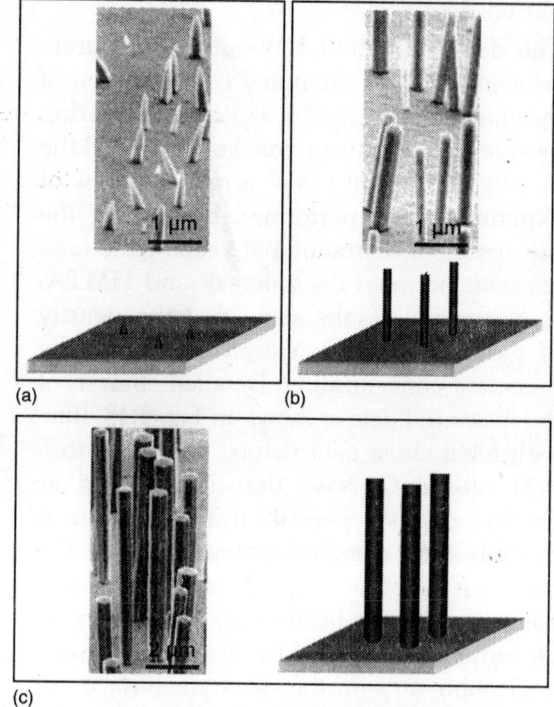

Figure 4.49 *Time-dependent of ZnO nanowire morphology evolution: (a) 0.5 h, (b) 6 h, and (c) 48 h.*

3. *Temperature*

Temperature is an important factor in maintaining a high aspect ratio of the hexagonal prism shape of ZnO NWs. 340 K is found to be the optimum temperature for obtaining a high aspect ratio and well-defined hexagonal prism shape of the ZnO NWs [Fig. 4.50(b)]. When the temperature is reduced for example to 330 K [Fig. 4.50(c)]. the aspect ratio becomes smaller than that at 340 K. The aspect ratio of the ZnO NWs is determined by the relative growth rate of the polar surfaces and non-polar surfaces. This means that the relative growth rate of polar surface to non-polar surface at low temperatures is smaller than that at high temperatures. By increasing the temperature for example to 365 K pyramid-shaped ZnO nanorods grow. [Fig. 4.50(a)]. The temperature-dependent growth behavior are understood from the basic theory of crystal nucleation and growth. When the temperature is low, the mobility and the diffusion length of the ions on the substrate are limited, which prohibit the ions to diffuse around, and therefore large size nuclei are formed. Thus, the NW density is low. At high temperature, the mobility and diffusion length of the ions are large enough to reach the sites of the firstly grown NWs, thus, no new nuclei are formed because the precursor ions have higher affinity to the already-formed seeds than they do to bare substrate, resulting in a lower density of NWs on the substrate. At an optimum temperature, such as 340 K, the mobility of the ions is moderate and their diffusion length is within a small range in the vicinity of the substrate; thus the accumulation of ions at local regions results in high density of NWs. However, their size is small because of the conservation of total ions that are able to migrate to near the substrate.

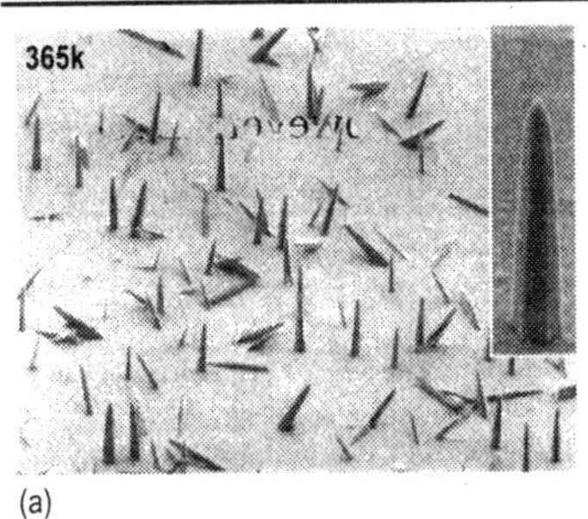

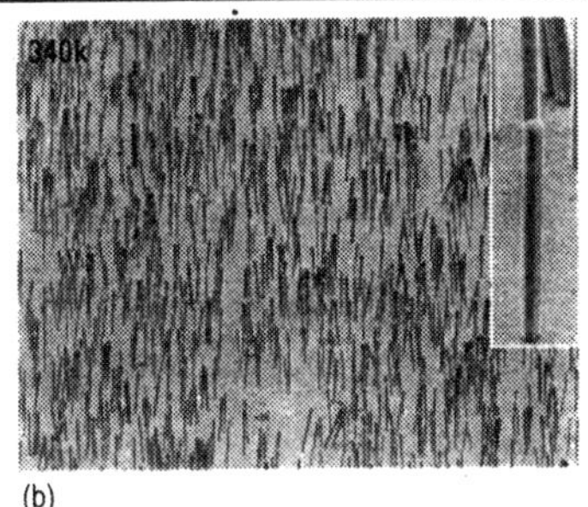

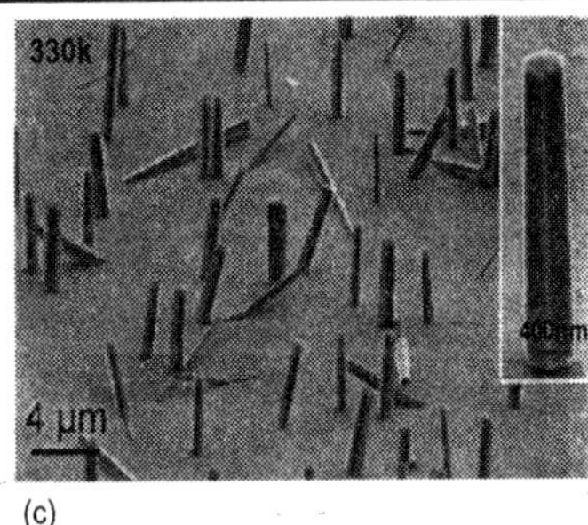

Figure 4.50 *ZnO nanowires grown at different temperatures: 60° tilt degree view of ZnO nanowires synthesized at (a) 365 K. (b) 340 K. and (c) 330 K. The three images are recorded at the same magnification, and the three inset images are at the same magnification.*

As discussed above HMTA functions as a slowly decomposing weak base, which maintains the weak basic environment in the solution and successively provides Zn^{2+} ions with OH^- ions. From reaction (1), 7 mol reactant produce 10 mol product. Thus, if the reaction temperature is increased, reaction (1) move forward by virtue of entropy increase, which means that HMTA decomposes more quickly at a high temperature than it would at a low temperature. That is to say, at the beginning, HMT A has already decomposed to a relatively large degree and has produced enough OH^-, resulting in a sufficient and thorough growth of ZnO. Therefore, the base of ZnO nanorods is thicker than it is at lower temperature. Furthermore, both the axial and lateral growth rates are greatly increased at higher temperature. As time goes by, the supply of Zn^{2+} is limited in the reaction container and gradually becomes exhausted, which leads to the formation of (c) incomplete shape of the NWs (pyramids or tapped nanorods).

So, the concentration controls the nucleation density, growth time controls the aspect ratio, and temperature controls the morphology as well as aspect ratio. By adjusting these three parameters, high aspect ratio ZnO NWs, is obtained as shown in Fig. 4.51.

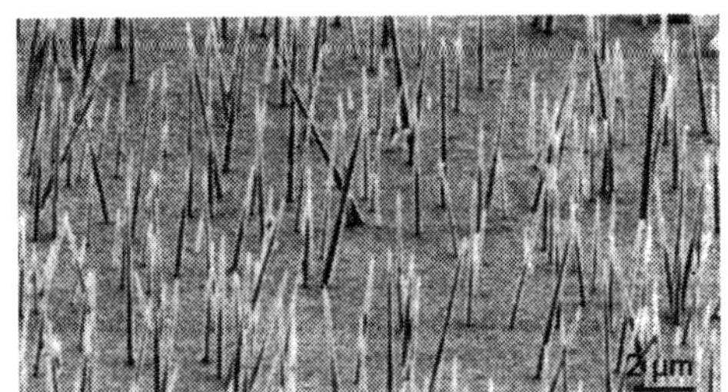

Figure 4.51 *ZnO nanowiores grown at 5 mM, 350 K for 48 h.*

4. Substrate

The effect of different substrates on the controlled growth of aligned ZnO NW arrays is investigated. The results show that the substrate substances have little influence on the growth of ZnO NW arrays as long as the substrates are treated with a standard fabricating procedure. Growth results on different substrates are shown in Fig. 4.52. ZnO NW arrays grow not only on single crystal Si wafers [Fig. 4.52(a)], but also on amo hous flexible polymer substrates [Fig. 4.52(b)]. The basic requirement for the aligned growth is that the substrate is smooth.

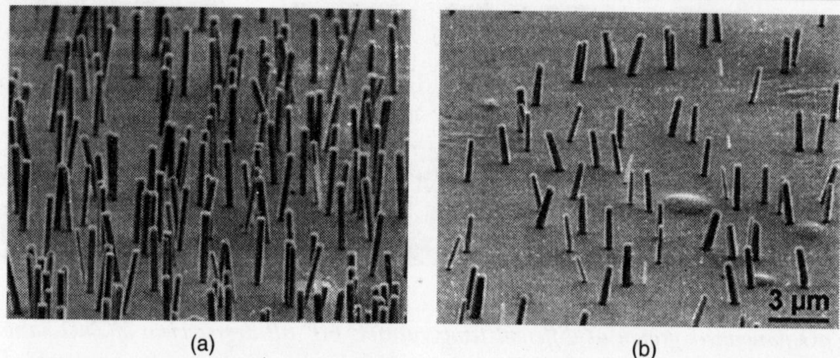

Figure 4.52 *ZnO nanowires grown on (a) stiff Si wafer and (b) flexible Kapton polymer substrate, which are at the same magnification.*

Structural Analysis

One-dimensional (1D) nanostructures, such as nanowires, nanobelt, and nanorods, are of nterest due to their unique properties and novel applications. The structure analysis of 1D nanostructures is done by achieving structural control i.e by controlling properties and device performances. Using ZnO, CdSe, and ZnS as examples and transmission electron microscopy TEM) the structure of nanowires and nanobelts, is analyzed for growth direction, side/top surface polar direction, surface reconstruction, point defects, dislocation, planar defects and win/bicrystal structures.

Thickness Fringes and Bending Contours in Imaging Nanowires and Nanobelts

For single crystal 1D nanostructures, electron diffraction by the crystal lattice produces various contrasts. Thickness fringes are created by a change in the projected thickness of the nanowire across its diameter (Figure 5.1(a)). To understand the thickness fringes, the two-beam diffracting ondition is used; one is the transmitted electron beam, and the other is the diffracted beam. An increase in crystal thickness results in an oscillation of the diffracted beam due to the multiple-scattering effect, producing the thickness fringes. The change in crystal thickness for he two adjacent thickness fringes is a few tens of nanometers. Using thickness fringes, the ymmetry of the nanowire is identified. For a straight and uniform nanowire, the thickness long each fringe is constant. But for nanowires with thicknesses smaller than a few tens of anometers, no thickness fringes are observed.

Bending contour is another contrast, which is frequently observed in bright-field TEM naging of nanobelts, see (figure 5.1(b)-(d)) The bending contour is due to a variation of atomic lane in respect to the direction of the incident electron beam (Figure 5.1(e)). Due to the crystal ending, the atomic planes have different orientations, thus the angle between the incident lectron beam and the local atomic plane changes across the specimen. If the beam is parallel the atomic plane at point A in Figure 5.1(e) for example, then there is little diffraction effect, nd the TEM image shows bright contrast. At point B, the incident beam strikes the atomic lane at the Bragg angle, strong diffraction is created, and thus, the TEM image of the local area shows darker contrast. The contrast is closely associated with the local symmetry, and in some cases, three dimensional (3D) bending can show "⋈" shape of contrast, see (Figure 5.1(b)) the center of which corresponds to the zone axis of the crystal. Bending contour is especially pronounced in oxide nanostructures.

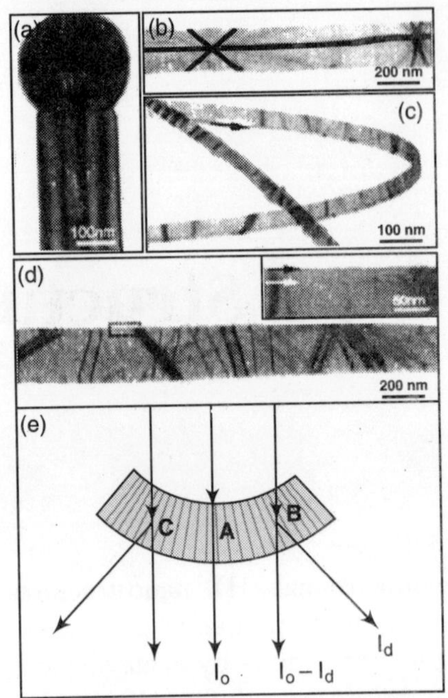

Figure 5.1 *(a) TEM image of Sn catalyzed ZnO nanowire, showing diffraction effect produced thickness fringes across the width of the nanowire. (b,c) TEM images of ZnO nanobelts, showing the bending' contours. After tilting the specimen, the up surface of the belt is seen in (d). (e) The formation of bending contour.*

Thickness of Nanobelt

To determine crystal thickness by TEM when the sample thickness is a few nanometer is done by depositing a nanostructure onto a flat solid substrate; atomic force microscopy is used. For nanobelts thicker than 50 nm, convergent beam electron diffraction (CBED) is applied to measure its thickness In some case, tilting a nanobelt the thickness is measured. The inset in Figure 5.1(d) is a ZnO nanobelt after being tilted for ~20^0. It shows the projected thickness side at the edge. The nanobelt thickness is calculated from the projection of TEM image and the sample tilting angle. For nanobelts or nanowires thinner than 20nm, quantitative comparison of the experimental high resolution TEM image with the simulated image gives the crystal thickness.

Figure 5.2 is a quantitative comparison of high-resolution TEM (HRTEM) image recorded from a ZnO nanobelt with the theoretically simulated images. The HRTEM image is recorded at 400kV from a c-plane dominated [01$\bar{1}$0] growth ZnO nanobelt. Besides a small distortion in the orientation of the nanobelt across its width, quantitative comparison between the observed image in the local region can be compared with the simulated image by changing crystal thickness, as grouped in Figure 5.2. The detailed contrast along the atomic columns in the HRTEM image in local areas change with the variation of the local thickness, which provide an effective technique for determining nanobelt thickness less than 20 nm. The HRTEM images from the five rectangles enclosed areas are enlarged in the insets of Figure 5.2 and the best matched images from simulation are displayed side-to-side. The simulation for spherical aberration Cs = 0.5 mm The determined thickness is potted at the lower part of Figure 5.2 as function of the distance from the edge of the nanobelt.

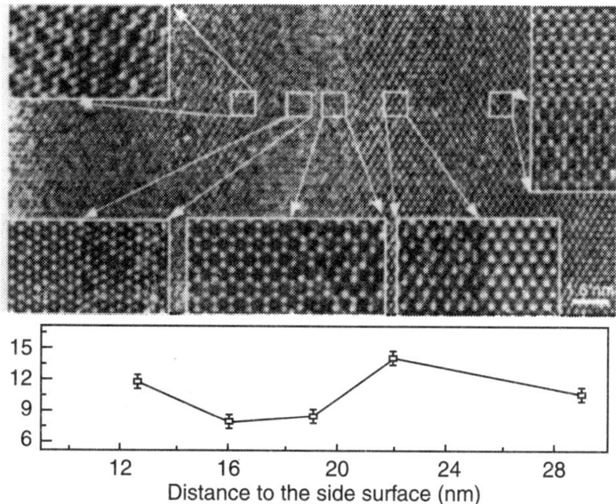

'igure 5.2 *Determination of local thickness by high-resolution TEM. HRTEM image of a ZnO nanobelts shows the dependence of image contrast onlocal sample thickness. The insets are th grouped experimental image together with the corresponding simulate dimage. A plot of the local thickness (nm) determined by the technique frmo several spots away from teh edge of the nanobelt.*

;rowth Direction of Nanowires and Nanobelts

;ombining TEM images with corresponding electron diffraction patterns is a key in structure nalysis. When determining the growth direction of a ID nanostructure, it is important to nsure that the incident electron beam is perpendicular to the growth direction. By depositing D nanostructures on a TEM grid, the nanowire may not be flat on the substrate, therefore at east dozens of nanobelts are measured to determine growth directions (unique and self-consistent). 'herefore, the data for more than one growth direction should be checked.

The operation of TEM relies on magnetic lenses, and Lorentz force results in an instrument etermined rotation between the recorded diffraction pattern and the image due to a change in ptical mode. Figure 5.3(a) is a bright-field TEM image of a ZnO nanowire at 400 kV. The lens f the TEM used has the rotation effect. After tilting the specimen perpendicular to the electron eam, the selected-area diffraction (SAED) pattern from the nanowire is recorded (See igure 5.3(b), without any correction in rotation angle. The diffraction pattern is [0001], but the anowire direction is not pointing to any low index diffraction spot in the pattern. This is aused by the magnetic rotation between the image and the diffraction pattern.

If the magnetic rotation angle between the image and the SAED pattern is known, then the rowth direction can be determined from the SAED pattern. If the rotation angle is unknown, ıen a simple technique is applied to determine its value. The principle relics on the shape ctor introduced by the sharp edge of the 1D nanostructure in the electron diffraction pattern. ;onsidering the shape effect of the nanowire, a streaking is produced in the diffraction pattern long a direction perpendicular to the nanobelt direction, as shown by the enlarged diffraction pot inserted in Figure 5.3(b). The rotation angle α is applied to define the growth direction of ıe nanowire, which is found be $[01\bar{1}0]$.

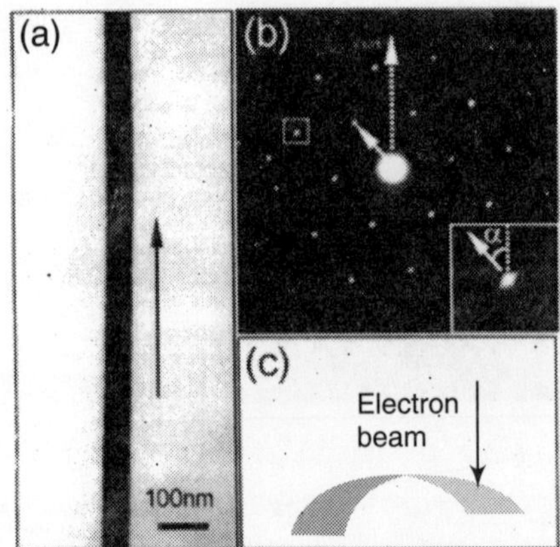

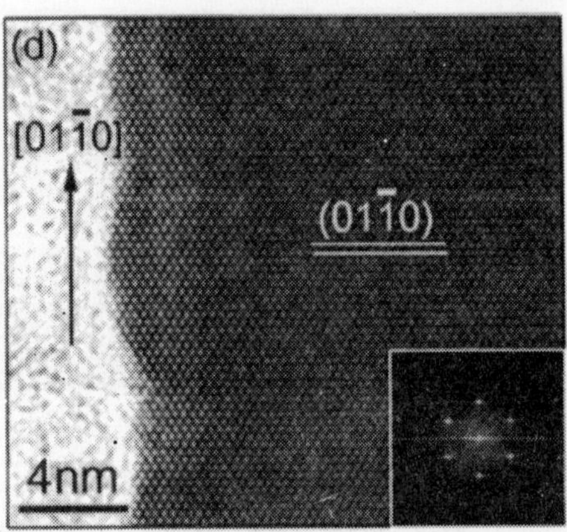

Figure 5.3 *Magnetic rotation angle and determination of the growth direction of a nanowire/nanobelt. (a TEM image of a ZnO nanobelt and (b) its corresponding selected area electron diffraction (SAEL pattern. The streaking of the spot in the inset of (b) is due to the sharp edge of the nanowire, whic can be used to correct the magnetic rotation angle between the image and the diffraction pattern. (c Schematic diagram showing the folded shape of a nanobelt and the incident electron beam, showin the possibility of incorrect determination of growth direction. (d) HRTEM image from the nanobe shown in (a), and its Fourier transform.*

The most direct method is using HRTEM image of the nanobelt and its Fourier transform for determining the nanowire growth direction. Figure 5.3(d) is a HRTEM image of the nanobel shown in Figure 5.3(a). The Fourier transform of the image gives the growth direction of th nanobelt [0110]

Shadow imaging is another technique for determining the nanowire growth direction. Usin a field emission TEM and converging the electron beam onto the nanobelt, the image and th converged beam diffraction pattern captured is in the diffraction mode (Figure 5.4(a)), b changing the beam convergence and using the brightness control knob so that the beam cros over is under/over the object Here the index of the diffraction spot g that is parallel to th nanobelt is the growth direction (provided the incident beam is perpendicular to the nanowire Figure 5.4(b) is an TEM image and the corresponding SAED pattern from a ZnO nanobelt. B increasing the beam convergence, the shadow image of the nanobelt is shown in the diffractio pattern (Figure 5.4(c)), which is pointing to the [2110] direction, the nanobelt growth directior

Another technique for directly determining the nanobelt growth direction is presented i Figure 5.5. The first step is to form a bright field TEM image using a parallel illumination bear and record the image at in-focus condition (Figure 5.5(a)). The second step is to converge th beam onto the specimen, then to underfocus the objective lens so much that the diffractio spots appear in the image mode (Figure 5.5(b)), The next step is to record the image onto th same film, The index of the diffraction spot parallel to the nanobelt is the growth direction.

Indexing Growth Direction and Growth Front Plane

After correctly determining the nanowire growth direction, it is necessary to align the imag with the diffraction pattern so that the information in real space is consistent to that provide

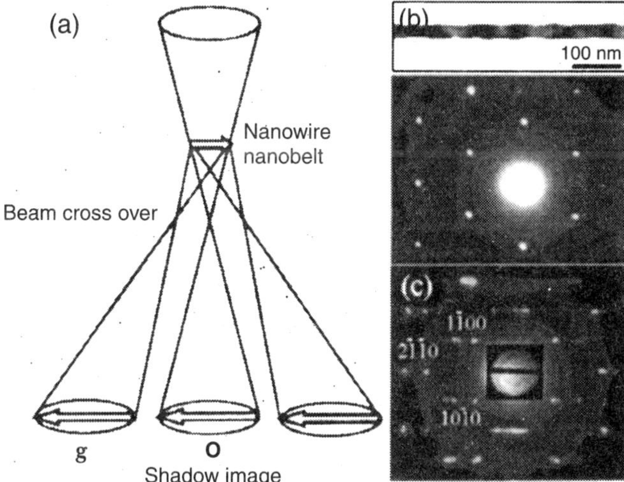

Figure 5.4 *Shadow image technique for determining the growth direction of a nanowire/nanobelt. (a) Optical diagram for shadow imaging. (b) TEM image and the SAED pattern from a nanobelt. (c) The corresponding shadow image recorded in diffraction mode.*

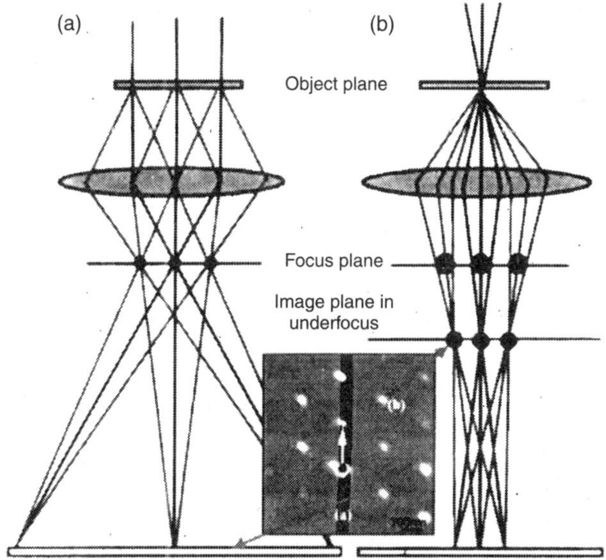

Figure 5.5 *(a) Optical diagram for receiving the imaging of a nanowire under parallel beam illumination. (b) Optical diagram for recording the diffraction information from the nanowire in the imaging mode by converging the electron beam and under-focusing the objective lens, so that the diffraction pattern appears in the image mode. The image shown is a double-exposed image and "diffraction pattern" recorded at the optical configurations of (a) and (b).*

by reciprocal space, For example, if the ZnO nanowire growth is along [000 1], it is necessary to point the (000 1) diffraction spot parallel to the nanowire growth direction when composing figures.

For cubic structures; the Miller index for a crystal plane is , the same as that for its norma direction, But for noncubic structure systems, the Miller index for nanowire growth directio may not be the same as the Miller index for the corresponding atomic plane perpendicular t the nanowire It structured SnO_2 nanobelt as an example, the normal direction of the wid surface of the ribbon is $[10\bar{2}]$; but the surface plane is close to $(10\bar{1})$ crystallographic plane (note SnOz has a tetragonal lattice structure), The axial growth plane is (201), but the growt direction is close to [101]

Side Surfaces of Nanowires and Nanobelts

The side surfaces of the as-grown nanowires that are imaged depend on the direction of th incident electron beam. Figure 5.6(a) is a TEM image of a α-Fe_2O_3 nanowire growing alon [110]. Three different patterns are for each direction of the incident electron beam i.e is [001 $[\bar{1}11]$ or $[2\bar{2}1]$ (Figure 5.6(b)-(e)). A common characteristic of the three patterns is that the (11(spot appears in all of them, because the incident beam is perpendicular to the nanowire.

The most direct way for determining the side surfaces is to prepare cross-section samples c nanowires and image them directly along the nanowire direction. The sample is prepared b embedding nanowires in a conductive epoxy, and cross-section sample can be made by slicin thin sections (using an ultramicrotone). Some of the nanowires with growth directions closel parallel to the incident electron beam are found in TEM, thus the side surface can be determine by either electron diffraction or HRTEM.

The determination of side surfaces needs a conjunction use of several techniques. Gol nanorods, for example, can have {110} type side surfaces, which are unusual for face-centere cubic metals. HRTEM, electron diffraction and dark-field TEM images with thickness fringe are co-applied to determine the surfaces.

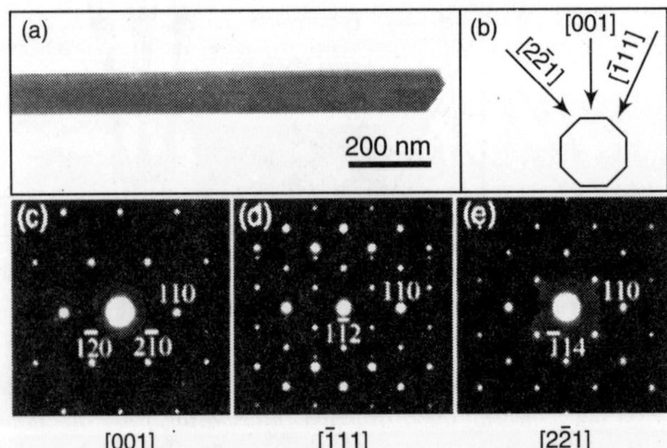

Figure 5.6 *Determination the side surfaces of a nanowire. (a) Low magnification TEM image of a α-Fe_2C nanowire. (b) Three possible incident electron beam directions for imaging the nanowire, and (c-e are the corresponding diffraction pattern along the three zone axes.*

Polar-Direction of Nanobelts

Some crystal surfaces have polarity that is produced by the atomic termination on the surface, for example the wurtzite structured ZnO can be viewed as a number of alternating planes composed of tetrahedrally coordinated O^{2-} and Zn^{2+} ions, stacked alternatively along the c-axis (as explained in polar surface induced growth phenomena). Since, the oppositely charged ions produce positively charged (0001)-Zn and negatively charged $(000\bar{1})$-O polar surfaces, it result in a normal dipole moment and spontaneous polarization along the c-axis. If the surface terminates with Zn or oxygen is important for understanding the polar surface dominated growth phenomena. From kinematic scattering theory, the polar surfaces because they cannot be determined because the diffraction intensities for the +g and -g diffraction spots because they are identical, so, the convergent beam electron diffraction (CBED) with strong dynamic diffraction effect is a unique tool for this.

CBED patterns are formed with a converged electron probe, illuminating the sample area, which one of a few nanometers. The convergence angle is equivalent to imaging the crystal through a range of directions, it is sensitive to the 3D crystal structure of the sample. The beam convergence, determines the size of the diffraction disks. The cross point of converged beam is at or below the specimen, depending on the quality of the specimen and the required convergence angle. The incident probe consists of many plane-wave components propagating along different directions, thus forming a converged conical electron probe (Figure 5.7) For an incident beam P, diffraction results in a complete point diffraction pattern consisting of P's due to Bragg reflection law. A similar set of point diffraction pattern is formed for another plane-wave component Q. Therefore, for cases where there are no disk overlaps, a perfect registration is retained between each incident beam direction and the diffracted beams. The intensity distribution in the disks in the CBED pattern contains the local symmetry and point group information of the electron beam illustrated area.

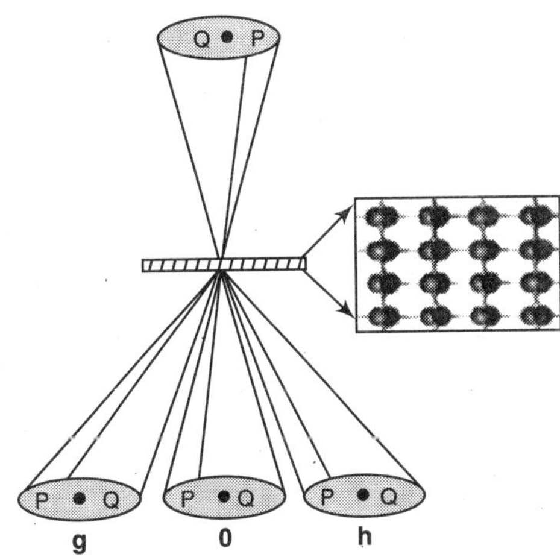

Figure 5.7 *Ray diagram of convergent beam electron diffraction for determining the polarization of ZnO.*

A key requirement for forming high quality CBED pattern is the sample thickness, It should be large enough to give a strong dynamic effect. Figure 5.8(a) shows a ZnO comb nanostructure. The SAED pattern in Figure 5.8(b) indicates the teeth of the comb grow asymmetrically along positive and negative *c* directions, But which direction is the positive direction remains to be determined. The CBED pattern recorded from the comb in Figure 5.7(c) shows asymmetry intensity distribution in the disks along the ±c directions. The CBED pattern is based on the well-established dynamic electron diffraction theories. Using the crystal structure of ZnO, by defining the incident beam geometry, and adjusting the crystal thickness to match the

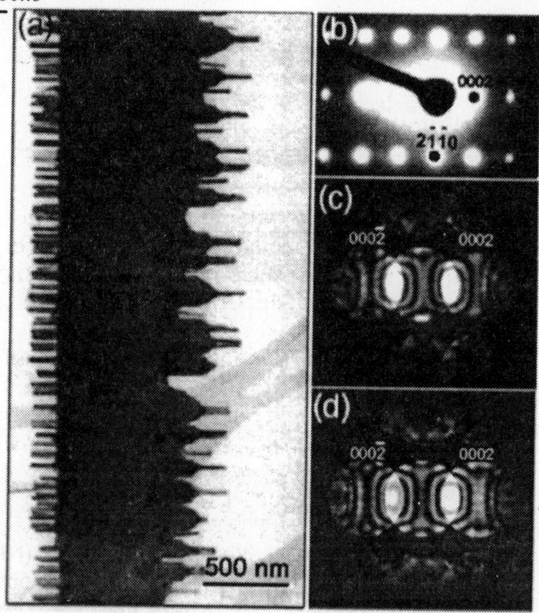

Figure 5.8 *Polar surface induced anisotropic growth. (a)TEM image of ZnO comb structure and (b) the corresponding SAED pattern. (c) and (d) are experimentla and simulated CBED pattern of the ZnO comb.*

experimental pattern, a theoretically calculated pattern is given in Figure 5.8(d), Comparing the simulated with the experimental patterns, the polar surfaces are determined Thus the teeth grow along [0001] on the Zn-terminated (0001) surface. The oxygen-terminated (0001) surface is chemically inert. The asymmetric catalytic behavior of the two oppositely charged surfaces results in the formation of comb structure.

Helical and Zigzag Structures

Now, TEM images provide two-dimensional (2D), thickness projected structure information of an object. Careful examination of the TEM images reveal a bit of 3D structure. Consider the helical and zigzag structures. Figure 5.9(a)-(c) shows three TEM images recorded from three different types of structures of ZnO: in-plane zigzag structure induced by a periodic change in growth directions; a helical structure of a nearly straight ZNO nanowire surrounded by an amorphous layer with a periodically "etched" helical concave during growth; and a spring structure. Among the three images, only the nanospring in Figure 5.9(c) is a 3D coiling of a thin nanowire. This is identified through the image contrast at the edge of the nanowire. Changing in contrast is possible by changing the focus of the objective lens. The segment of the nanowire lying on the focal plane shows minimum contrast (see the region indicated by a white arrowhead in Figure 5.9(d)) while that away from focus shows dark-white fringes at the edge of the nanowire (see the region indicated by a black arrowhead in Figure 5.9(d), reflecting the 3D shape of the coiling structure.

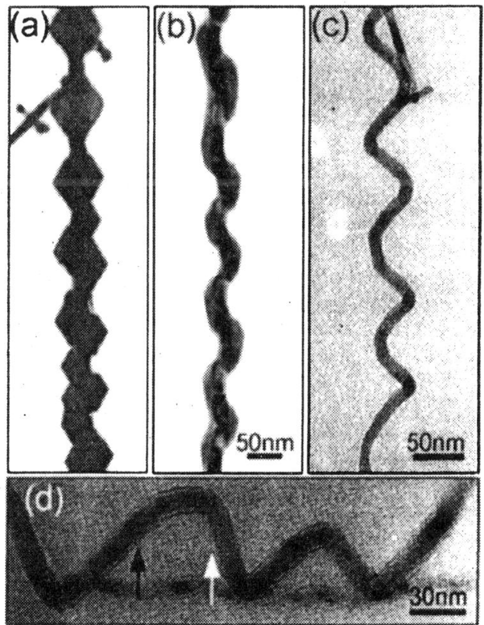

Figure 5.9 *Identification of 3D helical structures. TEM images of (a) a zigzag ZnO structure produced by swing of growth direction between [10$\bar{1}$ 0] and [01$\bar{1}$ 0], (b) a false spring structure and (c) ZnO helix. (d) An enlargement of a segment of the structure shown in (c)*

Surface Reconstruction

Large surface-to-volume ratio is the most typical characteristic of all of the nanomaterials. A larger percentage of dangling bonds on surfaces produce new surface structures, which are called reconstruction. For a crystalline material, the surfaces are usually dominated by some low index facets, such as {111} and {100} for gold. Direct imaging of surfaces of nanostructures at atomic resolution requires HRTEM. By imaging a nanostructure in profile, the reconstruction of the surfaces parallel to the incident electron beam is revealed.

Figure 5.10(a) is a low magnification TEM image of a ZnO nanowire that grows along [0001] with an Sn particle at its tip. The reconstruction of the (01$\bar{1}$0) side surfaces is presented in Figure 5.10(b). Besides the (01$\bar{1}$0) planes, (01$\bar{1}$1) and (01$\bar{1}\bar{1}$) fact are observed. The nanobelt shown in Figure 5.10(c) grows along [0$\bar{1}$10] direction, the top/bottom and side surfaces are ±(0001) and ±(2$\bar{1}\bar{1}$0) planes, respectively. While zooming in the upside surface, It is found that the (2$\bar{1}\bar{1}$0) plane are not smooth and reconstructed into a series of (1$\bar{1}$00) and (10$\bar{1}$0) facets.

Structures of Catalyst Particles in Vapor-Liquid-Solid (VLS) Growth

The nanowires obtained from the VLS process have a solid catalyst nanoparticle at the ends with sizes comparable to diameters of the connected nanowires. Metal particles such as Au and

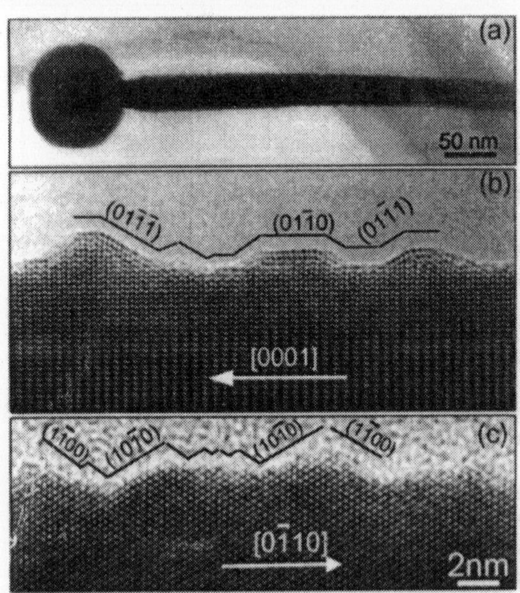

Figure 5.10 *Profile TEM imaging of reconstructed surfaces. (a) Low magnification and (b) high-resolution TEm image of a ZnO nanowire growing along [0001], and (c) HRTEM image of a nanobelt growing along $[0\bar{1}\,10]$, showsing surface reconstruction and tiny facets.*

Fe are effective for growing nanowires of Si, III-V compound, II-VI compound, and oxide: such as Au/ZnO,. Fe/SiO_2, Co/SiO_2,Ni/Ga_2O_3, Ga/SiO_2.

It is found that the catalyst particles have a fixed orientation relationship with their guided nanowires or nanobelts. The ZnO nanowire in Figure 5.11(a) is guided by an Au catalyst particle. The incident electron beam is $[01\bar{1}0]$ of ZnO. The thickness contrast indicates the surfaces of the nanowire are composed by six $\{01\bar{1}0\}$ facets. The SAED pattern including both the ZnO nanowire and the Au particle is displayed in Figure 5.11(b). The orientation, relationship is as $(0001)_{ZnO} \parallel (111)_{Au}$ and $[01\bar{1}0]_{ZnO} \parallel [\bar{1}\bar{1}2]_{Au}$. After examining the lattice mismatch between ZnO and Au. The orientation relationship is chosen to minimize interface lattice mismatch. This fixed orientation relationship is formed in the cooling process, during which the ZnO lattice decided the Au orientation when the Au changed from the liquid to solid state.

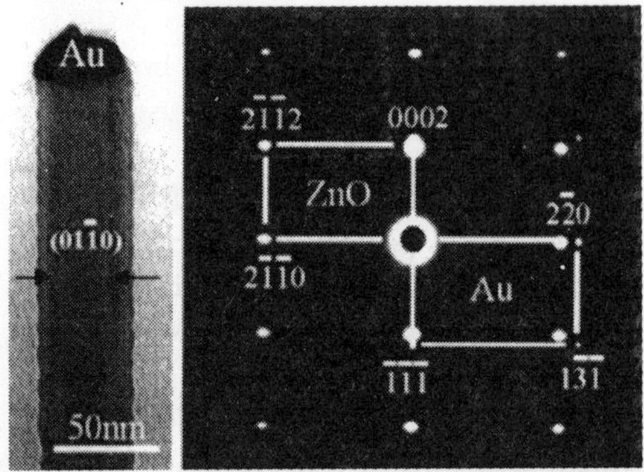

Figure 5.11 *Orientation relationship between catalyst particle and the grown nanowire. (a) TEM image of Au catalyzed ZnO nanowire, (b) the SAED pattern including both Au catalyst and ZnO nanowire.*

But Sn/ZnO show a different result. Using tin particle guided growth of ZnO nanostructures it is found that the interfacial region of the tin particle with the ZnO nanowire/nanobelt can be ordered (or partially cristalline) during the VLS growth, although the local growth temperature is much higher than the melting point of tin, and the crystallographic lattice structure at the interface is important in defining the structural characteristics of the grown nanowires/nanobelts. The interface takes the least lattice mismatch, thus, the crystalline orientation of the tin particle determine, the growth direction and the side surfaces of the nanowires/nanobelts. This has important impact on the understanding of the physical and, chemical process in VLS growth.

Dislocations in Nanowire/Nanobelts

Dislocations are rarely observed in nanowires and nanobelts, especially for oxides nanostructures. They are due to the size effect, which makes the dislocations unstable in nanowires and nanobelts.

Almost all of the oxide nanostructures are free of dislocations though recently dislocation in ultra-narrow ZnO nanobelts of 6 nm in width are observed figure 12(a) and (b) show bright-field and dark-field images of two fine ZnO nanobelts, respectively. The variation of the contrast (especially in the dark-field image) reveals the nonuniform strain field distribution along the nanobelt, which is created by the large deformation. The HRTEM image in Figure 5.12(c) reveals the existence of an edge dislocation with Burgers vector as $1/2[0001]+1/3[01\bar{1}0]$ at the twisted area indicated by an circle. At the left-hand side of the nanobelt, no dislocation is formed, but the inter-planar distance is expanded to ~5.4 Å at the exterior are and compressed to ~4.8 Å at the inner are to accommodate the local strain. Such a gigantic change of ~6% in interplanar distance without fracture for oxide is only possible at nanoscale. The nonuniform deformation along these fine nanobelts corresponds to a nonuniform strain field. When the accumulated strain in some local areas is beyond the structural endurance, dislocations are introduced. This metallic like structural characteristic for oxides is unusual.

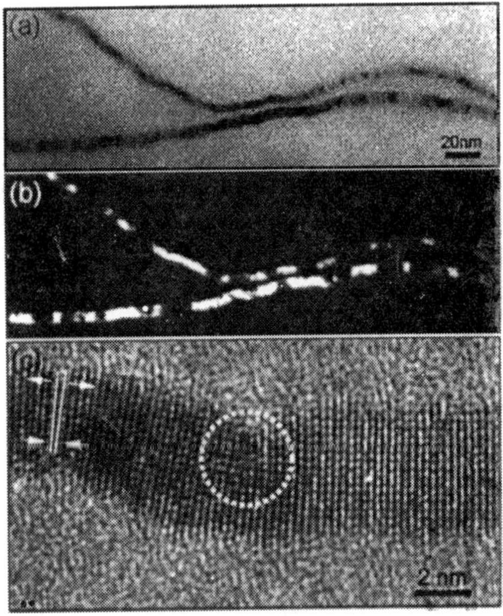

Figure 5.12 *Dislocation in ultra-narrow nanobelts. (a,b) Bright -field and dark-field TEM images of ultrafine ZnO nanobelts, (c) HRTEM image indicating the existence of a dislocation.*

Point Defects in Nanobelts

Imaging of point defects is challenging for TEM. TEM imaging relies on diffraction and interference and it is much less sensitive to point defects distributed randomly or with a short range ordering. Oxygen deficient oxide nanostructures are of great interest because their conductivity depends on oxygen content. TEM cannot be directly applied to determine oxygen deficiency in ZnO nanostructurcs.

TEM issued to directly Image defects created in ZnO nanobelts by Mn doping The Mn doping, is × 10^{15}cm^{-2} A higher density of mini-stacking faults is created (Figure 5.13(a)) Taking a Fourier transform of a HRTEM image and select only the ±(0002) diffraction spots by an filter, an inverse Fourier transform gives the distribution of the (0002) fringes in the image (Figure 5.13(b)) which clearly displays the mini-stacking fault.

Figure 5.13(c) is a HRTEM image recorded from an as-implanted ZnO nanobelt. A line scan along the dashed line gives an intensity profile across the atomic columns along $[01\bar{1}0]$. An ideal profile periodic with equal amplitude. The large scale change in the image contrast from column to column (see figure 5.13(d)) reflects the introduction of point defects as well as lattice distortion due to ion damage. The point defects can be oxygen deficiency and cation interstitials. After annealing the sample in air, the image contrast is very uniform (see Figures 5.13(e) and (f), because of the (disappearance of the point defects.

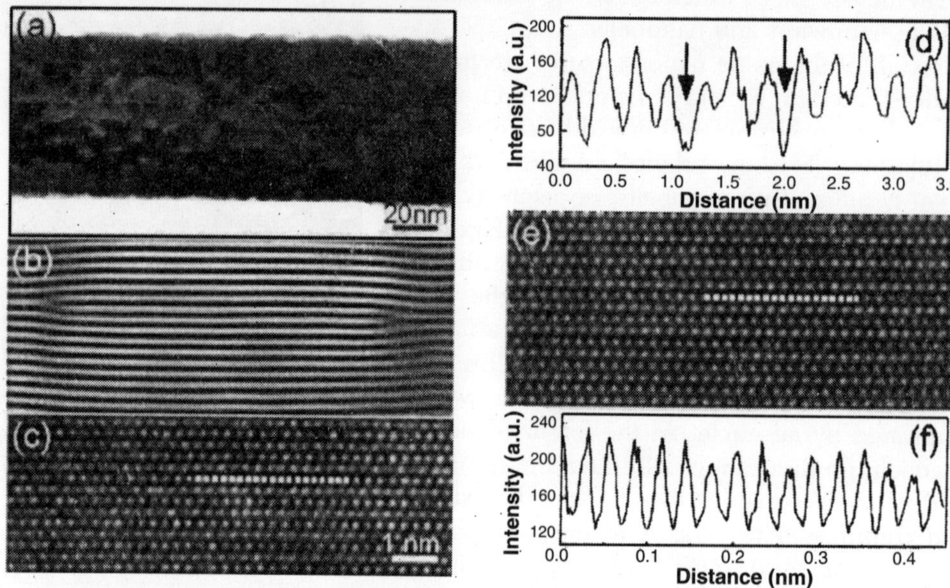

Figure 5.13 *Imaging point defects in nanobelts. (a) Low magnification TEM Image of a ZnO nanobelt doped with Mn, showing numerous mini-stacking faults produced byion implantation. (b) Fourier filtered image showing a dislocation loop created by the mini-stacking fault. (c) HRTEM image of the as-implanted sample, and (d) intensity line scan parallel to the (0001) plane, showing fluctuation in image contrast due to ion implantation induced point defects. (e) HRTEM image of the implanted ZnO nanobelt after being annealed in oxygen to compensate the lost oxygen and the displaced cations, and (f) intensity line scan parallel to the (0001) plane, showing the disappearance of the. fluctuation contrast in the image.*

Planar Defects in Nanowires and Nanobelts

Planar defects are the defects in nanobelts whose presence is essential for stabilizing the surfaces that exhibit energies. They lead anisotropic growth along a specific direction in the nanobelts. Planar defects can, be twins and bycrystals, stacking faults, or an interstitial stacking layer introduced by impurity. The former two can be easily distinguished from the last two by electron diffraction, Quantitative image simulations are required to identify the nature of the stacking faults.

(a) Stacking Faults

Stacking faults (Sf's) are planar defects in nanowire and nanobelts. They are formed by a change in stacking sequence of close packed atomic planes. Depending on the stacking faults lying in or perpendicular to their closed-packed plane, they are named basal-plane and prismatic-plane stacking faults separately.

In the wurtzite structure, there are three possible types of stacking faults, I_1 I_2, and E in bulk materials which are produced by extracting one layer, extracting two layers, and inserting a layer, respectively. The nature of stacking fault needs to be identified by combining HRTEM with image simulation or diffraction contrast extinguishing conditions. For Imaging the contrast of stacking faults: the diffraction beam g has to avoid the extinguishing condition as g.R=0. or n where R is the displacement crossing the defect and n is an integer. As we know, the R of stacking faults in wurtzite structures are $1/3[01\bar{1}0] + 1/2[0001]$ for type 1_1 case and $2/\ 3[01\bar{1}0)$ from type I_2 case.

A stacking fault in a ZnO nanobelt is displayed in Fig. 5.14(a). Its HRTEM image is show is Fig. 5.14(b). The synthesis process of the ZnO nanobelts are the same. An arrowhead highlights the c direction in the image. In spite of the perfect stacking sequence of wurtzite structure as ABABAB along the (0001) direction, a new consequence as identified as ...ABABACAC... in the defect area suggests the existence of an I_1 stacking fault. The displacement across the I_1 basal-plane stacking fault can be identified as $R=1/3(01\bar{1}0] + 1/2\ [0001] = 1/6[02\bar{2}\bar{3}]$.

Figure 5.15 a low magnification TEM image of another ZnO $(01\bar{1}0)$ growth nanobelt. There are two different basal-plane stacking faults exist in Fig. 5.15 (a), the top one is the I_1 stacking fault discussed in Fig. 5.14. A HRTEM image of the bottom defect is shown in Fig 5.15 (b). In the defect area, the stacking sequence described as .. .ABABCACAC .. It classified as an intrinsic type I_2 basal-plane stacking fault since it is equivalent to extracting, two periodic layers and a $2/3[01\bar{1}0]$ translation displacement is measured across the defect.

From crystal structural point of view, hexagonal wurtzite, structure takes space group as $P6_3mc$. It is described as a number of alternating planes composed of tetrahedrally coordinated cations and anions, stacking alternatively along the c-axis. The oppositely charged ions produce positively charged (0001) and negatively charged $(000\bar{1})$ polar surfaces, resulting in a normal dipole moment and sometimes even spontaneous polarization along the c-axis as well as a divergence in surface energy. In general, one-dimensional wurtzite nanostructures grow along the c-axis and their side surfaces are $\{01\bar{1}0\}$ and/or $\{2\bar{1}\bar{1}0\}$ due to their lower energies than that of (0001), resulting in almost neutralized deployment. I_1 type of basal-plane stacking faults are formed in [0001] growth nanowires. The 1D nanostructures growing normal to the [0001] direction, as shown in Figure 5.14 and 5.15 are accompanied by at least a stacking fault from the beginning to end. Noticing the surface kink at the intersection between the stacking fault and the growth front, it is a preferred position to adsorb the vapor molecules to accelerate the

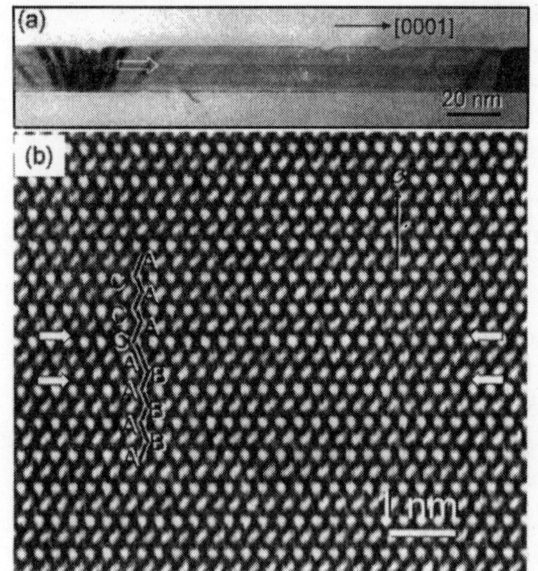

Figure 5.14 *Low-magnification (a) and HRTEM (b) image in ZnO nanobelt showing the type I_1 intrinsic basal-plane stacking fault. The incident electron beam in the HRTEM image is along [$2\bar{1}\ \bar{1}\ 0$] direction.*

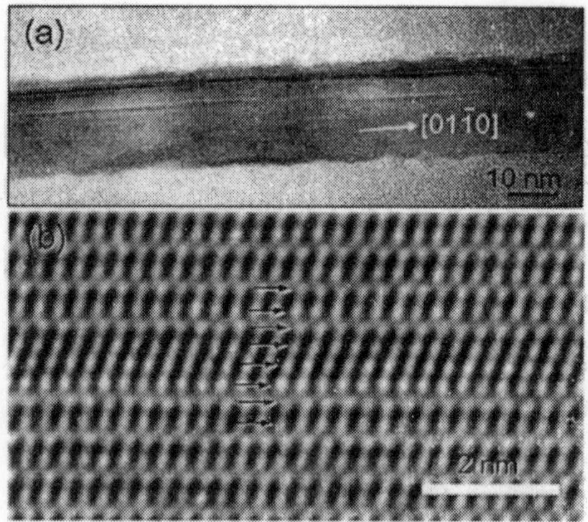

Figure 5.15 *Low-magnification (a) and HRTEM (b) images of a [$01\bar{1}\ 0$] growth ZnO nanobelt with incident electron beam along [$2\bar{1}\ \bar{1}\ 0$]J direction. The defect in (b) belongs to the type I_2 stacking fault.*

growth along such direction. As a result, the existence of basal plane stacking faults span the energy barrier set by the polar surfaces, and then lead the 1D nanostructures to grow along [$01\bar{1}0$) direction.

Besides basal-plane stacking faults also observed prismatic plane stacking faults in wider ZnO nanobelts (~1 μm). Prismatic plane stacking faults have two configurations classified by the displacements across them. The displacements across the defect, are 1/2 ($01\bar{1}\bar{1}$). and 1/6 ($02\bar{2}\ \bar{3}$) In order to determine the structural nature of the defect in the nanobelts, a series of dark-field images are recorded as shown in Fig. 5.16. The defect contrast can be seen clearly in Fig. 5.16(a) and (b), which are the corresponding bright-field and dark-field images under two-beam condition with the g = ($02\bar{2}0$). The diffraction condition is inserted in Fig. 5.16(b). After identifying the [$01\bar{1}0$] direction by the diffraction pattern, a part of the defect lies in the basal plane and part in prismatic plane. The extended contrast of defects while titling the belt suggests that they are planar defects and rules out the possibility of dislocation loops. Only one set of SAED pattern is made from the defects area, thus they cannot be twin structure. They must be stacking faults part in basal plane, and part in prismatic plane. By further tilting the nanobelt two extinguishing conditions of, the defects are obtained as shown in Fig. 5.16(c) and (d). They correspond to diffraction g of ($\bar{2}110$) and (0002) respectively. Both displacements 1/2($01\bar{1}\ \bar{1}$) and 1/6($02\bar{2}\ \bar{3}$) are perpendicular to (2110), or an integer when multiplied with (0002).

However, if the displacement is $1/2(01\bar{1}\bar{1})$, the stacking faults, lose contrast in the dark-field image formed by the $(02\bar{2}0)$ diffraction. As a conclusion, the stacking faults should take the displacement of $1/6(02\bar{2}\bar{3})$, which are the same as that of I_1 basal-plane stacking fault. HRTEM Fig. 5.17 confirms the deduction. The HRTEM images from the basal plane part of the defects are the same as that of I_1 stacking fault shown in Fig. 5.14(b).

Figure 5.17 (a) is a HRTEM image recorded from the prismatic plane part of the defect shown in Figure 5.16(a). Figure 5.17(b) gives a magnified defect area in (a). The displacement along the c-axis is easily identified. The projected width of the I_1 basal plane stacking fault along $[2\bar{1}\bar{1}0]$ direction, is just an atomic plane thickness as shown in Fig. 5.14. However the project of the prismatic-plane stacking fault in the same direction is nearly 3nm in Fig 5.17(b). A possible explanation is that the defect is not edgy on in Fig. 5.17. The prismatic stacking fault most likely lies in $(\bar{2}110)$ plane instead of the $(01\bar{1}0)$ plane. There is a $1/3[01\bar{1}0] + 1/2[000\bar{1}] = 1/6(202\bar{3})$ translation across the defect, which is self-consistent with the result received from deflation contrast analysis presented above.

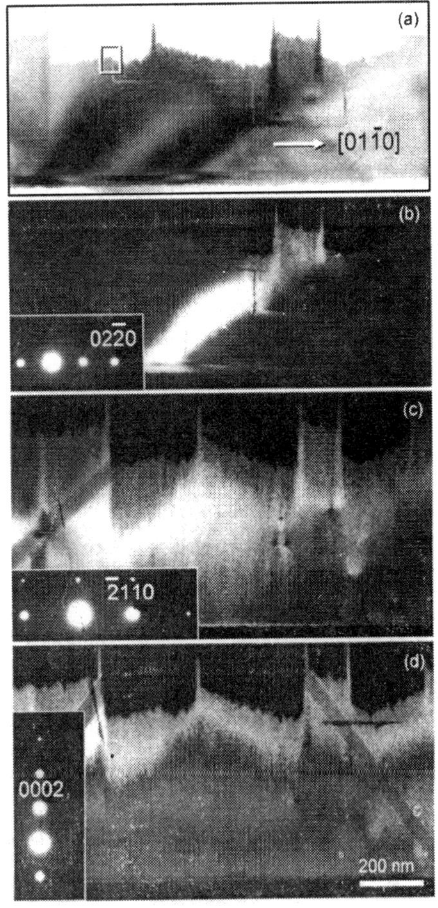

Figure 5.16 *Bright-field (a) and dark-field (b) images of a $[01\bar{1}0]$ growth ZnO nanobelt recorded under two-beam condition with $g=(02\bar{2}0)$. Prismatic and basal-plane stacking faults coexist in the belt. In the dark-field images recorded using $g=(\bar{2}110)$ (c) and 0002 (d), the contrast of the defect vanishes.*

For the difference between the basal-plane stacking fault and prismatic-plane one, an atomic model is built as shown in Fig 5.17(c) The horizontal lines localized the basal plane stacking fault, while the vertical line gives the prismatic plane stacking fault. The displacement between the left up line circled part and others is $1/6(202\bar{3})$. The ideal stacking sequences of Zn and oxygen ions can be describe as a ...AaBbAaBbAaBb... the capital and lower-case letters give the staking sequences of zinc and oxygen ions in [0001] direction, respectively, Now, consider the staking sequence of zinc ions as ABABABAB... while the sequences of oxygen ions are the same. When the defect remains in the basal plane, compared with perfect crystal case, the closest neighbors of an ion at defect area do not change and the bonding length between cations and anions remains the same. The only change occurs at the second closest neighbors, which is shifted form a hexagonal close-packed (hcp) system to a face center cubic (fcc) system. However, when the defect folds from basal plane to prismatic plane, (if the defect located in $(01\bar{1}0)$ plane), then at the vertical line located area, some of the bonds are compressed, some are expanded. The final configuration the system energy increases. This is why the prismatic-plan stacking fault chooses $\{2\bar{1}\bar{1}0\}$ planes.

The formation of the prismatic plane stacking faults is related to the secondary growth led by the self-catalysis effect. The basal-plane stacking faults are obtained in the bottom side of the nanobelt Fig 5.16 and they cross the whole nanobelt from the beginning to end and are presumed to guide the dominant growth of the nanobelt along $[01\bar{1}0]$. The exposed top and low surfaces of the nanobelt then are Zn-terminated (0001) and oxygen terminated $(000\bar{1})$ plane respectively. The self-catalysis growth at the Zn-terminated surface leads the secondary growth at (0001) surface, which is the up top of the nanobelt in Fig. 5.17. The domains growing from different nucleuses at (0001) surface merge together, at the end. However, the stacking, faults nucleated in each domain are hard to cross the boundary to elongate into nearby domain. Further the defect properties require that they do not terminate inside the body, therefore in order to minimize the system energy, the defects lying in the domain boundaries here those prismatic planes, are favorable. It corresponds to fold the stacking faults from basal plane to prismatic plane stacking fault to observed those prismatic-plane stacking faults.

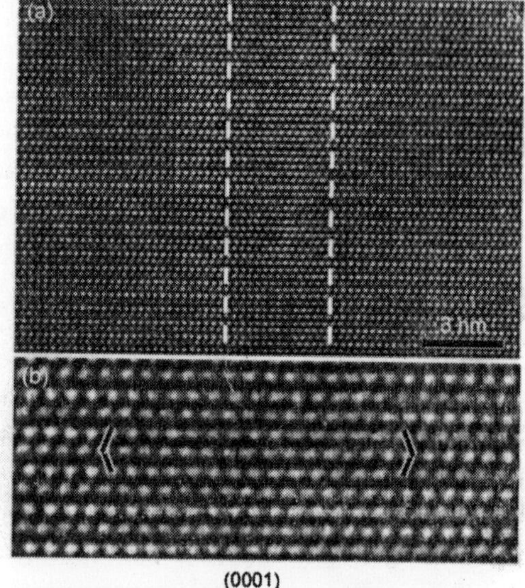

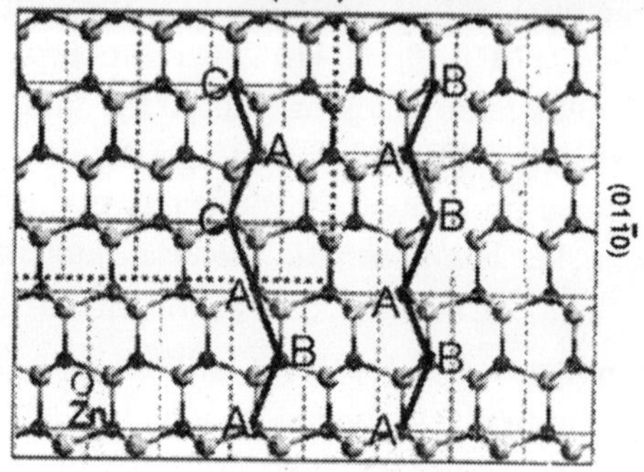

Figure 5.17 *(a) and (b) HRTEM images of ZnO nanobelts whit incident electron beam along$[2\bar{1}\bar{1}0]$ direction, showing the prismatic-plane stacking fault. Atomic model (c) shows the stacking fault folding from basal to prismatic plane.*

A low-magnification TEM image recorded from wurtzite-structured ZnS belt is given in Figure 5.18 displaying some planar defects. HRTEM images from the defect regions are recorded in figure 5.18(b) and (c). The inserted SAED pattern in pattern in Figure 5.18(b) came from the where the HRTEM image in Figure 5.18(b) is recorded. The c direction is marked in the image. The streaking along [0001] direction in the SAED pattern is due to the shape effect of the planae defects in c plane. Despite the perfect ZnS wurtzite structure, corresponding to the ABABAB stacking sequence parallel to the (0001) plane, a new sequence identified as to be ...ABABCACABAB... in the local area suggests the existence of I_1 and an I_2 stacking faults, as marked in the image. The ABABAB stacking sequence changes to ABCABC in some local area as shown in Figure 5.18(c), corresponding to the formation of the zinc blende phase in the wultzite-matrix. The structural models of the I_1 and I_2 stacking faults are illustrated in figure 5.19(a) and (b) respectively. There is $1/3[01\bar{1}0]+1/2[0001]$ translation across the I_1 stacking fault, and a translation of $2/3[01\bar{1}0]$ across the I_2 stacking fault.

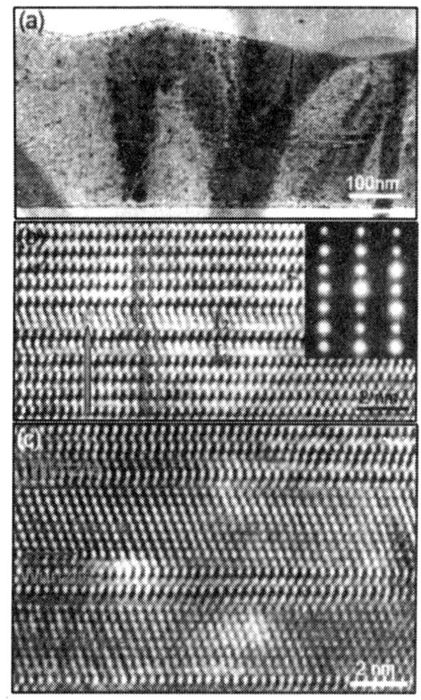

Figure 5.18 *Nature of stacking faults (SFs). (a) Low-magnification TEM image of a ZnS belt. (b) and (c) HRTEM image of local area in (a) showing the existence of stacking faults and zinc blende phase. The stacking sequence and types of SFs are indicated.*

(b) Twins and Bicrystals

Twin structure is most common in face-centered cubic structured metallic nanoparticles and silicon based nanowires. For wurtzite structures, there are possible twin boundaries defined by $(01\bar{1}1)$, $(01\bar{1}2)$ and $(01\bar{1}3)$. The theoretical calculations predict that $(01\bar{1}3)$ twin has the lowest energy.

When the size of materials shrinks into nanometer scale, not only $(01\bar{1}3)$ twin but also $(01\bar{1}1)$ and $(01\bar{1}2)$ twins are observed in ZnO 1D nanostructures. The population of $(01\bar{1}1)$ and $(01\bar{1}2)$ twins is much lower than that of the $(01\bar{1}3)$ twin, in agreement with theoretical energy calculation. Besides above three dominant twin structures, $(\bar{2}112)$ twin is also observed in 1D ZnO nanostructures.

Imaging twins by TEM prefers to carry out with incident electron beam along special orientation. The optimum orientation is parallel to the twin plan. For wurtzite structure, the beam direction is usely $[2\bar{1}\bar{1}0]$and/or$[01\bar{1}0]$. By controlling the deposition conditions, twin nanobelts of wurtzite structured ZnO and CdSe are synthesized. The most frequently observed twin structure takes the $(01\bar{1}3)$ twin plane. Nanowires/nanobelts with twin plane parallel to the growth direction are also called bicrystals.

Figure 5.20 Show a typical ZnO bicrystal. Its SAED pattern is displayed in Figure 5.20(c) which is composed of two sets of diffraction spots that have symmetrical geometrical distribution but a variable intensity. The two set patterns are labeled using subscripts Land R specifying the left-hand and right hand crystals, respectively. The common spot is the twin boundry plane $\bar{1}(013)$, and the incident beam direction is $[\bar{2}110]$. The existence of high density stacking faults in the dark-field image (Figure 5.15(c)) indicates the large local strain. The optimum orientation to image twins is parallel to the twin plane, so that the diffraction pattern shows mirror symmetry between the two sets of diffraction spots. Figure 5.21(a) is a HRTEM image from the twin boundary, which displays the mirror symmetry between the two crystals. HRTEM image

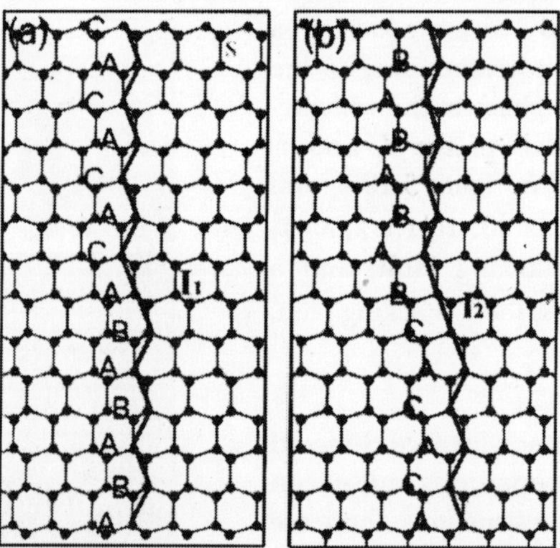

Figure 5.19 *(a) and (b) the structural models of the I_1 and I_2 types stacking faults observed in Figure 5.18(b) and (c) respectively.*

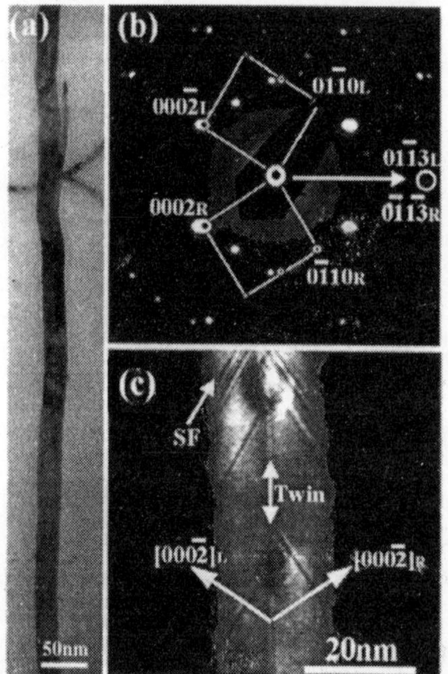

Figure 5.20 *Twinned/bicrystal nanobelt of ZnO. (a) and (b) TEM image and corresponding SAED pattern of a ZnO nanobelt that has a $(01\bar{1}3)$ twin parallel to the growth direction. (c) A dark-field TEM image showing the existence of stacking faults in the bicrystal nanobelt.*

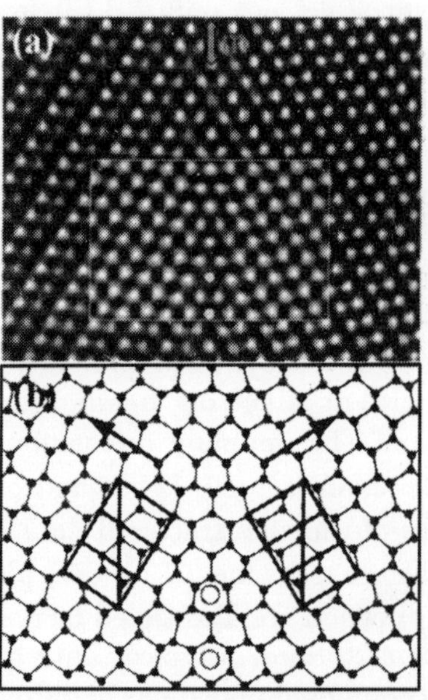

Figure 5.21 *Atomic model of the twinned/bicrystal noanobelt. (a)HRTEM image of the $(01\bar{1}3)$ twined ZnO nanobelt, the inserted is a simulation image based on the model in (b) The following parameters chosen for simulation: the sample thickness and defocus being 3.574 and 28.67 at 400 kV respectively.*

simulation, clarifies the Zn arrangement at the boundary. The best-matched structure model is shown in Figure 5.21. The simulated image based on the model is shown in Figure 5.16(a). In the structure model, the spheres are denoted by O and Zn ions respectively: The two arrowheads indicate the positive [0001] direction. When the [0001] is reversed on both side. the simulated image shows no difference, which means that HRTEM at this resolution is insufficient to distinguish the polar direction of ZnO. However, if the positive or negative [0001] direction is chosen, the ions arrangement is preserved.

Normally, ZnO nanomaterials take hexagonal wurtzite structure. However, they are cubic zinc-blende structure as well. For the atoms in both structures are tetrahedrally coordinated by atoms of the opposite species. The close packed planes stacking in ABCABC sequence will form zinc-blende structure, while stacking in ABABAB sequence gives the wurtzite structure. The formation of $(01\bar{1}3)$ twins seems relate to the zinc-blende nucleus in the growth.

It is considered that the surface contribution to the total energy becomes increasingly important as the size decreases. The favorable growth direction of 1D wurtzite nanostrucrures is along c-axis with exposed surfaces being $\{01\bar{1}0\}$ or $\{2\bar{1}\bar{1}0\}$ The c-axis in wurtzite structure corresponds to the (111) direction in zinc-blende. The exposed surface in zinc-blende structure will be {110} or {211}. Calculations of surface energies based on bond density level that the $\{01\bar{1}0\}$ planes in wurtzite structure is the energetically favorable surface in 1D nanostructure system. This is why ZnS nanowires take metastable wurtzite rather than stable zinc-blende.

However, the starting nucleus is in zinc-blende crystal structure for the limited high-energy surface areas. With the growth of the nucleus, the surface energy will drive the phase transition from zinc-blende to wurtzite, the eight (111) directions in zinc-blende can be separated into two classes, one is cation terminated another is anion terminated, Due to the self catalysis effect, normally only cation terminated (111) directions. corresponding to [0001] in wurtzite, grow further. The final structure introduced by the phase transition between zinc-blende and wurtzite is the well-known tetrapod structures. The angle between the two cation terminated (111) or [0001] directions is around 108° in tetrapod, structure, while the angles between the two [0001] direction in the (0113) twins are 116.7° in ZnO. There are no low-index planes between the two legs of the tetrapod. However, If the legs rotate 6-9°, the conjugated plane for the $(01\bar{1}3)$ twin is a favorable interface of choice.

Thus the growth of $(01\bar{1}3)$ twin is possibly, related to the formation of zinc-blende nucleus. Although $(01\bar{1}1)$ and $(01\bar{1}2)$ twins are not energetically favorable compared with $(01\bar{1}3)$ twin they are still observed in ZnO 1D system Fig. 5.22 (a) and (b) give the bright-field and dark-field TEM images of a $(01\bar{1}1)$ twin. The twin plane has been highlighted by an arrowhead. The SAED pattern is displayed in fig. 5.22(c). Solid and dashed rectangles separate the two set of diffraction spots. The shared plane is identified as $(01\bar{1}1)$ corresponding to the formation of $(01\bar{1}1)$ twin structure. The HRTEM image in fig. 5.22(d) gives clear mirror symmetry between the two parts of the twin.

The dark field image and SAED pattern of a $(01\bar{1}2)$ twin is displayed in fig 5.23 (a) and (b) respectively. The diffraction pattern indicates the existence of a $(01\bar{1}2)$ twin by the shared $(01\bar{1}2)$ diffraction of the two set of diffraction spots. Besides the diffraction spots coming from the twin, there are diffraction spots in weak intensity double diffraction. Based on the image in Fig. 5.23 a smooth interface between the two parts of the twined crystals is not possible this indicates the energy of $(01\bar{1}2)$ twin is high.

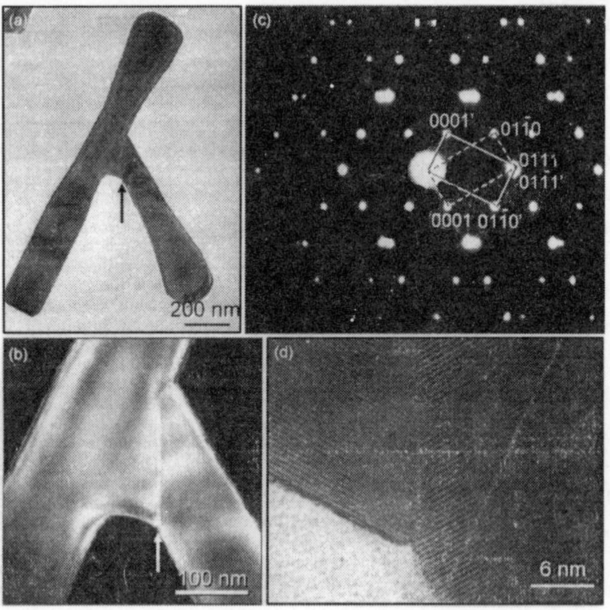

Figure 5.22 *(a and b) Bright-field and dark-field TEM images of (01$\bar{1}$ 1) twin structure in ZnO nanobelt. (c and d) the SAED pattern and HRTEM image of the (01$\bar{1}$ 1) twin with the incident electron beam along [2$\bar{1}$ $\bar{1}$ 0] direction.*

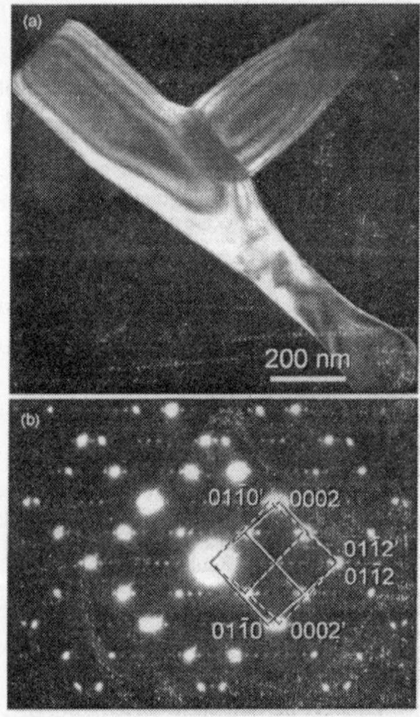

Figure 5.23 *(a) Dark field image of a (01$\bar{1}$ 2) twin ZnO nanobelt, (b) SAED pattern of (a)*

The $(01\bar{1}3)$ twin structure is also found in wurtzite structured CdSe bicrystals. An SEM image presented in Figure 5.24 a shows the bicrystal structure, with a sharp boundary at the center. The TEM image, is recorded parallel to the twin plane so that the two crystals are symmetrically imaged (Figure 5.24) The corresponding SAED pattern has two sets of diffraction spots and the fundamental units of the two are indicated. The common spot shared by the two sets of patterns is that from the twin plane, which is $(01\bar{1}3)$ (Figure 5.24(c)). The incident electron beam is along $[2\bar{1}\bar{1}0]$. Figure 5.24(d) is HRTEM image of the twin boundary, which presents the twin structure of the two sided crystals. High-density stacking faults also exist as indicated by arrowheads.

Besides $(01\bar{1}3)$ twin structure, the $(\bar{2}112)$ twin structure in ZnO bicrystal nanobelt is also observed Figure 5.25(a) and (b) shows bright field and dark-field images of the ZnO twinned nanobelts. The SAED pattern in Figure 5.25(c) indicate ,that the mirror plane of the twin structure is the $(\bar{2}112)$ plane. Only (0002) fringes are observed in the HRTEM image, as the

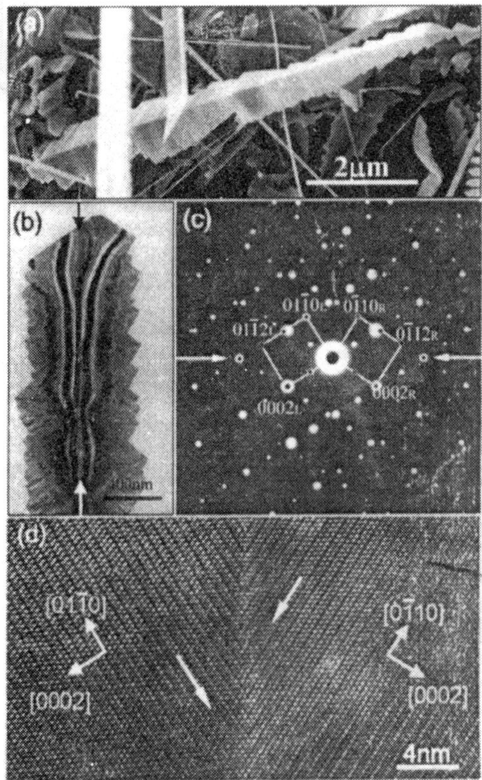

Figure 5.24 *Twinned/bicrystal nanowire of CdSe (a)SEM image CdSe $(01\bar{1}3)$ twined structure. (b)TEM image, (c) SAED pattern, and (d)the HRTEM image of the twin structure.*

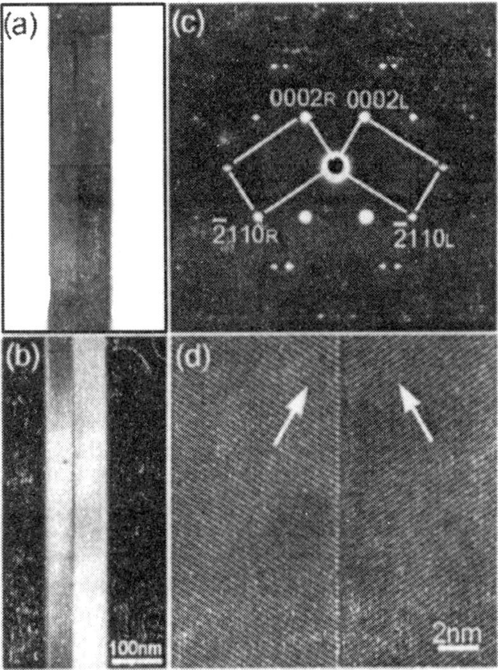

Figure 5.25 *ZnO $(\bar{2}112)$ twinned nanobelt. (a) and (b) are bright-field and dark-field TEM images, respectively. (a)SAED pattern, and (d) HRTEM image of the twin structure with the incidence electon beam along $[01\bar{1}0]$ direction.*

interplanar distance of $(\bar{2}110)$ is smaller than the resolution of microscope. The $(\bar{2}112\}$ type of twins is responsible for the formation of tetrapole/tetraleg structures of ZnO.. The diffraction contrast in Figure 5.25(a) and (b) indicates the large strain in the belt. Formation of stacking faults is an effective way to relax the strain.

Imaging twins are carried out at proper orientation. The optimum orientation is parallel to the twin plan. For wurtzite structure, the beam direction is usually $[2\bar{1}\bar{1}0]$ and/or $[01\bar{1}0]$. For imaging stacking faults, the beam direction is chosen so that the displacement of the lattice is not parallel to the beam. The $[01\bar{1}0]$ direction for wurtzite may not be a good choice because the translation vectors R of the stacking faults .are $1/3[01\bar{1}0] + 1/2[0001]$ for I_1 and $2/3[01\bar{1}0]$ for I_2.

(c) Interstitial Stacking Layer Related Planar Defects.

One-dimensional ZnO nanostructures without catalysts grow along the c-axis and the side surfaces are $\{01\bar{1}0\}$ and $\{2\bar{1}\bar{1}0\}$ due to their lower energies than that of (0001), resulting in a vanishable dipole moment and much reduced piezoelectricity. To maximize ,the effect of polar surface and the piezoelectricity, nanobelt grow along $[01\bar{1}0]$ or $[2\bar{1}\bar{1}0]$ are preferred. Now, Zn nanobelts that grow along [0110] are always accompanied with basal plane stacking faults. So the energy barrier set by the (0001) polar surfaces is balanced by introducing basal-plane planar defects To synthesize polar-surface dominated ZnO nanobelt impurity atoms such as In are introduced in the source material to form planar defects in the nanobelts. As a result the $[01\bar{1}0]$ growth nanobelts are formed. Some of the $[01\bar{1}0]$ growth nanobelts are self coiling and thus form a complete nanoring structure. Planar defects cerated by In doping are used as a mechanism for stabilizing the polar surface dominated nanobelts.

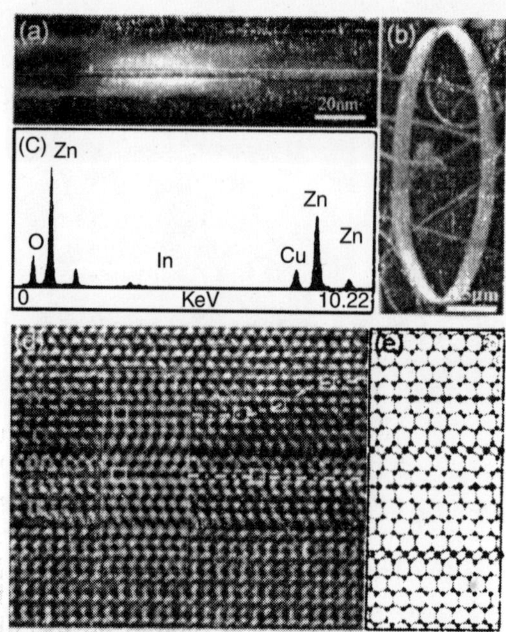

Figure 5.26 *Planar defects created by doping. (a) Dark-field TEM image of ZnO nanobelt with planar defects. (b) SEM image of a ZnO nanoring .formed by self-coiling of a nanobelt. (c) EDS showing the presence of In ions in the structure. (d) HRTEM image of the planar defects. and (e) the structural model. A simulated image is inserted in (d). The simulation is for a sample thickness of 2.924 nm, defocus -28.67 nm. at 400 kV.*

The doped impurity chosen is indium. As a result the $[01\bar{1}0]$ growth nanobelts are formed. Figure 5.26(a) shows a ZnO nanobelt with planar defects parallel to the (0001) polar surfaces. Self-coiling *of* the polar nanobelt, driven by the electrostatic interaction between the polar surface forms seamless, single-crystal nanorings (Figure 5.26(b)). Energy dispersive X-ray spectrum

(EDS) recorded in TEM from the nanoring shows the presence of minor indium besides majority *of* Zn and O (Cu and Si signals come from the copper grid and the substrate) (Figure 5.26(c)). Quantitative analysis indicate the atomic ratio of *In:Zn* ≈ 1.15, suggests that In atoms are doped in the ZnO lattice.

Hence, the planar defects in nanobelts and nanorings can be related to local segregation of in ions. The HTREM image in Figure 5.26(d) is a nanobelt with two sharp contrast planar defects, which is the building block of a nanoring. After constructing several possible structural models, detailed image simulations of experimental data is taken. If the two dark layers labeled I and II in the image are considerd as two In–I octaheldral layers, then structural model is the one displayed in Figure 5.26(e). If (0001) surface is defined as Zn-terminated and the $(000\bar{1})$ surface as oxygen terminated , so that the polarization is along c-axis, the two slabs of ZnO on both sides of the In-O octahedral layer has opposite polarization, which means that the In-O layer effectively induces a "head-to-head" polarization domain, so called inversion domain boundary (IDB). To configure the two sharply contrasted planar defects (I and II) there exisis another type of defect labeled as (III) between the I and II layers and it corresponds to a "tail-to-tail" IDB. The III layer exist in the image, the bright spots forms a rectangle pattern, as indicated between the I and II layers. On the other hand, in the structural model of In_2O_3, the two slabs of ZnO on either sides of In-O octahedral layer take not only head-to-head polarity as shown in figure 5.26(e). In the first case, the 4-fold symmetry axis of the In-O octahedral lies in the ZnO c-plane to form a head-to-head IDB. In the second case, the 4-fold symmetry of the In-O octahedron is parallel to the c-axis to form a tail-to-tail IDB. Such tail-to-tail layer also exists above the I layer, as labeled to be the IV layer This IV layer, moves across c-planes, as indicated by a dashed line. The transverse translation of the IV layer results from the relocation of the doped indium ions. It is worth noticing that the translation in the [0001] direction across the tail-to-tail IDB is very small, and there is no translation along the $[01\bar{1}0]$ direction.

Metal-oxide-metal Heterojunction Nanowires

Oxide nanowires (nanoribbons, nanobelts, or nanorods) are used in multifunctional bottom-up nanoelectronics, taking advantage of the size-dependent properties of oxides, such as dielectric, ferroelectric, piezoelectric, pyroelectric, chemical-sensing, bio-sensing, electro-optic, and magnetoresistance.

Metal-oxide-metal (MOM) nanowires, where a nanoscale segment of a functional oxide is sandwiched axially between two similar or dissimilar noble-metal nanowires, have distinct advantages over, all-oxide nanowires; examples of MOM nanowire are Au-SnO_2-Au and Au-NiO-Au. These systems, synthesized using a combination of template-based synthesis methods. All-metal Au-Sn-Au and Au-Ni-Au nanowires are electroplated inside the nanoholes of anodic aluminum oxide (AAO) templates, as shown in Fig. .5.27. These nanowires wire are then released by dissolving the AAO templates, and the dispersed nanowires are heat-treated to oxidize the Sn segment into SnO_2 and the Ni segment into NiO within each of the wires.

The AAO templates are prepared. High-purity Al foil 100 μm thick are anodized in 0.3 M oxalic acid under 40 V direct current (dc) at 290K The membranes are treated with a saturated solution of $HgCl_2$ to remove any Al metal. The membranes are then etched with 5wt% H_3PO_4 for 45 min to remove the barrier layer. The average diameter of the nanoholes in the resulting

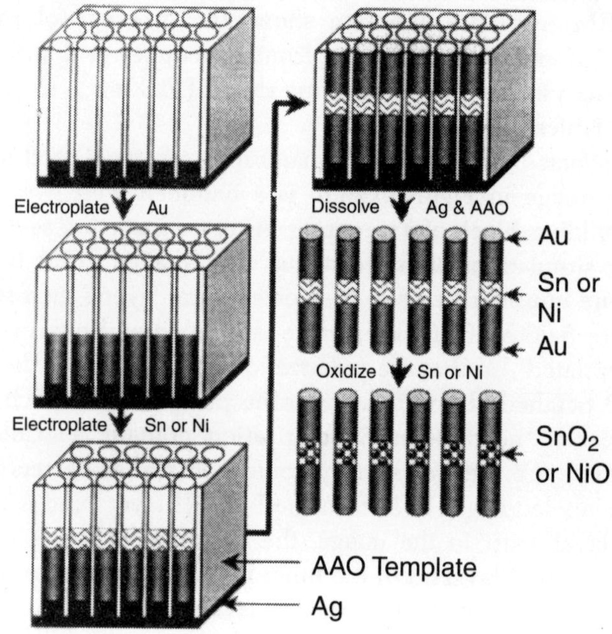

Figure 5.27 *The multi-step synthesis process used to synthesize Au-SnO_2-Au and Au-NiO-Au nanowires.*

AAO templates is ~60 nm. AAO templates with nanohole size of ~220 nm are also used. One side of the AAO templates is sealed by thermally evaporating a 0.5-μm Ag thin film.

Constant-current (0.5 mA) electroplating inside the nanoholes of the AAO templates is performed. With the Ag thin film serving the cathode and Ag foils as the anode, a small segment of Ag (~0.5 μm) is electroplated using Ag solution. The Au nanowire segment is then electroplated with a Pt foil as the anode. The Sn segment is electroplated using 3 M solution of tin (II) chloride with Sn foil as the anode. The Ni segment is electroplated using 0.5 M nickel (II) sulfate with a Ni foil as the anode. Finally, the Au nanowire segment is electroplated. In some cases, the AAO template filled with the all-metal nanowires is fractured, and the nanowires are imaged in a scanning electron microscope (SEM).

The Ag thin film and the Ag nanowire segment are dissolved in 2 M HNO_3. The AAO template is then dissolved in 3 M NaOH, while the solution is being diluted and ultrasonicated. The released nanowires are then concentrated using a centrifuge and rinsed, a process that is repeated 5 times.

The released all-metal nanowires are placed on high purity Pt foils and subjected to heat treatments in air. The Au-Sn-Au nanowires are subjected to a two-step heat-treatment; at 460K for 0.5 h and then at 920K for 0.5 h. The low-temperature heat treatment is done below the melting point of Sn (510K) and its lowest melting alloy (495K) with Au; for converting the Sn to SnO, while the high-temperature heat treatment is for converting the SnO to SnO_2. In the case of Au-Ni-Au nanowires, only a one-step heat treatment at 870K is used to oxidize Ni to NiO because melting points of Ni (1730K) and its alloys (1130K) with Au are much higher than the oxidation temperature.

The resulting Au-SnO_2-Au and Au-NiO-Au (MOM) nanowires are dispersed in deionized water or ethanol. Drops of solution containing the MOM nanowires are placed on 3-mm transmission electron microscope (TEM) grids (Al or Au) and dried. The nanowires are then observed in TEM operated-at 200 kV. The TEM is equipped with atmospheric thin-window energy dispersive spectroscopy (EDS) system. Some of the MOM nanowires are also observed in the SEM.

Figures 5.28(a) and (b) show SEM images of all-metal Au-Sn-Au and Au-Ni-Au nanowires inside a fractured AAO template, respectively. The uniformity of the different metal segments is clearly visible. The interfaces between Au and Ni show the rough surfaces of the tips of Au nanowires.

Figure 5.29 shows TEM bright-field images of an isolated Au-SnO_2-Au nanowire synthesized inside an AAO template, at low and high magnifications The insets in Fig. 5.29 shows selected-area electron diffraction patterns (SAEDPs), which confirm the metal part of the nanowire (dark) to be Au, and the oxide part of the nanowire (light) to be polycrystalline tetragonal SnO_2 (referred as cassiterite or rutile). EDS also confirms the presence of Sn and O in the light (SnO_2) segment , and Au in the dark (Au) segments. The overall length of the MOM nanowire is over 2 μm, and the diameter of the Au part of the nanowire is ~60 nm. The SnO_2 segment of the MOM nanowires is ~60 nm in diameter and ~70 nm long. Figure 5.30 is a high-resolution TEM image of the polycrystalline SnO_2 segment, showing grains ranging from 5 to 10 nm in size.

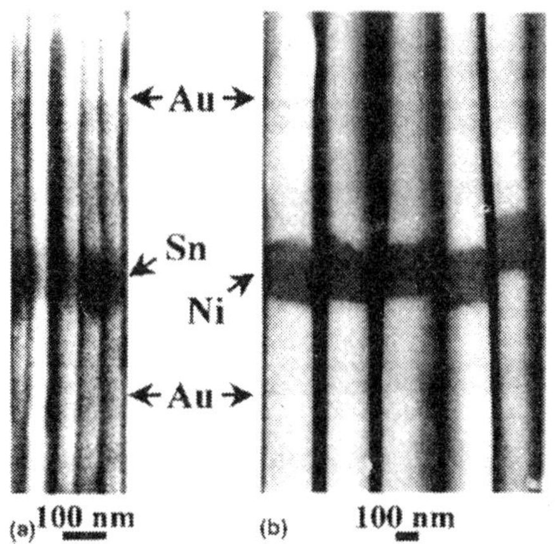

Figure 5.28 *SEM micrographs of all-metal nanowires: (a) Au-Sn-Au nanowires inside fractured in-house-prepared AAO template and (b) Au-Ni-Au nanowires inside fractured AAO template.*

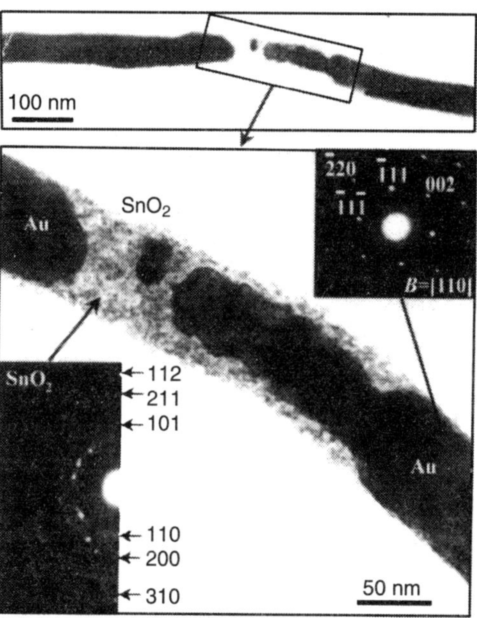

Figure 5.29 *TEM bright-field micrographs showing an isolated Au-SnO_2-Au nanowire synthesized inside an in-house prepared AAO template, at low (top) and high (bottom) magnifications, The dark regions are Au and the light region is SnO_2 as identified using the corresponding SAEDPs (insets).*

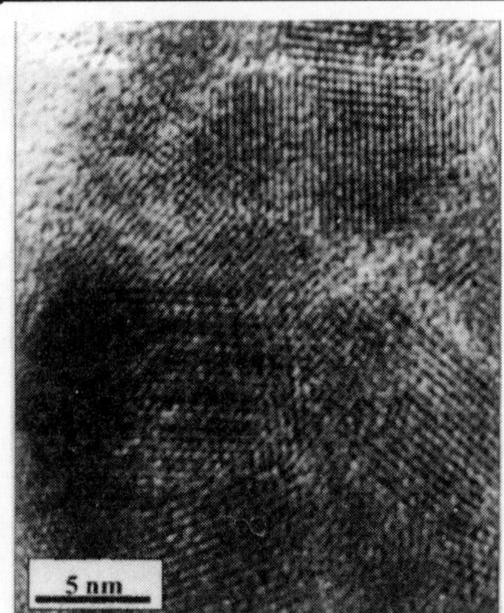

Figure 5.30 *High-resolution TEM micrograph of the SnO_2 region in the Au-SnO_2-Au nanowire.*

Figure 5.31 shows SEM micrographs of an isolated Au-SnO_2-Au nanowire synthesized inside an AAO template, The overall length of this MOM nanowire is over 10 μm, and the diameter of the Au part of the nanowire is ~220 nm. The SnO_2 segment of the MOM nanowires is ~220 nm in diameter and ~400 nm long.

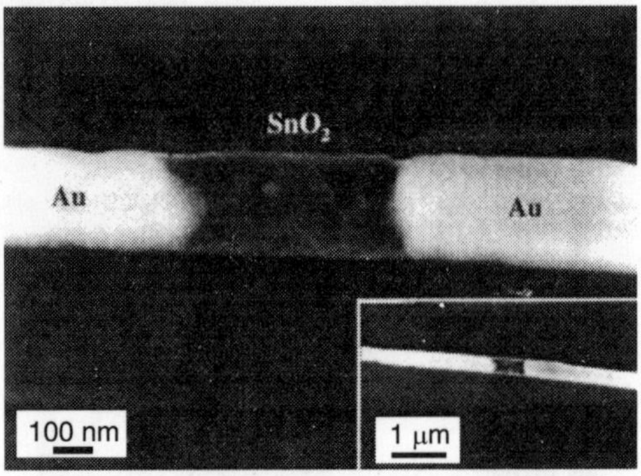

Figure 5.31 *SEM micrographs showing an isolated Au-SnO_2-Au nanowire synthesized inside an Anopore AAO template at high and low (inset) magnifications.*

Figure **5.32** shows TEM images of Au-NiO-Au nanowire synthesized inside an AAO template, at low and high magnifications. The SAEDP in **Fig 5.32** confirms the oxide part of the nanowire (light) to be polycrystalline cubic NiO (referred to as bunsenite). EDS also confirmed the presence of Ni and O in the light (NiO) segment, and Au in the dark (Au) segments. The overall length of the MOM nanowire is over 7 μm, and the diameter of the Au part of the nanowire is ~270 nm. The NiO segment of the MOM nanowires is ~300 nm in diameter and ~200 nm long.

The interface between Au and SnO_2 observed in Fig. 5.29 is due to the diffusion of Au in Sn prior to the oxidation of Sn, which is observed in Au-Sn-Au nanowires. At the oxidation heat-treatment 470K the diffusivity of Au in Sn is 1.9×10^{-12} m^2 s^{-1}, while the diffusivity of Sn in Au at the same temperature is 6.2×10^{-22} m^2 s^{-1}. In the thicker nanowires (Fig. 5.31) the Au-SnO_2 interface is not as jagged due to the longer diffusion lengths involved. The interfaces between Au and NiO in the thicker Au-NiO-Au nanowires (Fig. 5.32) are smoother, due to the long diffusion paths. Also, the diffusivity of Au in Ni at the oxidation temperature (870K) is much

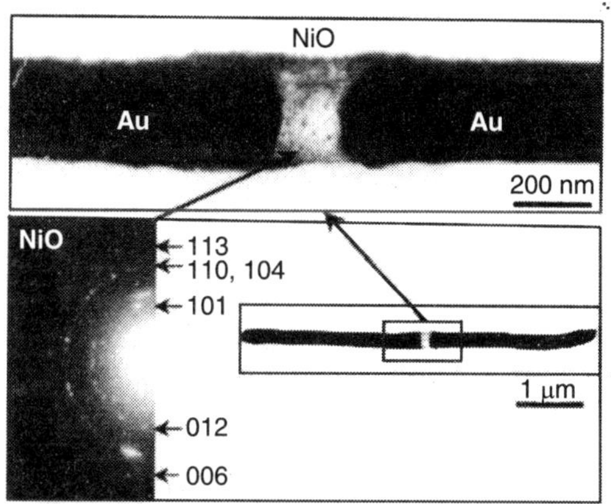

Figure 5.32 *TEM bright field microgrhs showing an isolated Au-NiO-Au nanowire synthesized inside an Anopore AAO template, at low (bottom right) and hight (top) magnification The dark regions are Au and the light region is NiO, the latter identified using corresponding SEADP (bottom left.)*

lower (1.0×10^{-20} m^2 s^{-1}) than that for Au in Sn at 470K. The diffusivity of Ni in Au at 870K is estimated at 8.9×10^{-17} m^2 s^{-1}) The jaggedness of the Au-NiO-Au in terraces is carried over from the inherent jaggedness of the interface in the as-deposited all-mental au-Ni-Au nanowires (Fig 5.28). In cases where inter diffusion in not desired slower diffusing Pt is used as the nobel metal part of the MOM naowires.

In addition to Sn and Ni a wide range of precursor metals are deposited by aqueous electroplating (e.g., Cr, Mn, Fe, Co, Cu, Zn, Y, Cd, In, Pb and Bi). Metals such as Ti and Al are electroplated using organics, while others cannot be deposited electrochemically due to thermodynamic limitations. Thus it is possible to prepare all-metal nanowires where the precursor-metal segment is any one of the metals listed above, which can then be oxidized selectively to result in the corresponding metal oxide. In order to produce doped single oxides or mixed oxides, it is possible to co-electoplate a combination of two or more precursor metals to form alloy segment, which can then be oxidized. Alternatively thin stripes of two or more types of precursor metals of predetermined thicknesses can be electroplated, followed by heat-treating the nanowires to cause interdiffusion of those metals before final oxidation.

The MOM heterojunction nanowires address three critical issues that are encountered when using all oxide nanowire building blocks.

First all-oxide nanowires are assembled across lithographically developed metal contact pad electrodes to create devices, where the length of the nanowire spanning the electrodes defines the active region. That length ranges from submicron to several tens of microns as it is limited by the 2D lithography technique. This results in large ohmic losses, requiring high device operating voltages, and functional properties of oxide nanowires that are highly temperature-sensitive. In MOM nanowire the dimension of the function oxide is controlled accurately and thus the active region of the assembled MOM nanowire is independent of the contace-pad electrodes geometery. Also in the case of nanowires of the Au-SnO_2 –Au type, heterojuncutions between Au and SnO_2 are useful in providing schottky contacts and catalysis sites in chemical sensing applications.

Second, the nanoscale nature of the all oxide nanowires calls for the use of electrical, chemical, or biological methods for their site specific assembly. But these methods are suited for metals compared to oxides. In MOM nanowires the noble metal part is amenable to electrical, chemical, or biological methods for the site-specific assembly of the MOM nanowires into circuits.

Third, majority of the oxide (and nano-oxide semiconductors) nanowires are synthesized using vapor phase or non-template chemical methods, where the resulting product is a tangled mixture of nanowires of various lengths, diameters and morphologies Although the vapor-phase methods, and some of the chemical methods, produce high-quality single-crystal oxide nanowires, large-scale, reproducible isolation of the desired nanowires from the mixture requires additional sorting steps, which are often intractable. So the synthesis method described above offer better control over the structure and the characteristics of the resulting oxide building blocks. The diameter of the nanowire is adjusted by tuning the AAO nanohole diameters, which ranges from 18 to 420 nm. The total length of the MOM nanowire and the length of the individual segments is controlled through electroplating conditions. However the interdiffusion between the precursor and noble metals before complete oxidation and the attendant instability of the interfaces pose a lower limit to thickness of oxode segment that is achieved using this method.

Heterostructured Nanowires

The success of semiconductor integrated circuits has largely hinged upon the ability to form heterostructures through carefully controlled doping and interfacing. The 2D semiconductor interface is used in optoelectronic devices such as LEDs, laser diodes, quantum cascade lasers, and transistors. For 1D nanostructures the capability of forming heterostructures is important for their potential applications as efficient light emitting sources and thermoelectric devices. While there are a number of techniques (e.g., MBE) for generating heterostructure and superlattices in thin films, a scheme for forming heterojunctions and superlattices in 1D nano structures with well defined coherent interfaces is demonstrated. Semiconductor nanowires and CNTs are used with homogeneous systems with a few exceptions that involve heterostructures; example heterojunction formed between CNTs and SiC nanowires and p-n junctions on individual CNTs or $GaAs/Ga_{1-x}In_x$ As nanowires, A electrochemical method is demonstrated for fabricating striped metal nanorods. This method, yields polycrystalline products with less than ideal interfaces. The use of a hybrid pulsed laser ablation/chemical vapor deposition (PLA-CVD) process is also made for generating semiconductor nanowires with periodic longitudinal heterostructures. Here Si and Ge vapor sources are independently controlled and alternately delivered into the VLS nanowire growth system. As a result single crystalline nanowire with Si/SiGe superlattice structure are obtained.

Figure 5.33 shows a scanning transmission electron microscopy (STEM) image of two such nanowires in the bright-field mode. Along the longitudinal axis of each wire, dark stripes appear periodically, reflecting the alternating deposition of the SiGe alloy and Si segments. Since the electron scattering cross section of Ge atoms is large than that of Si atoms the SiGe alloy block appears to be darker than the pure Si block The chemical composition of the dark region is also examined using EDX spectroscopy, and it shows a strong Si peak and apparent Ge doping (~12wt.-%Ge) The periodic modulation of Ge doping is confirmed by scanning a focused electron beam along the nanowire growth axis and tracking the change in X-ray signal from Si and Ge atoms in the wire (Figure 5.23(b)). Both Si and Ge X-ray signals show periodic modulation, with their intensities anti co-related: wherever the X-ray signal from Ge show a maximum, the signal from Si show a minium. This observation is also supports the formation of a Si/Si Ge superlattice along the wire axis.

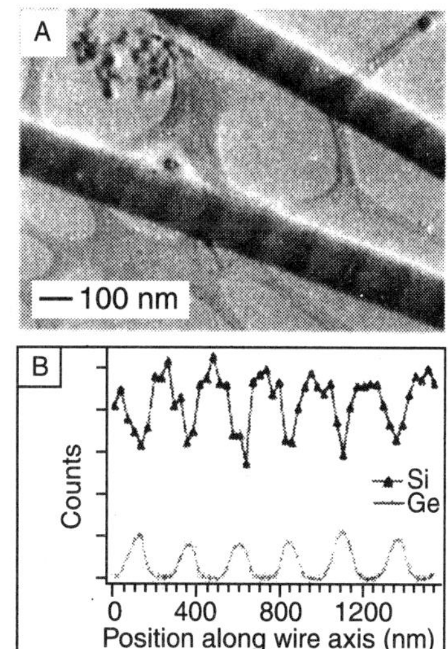

Figure 5.33 *TEM image of two typical Si/SiGe superlattice nanowire(B) Compositional profiles showing the spatial modulation in Si and Ge contents along the longitudinal axis of a nanowire*

Using a similar approach, GaAs/GaP. heterostructure nanowires respectively; are prepared. Since the supply of vapor sources is programmed, this VLS with process modulated source can be easily extended to prepare various other heterostructures on individual nanowires in a "custom-designed" fashion. It also enables the creation of various function devices (e.g., p-n junctions, coupled quantum dot structure and heterostructured bipolar transistors)on individual nanowires. These heterostructured nanowires are further used as important building blocks to construct nanoscale electronic circuits and light emitting devices, example: super nanowires with reduced phonon transport and high electron mobilities are promising candidates for improving the efficiency of thermoelectric devices.

LONGITUDINAL HETEROSTRUCTURES

The growth of longitudinal heterostuctured nanowires involves using a single one-dimensional growth mechanism that can be easily switched between different material mid-growth. In order to obtain useful heterostructures, the growth mechanism must be compatible with the desired materials and produce well-defined and coherent interface with good control. Because the VLS growth mechanism can provide such control, longitudinal heterostructure synthesis is performed using this approach.

The use of a hybrid pulsed laser ablation/chemical vapor deposition (PLA-CVD) process is made for generating semiconductor nanowires with periodic longitudinal heterostructure. In the process, Si and Ge vapor sources are independently controlled and alternately delivered into the VLS nanowire growth system. As a result, single crystalline nanowires containing the Si/SiGe superlattice structure are obtained.

Figure 5.34 shows a TEM image of two such nanowires in the bright-field mode. Dark stripes appear periodically along the longitudinal axis of each wire , reflecting the alternating domains of Si and SiGe alloy. Because the electron scattering cross section of Ge is a large than that of Si the SiGe alloy block appears darker than the pure Si block. The chemical composition of the dark region is examined using energy-dispersive X-ray spectroscopy (EDS) which show a strong Si peak and apparent Ge doping (~12 wt% Ge). The Si and Ge signals are periodically modulated with anticorrelated intensities. This observation supports the formation of a Si/SiGe superlattice along the wire axis.

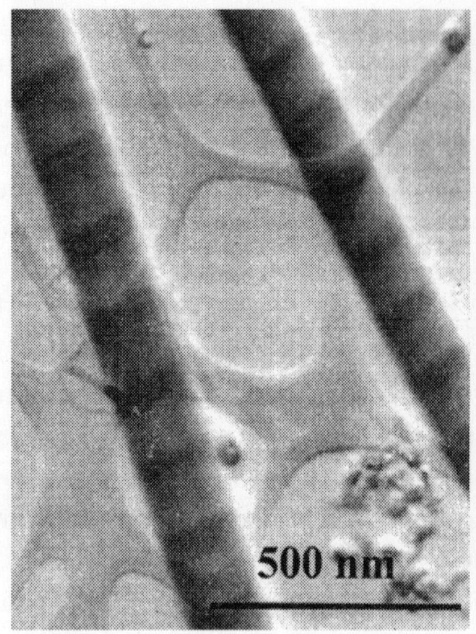

Figure 5.34 *Transmission electron microscopy (TEM) image of two Si/SiGe superlattice nanowires.*

Using a similar approach, GaAs/GaP and InAs/lnP heterostructured nanowires are prepared. Also produced are ZnSe/CdSe superlattice nanowires. Because the supply of vapor sources are readily programmed, the VLS process with modulated sources is useful for preparing a variety of heterostructures on individual nanowires in a custom-designed fashion. It also enables the creation of various functional devices (e.g., p-n junctions, coupled quantum dot structures, and heterostructured unipolar and bipolar transistors) on individual nanowires. These heterostructured nanowires are further used as important building blocks to construct nanoscale electronic circuits and light-emitting devices.

The synthesis of longitudinal nanowire heterostructures is done using a non-VLS mechanism. Striped Ag/Au and Au/Co nanowire superlattices are fabricated using a sequential electrochemical method inside anodic aluminum oxide templates. Also synthesized are GaN p-n junctions using a chemical vapor transport vapor-solid process by introducing the p-type dopant Cp_2Mg mid-growth.

Coaxial Nanowire Complexity and Functionality

The success of semiconductor integrated circuits has largely hinged on the capability of forming heterostructures through controlled doping and interfacing, as semiconductor heterostructures enable the confinement of electrons and holes, the guiding of light, and the modulation of phonon transport and carrier mobility. Based on understanding of nanowire growth different types of heterostructured nanowires are explored including, for example, coaxial heterostructured nanowires (COHNs), longitudinal heterostructured nanowires (LOHNs), and nanotapes (Figure. 5.35).

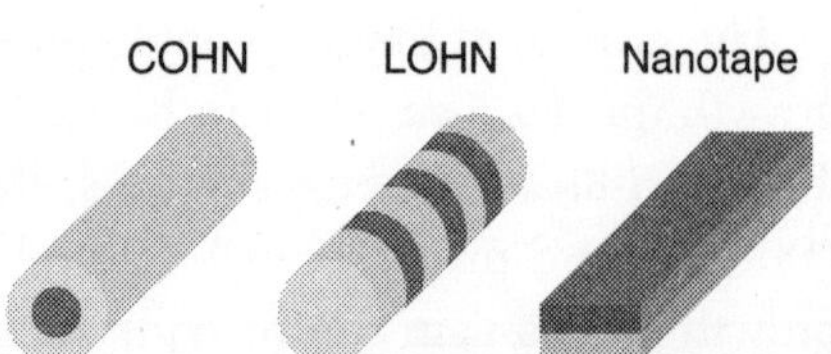

Figure 5.35 *Throe different types of heterostructured nanowires: a coaxial heterostructured nanowire (COHN), a longitudinal heterostructured nanowire (LOHN), and a nanotape.*

Coaxial nanowires are an important class of heterostructured nanowires that have significant technological potential. Coaxial structures can be fabricated by coating an array of nanowires with a conformal layer of a second material. The coating method chosen should allow excellent uniformity and control of the sheath thickness. Cladding nanowires with amorphous layers of SiO_2 or carbon is synthetically facile. A more exciting and difficult task, with greater technological importance, is to form heterostructures of two single-crystalline semiconductor materials. The synthesis of GaN / $Al_{0.75}Ga_{0.25}N$ core-sheath structures using a chemical vapor transport method (Figure 5.36) is achieved and ZnO/GaN core-sheath heterostructures are also grown using a MOCVD approach. Also, Si/Ge core-sheath wire are produce by chemical vapor deposition methods. It is important to point out that the choice of appropriate core and sheath materials with similar crystallographic symmetries and lattice constants is essential to achieve the deposition of single-crystalline epitaxial thin-film sheath structures and thereby produce high-quality materials. By choosing appropriate core and sheath material that have similar crystallographic symmetries and lattice constants is essential to achieve the deposition of single-crystalline epitaxial thin-film sheath structures, high quality materials are produced. The band structure of the different materials used in forming COHNs is chosen to achieve modulation doping. For example, the dopant atoms reside in the cladding, while the carriers are located in the core. This is the ID variation of the idea behind modulation doping in two-dimensional (2D) semiconductor heterostructures, which achieves very high electron mobilities by suppressing charged impurity scattering.

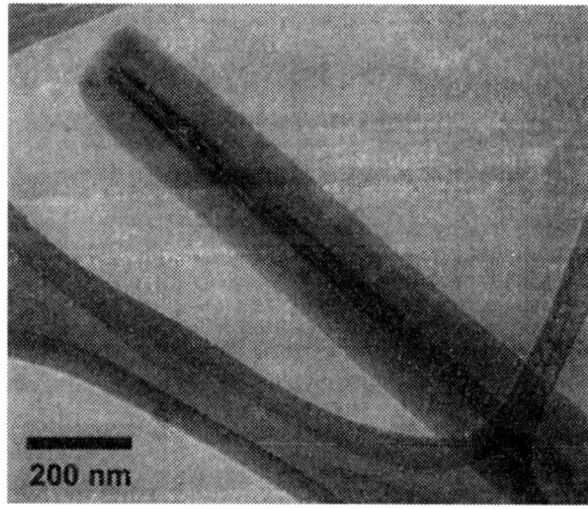

Figure 5.36 *Transmission electron microscopy image of a GaN/AIGaN core-sheath nanowire.*

Similar to the concept of creating a uniform sheath around a nanowire is the idea of coating 1D nanostructures (i.e., only along one side of the nanowire material). A Versatile approach to the synthesis of composite nanowire structures is developed in which the composition limitation is relaxed show that the resulting nanostructures have multiple functionalities such as luminescence, ferromagnetism, and ferroelectric or super-conducting properties.

In this process, tin dioxide nanoribbons are used as substrates for the thin-film growth of various oxides (e.g., TiO_2, transition-metal-doped TiO_2, and ZnO) using pulsed laser deposition (PLD). The energetic nature of the laser ablation process makes the plume highly directional and enables selective film deposition on one side of the nanoribbon substrate by means of the shadow effect (Figure 5.37(a)). Electron microscopy and x-ray diffraction demonstrate that these functional oxides grow epitaxially on the side surfaces of the substrate nanoribbons with sharp structural and compositional interfaces.

The creation of epitaxial core-sheath: structures imparts the ability to synthesize single-crystalline nanotube materials derived from three-dimensional (3D) crystal structure by dissolving the inner core. This synthetic approach requires that the core and sheath material; exist in epitaxial registry and possess different chemical stabilities. This "epitaxial casting" is used to synthesize, GaN nanotubes with inner diameters of 30–200 nm and wall thicknesses 5–50 nm Hexagonal ZnO nanowires are used as templates for the epitaxial overgrowth of thin GaN layers in a MOCVD system. The ZnO nanowire templates are removed by simple thermal reduction and evaporation in NH_3/H_2 mixtures, resulting in ordered arrays of GaN nanotubes on the substrate. This is an example of single crystalline GaN nanotubes and this templating process can be applied to many other semiconductor systems (Figure. 5.37(b))

The use of a hybrid pulsed laser ablation/ chemical vapor deposition process for generating semiconductor nanowires with periodic longitudinal heterostructures In this process, Si and Ge vapor sources are controlled and delivered into the VLS nanowire growth system. As a result, single-crystalline nanowires containing the Si/ SiGe superlattice structure are obtained. Figure 5.37(c) shows a scanning transmission electron microscopy image of two such nanowires in the bright-field mode. Dark stripes appear periodically along the longitudinal axis of each wire, reflecting the alternating domains of Si and SiGe alloy. Since the supply of vapor sources can be programmed, the VLS process with modulated sources is useful for preparing a variety of heterostructures on individual nanowires in a "custom-designed" fashion. The process also enable the creation of various functional devices (e.g., p-n junctions, coupled quantum-dot structures, and heterostructured unipolar and bipolar transistors) on individual nanowires. These heterostructured nanowires can be further used as important building blocks to construct nanoscale electronic circuits and light-emitting devices.

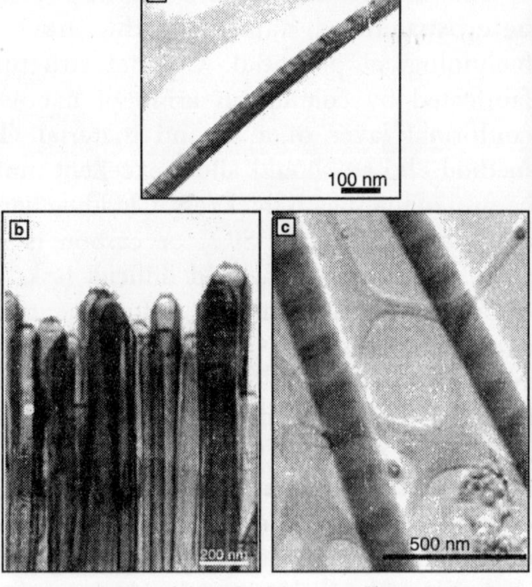

Figure 5.37 *(a) A highly crystalline SnO_2/TiO_2 composite nanoribbon showing the epitaxial growth of TiO_2 on the SnO_2 nanoribbon surface, (b) a cluster of single-crystalline GaN nanotubes prepared using the epitaxial casting methodology, and (c) two Si/SiGe superiattice nanowires.*

Functional Oxide Nanobelts

One of the most important contributions made by carbon nanotubes is having stimulated interest in one-dimensional nanostructures. Various techniques are practiced in order to invent a variety of quasi-one-dimensional nanostructures, the geometrical shapes of which include tubes, cages, cylindrical wires and rods, nails, coaxial and bi-axial cables, ribbons or belts sheets, diskettes and more. Exploration of nanostructures that exhibit functionality is the key to nanotechnology. Functional oxides are the fundamental ingredients of smart systems, because the physical and chemical properties of the oxides can be tuned and controlled through adjusting cation valence state and anion deficiency. The structures of functional oxides are very diverse and colorful, and there are endless new phenomena and applications. Such unique characteristics have made oxides the most diverse class of materials including semiconductivity, superconductivity, ferroelectricity, magnetism, and piezoelectricity.

Since the discovery of semiconducting oxide nanobelts oxide nanostructures have attracted considerable attention. The discovery of nanobelts is similar the discovery of nanotubes and it stimulates a vast interest in investigating nanobelt-based materials and applications.

Nanotechnology Based on Nanobelts

Semiconducting transition and rare-earth metal oxides are attracting significant attention as candidates for chemical and environmental sensors. This is because their electrical conductivity is highly dependent on the nature and concentration of the surface of adsorbed species. The key characteristics of these oxides are (a) cations with mixed valence and (b) oxygen vacancies. The latter are responsible for the observed high sensitivity of the electrical properties to the presence of adsorbed molecules and allow the tuning of the conductance of the oxide. The semiconducting oxide nanobelts grown are structurally perfect and geometrically uniform. These nanostructures are ideal materials for building nano-sized devices and sensors.

THE FAMILY OF NANOBELTS

Nanobelts Versus Nanowires

It is well known that a single-wall carbon nanotube can be metallic or semiconductor depending on the helical angle at which the graphitic sheet is rolled up. It is expected that the properties of a nanowire could be strongly affected by the structure of the side surfaces, analogous to the helical angle for carbon nanotubes, which determines its electronic structure. There are a few names being used for describing one-dimensionally elongated structures, such as nanorod, nanowire, nanoribbon, nanofiber, and nanobelts. The nanostructures "nanobelts" means that the nanostructure has specific growth direction and the top, bottom, and side surfaces are well-defined crystallographic facets. The requirements for nanowires are less restrictive than those for nanobelts because a wire has a specific growth direction, but its side surfaces may not be well defined, and its cross-section may not be uniform or specific in shape. Therefore it is believed that nanobelts are more structurally controlled objects than are nanowires; or simply, a nanobelt is a nanowire that has well-defined side surfaces. It is well known that the physical property of a carbon nanotube is determined by the helical angle at which the graphite layer was rolled up. It is expected that the physical and chemical properties of thin nanobelts and nanowires will depend on the nature of the side surfaces.

A few kinds of nanobelts (Table 6.1) are summarized the nanobelt structures of function oxides. Each type of nanobelt is defined by its crystallographic structure, growth direction, top surfaces, and side surfaces. Some of the materials can grow along two directions, but they can be controlled experimentally. Although these materials belong to different crystallographic families, they do have a common faceted structure, which is the nanobelt structure. In addition, nanobelts of $Cu(OH)_2$ MoO_3, MgO, and CuO have been successfully synthesized.

TABLE 6.1 *Crystallographic geometry of functional oxide nanobelts*

Nanobelt	*Crystal structure*	*Growth direction*	*Top surface*	*Side surface*
ZnO	Wurtzite	[0001] or $[01\bar{1}0]$	$\pm(2\bar{1}\bar{1}0)$ or $\pm(2\bar{1}\bar{1}0)$	$\pm(01\bar{1}0)$ or $\pm(0001)$
Ga_2O_3	Monoclinic	[001] or [010]	± (100) or ± (100)	± (010) or $\pm(10\bar{1})$
t-SnO_2	Rutile	[101]	$\pm(10\bar{1})$	± (010)
o-SnO_2 wire	Orthorhombic	[010]	± (100)	± (001)
In_2O_3	C-rare earth	[001]	± (100)	± (010)
CdO	NaCI	[001]	± (100)	± (010)
PbO_2	Rutile	[010]	± (201)	$\pm(10\bar{1})$

ZnO Nanobelts

For, nanobelts, ZnO is the most extensively used structure. Thermal evaporation of ZnO powders (purity: 99.99%; melting point:2245 K) at 1670 K result in ultralong ZnO nanobelts (Figure 6.1(a)). The typical lengths of the ZnO belts are in the range of several tens to several hundreds of micrometers; some of them even have lengths on the order of millimeters. Energy-dispersive X-ray spectroscopy and X-ray diffraction measurements show that the sample is Wurtzite (hexagonal)–structured ZnO with a lattice constant of a = 3.249 A and c = 5.206 Å. consistent with the standard values for bulk ZnO.

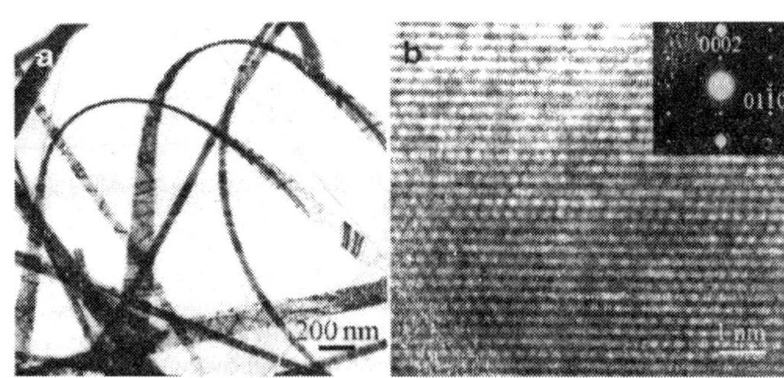

Figure 6.1 *(a) Transmission electron microscopy (TEM) image of the as-synthesized ZnO nanobelts. (b) High-resolution TEM image recorded with the incident electron perpendicular to the top surface of the nanobelt.*

Transmission electron microscopy (TEM) images reveal that the geometrical shape of the ZnO nanobelts is distinct in cross section from the nanotubes or nanowires. Each nanobelt has a uniform width along its entire length, and the typical widths of the nanobelts are in the range of 50–300 nm. No particle is observed at the ends of the nanobelts. A ripple-like contrast that appeared in the TEM image is due to strain resulting from the bending of the belt. High-resolution TEM and electron diffraction show that the ZnO nanobelts are structurally uniform and single crystalline (Figure 6.1(b)).

SnO_2 Nanobelts

Single crystallin SnO_2 nanobelts of rutile structure are consistently synthesized by thermal evaporation of either SnO_2 powders (purity: 99.9%, melting point: 1900 K) at 1620 K or SnO powders (purity: 99.9%, melting point 1350 K) at 1270 K. TEM images (Figure 6.2(a)) display the characteristic shape of the SnO_2 nanobelts. High-resolution TEM images (Figure 6.2(b)) reveal that the nanobelts are single crystalline and dislocation free. Electron diffraction patterns indicate that the SnO_2 nanobelt grows along [101], and it is enclosed by ± (010) and ± $(10\bar{1}1)$ crystallographic facets.

In addition to the normal rutile-structured SnO_2, it is possible to form an orthorhombic superlattice-like structure. The orthorhombic structure is from in a thin nanowire, coexist

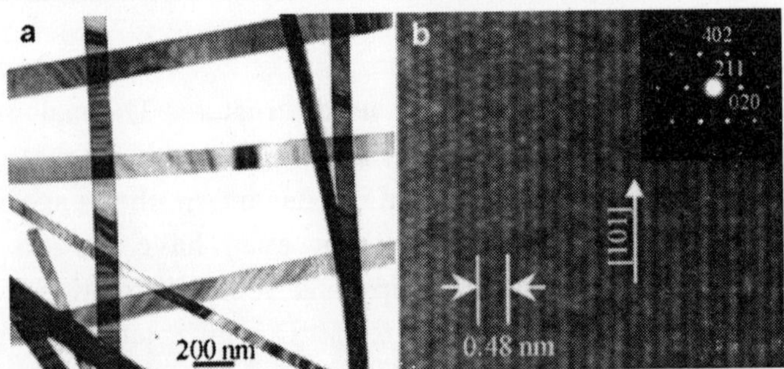

Figure 6.2 *(a) Transmission electron microscopy (TEM) image of the as-synthesized SnO_2 nanobelts. (b) High-resolution TEM image recorded with the incident electron perpendicular to the top surface of the nanobelt.*

with the normal rutile-structured SnO_2 in a sandwiched nanoribbon, or occur in the form of nanotubes. This result is distinct from that for bulk SnO_2, where pressures in excess of 150×10^5 kPa are required to form the orthorhombic form.

In_2O_3 Nanobelts

Nanobelts of In_2O_3 with C-rare earth crystal structure are also synthesized by the above synthesis method. Figure 6.3 shows TEM image of as synthesized In_2O_3 nanobelts. The evaporation of In_2O_3 powders (purity: 99.99%, melting point ~ 2290 K) at 1670 K yields In_2O_3 nanobelts. TEM observations are show that most of the In_2O_3 nanobelts have uniform width and thickness along their lengths. Typically, the In_2O_3 nanobelts have widths in the range of 50–150 nm and lengths of several tens to several hundreds of micrometers. Electron diffraction analysis shows that the In_2O_3 nanobelts are single crystalline and grow along ⟨100⟩, the surfaces being enclosed by {100}. In_2O_3 nanobelts growing along [120] type, with top and bottom surfaces (001) and side surface being $[2\bar{1}0]$ have also been observed. Transport properties of In_2O_3 nanobelts are also achieved.

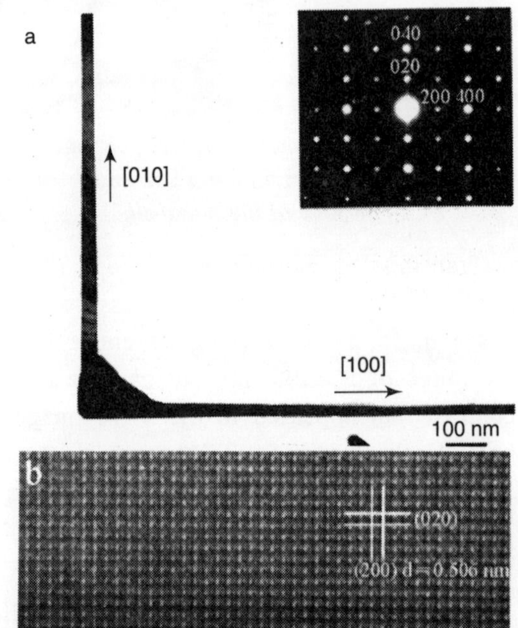

Figure 6.3 *(a) Transmission electron microscopy (TEM) image of the as-synthesized L-shaped In_2O_3 nanobelts. The inset is the electron diffraction pattern recorded from the nanobelt. (b) High-resolution TEM image recorded with the incident electron perpendicular to the top surface of the nanobelt.*

CdO Nanobelts

Nanobelts of CdO with NaCl cubic structure are also synthesized by evaporating CdO powders (purity: 99.998%, melting point: 1690 K) at 1270 K. Besides CdO nanobelts, many single crystalline CdO sheets with sizes on the order of several to several tens of micrometers are formed (Figure 6.4). These CdO sheets usually have shapes of rectangles, triangles, and parallelograms. The lengths of the CdO nanobelts are usually less than 100 μm, and their widths are typically 100–500 nm. Electron diffraction patterns show that the nanobelts grow along [100] and their surfaces are enclosed by ± (001) and ± (010) facets.

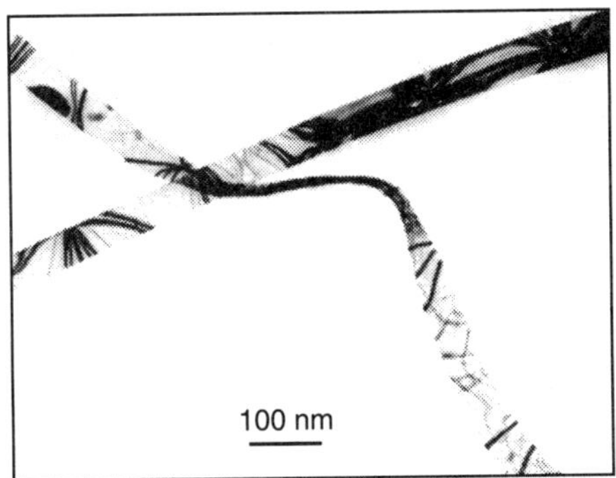

Figure 6.4 *TEM images of the as-synthesized CdO nanobelts.*

Ga_2O_3 Nanobelts

Nanoribbons and flat nanosheets of Ga_2O_3 are synthesized by evaporating GaN at high temperature in the presence of oxygen. The as-synthesized nanoribbons and nanosheets are pure, structurally uniform, single-crystalline and free from dislocations. The nanoribbons and nanosheets all have monoclinic β Ga_2O_3 structure ($C2/m$, a = 12.23, b = 3.04, c = 5.80 Å, and β = 103.7°). The flat top and bottom surfaces for both nanoribbons and nanosheets are ± (100) and the side surfaces are ± (010), and ± $(10\bar{1})$ for nanoribbons and ± (010), ± $(10\bar{1})$, and ± $(21\bar{2})$ for nanosheets. The axis direction of nanoribbon growth is along either [001] or [010].

PbO_2 Nanobelts

β-PbO_2 nanobelts, with a rectangular cross section, a typical length of 10–200 μm, a width of 50–300 nm, and a width-to-thickness ratio of 5–10, have been successfully synthesized by evaporation of commercial PbO powders at high temperature. The PbO_2 nanobelts are enclosed by top surfaces (201) and side surfaces ± $(10\bar{1})$, and their growth direction is [010]. Each PbO_2 nanobelt has a large polyhedral Pb tip at one of its ends, suggesting the growth is dominated by a vapor-liquid-solid mechanism. Electron beam irradiation of the PbO_2 nanobelts results in the phase transformation from PbO_2 to PbO and finally to Pb.

ZnS Nanobelts

ZnS (zinc sulfide) has two types of crystal structures: hexagonal Wurtzite ZnS (referred to as the hexagonal phase) and cubic zinc blend ZnS (referred to as the cubic phase). Typically, the stable structure at room temperature is zinc blend, with few observances of stable Wurtzite ZnS. Wurtzite structured nanobelts, nanocombs: and nanowindmill are synthesized using simple catalyst-free thermal evaporation techinque. The synthesis technique is based on thermal evaporation of ZnS powders at elevated temperature. The as-synthesized sample is composed of several types of structures. The most typical structure observed is ZnS nanobelts (Figure 6). The nanobelts have a uniform cross section along their length, with a typical width of 2-30 μm and extend to over 100 μm in length. Electron diffraction indicates that all three types of structures are single crystalline with the Wurtzite ZnS structure (a = 3.82 Å, c = 6.26 Å). The comb-like, saw-toothed belts and the regular belt structures were found in the same growth-temperature range. The nanobelt grows along [0001], with side surfaces $(01\bar{1}0)$ and the top surfaces $(2\bar{1}\bar{1}0)$. Silicon-based nanobelts have also been reported.

SYNTHESIS

The oxide nanobelts are synthesized by solid-vapor process. In principle, the thermal evaporation technique is a simple process in which condensed or powder source material is vaporized at elevating temperature, and then the resultant vapor phase condenses at certain conditions (temperature, pressure, atmosphere, substrate, etc.) to form the desired product. The processes are usually conducted in a horizontal tube furnace, as shown in Figure 6.5, which is composed of an alumina tube, a rotary pump system, and a gas supply and control system. A view window is set up at the left end of the alumina tube, which is used to monitor the growth process. The right-hand end of the alumina tube is connected to the rotary pump. Both ends are sealed by rubber O-rings. The ultimate vaccum for this configration is ~ 266 × 10^{-3} Pa. The carrying gas comes in from the left end of the alumina

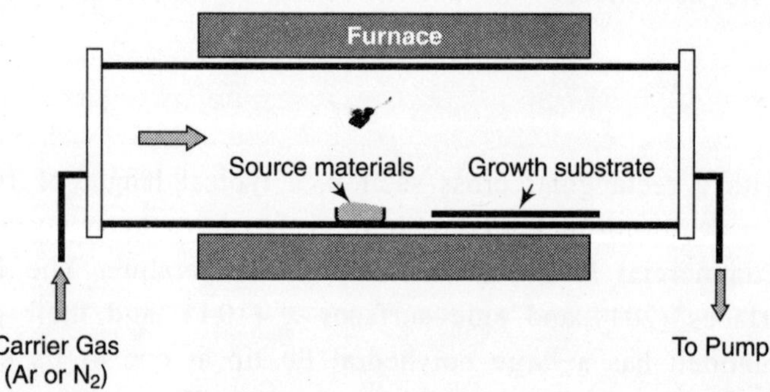

Figure 6.5 *Diagram of experimental apparatus for growth of oxide nanostructures.*

tube and is pumped out at the right end. The source material is loaded on an alumina boat and positioned at the center of the center of the alumina tube, where the temperature is the highest. Alumina substrates are placed downstream for collecting growth product. This simple setup can achieve high control of the final product.

There are several processing parameters such as temperature, pressure, carrier gas (including gas species and its flow rate), substrate, and evaporation time period that can be controlled and need to be selected properly before or during the thermal evaporation. The source temperature selection mainly depends on volatility of the source material. Usually, it is slightly lower than the melting point of the source material. The pressure is determined according to the evaporation rate or vapor pressure of source material. The substrate temperature usually drops with increasing distance from the position of source material. The local temperature determines the type of product to be received. It is also noted that the thermal evaporation process is very sensitive to the concentration of oxygen in the growth system. Oxygen influences not only the volatility of the source material and the stoichiometry of the vapor phase, but also the formation of product. After evacuating the alumina tube to $\sim 266 \times 10^{-3}$ Pa, thermal evaporation is conducted at a certain temperature for 2 h under a pressure of 2.66×10^4 to 7.98×10^4 Pa and an Ar carrier gas of 50 sccm (standard cubic centimeters per minute). The on and off time as well as the length of inletting the carrier gas could greatly affect the growth kinetics.

GROWTH OF OXIDE NANOBELTS

To understand the growth of a uniform belt structure without the presence of hetero-catalysis particles, a mechanism for nanobelt growth assumes that the source material vaporizes into molecular species at high temperature, and the molecules are composed of the stoichiometric cation-anion molecules (Figure 6.6(a)). When condensed onto the substrate at a lower temperature region, the cation-anion molecules are arranged in such a way that a proper cation-anion coordination is preserved to balance the local charge and structural symmetry (Figure 6.6(b)), forming a small nucleus. The newly arrived molecules continue to deposit on the formed nucleus, while the surfaces that have lower energy, such as the side surfaces, start to form. Because the growth temperature is in the range of

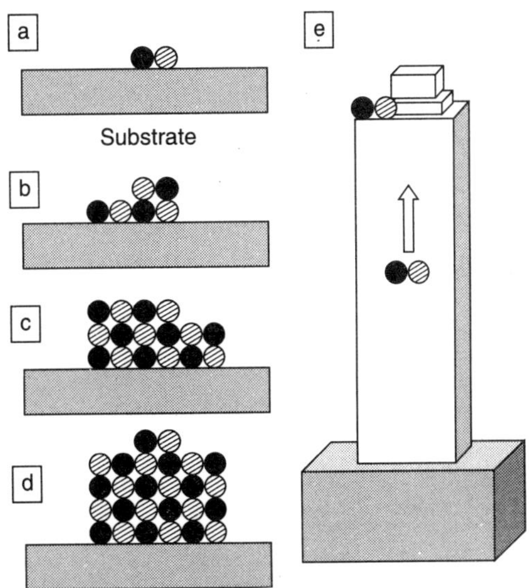

Figure 6.6 *A growth process of a nanobelt*

1070–1270 K the mobility of the atoms and molecules is high enough, and the low-energy surface tends to be flat, thus preventing the accumulation of the newly arrived molecules onto the surface, resulting in expansion in surface area as more molecules stick on the rough growth front (Figure 6.6(c)). The rough structure of the tip leads to a rapid accumulation of incoming molecules, resulting in the fast formation of a nanobelt (Figure 6.6(d)), and after some time a nanobelt is formed (Figure 6.6(e)). The newly arrived molecules can continue to stick on the growth front, or the side surfaces, but the smooth side surface and the high molecular mobility at the growth temperature prevents them from remaining on the surface. The molecules randomly diffuse on the surface and finally find the lower-energy sites at the growth front. The molecules do not stick to the edge of the nanobelts because of the unbalanced coordination and possibly higher energy. The size of the nanobelt cross section is determined by the growth temperature and supersaturation ratio in kinetics of crystal growth, as mentioned above.

SELF-CATALYZED GROWTH

Self-catalyzed growth is a process in the formation of nanobelts, in which a small thin layer or even an atom-thickness metal layer such as Zn presents at the growth front, leads to growth. The layer is quickly oxidized after the nanobelts are exposed to air after growth.

Figure 6.7(a) shows a $[01\bar{1}0]$ high-resolution TEM profile image (recorded from the growth front) of a nanobelt growing along [0001]. The side and front surfaces are rough, and there are tiny nanocrystals being formed at the growth front. These nanocrystals preserve the orientation of the nanofingers and possibly grow epitaxially, and their lattice spacing fits to ZnO, as presented in the enlarged image from the growth front (Figure 6.7(b)). The top layer of the nanocrystal has a darker contrast,

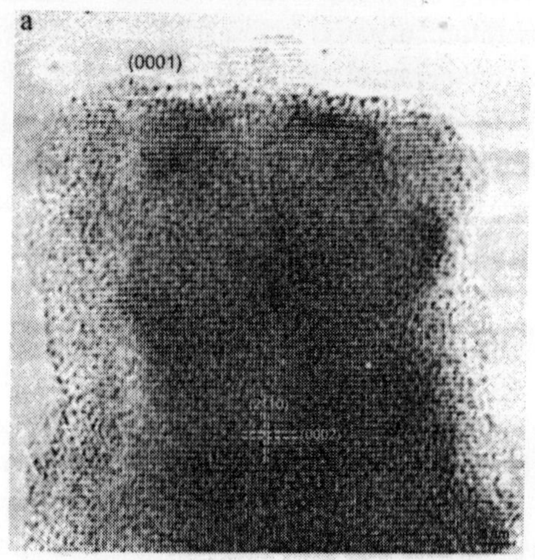

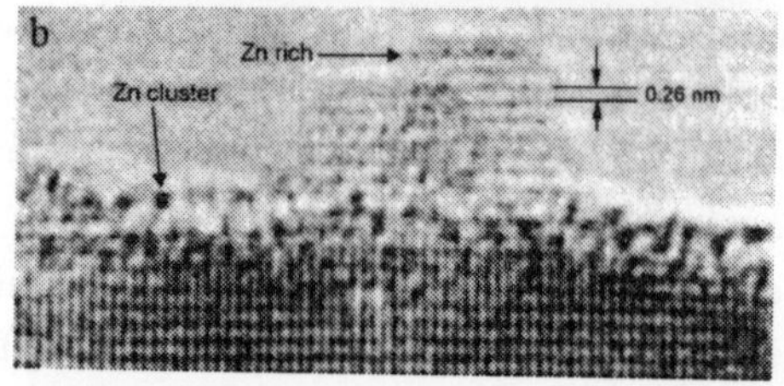

Figure 6.7 *(a) High-resolution TEM image recorded from a nanofinger growing along [0001], showing rougher surface and the presence of small ZnO nanocrystals as well as possible Zn clusters at the growth front.*
(b) An enlargement of the growth front to display the fine structures. The image suggests the presence of Zn clusters at the growth front, which may self-catalyze the process for the growth.

indicating an enrichment of Zn at the local region. A striking feature observed in the image is the presence of some dark spots at the growth front indicated by an arrowhead. These tiny clusters of sizes ~ 0.5 nm correspond to tiny Zn clusters.

HIERARCHICAL SUPERSTRUCTURES

Using a mixture of ZnO and SnO_2 powders in a weight ratio of 1:1 as the source material, a complex ZnO nanostructure is created. Figure 6.8(a) shows a low-magnification SEM image of the as-synthesized products with a uniform feature consisting of sets of central axial nanowires, surrounded by radial-oriented tadpole like nanostructures. The morphology of the string appears like a liana, and the axial nanowire is the rattan, which has a uniform cross-section with dimension in the range of a few tens of nanometers. The tadpole-like branches have spherical balls at the tips (Figure 6.8(b)), and the branches display a ribbon shape. Energy-dispersive X-ray spectroscopy analysis shows that the tadpole-like structure and the central nanowire are ZnO, while the ball at the tip is Sn. The ribbon branches have

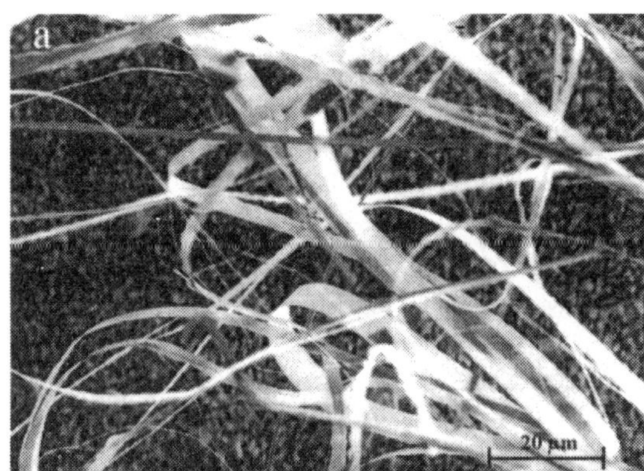

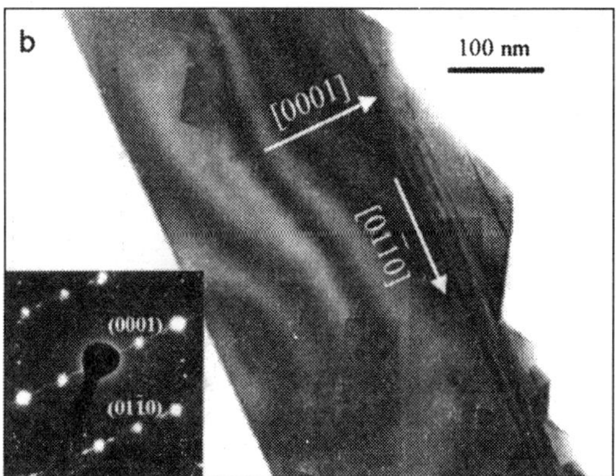

Figure 6.8 *(a) SEM image of the as-synthesized ZnS nanobelts. (b) TEM image of a ZnS nanobelt and the corresponding electron diffraction pattern (insert).*

a fairly uniform thickness, and their surfaces are rough with steps. The contact point between the nanoribbon with the axial nanowire is rather small, in the order of a few tens of nanometers, while far away from the contacting point the uanoribbon size is rather large, in the order of 100-200 nm.

In the synthesis, a mixture of SnO_2 and ZnO powder is used as the source material. SnO_2 can decompose into Sn and O_2 at high temperature; thus, the growth of the nanowire-nanoribbon junction arrays is the result of a vapor-liquid-solid growth process, in which the Sn catalyst particles are responsible for initiating and leading the growth of ZnO nanowires and nanoribbons. The growth of the novel structure presented is be separated into two stages. The first stage is the fast growth of the Zn axial nanowire along [0001]. The growth rate is so fast that a slow increase in the size of the Sn droplet has little influence on the diameter of the nanowire; thus, the axial nanowire has a fairly uniform shape along the growth direction. The second stage of growth is the nucleation and epitaxial growth of the nanoribbons owing to the arrival of tiny Sn droplets onto the ZnO nanowire surface. This stage is much slower than the first stage because the lengths of the nanoribbons are uniform and much shorter than those of the nanowire. Since Sn is in a liquid state at the growth temperature, it tends to adsorb the newly arriving Sn species and grow into a larger-sized particle (i.e., coalescing). Therefore, the width of the nanoribbon increases as the size of the Sn particle at the tip becomes larger, resulting in the formation of the tadpole-like structure observed in TEM (Figure 6.9(b)). The ZnO nanowire is likely to have a hexagonal cross section bounded by $\pm(10\bar{1}0)$, $\pm(01\bar{1}0)$ and $\pm(\bar{1}100)$, which are six crystallographic equivalent planes. The Sn liquid droplets deposited onto the ZnO nanowire lead to the simultaneous growth of the ZnO nanoribbons along the six growth directions:$\pm[10\bar{1}0]$, $\pm[0\bar{1}10]$, and $\pm[\bar{1}100]$ (Figure 6.9(c)). The angles between the two adjacent growth directions is 60°, resulting in the sixfold symmetric distribution of the nanoribbons around the nanowire.

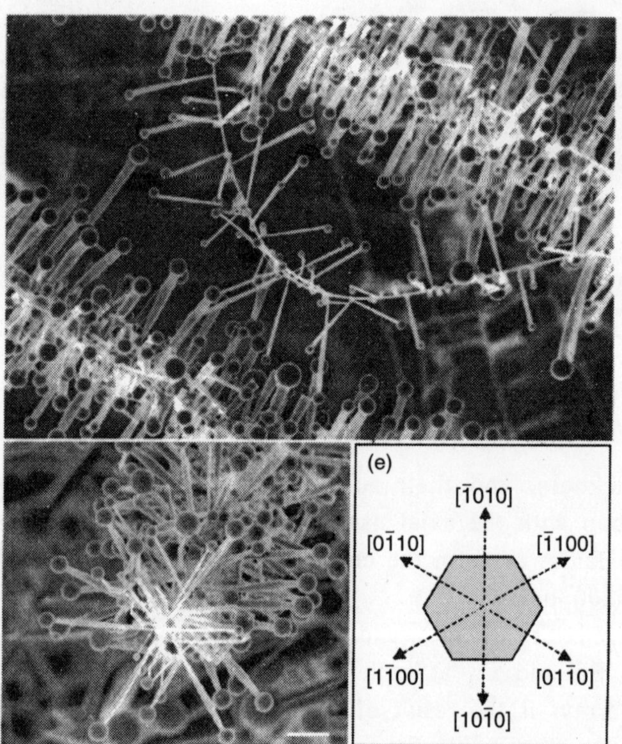

Figure 6.9 *(a,b) SEM images of the hierarchical ZnO nanowire-nanobelt junction arrays grown using Sn catalyst. (c) Growth model of the hierarchical structure.*

Hierarchical structures of $ZnO-In_2O_3$ have also been synthesized. Using the flat surface of a nanobelt, a composite layered structure of two oxides has been grown, forming oxide-oxide or metal-oxide tape structures.

METAL NANOBELT

Zn-ZnO core-shell nanobelts and nanotubes are synthesized with kinetic control. Pure ZnO powders are selected as the source material, which is loaded in an alumina boat and positioned at the center of a horizontal tube furnace. The silicon substrates for collecting the products are positioned at the low-temperature region of the furnace. The synthesis process has two steps. The first step is to control the evaporation and decomposition process, which is carried out by evacuating the alumina tube to ~ 266×10^{-3} Pa at 1620 K, keeping it there for 5 minutes, without introducing a carrier gas. The second step is to control the growth, which is also carried out at 1620 K for 30 min but with the Ar carrier gas at a pressure of 2.66×10^4 to 3.0×10^4 Pa and a flow rate of 25 sccm (standard cubic centimeters per minute). The substrate is placed downstream in the temperature region of 470–570 K.

TEM images illustrate that the Zn nanobelts have widths of ~ 50 nm (Figure 6.10(a)). Diffraction contrast across the structure suggests that it has a belt shape with a uniform

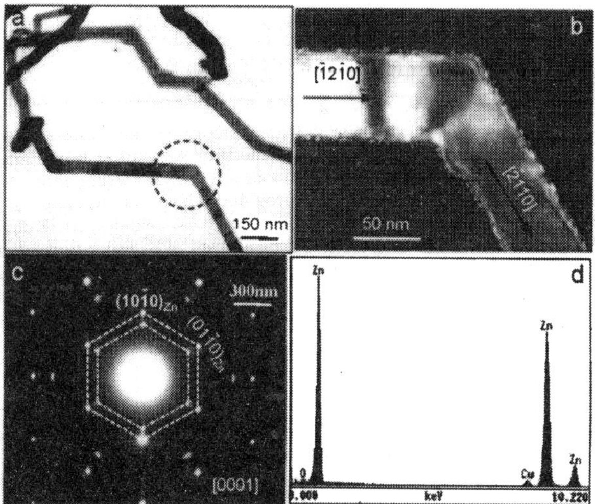

Figure 6.10 *(a) Low-magnification transmission electron microscopy (TEM) image of the as-synthesized Zn nanobelts with zigzag structure. (b) A dark-field TEM image from the circled region in (a), showing uniform contrast across its width and the surface ZnO layer. (c) The corresponding [0001] electron diffraction pattern recorded from the circled region in (a), displaying the presence of ZnO on the surface of the Zn nanobelt. (d) Energy-dispersive X-ray spectroscopy spectrum recorded from the nanobelt, showing that the dominant element is Zn. The tiny Cu peak came from the TEM grid.*

thickness (the contrast presented in Figure 6.10(a) is due to the bending of the nanobelts along its length). Zigzag structures are observed in the synthesized products. Dark-field images show that the ZnO covering on the top and bottom (0001) surfaces is of uniform thickness, while the ZnO on the side surfaces exhibits grainy structure (Figure 6.10(b)); however, the electron diffraction pattern (Figure 6.10(c)) indicates two sets of single-crystal diffraction spots, which are indexed as [0001] Zn and [0001] ZnO with an epitaxial orientation. This type of composite nanobelt grows along $\langle 2\bar{1}\bar{1}0\rangle$, its top and bottom surfaces being (0001) and side surfaces {0110}. The zigzag structure presented in Figure 6.10(a) is due to a change in the growth direction from [1210] to [2110], which are two crystallographically equivalent directions. A change in growth direction among the equivalent group of ± $[2\bar{1}\bar{1}0]$, ± $[1\bar{2}10]$, and ± $[11\bar{2}0]$ produces turning points in the growth directions. Energy dispersive X-ray spectroscopy shows that the nanobelt is dominated by Zn, with very little oxygen (Figure 6.10(d)).

In the synthesis experiment ZnO powder is used as the raw material, while the produced products are Zn-ZnO core-shell nanobelts and ZnO nanobelts distributed in distinct temperature regions. The decomposition of ZnO occurs when it is subject to high enough temperature in vacuum. The thermodyanamics of these processes can be presented as follows. The solid vapor process and the decomposition process are expressed as

$$ZnO(s) \leftrightarrow ZnO(g) \quad (1)$$

$$ZnO(g) \leftrightarrow Zn(g) + {}^{1}/{}_{2}O_2(g). \quad (2)$$

The Zn nanobelt is formed by deposition of the Zn vapor at the low-temperature region, while the surface oxide layer is formed after forming the Zn nanobelt due to surface oxidation. Through a chemical approach, Ag nanoparticles can also be transformed into nanobelts.

NANOBELTS AS TEMPLATES FOR SYNTHESIS OF NEW MATERIALS

Using the as-synthesized ZnO nanobelts as templates, nanostructured ZnS nanocables and nantubes have been synthesized by chemical reaction. Based on the geometrical shape of the nanobelt template, ZnS nanobelts are produced based on the reaction

$$ZnO + H_2S \rightarrow ZnS + H_2O. \quad (3)$$

Figure 6.11 shows a comparison of the SEM images recorded from ZnO nanobelts before and after reacting with H_2S. Besides some particle-shape reaction products, nanobelts of ZnS are formed, as indicated by arrowheads, but the nanobelts have pore sizes of ~ 30 nm. A rolled nanobelt is (see Figure 6.11(b)) produced by the surface tension introduced after the reaction. Such a shape is observed for the as-synthesized ZnO nanobelts. TEM analysis shows that the converted ZnS nanobelts have two types of structural configurations: rectangular ZnO-ZnS nanocables and ZnS nanotubes.

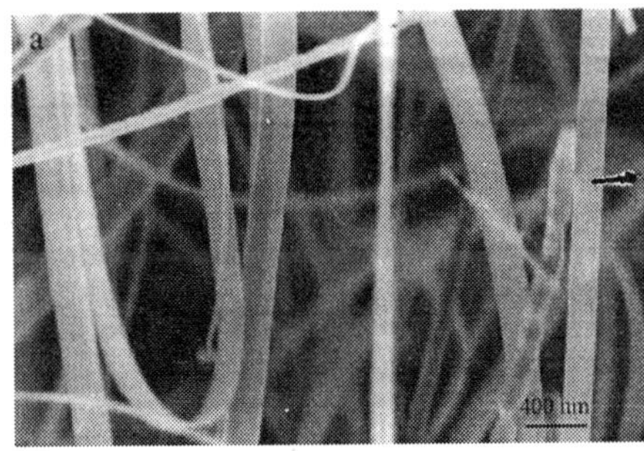

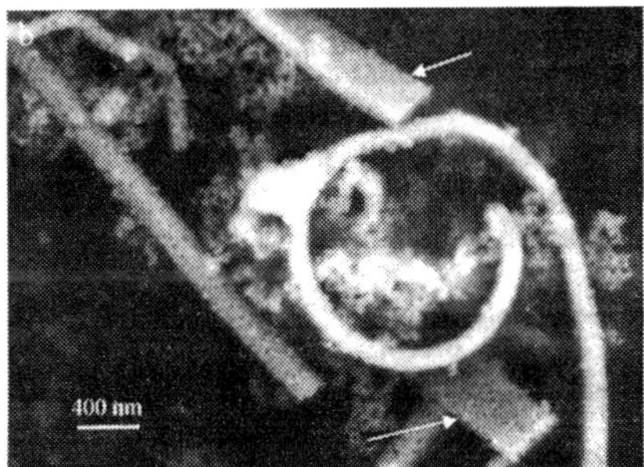

Figure 6.11 *SEM images of (a) ZnO nanobelts and (b) the ZnS nanobelts converted through chemical reaction with H_2S.*

The conversion of ZnO nanobelts into ZnS nanocrystallite-structured nanocables and nanotubes occurs in solution. Due to the limited solubility of ZnO in water, the nanocable still preserves the rectangular cross section. The pores in the structures are produced because of two factors. The excess H_2O produced in the reaction, is present in the structure and form the pores. From the structural point of view, ZnS has the zinc blende structure (cubic) with a = 0.54,109 nm, and ZnO has the Wurtzite structure (hexagonal) with a = 0.3249 nm and c = 0.52,065 nm; both are incompatible in structure. Thus, the substitution reaction is unlikely to produce single-crystalline ZnS. The formation of nanocrystallites is expected especially when the reaction temperature is at room temperature.

FIELD-EFFECT TRANSISTOR BASED ON A SINGLE NANOBELT

Field-effect transistors are fabricated using individual nanobelts. Large bundles of either SnO_2 or ZnO nanobelts are dispersed in ethanol by ultra-sonication until mostly individual nanobelts are isolated. These ethanol dispersions are dried onto a SiO_2/Si substrate for imaging by noncontact-mode atomic force microscopy. SnO_2 field-effect transistors are fabricated by depositing SnO_2 nanobelt dispersions onto SiO_2/Si (p^+) substrates, followed by treatment in an oxygen atmosphere at 1070 K for 2 h. The SiO_2/Si substrates are then spin-coated with poly(methyl methacrylate) (PMMA), baked, exposed to electron-beam lithography for the definition of electrode arrays, and developed. A 30-nm-thick layer of titanium is deposited by electron-beam evaporation to serve as the source and drain electrodes, and the remaining PMMA is lifted off in hot acetone. An atomic force microscopy image of the field-effect transistor (FET) and the diagram are given in Figure 6.12(a). The principle of this device is that controlling the gate voltage controls the current flowing from the source to the drain.

An alternative way of contacting the nanostructures is also applied to ZnO nanobelts. In this case FETs are fabricated by depositing dispersed ZnO nanobelts on predefined gold electrode arrays. In both cases the SiO_2 gate dielectric thickness is 120 nm, and the back

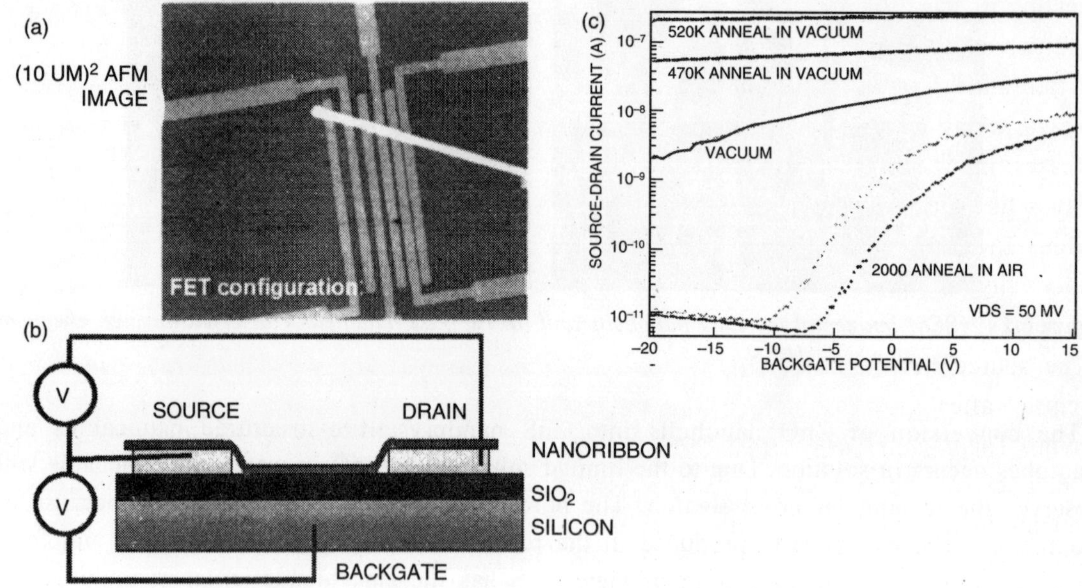

Figure 6.12 *(a) AFM image of a FET device made using a single nanobelt. (b) Working diagram of the device. (c) SnO_2 nanobelt FET oxygen sensitivity. Source-drain current versus gate bias for a SnO_2 FET after various treatments; measured in air, vacuum, 470 K vacuum anneal, 520 K vacuum anneal, 420 K air anneal.*

gate electrode is fabricated by evaporation of gold on the Si (p^+) side of the substrate. Also, in both fabrication schemes, the electrode arrays are variably spaced. They include electrode gaps as small as 100 nm and as large as 6 μm.

By forming metal electrode-nanostructure electrical contacts and capacitive coupling the nanostructure to a nearby gate electrode, an FET is produced using a nanobelt that allows the exploration of new aspects of the physical and chemical properties of the nanostructures.

The alternative way of contacting the nanobelts by simply depositing them on top of prefabricated gold electrodes led to very resistive contacts. A typical ZnO FET showed a gate threshold voltage of –15 V, a switching ratio of nearly 100, and a peak conductivity of 1.25 × 10^{-3} $(\Omega\text{-cm})^{-1}$. A completely analogous behavior is observed in the case of carbon nanotubes deposited on top of Au electrodes or covered by Ti electrodes.

ELECTRICAL PROPERTIES OF NANOBELTS

The conductivity of a nanobelt is tuned by controlling its surface and volume oxygen deficiency. Before electrical measurement, SnO_2 nanobelts are annealed in a 1-atm oxygen environment at 1070 K for 2 hours. Without this treatment, the as-produced nanobelts exhibit no measurable conductivity for source-drain biases from –10 V to 10 V but for gate biases from –20 V to 20 V, the SnO_2 nanobelts exhibit considerable conductivity. By further annealing of the devices at lower temperatures in vacuum, oxygen, or ambient, the electrical properties of the nanobelts are tuned.

After annealing of the SnO_2 devices in vacuum at 470 K the nanobelt conductivity increases along with an associated negative-shift in gate-threshold voltage. Smaller, additional increases in conductivity are observed after additional vacuum anneals. So, the nanobelt behaves like a metal, and with the gate field does not affect the current flowing through the device. In contrast, annealing nanobelt devices in ambient at 870 K decreases the conductivity, along with a shift in the gate-threshold voltage in the opposite, positive direction (Figure 6.12(b)).

The source-drain conductivity at zero gate bias spans 3 orders of magnitude from 0.09 $(\Omega\text{-cm})^{-1}$ after annealing at 870 K in ambient to 75.3 $(\Omega\text{-cm})^{-1}$ after annealing at 520 K in vacuum. The changes in conductivity with low-temperature annealing are due to variations in the number of oxygen species adsorbed on the SnO_2 surfaces or in the number of oxygen vacancies in the SnO_2 bulk with the amount of oxygen in the environment. The number of equilibrium surface and bulk oxygen defects are a function of the environmental oxygen partial pressure and temperature. Annealing in vacuum decreases the number of adsorbed oxygen species and increase the number of bulk oxygen vacancies, while annealing in oxygen or ambient should do the opposite. The bulk and surface oxygen vacancies in SnO_2 act as electron donors which increases SnO_2 conductivity and decreases the gate-threshold voltage.

The SnO_2 nanobelt conductivity increases and the gate threshold voltage decreases by taking a nanobelt device from ambient into vacuum without annealing (Figure 6.12(b)). Since the diffusion of bulk oxygen vacancies at room temperature are limited, this indicates that surface oxygen desorption takes place at this temperature. Because of their small dimensions, semiconducting oxide nanobelts have on the order of 10^{20} surface oxygen sites per cubic centimeter of material. Thus, even for partial changes in the concentration of adsorbed oxygen species, large changes in nanobelt conductivity are observed.

Annealing at 470 K and 520 K induces further changes in conductivity (Figure 6.12(b)). Therefore the surface or bulk nanobelt composition must change as a result of these annealing steps. For temperatures below 1170 K the equilibrium bulk nonstoichiometry of SnO_2 is insignificant in comparison to surface nonstoichiometry in samples of similar surface-to- volume ratios. So, the surface oxygen vacancies should control conductivity after 470 K and 520 K anneals. Annealing also desorbs species other than oxygen from the nanobelt surfaces. Water acts as a mask on the nanobelt surface, decreasing sensitivity to environmental oxygen until removed. The strong dependence of the conductance on the oxygen deficiency in nanobelts is an important characteristic of functional oxide, which is capable of tuning and controlling the electrical properties of the nanodevice.

PHOTOCONDUCTIVITY OF NANOBELTS

UV–light irradiation of the nanobelt diode of SnO_2 in air results in a significant increase of the conductivity (Figure 6.13). Light with a wavelength of 350 nm (E_λ = 3.54 eV) is used, exceeding the direct band-gap of SnO_2. The increase in the conductivity results from photogeneration of electron-hole pairs as well as doping by UV light-induced surface desorption. These processes are observed by introducing a shutter between the light source and the SnO_2 nanobelt so that the flux of UV photons could be turned on and off.

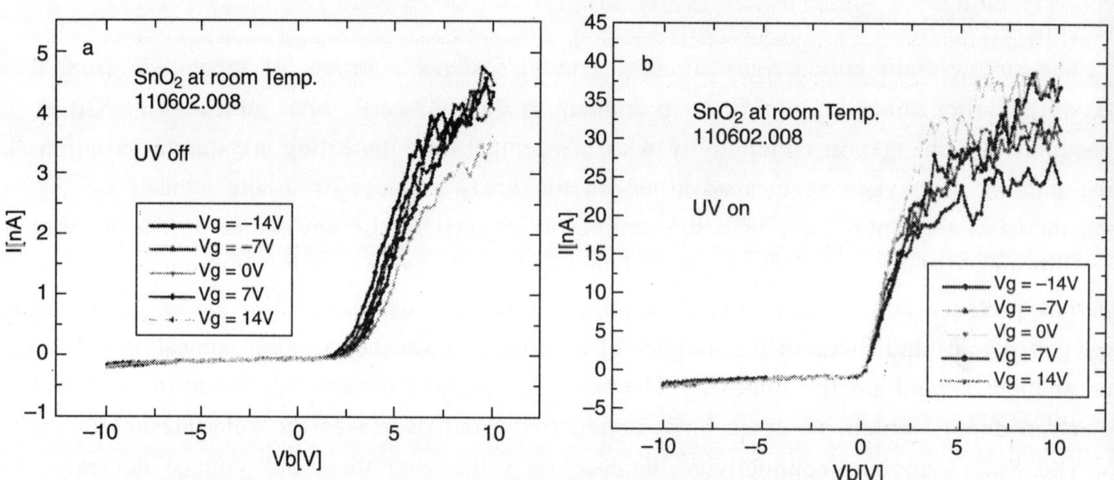

Figure 6.13 *Solid state diode of a single nanobelt. The current versus voltage (I–V) characteristic curves of a diode made using a single SnO_2 nanobelt, showing its response to gate voltage (a) without and (b) with UV radiation.*

The unique geometrical shape of nanobelts is ideal for field emission. MoO_3 Nanobelts and exhibit superior performance. The work function at the tips of individual ZnO nanobelts are measure by an another technique.

GAS SENSORS BASED ON NANOBELTS

Conductometric metal oxide semiconductor thin films are the most promising devices among solid state chemical sensors, due to their small dimension, low cost, low power consumption, on-line operation, and high compatibility with microelectronic processing. The fundamental sensing mechanism of metal oxide-based gas sensors relies on a change in electrical conductivity due to the interaction process between the surface complexes, such as O^-, O_2^-, H^+, and OH^-, reactive chemical species and the gas molecules to be detected.

Although many oxides have been investigated for their gas-sensing properties, mostly gas sensors are mainly made of SnO_2 in the form of thick films, porous pellets, or thin films.

The effects of the microstructure, namely, ratio of surface area to volume, grain size, and pore size of the metal oxide particles, as well as film thickness of the sensor are well recognized. Lack of long-term stability has prevented a wide range application of this type of sensor. Nanobelts of semiconducting oxide, with a rectangular cross section in a ribbon-like morphology, are very promising for sensors because their surface-to-volume ratio is very high, the oxide is single crystalline, the faces exposed to the gaseous environment are always the same, and the size produces a complete depletion of carriers inside the belt. The deposition technique is very simple and cheap and the size and shape is easily controlled. In polycrystalline and thick-film devices, only a small fraction of the species adsorbed near the grain boundaries is active in modifying the device electrical transport properties.

In the new sensors based on single-crystalline nanobelts, almost all of the adsorbed species are active in producing a surface depletion layer. Free carriers should cross the belt along the axis, in a FET channel–like way. In addition, the size of the depletion layer for tin oxide, (due to oxygen desorption), penetrates 50 nm or more through the bulk, the belts are almost as depleted of carriers as a pinched-off FET because the belt thickness is typically less than 50 nm. The presence of poisoning species should switch the structures from pinched-off to conductive channel, strongly modifying the electrical properties. A further reduction of belt size could bring the development of quantum-confined structures and nanodevices.

For the fabrication of sensors, a platinum interdigitative electrode structure is made using a lithography and metal-deposition technique on an alumina substrate. A platinum heater is attached to the back of the substrate to control the working temperature of the sensor. Then, several nanobelts are placed onto the electrodes for measuring their electric conductance, after ensuring the contact of the nanobelts with the electrodes. The flow-through technique is used to determine the gas-sensing properties of the thin films. A constant flux of synthetic air equal to 0.3 liters min^{-1} is mixed with the desired amount of gaseous species; flows through a stabilized sealed chamber at 290 K, atmospheric pressure and controlled humidity. Electrical characterization is carried out by a volt amperometric technique at constant bias of 1 V, and a picoammeter which measures the change of electrical current.

Figure 6.14 shows the isothermal response of the current flowing through the tin oxide nanobelts as two square concentration pulses of CO (250 and 500ppm, 30% respectively) being fed into the test chamber at a working temperature of 670 K 30% RH (relative humidity at 290 K). The electric current increases for about 60% and 100% with the introduction of 200 and 500 ppm CO, respectively. The sensor response, defined as the relative variation in conductance due to the introduction of gas, is about $\Delta G/G = 0.9$.

Figure 6.14(a) also shows the isothermal response of the current flowing through the nanobelts as a square concentration pulse of 0.5 ppm nitrogen dioxide is fed into the test chamber at a working temperature of 670 K and 30% RH. The sensor response is $\Delta G/G = -15.5 = -1550\%$, which is extremely high and sensitive. This means that the sensitivity of the sensor is on the level of a few parts per billion.

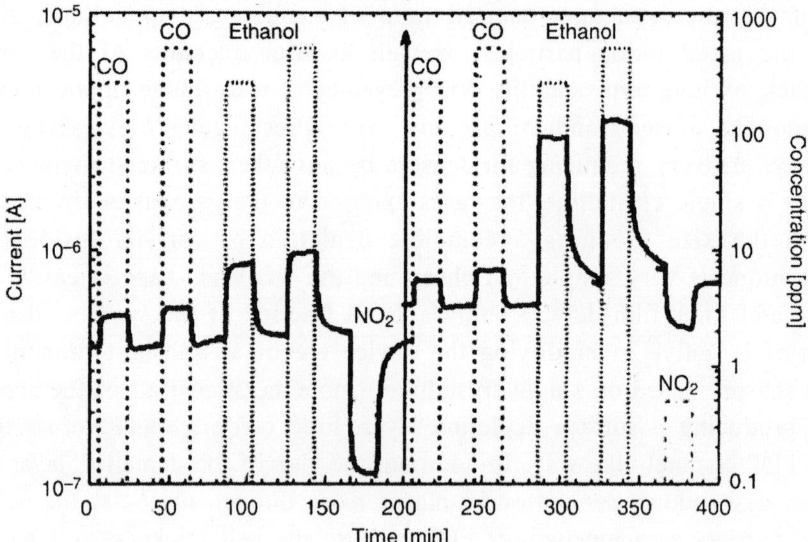

Figure 6.14 *Gas sensors made using nanobelts: Response of the conductance through SnO_2 nanobelts to the concentration of the surface-adsorbed CO, ethanol, and NO_2 gases at two different temperatures.*

In general, the selectivity of the oxide is important. It is improved by fabricating sensors using several types of nanobelts, or by functionalizing the surfaces of the nanobelts. It is, important to note that CO and ethanol increase the conductivity, while NO_2 decreases the conductivity of the SnO_2 nanobelts.

Note: The small size offered by the sensors based on individual nanobelts is a major advantage for application in biotechnology, because it gives the potential of implanting in biological systems. Nanosensors based on nanobelts may be unique in detecting single cancer cells and measuring pressure in biofluid.

HEAT TRANSPORT THROUGH A NANOBELT

Heat transport at nanoscale is very interesting and technologically important. With the reduction of object size, phonon modes and phonon density of states change drastically, resulting in unusual thermal transport phenomena in mesoscopic systems. Theoretical investigation of thermal conductance in a one-dimensional nanowire predict quantum effects at very low temperature: $G_{th} = \pi^2 K^2{}_B T/3h$. Experi-mental measurement has proved this prediction. Thermal transport along a single SnO_2 nanobelt is carried out (Figure 6.15(a)). Thermal contact micropads are fabricated using a lithography technique. The thermal conductance across the nanobelt measured as a function of the local temperature is given in Figure 6.15(b). The thermal conductivity of the nanobelts is significantly suppressed in comparison to bulk, owing to increased phonon-boundary scattering and modified phonon

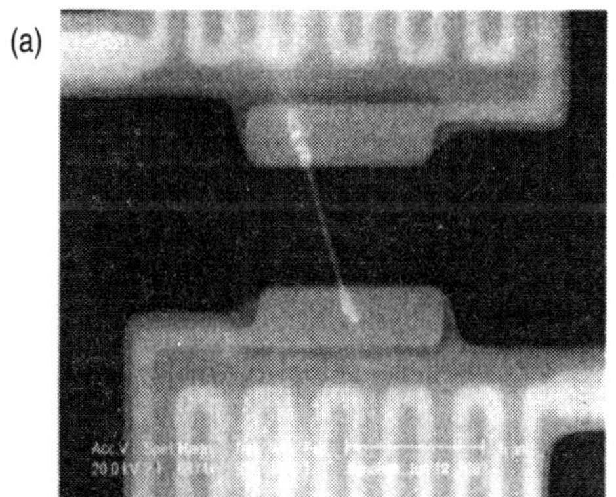

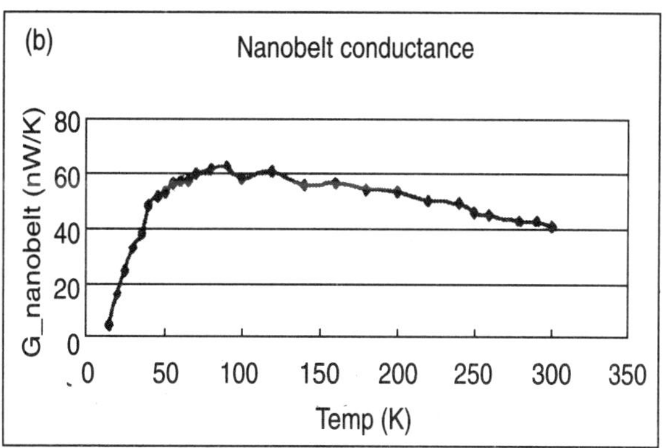

Figure 6.15 *Measurement of thermal transport through a single SnO_2 nanobelt.*

dispersion. This size effect can lead to localized heating in nanoelectronics but may find potential use for improving thermoelectric performance.

NANOBELTS AS NANORESONATORS

Another key quantity in the application of nanobelts is their bending modulus. This quantity is measured using a technique developed for carbon nanotubes. Based on the electric field–included resonant excitation, the mechanical properties of individual nanowires are measured by TEM. Using this method, mechanical properties of carbon nanotubes, silicon nanowires, and silicon carbide-silica composite nanowires are known.

To carry out the mechanical property measurements of a nanobelt, a specimen holder for TEM (200 kV) is built for applying a voltage across a nanobelt and its counter electrode. Mechanical resonance is induced if the applied frequency matches the natural resonance frequency of the nanobelt. Due to the mirror symmetry of the nanobelt, there are two distinct fundamental resonance frequencies corresponding to the vibration in the thickness and width directions. These are known from the classical elasticity theory as

$$v_x = \frac{\beta_1^2 T}{4\pi L^2}\sqrt{\frac{E_x}{3\rho}}, \tag{4}$$

$$v_y = \frac{\beta_1^2 W}{4\pi L^2}\sqrt{\frac{E_y}{3\rho}}, \tag{5}$$

where $\beta_1 = 1.875$ for the first harmonic resonance; E_x and E_y are the bending modulus if the vibration is along the x-axis (thickness direction) and y-direction (width direction), respectively; ρ is the density, L is the length, W is the width, and T is the thickness of the nanobelt. These two modes can be observed separately in experiments.

Changing the frequency of the applied voltage, it is found that there are two fundamental frequencies in two orthogonal directions transverse to the nanobelt. Figure 6.16(a) and (b) shows the harmonic resonance, with the vibration planes nearly perpendicular and parallel to the viewing direction, respectively. For calculating the bending modulus, it is critical to accurately measure the fundamental resonance frequency (v_1) and the dimensional sizes (L and T or W) of the investigated ZnO nanobelts. To determine v_1 the stability of resonance frequency is checked to ensure that one end of the nanobelt is tightly fixed and the resonant excitation is checked around the half value of the resonance frequency. The specimen holder is rotated about its axis so that the nanobelt is aligned perpendicular to the electron beam and so the real length (L) of the nanobelt can be measured. The projection direction along the beam is determined by the electron diffraction pattern, so that the true thickness and width is determined because the normal direction of the nanobelt is $[2\bar{1}\bar{1}0]$. Based on the experimental data, the bending moduli of ZnO nanobelts is calculated using Equation 1 or 2. The experimental results are summarized in Table 6.2. The bending modulus of the ZnO nanobelts is ~ 52 GPa. The experiments show that nanobelts are effective nanoresonators exhibiting two orthogonal resonance modes, which are used as probes for scanning probe microscopy (SPM) operated in tapping and scanning modes.

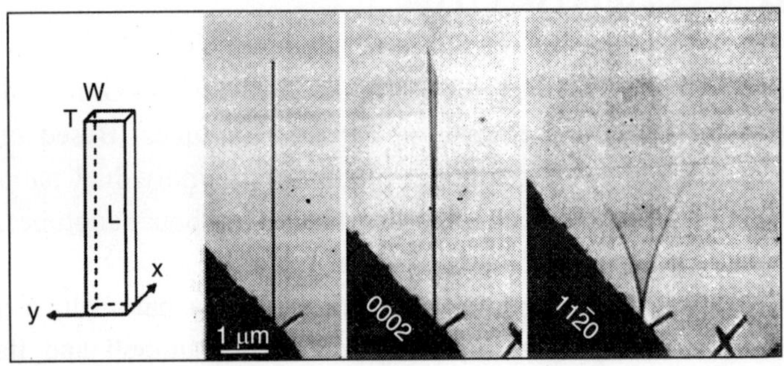

Figure 6.16 *Nanoresonators made of nanobelts. Measuring the bending modulus of a ZnO nanobelt by electric field-induced mechanical resonance in TEM. (a) Geometrical shape ora nanobelt. (b,c) Mechanical resonance ora nanobelt along the two orthogonal directions closely perpendicular to the viewing direction (v_x = 622 KHz) and nearly parallel to the viewing direction (v_y = 691 KHz), respectively.*

Table 6.2 *Bending modulus of the ZnO nanobelts (E_x and E_y represent the bending modulus corresponding to the resonance along the thickness and width directions, respectively)*

	Length *L (µm)*	Width *W (nm)*	Thickness *T (nm)*	Fundamental Frequency *(kHz)*				Bending Modulus *(GPa)*	
Nanobelt	(± 0.05)	(± 1)	(± 1)	*W/T*	v_{x1}	v_{g1}	v_{g1}/v_{x1}	E_x	E_y
1	8.25	55.	33	1.7	232	373	1.6	46.6 ± 0.6	50.1 ± 0.6
2	4.73	28	19	1.5	396	576	1.4	44.3 ± 1.3	45.5 ± 2.9
3	4.07	31	20	1.6	662	958	1.4	56.3 ± 0.9	64.6 ± 2.3
4	8.90	44	39	1.1	210	231	1.1	37.9 ± 0.6	39.9 ± 1.2

Using a technique developed for measuring the work function at the tip of a carbon nanotube, the work function at the tip of an oxide nanobelt is also measured.

NANOBELTS AS NANOCANTILEVERS

Cantilever-based SPM techniques are one of the most powerful approaches in imaging, manipulating, and measuring nanoscale properties and phenomena. The most conventional cantilever used for SPM is based on silicon, Si_3N_4 or SiC, which is fabricated by an e-beam or optical lithography technique and has typical dimensions of thickness ~ 100 nm, width ~5 μm, and length ~50 μm. Utilization of nanowire- and nanotube-based cantilevers have several advantages for SPM. Carbon nanotubes can be grown on the tip of a conventional cantilever and used for imaging surfaces with a large degree of abrupt variation in surface morphology. The manipulation of nanobelts is achieved by atomic force microscopy and can be used as nanocantilevers.

Combining microelectromechanical system technology with self-assembled nanobelts, cantilever are produce with much improved sensitivity for a range of devices and applications. Semiconducting nanobelts are used in cantilever applications. Structurally they are defect-free single crystals, providing excellent mechanical properties. The reduced dimensions of nanobelt cantilevers offer a significant increase in cantilever sensitivity Combining these techniques with micromanipulation the horizontal alignment of individual ZnO nanobelts onto silicon chips is made. The aligned ZnO cantilevers shown in Figure 6.17 have a range of lengths. So, the resonance frequency of each cantilever can be tuned and thus modify cantilevers for different applications such as contact, noncontact, and tapping mode atomic force microscopy. Periodic contrast of the ZnO cantilevers is observed as a result of electronic charge-induced vibrations during SEM operation. Such contrast is absent in regions where the nanobelts are in direct contact with the silicon substrate, suggesting adequate adhesion forces between the cantilevers and the silicon chip.

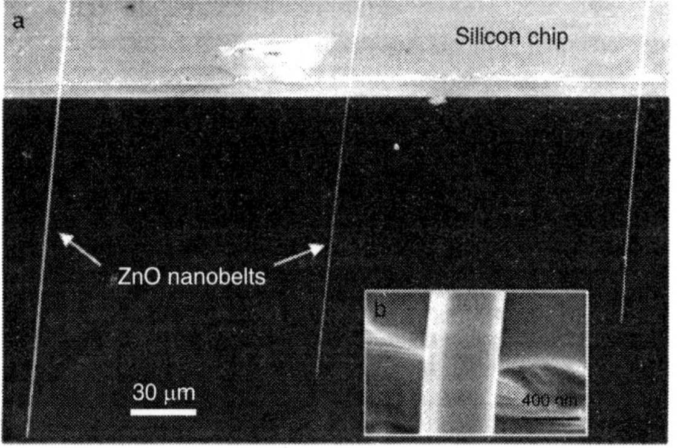

Figure 6.17 *(a) Nanobelts as ultrasmall nano-cantilever arrays aligned on a silicon chip. (b) An enlarged SEM image recorded from the nanobelt cantilever.*

PIEZOELECTRIC NANOBELTS WITH POLAR SURFACE

The noncentral symmetry and the tetrahedral coordinated $(ZnO4)^{6-}$ unit in ZnO results in anisotropic piezoelectric properties. Structurally, the Wurtzite-structured ZnO crystal is described as a number of alternating planes composed of fourfold coordinated O^{2-} and Zn^{2+} ions, stacked alternately along the x-axis.

The oppositely charged ions produce positively charged (0001)-Zn and negatively charged $(000\bar{1})$-O polar surfaces, resulting in a normal dipole moment and spontaneous polarization as well as a divergence in surface energy. To maintain a stable structure, the polar surfaces generally have facets or exhibit massive surface reconstructions, but ZnO ± (0001) is an exception, which is atomically flat, stable, and without reconstruction.

ZnO has a rich family of structures, which are mainly defined by the (0001), $\{01\bar{1}0\}$ and $\{2\bar{1}\bar{1}0\}$ facets as well as the anisotropic growth along [0001], $\langle 01\bar{1}0\rangle$, and $\langle 2\bar{1}\bar{1}0\rangle$. The most desirable morphology to maximize the piezoelectric effect is to create nanostructures that preserve large (0001) polar surfaces (110, 111). However, ZnO (0001) has a surface energy that diverges with sample size due to the surface polarization charge. Therefore, growth of (0001) surface-dominated freestanding nanostructures needs to overcome the barrier of surface energy.

ZnO nanobelts that are dominated by the (0001) polar surface have been synthersized. Hundreds of nanobelts laid down naturally onto a carbon film by electron diffraction are examined. More than 90% of them show the same orientation of [0001] with respect to the

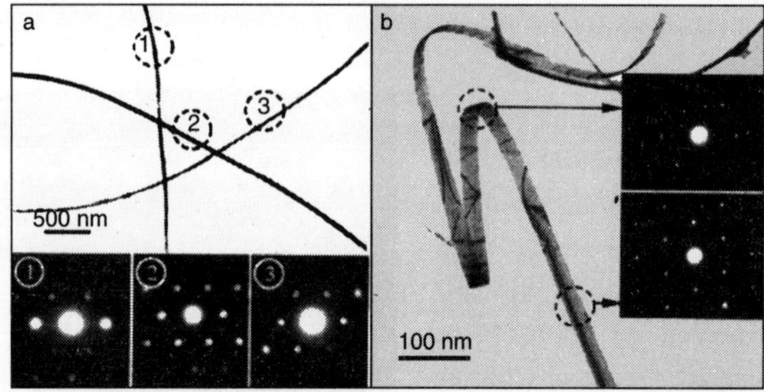

Figure 6.18 *Controlled growth of (0001) polar surface–dominated ZnO nanobelts. (a) Low-magnification TEM images and the corresponding electron diffraction patterns recorded from the areas as indicated by a sequence number from a TEM grid without tilting, showing their unanimous [0001] orientation on a flat carbon substrate. (b) Low-magnification TEM image and the corresponding electron diffraction patterns recorded from the circled regions, displaying the geometry of the nanobelt. The difference between the two electron diffraction patterns is due to the bending in the local regions. The contrast observed in the image is the bending contour in electron imaging produced by the deformation of atomic planes.*

incident electron beam (Figure 6.18(a)), indicating that the top flat surfaces of the nanobelts are the polar ± (0001) facets. This demonstrates the success of overcoming the surface energy barrier by growth kinetics in achieving structural control. The ZnO nanobelt has a Wurtzite structure with lattice constants of a = 0.325 nm and c = 0.521 nm. Indexing of the electron diffraction pattern shows that the nanobelt grows along $[2\bar{1}\bar{1}0]$ (the a-axis), with its top and bottom surfaces ± (0001) and the side surfaces ± $(01\bar{1}0)$. High-resolution TEM shows that the nanobelt is single crystalline without the presence of dislocations. Due to the small thickness of 5–20 nm and large aspect ratio of ~ 1:4, the flexibility and toughness of the nanobelts are extremely high, so they can be bent or twisted without fracture (Figure 6.18(b)).

Nanobelts with large polar surfaces are ideal systems for understanding piezoelectricity and polarization-induced ferroelectricity at nano-scale. The different polar surfaces can be used as selective catalysts. The piezoelectric and ferroelectric nanobelt structures could be nano-scale sensors, transducers, and resonators, which are functional components for a microelectromechanical system and nanoelectromechanical system.

ALIGNED GROWTH OF ZnO NANORODS

Tin is an excellent catalyst for growth of ZnO nanostructures. Using tin as a catalyst, we have grown aligned ZnO nanowires on a polycrystalline alumina substrate. The SEM image shown in Figure 6.19(a) clearly displays the reasonable alignment among the nanowires. Higher magnification SEM images show that the nanowire has a nonuniform cross section along its length and it becomes sharper toward the tip. The very tip has a Sn-rich head (Figures 6.19(b) and (c)). Gold is used for growth of aligned ZnO nanorods.

Figure 6.19 *(a,b,c) Low-magnification and high-magnification SEM images of ZnO nanorods grown on alumina substrate with Sn at the growth front. (d) Orientation-ordered epitaxial growth of ZnO nanorods on a large ZnO crystal, showing identical orientation.*

By using Sn particles reduced from SnO_2, well-aligned ZnO nanorods with identical crystallographic-orientation have been synthesized using a vapor-transport deposition process. Orientation-ordered nanorods grow normal to the c-planes of the as-deposited micron-sized ZnO rods on a polycrystalline Al_2O_3 substrate, and each nanorod is along [0001] and enclosed by $\{2\bar{1}\bar{1}0\}$ facet surfaces. The nanorods remain in identical crystal orientation with a homoepitaxial orientation relationship with the microrod. By controlling the growth time at high temperature, uniform lengths of aligned nanorods are achieved.

NANORINGS AND NANOHELIXES OF PIEZOELECTRIC NANOBELTS

A striking new phenomenon for the piezoelectric nanobelt presented in Figure 6.18 is the formation of helical structures by rolling up single-crystalline nanobelts (Figure 6.20(a) to (c)). A single crystal nanobelt form helical nanosprings and nanorings. The helical nanostructure has a uniform shape, with radii of ~500–800 nm and evenly distributed pitches. Each is made of a single-crystal ZnO nanobelt that is dominated by the (0001) polar surface. Electron diffraction indicates that the direction of the radius toward the center of the helical ring is always [0001] along the entire perimeter without significant twisting , and the circular plane of the nanoring is $[01\bar{1}0]$. A TEM image of a helical nanospring is shown in Figure 6.20(d). It shows the projected shape of the single-crystal twisted nanobelts. A model of the structure of the nanobelt is shown in Figure 6.20(e). These types of helical nanosprings and nanorings are observed only for the (0001) plane-dominated ZnO.

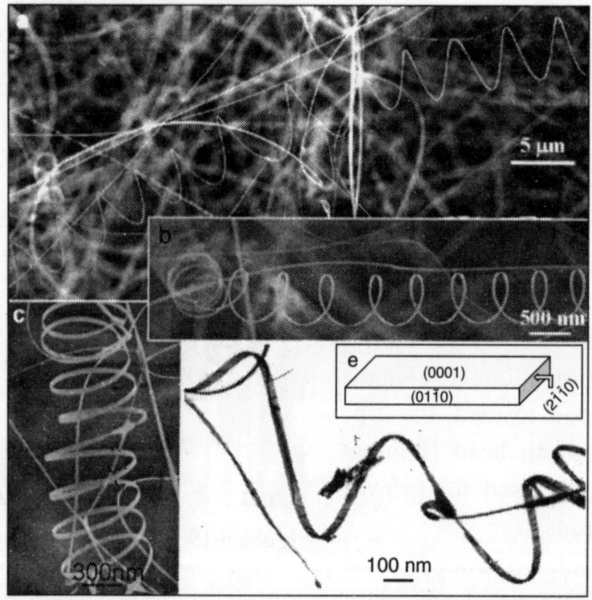

Figure 6.20 *Nanosprings and nanorings of piezoelectric nanobelts. (a-c) SEM images of the as-synthesized single-crystal ZnO nanobelts, showing helical nanosprings. The typical width of the nanobelt is ~30 nm, and pitch distance is rather uniform. (d) TEM image of a helical nanospring made of a single-crystal ZnO nanobelt. (e) The structure model of the ZnO nanobelt.*

The formation of the nanorings and nanohelicals are understood from the nature of the polar surfaces. The $(000\bar{1})$ plane is terminated with Zn [(0001)-Zn] or oxygen [$(000\bar{1})$-O], resulting in positively and negatively charged top and bottom surfaces, respectively. Physically, if the surface charges are uncompensated during growth, the net dipole moment tends to diverge and the electrostatic energy increases. For a thin nanobelt lying on a substrate, the spontaneous polarization induces electrostatic energy (due to the dipole moment), but rolling up to form a circular ring would minimize or neutralize the overall dipole moment thus reducing the electrostatic energy. However, bending of the nanobelt produces elastic energy. The stable shape of the nanobelt is determined by the minimization of the total energy contributed by spontaneous polarization and elasticity.

The growth of polar facet–dominated nanobelt surfaces shows the development of piezoelectric one-dimensional nanostructures. Since the ZnO $(2\bar{1}\bar{1}0)$ plane has surface energy lower than that of either (0001) or $(01\bar{1}0)$, a fast growth along $[2\bar{1}\bar{1}0]$ (a-axis) to form a nanobelt structure, it may be unfavorable from the energy point of view. However, the success of controlled growth of free-standing (0001) polar surface-dominated nanobelts along the a-axis demonstrates the feasibility of overcoming the energy barrier through growth kinetics, thus providing an approach for growing structurally controlled nanobelts.

Helical structure is the most fundamental structural configuration for DNA and many biological proteins and is due to van der Waals force and hydrogen bonding. For one-dimensional nanostructures, nanocoils and nanoring are observed for carbon nanotubes and SiC. The former are created by a periodic arrangement of the paired pentagon and heptagon carbon rings in the hexagonal carbon network, and the latter by a stacking of platelets around the growth axis. These helical structures are produced by the presence of point or planar defects (twins and stacking faults), which is a completely different mechanism from the process that we have proposed for the formation of single crystal helical nanosprings and nanorings. The striking new feature of the helical nanostructures for single-crystalline ZnO nanobelts is that they are spontaneous polarization-induced structures. The nanobelts and helical nanostructures are an ideal systems for piezoelectricity and polarization-induced ferroelectricity at nanoscale. The different polar surfaces could be used as selective catalysts. The piezoelectric and ferroelectric nanobelt structures could be nanoscale sensors, transducers, and resonators.

POLAR SURFACE-INDUCED ASYMMETRIC GROWTH

Comb structures of ZnO are achieved, but the mechanism that drives the growth is not elaborated. The comb structures (Figure 6.21) with the comb-teeth grow along [0001], the top and bottom surfaces being ± $(01\bar{1}0)$ and the side surfaces ± $(2\bar{1}\bar{1}0)$. As shown above, the ZnO (0001) surface is terminated with Zn or O. Using advanced TEM and a convergent beam electron-diffraction technique, it is found that the comb structure is due to the surface polarity. The positively charged ZnO (0001)-Zn surface is chemically active, and the negatively charged $(000\bar{1})$-Osurface is inert in the growth of nanocantilever arrays. Longer and wider nanofinger arrays are grown from the (0001)-Zn surface, which is a self-catalyzed process due to the enrichment of Zn at the growth front. The chemically inactive $(000\bar{1})$ surface typically does not grow nanobelt structures. The self-catalyzed process is likely a mechanism for the growth of nanobelts without the presence of foreign metallic catalysts. The nanocantilever arrays have potential application as nanoscale sensor arrays and tweezer arrays.

Figure 6.21 *SEM image of comb-like nanostructute of ZnO, which is the result of surface polarization–induced growth.*

Properties of Nanowires and Nanobelts

MECHANICAL AND THERMAL STABILITY

The small sizes and high surface-to-volume ratio of one-dimensional nanostructure endow them with a variety of interesting and useful mechanical properties. Their high stiffness and strength makes them applicable in tough composites and as nanoscale actuators, force sensors and calorimeters. One dimensional nanostructures show unique stability effects driven by the dominance of their surface and internal interfaces.

Understanding the mechanical properties of nanostructures is essential for the atomic-scale manipulation. Modifications of these materials behave qualitatively different when the dimensions are reduced from micro to nanoscale; for example the hardness and yield stress of a polycrystalline material increases with decreasing grain size on the micrometer scale. The change in grain size is due to effective disruption of dislocation motion by grain boundaries, a phenomenon known as the Hall-Petch effect. However on solids composed of nanoscale grain the Hall-Petch relation breaks down at a critical grain size, below which a material softens i.e. on the nanometers scale, an opposite behavior is observed. It can be explained in term of the piling up of dislocations at the boundaries when the crystal planes in individual grains are sheared. As the grains are downsized the area of their boundaries increases and thereby makes the material tougher by blocking dislocations more effectively; for example, the samples of nanocrystalline copper and palladium becomes softer by decreasing the grain size. This abnormal behavior is due to sliding motions at grain boundaries. As a result, the strength of a polycrystalline material first increases and then decreases with decreasing grain size, and there exists a characteristic length for each solid material to achieve the toughest strength. For copper and palladium nanomaterials, this characteristic length is around 19.3 and 11.2nm respectively.

The single-crystalline 1D nanostructures, are significantly stronger than their counterparts that have larger dimensions. This property is attributed to a reduction in the number of defects per unit length (Defects often lead to mechanical failure.)

As the scale of materials reduces to nanometers their surfaces minimize and so does their free energy which produces structural changes that propagate into the bulk. Surface-included global reconstruction is observed and modeled in free standing nanofilms of various metals particularly gold. They show that at a critical thickness, face centered cubic (f.c.c) metal films with {100} orientations restructure to a low energy {111} orientation to relieve the large tensile stress present in the {100} surfaces of these materials. This phenomenon is explored in gold nanowires using atomistic simulations. An f.c c. to body-centered tetragonal phase transition is observed in nanowire with a <100> initial crystal orientation and cross-sectional area below $4nm^2$. The transition is nucleated at the ends of the nanowire and propagates inward at a tenth the speed of sound in gold. No such effect is found in wires with <110> or <111> growth directions as these orientations feature surfaces that lack sufficient stress to overcome bulk stability.

Nanowire synthesis techniques yield single-crystalline structures with a much lower density of line defects than is typically found in bulk materials. As a result, one dimensional nanostructures often feature a mechanical strength, stiffness, and toughness approaching the theoretical limits of perfect crystals making them attractive for use in composites and as actuators in nanoelectomechanical systems (NEMS).

Characterzing the mechanical properties of nanowires/nanotubes/nanorods (NWs/NTs/NRS) is of great importance for their applications in electronics optoelectronics sensors and actuators. There are several techniques that have been developed for measuring the elastic properties of individual NTs. (see chapter 8). One of the technique is based on quantifying the deflection of a carbon NT that is affixed at one end and the other end is free to be deflected by an atomic force microscope tip. The NT is laid in parallel to a solid substrate, and the elastic modulus of a carbon NT is calculated from the force-deflection curve. Yet another technique uses two atomic force microscopes which stretch a carbon NT that is glued at both ends to the two tips respectively: the stretching force-displacement curve gives tensile strength and elastic modulus. In the third technique an atomic force microscope tip is used to bend a NT or a bundle of single-walled NTs lying across a hole in a solid substate. Quantifying the thermal vibration amplitude of a NT in TEM also yields its elastic modulus. AFM is also applied for the transversal elasticity of NTs. For all of these techniques, the NTs are removed from the substrate used in the growth and are manipulated for the measurements.

The bending force is measured as a function of displacement along the unpinned lengths. Continued bending of these nanorods lead to fracture. Based on the measurement taken, a Young's modulus of 610-660 GPa is estimated for these nanorods, which is in good agreement with the value of ~600 GPa predicted for [111] oriented SiC. The large Young's modulus associated with SiC nanorods implies that materials can be used as reinforcing elements in generating strong composites (with ceramics, metals, or polymers serving as the matrix) Thus atomic force microscopy is used to determine the mechanical properties (elasticity strength and toughness) of individual structurally isolated SiC nanorods that are pinned at one end to the surface of a solid substrate.

Another useful method is based on resonance vibration for measuring the mechanical properties of an individual 1D nanostructure of a NT/NW by insitu transmission electron microscope (TEM). One of the ends of a CNT is fixed to the surface of a TEM sample holder, and a nanoparticle is attached to the free end of this nanotube. The resonance is stimulated by applying ac voltage across two electrodes, one of which is a carbon NT that is glued to a metal tip affixed on a specimen holder The resonance frequency together with the geometrical parameters of the NT provided by TEM yields the elastic modulus. By measuring the resonance frequency the Young's modulus of this nanotube can be calculated. In principal this method can be further extended to measure the mechanical parameters of inorganic nanowires.

This method is also used for determining the elastic properties of one-dimensional nanostructures based on electric-field-induced resonant excitant of single SiC/SiO_2 wires and ZnO belts in a TEM. By applying an alternating electric field tuned to the natural vibration frequency of a ZnO belt pinned at one end to a TEM grid, it is found that the quasi rectangular belts exhibit dual fundamental frequencies and an average bending modulus of 52 GPa, close to the theoretical value.

AFM is also used to measure the change in length of Au nanowires during elongation compression cycles. The nanowires are formed between two gold structures (one of them is coated as a thin film on the AFM tip) by pushing them into physical contact . It is found that the Au nanowires are elongated under stretching in quantized steps up to three integer multiples of 0.176 nm and that they are spontaneously shortened in steps of 0.152nm when relaxed. This is due to the sliding of crystal planes within the Au nanowires created by stacking of crystal faults and thus change in the local structure from ccp to hcp. This approach is extended to examine the atomic events occurring during the plastic failure of various metals and their alloys. An STM supplemented with a force sensor is adopted to investigate the mechanical properties of a freely suspended chain of single Au atoms.

It is found that the bond strength of the nanowire is approximately twice that of a bulk metallic bond. Computation of this system indicates that the total effective stiffness of the nanowire has a strong dependence on the local arrangement of atoms at the chain bases (that is the contacts between the nanowires and the pads)

The Young's modulus of ZnO nanowires is found to decrease dramatically with increasing diameter, reaching the ZnO bulk value for diameters larger than 120nm. This behavior is due to a surface stiffening effect dominating at large surface-to-volume ratios. The Young's modulus of ZnO nanobelts with a wide range of lateral dimensions ranging from 20 to 230nm in thickness and 30 to 700 nm width is also measured. A wide spread of modulus values between 30 and 160 GPa, without a clear dependence on their thickness t width w,or surface-to-volume ratio is f. Molecular dynamics simulations show a noticeable size dependence of the elastic properties only for ZnO nanobelts with lateral dimensions below 4nm, for which the effects of surfaces stresses become significant.

In addition mechanical actuation based of the unique features of one-dimensional structures, for instance, flexible SiO_2 helices heated by electron beam show exapansion, contraction behavior similar to that of spring is also observed and entangled sheets of V_2O_5 nanowires are used as electromechanical actuators in liquid media. The large surface area and Young's modulus of the freestanding sheets are key to their ability to generate substantial forces (5.9 MPa) in response to reversible cation intercalation and double layer charging. The thermomechanical bending of bilayer ribbons is explored via the biametallic effect.

It is shown that epitaxial Cu-SnO_2 bilayers act as reversible thermal switches at temperatures <473K with tip displacements of several hundred nanometers as monitored by TEM. The use of detection schemes, opens application for these and related one dimensional structures as NEMS components and ultrasensitive force transducers.

Because of their very small size and weight nanomechanical resonators are capable of heat detection at the quantum limit and mass sensing at the level of individual molecules. The resonance frequency of a cantilever beam f_o scales linearly with the deometric factor t/L^2 (where t is thickness and L the length of the cantilever beam) whereas its mass sensitivity is roughly proportional to f_o^2 so that short light structures provide the highest sensitivity. For fabricating and detecting the motion of ultrahigh-frequency (1 GHz) resonators; lithographically produced nanowires are made. 5.5 fg mass detection is achieved using larger Si cantilevers in ambient conditions (100) which heralds the wide use of nanowires in force microscopy, high-frequency circuitry, and calorimetry in the quantum regime.

Thermal Stability

The thermal stability of 1D nanostructures is of critical important for their implementation as building blocks in nanoscale electronic and photonic devices. The melting point of a soild material is greatly reduced when it is processed as nanostructures. Also, the melting temperature of a crystal is inversely proportional to its effective radius for grains smaller than 20 to 40nm. This effect is due to a large fraction of atoms with low coordination numbers present in solids with high surface-to-volume ratios. The size dependent thermal melting, in thin one-dimensional nanostructures cannot be verified due to the difficulty of fabricating freestanding rods or wires <10 nm diameter.

Spectroscopic methods are used to investigate the photothermal melting and shape transformation of gold nanorods dispersed in micellar solution. It is found that the nanorods melt and get transformed into spherical particles when they are exposed to femtosecond laser pulses at moderate energies. At higher energies (for fs pulses) or when exposed to nanosecond laser pulses, these nanorods fragment and then transform into spherical particles with smaller dimensions. It is also found that an average energy of ~60fJ is required to melt a single gold nanorod (with an aspect ratio of ~4.1) dispersed in an aqueous medium.

High temperature TEM is used to investigate the melting and recrytallization of Ge nanowire encapsulated by carbon sheaths. The Ge nanowires (10-100nm in diameter) are prepared using the VLS process, and are coated with carbon sheaths 1-5nm thick to confine the molten Ge and prevent the formation of liquid droplets at elevated temperature. Two distinct features are observed in the melting-recrystallization cycle. One is the lowering of the melting points, which is inversely proportional to the diameter of the nanowire and the other is the large hysteresis loop associated with the melting-recrystallization cycle. For example a Ge nanowire of 55nm in diameter and ~1 μm in length start melting from both ends at 923K (the melting point of bulk Ge is 1200K) The melting then moves toward the center of this wire, and the entire wire is molten at ~1120K. Upon cooling the recrystallization process occurs at a temperature 830K much lower than the initial melting temperature. By tasking advantage of the relatively low melting points of nanowire encapsulated in carbon sheaths. The capability to manipulated individual nanowires is shown using techniques such as cutting, interconnecting, and welding. Figure 7.1 shows a set of TEM snapshots. It clearly demonstrates the welding of two Ge nanowires encapsulated within a carbon sheath. In this case, the two Ge nanowires melt and the liquid fronts move toward each other as the temperature is increased. They are eventually welded to from a single crystalline nanowire after recrystallization.

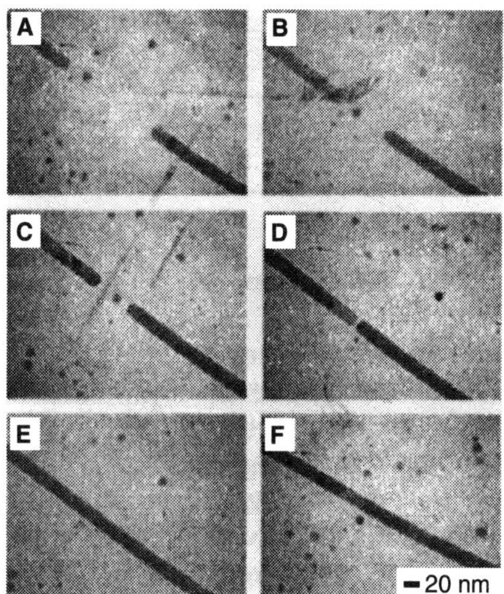

Figure 7.1 *TEM snapshots that show the melting, flowing and welding of two Ge nanowires (encapsulated by a carbon sheath) into a single nanowire.*

It is also observed that silicon nanostructures with various norphologies are formed at different temperature when a Si/SiO_2 mixture is thermally evaporated in a alumina tube. In addition to Si nanowires, many other types of Si nanostructure (e.g., those in the shape of an octopus pin tadpole, and linear chanin) are obtained with significant yields. These nanostructures evolve from Si nanowire through spheroidization, a process that involves material transport along individual nanowires. Both formation and annealing temperatures play important role in determining the relative ratios of these nanostructures in the products. It is possible to achieve a tight control over the morphology and crystallinity of these Si nanostructure by varying the temperature.

The reduction in melting point of nanowire has several implications Firstly, the annealing temperature necessary for the synthesis of defect-free nanowires may be a small fraction of the annealing temperature required for the bulk material. It is thus possible to perform zone refining to purify nanowires at a modest temperature. Secondly, a reduction in melting point enables to cut, interconnect and weld nanowires at a mild temperature. This capability provides a new tool to integrate 1D nanostructure into functional device and circuitry. Thirdly, as the thickness of nanowires is reduced to a smaller and smaller length scale , their stability becomes extremely sensitive to environmental changes such as temperature fluctuation and residual stress variation. Nanowires spontaneously undergo a a spheriodization process to break up into shorter segments at room temperature to reduce the relatively high free energy associated with a 1D system, when their diameter are sufficiently thin or the bonding between constituent atoms are too weak. Interface-driven instability is a key feature of one-dimensional nanostructure, but confined to nanowire substrates.

Now, considerable thermal properties of bilayer nanoribbons for nanoscale interfaces such as the silicon aluminum contact in nanoelectronics. By perturbing and monitoring individual bilayers using TEM, the details of interfacial processes such as diffusion electromigration, grain growth, melting and reaction between two well-defined as deposited materials are known. For example, the slow heating of Cu-SnO_2 bilayer nanoribbons causes interfacial stress that bends the structures elastically at temperature <473K. At intermediate temperatures the initially smooth Cu layer thickens and breaks up into three-dimensional islands as wetting state becomes thermally accessible. Finally reduction of the SnO_2 substrate by Cu at ~823K leads to etching of the interface and the appearance of several new phases.

Because the magnitude and decay time of the photoresponse is highly dependent on the presence of ambient O_2 the photocurrent in n-type oxide nanowires should be a product of electron-hole pair formation and electron doping (caused by the photo-induced desorption of oxidizing surface species, including oxygen). So, the sensitive dependence of nanowire conductivity on adsorbate molecules is used to fabricate single-crystalline nanowire gas sensor. A SnO_2 nanoribbon bathed in UV light is used to reversibly detect 3-100ppm NO_2 at room temperature. NO_2 adsorption on the ribbon surface traps free electrons and widens the region of depleted electron density near the surface, thereby causing the conductivity to drop, whereas the UV light continuously desorbs NO_2 to make the sensing dynamic. This is also used in nanowire based gas sensing ZnO belts, In_2O_3 wires, polycrystalline SnO_2 wires, and TiO_2 tubes arrays.

Nanowire chemical sensors operate via chemical gating induced by the surface adsorption of analyte molecules although other sensing mechanisms exist. The very high surface-to-volume ratios of thin one dimensional nanosturctutres endow them with inherently high sensitivity and short response time; however selectivity is a major problem, especially in the detection of gases. For example the reactive surfaces of oxide nanowires and carbon nanotubes (CNT) interact with most oxidizing and reducing vapors, which complicates many practical applications. The density functional theory (DFT) is used for calculations to understand the details of molecular adsorption on SnO_2 nanoribbon surfaces. It is found that (a) oxygen chemisorbs only on surfaces that contain oxygen vacancies; (b) adsorbed NO_2 exists primarily as tightly bound NO_3 species, is confirmed with X-ray absorption near-edge spectroscopy (XANES); and (c) many surface species are mobile at 300K and some can oxidize the SnO_2 lattice itself, potentially causing sensor signal drift. Selectivity in nanowire sensors is more easily addressed in liquid media, where ligand receptor binding (e.g., bitotin-streptavidin) and other surface functionalization schemes can provide molecular discrimination It is found that nanowire-and CNT based chemical sensing are among the first major applications for one-dimensional nanostructures.

Electron Transport Properties

CNTs and nanowires have been explored as building blocks to fabricate nanoscale electronic devices through self-assembly "bottom-up" approach example: field effect transistors (FETs), p-n junctions, bipolar junction transistors, complementary inverters, and resonant tunneling diodes. Some metal nanowires undergo a transition to become semiconducting as their diameters are reduced below certain values. For instance, two-probe measurements on arrays of single-crystalline Bi nanowires indicate that these nanowires under go a metal-to-semiconductor transition at a diameter of ~52nm, but of ~40nm in diameter these nanowires become semiconductors or insulators because their resistances increased with decreasing temperature.

So the external conduction sub-band and valence sub-band of this system moves in opposite directions to open up a bandgap.

Here, the carrier mobility is also suppressed by carrier confinement along the long axis of wire and by surface imperfection. Gold represents another metal whose electron-transport properties are used for short nanowires as thin as a single linear chain of atoms. Because these wires are extremely short in length (usually a few atoms across, sometimes also referred to as point contacts), their conductance is found to be in the ballistic regime with the transverse momentum of electron becoming discrete. The transport phenomena (e.g., conductance quantization in units of $2e^2h^{-1}$) observed in this kind of 1D system are found to be independent of material. As for semiconductor measurement on a set of nanoscale electronic devices (Fig. 7.2) indicate that GaN nanowire as thin as 17.6nm can function as a semiconductor.

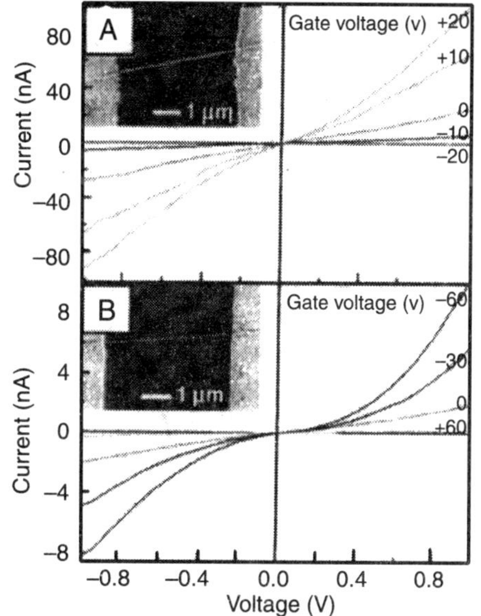

Figure 7.2 *Room temperature I-V behavior for a Te-doped (A) and Zn-doped (B) InP nanowire at various gate potentials. The inset show the nanowire sitting across two electrodes. The diameter of the nanowire is 47 and 45nm respectively.*

It is found that transport measurement of Si nanowires with a thickness of ~15nm become insulating. Nanorods containing diode junctions are synthesized and then assembled into arrayed systems. These semiconductor nanowires are assembled into cross-bar-p-n junctions and junction arrays with controllable electrical characteristics that yields as 95%. These junctions are further used to create integrated nanoscale FET arrays (Fig 7.3) with nanowires as both the conducting channel and gate electrode. In addition OR, AND, and NOR logic gate structures. In another example doped silicon nanowires are used to fabricate passive diodes, bipolar transistors and complementary inverters using self assembly.

There are several appealing features for this "bottom-up" approach to nanoelectronics. First, the size of the nanowire building blocks is tuned to sub-100nm and smaller, which lead to a high density of devices on a chip. Second, the material system for the nanowires are essentially unlimited which gives great flexibility to select the right materials for the desired device functionality. For example several GaN nanowires based nanodevices of interest for their high-power/high-temperature electrical applications.

Phonon-Transport Properties

Phonon transport is greatly impeded in thin (i.e.,d $< \Lambda$, where d is the diameter and Λ is the phonon mean free path) one-dimensional nanostructures as a result of increased boundary

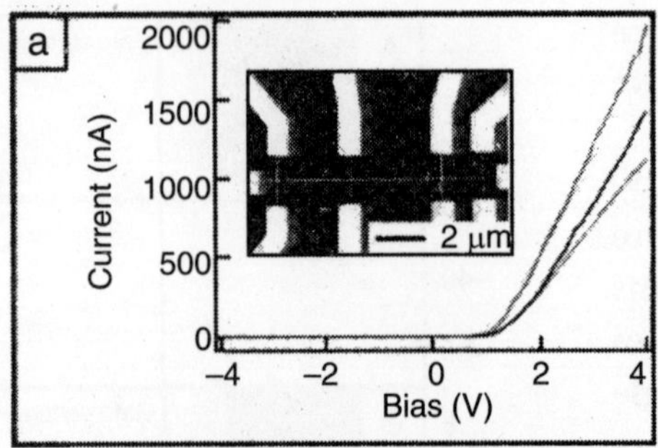

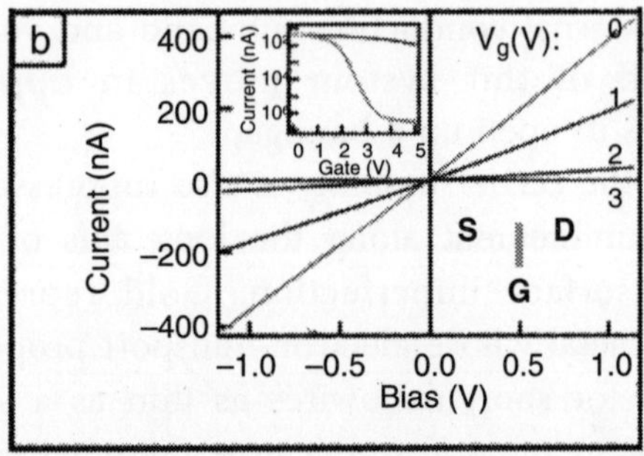

Figure 7.3 *(a) I-V behavior for a multiple junction array formed between four p-type Si nanowires and one n-type GaN nanowire The four curves represent the I-V response for each of the four junctions. Inset is a scaning electron microscopy image of the multiple junction device. (b) Gate-dependent I-V characteristics of a crossed nanowire field-effect transistor. The gate voltage (0,1,2, and 3V) for each I-V curve is indicated. The inset (top left) shows I vs V gate for n-type GaN nanowire (Bottom trace) and global back (top trace) gates when the bias is set at 1V. The transconductance for this device is 80 and 280 nS (V_{sd}=1V) when a global back gate and nanowire gate is used respectively. The inset (bottom right) shows the measurement configuration.*

scattering and reduced phonon group velocities stemming from phonon confinement. As the dimension of a 1D nanostructure is reduced to the range of phonon mean free paths (MFPs), the thermal conductivity is reduced due to scattering by boundaries, example; as the diameter of a silicon nanowire becomes smaller than 20 nm, the phonon dispersion relation is modified- (as a result of the phonon confinement) such that the phonon group velocities are reduced. Another example is heat conduction in cylindrical arid rectangular semiconducting nanowires that consider modified dispersion relations and all important scattering processes predict a large decrease (>90%) in the lattice thermal conductivity of wires tens of nanometers in diameter.

Molecular dynamics (MD) simulations also show, that the thermal conductivities of Si nanowires can be two orders of magnitude smaller than that of bulk silicon in the temperature range from 200 K to 500 K. The reduced thermal conductivity is desirable in applications such as thermoelectric cooling and power generation, but is not preferable for other applications such as electronics and photonics. Yet poor heat transport is advantageous for thermoelectric materials, which are characterized by a figure of merit {$ZT = a^2T/[\rho(K_p + K_e)]$, with a, T, ρ, K_p, and K_e the Seebeck coefficient, absolute temperature, electronic resistivity, lattice thermal conductivity, and electronic thermal conductivity, respectively} that improves as phonon transport, worsens. It is observed that *ZT* increases above bulk values in thin nanowires (by tailoring their diameters, compositions, and carrier concentrations).

Measurements of the overall thermal conductivity of Si/SiGe nanowires (Figure 7.4) are made as a function of temperature (20 to 300 K) and nanowire diameter using a suspended microdevice in vacuum. Individual Si/SiGe wires with a superlattice period of 100-150 nm exhibit a thermal conductivity substantially lower than that of Si/SiGe superlattice films with 30 nm periodicity (10-60% lower, depending on temperature). The broadness of the imbedded Si/SiGe interfaces and the moderate Ge concentration in these wires show that alloy (impurity) scattering is the dominant phonon scattering mechanism for short-wavelength phonons, whereas

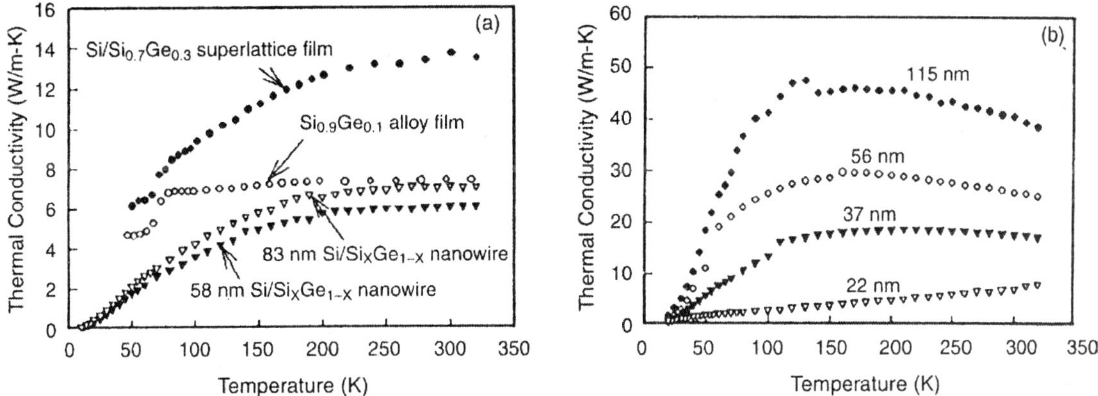

Figure 7.4 *(a) Thermal conductivities of 58 nm and 83 nm diameter single crystalline Si/Si_xGe_{1-x} superlattice nanowires. The value of x is ~0.9 to O.95 and the superlattice period is from 100 to 150 nm. Thermal conductivities of a 30 nm period two-dimensional $Si/Si_{0.7}Ge_{0.3}$ superlattice film and $Si_{0.9}Ge_{0.1}$ alloy film (3.5 μm thick) are also shown. (b) Thermal conductivities of different diameter single crystalline pure Si nanowires.The number beside-each curve denotes the corresponding wire diameter.*

boundary scattering plays a major role in disrupting phonons of all wavelengths. Comparison of superlattice wires with undoped Si nanowires of similar diameter (Figure 7.4) shows that the former have a conductivity roughly five times smaller at 300 K, or ~500 times less than the bulk conductivity for silicon.

4. Optical Properties

The optical properties of nanowires are determined by employing different optical characterization and analytical techniques. The complex dielectric function ($\varepsilon_1 + \varepsilon_2$) of the nanowires are deduced with the help of effective medium theories by considering the nanowires and the host matrix in which the nanowires are embedded to act as a single material. The refractive index(n) and the absorption coefficient (k) of the medium are related to ε_1 and ε_2, respectiely for the composite medium The complex dielectric function of the nanowires is determined directly with the help of standard reflection and transmission measurements combined with Maxwell's equations. The band gap and temperature variation of band gap of the nanowires is determined from the complex refractive index measurement which are considered to be important parameters for selection of materials for particular photonic applications. The plasmon frequency donor atom concentration and carrier concentration of nanowires is obtained by analyzing the infrared (IR) spectra of the nanowires.

Metallic nanowires exhibit interesting plasmon absorption effect. The energy of the surface plasmon band is sensitive to various factors such as particle size, shape, composition, surrounding media and inter particle interactions. The changeover from spherical to rod shaped nanostructured of Ag leads to splitting from an original single absorption band to two absorption bands that separate and become very prominent with increasing aspect ratio. It is found that single crystal Ag nanowire arrays with high aspect ratio exhibits two plasmon absorption bands as shown in figure 7.5. The first peak arises due to the transverse plasmon resonance and the second peak is attributed to the longitudinal plasmon resonance. The surface plasmon modes on Ag and Au nanowires show the occurrence of multipolar plasmon resonances, which can be explained by the standing plasmon wave considerations. Multiple peaks have appeared in the plasmon absorption spectra of the sulfide coated gold nanorods.

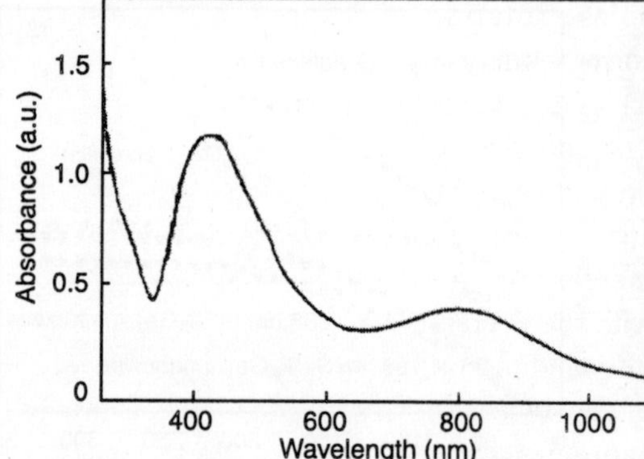

Figure 7.5 *Two plasmon absorption peaks for Ag nanowires*

The photoluminescence (PL) spectroscopic of semiconducting nanowires (InP, CdS ZnO nanowires) show that the PL energy peak and band gap increase with the decrease of the wire diameter. The above phenomena confirms the effect of quantum confinement in nanowires when the wire diameter is reduced.

Fluorescence study of nanowires provides information about the band gap, quantum confinement effect, strain in nanowires, oxygen vacancies, electron effective masses and Fermi energies. The number of subbands of the nanowires are measured with help of magnetooptic techniques. The determination of the number of subbands in nanowires is very important for determining the electron transport properties in them. It is to be noted that magnetic properties of the nanowires can also be determined by the above techniques.

Figure 7.6 (a) and (b) show the magneto-optical (MO) spectra of different diameter Ni nanowire The magneto-optical polar rotation (see Figure 7.6 (a)) reveals that arrays of 35 nm diameter Ni nanowire exhibit sharp peak like Ni around 3.4eV due to an enhancement in the

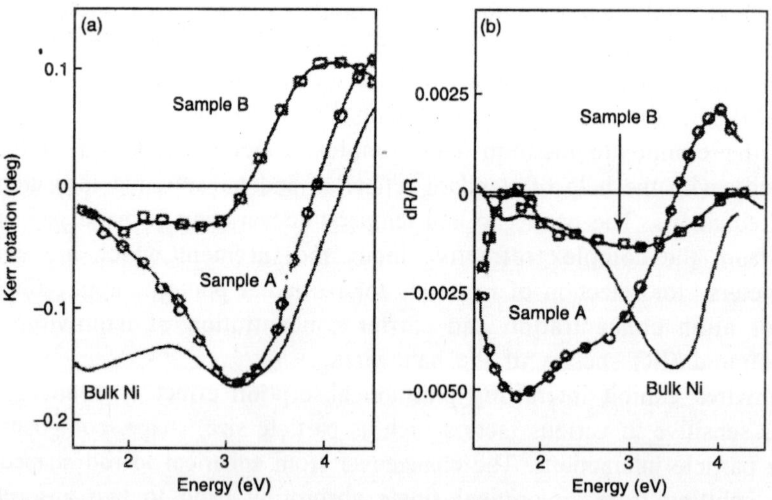

Figure 7.6 *MO polar (a) Kerr spectra and (b) reflectivity comparison between Ni nanowires array and bulk Ni. A and B represent nanowires wire diameter 35nm and 180nm respectively.*

polar rotation Figure 7.6 (b) shows the corresponding zero magnetic field reflectivity (ΔR/R) of the nanowires.

Nanowire arrays exhibits non-linear optical properties, which make them very attractive for application as photonic material. The sharp increase in the energy of the absorption peak with the decrease of the wire diameter is attributed to quantum confinement effect and the occurrence of the blue shift. The extinction spectra of silver nanowires with different diameters shown in Figure 7.7 reveal that for fixed length (*h*) of the nanowires when the diameter (*d*) decreases, λ shift towards the shorter wavelength side because of the quantum confinement effect.

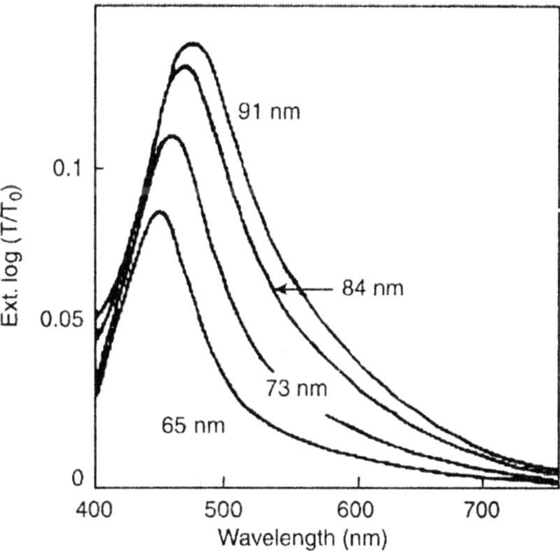

Figure 7.7 *The extinction spectra of different diameter Ag nanowires. The electric field of the incident light is perpendicular to the wire axes. Here, d varies from 65-91nm and A = 840nm.*

Applications

Uniform morphology and optical properties of nanowires have potential for various optical applications The n-p junction of nanowires is found to be capable of light emission by virtue of their photoluminescence (PL) or electrolumine science (EL) properties. The use of *p-n* junction nanowires has been contemplated for laser applications. It is a established that ZnO nanowires of wire diameter smaller than the wavelength of emitted light exhibits lasing actions at lower threshold energy compared to their bulk counterpart. This is due to the exciton confinement effect in laser action which decreases the threshold lasing energy in nanowire. This effect is observed in small diameter ZnO (385 nm Diameter) and GaN nanowires.

The *n-p* junction nanowires or superlattice nanowires with p-n junctions are used as light emitting diodes. The huge surface area and high conductivity along the length of nanowires are suitable for inorganic-organic solar cells. The solar cell made of CdSe nanowires has high efficiency. The subwavelength diameter Si nanowire is used as lowloss optical waveguides within visible to IR range of spectrum. The optical losses of Si sub-wavelength diameter nanowires are much lower compared to that of other sub-wavelength diameter metallic plasmon waveguides. The optical loss of different diameter Si nanowires transmitting the wavelength of 633 nm and 1550nm are shown in Figure 7.8 (a) and Figure 7.8 (b) represents a SEM image of a coiled 260nm diameter Si nanowire of nearly 4nm in length.

Nanowires made of various metal segments like Ag, Au Ni Pd etc are used as barcode tags for different optical read outs. When the intensity of the incident photons are increased the electron density of the subband edges also increases. Due to this these quantum wire develop strong nonlinearity. Therefore, nanowires are used to develop optical switches. These optical switches operate at lower energy and with enchanced switching speed compared to the know switches.

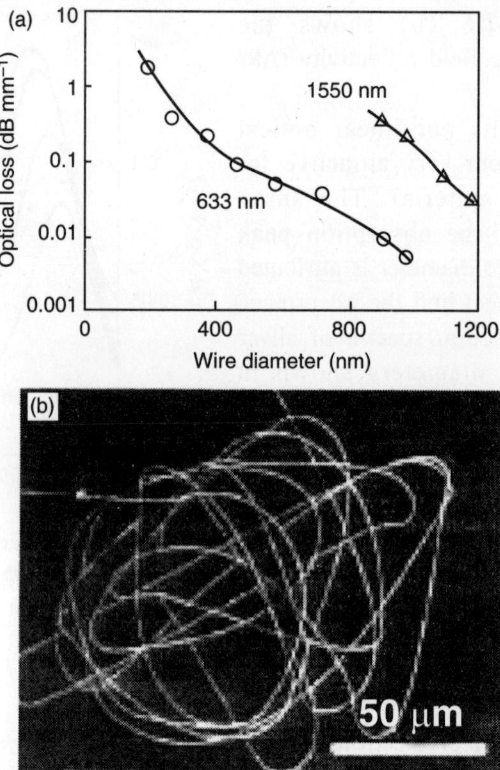

Figure 7.8 *(a) optical loss of silica wires measured at a wavelength of 633nm (field white circle) and (b) SEM image of a coild 260nm diameter Si nanowires of length of nearly 4nm.*

As for quantum dots, size-confinement is used in determining the energy levels of a nanowire once its diameter is reduced below a critical value (the Bohr radius). It is found that the absorption edge of Si nanowires (synthesized with hexane supercritical fluid as the solvent) is significantly blue-shifted as compared with the indirect bandgap (~1.1 e V) of bulk silicon.

Also observed are sharp, discrete features in the absorption spectra and relatively strong "band-edge" photoluminescence (PL). These optical features originate from quantum-confinement effects, although surface states also make additional contributions. In addition, the variation in growth direction for these Si nanowires lead to different optical signatures. For example, the <100> oriented nanowires display a strong feature reminiscent of the $L \rightarrow L$ critical point in the Si band structure with a slowly rising phonon-assisted optical transition, whereas the <110> oriented Si nanowires exhibit distinctly molecular-type transitions. The <100> oriented wires also exhibit a significantly higher exciton energy relative to the <110> oriented ones.

The light emitted from nanowires is highly polarized along their longitudinal axes. There exists a striking anisotropy in the PL intensities recorded in the direction parallel and perpendicular to the long axis of an individual, isolated indium phosphide (InP) nanowire

(Fig. 7.9). The magnitude of polarization anisotropy can be quantitatively explained in terms of the large dielectric contrast between the nanowire and the surrounding environment, as opposed to quantum mechanical effects such as mixing of valence bands. This large polarization response is used to fabricate polarization-sensitive nanoscale photodetectors that find use in integrated photonic circuits, optical switches and interconnects, near-field imaging and high-resolution detection.

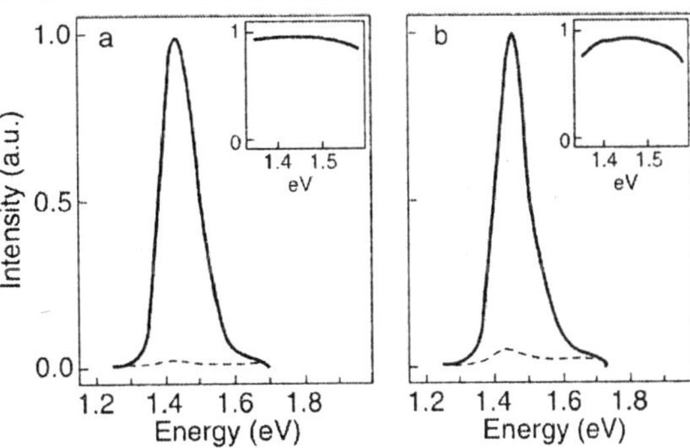

Figure 7.9 *(a) Excitation and (b) emission spectra recorded from an individual InP nanowire of 15 nm diameter. The polarization of the exciting laser is aligned paraliel (solid line) and perpendicular (dashed line) to the long axis of this nanowire respectively. The inset plots the polarization ratio as a function of energy.*

Semiconductor nanorods can be used to enhance the processibility and efficiency of solar cells. Thin-film photovoltaic devices are fabricated by blending CdSe nanorods with polythiophenes to obtain hybrid materials. The performance of such a device is tuned by controlling the aspect ratios of the nanorods. It is found that nanorods are superior to quantum dots in photovoltaic applications, because they provide a direact path for electrical transport at much lower loadings. Under Air Mass (AM) 1.5 Global solar conditions, a power conversion efficiency of as high as 1.7 % is achieved.

For nanorods made of noble metals such as Au and Ag, their surface plasmon resonance (SPR) properties exhibit two SPR modes, corresponding to the transverse and longitudinal excitations. While the wavelength of transverse mode is essentially fixed around 520 nm for Au and 410 nm for Ag, their longitudinal modes are tuned to the span across the spectral region from visible to near-infrand (NIR) by controlling their aspect ratios. It is demonstrated to at gold nanorods with an aspect ratio of 2.0-5.4 can fluoresce with a quantum yield more than one million times that of the metal. These properties, coupled with biological inertness make Au and Ag nanorods ideal candidates for use as colorimetric markers or sensors, and contrast-enhancing reagents for in-vivo optical imaging.

Nonlinear Optical Properties

Semiconductor nanowires are also used as frequency converters or logic/ routing elements in nanoscale optoelectronic circuitry. Coherent NLO phenomena, such as second and third-harmonic generations (SHG and THG, respectively), depend explicitly on the crystallographic structure of a medium, as well as the polarization scheme. The temporal response of non-resonant harmonic generation is similar to the pulse width of the incident laser (in some cases, as short as 20 fs), while incoherent processes are at least 2–4 orders of magnitude slower. Moreover, non-resonant SHG is essentially independent of wavelength below the energy bandgap of a semiconductor, most often including the 1.3–1.5 μm wavelength region typically used in optical-fiber communications. Microcrystalline ZnO thin films show a large second-order nonlinearity a parameter that determines the efficiency of a material as a nonlinear optical converter at optical frequencies.

An NSOM operated in the oblique collection mode is used to measure the NLO properties of individual ZnO nanowires. Here sample is illuminated in the far-field, as it is preferred for nonlinear near-field imaging because of its suitability for high incident pulse intensity experiments. Fig 7.10 depicts two SHG images, showing the dependence of SHG signal intensity on the polarization of input pulse and the orientation of ZnO nanowire. The dependence arises from two independent, non-vanishing components of $\chi^{(2)}$ observed in SHG for ZnO, $\chi^{(2)}_{zzz}$ and $\chi^{(2)}_{zxx}$ two wires are situated approximately normal to each other, with an s- and p-polarized incident beam in Figure 7.10 (a) and (b) respectively The polarization ratio, $SHG_{s\text{-}inc}/(SHG_{p\text{-}inc} + SHG_{s\text{-}inc})$, is 0.90 for wire No. 1. The wire No.2 has an average signal that, is 2.5 times that from wire No.1. The SHG polarization effect, is also quantitatively analyzed by taking polarization traces from several wires. In doing so, the near-field probe is maintained above each wire, and the input polarization is rotated as the SHG signal is monitored (Fig 7.10(c). The theoretical traces are computed to fit the polarization data. Nearly all the wires tested (80-100 nm in diameter) exhibit a ratio $\chi^{(2)}_{zzz} / \chi^{(2)}_{zxx}$ of approximately 2.0-2.3, while one larger wire (~.125 nm in diameter) yielded $\chi^{(2)}_{zzz} / \chi^{(2)}_{zxx} = 4.2$. These ratios when compared to the value (3.0) for a ZnO bulk crystal and 6.0 times higher than that of polycrystalline thin films.

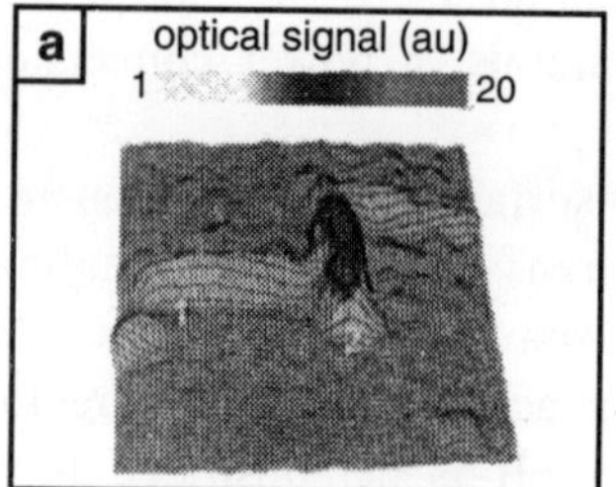

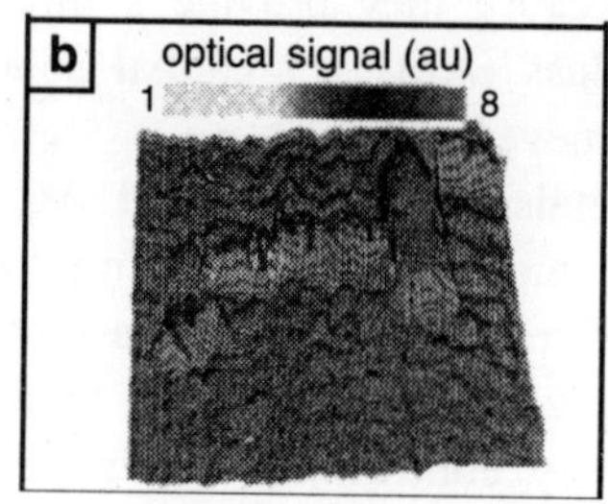

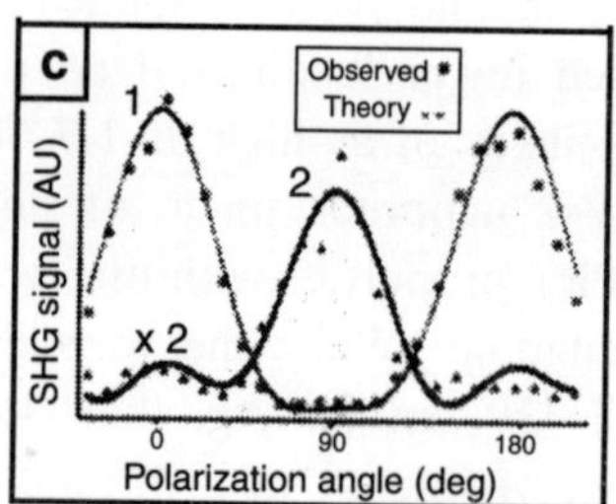

Figure 7.10 *Dependence of nanowire single harmonic generation (SHG) signals on the polarization a) A combined topographical and SHG image of two wires at angle of approximately 90⁰. The image size ~13 × 13 μm² and the maximum topographic height 130nm. The beam is s-polarized, and incident from the right (b) An SHG image recorded from same region as in (a) with a p-polarized incident beam. (c) Polarization dependent SHG data and theoretical predictions taken from the wires labeled in (a) the theoretical curves are calculated for the SHG signal and $\chi^{(2)}_{eff}$ for a hexagonal crystal.*

The absolute magnitude of each component of $\chi^{(2)}$ for individual ZnO nanowires $\chi^{(2)}_{zxx}$ and $X^{(2)}_{zzz}$ are determined by measuring the nanowire SHG with respect to a ZnSe disk (at 1.4 μm excitation). $\chi^{(2)}_{zzz}$ for an individual ZnO wire is found to be 5.5pm V^{-1} and for $\chi^{(2)}_{zzz}$ is 2.5pm V^{-1}. $\chi^{(2)}_{zzz}$ value is considerably lower than the value (18pm V^{-1}) for the bulk crystal but agrees with values for ZnO thin films (4-10pm V^{-1}). Also, the SHG signal for an individual nanowire is wavelength independent ($\lambda_{SHG} > 400$ nm) and relatively efficient with a $\chi^{(2)}_{eff}$ ($\geq 5.5 pmV^{-1}$) larger than that of β-barium borate (BBO, $\chi^{(2)}_{eff} \approx 2.0$ pmV^{-1}). These measurements indicate that ZnO nanowires are potentially useful as an effective frequency converter in the UV region. They are also used as logic components in nanoscale optoelectronics.

Photoconductivity Optical Switching and Chemical Sensing Properties

Electronic conductivity in semiconductor nanowires belts and tubes is substantially enhanced by exposing these structures to photons of energy greater than their bandgaps.

Among all nanoscale devices switches are critical for important applications like memory and logic. Electrical switching on the nanometer and molecular scales has been achieved through the application of a gate potential by nanotube transistor. It is also possible to create highly sensitive electrical switches by controlling the photoconductance of individual semiconductor nanowires. For example the conductance of ZnO nanowire is founded to be extremely sensitive to ultraviolet light exposure. The light induced insulator to conductor transition enables them to reversibly switch a nanowire between OFF and ON state. In a typical experiment four probe measurements on individual ZnO nanowires indicate that they essentially insulate in the dark, with a resistivity >3.5 MΩ cm^{-1}. When these nanowires are exposed to a UV light source with wavelengths below 400nm their resistivity is instantly reduced by 4 to 6 orders in magnitude. In addition to their high sensitivities these photoconductive nanowires exhibit an excellent wavelength selectivity. Figure 7.11(a) shows the evolution of photocurrent when a ZnO nanowire is exposed first to very intense visible light at 532nm (Nd:YAG,the second harmonic) for 200s and then to UV light at 365nm. There is no photoresponse at all to the green light while exposure to the less intense UV light a to change in conductivity by 4 orders of magnitude. Measurements on the spectral response show that the ZnO nanowires have a cut-off wavelength at 385 nm, which is same as the bandgap of ZnO.

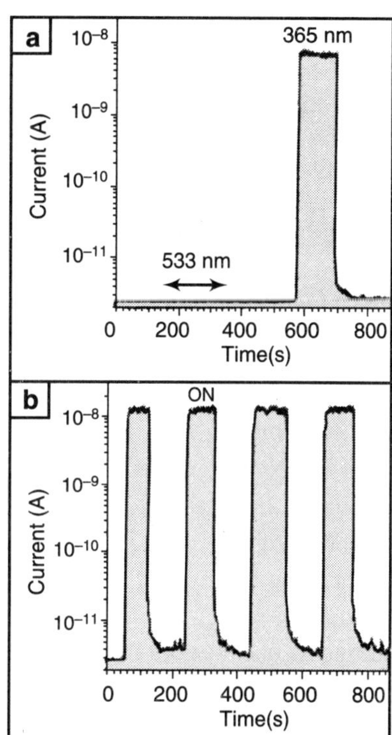

Figure 7.11 *(a) The photoresponse (change in current under a constant bias) of a ZnO nanowire to light exposure at two different wavelengths:532 and 365nm. (b) Reversible switching of a ZnO nanowire between its low and high conductivity states.*

It is established that oxygen chemisorption enhanes the photosensitivity of bulk or thin-film ZnO. A similar photoresponse mechanism can be applied to the nanowire system with nanowire which further enhances the sensitivity of the device. As, the photoresponse of ZnO consist of two parts: a solid state process where an electron and a hole are created ($hv \rightarrow h^+ + e^-$) and a two-step process involving oxygen species adsorbed on the surface. In the dark, oxygen molecules adsorb on the oxide surface as negatively charged ions by capturing free electrons of the n-type oxide semiconductor ($O_{2(g)} + e^- \rightarrow O^-_{2(ad)}$) there by creating a depletion layer with low conductivity $\chi^{(2)}_{eff}$ in the vicinity of the nanowire surface. Upon exposure to UV light, photogenerated holes migrate to the surface and discharge the adsorbed oxygen ions through surface electron-hole recombination ($h^+ + O^-_{2(ad)} \rightarrow O_{2(g)}$). At the same time, photogenerated electrons destroy the depletion layer, and thus significantly increase the conductivity of the nanowire.

Thus, ZnO nanowires have the potential as candidates for optoelectronic switches with the insulating state as "OFF" in the dark and the conducting state as "ON" when exposed to UV light. Figure 7.11(b) shows the photoresponse of a ZnO nanowire as a function of time when the UV-lamp is switched on and off. It shows that this nanowire can be reversibly and rapidly switched between the low and high conductivity states. With some further optimization on the nanowire composition (e.g., through proper doping)the photocurrent decay time (or the response time) can be reduced to the microsecond scale. So, the nanowires can serve as very sensitive UV-light detectors in many applications such as microanalysis and missile plume detection and as fast switching devices for nanoscale optoelectronics applications where ON and OFF states can be addressed optically.

Nanowire Lasing

Nanowires with flat end facets can be exploited as optical resonance cavities to generate coherent light on the nanoscale. Room temperature UV lasing is demonstrated in laboratories for the ZnO and GaN nanowire systems with epitaxial arrays, combs and single nanowires. ZnO and GaN are wide bandgap semiconductors (3.37.3.42 eV) suitable for UV blue optoelectronic applications. In addition ZnO has an excitation binding energy as high as 60 meV which is significantly larger than that of ZnSe (22meV) and GaN (25 meV) This indicates that the excitations in ZnO are thermally stable at room temperature.

The large binding energy for excition ZnO (~60meV) permits lasing via excition-excition recombination at low excitations at low excitation conditions, whereas GaN supports an electron hole plasma (EHP) lasing mechanism. Far field imaging and near-field scanning optical microscopy (NSOM) is used for photon confinement in these small ($d \leq \lambda$; where d is the nanowire diameter and λ is the wavelength) cavities.

Well faceted nanowires with diameters from 100 to 500nm support predominantly axial Fabry-Perot waveguide modes (separated by $\Delta\lambda = \lambda^{2l}[2Ln(\lambda)]$ where L is the cavity length and $n(\lambda)$ is the group index of refraction owing to the large diffraction losses suffered by transverse trajectories. Diffraction prevents smaller wires form lasing: PL is lost instead to the surrounding radiation field ZnO and GaN nanowires produced by VLS growth are cavities with low intrinsic finesse (F) owing to the low reflectivity (R) of their end faces (102)(~19%) [where F =

$\pi R^{1/2}/(1-R)$], such that the confinement time for photons is short and photons travel an average of one to three half passes before escaping from the cavity. Far field imaging indicates that PL and lasing emission are localized at the ends of nanowires This shows strong waveguiding behavior which is consistent with axial Fabry-Perot modes.

The transition from spontaneous PL to optical gain is achieved by exciting a high density of carriers via pulsed UV illumination The dependence of nanowire emission on pump power (Figure 7.12) shows three regimes, corresponding to (a) spontaneous emission followed by (b) stimulated emission (lasing) above a certain threshold fluency and (c) saturation through gain pinning at high pump power The lasing thresholds observed in nanowires vary across several orders of magnitude as a consequence of differing nanowire of dimensions quality of the particular nanowire and coupling to the substrate (the lowest threshold observed of ZnO is ~70 nJ cm^{-2}:for GaN ~500nJ cm^{-2}) The appearance of narrow cavity modes (line width 0.25-1.0nm) spaced in agreement with cavity dimensions confirms the lasing behavior The spectral position of the ZnO gain profile is typically nearly independent of pump power at the moderate pumping intensities that correspond to excition-excition lasing but exhibits significant red shift near saturation as band filling and charge screening induce an excition EHP transition GaN nanowires on the other hand show a consistent red-shift from threshold to saturation owing to band gap renormalization.

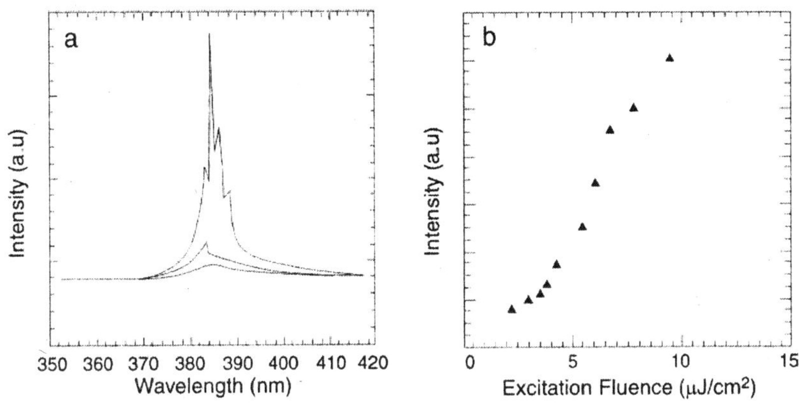

Figure 7.12 *(a) Spectra of light emission from GaN/AlGaN core-sheath nanowires below near and above threshold (about 2-3 μJcm^{-2}) (b) The power dependence of output integrated emission intensity.*

In a typical set-up for lasing experiments (Fig 7.13(a)) one end of the nanowire is terminated in the epithetical interface between the sapphire and ZnO and the other end in the flat (0001) plane of hexagonal ZnO Fig 7.13 (b). Considering the refractive indices of sapphire (1.8), ZnO (2.5), and air (1.0) both ends of each nanowire serve as good mirrors to construct an optical cavity. Such a natural cavity/waveguide configuration suggests a simple approach to the fabrication of nanoscale optical resonance cavities without cleavage and etching. In an experiment the nanowires are optically pumped by the fourth harmonic of a ND:YAG (YAG: yttrium aluminium-garnet) laser with various intensities. Light emissions are collected in the direction either normal to the end surface plane or along the longitudinal axis of the nanowire. Figure 7.13(c) shows the emission spectra at different pumping powers. When the excitation

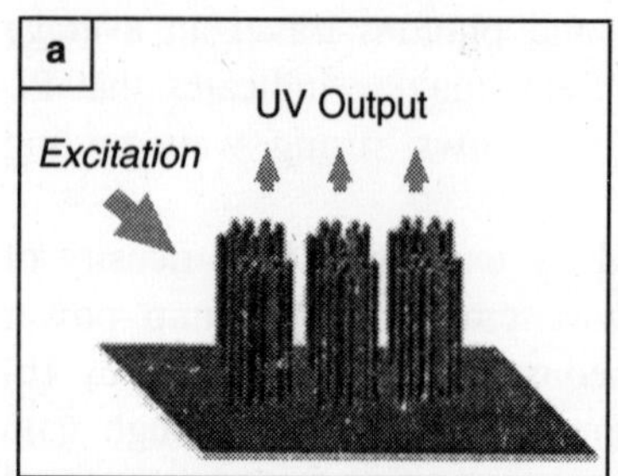

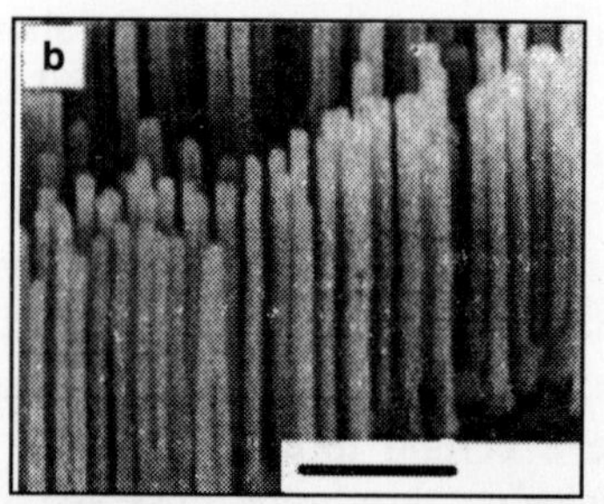

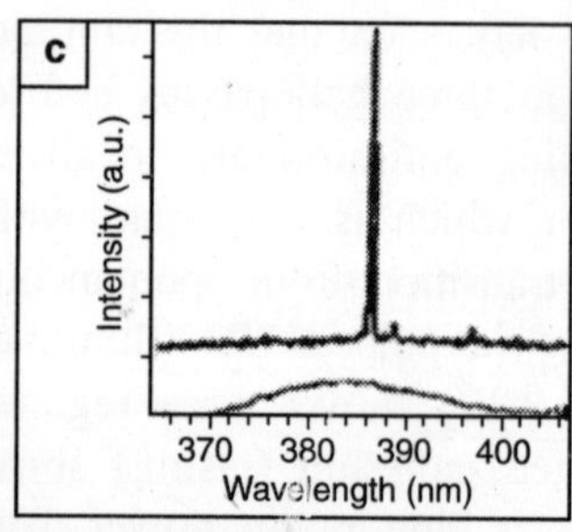

Figure 7.13 *(a) Excitation and detelection configurations used for the lasing. (b) SEM image of a 2D array of ZnO nanowires grown as uniaxial crystals on the surface of a sapphire substrate. The power dependent emission spectra recorded from a 2D array of ZnO nanowires with the excitation energy being below (below trace) and above (top trace) the threshold.*

power is below the threshold, the spectrum (Fig 7.13(c)) bottom trace consists of a broad spontaneous emission peak with a full width at half maximum (FWHM) of ~17nm. This spontaneous emission originates from the recombination of excitations through an excition collision process, in which one of the excitations radiatively recombines to generate a photon. The emission peak is narrowed as the excitation power is increased, owing to the preferential amplification of frequencies close to the maximum of the gain spectrum. Once the power exceeds the threshold (~40 $KWcm^{-2}$) a sharp peak emerges in the emission spectrum (Fig 7.13(c). Top trace) the line width of this peak can be more 50 time smaller than that of the spontaneous emission peak. Above the threshold, the integrated emission intensity also increases rapidly with the pumping power. The narrowing of the line width and rapid increase in intensity support stimulated emissions by these arrayed nanorods.

The lasing action in arrayed ZnO nanowires without mirrors suggests that single crystalline well-facetted nanowires can function as effective resonance cavities. So, well cleaved nanowire can be used as natural optical cavities. It can be extended to many other different semiconductor systems. The lasing effects in GaN nanowire create p-n junction in these individual nanowires, so it is possible to make electrically pumped UV and blue lasers out of individual nanowires. The chemical flexibility as well as the one dimensionality of these nanowires makes them ideal candidates for fabricating miniaturized laser light sources. These miniaturized nanolasers find applications in nanophotonics and microanaysis.

Figure 7.14 *Near-field optical image of an individual lasting ZnO nanowire. The lasing emission is only seen at the two ends of this nanowire.*

The lasing effect in nanowires is also confirmed by the optically characterizing individual ZnO nanowires using near field scanning microscopy (NSOM). Figure 7.14 shows a typical example where the nanowire are excited with short pluses (<1ps) of

4.3eV photons (285nm) in order to observe PL and lasing. The emission from an individual nanowire is collected using a chemically etched fiber optic probe held in the constant gap mode both topographic (based on the shear-force feedback signal) and optical information is collected during forward and reverse scans.

As show in Figure 7.14 lasing is observed form a nanowire that is situated at an angle to the quartz substrate the intense signal collected near the end confinement of the emitted photons to a cone-shaped region near the end of the nanowire It is found that silver nanowires can also serve as waveguides to direct the propagation of photons through surface plasma coupling.

Two aspects of lasing in one dimensional ZnO structures observed are its ultra fast dynamics in nanowires and its manipulation in nanoribbons. Time resolved second harmonic generation (TR-SHG) and transient photoluminescence spectroscopy are used to probe carrier relaxation dynamics near the lasing threshold, as well as under gain saturation conditions. Above the lasing threshold, a bi-exponential decay of the PL is observed with a fast component (~10ps) corresponding to exciton exciton lasing and slow component (~70ps) owing to free excition spontaneous emission. The fast process shifts to shorter times with increase influence of EHP dynamics at higher carrier densities (Figure 7.15). The SHG transient which moniter the overall repopulation of the valence bands after excitation shows a fast component with a decay time that decreases from 5 to ~1ps from threshold to saturation through a multi-body scattering process consisting of both radiative and nanoradiative events. For the anoribbon the dependence of the lasing threshold and spectrum on the ribbon length is determined by successively etching isolated ribbons using a focused ion beam (FIB). The threshold pump fluency and nanoribbon length are found to be inversely propotional for lengths greater than 10 μm whereas most ribbons shorter than 5 μm fail to lase at any pump intensity because gain volume is lost.

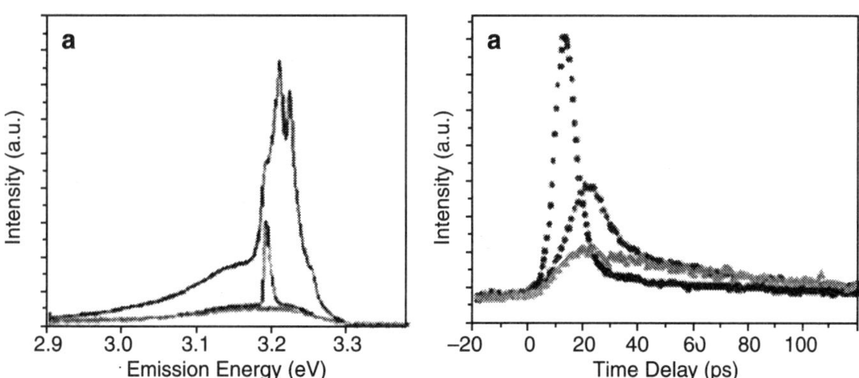

Figure 7.15 *(a) PL/lasing spectra of single ZnO nanowire near the lasing threshold (excitation ~1 μJ/cm^{-2}) and (b) transient PL response. Long decay component is 70 ± 7ps and short component is 9 ± 0.8ps(red) and 4.0 ± 0.3ps (black)*

MAGNETIC PROPERTIES

Magnetic nanowire arrays consisting of isolated needle-like magnetic nanowires are used for perpendicular magnetic recording. The nanowires that are introduced into the pores of alumina template by electodeposition produce bit densities in excess of 100 Gbit m^2. When the magnetic field is applied parallel to long axis of the magnetic nanowire, it exhibits a coercive field, which in inversely proportional to the pore diameter. The squareness of the hysteris loop can be increased from 30% up to nearly 100% by decreasing the wire diameter.

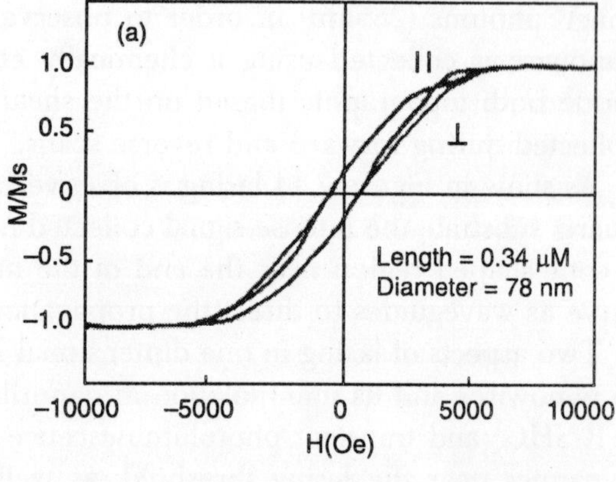

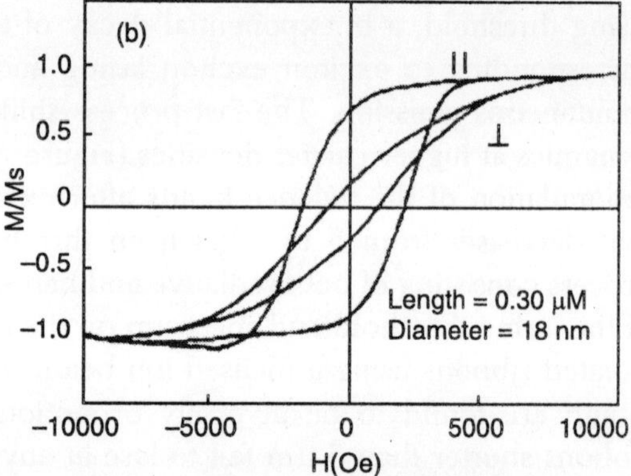

Figure 7.16 *The variation of squareness of hysteresis loops of Co nanowire arrays with applied magnetic field (H) parallel (||) and perpendicular (⊥) to the wire axes: (a) nanowire diameter, 78nm and (b) nanowire diameter 18nm*

Figure 7.16 (a) and (b) show the variation of squareness of the hysteresis loop as a function of the applied magnetic field (H) parallel and perpendicular to the wire axes for different nanowire diameters. A gradual increase of coercivity is observed with increasing aspect ratio (length) *(l)*/diameter *(d)*, but little charge is observer when *1/d*>10. Moreover, depending on the geometry, different sites on the surface, differ in local coordination number which leads to the variation of moment with local environment. The changes in the diameter of the nanowire influence the surface energies of the different crystallographic planes. It is found that in the case of *bcc* α-Fe, the order of surface free energy (γ) for (110), (100) and (111)Planes is $\gamma_{110} < \gamma_{100} < \gamma_{111}$, where the values of γ_{110} and γ_{100} are close to each other However, it is found that as the diameter of the nanowire increases, the differences in the surface free energy between the crystallographic planes become more significant. Therefore, when nanowires are processed at low temperatures, growth of (110) and (100) planes are nearly equal giving rise to octagonal cross section of the nanowires. At higher temperatures, growth of (110) plane predominates and square shaped nanowires are formed. This change in morphology causes a change in the magnetization characteristic of the material. Therefore the magnetic properties of the nanowires can be tuned by changing their diameter, which also change coercivity, magnetization, remanace and squareness of the hysteresis loop.

It is found that the total magnetic anisotropy of these nanowires is significantly influenced by the magnetic anisotropy, from of the nanowire and the demagnetization fields between the nanowires.

Hexagonally arranged Ni-nanowires embedded in anodic alumina templates exhibit a strong enhancement in their magnetoopical (MO) response. The magnetic nanowires of Fe, Co, Ni show much enhanced magnetic coercivity than that of their bulk counterpart. The coercivity is strongly influenced by annealing of the wire at different temperatures, the aspect ratio and the wire diameter. Figures 7.17(a) and (b) show coercitivity of the magnetic Fe, Co and Ni nanowires with their diameters at room temperature and at zero temperature. The temperature dependence of coercivity as function of wire diameter of Fe nanowire is shown in Figure 7.18(a) and the variation of saturation magnetization as a function of temperature for different diameters of Fe, Co and Ni nanowires are shown in Figure 7.18 (b)]

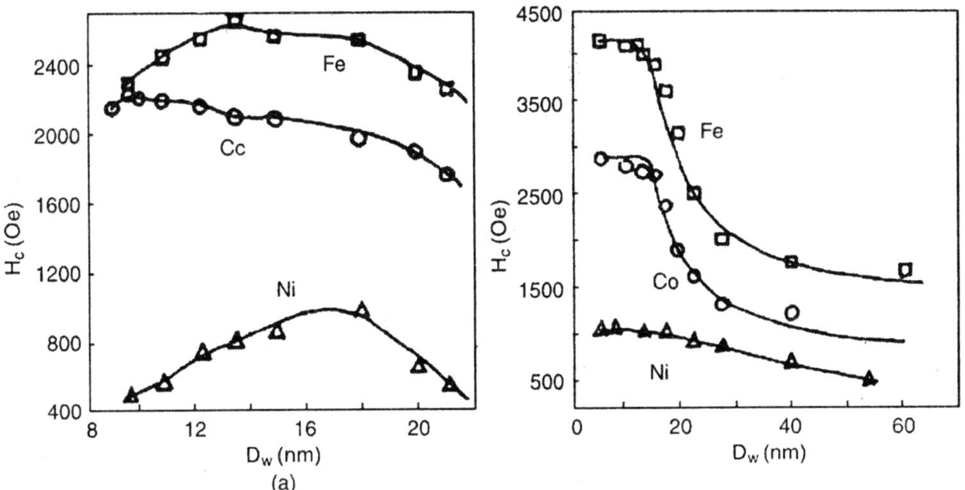

Figure 7.17 *Coercivity as a function of nanowire diameter (D_w) for magnetic nanowires of Fe, Co and Ni at (a) room temperature and (b) zero temperature.*

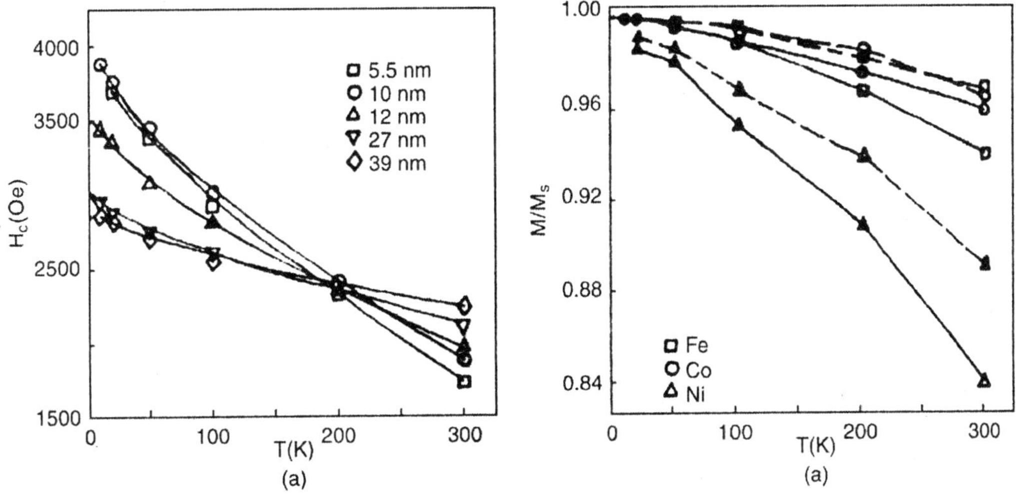

Figure 7.18 *Variation of (a) coercivity as a function of temperature with different diameters of Fe nanowires and (b) magnetization as a function of temperature for different diameters of Fe, Co and Ni nanowires (soild lines: 5.5nm diameter nanowires, dashed line: 10nm diameter nanowires)*

The magnetoresistance (MR) property of nanowires provide information on the quantization effect, wire boundary scattring of electrons and the effect of doping and annealing on scattering. The longitudinal MR where the magnetic field is applied along the wire axes, the MR peak maxima at a particular applied field (B) varies linearly with the reciprocal of the wire diameter for a specified temperature. The appearance of the maxima is attributed to the classical size effect. In transverse MR of Bi nanowires, where the magnetic field is applied perpendicular to the wire axis, the temperature follows $T \propto B^2$ in the range of $0 \leq B \leq 5.5T$.

It is found that the MR of nanowires is highly diameter dependent. The longitudinal MR as a function of magnetic field for Bi $_{0.85}SB_{0.15}$ alloy nanowires with 40nm diameter at different temperatures are linear as shown in Figure 7.19 (a) On the other hand 65nm diameter $Bi_{1-x}SB_x$ nanowires exhibit different types of longitudinal MR as a function of the applied field as is seen in Figure 7.19 (b) The maxima in MR are attributed to the bondary scattering effect.

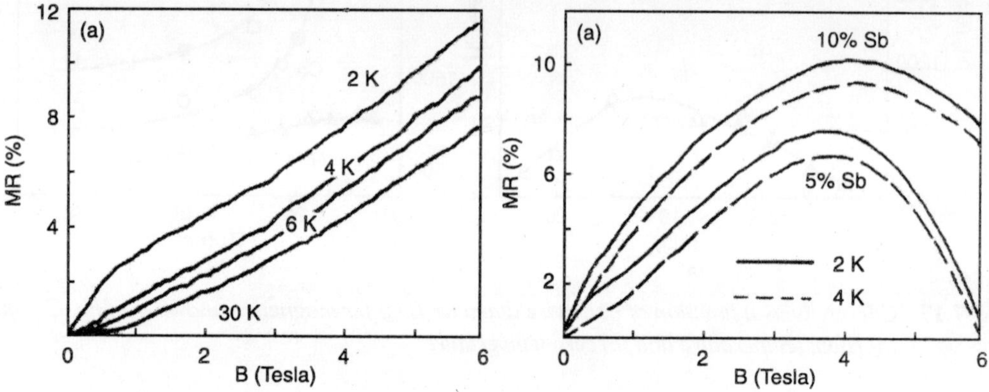

Figure 7.19 *Longitudianl MR of (a) 40nm diameter $Bi_{0.85}$ $Sb_{0.15}$ alloy nanowires and (b) 65nm diameter Bi-Sb alloy nanowires as a function of magnetic field, B at different temperatures.*

Also, the 1-D confined system at low magnetic field can be transformed into 3-D confined system by applying a longitudinal magnetic field to the nanowire arrays at very low temperatures. The Bi nanowires of diameters up to 200nm show oscillating effect, which is due to the field strength variation when the quantized Landau level passes through the Fermi energy level.

Applications

The potential applications of nanowires lie in the magnetic information storage medium. Periodic arrays of magnetic nanowire arrays possess the capability of storing 10^{12} bits/in^2 of information per square inch of area. The small diameter, single domain nanowires of Ni, Co fabricated into pores of porous anodic alumina are most suitable The high aspect ratio of the nanowires results in enchanced coercivity and suppresses the onset of the superparamagnetic limit, which is important for preventing the loss of magnetically recorded information between the nanowires. Suitable separation between the nanowires is maintained to avoid the inter-wire interaction and magnetic dipolar coupling. It is found that nanowires can be used to fabricate stable magnetic medium with packing density>10^{11} wire cm^{-2}.

Thermoelectric properties

Nanowires exhibit interesting thermoelectric properties. The unique electronic band structure and diameter dependent variation in electron density of state make nanowires promising materials for various thermo-electronic applications Metal nanowires exhibit many fold increase in Seebeck coefficient due to their enchance density of electronic states at the one-dimensional sub-band edges, which is attributed to the quantum confinement effect. The thermopower of the metallic nanowire also increases because of the enhanced electronic density of state near the Fermi energy level. The seebeck coefficient and thermopower of metallic nanowires are influenced by the wire diameter and the alloy phase concentration. It is observed that the thermal conductivity of nanowires is influenced by the wire diameter. The variations of thermal conductivity *(k)* as a function of temperature for Si nanowires of diameter are shown in Figure 7.20.

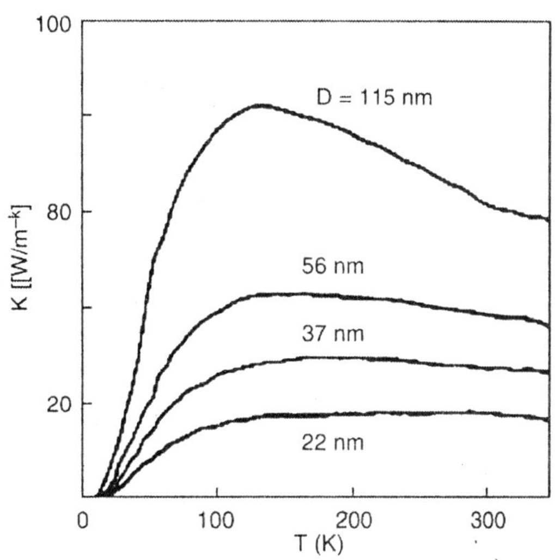

Figure 7.20 *Thermal conductivity (k) as a function of temperature for Si nanowires of different diameters (D)*

In Sb and Si doped Bi nanowires, the thermopower increases on decreasing the wire diameter. A similar phenomenon is observed in case of alumina doped Zn nanowires. Bi nanowires also exhibit significant temperature (T) dependence of resistance, R(T) and the phenomena is strongly influenced by the variation of diameter of the nanowires. They also exhibit quantum confinement effect below certain critical diameter(~48nm) and the semimetallic phase of Bi transforms into the semiconducting phase . This phenomenon implies that at elevated temperature or at room temperature, the semimetallic phase transformation of Bi nanowires is possible only by reducing the wire diameter. The crystalline structure of the material also influences the quantum size effect of nanowires. In polycrystalline nanowires the resistance maxima is not observed in R(T) vs T curve because of less carrier mobility due to the boundary scattering effect. Plots of R (T)/R (270K) ratio measured as a function of temperature, T (K) for 40nm Bi and Te doped Bi nanowires are shown in Figure 7.21(a). The average carrier mobility ratio, μ(T)/μ(270 K) for the same nanowire are also calculated from the R (T) value and the T-dependent carrier density measurement. Figure 17(b) shows the temperature dependence of mobility ratio, μ (270)/μ(T) where μ stands for the doped Te ion concentration (ions cm^{-3}).

The electronic wave function becomes more localized with the reduction of diameter of the nanowires. This phenomenon is considered to be the key factor that causes nanowires to exhibit interesting thermoelectric properties.

Applications

The enhanced thermopower and may fold increase in the seebeck coefficient of nanowires make them very attractive for thermoelectric cooling system and energy conversion devices.

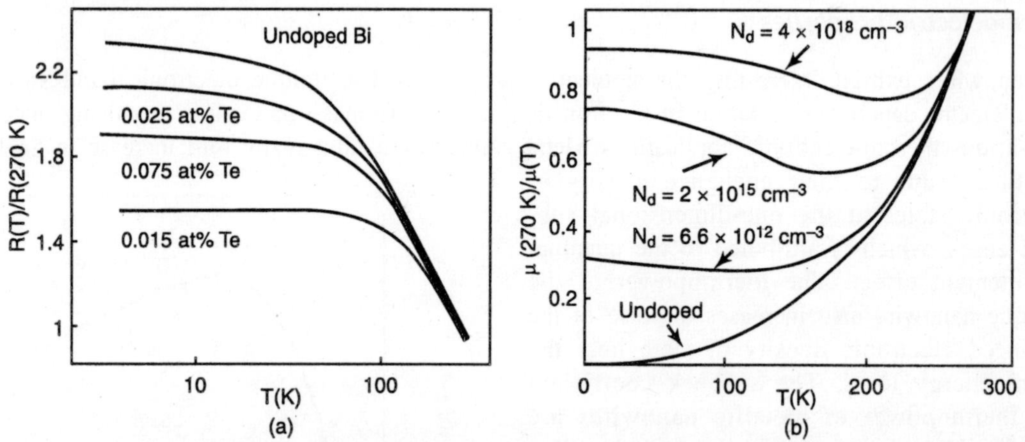

Figure 7.21 *(a) The R(T)/R (270K) of 40nm diameter Bi nanowire different amounts of Te alloy concentrations (b) carrier mobility ratio as a function temperature of 40nm Bi nanowire with different doped Te atom concentration.*

Electrical properties (transport properties)

Electron transport properties of nanowires are very important for electrical and electronic applications as well as for understanding the unique one dimensional carrier transport mechanism. It is observed that the wire diameter, wire surface condition, crystal structure and its quality, chemical composition, crystallographic orientation along the wire axis etc are important parameters which influence the electron transport mechanism of nanowires. Figure 7.22 (a) shows the I-V characteristic of the Cu nanowires both at room temperature and at 4.2K. The plot shows the linear ohmic behavoiur. However, by oxidation when the metallic Cu nanowires are transformed into semiconducting Cu_2O nanowires and placed between two electrodes, it forms two Schottky diodes in series and double diode like I-V characteristic curve is obtained as shown in Figure 7.22(b).

Also quasi one dimensional nanowires exhibit both ballistic and diffusive type electron transport mechanism, which depends upon the wire length and diameter. Ballistic type transport phenomena is associated with predominant carrier flow without scattering which is due to the fact that the carrier mean free path is longer than that of the wire length. Ballistic type transport mechanism is normally observed at the contact junction of nanowire and other external circuits, where the conductance is quantized into a integral multiple of $2e^2/h$, called the universal conductance unit (*e* is the electronic charge and *h* the plank's constant)

In the case of ballistic type transport in metallic nanowires, the conductance quantization phenomena takes place because of the fact that the nanowire diameter be comes comparable to the electron Fermi wavelength of that metal. As compared to the ballistic transport mechanism, in diffusive type conduction mechanism, the carrier mean free path is smaller than that of the wire length. Therefore various types of scattering affect the carriers and the carrier transport mechanism of nanowires becomes quite similar to that of bulk counterpart.

The electron transport mechanism of the metallic and semiconducting nanowires is obtained by considering three parameters viz. wire diameter (d) carrier mean free path (l_c) and the de Broglie wavelength of electron (λ_e) When the wire diameter (d) is larger than the carrier mean

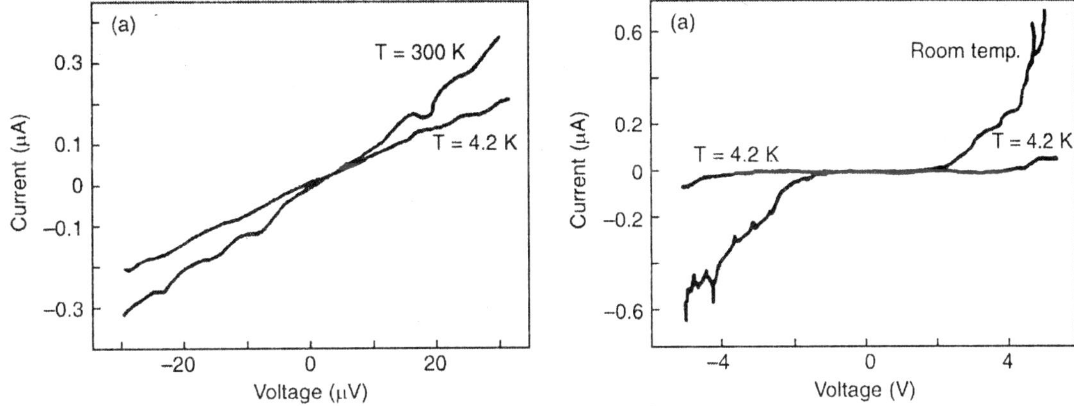

Figure 7.22 *(a) I-V characteristics of the metallic Cu nanowire at 4.2 K and at room temperature and (b) I-V curves of the wire at 4.2 K and room temperature recorded after oxidation of Cu nanowires.*

free path (λ_e) and the de Broglie wavelength of electron (λ_e) i.e $d >> l_c$ and $d >> \lambda_e$, the electron transport mechanism of nanowire remains similar to that of bulk counterpart. However if the wire diameter (d) become smaller or comparable to that of the carrier mean free path (l_c) and much larger than the de Broglie wavelength of electron (λ_e) i.e. $d \leq l_c$ and $d >> \lambda_e$, the electron transport mechanism of nanowires follows the classical laws of physics. But, when the carrier mean free path (l_c) and the de Broglie wave length of electron (λ_e) both become comparable to the wire diameter i.e. $d \leq l_c$ and $d \sim \lambda_e$, the phenomena of quantum confinement occurs in nanowires which includes significant change in the density of electron states. The electronic density of states (DOSs) of nanowires is diameter dependent. For example, electronic DOSs to Ti nanowires of diameter < 1nm exhibit some discrete peaks like molecules as shown in Figures 7.23 (a) and (b). Thicker Ti nanowires (diameter > 1 nm) show bulk Ti like electronic DOSs, because by the overlapping of discrete molecular levels with each other as shown in Figures 7.23 (c) and (d).

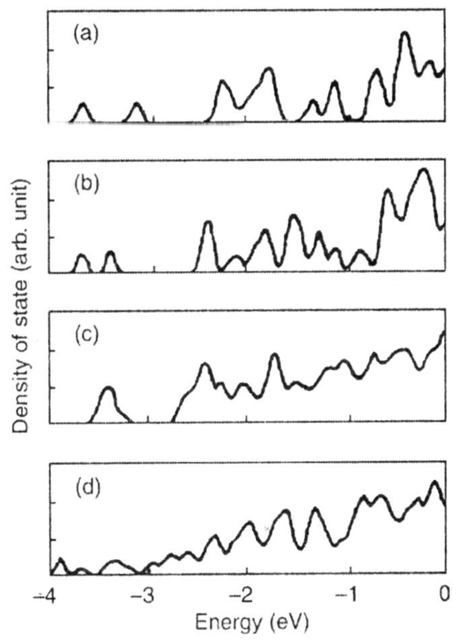

Figure 7.23 *Electronic densities of states (DOSs) of Ti nanowires (a) and (b) wire diameter<1nm and (c) and (d) wire diameter>1nm with 0.05 eV Gaussian broadening. The Fermi level is set as zero on the energy axis.*

However it is found that the conductance of nanowires strongly depends on their crystalline structure. For example in the case of perfect crystalline Si nanowires having four atoms per unit cell, generally three conductance channels are found. One or two atom defect either by addition or removal of one two atoms may disrupt the number of such conductance channel and may cause variation in the conductance. Variation in conductance due to the above reason is shown in Figure 7.24 where conductance is plotted as a function of energy for perfect Si nanowires and single atom addition defect nanowire.

The change in the surface conditions of the nanowires causes remarkable change in the transport behaviour. As illustrated in Figure 7.25 it is seen that change in electrical conductivity is caused by the variation of the surface scattering phenomena of carriers in nanowires when the diameters of the nanowires are changed.

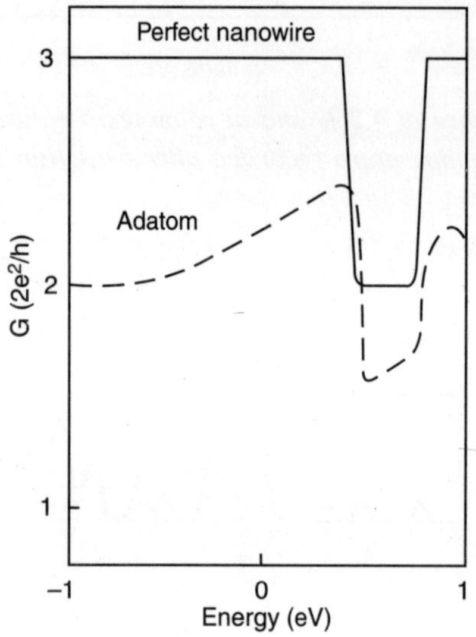

Figure 7.24 *Conductance, $G(2e^2/h)$ of silver nanowires with single-atom defects: Adatom (dotted line) and for perfect nanowires (solid curve)*

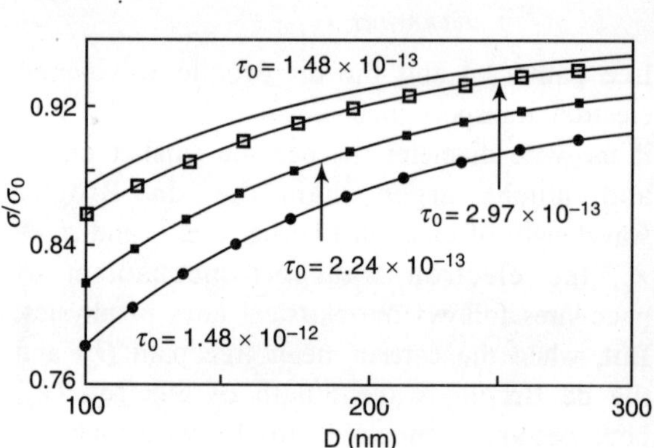

Figure 7.25 *Relative conductivity vs nanowire radius (D) at 300K where σ_0 is the bulk conductivity to scattering time τ_0 (in s).*

It is also found that the redox molecules like cobalt phthalocyanine layer coating on the n-InP nanowires change their conductance. The resistance of Pd nanowire changes due to the adsorption of hydrogen gas molecules on the wire surface.

Electron transport mechanism of superlattice nanowires have application in thermoelectronics nanobarcodes, nanolasers, one dimensional waveguide and resonant tunneling. Superlattice nanowires consist of periodic modulation of chemical composition and crystal structure in the form of quantum dots along the wire axes. The electron transport phenomena between the quantum dots are governed by the quantum tunneling mechanism in which specific regions of the nanowires with a particular composition act as the potential barrier. As for example in case of the InAs-InP superlattice nanowires the InP dots as the potential barrier. It is to be noted that the interfaces between the quantum dots have the capability of blocking the phonon conduction along the wire axes without affecting the electrical conduction.

Applications

Nanowires possess the potential for use in numbers electronic applications. Junctions of semiconductor nanowires such as GaAs and GaP have shown good rectifying characteristics. Several semiconductor devices such as junction diodes memory cells and transistors, FETs LEDs and inverter etc have already been fabricated using nanowire junctions. Figure 7.26(a) shows a diagram of FET made of nanowires. The LED fabricated by a crossed junction of p-Si and n-GaP with their-V characteristics and EL spectra are shown in Figures 7.26 (b) and (c) respectively. These basic building blocks of semiconductor devices can be used to fabricate complex IC chips of much reduced dimension.

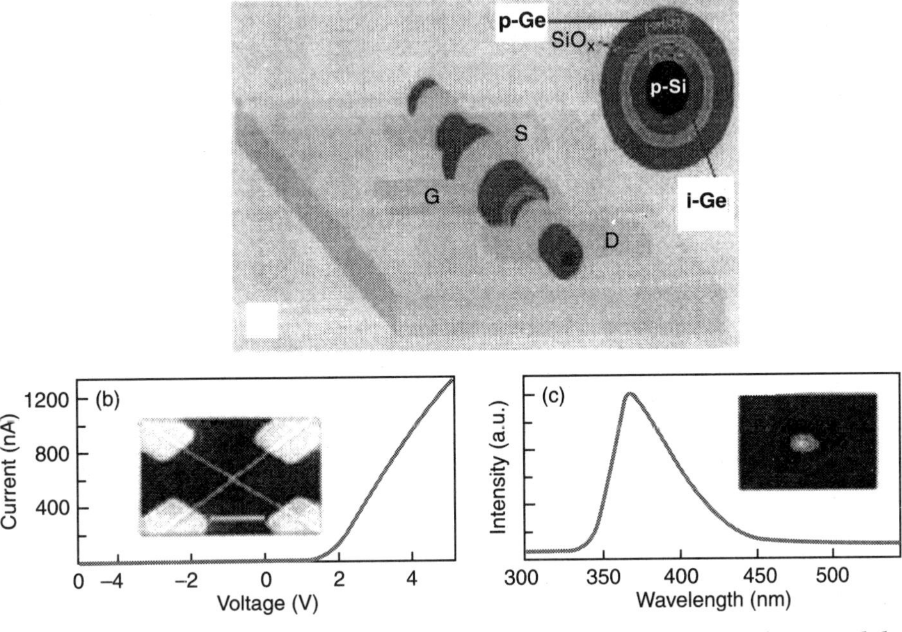

Figure 7.26 *(a) A coaxially fabricated nanowire FET. Inset: Cross sectional view of FET with layers of i-Ge (10nm) SiO_2 (4 nm) and p-Ge (5nm). The source (S) drain (D) and gate (G) electrodes are also shown, (b) I-V characteristics of a p-Si/n-GaN crossed junction LED. inset: Scanning electron microscopic (SEM) image of the junction LED (c) EL spectra of the LED. Inset: EL image representing the spatial of intensity having a maximum at the crossed junction.*

These junction devices exhibit transport and current rectifying properties similar to their parent counterpart. Several attempts are made to fabricate nanowire diode. It is demonstrated that the crossings of phosphorus (n) and boron (p) doped Si nanowires form p-n junction diodes. Similar type of junction diode are formed by n and p type InP nanowires Figures 7.27(a) and (b) show an in plane gate (IPG) quantum wire transistor (QWRTr) and a warp gate (WPG) quantum wire transistor made of GaAs nanowires, respectively. A WPG QWR based binary decision diagram (BDD) device is shown in Figure 7.27 (c) and Figure 7.27 (d) represents the SEM image of hexagonal BDD 2 bit digital Adder circuit.

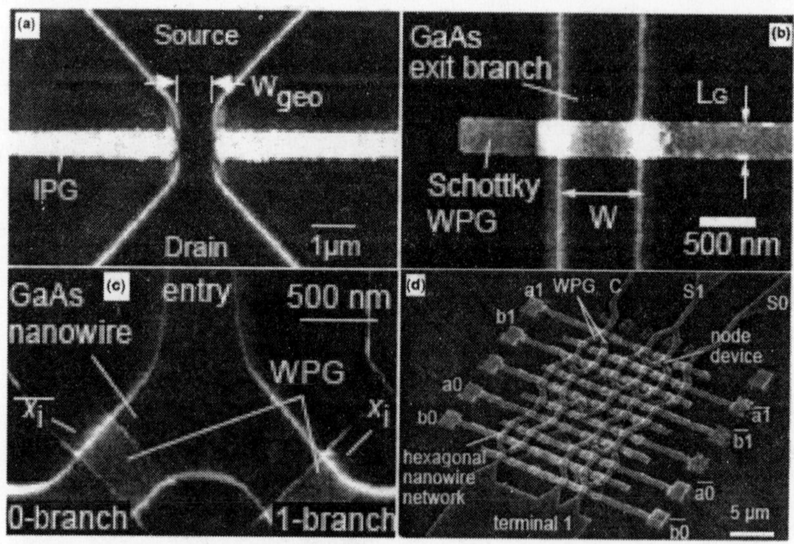

Figure 7.27 *SEM image of (a) IPG QWRTr, (b) WPG QWRTr (c) WPG QWR based BDD device and (d) hexagonal BDD2 bit digital adder.*

The field effect transistors made of nanowires exhibit remarkably modified conductance behaviour and they are very attractive because of their morphologic advantages. The operational speed of FETs made of nanowires is much faster compared to that of bulk FETs.

Several heterogeneous junctions are formed inside single superlattice nanowire by varying the materials compositions along the wire axes perpendicular to the wire axes. Nanowire junctions also perform certain logical operations and are used as logic gates (see Figure 7.28). The electroluminescence (EL) properties of nanowires are used for different opto-electronic applications For example Figure 7.29(a) represents three crossed junctions made of p-Si nanowire with n-GaN, n-CdS and n-CdSe nanowire and their corresponding EL spectral peak (Figure 7.29 (b)).

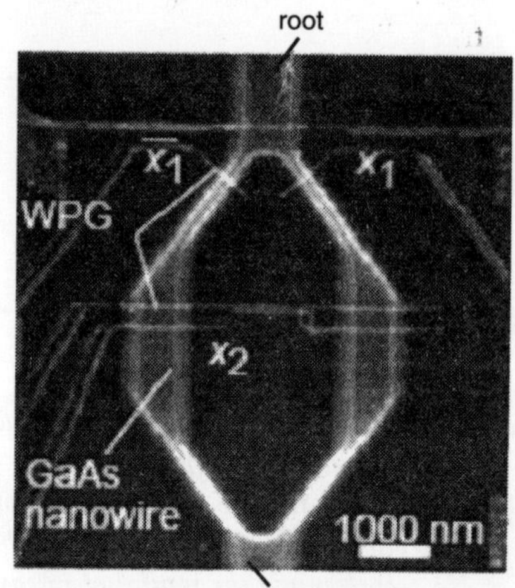

Figure 7.28 *SEM image of WPG single electron BDD OR logic gate.*

Nanowire arrays are used in field emission devices such as flat panel displays because of the significant drop in the work function of the surface electrons in those small diameter and high curvature tips of the nanowires

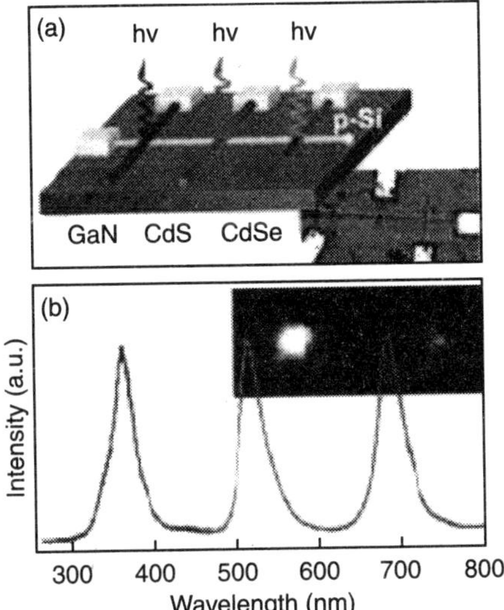

Figure 7.29 *(a) Three crossed junction diodes made of different nanowires. Inset: SEM image of the junction is shown and (b) corresponding EL spectrum from the three junctions are shown.*

CHEMICAL PROPERTIES

Nanowires exhibit chemical properties because of their enhanced surface to volume ratio, high aspect ratio, large curvature at the nanowire tips and huge number of surface atoms. The high chemical reactivity interesting electrical and electronic properties of nanowires make them attractive for sensor devices applications. Figure 7.30 shows an Au/Pi/Au nanowire with a

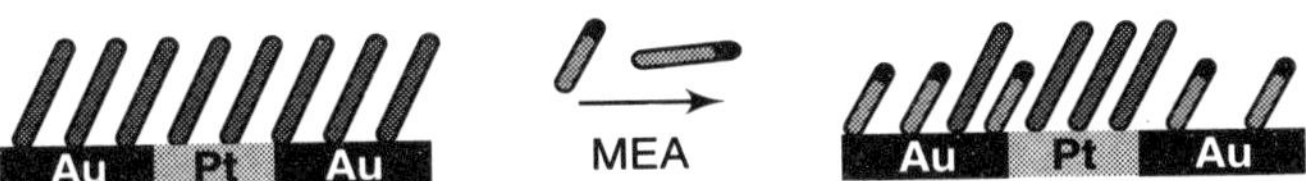

Figure 7.30 *An Au/Pt/Au nanowire with a butane-isonitrile monolayer replaced by MEA from all the Au part of the nanowire.*

butaneisonitrile monolayer adsorbed on the wire surface. The butaneiso nitrile molecules are adsorbed to the Au part of the nanowire and replaced by 2-mercaptoethylamine (MEA) when the wire is exposed to MEA because of the enhanced chemical reactivity of Au over Pt. Chemical and biological sensors made of nanowires as sensing probe exhibit enhanced sensitivity and fast responsivity compared to the conventional sensors, as they need less electrical power to work. The capability of providing real space time distribution of a particular species improves significantly if arrays of nanowires are used in the sensing probe. These types of nanowires sensors are used estimate the real concentration of particular substance quickly the function of nanowire based conductivity which arises due to the adsorption of the molecules that are required to be detected. This is shown 7.31 figure where the electrical current as a function of time, plotted for In_2O_3 nanowires and shows enhancement in current when the protein specific antigen (PSA) molecules are adsorbed on the wire surface. As the fast and accurate detection of

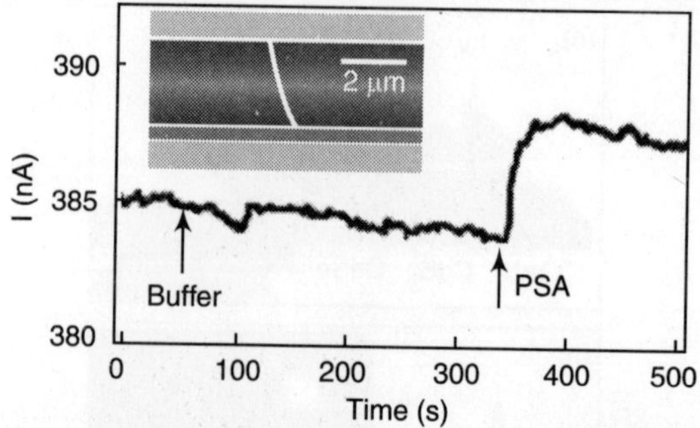

Figure 7.31 *The curve represent the change in current as a function of time for an individual In_2O_3 nanowire when exposed to buffer and PSA. Inset SEM image of the single In_2O_3 nanowire sensor devices.*

various biological compounds is a primary necessity for medical diagnosis, nanowire based sensors hold a lot of promise in this area.

Applications

The nanowire based pH sensors and Pb nanowire based hydrogen gas sensors have been developed.

DOPING

Considerable efforts have been devoted to energy-band engineering by doping for controlling their electrical properties and assembling NWs into increasingly complex structures.

Carbon nanotubes (CNTs), sillicon nanowires, and ZnO nanowires/nanobelts are most outstanding examples of one dimensional (1D) nanostructures. It is known that conductivity of these 1D nanostructures is strongly affected by doping, such as CNT can be doped with nitrogen, Si can be doped with boron, and ZnO can be doped with Mg. In comparison to both carbon and silicon, the conductivity of ZnO can be tuned by changing the oxygen deficiency of vacancies. Controlling oxygen diffusion in ZnO is important for understanding its electronic and optical properties.

Numerous studies based on ZnO nanostructures have demonstrated novel applications due to their semiconducting and piezoelecric properties . For ZnO n-type conductivity is relatively easy to realize via excess Zn, or with Al, Ga, or In doping, but p-type doping has also been achieved. There are many possible strategies for doping ZnO in order to make p-n junctions for advancing the technological uses of ZnO based electronic and optoelectronic devices. The rectifier, a fundamental device for electronics, normally consists of a p-n unction diode. The role of the dopants is the formation of a p-n junction and creation of an electrostatic potential energy barrier at the junction.

Being a key functional material with versatile properties, such as dual semiconducting and piezoelectric properties, ZnO has important applications in optoelectronic devices, sensors, lasers, transducers and photovoltaic devices. In addition, the morphology and the dopant concentration of ZnO nanostructures can be controlled by tuning the growth conditions, which further broadens their applications. ZnO nanoparticles are nanotoxic, biosafe and biocompatible, and are used in many applications such as drug carriers and cosmetics.

Elastic Properties of Nanowires & Nanobelts

The elastic modulus of individual NWs is determined by an AFM based technique, by simultaneously recording the topography and lateral force image in AFM contact mode when the AFM tip scans across the aligned nanowire arrays.

This technique allows a measurement of the mechanical properties of individual NWs of different lengths in an aligned array without destructing or manipulating the sample.

The aligned ZnO NW arrays are grown using gold as catalyst by a vapor-liquid-solid method (VSL) process. By choice of an appropriate susbstrate, the epitaxial growth of ZnO on the single-crystal substrate, such as α-Al_2O_3, GaN, AIN, or $Al_{0.5}$ $Ga_{0.5}N$, yield aligned ZnO nanowires. Figure 8.1(a) shows a scanning electron microscopy (SEM) image of the as-grown ZnO nanowires on a α-Al_2O_3 sustrate, showing well-aligned distribution but a large variation in height/length. If the density of the NWs is too high and/or the NW is too long, the AFM tip is probably too big to reach the surface of the substrate without simultaneously touching two nanowires. A NW array is grown that has relatively less density or shorter length (Figure 8.1(b)), and the AFM tip can exclusively reach one NW and the growth substrate without touching another nanowire. Thus, our calculation can be carried out with considering only one nanowire. Each NW is a single crystal and grows along [0001] and is enclosed by six $\{01\bar{1}0\}$ sides facets (Figure 8.1(c)). The NWs exhibit a uniform shape, and their size is measured from the TEM image to have an average diameter of 45 nm (Figure 8.1(d)).

The elastic measurements are performed by an AFM operated in contact mode. A Si tip with 20° cone angle is used for the measurements. The rectangular cantilever has a calibrated normal spring constant of 4.5 Nm^{-1}. The set point (constant normal force) is 21 nN. The scanning velocity is kept at 7.8 at $\mu m/s^{-1}$.

The principle for the AFM measurement is illustrated in Figure 8.2. In AFM contact mode, a constant normal force is kept between the tip and sample surface. The tip scans over the top of the ZnO NW and the tip's height is adjusted according to the surface morphology and local

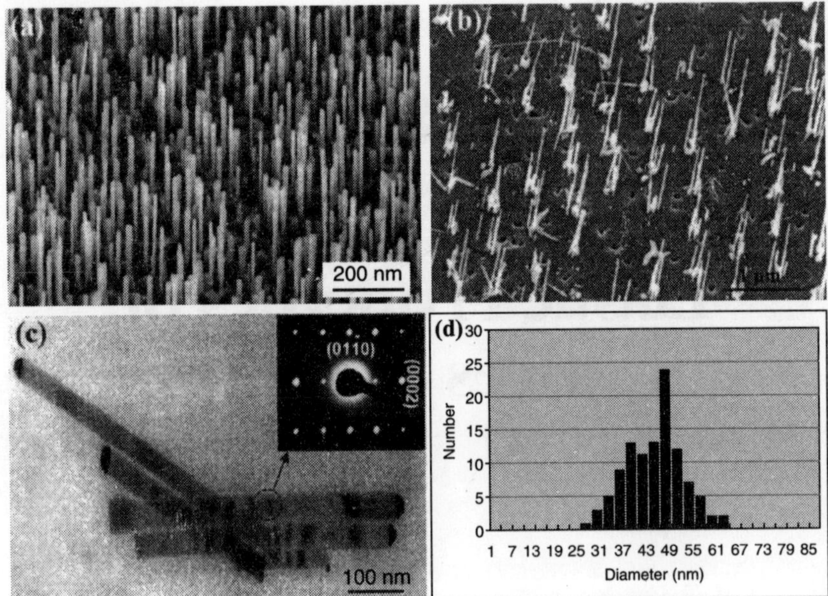

Figure 8.1 *(a, b) SEM images of ZnO nanowire arrays grown on sapphire surfaces. (c) TEM image of the nanowires with gold catalyst at their tips. Inset is an electron diffraction pattern of a nanowire. (d) Statistical distribution in nanowire diameter measured from TEM images.*

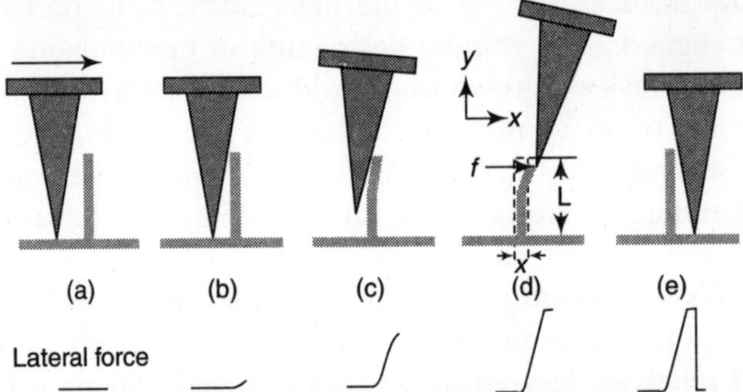

Figure 8.2 *Procedures for measuring the elastic modulus of a nanowire in the AFM contacting mode.*

contacting force. Before the tip meets a NW, a small lateral force is observed (Figure 8.2(a)). When the tip comes in contact with a NW, the lateral force increases almost linearly as the NW elastically bends form its equilibrium position (Figure 8.2(b) and (c)). At the largest bending position, as illustrated in Figure 8.2(d), the tip crosses the top of the NW, then the NW is released; the lateral force drops suddenly and reaches the ordinary level (Figure 8.2(e)). Considering the size of the NW, the thermal vibration of the NW at room temperature can be ignored.

During the scanning process, the AFM works in the following way: When the tip contacts the NW, the scanner (or cantilever) retracts at the same time in order to maintain a constant normal force, the cantilever is twisted, and a rapid change in lateral signal is detected by the photodetector. At the largest bending position, the scanner retracts to the highest position and the lateral signal reaches the maximal value. As the tip keeps moving at a low scanning speed, the bent NW is released by the tip, then the scanner extends quickly to touch the substrate, and the twisted cantilever immediately recovers to its equilibrium position. From the torsion of the cantilever, the lateral force is measured.

The as-synthesized sample is loaded upward on a specimen holder, and the tip of the AFM is scanned across the aligned NW arrays. Both the topography image (feed back signal from the scanner) and the lateral force image are recorded simultaneously. The scanning area varies from 25 to 100 μm^2. In contact mode, the AFM tip keeps in touch with the sample surface at a constant force (set point). In contact mode, as the tip scans over the vertically aligned NWs, the NWs are bent consecutively if their density is reasonably low. The bending force is captured in the lateral force image (Figure 8.3(b)), and the bending distance is directly recorded in the topography image (Figure 8.3(a)). The elastic modulus is derived from the relationship between maximal lateral force and maximal bending distance.

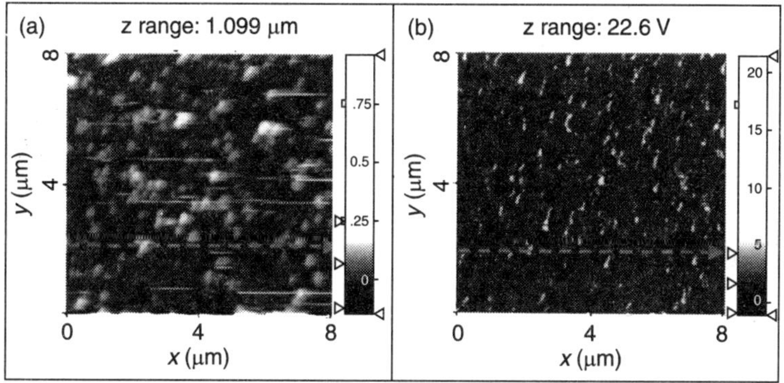

Figure 8.3 *(a) Topography image and (b) lateral force image of the aligned ZnO nanowire arrays received in AFM contacting mode. The elastic modulus of each nanowire in the scanning range is derived from these images.*

The elastic modulus is derived based on the following calculation. As the AFM tip swipes over the NW with its side and top surfaces in contact with the NW, a lateral force is applied on the top of the NW. At the largest bending position (Figure 8.2(d)), the applied lateral force reaches the maximum, so does the bending distance. The lateral force is directly read from the lateral friction force image, and the corresponding bending distance is read from the topography image. From the geometrical relationship illustrated in Figure 8.2(d), when a vertical NW experiences a lateral force f parallel to the scanning direction, the displacement of the NW under the small deflection approximation is expressed as.

$$EI\frac{d^4x}{dy^4} = (f_0 + f)\delta(y - L) \tag{1}$$

where f_0 is the projected component of the friction force between the tip and the NW in parallel to the tip scanning direction, E and I are the elastic modulus and momentum of inertia of the NW, x is the lateral displacement perpendicular to the NW, y is the height from the fixed end (root) of the NW to the point where the lateral force is applied, which is approximately the tip of the NW (y = L), and the contact is assumed to be a point, and L is the length of the NW. f_0 is much smaller than the bending force f especially when the scanning speed is low; it thus is dropped out in equation 1. The applied lateral force f is expressed as

$$f = 3EI\frac{x}{L^3} \tag{2}$$

From Hook's law, the spring constant is $K = f/x$; thus, the elastic modulus is expressed as a function or the spring constant K, the length of the NW, and the momentum of inertia: $E = KL^3/3I$. The ZnO nanowire growing along [0001] usually has a hexagonal cross section with side length of a (a is the radius of the NW), for which the momentum of inertia is $I = (5(3^{1/2})/16)\ a^4$. From the SEM and TEM images, the aligned ZnO NWs have a uniform diameter of 45 nm (Figure 8.1(d)) and heights varying from 200 to 800 nm (Figure 8.1(d)). The elastic modulus is given by

$$E = \frac{16L^3K}{15(e^{1/2})a^4} \tag{3}$$

From the topography image (Figure 8.3(a)), the bright spots with tails are from the NWs, and the tails are due to the deflection of the NWs along the scanning direction of the tip. The line scan speed of the tip is 7.8 $\mu m/s^{-1}$, and the gain is set at a medium range. At the same points in the corresponding lateral force image (Figure 8.3(b)), there are also spots. To ensure that the center of the conical tip touches the center of the NW, both curves are read from the center of the NW as indicated in the images by dashed lines. Taking a line scan across the middle point of a spot in the topography image, a curve for the scanner retracting distance verses the NW lateral displacement is obtained, as shown in Figure 8.4(a). A change in the profile occurs when the tip scans over the NW; the relatively flat part is when the tip scans on the substrate where there are no NWs. Taking a scanning profile at the corresponding line in the lateral force image (as indicated in Figure 8.3(b), the maximal lateral force for bending the NW is measured. A sudden change in the curve occurs when the tip meets the NW and the friction force increases rapidly. Parts (c) and (d) of Figure 8.4 are the enlarged portions of the curves enclosed by rectangles in parts (a) and (b) of Figure 8.4 respectively. As in Figure 8.4(c), the scanner retracting distance and the lateral displacement are the differences in y and x coordinates, respectively, between points A and B. In the corresponding lateral force curve in Figure 8.4(d), the difference in heights between points C and D is the maximum bending force of the NW. It is noticed that the profile of lateral force is not as asymmetric as shown in Figure 8.2. This is because the gain of the AFM has to be set very high to observe the ideal profile, but a high gain results in an ultrasensitivity of the measurement to system noise. A choice of medium gain produces a lateral force profile that is not as asymmetric. In addition, the nanowire vibrates after being released by the tip, and a subsequent contact of the vibrating tip with the body of the cone-shape tip also reduces the asymmetry of the profile.

The length of the nanowire is derived from the AFM image, as demonstrated in Figure 8.5. When the tip meets a NW the scanner/cantilever retracts and at the same time the tip is

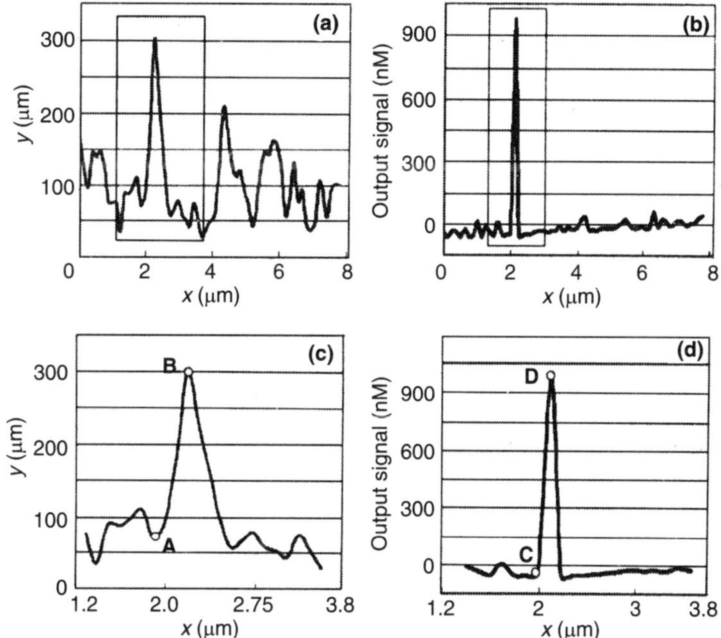

Figure 8.4 *(a, b) Line profiles along the dashed lines in Figure 3 (a, b) showing the curves of scanner retracting distance versus the nanowire lateral displacement and the lateral force versus the lateral displacement, respectively. (c, d) Enlarged portions of the areas enclosed by rectangles in (a, b), respectively. The set point of the AFM was 0.15 V, while the photodetector output was ~ 6 V at the maximum lateral force.*

deflected. From the point at which the scanner begins to retract to the largest retracting position, the vertical and lateral displacements are represented by h and d, respectively, which are measured from the topography image (Figure 8.4(a)). The true lateral displacement of the NW is considered for the bending of the cantilever. For the same force applying to the cantilever and the NW, the lateral displacement of the cantilever is 4 orders of magnitude smaller than that of the NW; thus, it can be ignored in the calculation. Therefore, the length of a NW is given by $L \approx (h^2 + x_m^2)^{1/2}$, and the bending displacement of the NW is given by $x_m = d - h \tan \theta$, where 2θ is the apex angle of the AFM tip ($2\theta = 20°$). From the same position in the lateral force image, the lateral force f_m, that caused the maximal NW bending is obtained. Combining the measured x_m and f_m from the two line profiles, the spring constant $K = f_m/x_m$ is obtained.

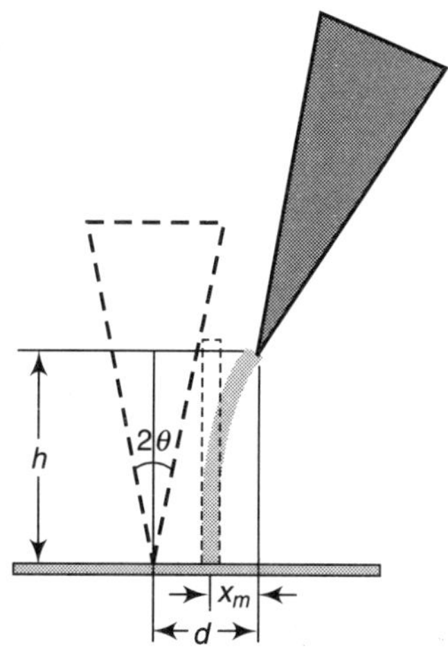

Figure 8.5 *Tip-nanowire geometry for calculating the length/height of the nanowire.*

The elastic modulus of several well aligned ZnO NWs is presented in Table 8.1. The lengths of the nanowires vary from 167.9 to 683.4 nm, the corresponding elastic modulus varies from 15 to 47 GPa. The averaged elastic modulus of ZnO NW is 29 GPa. This is consistent with the results measured by TEM. From equation 3, the estimated error in E measurement is

$$\Delta E/E = 3\,|\Delta L/L| + 4|\Delta a/a| + |\Delta f/f| + |\Delta x/x| \tag{4}$$

Table 8.1 *Measured Elastic Modulus of Aligned ZnO Nanowires*

nanowire	*length (μm)*	*elastic modulus (GPa)*
1	0.277	20
2	0.281	22
3	0.452	20
4	0.211	36
5	0.301	42
6	0.640	36
7	0.683	47
8	0.209	35
9	0.168	15
10	0.209	35
11	0.194	22
12	0.285	28
13	0.631	21
14	0.328	31
15	0.641	20

For typical values of $\Delta L = 10$ nm. $L = 300$ nm, $\Delta a = 0.5$ nm. a = 22.5 nm, $\Delta f/f = 3\%$ and $\Delta x/x = 3\%$. Equation 4 yields $\Delta E/E = 26\%$. which gives $E = 29 \pm 8$ GPa. The spread of data in Table 8.1 cannot be totally attributed to experimental error because the elastic modulus depends on the size and length of the NWs. It is pointed out that the variation in diameter as presented in Figure 8.1(d) is one of the major errors in this type of measurement because the size of the nanowire cannot be accurately measured by AFM without destructing the sample.

So, by this method for measuring the elastic modulus of individual nanowires aligned on a solid substrate using AFM has two advantages. One, it is feasible to measure the elastic modulus of an as-grown nanowire without destructing or manipulating the sample. Second, the measurements are carried out individually, systematically, and almost simultaneously for all of the nanowires aligned in the scanning area. The lengths of the NWs can be different. The topography image and the lateral force image simultaneously capture the geometrical profile and mechanical properties of all the nanowires in the area. An analysis of the images gives the elastic modulus of individual nanowires. The disadvantage of the technique is its inaccuracy in evaluating the size of individual nanowires without destructing the sample. For the ZnO nanowires grown on sapphire surfaces with an average diameter of 45 nm, the elastic modulus is measured to be 29 ± 8 GPa. This technique can be applied to one-dimensional nanostructures of any material as long as they are aligned on a solid substrate and their density is relatively low and/or heights are relatively short.

Diameter-Dependent Radial and. Tangential Elastic Moduli of ZnO Nanowires

The size-dependent elasticity of ZnO nanowires is measured using contact resonance atomic force microscopy (CR-AFM) and friction-type measurements. (Figure 8.6) Elastic properties of ZnO NWs are also measured by using CR-AFM. The indentation modulus and the Young's modulus in the radial direction of ZnO NWs that are laid down on a substrate is determined through the axial modulus and also from measurements of the deflections of NWs in various cantilever or suspended-beam configurations. Below is given a method for determining the mechanical properties of NWs as an alternative which is useful for NWs that are supported, formed, or bound longitudinally to a substrate realistic sphere-on-cylinder geometry is used for the contact between the AFM probe tip and the nanowire. The contact geometry for extracting elastic properties from measured date is a key factor for accurate determinations of the elastic moduli of ZnO NWs, with diameters that are comparable with the dimensions of the probe tip. In addition to the radial Young's modulus, another elastic property, the lateral shear modulus is determined from friction-type measurements. It is found that both the radial Young's modulus and the lateral shear modulus are size dependent for diameters smaller than 80 nm. The methods are described below:

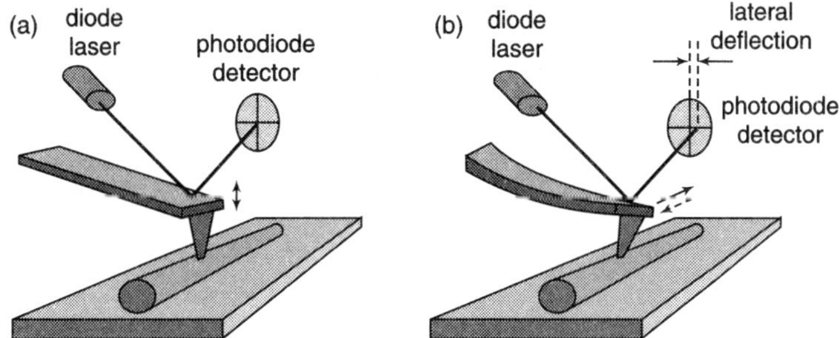

Figure 8.16 *(a) Contact resonance atomic force microscopy (CR-AFM) experiments on a ZnO NW. The radial indentation modulus of the NW is determined from the change in resonance frequency that occurs when the probe tip is brought from air into contact with the wire. (b) Friction-type measurements on top of a nanowire. The lateral deflection, which is proportional to the lateral force on the cantilever, is used to determine the tangential shear modulus of the ZnO NW.*

For, the synthesis and characterization, ZnO NWs are grown on Au catalyst layers deposited on SiO_2 substrates through thermal evaporation of Zn powder at 820 K in a N_2 and O_2 gas mixture. The NWs are then cleaved from the substrate, suspended in ethanol, ultrasonicated, and dispersed on clean Si(111) wafers.

Scanning electron microscopy (SEM) and AFM images of individual NWs before and after their removal from the substrate are shown in Figure 8.7 (a) and (b). High-resolution transmission electron microscopy (HRTEM) and electron diffraction analysis reveal that the ZnO NWs are single crystals grown along [0001] with an interplanar spacing of 2.55 ± 0.05 Å (Figure 8.7(c)), which corresponds to the known spacing between the adjacent (0002) planes of ZnO wurtzite.

The SEM and the AFM images reveal smooth wires with lengths of several microns. Due to the growth procedure and conditions, the NWs have rounded cross sections (Figure 8.7(a))

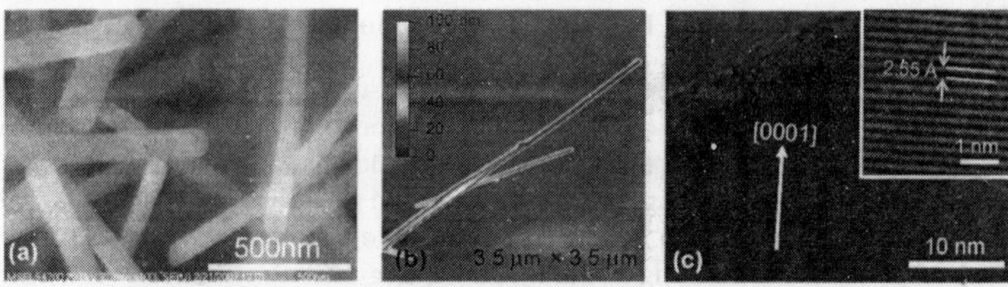

Figure 8.7 *Morphological characterization of the ZnO NWs. (a) SEM image of as-grown nanowires. (b) AFM image of a nanowire that makes contact (parallel to its axis) with the Si(111) substrate. (c) HRTEM images of a ZnO NW showing the spatial periodicity of the (0002) atomic planes; the image was taken in the vicinity of a wire end and shows the presence of a polycrystalline surface layer (upper left comer). The inset shows a magnification of the (0002) planes, identifying the interplanar spacing of 2.55 Å.*

rather than hexagonal ones. AFM profiles of the ZnO NWs are also consistent with circular cross selection as seen in Figure 8.8(a). The radius of the nanowire and that of the probe tip (approximated as a spherical surface) as calculated by deconvoluting the contributions of the tip and the ZnO NW to the average cross-sectional AFM profile (Figure 8.8(b)).

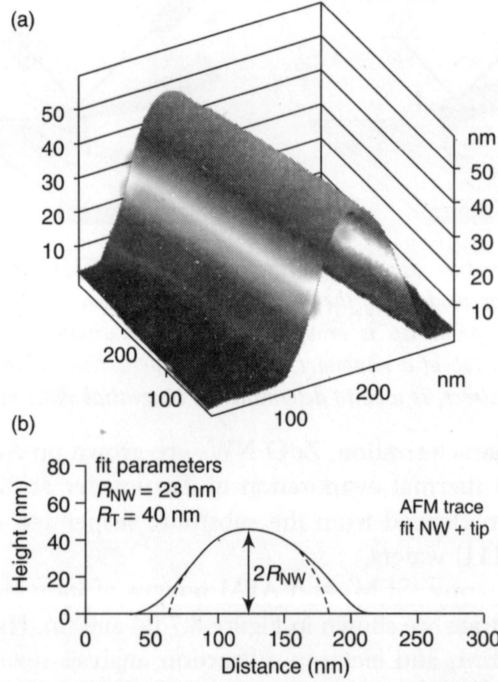

Figure 8.8 *The contact between the probe tip and a 46 nm diameter ZnO NW. (a) AFM profile of a 300 nm segment of the nanowire. (b) cross section of the profile shown in panel (a), which is used to calculate the radius of the spherical AFM tip via an erosional fit of the cross section.*

An AFM is used to find the radial and tangential stiffnesses of the contact between an AFM probe and the tested materials. The AFM probes pre single-crystal Si cantilevers with integrated Si(100) tips. The spring constants k_c of the cantilevers is in the range of $8 < k_c < 12$ Nm^{-1}, as determined from both the thermal-noise and the Sader methods. For the CR-AFM measurements, a lock-in amplifier is used to induce vibrations in the cantilever and to collect the signal on the photodiode detector. The applied load of 250 nN is kept constant throughout the measurement and is much larger than the adhesion force while it is small enough so that the contact deformations are in the linear elastic regime.

The contact between the probe tip and the tested material behaves as a spring, characterized b a contact stiffness that is specific to the direction of the movement of the tip (radial or lateral see Figure 8.6). The mathematical details necessary to calculate elastic moduli from the measured contact stiffnesses is given below, with special focus on the geometry of the contact between the nanowire and the AFM tip.

For, the contact geometry, the probe tip is assumed as a spherical surface of radius R_T. The validity of this is confirmed by taking SEM images of the tip during the course of the measurements. In a classical Hertzian model, the contact area between the spherical tip and the cylindrical nanowire of radius R_{NW} is an ellipse of eccentricity e which is calculated (numerically) from

$$\frac{R_{NW}}{R_T + R_{NW}} = (1 - e^2)\frac{K(e) - E(e)}{E(e) - (1 - e^2)K(e)} \tag{1}$$

where K (e) and E (e) are the complete elliptic integrals of first and second kind, respectively. The major semiaxis a of the contact ellipse depends (in addition to R_T and R_{NW}) on the normal force P and the reduced elastic modulus E^* of the materials in contact, via

$$a = \frac{K(e)}{\pi}\left(\frac{6PR_T}{\gamma E^*}\right)^{1/3} \tag{2}$$

where $\gamma = 2(1 - e^2)K^3(e)(2R_{NW} + R_T)/\pi^2 E(e)R_{NW}$. In the isotropic approximation, the reduced elastic modulus E^* in equation 2 is defined as

$$\frac{1}{E^*} = \frac{1}{M_T} + \frac{1}{M_1} = \frac{1 - \nu_T^2}{E_T} + \frac{1 - \nu_1^2}{E_1} \tag{3}$$

Where M_T, E_T and ν_T are the indentation modulus, Young's modulus, and Poisson ratio of the tip and M_1, E_1, and ν_1 are the same quantities for the material with which the tip is brought in contact.

In the regime of small deformations, the contact is described by a radial stiffness k^* defined as the derivative of the normal force P with respect to the change in the distance δ between the center of the sphere (tip) and the axis of the nanowire

$$k^* = \frac{\partial P}{\partial \delta} = \frac{\pi a E^*}{K(e)} \tag{4}$$

where $\delta = (9\gamma P^2/16 R_T E^{*2})^{1/3}$ for the contact geometry. Contact stiffness measurements on a reference material (the Si(111) substrate) and on the nanowire are made, and then the ratio

k^*_{ref}/k^*_{NW} is determined in order to eliminate the force P that appears in equation 4. Making use of equation 3, the radial indentation modulus M_{NW} of the nanowire is determined from

$$\frac{1}{M_{NW}} = \gamma^{-1/2}\left(\frac{k^*_{ref}}{k^*_{NW}}\right)^{3/2}\left(\frac{1}{M_{ref}}+\frac{1}{M_T}\right)-\frac{1}{M_T} \tag{5}$$

where M_T and M_{ref} are the indentation moduli of the tip and the reference Si (111) substrate respectively.

The ratio k^*_{ref}/k^*_{NW} in equation 5 is calculated using CR-AFM measurements on each ZnO NW and on Si (111). In these measurements, the resonance frequency of the probe changes, is when it is brought from air into contact with a material, with the frequency shift determined by the geometry of the contact and the elastic properties of the materials involved. Figure 8.9(a) shows histograms of the lowest resonance frequency f_1 for contact with a 46 nm diameter NW and the Si(111) substrate. No significant change is detected in the resonance f_1 measured on the substrate before and after testing the ZnO NW. With the resonance frequencies identified, the contact stiffness on each tested material is calculated using a clamped-coupled cantilever model system. The lowest resonance frequencies f_1 and f_2 are expressed in terms of the fundamental resonance in air, f, and plotte them as functions of the stiffness ratio k^*/k_c (refer to Figure 8.9(b). Once the curves .f_1/f and f_2/f are fully determined by using the measured f_1 and f_2 values for the contact with Si(111), it is sufficient to measure just one resonance (e.g., f_1) for the contact with the ZnO NW, and the contact stiffness k^* is found from the corresponding frequency curve. To verify that the contact stiffness determinations are reliable, both f_1 and f_2 are measured for ZnO NWs and it is found that both give the same value for the contact stiffness k^* (see Figure 8.9(b)).

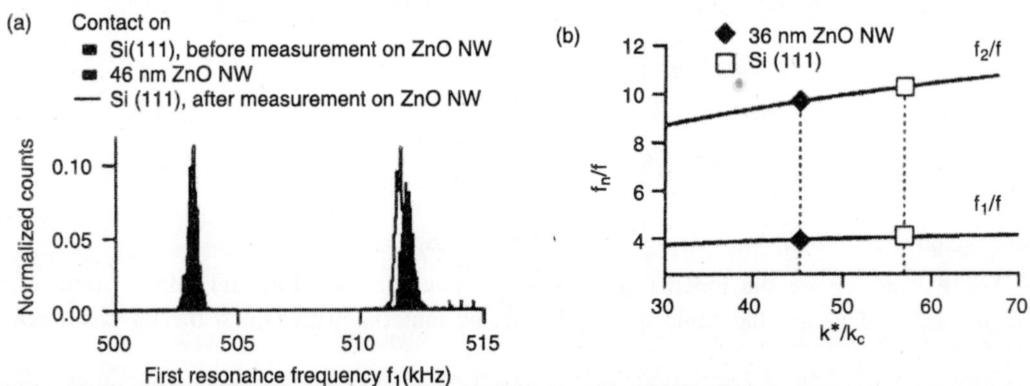

Figure 8.9 *(a) Histograms of the first resonance frequency f_1 for an AFM probe on the Si (111) substrate, on a 46 nm diameter ZnO NW, and again on the substrate. (b) Contact resonance frequencies f_1,f_2 normalized to the fundamental resonance in air land plotted as functions of the normalized contact stiffness k^*/k_c. These resonance frequencies depend on probe parameters that are determined by requiring that the experimental f_n/f (n = 1,2) values for the contact with Si(111) lie on the corresponding curves. The contact stiffness k^* for any given nanowire is determined by using experimental values of the frequency ratios on either of the two curves, f_1/f or f_2/f.*

Now, consider the measurements used for determining the tangential shear modulus of the NWs. These friction-type measurements are performed with a nonvibrating probe tip dragged back and forth along the surface of the nanowire (Figure 8.6(b), with a speed of 10 nm s^{-1} and under a constant normal force of 250 nN. The lateral contact stiffness is characterized in the no-slip linear elastic regime. In this regime, the variation of the lateral force is proportional to the lateral voltage signal V from the photodiode (Figure 8.6(b)) and is recorded as a function of the displacement along the materials (Figure 8.10). The measurements yield the lateral stiffness κ, defined as the change in lateral force F on the probe tip with displacement, $\kappa = \partial F / \partial d$. The lateral stiffness is expressed as $\kappa = \alpha s$, where α is a calibration coefficient that describes the proportionality between the lateral force and the deflection signal from the photo diode, and $s = \partial V / \partial d$ is the slope of the deflection curves (See Figure 8.10). As measured, κ includes not only a contribution κ^* from the contact between the tip and the material but also a contribution K_p that comes from twisting the AFM probe

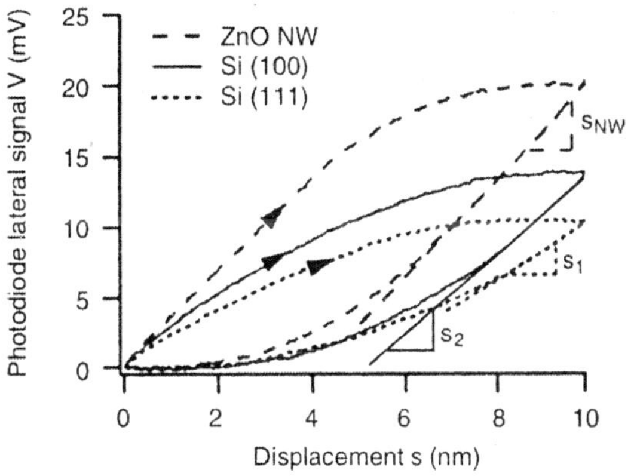

Figure 8.10 *Lateral force-displacement curves measured on top of a 36 nm diameter ZnO NW on Si(111), and on Si(100). The arrows indicate the scan directions on the friction loops. The lateral force F in each case is proportional to the lateral signal V from the photodiode (refer to Figure 1b). Instead of working with an instrument-specitic calibration between F and V the lateral contact stiffnesses s_l, s_2 and s_{NW} are determined and used; i.e., the slopes of the tangents to the friction loops in the linear non-slip regime.*

$$\frac{1}{\kappa} = \frac{1}{\kappa^*} + \frac{1}{\kappa_p}$$

(6)

The tangential shear modulus G_{NW} is calculated from the measured lateral contact stiffness κ as described below.

Using again the Hertzian model, the lateral contact stiffness κ^* is written in terms of lateral shear moduliG$_T$ and G_{NW} of the two solids in contact, that is, the tip and the nanowire

$$\frac{1}{\kappa^*} = \frac{1}{2\pi a}\left[\frac{K(e) - \nu_T B(e)}{G_T} + \frac{K(e) - \nu_{NW} B(e)}{G_{NW}}\right] \tag{7}$$

where $B(e) = [E(e) - (1 - e^2)K(e)]/e^2$ and a is given by equation 2. Equation 6 contains two instrument-specific parameters α and κ_p, with the former requiring tedious calibration. However, these parameters are eliminated by performing additional measurements on two reference materials, which is chosen to be the Si(001) and Si (111) surfaces. Equations 6 and 7 (for the contact with the ZnO NW), in combination with their analogues for the measurements on the two reference surfaces,- allow for the determination of G_{NW} in terms of the slopes s_l, s_2 [for Si(111) and Si(001), respectively], and s_{NW} defined in the linear elastic regime (Figure 8.10)

The radial indentation modulus M_{NW} is computed via equation 5 and is listed in Table 8.2 for several nanowires with diameters below 150 nm. The table includes the radial Young's modulus E_{NW}, calculated in the isotropic approximation (equation 3) and using an average Poisson ratio of $v_{NW} = 0.3$. The lateral shear modulus G_{NW} is also given in Table 8.2. The elastic moduli for ZnO is nanobelts of large cross sections is determined in order to have a clear idea of their bulk limits. The Young's modulus for the nanobelt is 112 GPa which is as observed for single-crystal ZnO bulk oriented along [0001]. The data shows that the size effects for [0001] ZnO NWs are significant for wire diameters smaller than 80 nm. It is found that all three moduli (M_{NW}, E_{NW}, and G_{NW}) approximately double their values (with respect to the bulk limits) when the NW diameter is decreased to 25 nm (Table 8.2).

Table 8.2 *The Radial Indentation Modulus M_{NW}, Young's Modulus E_{NW}, and Lateral Shear Modulus G_{NW} for Several ZnO NWs**

NW diameter (nm)	M_{NW} *(GPa)*	E_{NW}	G_{NW}
25.5	217.5	198.0	109.4
36.2	170.0	154.7	97.5
36.6	170.4	155.1	98.0
46.0	122.6	111.5	–
61.0	140.2	127.6	75.4
80.0	112.8	102.6	–
134.4	114.6	104.3	50.7
Nanobelt	109.8	100.0	51

*The elastic moduli are determined for a ZnO nanobelt with a 250 nm × 450 nm rectangular cross section (last row). The maximum uncertainty for each modulus is ± 15.0 GPa.

The elastic moduli are calculated using the sphere-on-cylinder (SOC) model for the contact between the tip and the ZnO NW, as opposed to the simpler and more widespread sphere-on-elastic halfspace (SOH) model. Despite the complexity of the SOC model, when the wire diameters are compared to the diameter of the probe tip (see Figure 8.8(b), the SOC model is used; since it yields more accurate and reliable elastic properties. A recalculation of the ZnO NW elastic moduli based on the simpler SOH geometry gives different values than those in Table 8.1. Still, the results of the sphere-on-halfspace and sphere-on-cylinder analyses converge in the limit of thick nanowires. The indentation modulus (114.6 GPa) for the largest wire computed using the SOC geometry is only slightly greater than the indentation modulus of the 250 nm × 450 nm nanobelt (~ 110 GPa), which is calculated using the SOH model.

The size dependence of the elastic properties of ZnO NWs is known by recognizing that the thinner nanowires have a relatively larger number of atomic bonds in the surface and near-surface layers and that these bonds are shorter and stiffer than their bulk counterparts. This is supported by density functional theory calculations which consistently show that the interlayer distance and the bond lengths decrease for the relaxed ZnO$(1\bar{1}20)$ and ZnO$(10\bar{1}0)$ surfaces, which are both parallel to the wire orientation. A more quantitative approach is from modeling the nanowire as a cylindrical core (bulk) and a shell (surface region) that have different radial elastic moduli E_c and E_s, respectively (for simplicity, consider the Poisson ratio to have the same value of 0.3 in the core as it has in the shell). For a core-shell nanowire with a shell thickness t and core radius $R_c = R_{NW} - t$, estimate of the radial Young's modulus E_{NW}. It is found from the analysis of strain under conditions of uniform radial stress, which yields

$$\frac{R_{NW}}{E_{NW}} = \frac{t}{E_s} + \frac{R_c}{E_c} \tag{8}$$

The strain analysis for a NW subjected to a shear load distributed uniformly on its lateral surface also yields a equation for the shear modulus G_{NW};

$$\frac{R_{NW}}{G_{NW}} = \frac{t}{G_s} + \frac{R_c}{G_c} \tag{9}$$

Substituting the Young's modulus data from Table 8.2 to equation 8, the parameters of the core-shell model are estimated to be E_c = 95 GPa, E_s = 190GPa, and t = 12 nm. The fit curve is depicted in Figure 8.11 and shows that in the bulk limit, the radial Young's modulus tends to the fitting parameter E_c. The value t = 12 nm is consistent with estimations of the shell thickness based on HRTEM images (Figure 8.10) which show a polycrystalline surface layer that is at least 5 nm thick. The shell thickness t = 12 nm is obtained which also agrees with an assessment based on a core-shell model for the axial Young's modulus. But the assessment, places the shell thickness at 4.4 nm is the result of extracting the effective axial Young's modulus of the nanowire from the sum of the flexural rigidities of a core and a shell that have different axial moduli ($E_c^{||}$ and $E_s^{||}$ respectively). It is found that the core-shell analysis for the axial Young's modulus is not applicable here because the expressions of core-shell models for elastic properties of nanowires depend on the particular elastic property being analyzed and on the type of deformation to which the nanowire is subjected. However, for both the axial Young's modulus and the radial modulus, suitably designed core-shell models appear to capture the increase of the corresponding elastic modulus with decreasing nanowire diameter.

With the shell thickness determined from fitting the E_{NW} values, the parameter t is found to be in consistent with the shear modulus data. Using t = 12 nm in eq 9, it is found that $G_c = G_s/2$ = 55 GPa. The curve (equation 9) shown in, Figure 8.11, tends to the bulk limit value G_c when the wire radius increases. Therefore, the shell thickness (t = 12nm) estimated from the Young's modulus data is also consistent with a core-shell model that describes the shear modulus (equation 9).

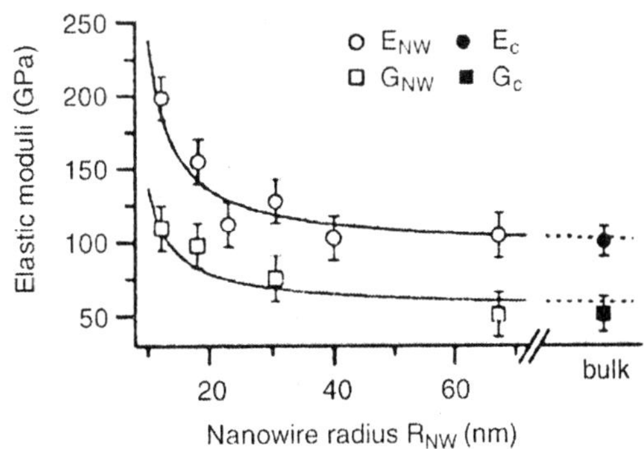

Figure 8.11 *Elasic moduli E_{NW} and G_{NW} obtained form measurements (data points)and filtted as functions of the nanowire radius according to equation 8 for E_{NW} and equation 9 for G_{NW}.*

A closer look at the parameters (t = 12 nm, $E_s = 2E_c$ = 190 GPa) for core-shell model (equation 8) and for axial deformations (t = 4.4 nm, $E_s^{||} = 1.5\,E_c^{||}$ = 210 GPa) show that, both core-shell models remain qualitative because, in both of them, the thickness is considered as the radius of the smallest nanowire and the value E_s closely corresponds to the thinnest nanowire. In neither of the models the Young's modulus, saturate at a constant E_s as the diameter is decreased beyond a certain value, which means that the fits only hold within the diameter range.

As the ZnO NWs have a polycrystalline shell (Figure 8.7) it cannot be in epitaxial relation with the core of the nanowire, which means that the stress created at the interface effectively modifies the elastic properties of the core. Even in the absence of a polycrystalline shell, the strain experienced by the surface layers due to reconstruction and relaxation depends on the diameter of ZnO NWs, and thus, the concept of a shell with constant elastic modulus hold as an approximation.

The above approach is for investigating the mechanical response of NWs in the following ways: (a) no macroscopic displacement of the wires (or segments thereof) is involved, (b) when nanowires are used as substrate-supported parts of piezoelectric or electromechanical devices, the investigations can be performed that is, by taking the substrate (as is) into the AFM, and (c) detection of defects in the NWs can be readily made from the elastic response at various locations on the wire surface. Using the above methodology for ZnO NWs oriented along the [0001] direction, it is found that both the lateral shear modulus and radial indentation modulus increase significantly with decreasing nanowire diameter, which is due to a surface stiffening effect. Thus the shows core-shell model of the nanowire shows that these results can be understood by a comparison between the elastic properties of the bulk core and those of a (stiffer) surface shell of roughly constant thickness, with the elastic properties of the shell becoming predominant in the limit of small ZnO NW diameters.

PROPERTIES OF ZnO NANOBELTS

Size dependence

The Young's modulus of ZnO nanobelts is measured using an atomic force microscope following the modulated nanoindentation method.

Experiment

Their young's modulus is found to decrease significantly from about 100 to 10 GPa, as the width-to-thickness ratio increases from 1.2 to 10.3. This behavior is explained by a growth-direction-dependent aspect radio and the presence of stacking faults and nanobelts growing along particular directions.

In the modulated nanoindentation technique, a modulated signal induces a normal oscillation of the scanner supporting the sample, while the normal force is monitored by the deflection of the cantilever (Figure 8.12(a)) The frequency of the oscillations is set at 1.348 kHz.

The oscillations are small, about 1Å a, so that the tip sticks on the nanostructure. During the stick regime, the normal force F necessary to vertically move the nanostructure by d_{tot} with respect to the cantilever support coincides with the force needed to elastically stretch two springs in series: the cantilever, with normal stiffness k_{lever}. and the tip-sample contact with normal stiffness $k_{contact}$ The total stiffness k_{total} is given by:

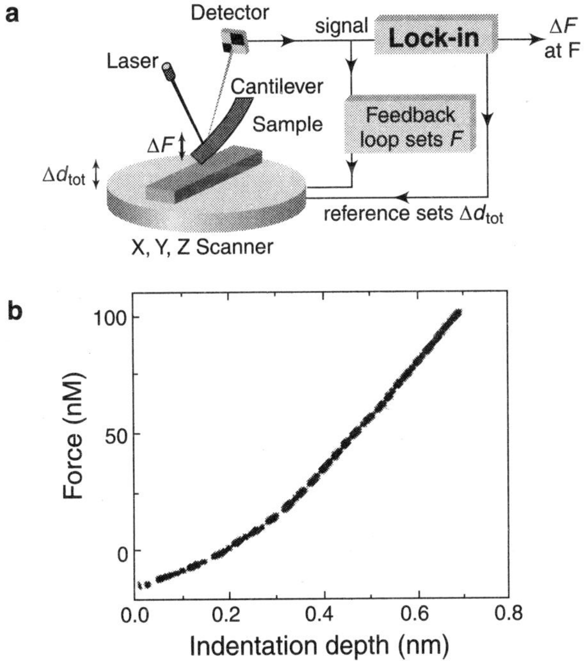

Figure 8.12 *(a) Experimental setup for the modulated nanoindentation method. (b) Normal force as a function of indentation depth for a nanobelt of width-to thickness ratio 2.9.*

$$\frac{\partial F}{\partial d_{tot}} = k_{total} = \left(\frac{1}{k_{lever}} + \frac{1}{k_{contact}}\right)^{-1} \tag{1}$$

d_{tot} is the sum of the contact deformation and the cantilever bending. Since k_{lever} is known, a measure of $\partial F/\partial d_{tot}$ at different normal loads F_o will yield k_{tolal} as a function of $\underline{F}_o$.

During the experiment, the value of $\partial F/\partial d_{tot}$ is measured by means of a lock-in amplifier. The lock-in amplifier vertically modulates the sample position by Δd_{tot} by exciting the piezo-tube of the AFM scanner with a sinusoidal signal of given amplitude and frequency, and it simultaneously measures the signal ΔF extracted from the AFM photodiode (PSD) via a signal-access module (Figure 8.13).

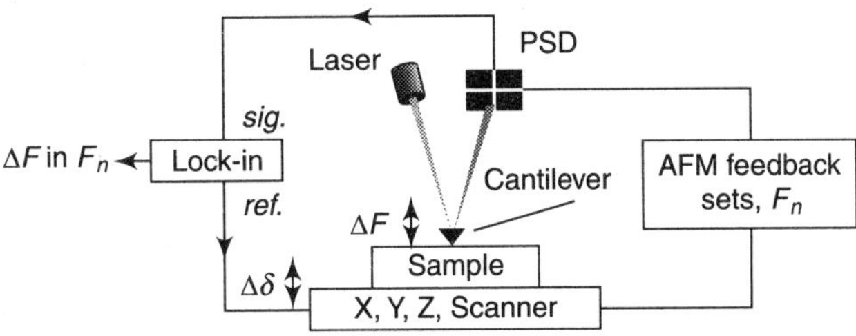

Figure 8.13 *Experimental setup for the modulated nanoindependent method.*

The elastic modulus is deduced from the k_{total} as a function of F_o by modelling the contact between the AFM tip and the nanobelt with the Hertz model. Since the nanobelt width is generally much larger than the AFM tip radius, the system is modelled as a sphere (tip) indenting a flat surface, in which case, the normal contact stiffness is proportional to the contact radius a.

$$k_{contact} = 2E^*a \tag{2}$$

where E^* is the reduced modulus of elasticity defined by $E^* = [(1 - \upsilon^2_1)/E_1 + (1 - \upsilon^2_2)/E_2]^{-1}$ with, υ_i and E_i the Poisson's ratio and Young's modulus of the indenter ($i = 1$) and sample ($i = 2$) respectively. The following values are used; $\upsilon_i = 0.27$, $E_1 = 169$ Gpa for the silicon tip, and $\upsilon_2 = 0.3$, an average value from the elastic constants tables. The contact radius a is given by

$$a^3 = \frac{3}{4}\frac{R\cdot(F + F_{adh})}{E^*} \tag{3}$$

where R is the tip radius and F_{adh} is the tip-sample adhesion force which is experimentally determined The indentation depth z is obtained by integrating $k_{contact}$ obtained at different values F of the normal force:

$$z(F) = \int_{F(z=0)}^{V_{max}} \frac{dF^*}{k_{contact}(F^*)} \tag{4}$$

where F_{max} is the maximum normal force used in the experiments. F (z = 0) can be negative due to the presence of adhesive force (figure 8.12(b)). Combining equations (1)–(3), the data k_{total} against F are fitted with:

$$k_{total} = \left\{\frac{1}{k_{lever}} + \frac{1}{2E^*[3/4\cdot R\cdot(F + F_{adh})/E^*]^{1/3}}\right\}^{-1} \tag{5}$$

where E^* is the only unknown fit parameter.

This modulated nanoindentation technique is tested on a silicon substrate) and it yields an average value of 143 GPa. The same silicon tip of radius around 60 nm, is used for the images and the modulated nanoindentation experiments. The normal cantilever spring constat 3 $N^{-1}m$ is calibrated.

The ZnO nanobelts used are prepared by vapour deposition, and deposited on a flat silicon substrate. The nanobelt samples are characterized by scanning electron microscopy and AFM operating in non-contact mode in air. The normal cantilever spring constant, 48 N/m is calibrated by using 'Sader' method. The amplitude of the oscillations for the modulated nanoindentation is 1.5 Å and the frequency is 14.018 kHz.

The scanning electron microscopy and AFM images show that the nanobelts have a rectangular cross-section, with lateral dimensions of several tens of nanometres to a few microns an lengths up to a few millimetres (Figure 8.14). As two different nanobelts are used one nanobelt is 450 nm thick and 1000 nm wide (labelled NB1), and the other is 820 nm thick and 1100 nm wide (NB2). Both are several hundreds of microns long.

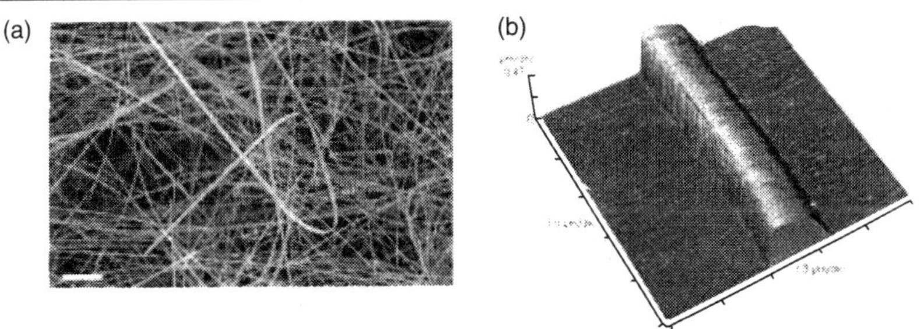

Figure 8.14 *(a) SEM image of the ZnO naobelts. The scale bar is 20 μm. (b) AFM image (5 × 5 μm) of a ZnO nanobelt.*

Figure 8.15 Shows the total stiffness against normal indentation force for NB1 (Figure 8.15(a)) and NB2 (Figure 8.15(b)). Using $v_1 = 0.27$, $E_I = 169$ GPa for the silicon tip, and $v_2 = 0.3$, an average value is calculated using the elastic constants from tables and the data gives Young's modulus values of 55 and 108GPa for NB1 and NB2, respectively, (which are consistent with those obtained from the mechanical resonance technique or other AFM methods.)

The significant difference in Young's modulus between the two nanobelts of distinct width-to-thickness ratio in the range 1–2 indicates that there is a size dependence of the elastic properties of the ZnO nanobelts. From other reported data, where the highest Young's moduli are measured on the nanobelts with the smallest width-to-thickness ratio they are usually lower than 1.5 (Figure 8.16); it is concluded that the modulus does not depend on the width-to-thickness ratio.

TEM images and electron diffraction patterns show that the ZnO nanobelts have a wurtzite structure and grow mainly along [0001] without defects or dislocations (Figure 8.17 (a) and (b)). There are indications that a few of these nanobelts are actually, nanowires, i.e., no

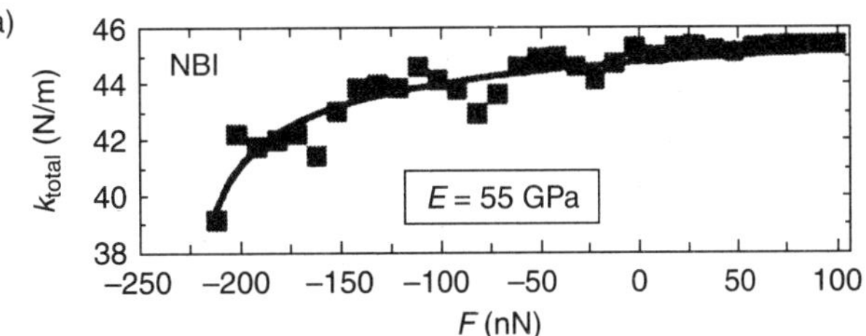

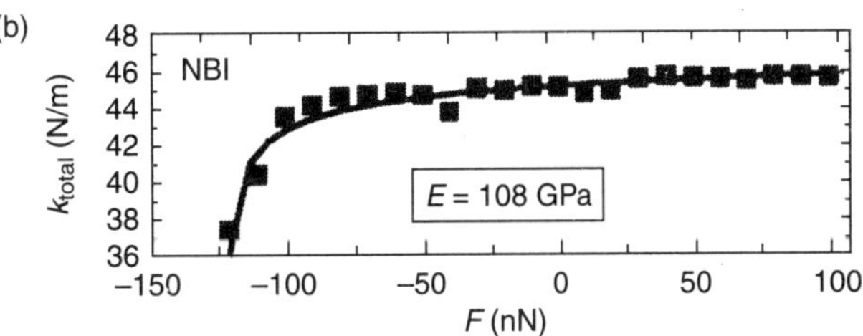

Figure 8.15 *Total stiffness versus normal identation force for (a) 1000 × 450 nm NB1 (b) 110 × 820 nm NB2.*

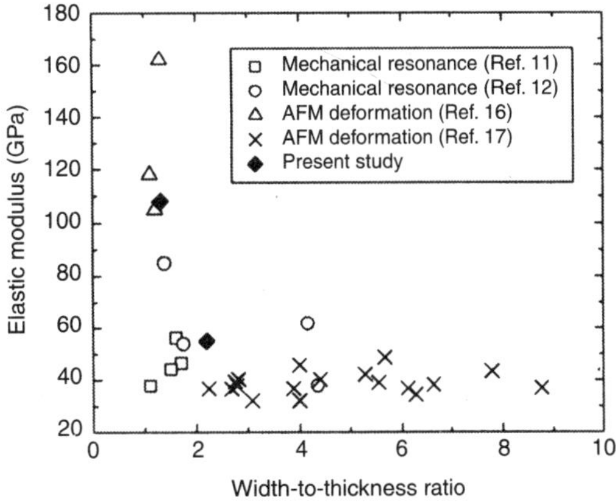

Figure 8.16 *Elastic modulus of ZnO nanobelts as a function of the width to thickness ratio.*

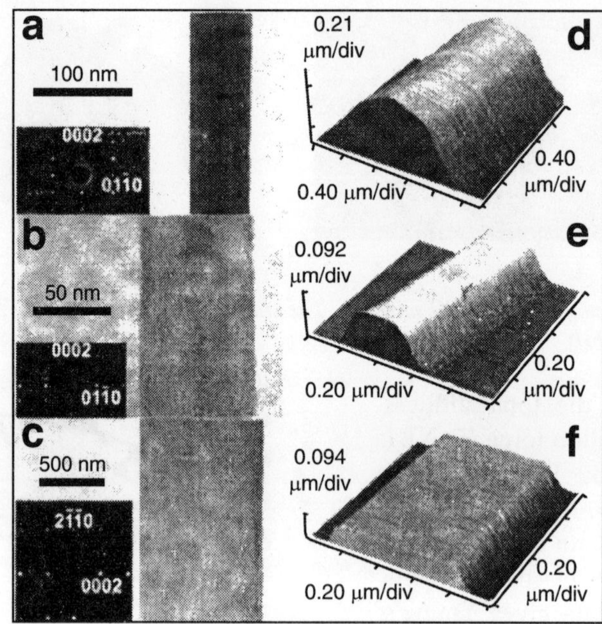

Figure 8.17 *TEM images of (a) nanowire; (b) nanobelt grown along the [0001] direction; (c) nanobelt grown along the [2$\bar{1}$ $\bar{1}$ 0] direction with a stacking fault parallel to the (0001) surface and present over the entire length of the nanobelt. Insets show the corresponding electron diffraction pattern. Note that the nanobelt in (c) is wider than 500 nm, From figure 3 in this nanobelt the narrow side surface is the (0001). AFM images of (d) nanowire (2 μm × 2 μm image); (e) nanobelt with a width-to-thickness ratio of 1.9 (1 μm × 1 μm image); (f) nanobelt with a width-to-thickness ratio of 9.5 (1 μm × 1 × m image).*

rectangular or square cross section (Figure 8.17(a)). Less common growth directions are [01$\bar{1}$0] and [2$\bar{1}$$\bar{1}$0]. Nanobelts grown along these directions present the polar (0001) surface at the side surfaces. Figure 8.17(c) shows a TEM image of a nanobelt grown along the [2$\bar{1}$$\bar{1}$0] direction with the polar (0001) surface on the narrower side. This nanobelt has a single stacking fault running parallel to the (0001) surface over the entire length of the nanobelt (Figure 8.17(c)) consistent with previous observations of stacking faults parallel to the (0001) surface.

The distribution of the nanobelts growth direction is reflected in the size and shape distribution of the nanobelts as observed by AFM (see Figures 8.17(d) to (f) and 8.17(a) to (c). It is shown that 85% of the nanobelts have a width-to-thickness ratio(w/t), smaller than 3 (Figures 8.18(a), (d) and (e). Among these nanobelts. 27% present a circular/polygonal cross section characteristic of nanowires. For these nanowires, the average w/t is 1.4 ± 0.2 (see Figures 8.18(a) and (d). Finally, only 15% of the nanobelts have 3 < w/t < 10 (see Figures 8.18(a) and 8.17(f).

To investigate the size dependence of the elastic properties of ZnO nanobelts, the Young's modulus, E_{NB} of 14 different nanostructures (nanobelts or nanowires) is measured as a function of surface-to-volume ratio, w, and t (see Figures 8.18(d) to (f)). No clear correlation is found between E_{NB} and surface-to-volume ratio, w, and t. The surface-to-volume ratio is defined here

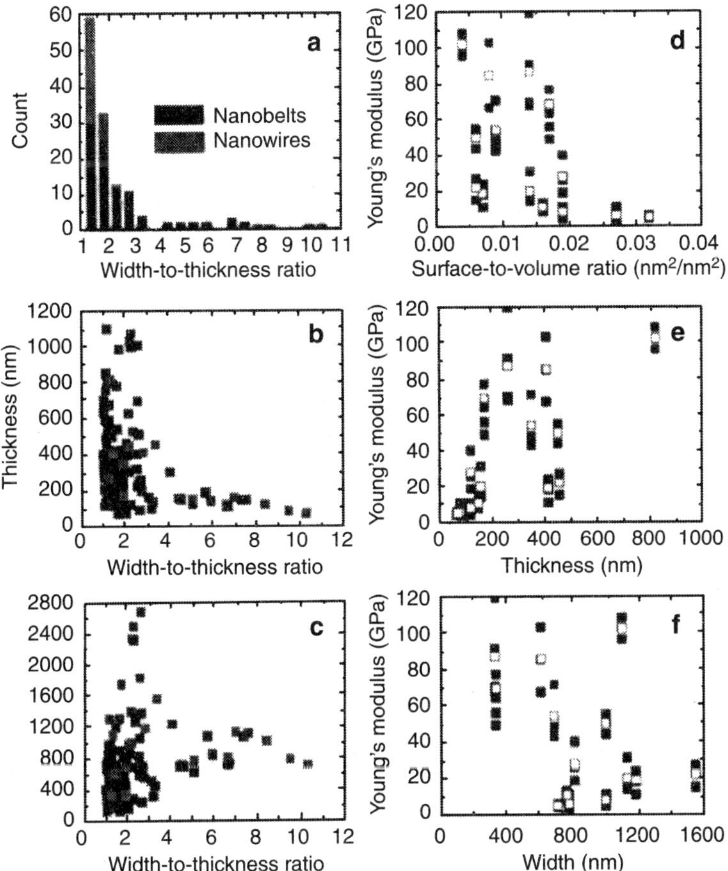

Figure 8.18 *Characteristics of ZnO nanostructures determined by AFM: (a) width-to-thickness ratio distribution. (b) thickness against width-to-thickness ratio, and (c) width against width-to-thickness ratio. The Young's modulus of 14 of the nanobelts analyzed in (a), (b), and (c) is plotted as a function of (d) surface-to-volume ratio. (e) thickness, and (f) width. In (d–f), the solid symbols are data points, and the open symbols are average values of the measurements from the same nanobclt.*

as the ratio between the perimeter and the area of the cross section. Most of the nanobelts are greater than 100 μm in length (the smallest length being 40, μm), the error due to end effects is less than 1 %. Among the nanobelts a few have similar w or t but different w/t. For example, two nanobelts with comparable thicknesses t = 174 nm and t = 161 nm but different w/t of 1.94 and 7.01 have average moduli of 69 and 20 GPa, respectively. Similarly, two nanobelts with w ~ 1000 nm but w/t = 2.22 and 8.21, have average moduli of 50 and 8 GPa. respectively. This show that the key parameter controlling the elastic properties of the nanobelts is the width-to-thickness ratio.

Figure 8.19 shows E_{NB} against w/t for ZnO nanobelts. A clear tendency emerges from this graph, showing that the Young's modulus decreases from about 100 GPa to about 10 GPa as w/t increases from 1.2 to 10.3. A sharp decrease is observed for ratios between 2 and 3, while for ratios over 3, the Young's modulus remains constant. The w/t dependence of the Young's modulus is shown in Figure 8.19. It is also found that the highest Young's moduli, above 100 GPa, is measured for nanobelts with the smallest w/t usually lower than 1.5. Also measured is the elastic modulus of four ZnO nanobelts with 1.4 < w/t < 4.4 and 85 > E_{NB} > 38 GPa, is obtained with E_{NB} = 85 GPa corresponding to w/t = 1.4.

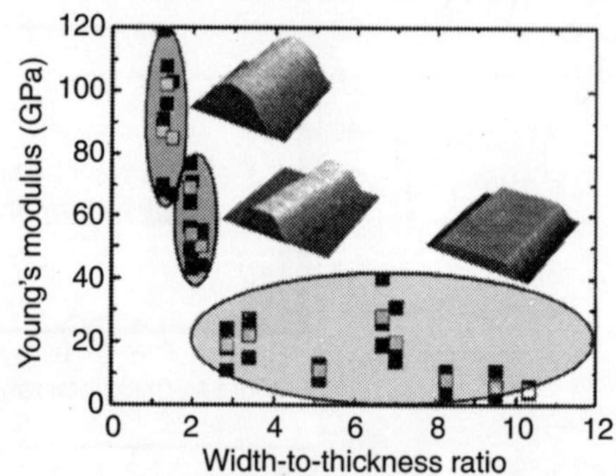

Figure 8.19 *The Young's modulus of ZnO nanobelts measured with an atomic force microscope by means of the modulated nanoindentation method. The elastic modulus is found to depend strongly on the width-to-thickness ratio of the nanobelt, decreasing from about 100 to 10 GPa, as the width-to-thickness ratio increases from 1.2 to 10.3.*

The origin of the low elastic modulus of nanobelts with high w/t; surface effects and the possibility of a structural phase transition shows that the size dependence of ZnO nanowires moduli can be explained by considering the nanowire as a composite wire composed of a core with a modulus similar to the bulk and a shell with a higher modulus.. The increase of the modulus with a decreasing diameter of the wires is due to composite modulus, where the higher shell modulus dominates at large surface-to-volume ratios. However, the surface-to-volume ratio is more than 1 order of magnitude smaller than that required for surface effects (0.004–0.032 compared with 0.08 nm^{-1}).

A structural phase transition leads to a sudden stress drop and therefore an apparently low Young's modulus in the nanobelt. At a pressure around 9 GPa, a phase transition is observed in ZnO crystals and this phase transition pressure can be as low as 6 GPa. Considering a maximum load F of 200 nN, E* = 75 GPa, a tip radius of 60 nm, (from equation 3) gives a contact radius of 5 nm, and a pressure of 3 GPa. No hysteresis or force variations are observed when the tip is approaching or retracting during the indentation process. The force-indentation curve yields an exponent of 1.50 ± 0.05.

Due to no surface effects or structural phase transitions it is necessary to know the relationship of w/t to any structural property of the nanobelts. X-ray diffraction on single ZnO nanobelts give a relationship between w/t and the growth direction. Nanobelts with w/t = 8.5 and 8.7 grew along [$01\bar{1}0$], i.e., perpendicular to [0001] and with a polar narrow side surface, while nanobelts with w/t = 1.5 and 2.1 grew along [0001]. These are consistent with the TEM and AFM measurements shown in Figures 8.17 and 8.18 and that high w/t is less common and is associated to a less common growth direction, i.e., [$2\bar{1}\bar{1}0$] or [$01\bar{1}0$].

However, in bulk ZnO, the Young's modulus along [0001] is close to the one along a perpendicular direction, i.e., [$2\bar{1}\bar{1}0$] or [$01\bar{1}0$). By means of indentation system with a 1.5 μm diamond tip, the Young's moduli along the [0001] and [$01\bar{1}0$] directions are found to be 163 and 143 GPa, respectively. This agrees with the measured moduli which are E = 180 GPa for the sample oriented (0001) and E = 153 GPa for the sample oriented ($01\bar{1}0$).

Figure 8.17 and TEM measurements. indicate that nanobelts or nanowires grown along [0001] are free of dislocations and stacking faults, while nanobelts grown along [01$\bar{1}$0] and [2$\bar{1}\bar{1}$0] with the polar (0001) surface as the narrower side surface exhibit stacking faults parallel to the (0001) surface over the entire length of the nanobelts. So the low Young's modulus in nanobelts with high w/t is due to the presence of planar defects in nanobelts grown along [01$\bar{1}$0] or [2$\bar{1}\bar{1}$0] and identified as high w/t nanobelts. A similar planar defects is observed in WO_3 nanowires, where the Young's modulus decreases from about 300 GPa (bulk value) to 100 GPa, as the, diameter of the nanowire increases from 16 to 53 nm. This is due to size dependent defect concentration.

The origin of a larger Young's modulus for nanobelts with w/t = 1 as compared with nanobelts with w/t = 2 is unclear. However, the majority of the nanobelts present, the highest modulus values, are identified as nanowires from AFM topography measurements. The symmetry of the structure has thus a role.

Resonance of a Nanobelt

Based on the electric field-induced resonant excitation, an technique is developed for measuring the mechanical properties of individual nanowires by transmission electron microscopy (TEM). By this method, mechanical properties of carbon nanotubes, silicon nanowires and silicon carbide-silica composite nanowires are known.

The mechanical property of a nanobelts is measured with TEM by applying a voltage (200kV) across a nanobelt, and its counter electrode. Mechanical resonance is induced if the applied frequency matches the natural resonance frequency of the nanobelt Due to the mirror symmetry of the nanobelt (Fig. 8.20(a)). there are two distinct fundamental resonance frequencies corresponding to the vibration in the thickness and width directions, which are given from the classical elasticity theory as:

$$v_x = \frac{\beta_1^2 T}{4\pi L^2}\sqrt{\frac{E_x}{3\rho}} \tag{11}$$

$$v_y = \frac{\beta_1^2 T}{4\pi L^2}\sqrt{\frac{E_y}{3\rho}} \tag{12}$$

where β_1 = 1.875 for the first harmonic resonance; E_x, and E_y are the bending modulus if the vibration is along x-axis (thickness direction) and y direction (width direction), respectively; ρ is the density, L is the length, W is the width and T is the thickness of the nanobelt. The two mode's are decoupled and they can be observed separately.

Changing the frequency of the applied voltage, two fundamental frequencies are observed in two orthogonal directions transverse to the nanobelt. Figure 8.20 (b) and (c) show the harmonic resonance with the vibration planes nearly perpendicular and parallel to the viewing direction, respectively. For calculating the bending modulus, the fundamental resonance frequency (v_1) and the dimensional sizes (L and T or W) of ZnO nanobelts are measured accurately. To determine v_1, the stability of resonance frequency is checked to ensure that one end of nanobelt

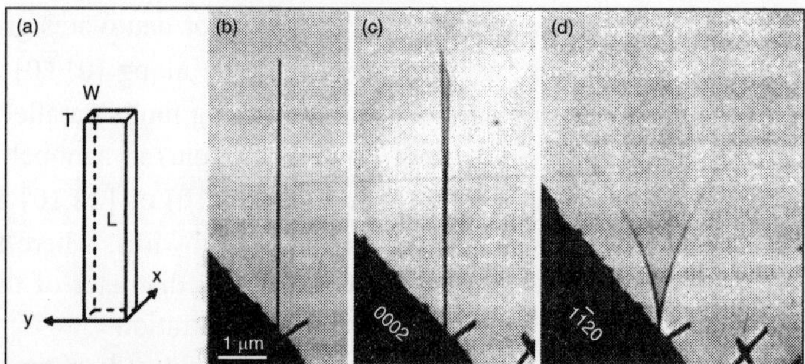

Figure 8.20 *Measuring the bending modulus of a nanobelt. (a) Geometrical shape of a nanobelt. (b) A ZnO nanobelt at stationary. (c, d) Mechanical resonance of the nanobelt along two orthogonal directions, respectively, closely perpendicular to the viewing direction (v_x = 622 KHz), and nearly parallel to the viewing direction (v_y.=_691 kHz)*

is tightly fixed, and the resonant excitation is checked around the half value of the resonance frequency. The specimen holder is rotated about its axis so that the nanobelt is aligned perpendicular to the electron beam, so the real length (L) of the nanobelt can be measured. The projection direction along the beam is determined by electron diffraction pattern to determine the true thickness and width because the normal direction of the nanobelt is [2110]. Bases on the data, the bending moduli of the ZnO nanobelts is calculated using equation (11) or (12). The bending modulus of the ZnO nanobels is 52 GPa. It is found that nanobelt is effective nanoresonators, exhibiting two orthogonal resonance modes, which can be used as probes for SPM operated in tapping and scanning modes.

ELASTIC DEFORMATION BEHAVIOR OF BRIDGED NANOBELTS

The elastic deformation behavior of one-dimensional nanostructures is measured by fitting the image profile using atomic force microscopy in contact mode along the entire length of a bridged/suspended nanobelt /nanowire/nanotube under different load force.

A continuous scan of a ZnO nanobelt (NB) that is supported at the two ends by an AFM tip in contact mode; a quantitative fitting of the elastic bending shape of the NB as a function of the bending force provides a reliable and accurate method for measuring the elastic modulus of the NB.

The ZnO NBs used are prepared by physical vapor deposition, A silicon substrate is prepared with long and parallel trenches caved at its surface by nanofabrication. The trenches are about 200 nm deep and 1.25 μm wide [Fig. 8.21(a)]. Long ZnO NBs are manipulated across the trenches. The morphology and dimensions of the NB are captured-by SEM and AFM. The SEM image gives the width of the trench and the length and width of the NB, and the AFM image gives the thickness of the NB [Figs. 8.21.(b) and (c)]. For the measurements, the NBs are scanned along its length direction using an AFM tip in contact mode at a constant applied force; and bending images of a series NBs are recorded by changing the magnitude of the contact force. The elastic deformation behavior of the NBs is calculated. The cantilevers (spring constant of ~2N^{-1}m) is used after calibrating them so as to calculate the AFM contact forces.

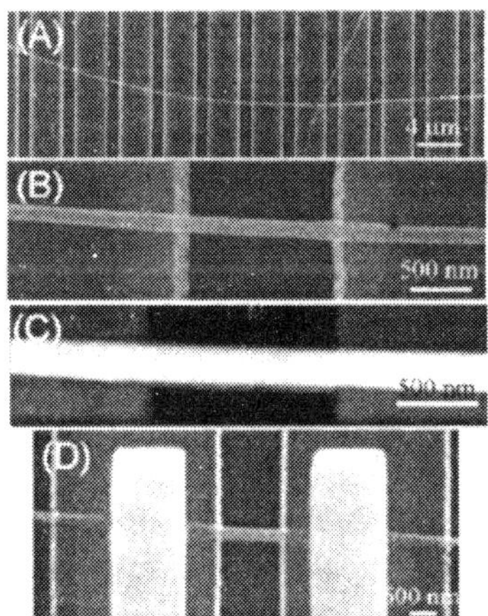

Figure 8.21 *(a) Low-magnification SEM image of a silicon substrate with parallel trenches. Long nanobelts (NBs) are lying on the trenches. (b) SEM image of one NB bridged over a trench (c) and the corresponding AFM image of the bridged NB over the trench. (d) The coraesponding SEM image of the same NB in (b) after depositing Pt pads at the two ends.*

The profiles of a suspend NB along the length direction under different contact forces are shown in Fig. 8.22(a). Each curve is received by an average of ten consecutive measurements along its length under the same loading force. Due to a small surface roughness (~ 1 nm) of the NB, the curves are not perfectly smooth. In addition, the as-attached NB on the trenches is not perfectly straight, due to initial bending during the sample manipulation. So, to eliminate the effect of the surface roughness and initial bending of the NB on force curve quantification, deflection curves are calibrated by subtracting the profile measured under low applied force (~ 100 nN) from those measured at higher applied forces. Due to the presence of the trench, a reasonable force is applied to make the tip in contact with the NB. To obtain a good image under common AFM, the set point is chosen to be ~ 0.5 V, which correspond ~ 100 nN for the AFM. As a result a normalized force is also defined by subtracting the 100 nN from the applied force. The final results are as shown in Fig. 8.22 (b). Some small ripples appear at the middle of

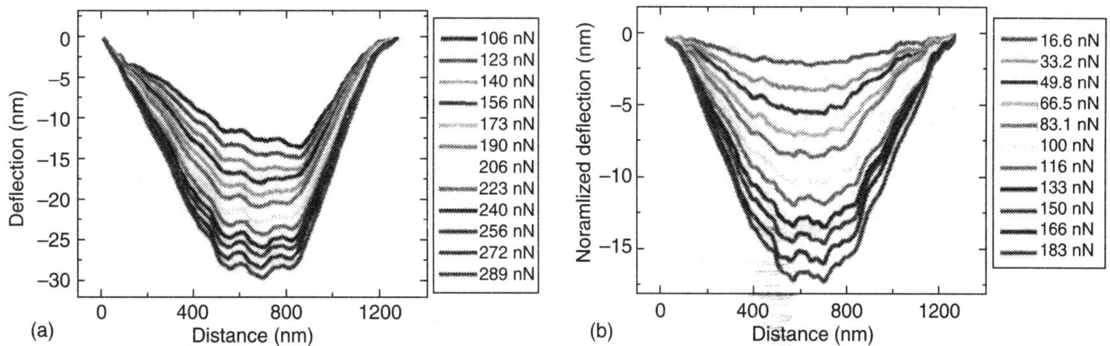

Figure 8.22 *(a) AFM image profiles of one suspended NB under different load forces in contact mode. (b) The normalized AFM image profile after removing the surface roughness by subtracting the image acquired at 106 nN from the data in (a).*

the NB and become pronounced with the increase in contact force. This is due to the rippling effect of the NB with a compressed top surface in contacting with the AFM tip, because the ZnO NB cannot create and preserve edge dislocations to accommodate the deformation.

For the measurements keep the load small so that the maximum deflection of the NB is less than the half thickness of the NB and assume that the deformation process is elastic. Also keep the AFM scanning rate at 0.5 ~ I Hz for the measurement to be static. Repeat these measurements by increasing and decreasing the load to ensure the reproducibility in the profile (within 1 nm) and negligible hysteresis.

The profile images of the NBs record the deformation of all of the points along its length under different applied forces. One profile contains up to 650 points. Every point on the suspended portion of NB in the images should be regarded as a mechanical measurement done. By filling the shapes of all of the curves measured at different applied forces, the elastic modulus can be derived.

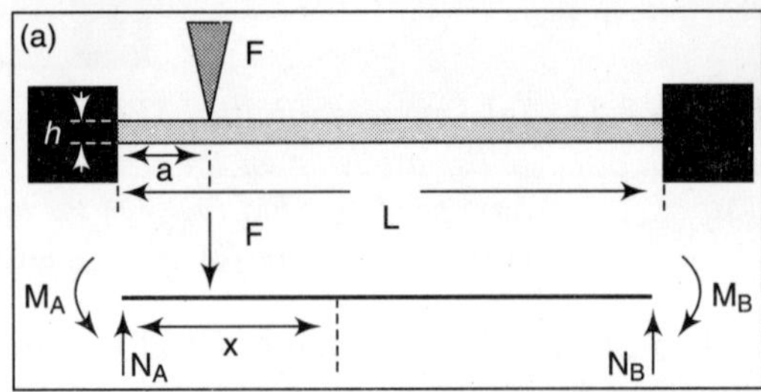

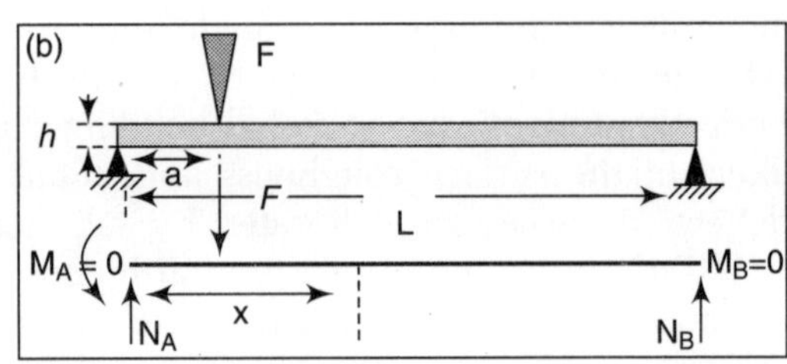

Figure 8.23 *Diagrams of the (a) clamped-clamped beam model (CCBM) and (b) the free-free beam model (FFBM).*

Now, there can be two models derived under different boundary conditions. One is the clamped-clamped beam model (CCBM) [Fig. 8.23(a)]. in which the two ends of the NB are affixed, so that the deflection v and its first derivative dv/dx are both 0 at $x = 0$ and $x = L$, where L is the width of the trench. The other is the free-free beam model (FFBM). [Fig. 8.23(b)], in which the two ends of the NB can freely slide, which means that only the support force exists and there is no force moment; thus, only $v = 0$ at $x = 0$ and $x = L$. When a concentrated load F is applied at point a away from the A end, the differential equation that determines the deflection of the entire beam under small angle deformation is

$$EI\frac{d^2v}{dx^2} = -N_A x + F[x - a] + M_A, \quad (1)$$

where E is the bending modulus, I the moment of inertia which is $wh^3/12$ for the rectangular beam, N_A the support force and MA the force moment at the A end, and $[x-a]$ is a step function, which means that $[x-a] = 1$ when $x \geq a$ and $[x-a] = 0$ when $x < a$. In the CCBM, the solution of equation (1) is

$$v = \frac{Fa^3(L-a)^3}{3EIL^3} \quad (2)$$

This equation gives the deflection of the NB at the contact point a of the AFM tip under a constant applied force F. In the FFBM, the deflection of the NB at the contact point a is

$$v = \frac{Fa^2(L-a)^2}{3EIL^3} \quad (3)$$

Also, it is also possible to have clamped-free mixed case, and v will have an asymmetric form.

The CCBM theory is assumed because it is assumed that the adhesion force between the 1D nanostructure and the substrate is strong enough to clamp the two ends of the nanostructure. But the nanostructure can bear a relatively large lateral force without any observable movement. Therefore, some metal pads are deposited for the CCB conditions.

The two different models give totally different results. For example, at the middle point $a = L/2$, $E = FL^3/192Iv$ for CCBM, and $E = FL^3/48Iv$ for the FFBM, which are different by a factor of 4 in the measured elastic modulus. But it is not possible to identify model for data analysis for the measurement made at one contact point, such as the middle point, of the 1D nanostructure.

Equations (2) and (3), show that the shapes of two deformation curves and their slopes are different. By fitting the shape of the curve measured under different applied forces, it is unique to determine which model is more precise to quantify the elastic deformation behavior of the NB. Figure 8.24 (a) shows an example of curve fitting based on the two models. The FFBM is ound to fit much better than the CCBM, especially when the applied force is large (> 190 nN). Under lower applied force, the surface roughness of the NB and the noise from the AFM system make it more difficult.

To examine the validity of the FFBM (see Fig 8.24(a)), elastic deformation is made from a NB, before and after affixing the two ends. Figure 8.21(d) shows a SEM image of a NB after depositing two Pt pads at the two ends by focus ion beam microscopy. Before Pt deposition, the maximum relative deflection of the NB at the middle point is measured to be 5.4 nm at a normalized force of 117 nN. After Pt deposition, the maximum relative deflection of the NB decreases to 1–1.5 nm at a normalized force of 119 nN. The ratio is ~ 4. This indicates that the elastic deformation behavior of a bridged NB with two ends "free" can be modeled using FFBM.

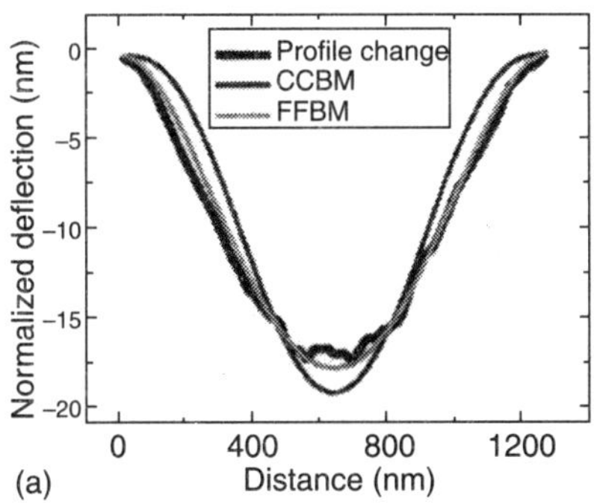

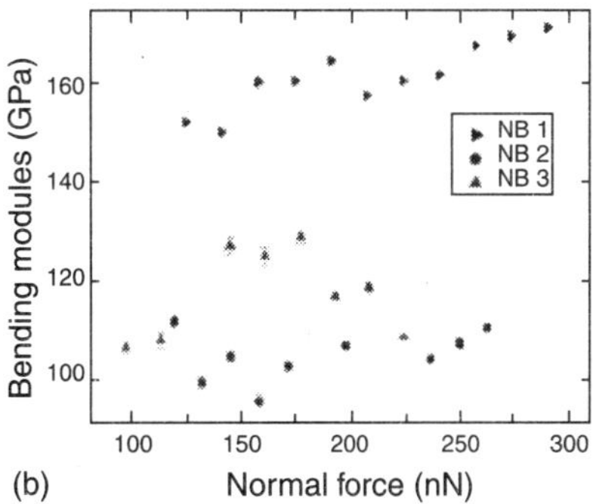

Figure 8.24 *(a) Curve fitting using the CCBM and FFBM for the image profiles of NB 1 acquired under normalized force of 183 nN, (b) The bending modulus from the FFBM fitted curves under different load forces. The error bars are introduced with consideration of the uncertainty in cum: fitting. The dimensions of NB 1 are 1.270 μm in length. 90 nm in width. and 70 nm in thickness; the dimensions of NB 2 are 1.253 μm in length. 115 nm in width, and 95 nm in thickness; and the dimensions of NB 3 are 1.232 μm in length. 125 nm in width, and 115 nm in thickness.*

At a large load force, as the friction between the NB and the edge of the trench is large enough to prevent the sliding of the NB, therefore the CCBM model should be considered, although the two ends are unfixed. It is found that the adhesion between the ZnO NB and the silicon substrate is weak, due to the incompatible crystal structure systems and the large lattice mismatch. So, ZnO NBs can be displaced by AFM tip during imaging, although the scanning direction is along the NB for minimizing the possibility of large impact from the AFM tip. Thus, the adhesion force and the force moment at each side of the NB should be regarded as zero, so that the FFBM is the most reasonable model.

Figure 8.24 (b) shows the bending modulus derived from the curve fitting at different loads for three different NBs. The elastic moduli for NBs 1, 2, and 3 remain consistent, respectively, under different loaded forces, indicating that the deformation is elastic although ripples are observed from the deformation curve [Fig. 8.22(b). The elastic moduli for NBs 1, 2, and 3 are 162 ± 12, 105 ± 10, 118 ± 14 GPa respectively. The difference for the three NBs is related to their sizes. The bending modulus is larger than that measured by mechanical resonance measured by TEM.

9

Field Emission Properties

FIELD EMISSION OF DENSITY-CONTROLLED ZnO NANOWIRE

Field emission (FE) is one of the most important applications of ZnO since the emitting efficiency can be highly improved by the alignment, as the emitter density is very critical to the emission efficiency.

In a FE measurement, the distance between the sample and counter electrode is an important parameter to determine the field-enhancement factor. However, this distance is usually difficult to determine inside a vacuum-testing chamber. So, measuring system is developed inside a scanning electron microscope chamber, in which the emitting distance can be directly observed. The working vacuum of the chamber at 6×10^{-4} Pa provides a reasonable vacuum condition for emission testing. This system provides a novel and reliable way to measure the FE property with knowledge of the exact emitting distance, nanowire density, and the region being tested.

For, measurements on a series of aligned ZnO NWs with controlled density and size, it is found that the NWs with a density between 60 and 80 μm^{-2} and a length of 1 μm give the highest emitting current, which is 20 μA at a field of 10 V μm, detected by using a 0.03 mm^2 W counter electrode.

Experiment

Density-controlled aligned ZnO NWs are grown on an expitaxial layer of AlGaN coated on a *c*-parallel orientated sapphire substrate. The density is controlled by varying the thickness of a gold catalyst from 1 to 8 nm on a 1 cm × 2 cm substrate. The growth of ZnO NWs is through a vapor-solid-liquid (VLS) process. A mixture of equal amounts (by weight) of ZnO and graphite powders are used as source materials and loaded in an alumina boat. Which is located at the center of an alumina tube. Ar mixed with 2 % O_2 is used as the carrier gas at a total flow rate of 50 sccm. The substrate is placed down stream in a temperature zone of 1120 K. A horizontal tube furnace is used to heat the center of the alumina tube to 1270 K at a rate of 50 K min^{-1}, and the temperature is held at the peak temperature for 30 min under a pressure of 30 m-mmHg. After the growth, the system is slowly cooled down to room temperature under the flow of the mixing carrier gas.

The measuring systems is set up inside a SEM chamber. As shown in Figure 9.1(a), under the electron gun, the NW sample is perpendicularly mounted on a cliff side of a triangular-shaped scanning electron microscope sample holder, which is attached to the SEM stage with a moving resolution of a few nanometers. A W needle with a flat top is used as the counter electrode and held by a stainless steel mini post system. In order to achieve independent movements of the W electrode, the post system is attached to another *x–y* movement stage (20 nm moving resolution) that is screwed on the side of the SEM stage. This *x-y* stage is controlled by an outside controller through the electronic interface built on the door of the scanning electron microscope chamber. Meanwhile, the NW sample and the W needle are connected to the negative and positive electrodes, respectively, of an outside high DC voltage source. The voltage is read out by the meter on the voltage source and the current is measured by using a pico-ampere meter.

Figure 9.1(b) shows an optical photo of the scanning electron microscope chamber equipped with this measuring system. Before the measurement, the W needle and the NWs are aligned under the electron beam so that the scanning electron microscope sees both of them. Since the *x–y* mechanical stage has a lower movement resolution, the W needle, is located first until the flat tip is observed at the center of the microscope's imaging screen. Then, the NW sample is moved towards the W tip, and focused both of them by the electron beam at the same time so that they are aligned at the same height, as shown in the inset of Figure 9.1(b).

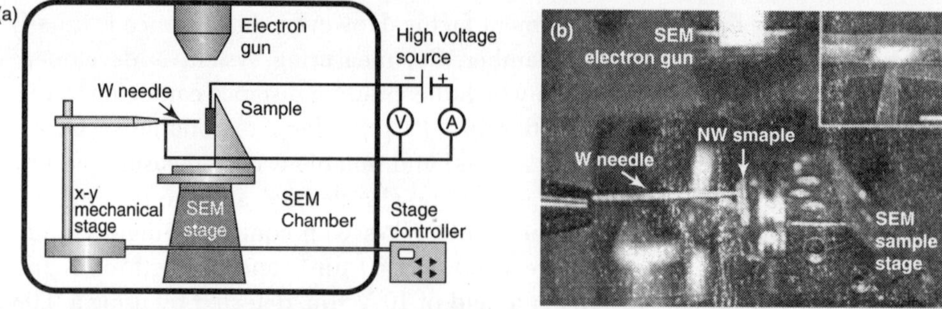

Figure 9.1 *Setup of the scanning electron microscope in situ FE measurement system. (a) Schematic of the setup. (b) Photo of the system inside the scanning electron microscope chamber; inset: a scanning electron microscopy (SEM) image of the W needle tip pointing to the nanowire arrays, the scale bar represents 500 mm.*

The typical morphology of the density-controlled aligned ZnO NWs is shown in the scanning electron microscopy (SEM) image in Figure 9.2(a). This shows the NW sample with the highest density (110 μm^{-2}). Every single NYW is self-aligned perpendicular to the substrate and there is no bending or interconnects between the NWs. The length of the NWs vary from. 0.7 to 1 μm and their widths are 30-40 nm, as shown in the transmission electron microscopy (TEM) image in Figure 9.2(b). A, gold particle (10–20 nm) is observed at the tip of most of the NWs. All of the NWs are single crystalline and their growth direction is along the [0001] direction (Fig. 9.2(c)).

During the FE measurement, the vacuum of this scanning electron microscope chamber is at 6×10^{-4} Pa. The distance between the NW array and the W tip is controlled by fine movement of the scanning electron microscope stage and measured directly from the SEM image, as shown in the inset of Figure 9.2(a). This distance is kept at 20 μm for all of the measurements. The NWs are also examined by using SEM after FE tests. Most of the gold catalyst tips a gone after several electron emission steps. Notably, after the emission, the highest density NWs are attracted together to form bundles, as shown in Figure 9.2(d). This is due to the potential-induced self-attracting effect, upon the application of a high voltage. Also, the magnetic field created by one stream of FE current by a few NWs creates an attractive Lorentz force for the next elf- stream of NWs, similar to the self-attraction of two parallel conducting wires. Moreover, during the emission, the Joule heating effect sinter them and some of them eventually merge together (inset of Fig. 9.2(d)). This phenomenon is not found for aligned NW arrays with lower densities.

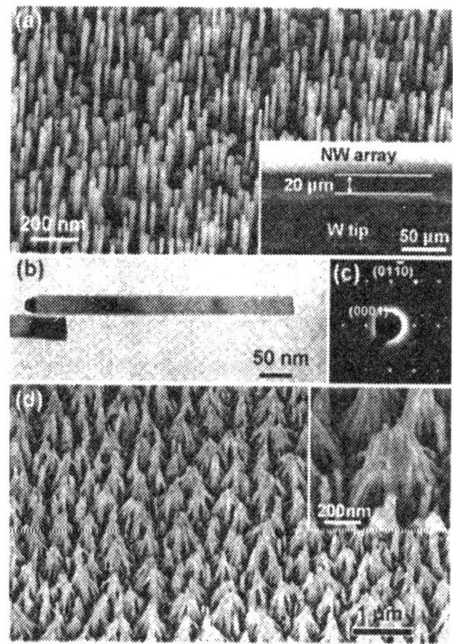

Figure 9.2 *(a) A representative SEM image of aligned ZnO NWs; inset: an SEM image of the FE testing condition. (b) A transmission electron microscopy (TEM) image of a typical ZnO nanowire. (c) A corresponding electron diffraction pattern. (d) An SEM image of the highest density ZnO NWs after FE testing; inset: an enlarged SEM image of one nanowire bundle.*

The FE properties are measured on the aligned ZnO NWs with five different densities. All of them are grown on the same substrate and measured one after the other under the exact same conditions. Typical top-view SEM images of the five samples are shown in Figure 9.3(a), where they are identified from I to V with a density of 108,86,64,45 and 28 NWs/μm^{-2}, respectively. The FE *I–V* curves are shown in Figure 9.3(b). The counter electrode (the W needle tip) has an area of only 0.03 mm^2, which is much smaller than the size of the sample. However, considering the density of the aligned NWs, the number of NWs covered under the counter electrode are of the order of 10^6 to 10^7, which is big enough to show a representative universal property. Therefore, the total current received is determined by the size of the W tip so that the measured emission current for samples of different grown areas is compared directly. Similar to CNTs, the highest density NWs do not prove to be a very effective emission source (the curve composed by diamond dots). The emission is turned on by the electric field around 20 V μm^{-1} and the emission current increases to 20μA when the electric field ramped up to 25 μm^{-1}. This is caused by a screening effect due to the very small space between the NWs. However, from the SEM observation, the agglomeration of the NWs largely reduces the sharp emitting tips. Therefore, the self-attracting phenomenon in high-density NWs is responsible for their low emission efficiency.

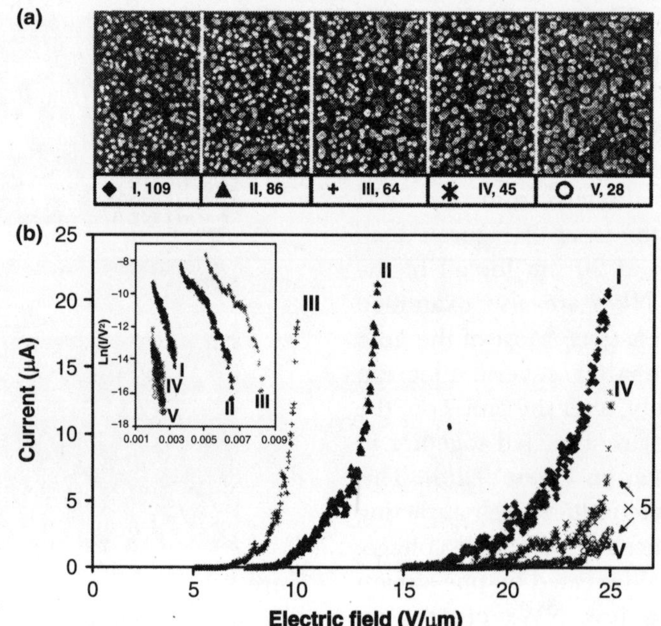

Figure 9.3 *Top-view SEM images of the aligned ZnO nanowire arrays with five different densities (NWs/mm^2). Curves IV and V are magnified five times (for better illustration). (b) Corresponding FE I–V characteristics and the converted F–N curves.*

The highest emission efficiency is observed on samples II and III, which have medium densities (86 and 64 μm^{-2}) and are marked by triangles and crosses, respectively, in Figure 9.3(b). Their turn-on electric field is 8–10 V μm^{-1} and the emission current sharply jumps to 20 μA at a field of 10^{-13} V μm^{-1}. When the NW density is decreased to 45 μm^{-2} and lower, the emission efficiency becomes very low, as shown by the star and circle points in Figure 9.3(b). Their highest emission current reaches 1–2 μA under an electric field of 25 V μm^{-1} (500 V).

The emission *I–V* curves are converted by using the Fowler-Nordheim (F-N) equation and plotted as ln(I/V^2) versus 1/V, as shown in the inset of Figure 9.3(b). The five curves exhibit almost straight lines and the field-enhancement factors (β) are calculated from the slopes (k_{F-N}) according to the equation

$$k = B\phi^{3/2}d/\beta \tag{1}$$

where φ is the work function of ZnO, d is the emitting distance, and $B = 6.83 \times 10^9$ VeV$^{3/2}$ m^{-1}. By taking 5.2 eV as a fixed work-function value for ZnO NWs and keeping the distance at 20 μm, the field-enhancement factors are calculated and listed in Table 9.1. In CNT the distance between the emitters is the key parameter that determines the field-enhancement factor. So, the average distance between the NWs is estimated by assuming that they are evenly distributed on the substrate. The best field-enhancement factor of aligned ZnO NWs is found to be >800, which is a reasonable value for the application as an electron-emitting source. Thus, the optimum interspacing of ZnO NWs should be 110–130 nm, which is about 10 times smaller than the theoretical calculation for CNTs. The value of β dramatically drops when the interspacing increase to 150 nm. The small number of emitting tips (NWs) on the substrate is considered as the reason for this low emitting ability. However, for the low-density NW arrays, there are many gold/ZnO particles left on the substrate during the growth. The electric-field distribution around the NW tips is disturbed by the neighboring particles so that the field-enhancement factors are further reduced.

Table 9.1 *Field-enhancement factors of ZnO NWs with different densities.*

Sample	*NW density [mm^{-2}]*	*Average distance between NWs [nm]*	$k_{F\text{–}N}$	β
I	109	96	–3252	498
II	86	108	–2019	802
III	64	125	–1883	860
IV	45	150	–4371	370
V	28	190	–5185	312

In reality, the NW distributions are very random, while the theoretical prediction is based on a uniform arrangement. It is necessary to reveal how close the assumption is made to real cases. Therefore, additional FE tests are performed on a patterned ZnO NW array with almost the same density as the unpatterned low-density NWs. Both samples are shown in Figure 9.4(a). The distance between the NWs in the patterned sample is much more uniform than the unpatterned one. Since the density is controlled by the prepatterned catalyst dots, no redundant gold particles are left on the substrate, while in the unpatterned sample, a lot of gold/ZnO particles are found lying around the NWs. The corresponding FE *I–V* characteristics and their F–N curves are shown in Figure 9.4(b). The patterned sample shows a low turn-on electric field at 9 V μm^{-1}, which is less than half of that given by the unpatterned sample (21 V μm^{-1}). Both F–N curves exhibit an almost linear shape. From their slopes, the field-enhancement factors are calculated and are listed in Table 9.2, where the β of the patterned sample is 716, almost two times larger than the unpatterned sample. This high emitting efficiency of the patterned NWs is due to the uniform NW distribution and the clean substrate. First, NWs in the unpatterned sample are unevenly distributed, especially for the low-density ones. As shown in the right-hand-side image of Figure 9.4(a), several NWs grow very close together but are far away from the others. So, the screening effect is applied on these NWs to reduce the emitting efficiency. On the other hand, the large gold particles disturb the electric-field distribution during FE. Therefore, the evenly distributed aligned NWs without disturbance from other metal particles give a much higher emitting efficiency.

Table 9.2 *Summary of the field-enhancement factors of patterned/unpatterned ZnO NWs.*

Sample	*NW density [mm^{-2}]*	*Average distance between NWs [nm]*	$k_{F\text{–}N}$	β
Patterned	30	182	–2262	716
Un patterned	32	177	–3879	417

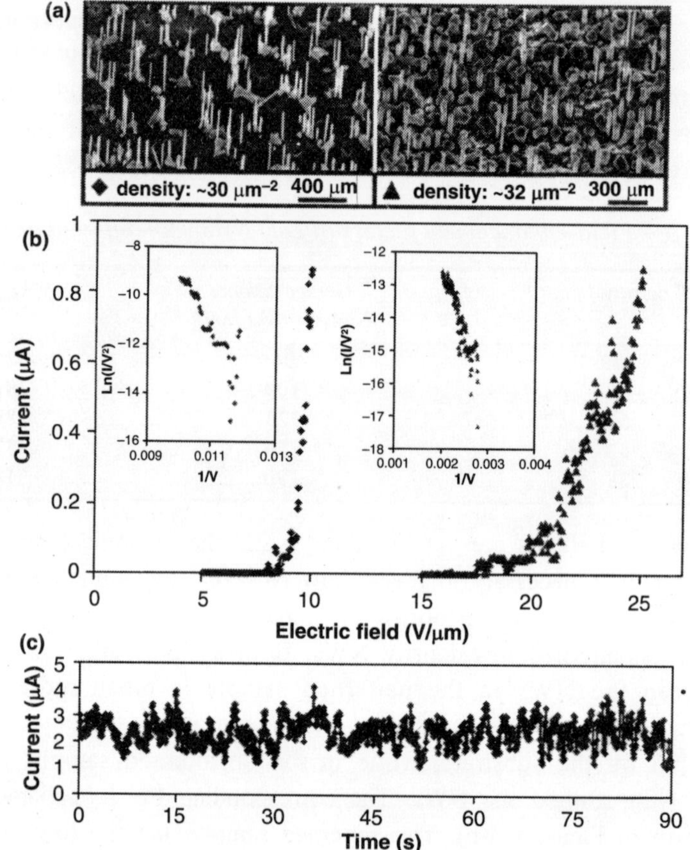

Figure 9.4 *(a) SEM images of patterned (left) and unpatterned (right) ZnO NWs with very close densities. (b) Corresponding FE I–V characteristics and the converted F–N curves. (c) Emission current of the patterned ZnO NWs showing the stability.*

A FE stability measurement is performed on the patterned ZnO NW sample by keeping the electric field at 12 V μm^{-1}. The emission current is recorded for 90 s, as shown in Figure 9.4(c). The current is fluctuated between 2 to 4 μA and remains in this region during the entire measurement.

Work Function at the Tip of a Nanowire/Nanobelt

The workfunctions are for bulk surfaces that do not have a curvature. For nanowire based materials, such as ZnO, it is the workfunction at the tip of the nanowire that matters to the field emission. A technique for measuring the workfunction at the tip of a single nanotube/nanowire is described below. The measurement is done by TEM. Figure 9.5(a) is a low magnification TEM image used in the experimental technique for the measurement of the workfunction at the tip of a ZnO nanobelt. One end of a nanobelt, is electrically attached to a gold wire, and the other end faces directly against a gold ball. Because of the difference in the surface work function between the ZnO nanobelt and the counter Au ball, a static charge Q_0 exists at the tip of the nanobelt to balance this potential difference. The magnitude of Q_0 is proportional to the difference between work function of the nanobelt tip (NBT) and the Au electrode, $Q_0 = \alpha(\phi_{Au} - \phi_{NBT})$ where α is related to the geometry and distance between the nanobelt and the electrode.

The measurement is based on the mechanical resonance of the nanobelt induced by an externally alternative electric field. ZnO nanobelt can be regarded as a vibration cantilever clamped at one end. Experimentally, a constant voltage V_{dc} and an oscillating voltage V_{ac} cos $2\pi ft$ are applied onto the nanobelt (Fig 9.5(a)), where f is the frequency and V_{ac} is the amplitude. Thus, the total induced charge on the nanobelt is

$$Q = Q_0 + \alpha e(V_{dc} + V_{ac} \cos 2\pi ft) \quad (1)$$

The force acting on the nanobelt is proportional to the square of the total charge on the nanobelt

$$F = \beta[Q_0 + \alpha e(V_{dc} + V_{ac} \cos 2\pi ft)]^2$$

$$= \alpha^2\beta\{[(\phi_{Au} - \phi_{NBT} + eV_{dc})^2 + e^2V_{ac}^2/2] + 2eV_{ac} (\phi_{Au} - \phi_{NBT} + eV_{dc}) \cos 2\pi ft + e^2V_{sc}^2/2 \cos 4\pi ft \quad (2)$$

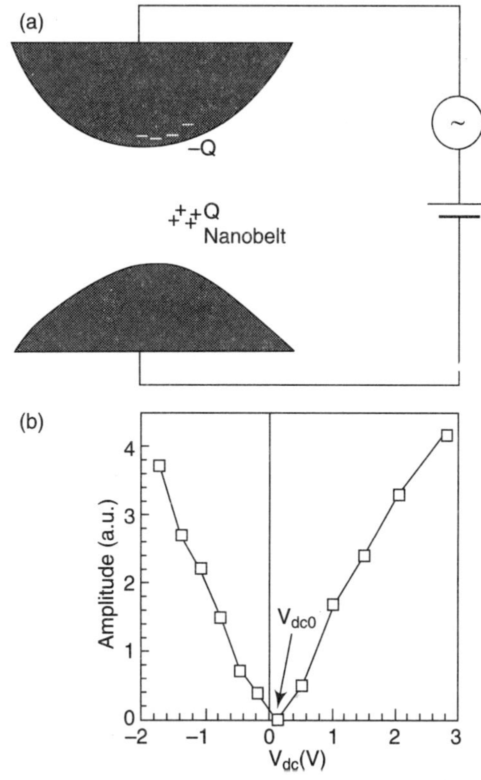

Figure 9.5 *(a) Set up for measuring the work function at the tip of a ZnO nanobelt. (b) A plot of vibration amplitude of a ZnO nanobelt as a function of the applied direct current voltage, from which, the offset voltage V_{dc0} = 0.12 V.*

where β is a proportional constant. In equation 1 the first term is constant and it causes a static deflection of the ZnO nanobelt. The second term is a linear term, and the resonance occurs if the applied frequency f is tuned to the intrinsic mechanical resonance frequency f_0 of the ZnO nanobelt. According to vibration theory and with consideration the rectangular cross-section of the nanobelt, mechanical resonance is along the thickness or width direction. The last term in equation 2 have is the second harmonics. The most important result of equation 2 is that, for the linear term, the resonance amplitude A of the nanobelt is proportional to $V_{ac}(\phi_{Au} - \phi_{NBT} + eV_{dc})$.

The principle of the measurement is as follows. First set $V_{dc} = 0$ and tune the frequency f to get the mechanical resonance induced by the applied oscillating field. Second, under the resonance condition of keeping $f = f_0$ and V_{ac} constant, slowly change the magnitude of V_{dc} from zero to a value V_{dc0} that satisfies ϕ_{Au} – ϕNBT + $eV_{dc0} = 0$. At this moment, the resonance amplitude A becomes zero although the external AC voltage is still in effect. Therefore, the work function at the tip of ZnO nanobelt is $\phi_{NBT} = \phi_{Au} + eV_{dc0}$, while ϕ_{Au} = 5.1 eV. Figure

9.5(b) is a plot of the vibration amplitude A of the nanobelt as a function of the applied direct current voltage V_{dc}, from which the value for V_{dc0} is determined. The dual-mode resonance of the nanobelts observed experimentally. It is important to point out that it does not matter which resonance direction/frequency one chooses, either along the thickness or width direction, the measured V_{dc0} remains the same.

To investigate the dependence of the work function of ZnO nanobelts on the size of their cross-section, i.e., the width W and thickness T of the nanobelts, the geometrical size of the nanobelts is measured from which the work function is measured. Using the projected dimension measured from the TEM image and the orientated angle determined from the corresponding electron diffraction pattern, the geometrical parameters of the nanobelts are obtained. The work function at the tip of ZnO nanobelt is around 5.2 eV, and shows no significant dependence on the cross-section size of the ZnO is nanobelts. The technique demonstrated here is applied to any type of nanowires/nanotubes.

Emission Characteristics of Aligned Nanowires (Mo, MoO_2 and MoO_3)

The sharp geometrical shape of nanotubes and nanowires indicates that they are ideal for superior field emission. The growth of aligned nanotubes and nanowires stimulate a great interest in exploring the applications of 1D nanostructure for field emission. The unique geometrical shape of nanobelts is ideal for field emission. Nanowires of refractory materials are potential candidates for field emission applications.

Aligned Mo, MoO_2 and MoO_3 nanowires are used to illustrate the field emission characteristics of NWs. The NW arrays are grown by a solid vaporization of metal powders (see the SEM images in Fig. 9.6). The field emission measurements are carried out in a vacuum chamber of ~2.6×10^{-5} Pa Hg at room temperature. The nanowire arrays are first adhered using silver paint to the surface of an oxygen free high conductivity copper. A transparent anode consisting of a quartz plate of 4 cm in diameter coated with conducting tin oxide film is placed in front of and parallel to the surface of the sample cathode. Figure 9.6 shows the plots of field emission current versus applied electric field of the metallic Mo, MoO_2 and MoO_3 nanowire arrays. From the plots, the field emission turn-on field (E_{to}) and threshold field (E_{thr}), are determined for these arrays, which are defined to be the macroscopic fields required to produce a current density of 10 μA cm^{-2} and 10 mA cm^{-2}, respectively. The E_{thr} of Mo, MoO_2 and MoO_3 nanowire arrays are 6.24 MVm^{-1}, 5.60 MVm^{-1} and 7.65 MVm^{-1}, respectively. Nanowires of Mo oxides have better field emission performance than the metallic Mo nanowires.

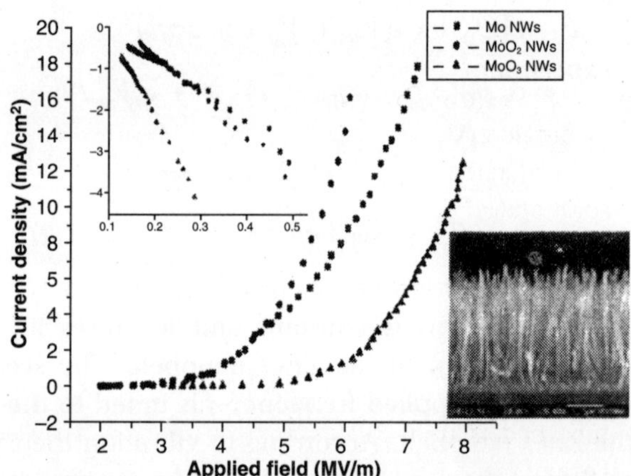

Figure 9.6 *Field emission current versus electric field (I–E) plots of the Mo, MoO_2 and MoO_3 nanowire arrays and their corresponding F-N plots (inset). The SEM image on right-hand side is aligned Mo nanowires.*

The emission characteristics are analyzed using the Fowler-Nordheim (F-N) theory: $J \alpha (E^2\beta^2/\Phi) \times \exp(-B\Phi 1.5/E\beta)$, where J is the emission current density, E *is* the applied field ($E = V/d$), Φ is the work function, β is the enhanced factor, and $B = 6.83 \times 10^9$ ($VeV^{-3/2}Vm^{-1}$). By plotting $\ln(J/E^2)$ versus $1/E$, a straight line is obtained (insertion in Fig. 9.6). The linearity of these curves implies that the field emission from these nanowire arrays follow the F-N theory. By using the work function of bulk Mo (4.24 eV), the β of Mo nanowire arrays is calculated to be about 4400, which is high enough.

Growth and Field-Emission Properties–Aligned AlN Nanorods

Aluminum nitride, an important member of the group III nitrides with the highest bandgap of about 6.2 e V, has excellent thermal conductivity, good electrical resistance, low dielectric loss, high piezoelectric response, and ideal thermal expansion, matching that of silicon. The interest in field-emission (FE) applications of AlN materials has grown because they exhibit a negative electron affinity. Exhibiting a negative electron affinity means that electrons excited into the conduction band can be freely emitted into vacuum. In addition, high-current emission at a relatively low field is most attractive for FE applications. As a result, significant effort is being devoted to reducing the tip size and increasing the density of the emitting sites by using hierarchical nanostructures. Therefore, the synthesis of AlN nanostructures, such as nanowires, nanotubes, nanocones, nanotips, hierarchical comb-like structures, and nanobelts, with controlled high-aspect-ratio shapes and sizes is important. The growth of well-aligned AlN nanorods with hairy surfaces by a vapor-solid (VS) process is given below.

Experiment

Single-crystalline, 3–5 Ωcm, phosphorous-doped Si(001) wafers are used. The silicon wafers are cleaned chemically by the standard RCA cleaning process. The wafers are dipped in a dilute HF solution (HF/H_2O = 1:50) for 2 min immediately before loading into a thermal-evaporation system. An ultrathin Ni film (3 nm thick) is deposited at $\sim 2.6 \times 10^{-5}$ Pa onto a Si substrate at room temperature by thermal evaporation at the rate of 0.1 nms^{-1}. This Ni thin film serves as the catalyst for heterogeneous growth of the AlN nanostructures. The processes are carried out in a horizontal tube furnace, which is composed of a horizontal quartz-tube furnace, an alumina tube, a rotary-pump system, and a gas supply and control system. A viewing window is set up at the end of the alumina tube and is used to monitor the growth process. The opposite end of the tube is connected to the rotary pump. Both ends are sealed by rubber O-rings. The ultimate vacuum for this configuration is $\sim 2.6 \times 10^{-5}$ Pa. $AlCl_3$ and $(NH_4)_2CO_3$ are used as aluminum and nitrogen sources, respectively. 0.004 mol of $AlCl_3$ powder is placed at the center of the horizontal alumina-tube furnace, where the temperature is the highest. 0.004 mol of $(NH_4)_2CO_3$ powder is positioned at the front part of the horizontal alumina-tube furnace upstream of $AlCl_3$. The substrates are placed in a lower temperature zone, downstream from $AlCl_3$. In a typical run, after being evacuated to a pressure of $\sim 2.6 \times 10^{-5}$ Pa, the furnace is heated from room temperature to 1270 K at a heating rate of 60 k min^{-1} under H_2 flow at 25 sccm (standard cubic centimeters per minute). The samples are held at 1270 K for 30 min, and the chamber pressure is kept at 2×10^{-5} Pa.

After the growth process, the product is identified by XRD with Cu Kα radiation. The substrate-bound nanorods are mechanically scraped off, sonicated in ethanol, and deposited on carbon-coated copper grids for TEM characterization. Morphology of grown AlN nanostructures is performed with a TEM with a point-to-point resolution of 0.18 nm, operating at 400 kV, and with a SEM.

The CL spectrum, is measured in a scanning electron microscope with an electron-probe microanalyzer. CL spectra are accumulated in the single-shot mode in 1s. The FE measurements of the AlN nanostructures are carried out at a vacuum of 1.3×10^{-5} Pa using a spherical stainless-steel probe (1 mm in diameter) as the anode. The emission current is recorded on the level of nanoamperes. The distance between the anode and the emitting sample's (cathode) surface is fixed at 50 μm.

The well-aligned AlN nanorods with hairy surfaces provide a new hierarchical nanostructure, and serve as a promising candidate for FE emitters because of their low electron affinity and the geometry of the multiple-nanotip surfaces.

The structure of the as-grown products of AlN nanorods is determined by X-ray diffraction XRD). As shown in Figure 9.7 all of the diffraction peaks in the XRD pattern they correspond to a hexagonal wurtzite-structured AlN crystal with lattice constants of $a = 0.3114$ nm and $c = 0.4986$ nm.

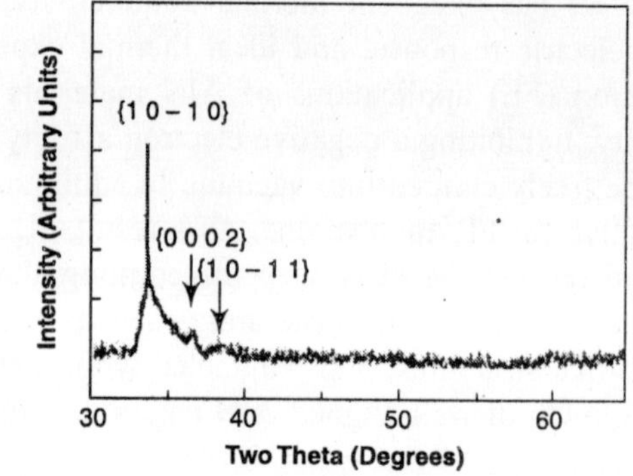

Figure 9.7 *XRD pattern of the as-synthesized products of AlN nanorods.*

No other impurity phases are found in the XRD pattern. The morphology of the as-synthesized products is characterized by scanning electron microscopy (SEM). As shown in Figure 9.8(a), a typical low-magnification SEM image indicates that the as-synthesized product consists of a large quantity of uniform nanostructures. A representative cross-sectional SEM image (Fig. 9.8b) reveals that AlN nanorods grow vertically from the substrate to form arrays with high density. These well-aligned, columnar AlN nanorods have diameters of about 200–400 nm and lengths of up to ~3.4 μm, resulting in an aspect ratio as high as 17. Figure 9.8(c) shows the morphologies of AlN nanostructures. On each columnar nanorod, there are a number of uniform crystallites radially oriented along the main trunk. From the high-magnification SEM images shown in Figure 9.8(d), it is observed that each columnar nanorod is composed of high-density AlN nanotips grown radially along the main trunk. The length and diameter of the numerous AlN nanotips on a columnar trunk of AlN are about ~35 nm and ~3–15 nm, respectively, resulting in average aspect ratios of ~2.3–11.6.

The AIN nanostructures are further characterized by transmission electron microscopy (TEM) and high-resolution TEM (HRTEM). Figures 9.9(a) and (b) show bright-field and dark-field TEM images of two nanorods. It also reveal that each columnar nanostructure consists of a main trunk (nanorod core) and packed secondary branches (nanotips). The nanorod cores extend throughout the entire length, whereas the nanotips grow on the surface of the nanorod cores in the radial direction. Each nanorod is composed of tiny nanocrystallites of AlN. The

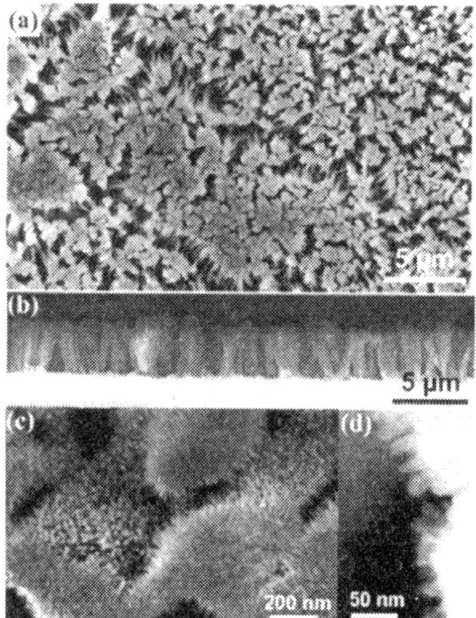

Figure 9.8 *SEM images of the morphologies of the AlN product. (a) Low-magnification plan-view image; (b) low-magnification cross-sectional image; (c) medium-magnification image of AlN nanorods with multiple-nanotip surfaces; (d) high-magnification image of the surface of an AlN nanorod.*

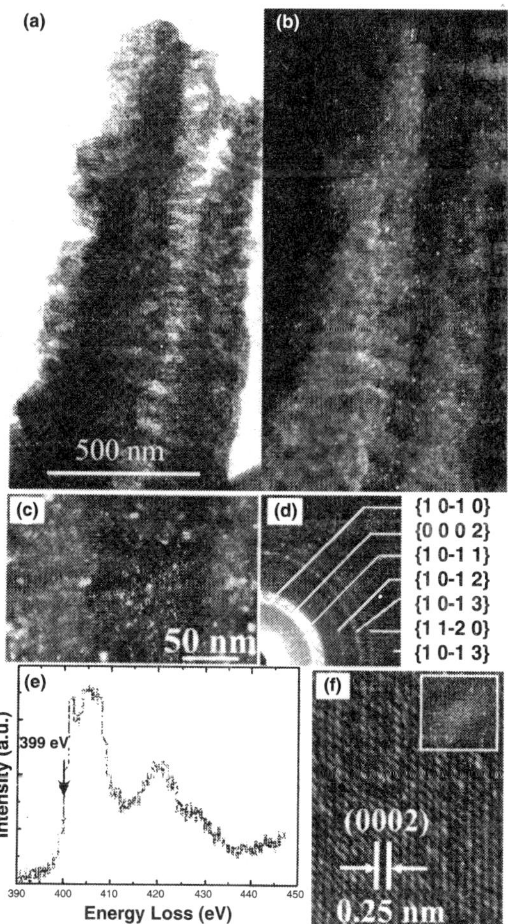

Figure 9.9 *Characterization of the AlN nanostructures. (a) Bright-field and (b) dark-field TEM images of two columnar AlN nanorods; (c) enlarged dark-field TEM image showing the AlN nanocrystallites, 3-5 nm in diameter; (d) selected-area electron diffraction pattern of the nanorod, displaying a polycrystalline structure of each rod; (e) nitrogen K-edge energy-loss near-edge structure recorded from the sample; (f) HRTEM image of a single nanotip. The inset shows a fast-fourier-transform pattern.*

bright dots in the dark-field TEM images (Figs. 9.9(b) and (c) correspond to the nanocrystallites, which have diameters of 3-15 nm. Selected-area electron diffraction (SAED) is utilized to investigate the phase of the synthesized nanostructures. The rings recorded in the diffraction pattern correspond to the wurtzite AlN crystal, as shown in Figure 9.9(d). Quantifying the nitrogen content is essential to identify the existence of AIN. Electron energy-loss spectroscopy (EELS) measurements are made to analyze the composition of the synthesized nanorods. Since EELS has a high sensitivity to light elements, it is well suited for chemical analysis of AlN a structures. It is known that the energy-loss near-edge structure (ELNES) of an EELS spectrum represents a "fingerprint" of the local electronic structure, i.e., the observed ELNES provides information concerning the structure and chemical properties of the atoms that have undergone excitation. Figure 9(e) shows a representative EELS spectrum after subtraction of the background, showing nitrogen K-edge ELNES. The shape of the ELNES from the nitrogen K-edge is very similar to that of unoxidized AlN thin films. No doubt nitrogen is one of the primary elements in the assynthesized nanostructures. Figure 9.9(f) shows an HRTEM image of the nanotip growing along [0001]. The inset to Figure 9.9(f) shows a fast Fourier transform (FFf) of the image.

Mostly, the AlN nanotips grow on the surface of columnar AlN nanorods. However, they also grew on the surface of wetting layers (or the bottom of AlN columnar nanorods), forming an entirely nanotip-covered structure (Fig. 9.10). This corresponds to the morphology of nanostructures grown by the VS mechanism.

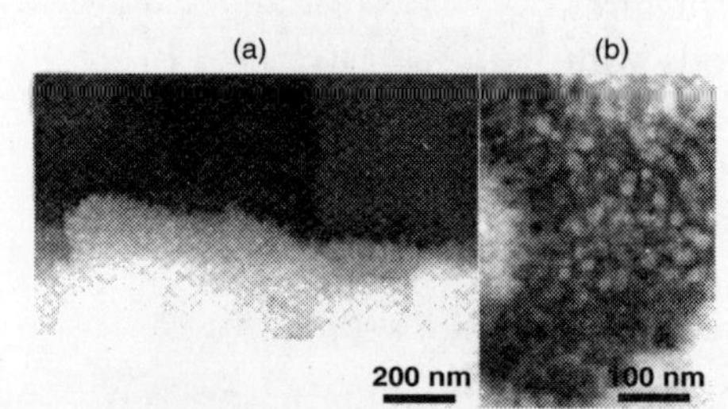

Figure 9.10 *High-magnification SEM images of wetting layers of AlN nanostructures: (a) cross-sectional image; (b) plan-view image.*

Consider, the chemical process for forming AlN, $AlCl_3$ and $(NH_4)_2CO_3$ are used as aluminum and nitrogen sources, respectively. The sublimation temperature of $AlCl_3$ is 450 K $(NH_4)_2CO_3$, on exposure to air or heating, undergoes decomposition as

$$(NH_4)_2CO_3 \rightarrow 2NH_3(g) + CO_2(g) + H_2O(g) \tag{1}$$

with liberation of ammonia. The boiling point of NH_3 is 237 K Because of the low sublimation temperature of $AlCl_3$ and the low boiling point of NH_3, AlN is observed to form even at 870 K or lower via the gas-phase reaction of $AlCl_3$ and NH_3.

$$AlCl_3(g) + 4NH_3(g) \rightarrow AlN(s) + 3NH_4Cl(g) \tag{2}$$

Based on observations, a possible mechanism for the formation of the hierarchical AlN nanostructures is likely to consist of the following consecutive steps: 1) As the processing temperature increases, thermal decomposition of $AlCl_3$ and $(NH_4)_2CO_3$ powders results in the formation of Al and N vapors. The ultrathin films of Ni catalyst provide an energetically favored site for the absorption of incoming AlN vapor (or clusters), resulting in the formation of a solid-solution compound, i.e., heterogeneous growth. With the continuing adsorption of AlN vapors, the solid-solution compound becomes supersaturated. The AlN compound then precipitates from the solution, yielding the 1D AlN nuclei at the interface between the substrate and the supersaturated compound solution, resulting in continuing growth of 10 columnar AlN nanorods. In addition, no compounds are found to deposit on the Si substrate without the presence of a Ni catalyst. Therefore, the Ni catalyst is critical for growing the well-aligned AlN nanorods. Similarly, cone-like AIN arrays are found to grow in the reaction between $AlCl_3$ and NH_3 catalyzed by a Ni thin film. (2) The columnar AlN nanorods are also used as templates for the adsorption of the AlN vapors during the process, resulting in the formation of AIN nuclei on the side surfaces of growing AlN columnar nanorods. The AlN nanotips grow on the columnar AlN nanorods with a continuous supply of AIN vapor, i.e., the hairy structure is formed by a homogeneous growth process. Finally, the AIN hierarchical nanostructures are obtained. Secondary AlN nanorods grown on the stems of AlN nanorods/nanobelts via condensation of AlN vapor are also observed.

Nanorods are ideal objects for electron FE. Figure 9.11 shows the FE current density as a function of the applied field as a current density versus electric field (*J*–*E*) plot, and the inset shows a ln(J/E^2) $-1/E$ plot of the AlN nanostructures. The well-aligned columnar AlN nanorods with multiple-nanotip surfaces have excellent turn-on-field values. The turn-on field (defined to be the electric field required to generate a current density of 0.01 mA cm^{-2}) and threshold field (defined to be the electric field required to generate a current density of 1 mA cm^{-2}) for the AlN nanostructures are found to be about 3.8 and 7 V μm^{-1}, respectively. Although the value of the turn-on field from the AlN nanostructure on a sapphire substrate is higher than the best value found for carbon nanotubes and SiC, they are much lower than those of many other types of emitters such as carbon nitride, Si nanostructures, MoO_3 nanobelts, and ZnO nanowires. The selection of the substrate is found to be critical to the FE properties of AlN nanostructures. Si is selected as the substrate since it is more compatible with the processing technology of Si-based microelectronic devices.

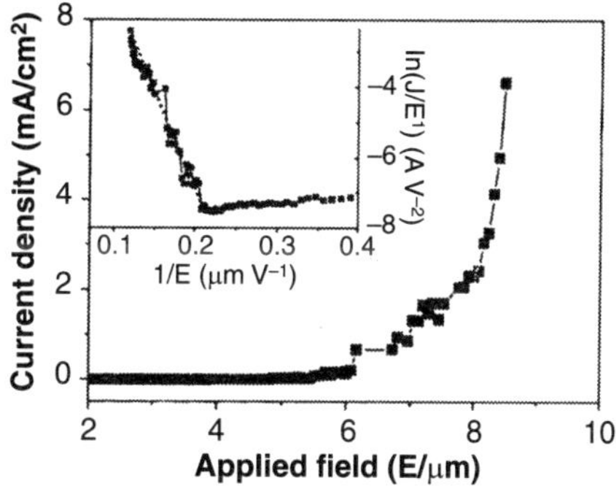

Figure 9.11 *Field emission current density versus electric field (J-E) for hierarchical AlN nanostructures. The inset shows the corresponding Fowler-Nordheim relationship (In(J/E²)–1/E plot).*

Based on the Folwer-Nordheim (FN) model, the FE current from a metal or semiconductor is attributed to the tunneling of electrons through the potential barrier from the material into vacuum under the influence of an electric field. In order to understand the FE behavior, the *J*–*E* data is analyzed by applying the FN equation

$$J = (A\beta^2 E^2/\Phi)\exp[-B\Phi^{3/2}(\beta E)^{-1}] \tag{3}$$

where *J* is the current density, *E* is the applied electric field, and Φ is the work function. *A* and *B* are constants, corresponding to 1.56×10^{-10} AeVV^{-2} and 6.83×10^3 $eV^{-3/2}$ μm^{-1}, respectively. The FN plot is shown in the inset to Figure 9.11. By determining the slope of the ln(J/E^2)$-1/E$ plot using the work-function value of AlN (3.7 eV), the field-enhancement factor, *β*, is calculated to be about 950 for well-aligned, columnar AlN nanorods with multiple nanotip surfaces. The *β*-value is highly dependent on the geometrical features of the nanostructures. Efficient electron emission from well-aligned, columnar AlN nanorods is due to the small radii of curvature of their multiple-nanotip surfaces, resulting in a high *β*-value.

Figure 9.12 show the room-temperature cathodolummescence (CL) emission spectrum, demonstrating a broad emission ranging from 350 to 700 nm with a peak at about 519 nm in the yellow-light range. The emission does not result from band-edge emission, but is from

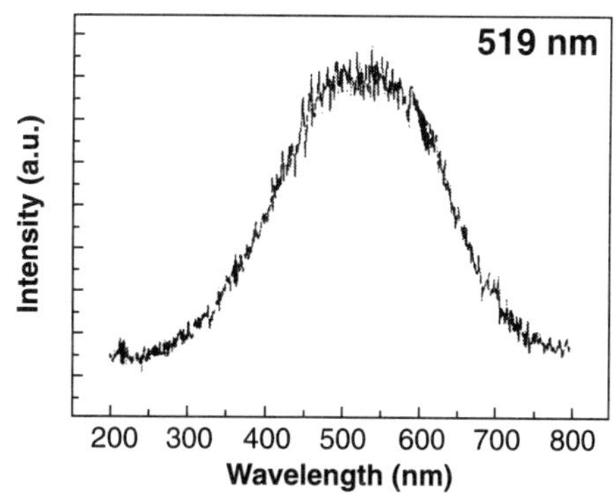

Figure 9.12 *The room-temperature CL emission spectrum of hierarchical AlN nanostructures.*

deep-level or trap-level emission. The intensive light emission from AlN materials is due to the nitrogen deficiency and the radiative recombination of a photon- (or electron-) generated hole with an electron occupying the nitrogen deficiency; this phenomenon is also observed in AlN thin-film and nanostructures. The intensive CL emission indicates that, well-aligned, columnar AlN nanorods with multiple-nanotip surfaces have potential application in light-emitting nanodevices.

Silicon Nanowires

Semiconductor nanostructures, nanoagglomerates, and nanowires have potential applications for the development of nanodevices. The vapor-liquid-solid (VLS) process where gold particles act as a mediating solvent on a silicon substrate forming a molten alloy is applied to the generation of silicon whiskers. The VLS reaction generally leads to the growth of silicon whiskers epitaxially in the ⟨111⟩ direction on signal crystal silicon ⟨111⟩ substrates. More recently, the ideas entailed in the VLS technique apply to develop laser ablation of metal containing silicon targets, obtaining bulk quantitative of silicon nanowire The apparatus used is depicted: in Fig. 10.1. Double concentric alumina tubes are heated to the desired temperature in a tube furnace. The inner tube is vacuum sealed by two water cooled stainless steel end pieces attached and tightly lock-press fit against custom viton o-rings. At one end of the furnace, UHP argon enters through the upstream stainless steel end piece, passes through a matched set of zirconia insulators to the central region of the inner tube oven, and flows over a

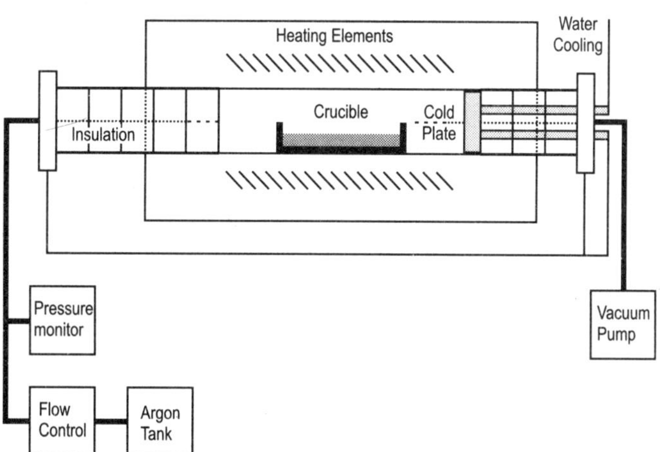

Figure 10.1 *A high temperature oven system used for the synthesis of crystalline silicon nanowire and silica nanospheres.*

crucible containing the silicon-silica (Si/SiO_2) mixture or powdered silicon monoxide at a flow rate of 100 sccm.

The total tube pressure in the inner tube is 3×10^4 Pa controlled by a mechanical pump attached to the tube through the downstream stainless steel end piece. This end piece is mechanically attached to a "water cooled" cold plate whose temperature is adjustable through a matching set of insulating zirconia blocks. Depending on the desired temperature range of operation, the crucibles containing the silicon/silicon oxide based mixtures are either quartz, alumina, or low porosity carbon. The condensation in of nanowire is observed as dark brown deposits in a narrow region on the wall of the inner tube close to the defining end points of the oven shell and corresponding to a temperature in the range 1170-1270K tities of SiO_2 nanospheres are deposited on the temperature controlled cold plate.

Figure 10.2 corresponds to transmission electron micrographs (TEM) of the exemplary virtually uniform and straight nanowires genemted from a 50/50 Si/SiO_2 equimolar mixture heated to a temperature of 1670 K at a total pressure of 3×10^4 Pa for 12h. The central crystalline silicon core is ~30 nm in diameter whereas the outer SiO_2 sheathing is 15 nm in thickness. The high-resolution transmission electron microscopy (HRTEM) views in Figs. 10.2(b) and 10.2(c) demonstrate a number of distinguishing characteristics. 10.2(c) demonstrates that the axes of the SiO_2 clad crystalline silicon nanowires are parallel to <111>. NWs having their axes parallel to <112> display twinning, high order grain boundaries, and stacking faults. At the Si-SiO_2 interface [Fig. 10.2(c)] the crystal planes are described as {211}. The wire which is depicted in Fig. 10.2 is virtually defect free. Fig. 10.2(b) shows that the inner crystalline silicon core undulates slightly. However, the fluctuations in the shading that are apparent in the HRTEM micrograph indicate that the wires are of sufficient quality that the detailed strain due to slight bending above the TEM mount can be observed in the TEM.

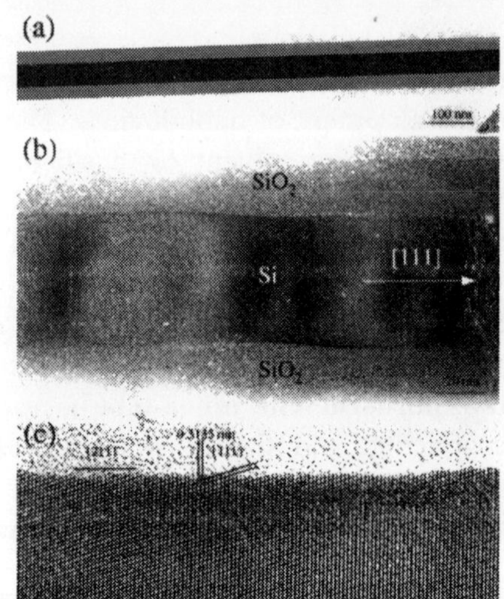

Figure 10.2 *TEM of (a) SiO_2 sheathed crystalline silicon nanowire, (b) closer view showing (1) slight wire undulations and stress pattems which are apparent in a virtually defect free nanowire, and (c) closer view showing crystalline core axes parallel to <111> direction. Synthesized at 1670 K from 50/50 Si/SiO^2 mix ($P_{Total} = 3 \times 10^4$ Pa. flow rate 100 secm of UHP argon) .*

Experiment I

Fabrication of Si NWs: A thermal evaporation technique is employed to fabricate the Si nanostructures. The evaporation source material is SiO powder. SiO is ground to powder, then put into a closed chamber in which the Al_2O_3 substrate is located above the powder. The chamber is pumped to lower then 10Pa. After this at 5.3×10^4 Pa, mixed Ar (90%) and H_2 (10%) gas is filled into the reaction chamber as the ambient gas The graphite plate is selected as the heating material. The SiO powder is heated to about 1470K. The temperature of the Al_2O_3 substrate is about 1170K. After the temperature is kept at 1470K for 20 min, the powder supply is cut off and the temperature is allowed to decrease to room temperature. Yellow sediments are found on the substrate. These yellow sediments, which are Si NWs with a thick SiO_x layer, are thoroughly washed with HF at room temperature. The thickness of SiOx surface layer is effectively reduced or totally removed. Freshly made Si NWs are distributed on a TEM copper grid with colloidal/carbon supporting thin film. The general morphologies of the Si NWs are shown in Figure 10.3(a) and (b). The electrons diffraction analysis shown in Figure 10.3(c) demonstrates high equal quality Si NWs with [110] longitudinal orientation. Some of the morphologies and microstructural details of the Si NWs, are shown in (Fig 10.4 to 10.6).

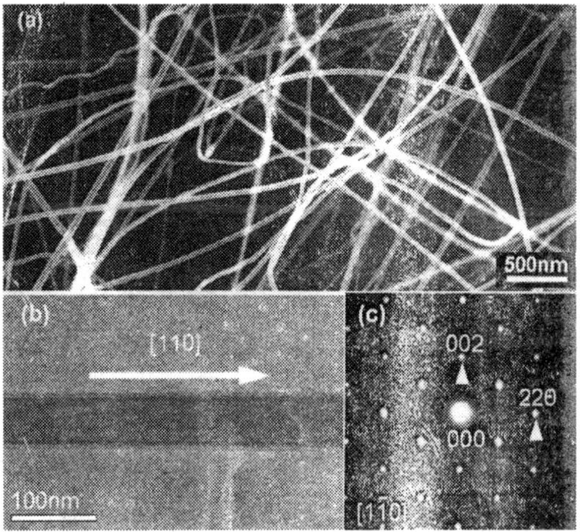

Figure 10.3 *The morphologies of Si nanowires and the structure as well as the orientation of Si nanowires are shown in (a) to (c). (a) is a low magnification image and (b) is an enlarged one: (c)is the SAEDP taken from. the framed region in (b) indicating the longitude direction of the nanowire is along.*

Axial Tensile Testing of the NWs: As shown in Figure 10.4 the method used conducts axial tensile testing of Si NWs by an ultrahigh-resolution transmission electron microscope along a particular crystallographic orientation. A copper TEM grid with a colloidal/carbon thin film is used to apply an external mechanical driving force to deform the nanowires. The Si NW tensile test are done through *controllable* electron irradiation/heating by a specially designed celloidin/carbon supporting film on a TEM Cu grid (the colloidal/carbon supporting film must have the properthickenss). Randomly distributed nanowires are bridged for two (designed) broken parts of the colloidal/carbon supporting film. Upon controlled irradiation/heating by the electron beam at particular designed positions, the celloidin/carbon thin film under goes polymerization shrinkage by linear contraction of 4-5 % and pulls the Si NWs axially, The elastic modulus of the celloidin/carbon thin film is about 1-2 GPa and the Poisson ratio is 0.25-0.35.

The axial-extensile experiment is conducted with two high-resolution electron microscopes. One is a regular high-resolution transmission electron microscope (HRTEM) and the second is a ultrahigh-resolution scanning transmission electron microscope with a field emission gun. The point resolution is 0.19 nm. Both transmission electron microscopes have a 20° double tilt specimen holder and an energy dispersive X-ray spectrometer. The electron energy losses spectrometer is attached to HRTEM. Most of the axial tensile tests are conducted with the regular high-resolution transmission electron microscope. Some of the UHRTEM tensile tests

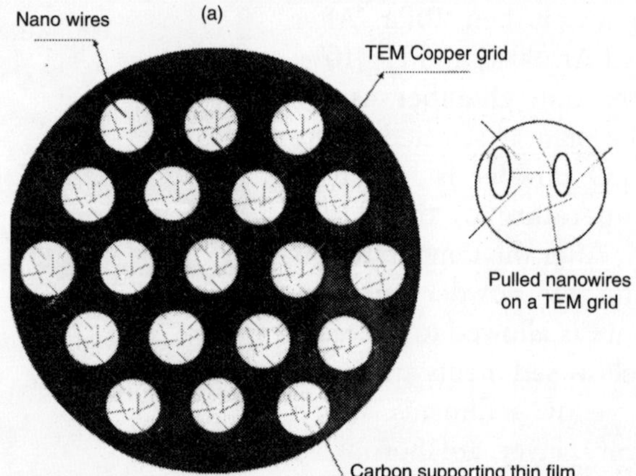

Figure 10.4 *The experimental methods for conducting the transmission electron microscopy tensile experiment on Si nanowires; (a) shows the Si nanowires being scattered on a Cu grid with carbon supporting thin film and (b) shows the broken carbon thin film pull the. Si nanowires axially under the electron beam heating.*

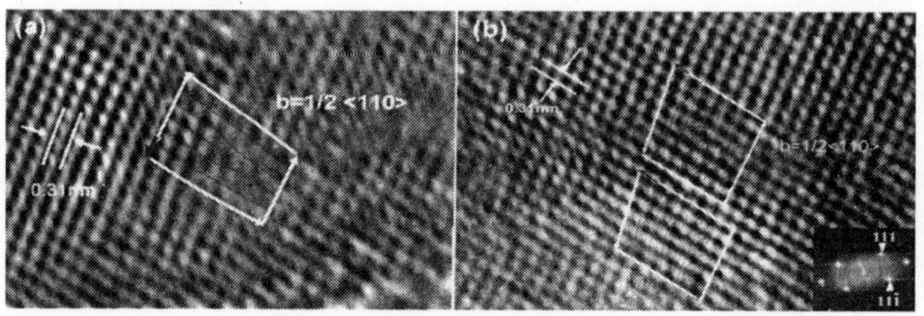

Figure 10.5 *The disloactions generated on two {111} planes have equivalent Schmidt factor values with respective to the loading force direction and the NW's orientation. (a) A dislocation emitted on (11-1) plane and (b) two dislocations on the (111) plane. The Burgers vector circuits are drawn in the figure and the Burgers vector projections are indicated by arrows.*

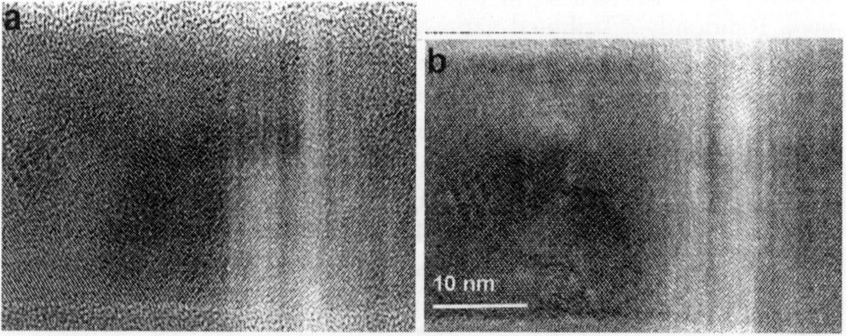

Figure 10.6 *Electron irradiation effects on a Si NW revealed by HREM observation. (a) is the HREM image along [110] direction prior to the electron irraidation shower and (b) the HREM image of the same Si NW with an electron irraidation shower by 1×10^{19} e cm^{-2} s^{-1} for 7200 s.*

on the[110]-oriented Si nanowire are conducted with the field emission gun, to achieve atomic level resolution imaging simultaneously with the axial tensile tests to elucidate the atomistic mechanisms. During the tensile tests, the electron beam is spread to about 20-50 μm to avoid the effects of electron irradiation on the Si NW/s. The operating voltage for both HRTEM observations is 200 kV and with a low electron flux dose rate around 5×10^{19} e cm^{-2} S^{-1}. The increased temperature of the Si NW specimens due to electron beam heating is a few degrees above room temperature.

Simulation of the HRTEM Images The cubic structured Si atomic model is built with a software and the HRTEM simulation is conducted. The multi-slice approximation is used in simulating the HRTEM images. The simulated HRTEM image is obtained under the conditions of 250 A thickness and defocus with 100 Å.

EELS and EDS Analysis: The electron energy loss spectroscopy (EELS) line scan experiments are obtained using a scanning transmission electron microscope (STEM). The EELS spectra are taken. The energy resolution is about 1.0 eV and the spatial resolution is 1.4 Å. The EELS spectra are taken across the Si nanowire diameter with a step size of 1 nm the collection time for each spectrum is 5 s. Specimen drifting is compensated automatically. The collected raw EELS spectra are background subtracted. EDS line scan analysis is conducted with the HRTEM imaging and the axial tensile tests on the Si NWs.

It is shown that large (5 to 50 μm diameter), low melting point gallium droplets can be used as an effective catalyst for the large-scale growth of highly aligned, closely packed silica nanowire bunches (see Fig. 11.1(a) and (b). The silica nanowires grow batch by batch. For each batch, numerous nanowires simultaneously nucleate, row at nearly the same rate and direction, and simultaneously stop growing. Tubes whose wall is composed of highly aligned silica nanowires with diameters of 15 to 30 nanometers and length of 10 to 40 micrometers are obtained (see Figure 11.1(b)). One of the most amazing phenomenon observed is the self-splitting of the silica nanowire, (Figure 11.1(c)), which is not observed in crystalline nanowires. To offer a simple interpretation about the formation of wire splitting, the five Zachariasens' rules for oxide glass formation are examined which are:

1. Oxide glass networks are composed of oxygen polyhedra.
2. The coordination number of each oxygen atom *is 2.*
3. The coordination number of each metal atom is 3 or 4.
4. Oxygen polyhedra share corners, not edges or faces.
5. Each polyhedron must share at least three corner.

The polyhedron for silica is $(SiO_4)^{4-}$, which is a tetrahedron. The glass is formed by sharing corners of the tetrahedral, but with out translation or orientation symmetry (see figure 11.1(d)) for example, take tetrahedron B, it has four corners to share. If this tetrahedron is on the surface of the nanowire, another tetrahedron *C* can be linked to B via corner sharing, a continuous growth along C' and D' leads to the formation of another nanowire. Thus, the splitting of the nanowire offers a new approach at nanoscale for splitting or converging optical signals using the silica nanowires. This is because the density of the silica nanowires can be adjusted according to the geometry and available space to create a *wide* range of nanostructures, such as the diatoms found in nature.

Plasticity of Silicon Nanowires

Elastic-plastic and fracture properties are determined for characterizing materials' mechanical behavior. Extensively Silicon nanowires (NWs) are one of the most important nanostructures used for fabricating various electronic and optoelectronic nanodevices, and they are a building block for the construction and assembly of functional nanometer-scale systems. Although the electrical and optical properties of Si NWs are known only limited information is available about the structure-mechanical property correlations of Si NWs. This is due to the difficulty of carrying out tensile or bending measurements on individual NWs, The elastic-plastic strains retained in NWs significantly affects their electronic properties by perturbing the band structure or changing the Fermi energy of the nanostructures, For example, the applied strains of continuous torsion on carbon NTs results in chirality variation and therefore introduce a distinct conductance oscillation from metallic to semiconductor. A strain induced giant piezoresistance effect is also observed for Si NWs.

TEM observation of the elastic-plastic-fracture processes of a single Si NWs recorded at atomic resolution produced by experiment I above is given below (see figure 10.9). It shows the strain-induced structural evolution process of Si NWs and its large strain plasticity (LSP). The LSP of Si NWs via a brittle-ductile transition originates from a dislocation-initiated amorphization. This is in contrast to them mechanical behavior of bulk Si.

The Si NW elastic elastic-plastic, and plastic deformation investigations are carried out by ultrahigh-resolution TEM (UHRTEM) by axially extending the Si NWs by means of a mechanical force created by a TEM specimen-supporting grid under electron beam irradiation (see figure 10.3). Figure 10.7(a) to (d) show a series of images' of a single Si NW that is extended by the force created by the skrinkage, of the broken colloidal thin film A clear plastic deformation is observed at the center of the NW none of crack is visible instead, the NW transforms from crystalline structure to amorphous at the central region of the axially extended NW. Of the nine Si NW samples (ranging from 15 to 70 nm in diameter), all show a large elastic strain if the NW diameter is below 70 nm, and a large plastic deformation rate is achieved. The strain rate is kept at a level of $10^{-5}s^{-1}$ A reduction in diameter as large as 426 % (the ratio of initial diameter over breakpoint diameter as a percentage) and 125 % elongation ratio are obtained at ambient temperature for a NW with an original diameter of 26 nm, as shown in Figure 10.7(a) to (d). The series of images in Figure 10.7(e) to (h) shows another example of an axial tensile event of a Si NW, in which buckling of the extended NW is observed, which is similar to the case NTs. The diameters of the axially extended NWs are measured prior to and after the plastic elongation. The axial elongation of the Si single crystalline NWs is revealed by comparing the corresponding lengths l_0 and l_f in Figure 10.7

Table 10.1 summarizes the diameters of the nine extended NWs together with the NWs' orientation. All of .the Si NWs have [110] growth direction. The Si NWs plasticity is evaluated by the diameter reduction ratio r_o/r_f as a percentage, where r_o and r_f, are the initial and final diatmeters of the Si NW, respectively, For the NW with r_o=26 nm and r_f=6.1 nm, a diameter reduction ratio of 426 % and an elongation of 125 % are obtained. This reveals the nature of the LSP. Figure 10.8 shows a plot of the diameter reduction ratio versus the Si NW diameter. Extrapolation of the plot gives brittle-ductile transition diameter of around 60 nm for ambient

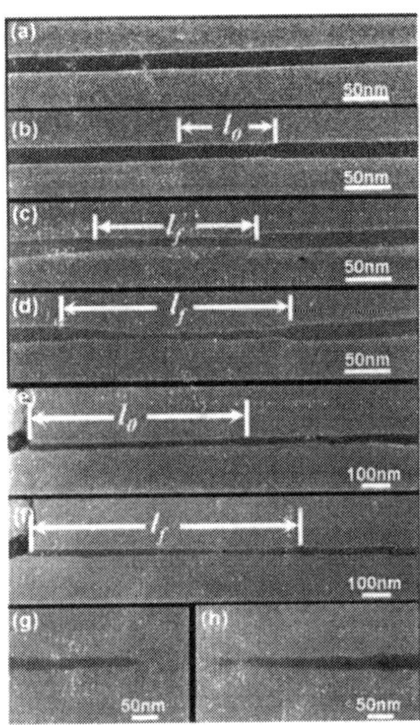

Figure 10.7 *(a-d) A Si NW during the application of axial tension.The arrows indicate the corresponding positions of the extended nanowire throughout the procedure. The original diameter of the Si NW is 26 nm and after plastic elongation the diameter reduced to 6.1 nm. (e-h) Extending a Si NW with buckling character. The axial elongation of the Si single-crystalline NW is obtained by comparing the corresponding l_0 and l_f*

Table 10.1 Details of the 9 axial-extended Si NWs

	Growth direction	*incipient diameter of nanowires (nm)*	*breakpoint of diameter(nm)*	*The ratio of incipiet diameter and breakpoint's diameter(%)*
1	<110>	15	2.3	652
2	<110>	17	2.8	607
3	<110>	23	4.7	490
4	<110>	26	6.1	426
5	<110>	33	10	330
6	<110>	34	10.6	321
7	<110>	44	16.5	267
8	<110>	50	27	185
9	<110>	70	66	106

temperature. The size dependent Young's modulus of Si are also shown in Figure 10.8 When the NWs diameter is large than 60 nm an abrupt brittle fracture occurs.

Consider the mechanism of the superplastic deformation of Si N through an axial tensile experiment using TEM. Figures 10.8(a) and (b) show low-magnification images of a Si NW, extended at the initial elastic-plastic transition and in the final broken state, respectively. Figure 10.9(c) is a HRTEM image of the axially extended Si NW at the initial elastic-plastic transition. Figure 10.9(d) is an HRTEM image of the extended Si NW at a later lime. Figure 10.9(e) is an enlarged HRTEM image taken from the framed region indicated in Figure 10.9(c). Based on the first-order approximation, assume that the white dots in the HRTEM images arc the projections of Si-Si "dumbbell" atomic pairs, as in conventional lattice images. The

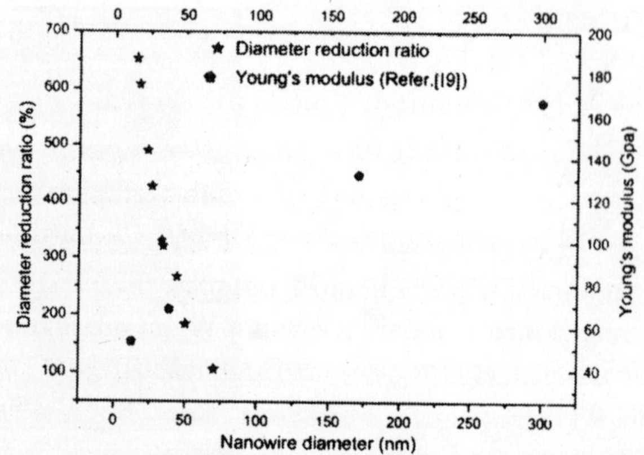

Figure 10.8 *The diameter reduction ratio as a function of nanowire diameter when axial tension is at a low temperature (room temperature). Young's moduius values as a function of diameter are plotted in the same figure for com0parison.*

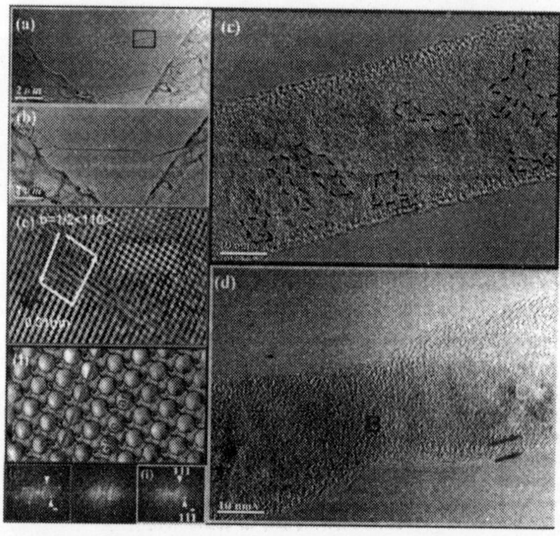

Figure 10.9 *High resolution electron microscopy of an axially extended Si NW. a-b) Low magnification images prior to and after the Si NW is plastically broken. c) HRTEM image of the Si NW in a status of elastic-plastic transition and prior to severe plastic deformation. d) The same Si NW after large-strain plastic deformation with a neck in the middle of the NW. g-i) The FFT electron diffraction patterns taken from areas A-C in (c), respectively. e) Enlarged HRTEM image taken from the framed region in (c). The dislocation structure is highlighted by dotted lines. A Burgers vector circuit is drawn, indicating a perfect edge dislocation. The simulated HRTEM images of the perfect crystalline lattices and the head-tail connection are shown as insets. f) Atomic structure of Si atomic lattice projected along the [110] orientation.*

HRTEM images is simulated by multiple-slice theory and the simulated HRTEM image for the perfect lattice is inserted in Figure 10.9(e), in which every white dot is regarded as a Si-Si dumbbell atomic pair. Tracking the extended Si NW HRTEM image it is found that the initial elastic-plastic transition is characterized by activation of plastic dislocations. In Figure 10.9(e), a Burgers vector circuit is drawn around a dislocation and the Burgers vector projection is directly determined. The dislocation Burgers vector is determined as (1/2)<110> which is due to a projection of <112>, as shown in Figure 10.9(e), a 60^0-type dislocation. This type of dislocation is dominant in this low-temperature plastically deformed Si NW so this dislocation is activated on the {111} shuffle planes. Figure 10.9(f) shows the two-dimensional (2D) projection of the diamond-type crystal line structure of the Si lattice along [110]. The two types of [111] shuffle and glide planes are indicated in Figure 10.9(f). Theoretically. the activation of dislocations on the {111} shuffle planes takes less energy (0.1 G) than on glide planes (0.35G) at low temperatures where G is the shear modulus The dislocations are emitted equivalently on the two $(1\bar{1}1)$ and $(\bar{1}\bar{1}1)$ planes with respect to the external loading force direction and the NW longitude direction (see Figure 10.5) Although the dislocation movement from one potential to another is directly revealed in a high temperature in a Si thin film. The generation of dislocations has of triggered the initiation of plasticity of Si NWs at room temperature. The high-pressure-induced plasticity, phase transformations of a β-Sn structure can induce the occurance of phase transformation. The availability of high shearing stress and slip systems on the (111) shuffle planes initiateplastic dislocations.

Further, following the initiation of dislocation plasticity, with increased strain, LSP of the axially extended Si NWs at low temperatures is observed. Throughout the whole procedure cracks are observed even at the atomic scale. The LSP of Si single-crystalline NWs is characterized by the continuous development of disorder in the lattice through emission of dislocations and disordering of crystalline structures and amorphization and plastic necking, as shown in Figure 10.9(d). Three fast Fourier transform (FFT) diffraction patterns are shown in Figure 10.9(g) corresponding to the areas A-C (Fig. 10.9(d)) of the NW, respectively. These indicate that areas A and C are crystalline and B has transformed to a disordered structure. The fraction of the disordered structure with respect to the original crystalline structure varies as a function of distance from the most strained region of the narrowest location of the superplastically deformed Si NW. This indicates that LSP of the axially extended Si NW is a continuous process involving creation of dislocations and disordering of the crystalline lattice.

The thickness of the NW surface with disordered structure before Fig. 10.9(c) and after (Fig. 10.9(d)) axial extension. This reveals that, during the process of axially extending the Si NW, the surface disordered lattice area of the Si NW expand along the radial direction from the original thickness of 3 nm to 7 nm. At the thinnest (most necked) region, the crystalline Si lattice completely transforms to a disordered structure. This observation show that the plasticity and LSP are surface-effect-triggered phenomena. It is known that the surface of a Si NW can be covered by a thin oxide layer. The oxygen diffusion into the lattice especially in the strained area, during plastic deformation is carried out using electron energy loss spectroscopy (EELS) in an attachment to the transmission electron microscope). Data is acquired simultaneously while conducting the axial tensile tests on the Si NW. Figure 10.10(a) and (b) shows two sets of

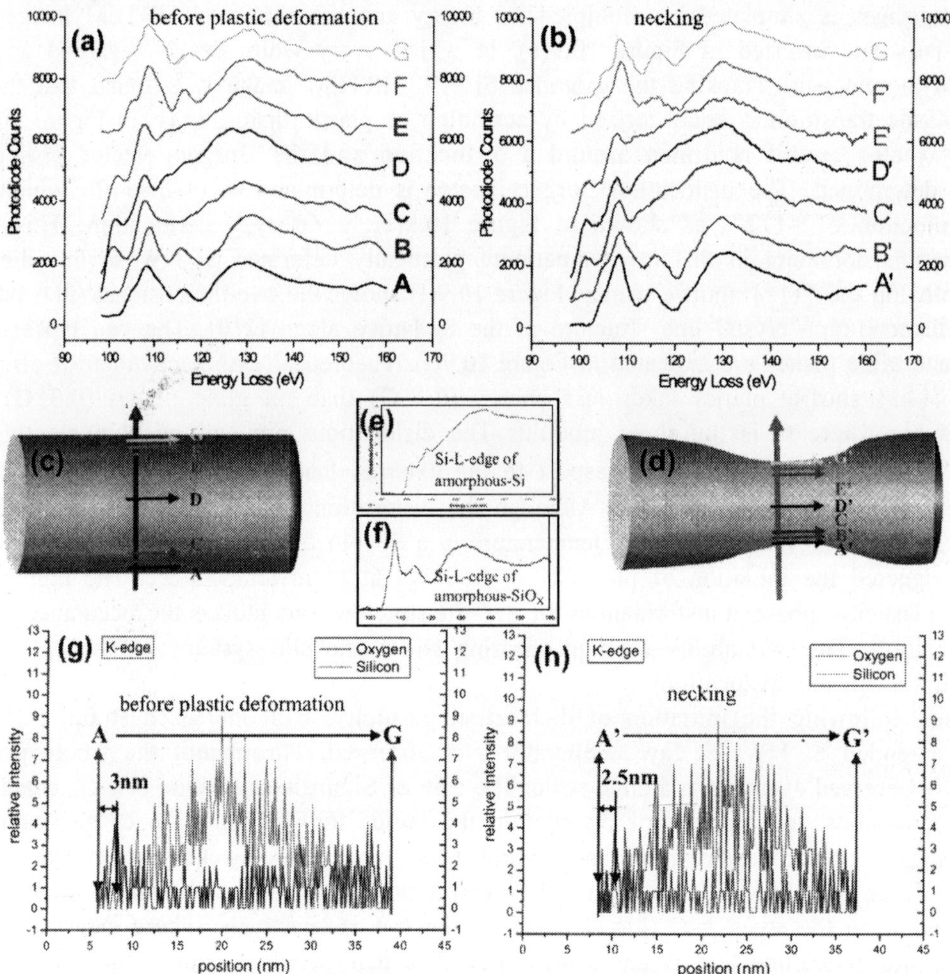

Figure 10.10 *Series of EELS spectra and EDS line scans taken from a Si NW prior to and after large-strain plastic deformation. Both series of EELS spectra are taken across the NW width. a) A series of spectra taken from the Si NW prior to plastic deformation. b) A series corresponding to the NW with a superplastic necking feature. c-d) the locations at which the spectra are taken for (a) and (b). respectively. e-f) EELS spectra for the Si L edge of Si and amorphous SiO., respectively. g-h) Line scan EDS spectra of elemental oxygen and silicon for the Si NW before and after neck formation. The surface oxygen layer thickness is measured as the distance from the starting position A or A′ to the first peak position of oxygen in both oxygen spectra.*

EELS spectra taken from a Si NW before and after superplastic deformation. For each set of EELS analysis, 32 spectra are taken across the diameter of the SiNW and seven of them are chosen to demonstrate that neither prior to nor after the superplastic deformation, the SiO*x* surface layer diffuses radially into the inner layers of the Si NW (especially in the highly strained area) Figure 10.10(a) shows a series of EELS spectra taken from the Si NW prior to plastic deformation. Figure 10.10(b) shows the corresponding EELS spectra taken from the superplastically deformed Si NW with a neck. The corresponding locations at which the EELS spectra are taken shown in Figure 10.10(c) and (d). From comparison of the fine structure of the Si L edge with standard samples, a thin, 3.5 nm thick SiO_x. surface layer is revealed, identified by the EELS spectra shown in Figure 10.10(a). This is consistent with the result directly derived HRTEM observation as shown in Figure 10.9(c). The Images in Figure 10.10(a) and (b) are taken along the same line across the NW at a time interval of 1 h. The Si $L_{2\text{-}3}$ edge characters of amorphous Si and Si*Ox* are also shown in Figure 10.10(e) and (f), respectively, The EELS spectra of the surface areas show typical Si*Ox* amorphous features with an $L_{2\text{-}3}$ edge at 103 eV for both Figure 10.10(a) and (b). The thickness of the surface Si*Ox* layer, is reduced to 3.0 nm after superplastic deformation. This is in agreement with the diameter reduction due to plastic necking and, attributed to the surface atomic diffusion process. The shapes of the Si $L_{2\text{-}3}$ edge of the EELS spectra acquired from position C, D and E (Fig. 10.10(a) and (c)) have distinctive feature in çomparison to the spectra acquired from locations A, B and F, G. The Si $L_{2\text{-}3}$ edge of spectra C, D, and E (Fig. 10.10(a)) starts at 98 eV, which is a feature of crystalline Si while the spectra C′, D′ and E′ (Fig. 10.10(b) and (d)) show features of disordered Si. As revealed in the reference EELS spectra the difference between crystalline Si and amorphous Si is not pronounced. This confirms that Sp^3 bonding is preserved in the plastic-deformation-induced disordering of Si structure β-Sn-type phase transform is not from either the electron diffraction or the EELS analysis. Energy dispersive X-ray spectrometry (EDS) line scans on the Si NWs prior to and after deformation i.e Figure 10.10(g) to (h), shows the K-edge spectra from the EDS line scans of silicon and elemental oxygen across the Si NW prior to and after necking deformation. There is an oxygen-rich layer in the oxygen spectra both prior to and after plastic deformation. The oxygen-rich layer is about 3 nm prior to plastic deformation, which is close to the EELS analysis results and the HRTEM observation. With plastic necking, the oxygen layer *is* reduced to about 2.5 nm, remaining at the same level, as shown in Figure 10.10(h). The oxygen in the thin Si*Ox* layer does not diffuse along the radial direction to contribute to the Initiation of dislocation plasticity and LSP of Si NWs.

From the direct atomic-scale imaging of emitted dislocations at ambient temperature in Si NWs during axial tensile experiments, the plasticity and LSP of Si NWs with dislocation velocity and the Young's modulus is obtained. The plasticity of a particular material (i.e.. whether it is brittle or ductile) is controlled by the dislocation velocity, which is regarded as the rate of emission of dislocations and/or dislocation motion. It is given by.

$$v = A\tau^{m}\exp\left[\frac{-U}{k_B T}\right] \tag{1}$$

where v is dislocation velocity, A and m are constants, τ is shear stress, k_B is the Boltzmann constant, T is absolute temperature, and U is the activation energy of a dislocation. The activation energy is regarded as a constant that reflects the atomic-atomic interactions and bonding characters and is correlated to shear modulus G and elastic modulus E by $U \propto Ga_o$ and $U \propto Ea_0/[2(1 + r)]$, where a_0 is the lattice constant and y is the Poisson ratio. The NW's surface structure differs from that of the bulk material and it strongly affects the NW's physical, chemical, and mechanical properties due to pronounced surface effects. The elastic modulus of Si is related to the size of the NW, and it decreases with decreasing NW diameter and film thickness. By assuming that a_0 and r are constants, the simple relation is:

$$U_N = \frac{E_N}{E_B} U_B \tag{2}$$

where U_B and E_B are the dislocation activation energy and Young's modulus of bulk silicon, respectively, and U_N and E_N are those of the NW. Then dislocation velocity in the NW is obtained as:

$$v_N = \left[\frac{-E_N U_B}{K_B k_B T}\right] \tag{3}$$

Using $U_B = 2.2$ eV. and $E_B = 169$ Gpa, the correlation of dislocation activation energy for the NWs, U_N with the diameter of the NWs, as shown in Figure 10.11(a) is obtained. The extrapolated Si NW's Young's modulus is plotted in the same figure. The dislocation activation energy U_N

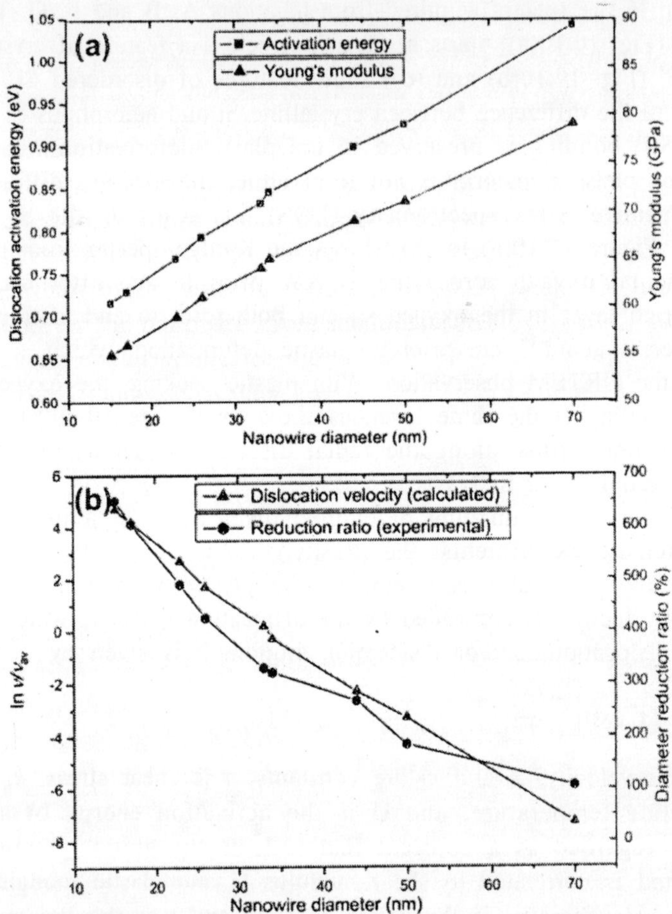

Figure 10.11 *(a) The dislocation activation energy of nanowires U_N with respect to the diameter of the nanowires. The extrapolated Si NW Young's modulus is also plotted. b) The normalized dislocation velocity ratio v/v_{avg} as a function of the diameter of the Si nanowires. The diameter reduction ratio is plotted in the same figure for comparison.*

decreases with decreasing NW diameter, reaching a value of 0.716 eV at a diameter of 15 nm, significantly lower than the 22 eV of bulk materials. The dislocation velocity for Si NWs is compared as a function NW diameter at room temperature. In Figure 10.11(b) the normalized dislocation velocity ratio v/v_{av} is plotted as a function of the Si NW diameter. Where v_{avg} is the average dislocation velocity for Si NWs. Now v increases rapidly with decreasing NW diameter. The plasticity evaluation parameter i.e. the diameter reduction ratio (Fig. 10.8 the star curve), is plotted in the same figure. Also v/v_{avg} shows a similar tendency to the diameter reduction ratio. This tendency demonstrates that the plasticity, that is, the diameter reduction ratio of Si NWs, is correlated with the dislocation velocity.

Experiment-II (Another method)

Synthesis of the SiNWs: SiNWs are synthesized by a vapor transport and condensation process using Au as a catalyst. The quartz tube inside the three zone furnace is evacuated to a pressure below $133\text{x}10^{-5}$ Pa using a diffusion pump. During the growth procedure a constant flow of 60 sccm Ar and 15 sccm H_2 is introduced as the carrier gas and the pressure inside the quartz tube is kept constant at 133Pa. The three zones are heated from room temperature to 1370,1170 and 970K respectively, at the same time. The Si source is placed in an alumina boat, in an upstream, 1370K zone. Au-coated Si substrates are heated downstream in the middle of the 1170K and 970K zones. After the samples are kept at the desired temperature for 90 min, the furnace ,is cooled down to room temperature.

Sample Preparation: SiNWs are dispersed uniformly in ethanol solution. Immersing a tungsten probe in the solution for several seconds causes a SiNW to be attached to the tip, and can be checked with an optical microscope. The samples are annealed in a vacuum furnace at 570K, at 50sccm Ar flow under 1.3×10^3 Pa.

Mechanical Property Measurement: A tungsten probe with a SiNW is located on a manipulator, which is remotely controlled in three dimensions. The length of the NW is estimated by rotating the NW in a scanning electron microscope. The cross section is assumed to be round. The AFM cantilever is made of Si_3N_4 with a spring constant of 0.006 N m^{-1} and loaded on a rotator that can be rotated through 360^0.

Elastic Properties and Buckling of Silicon Nanowires

The SiNWs used are fabricated by chemical vapor deposition by the vapor liquid solid (VLS) growth process. Figure 10.12 shows a transmission electron microscopy (TEM) image of an as-grown SiNW, revealing its core/shell structure. The diameters of the NWs are 40-90nm. The thickness of the outer oxide layer is about 5 nm. The inset shows a diffraction pattern, revealing the single-crystalline structure of the NWs.

Figure 10.12 *TEM image of the SiNWs with 5-10nm native oxide layer. Inset: Selected area electron diffraction pattern.*

The SiNW is manipulated using a tungsten probe and the AFM cantilever is used to measure the force-displacement response of a single SiNW. Figures 10.13(a) and (b) illustrate the principle of the mechanical manipulation setup.
First (see Figure 14*a*) the SiNW is attached to the probe. Then the SiNW is bent by pushing the probe against the AFM cantilever. The AFM cantilever is deflected, as shown in Figure 10.13(b), from which the displacement and the applied force is quantified, provided the mechanical behavior of the cantilever is characterized. The displacement is characterized by the distance between the two ends of the NW.

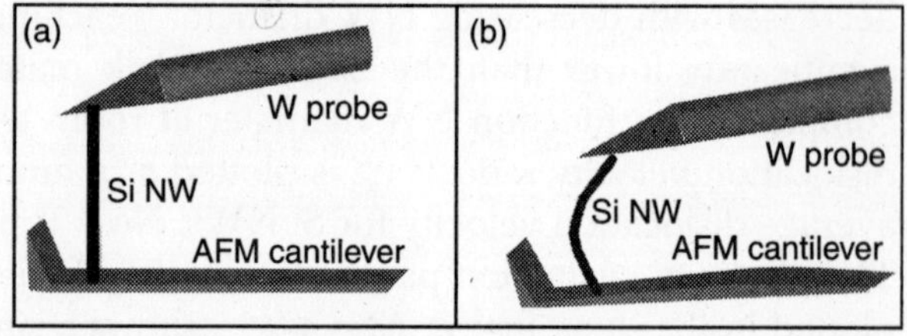

Figure 10.13 *The experimental steup before (a) and after (b) the maniputation*

Figure 10.14 shows images and the corresponding bending curves measured from a single SiNW when it is buckled by the probe. The NW is oriented perpendicular to the direction of the electron beam to ensure that there is no "hidden" displacement along the electron beam direction, in order to measure the displacement accurately. Figures 10.14(a) to (f) show that the SiNW is being pushed against the AFM cantilever and the deflection of the cantilever is presented in the images with reference to a stationary feature at the top of the images which is not connected to the cantilever and is out of focus in the images due to the difference in depth. With this stationary feature as a reference point, the NW deflection as well as the bending force is quantified. The NW is affixed at its two ends to the tip of the cantilever. The straight distance between the two ends of the NW (L) is used to characterize its buckling effect. Figure 10.14(g) shows a curve of the applied force versus the absolute value of the difference of the chord. i.e., $|L-L_0|$. The curve present the standard buckling behavior of a slender column under the action of an axial load.

Figure 10.14(h) shows the curve for the stress and strain based on Figure 10.14(g). Under the approximation that the strain is homogeneous and the NW become semicircular at the maximum deformation, the equations $R\alpha = L_o$ and $[R \sin(\alpha/2)] = L/2$ is solved as depicted in the inset of Figure 10.14(h). The stress is expressed as a function of the applied force F and the cross-sectional area of the NW,A: stress = F/A. The strain is expressed as a function of diameter of the NW,d and the radius of curvature R, strain = d/2R. At the maximum elastic point the strain of the NW is 1.5%, which is much higher than the 0.2% for most metallic materiais.

Figure 10.15 shows the experimental and calculated data of the strain versus $|L-L_0|$ during the buckling. In the calculation, the relationship between the buckled chord length L and the strain ε is calculated with a thin rod model. Now suppose that the buckling force applied by the AFM cantilever is parallel to the original unbuckled nanowire. The shape of the bent nanowire is shown in Fig. 10.15 which is given in the following form:

$$x = \sqrt{2}\lambda \times [\sqrt{1-\cos\theta_0} - \sqrt{\cos\theta - \cos\theta_0}] \tag{1}$$

$$y = \frac{1}{\sqrt{2}}\lambda \int_0^\theta \frac{\cos\theta}{\sqrt{\cos\theta - \cos\theta_0}}$$

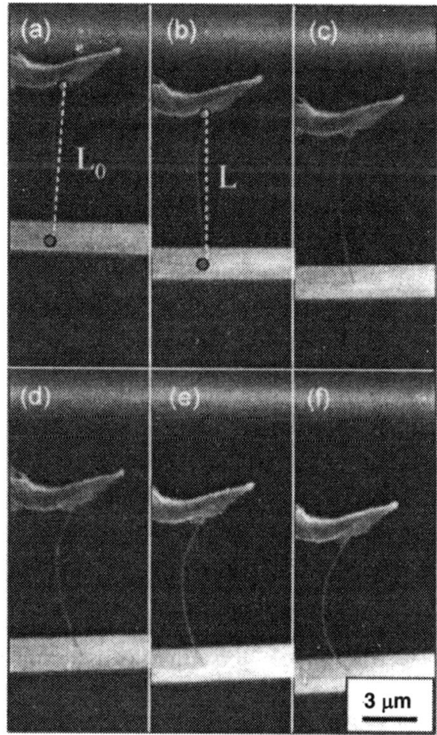

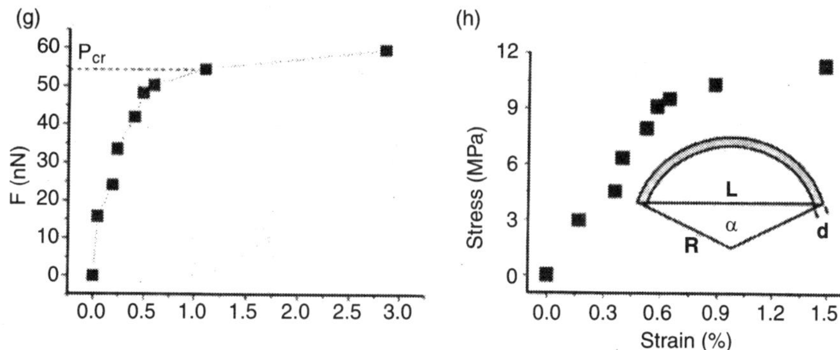

Figure 10.14 *(a-f) A series of SEM images showing the continuous buckling of the SiNW. g) Corresponding curve of the applied force F vs change in chord length $|L - L_0|$when the NW is buckled. h) Calculated stress-strain curve of the buckling of the NW. Inset diagram of the deformation.*

where θ is the local orientation of the tangent to the nanowire. λ is a characteristic length, defined as $\lambda = (EI_0/f)^{1/2}$, where E is Young's modulus, I_o is the moment of inertia of the cross section, and f is the magnitude of the buckling force. $\theta_o = \theta(L_0)$ is the local orientation at the top of the nanowire, where L_o is the chord length of the unbuckled wire, that is, the length of the nanowire. The maximum strain that occurs at the surface of the nanowire is

$$\varepsilon_{max} = r\frac{d\theta}{dl} = r\frac{1}{dl/d\theta} \qquad (2)$$

where r is the radius of the nanowire and

$$l = l(\theta) = \frac{1}{\sqrt{2\lambda}}\int_0^\theta \frac{\cos\theta}{\sqrt{\cos\theta - \cos\theta_0}}$$

is the arc length from the bottom of the wire to the point *(x,y)*.

The maximum strain a function of l is *calculated* at the position $l = Lo/2$ i.e at the middle of the wire. The strain versus chord length plot is given in Figure 10.15. The curve matches the experimental data which indicates that the assumption fits well for Figure 10.14(h). Thus the strain of the NW is larger than the conventional value for bulk materials.

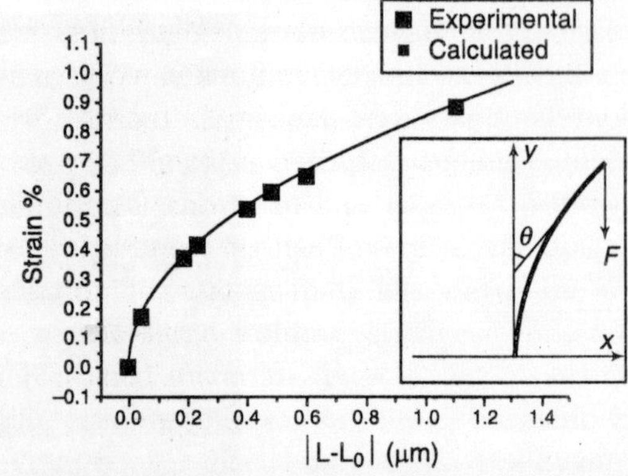

Figure 10.15 *A comparison of the experimental results and simulated curve of strain vs. change in chord length |L – Lo| Inset: The nanowire under buckling. The calculation is carried out for the middle point of the chord: l= Lo/2.*

The critical load P_{cr}, is marked in Fig. 10.14. After reaching the critical force, a slight increase of the force causes the NW to buckle. The NW returns to its original shape when the load is removed. The force deflection curves obtained show a high tendency for buckling effect, which is explained by Euler's formula. From Euler's formula, which is expressed as a function of the elastic modulus E, moment of inertia I_0, length of the NW L_0, and an effective factor K such that $P_{cr} = \tau E I_o/(KLo)^2$, the critical loads are used to evaluate the elastic modulus of the NW (assuming the validity of Euler's formula). Because both ends of the NW are pinned, the effective factor K is equal to 1. Figure 16 shows the P_{rc} versus I_o/L^2 curve for a few NWs with different diameters and lengths. Euler's formula indicates that the slope of the curve is equal to τ E, giving the mean elastic modulus of the NWs E = 175 Pa, which is consistent with the bulk value. This means that the elastic modulus of the NW is the same as the bulk value and is not changed by the reduction to a nanometer scale.

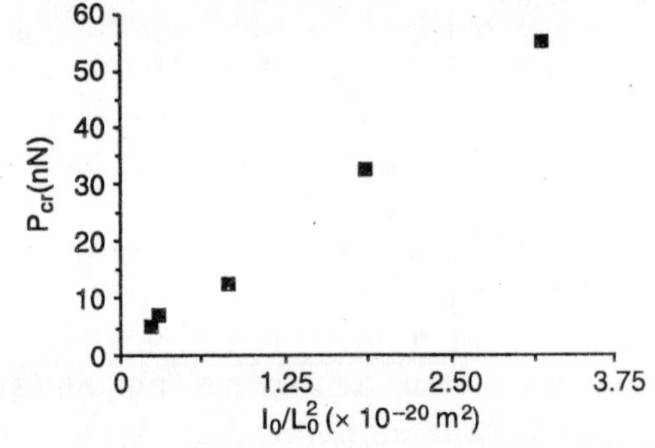

Figure 10.16 *Plot of the critical Pcr vs. I_o/L^2 for several NWs that have been manipulated.*

The elastic modulus of the NW is also measured by applying the force along different directions Figure 10.17(a) is a diagram of the manipulation process. First, a SiNW is attached to the probe and the head of the AFM. Then the SiNW is bent by moving the probe parallel to the AFM cantilever (vertically in Fig. 10.17(a), which causes the AFM cantilever to bend because of a small horizontal force created by the NW. Using the stationary feature at the bottom of the images (Figs. 10.17(b) to (f) as a reference, the tangential force F_t along the vertical direction (Figs 10.17(b) to (f) and the bending of the NW in the vertical direction is calculated.

From the Figures 10.17(b) to (f) of a SiNW, when it is pushed against the head of the AFM cantilever; the deflection of the cantilever is shown in the image with reference to the static

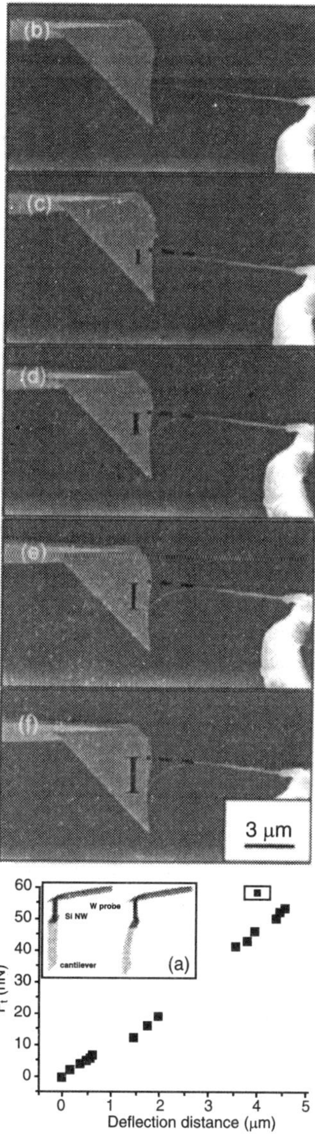

Figure 10.17 *(a) Diagram illustrating the principle of bending and the setup. b-f) A series of SEM images showing the manipulation and deflection of a SiNW. (g) Corresponding force-deflection distance curve of the NW.*

feature. The elastic modulus E is expressed a $E = kL^3{}_0/3l_0$ (where k is the spring constant). Figure 10.17(g) shows a curve from which the tangential force, as measured from the deflection of the cantilever, versus the deflection distance is obtained. From the curve, the elastic modulus is found to be 200 GPa. For large deflection distances, the mechanical behavior of the NW is about the same as in the small-deflection linear part. This observation shows that the linear regime of the NW is large. This helps in the prediction of mechanical response and the design of the mechanical devices in NEMS.

For all of the SiNWs, the compressive load buckles the NW but do not fracture it even under extremely large deformation. The increase in flexibility is due to decrease in thickness. For a slab of a material that has a thickness d, the elongation of the interatomic bond at the outer surface of the NW is $d\alpha/2$. If the critical bonding length before a bond breaks is s, under the condition of no generation of a dislocation, the maximum bending angle of the slab is $\alpha = 2s/d$. This means that the degree of bending is determined by the thickness of the object rather than its aspect ratio. For a NW of diameter 20 nm, its bending angle is 1000 times of that of a silicon wafer of thickness 20 μm. At the maximum elastic point the strain of the NW is 1.5%, which is larger than the conventional value of 0.2%. In addition, a change in crystallinity for nanowires contributes-to Its enhanced flexibility.

EXCITATION OF LOCAL FIELD ENHANCEMENT ON SILICON NANOWIRES

The large enhancement of the optical near-field close to illuminated particles when the incident wavelength is resonant with natural electromagnetic eigenmodes of the particle is recognized and has numerous applications. It is used in optical antennas for high-bandwidth intrachip and interchip connections, and has potential to improve the efficiency in harvesting energy from solar radiation. The excitation of the surface plasmon resonance (SPR) on metallic nanostructures is shown by Au, Ag, and Cu. The response of nonmetallic nanostructures to electromagnetic radiation and its applications are of interest, due to the fact that ,dielectrics have weaker damping, which result in a higher efficiency in radiative transfer.

This is demonstrated for Si nanowires (SiNWs) by calculating the spatial dependence of the optical field outside the surface of a cylindrical nanowire of radius R illuminated by a monochromatic plane wave of wavelength λ. with arbitrary electric field polarization (see Figure 10.18 below).

Scattering of a plane wave by an infinitely long cylindrical nanowire

The scattering geometry of a homogeneous and infinitely long, cylindrical nanowire irradiated by a plane wave is illustrated in Fig. 10.18. The nanowire of radius R is centered on the origin with its axis along the z-axis, and the incident direction is in the x-z plane and

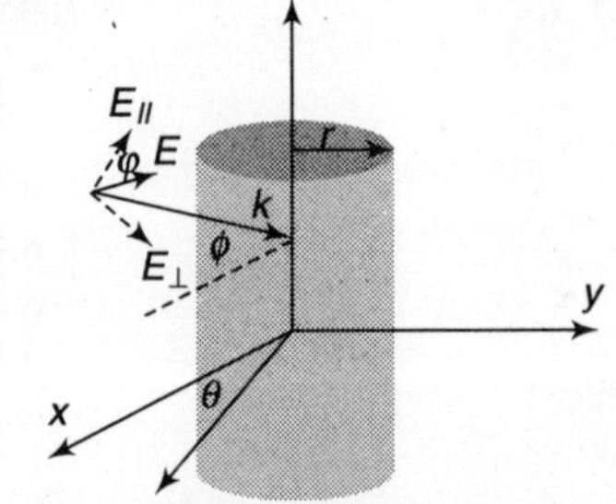

Figure 10.18 *The scattering geometry for the infinitely long cylindrical nanowire. $E_\perp$ and $E_\parallel$ are perpendicular and parallel to the plane of incidence (x-z).*

forms an angle ϕ with respect to the nanowire axis. The electric vector is polarized to from angle φ with the in plane component of the field $E_{\|}$ as shown.

The expression for the external field is given by

$$E_{ext}^{NW} = \frac{1}{\sqrt{k_0^2 - h^2}} \sum_{n=-\infty}^{\infty} (-i)^n e^{in\theta - ihz} (E_r \hat{r} + E_\theta \hat{\theta} + E_z \hat{z}) \tag{1}$$

$$E_r = \frac{in}{r}\left(-H_n(\sqrt{k_0^2 - h^2 r})a_r^2 + j_n(\sqrt{k_0^2 - h^2 r})\sin\varphi\right)$$

$$+ \frac{ih\sqrt{k_0^2 - h^2 r}}{k_0}\left(H_n'(\sqrt{k_0^2 - h^2 r})b_n^s - iJ_n'(\sqrt{k_0^2 - h^2 r})\cos\varphi\right)$$

$$E_\theta = \frac{nh}{k_0 r}\left(iJ_n(\sqrt{k_0^2 - h^2 r})\cos\varphi - H_n(\sqrt{k_0^2 - h^2 r})b_n^s\right)$$

$$+ \sqrt{k_0^2 - h^2 r}\left(H_n'(\sqrt{k_0^2 - h^2 r})a_n^2 - J_n'(\sqrt{k_0^2 - h^2 r})\sin\varphi\right)$$

$$E_z = \frac{k_0^2 - h^2}{k_0}\left(iJ_n(\sqrt{k_0^2 - h^2 r})\cos\varphi - H_n(\sqrt{k_0^2 - h^2 r})b_n^s\right)$$

where k_o $(= 2\pi/\lambda)$ is the propagation constant of the light in the medium, $h = k \cos \phi$, m is the complex refractive index of the cylinder, and $J_n(\sqrt{K_0^2 - h^2 r})$, $H_n(\sqrt{K_0^2 - h^2 r})$ are Bessel functions of the first kind and Hankel functions of the second kind, respectively. For normal incidence, $(\phi = 0$ and $h = 0)$, the expansion coefficients a_n^s and b_n^s: for TM mode excitation $(\phi = 0°)$ are

$$a_n^s = 0 \tag{2}$$

$$b_n^s = -ib_n = -i\frac{J_n(mk_0R)J_n'(k_0R) - mJ_n'(mk_0R)J_n(k_0R)}{H_n'(k_0R)J_n(mk_0R) - mH_n(k_0R)J_n'(mk_0R)} \tag{3}$$

and for TE mode excitation $(\varphi = 90^0)$

$$a_n^s = -ia_n = -\frac{J_n'(mk_0R)J_n(k_0R) - mJ_n(mk_0R)J_n'(k_0R)}{H_n(k_0R)J_n'(mk_0R) - mH_n'(k_0R)J_n(mk_0R)} \tag{4}$$

$$b_n^s = 0 \tag{5}$$

The electric field intensity I_{ext} is defined as $I_{ext}^{NW} = E_{ext}^{NW} \cdot E_{ext}^{NW\,*}$. For normal incidence, TM mode
$(\varphi = 0^0)$

$$I_{ext}^{TM} = \sum_{n=-\infty}^{\infty} (-i)^n e^{in\theta}(J_n(k_0 r) + b_n H_n(k_0 r))\hat{z} \cdot \sum_{n=-\infty}^{\infty} \left[(-i)^n e^{in\theta}(J_n(k_0 r) + b_n H_n(k_0 r))\hat{z}\right]^* \tag{6}$$

and for the TE mode *($\varphi = 90°$)*

$$\therefore \quad I_{ext}^{TM} = \sum_{n=-\infty}^{\infty} (-i)^n e^{in\theta}\left(\frac{in[J_n(k_0 r)\sin\varphi - a_n^s H_n(k_0 r)]}{rk_0}\hat{r} + (J_n'(k_0 r)\sin\varphi - a_n^s H_n'(k_0 r))\hat{\theta}\right) \times$$

$$\sum_{n=-\infty}^{\infty}\left[(-i)^n e^{in\theta}\left(\frac{in[J_n(k_0 r)\sin\varphi - a_n^s H_n(k_0 r)]}{rk_0}\hat{r} + (J_n'(k_0 r)\sin\varphi - a_n^s H_n'(k_0 r))\hat{\theta}\right)\right]^* \tag{7}$$

The average intensity $\overline{I_{ext}^{NW}}$ is defined as $\overline{I_{ext}^{NW}} = (2\pi)^{-1}\int_0^{2\pi}\left|E_{ext}^{NW}\right|^2 d\theta$. For normal incidence TM mode ($\varphi = 0^0$)

$$\overline{I_{ext}^{TM}} = \sum_{n=-\infty}^{\infty} |H_n(k_0 R)b_n + J_n(k_0 R)|^2 \tag{8}$$

and for TE mode ($\varphi = 90^0$)

$$\overline{I_{ext}^{TE}} = \sum_{n=-\infty}^{\infty} |J_n'(k_0 R) + a_n H_n'(k_0 R)|^2 + \frac{n^2}{k_0^2 R^2}|J_n(k_0 R) + H_n(k_0 R)a_n|^2 . \tag{9}$$

The Solution of Maxwell's equations and application of appropriate boundary, conditions at the dielectric nanowire-vacuum interface (equation 1) allows the local optical field $E_{\text{ext}}^{\text{NW}}$ to be expressed using the Lorenz-Mie formalism in cylindrical coordinates *(r.θ. z)*. Resonance occurs when λ. approaches one of the natural electromagnetic modes of the nanowire. Considering a SiNW under monochromatic plane-wave excitation (without loss of generality) with the incident electric field-polarized parallel (TM) or perpendicular (TE) to the axis of the nanowire, the spatial distributions of the calculated near-field intensity $I_{\text{ext}}^{\text{NW}} = |\text{E}_{\text{ext}}^{\text{NW}}|^2$ (see equation 6, 7) for the cases of resonant and nonresonant conditions (Figure 10.19) clearly demonstrate that the local field at resonance (Figure 10.19(a) is strongly confined to near the surface of the SiNW with the intensity enhanced by ~20-fold, while the nonresonant field (Figure 10.19(b) is weaker by 1 order of magnitude and is delocalized over a range of hundreds of nanometers away from the surface. Compared to the surface plasmon resonance enhanced field, that is significantly attenuated beyond a molecular-scale distance away from the metallic substrate, the resonant local field at the dielectric surface is expected to be much longer ranging in enhancing optical emission, absorption, optical force, and nonlinear optical response of mesoscopic systems as it extends over tens of nanometers (Figure 10.19(a)),

Now, the average intensity of the local field obtained by integration over the surface of the nanowire ($\overline{I_{\text{ext}}^{\text{NW}}} = (2\pi)^{-1}\int_0^{2\pi} |\text{E}_{\text{ext}}^{\text{NW}}|^2\ d\theta$ (equation 8, 9). The local light-matter interaction is at the surface of the nanowire is evaluated experimentally by monitoring selected optical signals such as the Raman scattering from chemisorbed molecules. Using the wavelength-dependent complex dielectric permittivity for Si, calculated values of $I_{\text{ext}}^{\text{NW}}$ as functions of the SiNW radius R and of λ. (Figure 10.20) reveals series of sharp peaks that arise from resonant scattering.

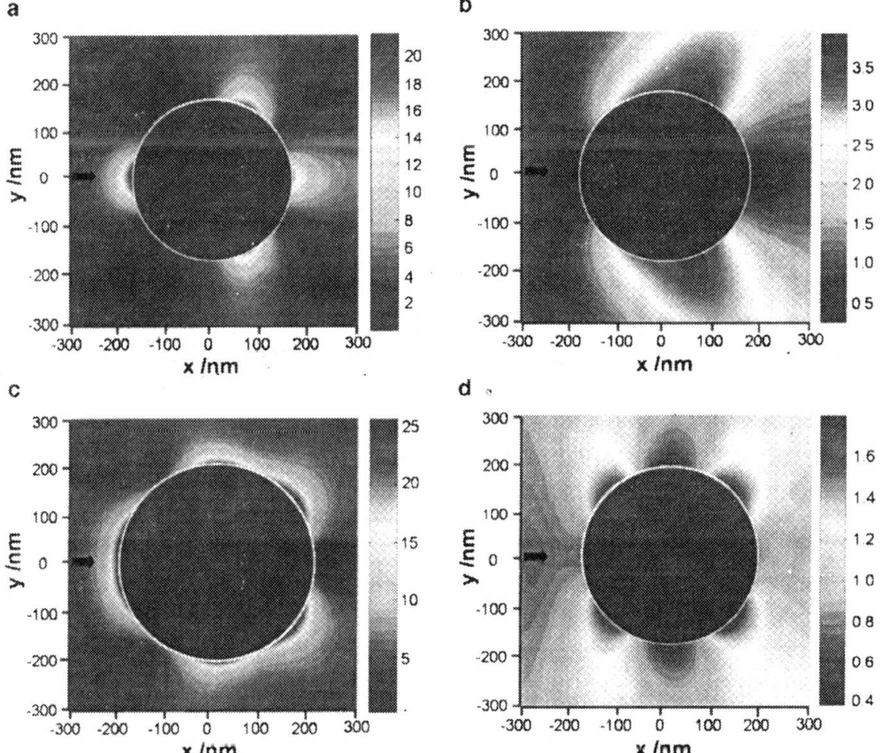

Figure 10.19 *Spatial distribution of the optical near-field in thc vicinity of a SiNW (cross-sectional view). Normally incident, monochromatic (λ = 785 nm) plane-wave radiation of unity intensity is denoted by the black arrow. (a) Resonant scattering and (b) nonresonant scattering for TM-mode excitation of SiNWs possessing radii of 169.3 and 179.8 mn, respectively. (c) Resonant scattering and (d) nonresonant scattering for TE mode excitation corresponding to radii of 180.2 and 208.7 nm, respectively The surface of each SiNW is denoted by a white circle. and the intensity of the internal field is intentionally set to zero for clarity. The nonresonant near-field extends to more than 300 nm from the surface, exceeding the area of the figure. The employed complex refractive index of silicon is 3.691-0.005i. The calculated resonant responses in panels a and c are shown for n = 3, where n is subscript in the scattering coefficients a_n and b_n as defined in equation (2), (3).*

The local field enhancement for SiNWs is carried out by collection of the Raman scattering from p-aminothiophenol (PATP) molecules chemisorbed onto Au nanoclusters (AuNCs)) which are formed following vacuum thermal evaporation of Au onto the SiNWs-(Figure 10.21(a)). Tapered SiNWs are prepared by metal nanostructure-catalyzed chemical vapor deposition. Au nanoclusters are formed on as-prepared tapered SiNWs (and on 300 nm SiO_2-coated Si(100) wafers without SiNWs as blank control substrates) via thermal evaporation of Au. The adsorption of organic molecules is carried out by immersing AuNC-coated tapered SiNWs (as well as control substrates) in 1 wt % ethanol solution of PATP for 24 h. Raman scattering spectra of the self-assembled PATP monolayers are collected in the backscattering configuration with 632.8

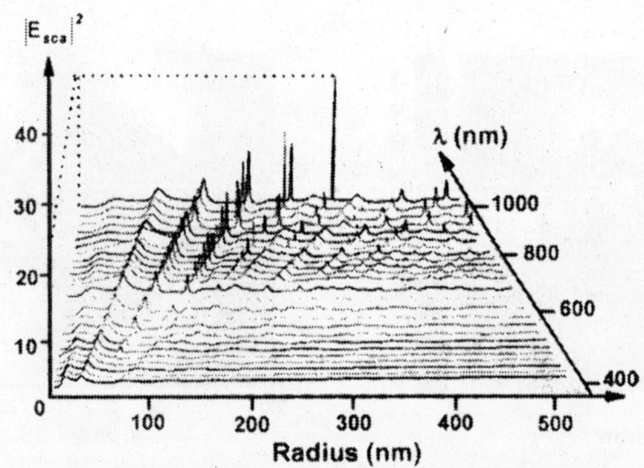

Figure 10.20 *Evolution of the average intensity of the local surface field of a SiNW as a function of the radius and the incident wavelength λ (400 to 1000 nm). The maximum integrated intensity of the near-field increase from ~3.5 at λ = 400 nm to ~26 at λ = 1000 nm for corresponding radii of ~25 and ~320 nm. respectively. Variations in the intensity of the peaks depend in part on the wavelength-dependent complex dielectric function of Si. as the stronger absorption of silicon for shorter wavelength tends to dampen the resonant character.*

nm HeNe (0.8 mW) and alternatively, 785 nm diode (0.08mW) laser excitation focused to Gaussian spot diameters of ~1.2 and~ 1.5 μm, respectively. The incident electric field is polarized perpendicular (TE) and parallel (TM) to the long axis of the tapered SiNWs. Spectra are collected from a number of locations from the base to near the tip of each tapered SiNW, where the diffraction-limited spot size limit the placement of the incident beam to ~ $\lambda/2$. For each λ. and each nanostructure, the ratio of Stokes-to-anti-Stokes-scattered intensity is evaluated to ensure that no laser-induced heating occurs. All integrated intensities are due to Lonrentzian fittings of the measured lineshapes. All values of enhancement are obtained from spectra collected from different locations along an individual tapered SiNW. The tapered SiNW are representative of all tapered SiNWs.

Raman-scattering signals from bulk PATP with a thickness of 300 μm are collected using identical conditions.

In the experiments, the role of the AuNCs as signal carriers: their SPRs along with the light-scattering resonant field together provides an enhancement that enables probing of the local light-matter interaction at the SiNW. The total field enhancement for this AuNCs-SiNW structure Q_{tot}^{NW} is approximated as the product of the field enhancement from AuNCs Q_{Au} and that from the SiNW Q_{Si}^{NW}, that is, $Q_{tot}^{NW} = Q^{Au}(Q_{Si}^{NW})^2$ with the SiNW enhancingm both the excitation of and the scattered radiation from the AuNCs (see below)

Model of the electromagnetic interaction between the SiNW and Au nanocluster

The near field in the vicinity of a Au nanocluster (AuNC) on a SiNW is expressed in terms of the superposition of several components $E_{101} = E_{inc} + E_{sca}^{NW} + E_{sca}^{Au} + E_{sca}^{Au-NW}$ with E_{inc}, E_{sca}^{NW}, E_{sca}^{Au} and E_{sca}^{Au-NW}, the amplitudes of the incident wave, the wave scattered from the

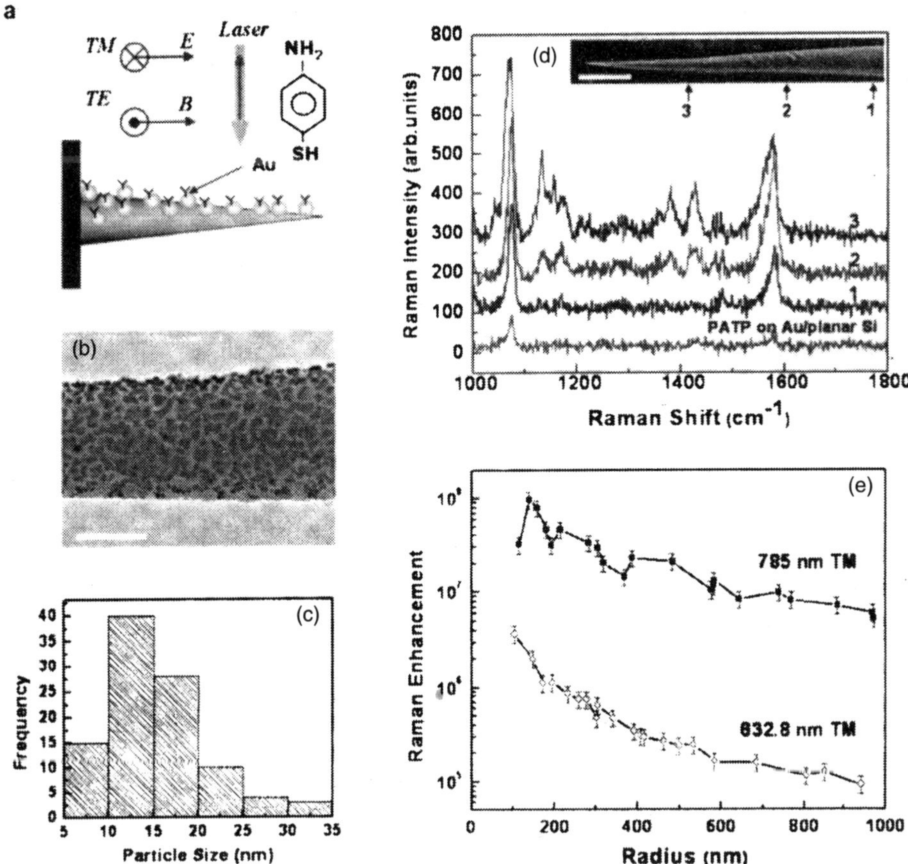

Figure 10.21 *Raman-scattering measurements. (a) experimental setup. The Raman spectra of PATP molecules that self-assemble onto Au nanocluster-coated and tapered SiNWs are collected in the backscattering geometry with TE and alternatively TM-polarized incident radiation. (b) A representative transmission electron micrograph of Au nanocluster-coated tapered SiNW; (c) Histogram of the size of the Au nanoclusters. (d) Raman spectra of PATP collected from the planar substrate and various locations along the tapered SiNW as denoted in the figure using 785 nm laser excitation. The inset is a scanning electron micrograph of a portion of the Au nanocluster-coated tapered SiNW. (e) Evolution of the Raman enhancement for the PATP molecules chemisorbed onto Au nanocluster (Au NC)-modified SiNWs as a function of the radius of the SiNWs collected using 632.8 (open circle) and 785 nm (solid square) with TM-polarized incident laser excitation.*

SiNW, the wave scattered from the AuNC, and the wave scattered first from the AuNC and then the SiNW, respectively. The total field enhancement is:

$$Q_{\text{tot}}^{\text{NW}} \approx \frac{|E_{inc} + E_{sca}^{NWi} + E_{sca}^{Au} + E_{sca}^{Au-NW}|^2}{|E_{\text{inc}}|^2} = \frac{|E_{inc} + E_{sca}^{NW} + E_{sca}^{Au} + E_{sca}^{Au-NW}|^2}{|E_{\text{inc}} + E_{sca}^{NW} + E_{sca}^{Au}|^2}$$

$$\times \frac{|E_{inc} + E_{sca}^{NWi} + E_{sca}^{Au}|^2}{|E_{inc} + E_{sca}^{NW}|^2} \times \frac{|E_{inc} + E_{sca}^{NW}|^2}{|E_{inc}|^2} \quad (10)$$

The enhancement contributed by the SiNW to the incident field is $Q_{Si}^{NW} = |E_{inc} + E_{sca}^{NW}|^2 / |E_{inc}|^2$. The scattered wave from SiNW, along with the incident light, contributes to the excitation field for the AuNC. If the phase difference (on account of the small size of the AuNCs) is neglected, the contribution from the AuNCs can be approximated as $Q_{Au} = |E_{inc} + E_{sca}^{NW} + E_{sca}^{Au}|^2 / |E_{inc} + E_{sca}^{NW}|^2$. Assuming the wave scattered from the AuNC interacts with the SiNW in the same way as the incident wave, the SiNW induces a second enhancement for the scattered radiation of the same magnitude as for the incident wave, $Q_{Si}^{NW} \approx |E_{inc} + E_{sca}^{NW} + E_{sca}^{Au} + E_{sca}^{AU-NW}|^2 / |E_{inc} + E_{sca}^{NW} + E_{sca}^{Au}|^2$ and therefore, the total field enhancement in the AuNC-SiNWs system is expressed as $Q_{Si}^{NW} \approx Q_{Au} \cdot (Q_{si}^{NW})^2$.

The description of the response from the AuNC as a metallic sphere on a planar substrate also includes the effect of its image sphere. Similar to the analysis for the AuNC-coated SiNW, the incident wave will strike the spheres either directly or after scattering at the surface, in which case it is an image wave that effectively emanates from the image sphere. Since the phase difference between the incident wave and that reflected from the surface is negligibly small, the total incident radiation E_{inc}^{Pl} striking the spheres on a planar substrate can is written as $E_{inc}^{Pl} = (1 - R)\ E_{inc}$, where E_{inc} is a function of the original incident field, and R is the reflection coefficient, which for normal incidence depends on the refractive index of the sphere, m (the medium here is taken to be vacuum) as $R = (1 - m) / (1 + m)$. Therefore, the existence of the planar silicon substrate enhances the intensity of the incident electromagnetic wave by $Q_{inc}^{Pl} = I_{inc}^{Pl} / I_{inc} = |1 - R|^2$. The Si substrate enhance the light scattered from AuNC as it does for the incident light, and taking into account the enhancement from the AuNCs Q_{Au}' the total estimated field enhancement for the AuNCs-planar substrate model is $Q_{inc}^{Pl} = Q_{Au} \cdot (Q_{inc}^{Pl})^2$.

Since both incident field and Raman scattered signal is enhanced, the Raman enhancement *(RE)* for the probing molecules (PATP) is equal to square of the field enhancement $RE^{Pl} = (Q_{Si}^{Pl})^2 = (Q_{Au})^2 \cdot (Q_{Si}^{Pl})^4$ for the planar substrate, and $RE^{NW} = (Q_{tot}^{NW})^2 = (Q_{Au})^2 \cdot (Q_{Si}^{NW})^4$ for the SiNW. THus, the local field enhancement of the SiNW is estimated by normalizing the two Raman enhancements

$$\frac{RE^{NW}}{RE^{Pl}} = \frac{(Q_{Si})^4}{(Q_{Si}^{Pl})^4} \quad \text{(xii)}$$

The values of RE^{NW} and RE^{Pl} are obtained from experiments by normalizing the Raman scattering intensities of PATP per unit volume from the two substrates to that from bulk PATP power $RE^{NW} = \left[I_{Ram}^{NW} / I_{PATP}^{NW}\right] \cdot \left[I_{Ram}^{Bulk} / N_{PATP}^{NW}\right]^{-1}$ and $RE^{Pl} = \left[I_{Ram}^{Pl} / I_{PATP}^{Pl}\right] \cdot \left[I_{Ram}^{Bulk} / N_{PATP}^{NW}\right]^{-1}$ where I_{Ram} is the integrated intensity of Raman peak of PATP (here, using the 1079 cm^{-1} peak), and N_{PATP} is the number of PATP molecules in the illuminated area.

Because an enhanced field results in an increased polarization for the Raman-shifted frequency, the enhancement for the Raman-scattering intensity of PATP is proportional to the square of the field enhancement, $RE^{NW} = (Q_{tot}^{NW})^2 = (Q_{Au})^2(Q_{Si}^{NW})^4$. Therefore, as the contribution $(Q_{Si}^{NW})^4$ the field enhancement Q_{Si}^{NW} from the SiNW. Thus, the intensities of the measured Raman scattering from PATP chemisorbed onto AuNCs evaporated onto the tapered SiNW are normalized to that from PATP on identically prepared AuNCs evaporated onto a bulk crystal Si wafer as shown in the scatter points in Figure 10.22. The samples used are as-grown tapered SINWs Po which possess an average apex angle of ~6° and length of ~25μm (Figure 10.23(d)) coated by thermally evaporated AuNCs of average size 15.5 ± 5.0 with a nanoscuster number density on the order of ~10^{11}cm^{-2} (Figure 10.21(d)) onto which a monolayer of PATP is deposited, resulting in an estimated molecular number density of ~2.56×10^{18} cm^{-2}.

The Raman scattered intensities for PATP chemisorbed on the AuNC deposited on the

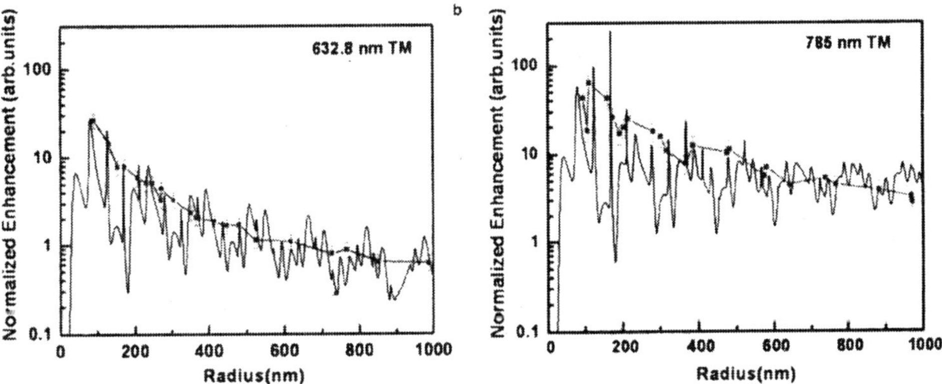

Figure 10.22 *Comparison of calculated and experimentally observed enhancement. Result shown are for TM-polarized excitation of wavelengths (a) 632.8 and (b)785nm.*

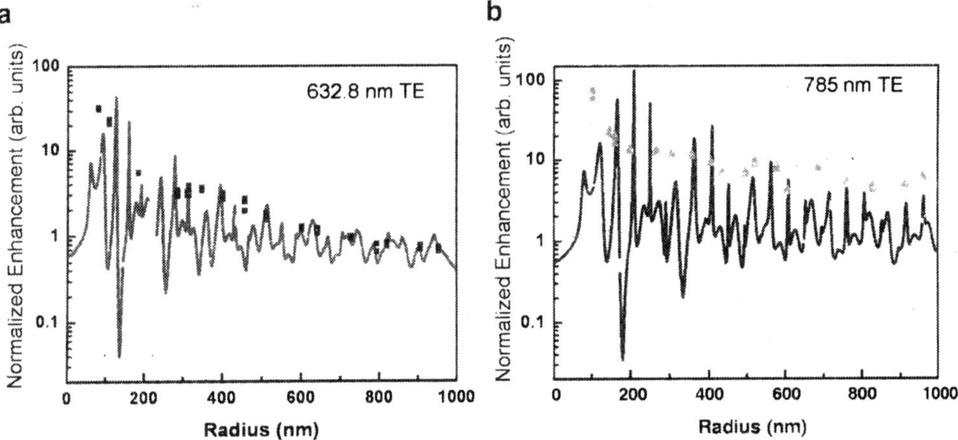

Figure 10.23 *Comparison of calculated and experimentally observed enhancement. Results shown in a and b are for TE-polarized, 632.8-nm and 785-nm excitation, respectively.*

SiNW, are much larger than those for the molecules on the planar substrate (Figure 10.22(d)) The effective values of Raman enhancement for the 1079 cm^{-1} peak are ~1.4 × 10^6, ~5.83.0 × 10^6, ~3.0 × 10^7, and ~1.0 × 10^8 for the planar substrate spectrum and for different locations along the tapered SiNW, respectively, as denoted by arrows in Figure 10.22(d). This diameter dependent variation in the Raman enhancement arises from the SiNW, and it allows to establish the contribution $(Q_{Si}^{NW})^4$ or the local field enhancement Q_{Si}^{NW} from the SiNW for light scattering, a gently tapered SiNW is considered. A chain of uniform diameter SiNWs with continuous change in size, facilitate the size-dependence of the optical near-field. Raman spectra is collected at selected probing locations from the base to near the tip of a singe tapered SiNW and are plotted as a function of R in Figure 10.21(e). The variations for two different excitation wavelengths (λ= 785 and 632.8 nm, Figure 10.21(e) are consistent with the intensity of the Raman scattering by optical phonons (520 cm^{-1}) in the SiNW alone.

The predicted (Figure 10.20) calculated from $\overline{I_{ext}^{NW}}$ $(= Q_{Si}^{NW})$ and the contributions to the calculated RE due to the SiNW is functions of R are plotted along with the corresponding experimental results, for the two excitation wavelengths, 632.8 and 785 nm TM-polarized excitation (Figure 10.22) and TE-polarized excitation (see Figure 10.22) are obtained Because the experimental results involve a normalization of the RE for PATP from Au modified SiNWs for the molecules from Au-modified planar silicon RE^{NW}/RE^{Pl}, the analysis account for the imaging effect of the planar substrate $(Q_{Si}^{NW}/Q_{Si}^{Pl})^4$. Figure 10.22 show that diameter and wavelength-dependences calculations agree with the experimental data.

Large Area, Dense Silicon Nanoware Array Chemical Sensors

Nanowire sensors are fabricated by-a labor intensive sequence of steps that included NW synthesis and dispersions followed by direct-write lithography and metallization. It results in one or a few selected specimens. Top down techniques based on electron beam lithography to define individual SiNWs on silicon-on-insulator (SOI) substrates are demonstrated. Electric field directed deposition or fluidic assembly are promising, but still present challenges. Furthermore, synthesized NWs often exhibit a variation in diameter, which introduce a spread in the electrical response of nanowires to analyte gases.

Below is given a simple, top-down technique based on nanoimprint lithography (NIL) to define SiNWs over a large area and with high density and uniformity. The transport and transistor characteristics of these NWs are measured and their application to chemical sensing using ammonia gas and liquid solutions of nitrobenzene and phenol in cyclohexane are demonstrated. The sensing mechanism is charge transfer between the analytes and the NWs and scales with the value of the Hammett parameter and the concentration of the organic solutions. Furthermore, the sensing of the SiNWs is enhanced as a result of an increased surface to volume ratio and decreased capacitance.

Now take boron doped (10- 20Ωcm) SOL wafers, with a 100 nm thick device layer and a 155 nm buried oxide layer and thermally imprint one wafer using a 200 nm dense pitch (100 nm line/space) Si grating mold. The mold itself is fabricated using laser interference lithography.

Transfer the pattern obtained into the underlying SOI device layer using reactive ion etching with Cl_2/ HBr plasma. Now obtain SEM images. The Scanning electron Microscopy (SEM) images of the individually addressable Al electrodes over the SiNWs are shown in Figs. 10.24 (a) and (b), while cross sections at 60° and 90° with respect to the plane of the figure are shown in Figs. 10.24 (c) and (d). The SEM images demonstrate the uniformity of the SiNWs which have an average diameter of 76±5 nm with a LER of 5 nm.

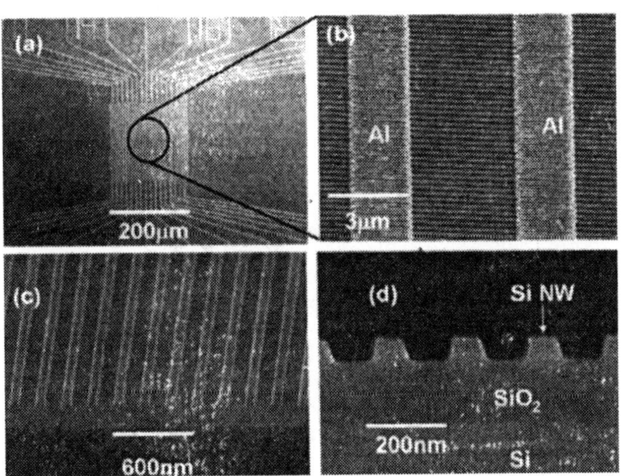

Figure 10.24 *Top-down image of one SiNW device showing. the interdigitated Al source-drain electrode. (b) Close-up image of SiNWs between two Al electrodes. [(c) and (d)] Cross section SEM images at 60° and 90° tilt.*

To form source and drain contacts arrays of interdicted Al electrodes 180 μm long and 4 μm apart, are defined over the SiNWs using optical lithography followed by electron beam evaporation and lift-off. Each Al electrode array consists of 32 individually addressable parallel contacts, with 900 SiNWs spanning the gap between any pair of electrodes. Anneal the metallized wafer in vacuum at 720K for 20 min to render the Al contacts Ohmic and expose to O_2 plasma to remove residual organics which also ensures that the nanowire surface is clean an oxidized. Also pattern and anneal Al contacts, of an additional SOI wafer from the same batch, but which is not patterned using NIL. For both the NW and thin film devices, the p-doped Si substrate of the SOI wafer is used as the gate, with an In Ohmic bottom contact.

A family of gate voltage sweeps, I_{SD}-V_g collected at V_{SD} =1.0 V for different Al electrode pairs for a selected interdigitated array is shown in Figure 10.25. The I_{SD}-$V_g$curves correspond to three regimes: hole accumulation (I), intrinsic or "off" (II), and electron accumulation (III), as indicated by the three band diagrams in the figure. The contacts are heavily *p* doped as a result of Al diffusion during the contact formation process. The I_{SD}-V_{SD} behavior of the device in the hole accumulation regime shows behavior similar to p-channel metal-oxide-semiconductors field-effect transistors, as indicated in Fig. 10.25.(b)

Figure 10.25(c) plots the resistance of the device of Fig. 10.25(a) as a function of nanowire length (i.e.. for different pairs of electrodes) in the hole accumulation regime, indicating that the resistance scales linearly with nanowire length, and thus that the transport is diffusive. Analysis

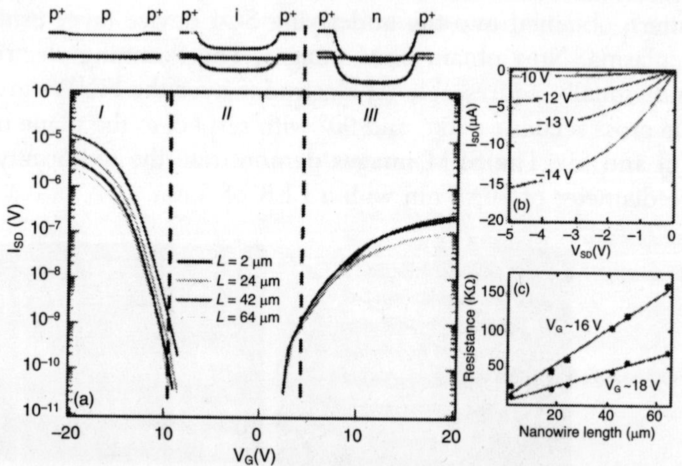

Figure 10.25 *(a) I_{SD} vs V_g. for one SiNW device collected for different AI electrode pairs with V_{SD} set at 0.1 V. (b) I_{SD} vs or V_{SD} for the same SiNW device collected at different gate voltages. (c) Resistance of the nanowire array of (a) as a function of nanowire length.*

of the I_{SD}-V_g characteristics using a standard transistor diffusion model gives a mobility or 81 $cm^2\ V^{-1}s^{-1}$ where an expression for the capacitance which accounts for the NW array geometry is used. Similar analysis for the thin film device yields a mobility of 78 cm $V^{-1}s^{-1}$. While these mobilities are lower than those typically reported for p-channel SOl transistors there is no significant degradation for the nanowire cross sections used.

The chemical sensing characteristics of the NIL-based SiNWs the device sensitivity to ammonia vapor is tested. Ammonia is a strong reducing agent and its effect on carbon nanotube transistors is to make the threshold voltage more negative. Figure 10.26 indicates that similar behavior is observed for SiNW sensors, where the I_{SD}-V_g curves are shifted to the left by 5.4 V. Also shown in Fig. 10.26 is the response for the thin film device, which shows a much smaller shift.

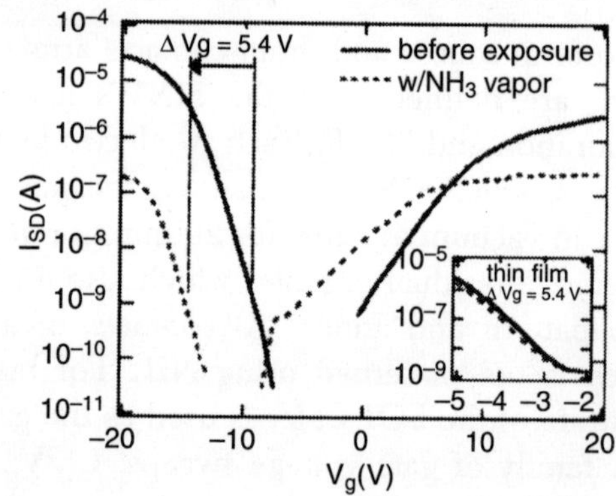

Figure 10.26 *Variation in I_{SD} vs V_G for a SiNW array exposed to ammonia vapor. Inset shows response of SOl thin film device to similar exposure.*

The enhanced response of the SiNWs is found by considering the charge in the channel with and without analytes, env=AC(V_t-Vg) and env+$e\alpha a^{-1}$ θ_s = AC(V'_t-Vg).where n is the number of carriers per unit volume, a is the area that a molecule occupied on the surface, $e\alpha$ is the charge transfer per molecule, θ is the surface coverage, C is the cpacitance per unit area, V_t is the threshold voltage, A is the device footprint area, and s is the area exposed to analytes [equal to (2t+ω)ML for M nanowires of length L and equal to A for the thin film].

Subtracting these two equations an expression for the shift in threshold voltage is

$$\Delta V_t = e\alpha a^{-1}\theta C^{-1}(s/A) \quad (1)$$

So the capacitance is proportional to $(1+l)/(1-l)$, where l is the separation between nanowires and K is the gate oxidise relative dielectric constant. The shift in threshold voltage become

$$\Delta V_t \propto 2t + \omega)/(l + \kappa\omega), \quad (2)$$

where the thin film limit is given by $\omega \rightarrow \infty$. Equation (2) shows that larger sensitivity is reached with small l or ω, highlighting the need for dense arrays of small diameter nanowires. The benefit to using small diameter nanowires is realized when $2t/l > 1/\kappa$. For SiNW device the s/A ratio is unchanged from the thin film device. This shows that it is lower capacitance (× 0.5) for SiNW array which is responsible for the increased sensitivity by a factor of ~2. Another factor that contributes to further increase in the sensitivity in the sensitivity of the NW device, is the LER, which gives a large s/A ratio and additional binding sites for the analytes created by the reactive ion etch.

For the sensing characteristics of the SiNWs the effect of different concentrations of nitrobenzene and phenol is measured in cyclohex ane solutions on the device electrical characteristics. Nitrobenzene and phenol are electron withdrawing and electron donating molecules, respectively, and their solution in cyclohexane is used to change the I_{SD}--V_g characteristics of carbon nanotube transistors. Solutions of nitrobenzene and phenol in cyclohexane with concentrations ranging from 0.001M to 0.25M are applied using a micropipette directly over one of the SiNW devices, while I_{SD}-V_g curves are measured just before and just after the solution application. The resulting curves for one of the concentrations of each chemical are shown in Fig. 10.27 where it is seen that in the increase

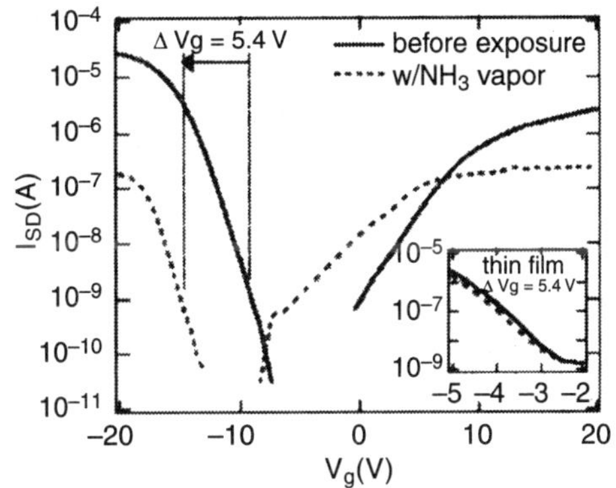

Figure 10.26 *Variation in I_{SD} vs V_G for a SiNW array exposed to ammonia vapor. Inset shows response of SOl thin film device to similar exposure.*

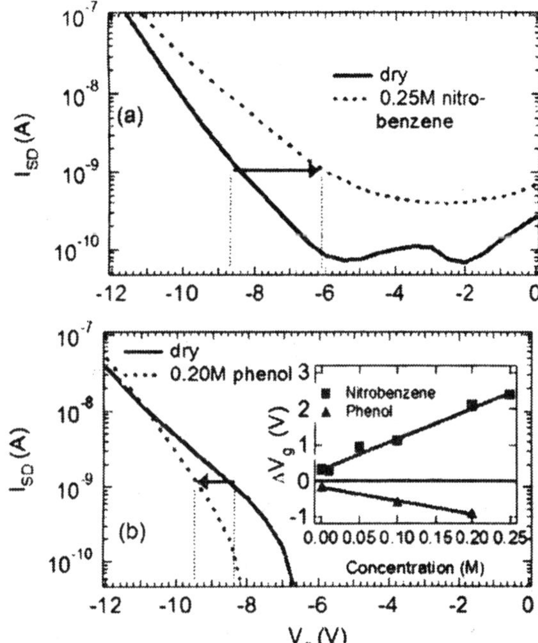

Figure 10.27 *Variation in I_{SD} vs V_G for (a) 0.25M of nitrobenzene and (b) 0.2M of phenol. dissolved in cyclohexane and applied directly over the SiNW array sensor. Inset shows shift in V G necessary to maintain I_{SD} =1 nA in accumulation mode as a function of the solute concentration.*

(decrease) in I_{SD} is consistent with the electron withdrawing (donating) character of nitrobenzene (phenol), as it effectively dopes the SiNWs p type (n type) Fig. 10.27 (b) (inset) shows the change in gate bias corresponding to I_{SD} of 1 nA. It is plotted as a function of the concentration, indicating a linear behavior. No measurable shift for the thin film device is observed when exposed to similar concentrations of nitrobenzene.

Now consider the adsorption of the molecules on the nanowire surface with the goal of extracting the binding energy. To relate θ to the analyte concentration in the liquid, also consider equilibrium between the liquid and the surface, with the partition function

$$Z = 1 + z_{vib} e^{(\mu - E_b)/kT}, \tag{3}$$

where μ is the chemical potential of the, analyte in the liquid, E_b is the binding energy of the molecule to the surface, and Z_{vib} is the contribution to the partition function due to vibrations. The chemical potential is $\mu = \mu_0 + KTInx$ where μ_0 is the chemical potential of the pure substance and x is the mole fraction. For a pure substance $\mu_0 = \Delta H_f - TS$, where ΔH_f is the enthalpy of formation and S is the entropy.

Assuming that the mole fraction of the solute is much less than the solvent, the equilibrium surface coverage as a function of the solute concentration is obtained as:

$$\theta = c/(c + c_0), \tag{4}$$

where

$$c_0 = \bar{c} z_{vib}^{-1} e^{-\mu_0/kT} e^{E_b/kT} \tag{5}$$

with $\bar{c}$ the solvent concentration. For $c << c_0$, equations. (1) and (4) give

$$\Delta V_g = (2t + w) Le\alpha a^{-1} C^{-1} (c/c_0) \tag{6}$$

A least-squares fit to the data of Fig. 4 (inset) for nitrobenzene –gives a value for the prefactor of 8.47 ± 53 V/M Using equations (5) and (6), a value of μ of –0.56 eV, and assuming $\alpha = 1$ and $z_{vib} = 1$, a value for the binding energy of 440 meV is obtained This value indicates strong physisorption of the nitrobenzene on the NW surface. As, $E_b >> kT$ it shows why the devices are very stable once exposed to the analytes. The binding energy is much less than chemisorption binding energies, and so it and refreshes the device, as does pure cyclohexane rinse. From these calculations, the density of molecules on the NW surface at 0.25M is found to be 0.3 molecules/nm/NW. This small density is consistent with linear behavior of Fig. 10.27(d).

As in the sensing properties of carbon nanotubes it is found at the voltage shift is proportional to the Hammett parameter σ_p. For nitrobenzene, $\sigma = 0.78$ while for phenol $\sigma_p = -0.4$; the ratio ~–2 is in reasonable agreement with the ratio of the slopes from Fig. 10.27 equal to –2.4.

HIGH-PERFORMANCE LITHIUM BATTERY ANODES USING SILICON NANOWIRES

There is great interest in developing rechargeable lithium batteries with higher energy capacity and longer cycle life for applications in portable electronic devices, electric vehicles and implantable medical devices. Silicon is an attractive anode material for lithium batteries because it has a low discharge potential and the highest known theoretical charge capacity (4,200 mAh g^{-1}). Although this is more than ten times higher than existing graphite anodes and much larger than various nitride and oxide materials, silicon anodes have limited applications because silicon's volume changes by 400% upon insertion and extraction of lithium which results in pulverization and capacity fading. Below is shown that silicon nanowire battery electrodes circumvent these issues as they can accommodate large strain without pulverization, provide good electronic contact and conduction, and display short lithium insertion distances.

Si NWs are synthesized using the VLS process on stainless steel substrates using Au catalyst. The electrochemical properties are evaluated under an argon atmosphere by both cyclic voltammetry and galvanostatic cycling in a three electrode configuration with the Si NWs on the stainless steel substrate as the working electrode and Li foil as both reference and counter-electrodes. No binders or conducting carbon is used. The charge capacity is the total charge inserted into the Si NW per mass unit during Li insertion whereas the discharge capacity is the total charge removed during Li extraction. For electrical characterization, single Si NW devices are contacted with metal electrodes by electron-beam lithography or focused-ion beam deposition.

Si bulk films and micrometre-sized particles are used as electrodes in lithium batteries Now it has shown capacity fading and short battery lifetime due to pulverization and loss of electrical contact between the active material and the current collector (Fig. 10.28(a)). The use of sub-

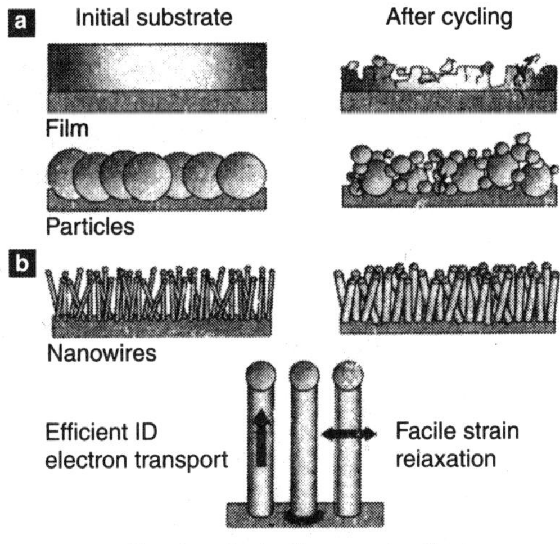

Figure 10.28 *Morphological changes that occur in Si during electrochemical cycling. (a) The volume of silicon anodes changes by about 400% during cycling. As a result, Si films and particles tend to pulverize during cycling. Much of the material loses contact with the current collector, resulting in poor transport of electrons, as indicated by the arrow. (b) NWs grown directly on the current collector do not pulverize or break into smaller particles after cycling. Rather. facile strain relaxation in the NWs allows them to increase in diameter and length without breaking. This NW anode design has each NW connecting with the current collector, allowing for efficient 1 D electron transport down the length of every NW.*

micrometre pillars and micro and nanocomposite anodes lead to improvement. Electrodes made of amorphous Si thin films have a stable capacity over many cycles but have insufficient material for a viable battery. The concept of using one-dimensional, (lD) nanomaterials is achieved with carbon, Co_3O_4, SnO_2 and TiO_2 anodes, and has shown improvements compared to bulk materials. Si nanowire (NW) anode configuration is shown in Fig. 10.28(b). Nanowires are grown directly on the metallic current collector substrate. This has several advantages and has lead to improvements in rate capabilities in metal oxide cathode materials. Firstly the small NW diameter allows for better accommodation of the large volume changes without the initiation of fracture that can occur in bulk or micron-sized materials (Fig. 10.28(a)). Secondly, each Si NWs is electrically connected to the metallic current collector so that all the nanowires contribute to the capacity. Thirdly, the Si NWs have direct 1D electronic pathways allowing for efficient charge transport. In electrode microstructures based on particles, electronic charge carriers must move through small interparticle contact areas. In addition, as every NW is connected to the current-carrying electrode, the need for binders or conducting additives, which add extra weight, is eliminated. Furthermore, the Si NW battery electrode are realized using the vapour-liquid-solid (VLS) or vapour-solid (VS) template-free growth methods to produce NWs directly onto stainless steel current collectors.

A cyclic voltammogram of the Si NW electrode is shown in Figure 10.29(a). The charge current associated with the formation of the Li-Si alloy began at a potential of ~330 mV and became quite large below 100 mV. Upon discharge, current peaks appeared at about 370 and 620 mV. The current-potential characteristics are consistent with experiments on microstructured Si anodes. The magnitude of the current peaks increases with cycling due to activation of more material to react with Li in each scan. The small peak at 150-180 mV is due to reaction of the Li with the gold catalyst, which makes a negligible contribution to the charge capacity.

Si NWs are found to exhibit a higher capacity than other forms of Si. Figure 10.29(b) shows the first and second cycles at the C/20 rate (20 h per half cycle). The voltage profile observed is consistent with a long flat plateau during the first charge, during which crystalline Si reacts with Li to from amorphous Li_xSi. Subsequent discharge and charge cycles have different voltage profiles characteristic of amorphous Si. Significantly, the observed capacity during this first charging operation is 4,277 mAh g^{-1}. The first discharge capacity is 3,124 mAh g^{-1}, indicating a coulombic efficiency of 73%. The second charge capacity decreases by 17% to 3,541 mAh g^{-1}, although the second discharge capacity increases slightly to 3,193 mAh g^{-1}, giving a coulombic efficiency of 90%. Both charge and discharge capacities remain constant for subsequent cycles, with little fading up to 10 cycles (Fig. 10.29(d)). As the charge-discharge, data are shown along with the theoretical capacity (372 mAh g^{-1}) for the graphite, currently used in lithium batter anodes; the charge data is reported for thin films containing 12-nm Si nanocrystals (NCs) in Fig. 10.29(d). This improves capacity and cycle life in the Si NWs.

The Si NWs also display high capacities at higher currents. Figure 10.29(c) shows the charge and discharge curves observed at the C/20, C/10, C/5, C/2 and 1C rates. Even at the 1C rate, the capacities remain >2,100 mAh g^{-1}, which is five times larger than that of graphite. The cyclability of the Si NWs at the faster rates is excellent. Using the C/5 rate, the capacity is stable at ~ 3,500 mAh g^{-1} for 20 cycles in another device Despite the improved performance, the Si NW anode shows an irreversible capacity loss in the first cycle. Although solid electrolyte

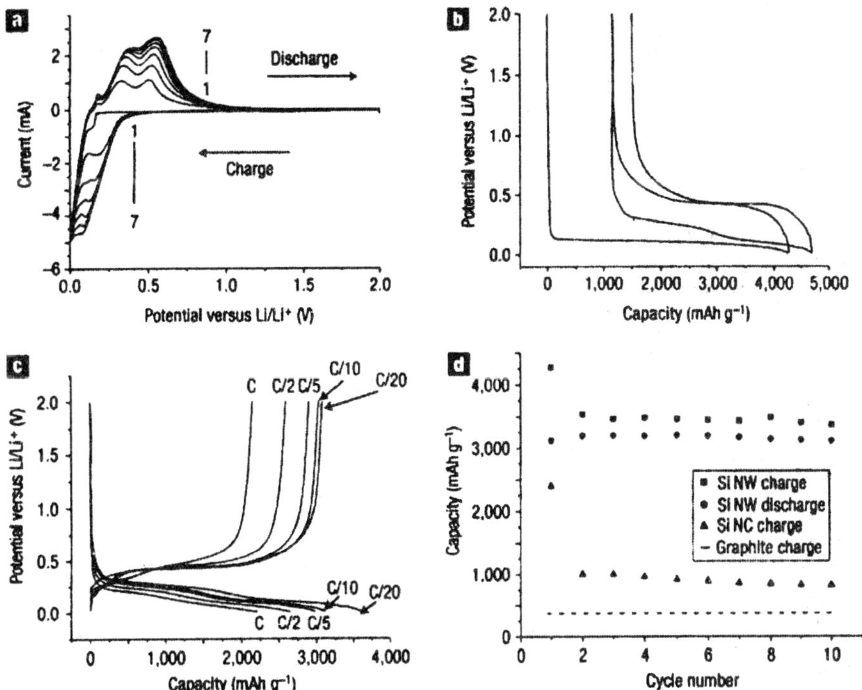

Figure 10.29 *(a) Cyclic voltammogram for Si NWs from 2.0 V to 0.01 V versus Li/Li$^+$ at 1 mV s^{-1} scan rate. The first seven cycles are shown. (b) Voltage profiles for the first and second galvanostatic cycles of the Si NWs at the C/20 rate. The first charge achieved the theoretical capacity of 4,200 mAh g^{-1} for $Li_{4.4}Si$. (c) The voltage profiles for the Si NWs cycled at different power rates. The C/20 profile is from the second cycle. (d) Capacity versus cycle number for the Si NWs at the C/20 rate showing the charge (squares) and discharge capacity (circles). The charge data for Si nanocrystals (triangles) and the theoretical capacity for lithiated graphite (dashed line) are shown as a comparison to show the improvement when using Si NWs.*

interphase (SEI) formation is observed in Si, this is not the cause for initial irreversible capacity loss, because there is no appreciable capacity in the voltage range of the SEI layer formation (0.5 – 0.7 V) during the first charge (Fig. 10.29(b)) Although SEI formation occurs the capacity involved in SEI formation is very small compared to the high observed charge capacity. The structural morphology changes during Li insertion are needed to understand the high capacity and good cyclability of Si NW electrodes. Pristine, unreacted Si NWs is crystalline with smooth sidewalls (Fig 10.30(a)) and has an average diameter of ~89 nm (s.d., 45 nm) (Fig. 10.30(e)). Cross-sectional scanning electron microscopy (SEM) shows that the Si NWs grow off the substrate and have good contact with the stainless steel current collector (Fig. 30(a) inset). After charging with Li, the Si NWs have textured sidewall (Fig. 10.30(b)), and the average diameter increases to ~141 nm (s.d., 64 nm). Despite the large volume change, the Si NWs remain intact and do not break into smaller particles. They also remain in contact with the current collector, suggesting minimal capacity fade due to electrically disconnected material during cycling.

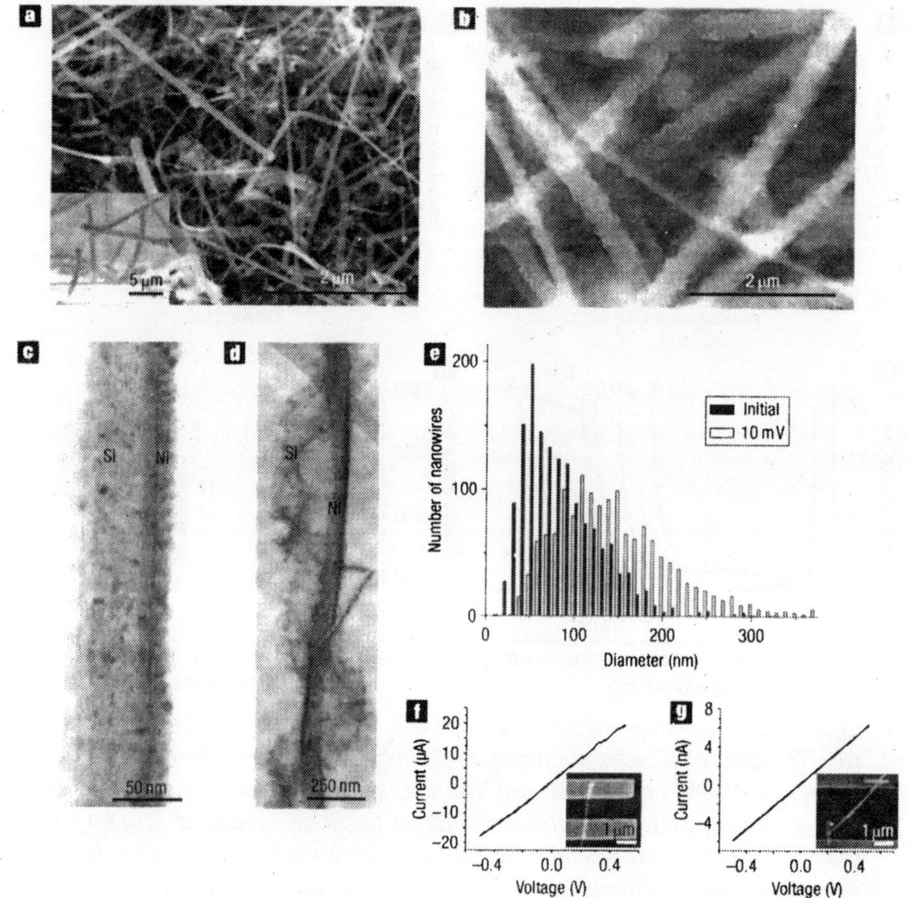

Figure 10.30 *(a-b) SEM image of pristine Si NWs before (a) and after (b) electrochemical cycling (same magnification). The inset in a is a cross-sectional image showing that the NWs are directly contacting the stainless steel current collector. (c,d) TEM image of a pristine Si NW with a partial Ni coating before (c) and after (d) Li cycling. (e) Size distribution of NWs before and after charging to 10 mV (bin width 10 nm). The average diameter of the NWs increased from 89 to 141 nm after lithiation. (f) I-V curve for a single NW device (SEM image, inset) constructed from a pristine Si NW. (g) I-V curve for a single NW device (SEM image, inset), constructed from a NW that had been charged and discharged once at the C/20 rate.*

The Si NWs also change their length during the change in volume. To evaluate this, 25-nm Ni, is evaporated onto as-grown Si NWs using electron beam evaporation. Because of the shadow effect of the Si NWs, the Ni only covers part of the NW surface (Fig. 10.30c), as confirmed by energy dispersed X-ray spectroscopy (EDS) mapping. The Ni is inert to Li and acts as a rigid backbone on the Si NWs. After lithation (Fig. 10.30(d), the Si NWs change shape and wrap around the Ni backbone in a three-dimensionally helical manner. This is due to an expansion in the length of the NW, which causes strain because the NW is attached to the Ni and can not freely expand but rather is buckled into a helical shape. Although the NW length increases after lithiation, the NWs remain continuous and without fractures, maintaining a pathway for electrons all the way from the collector to the NW tips. With both a diameter and length increase, the Si NW volume change after Li insertion appears to be about 400%.

Efficient electron transport from the current collector to the Si NWs is necessary for good battery cycling. For this electron transport measurements are conducted on single Si NWs before and after lithiation. The current versus voltage curve on a pristine Si NW is linear with a 25 kΩ resistance (resistivity of 0.02 Ω-cm) (Fig. 10.30(f)) After one cycle, the NWs become amorphous, but still exhibit a current that is linear with voltage with an 8 MΩ resistance (resistivity of 3 Ω-cm) (Fig. 10.30(g)). The good conductivity of pristine and cycled NWs ensure a efficient electron transport for charge and discharge.

The large volume increase in the Si NWs is driven by the dramatic atomic structure change during lithiation. To understand the structural evolution of NWs, the NW electrodes are characterized at different charging potentials. The X-ray diffraction (XRD) patterns are taken for initial pristine Si NWs, Si NWs charged to 150 mV, 100 mV; 50 mV and 10 mV, as well as after 5 cycles (Fig. 10.31(a)). XRD patterns of the as-grown Si NWs show diffraction peaks associated with Si, α-$FeSi_2$, Au (the Si NW catalyst) and stainless steel (SS). The α-$FeSi_2$ forms at the interface between the SS and the Si wires during the high temperature (800K), NW growth process. The α-$FeSi_2$ is not found to appreciably react with Li during electrochemical cycling, although a small amount of reaction takes place. After Si NWs are charged to 150 mV, the higher angle Si peaks disappear. Only the Si(111) peak is visible, but its intensity is greatly decreased. This is consistent with the disappearance of the initial crystalline Si and the start of

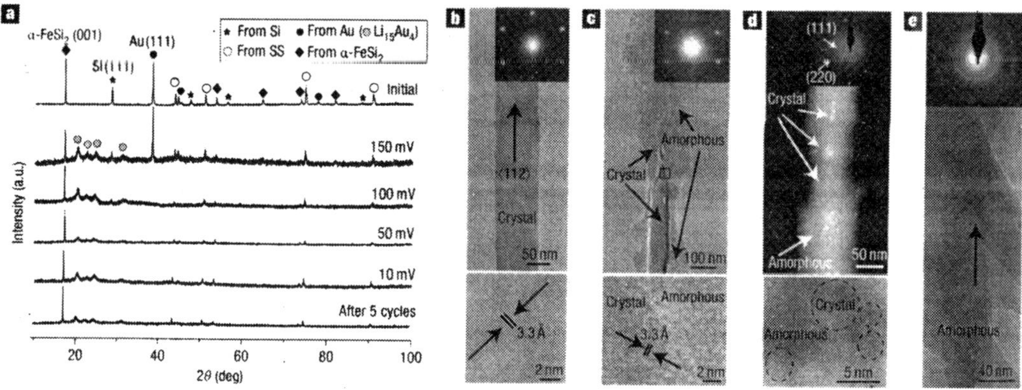

Figure 10.31 *(a) XRD patterns of Si NWs before electrochemical cycling (initial), at different potentials during the first charge, and after five cycles. (b-e) TEM data for Si NWs at different stages of the first charge. (b) A single-crystalline, pristine Si NW before electrochemical cycling. The SAED spots (inset) and HRTEM lattice fringes (bottom) are from the Si 1/3(224) planes. (c) NW charged to 100 mV showing a Si crystalline core and the beginning of the formation of a Li_xsi amorphous shell. The HREM (bottom) shows an enlarged view of the region inside the box. (d) Dark-field image of a NW charged to 50 mV showing an amorphous Li_xsi wire with crystalline Si grains (bright regions) in the core. The spotty rings in the SAED (inset) are from crystalline Si. The HRTEM (bottom) shows some Si crystal grains embedded in the amorphous wire. e, A NW charged to 10 mV is completely amorphous $Li_{4.4}$Si. The SAED (top) shows diffuse rings characteristic of an amorphous material.*

the formation of amorphous Li_xSi. The four broad peaks that appear in the lower angles are due to the formation of $Li_{15}Au_4$. At 100 mV, the pure Au peaks disappear, indicating that the Au has completely reacted with Li. The Si(111) peak is very weak at 100 mV, and disappears completely at 50 mV. It appears that Si NWs remain amorphous after the first charge, consistent with the non-flat voltage charging /discharging curve in Fig. 10.29(b). This contrasts, however, with other Si electrodes as, $Li_{3\cdot75}Si$ at potentials less than 30-60 mV. XRD show that this crystalline phase only forms at <50 mV for films thicker than ~2 μ m. This is not observed to be in Si NWs, because of their shape and small dimensions.

The local structural features of Si NWs during the first Li insertion with transmission electron microscopy (TEM) shows the selected area electron diffraction (SAED). The as grown Si NWs are single-crystalline. Figure 10.31(b) shows an example of a typical Si NW with a <112> growth direction. Figure 10.31(c) shows a Si NW with a <112> growth direction that is charged to 100 mV. There are two phases present, as expected from the voltage profile i.e. crystalline and amorphous phases The distribution of the two phases is observed both across the diameter (a crystalline core and an amorphous shell) and along the length. The SAED show the spot pattern for the crystalline phase (Si), but weak diffuse rings from the amorphous phase (Li_xSi alloy) are also observed. Li ions diffuse radially into the NW from the electrolyte, resulting in the core-shell phase distribution. The reason for phase distribution along the length is unknown At 50 mV, the Si NW become mostly amorphous with some crystalline Si regions embedded inside the core, as seen from the dark-field image and HRTEM (Fig. 10.31(d)) The SAED shows spotty rings representative of a polycrystalline sample and diffuse rings for the amorphous phase. At 10 mV (Fig. 10.31(e)), all of the Si have changed to amorphous $Li_{4.4}Si$, as indicated by the amorphous rings in the SAED. These TEM observations are in consistent with the XRD results (Fig. 10.31a) and voltage charging curves (Fig. 10.29(b)).

SiO_2 and SiC Nanowires

SiO_2 NANOWIRES GROWN FROM LARGE GALLIUM SURFACE

Generally, the nanoparticle growth is through the VLS process, in which one particle nucleates one nanowire and the size of the particle determines the diameter of the nanowire. But it is shown that a large size (5 to 50 μm in diameter), low melting point gallium droplet can be used as an effective catalyst for the large-scale growth of highly aligned, closely packed silica nanowire bunches (Figs. 11.1(a) and (b) because Ga has the longest temperature range in the liquid range in the liquid phase (from 300-4675K) In addition, small molten Ga droplets tend to agglomerate due to its high surface tension.

The growth of nanowires is carried out using following experiments:

Materials and Substrates. GaN powders (99,99%); Ultrahigh-purity argon (99.999% of purity with O_2 and H_2O contents less than 4 molar ppm and 3.5 molar ppm respectively) is used as the carrier gas; substrates of silicon (100) wafers and alumina plates are ultrasonically cleaned in acetone for 30 min before use.

Experiment

(1) Growth of SiO_2 Nanowire

The experimental setup for SiO_2 nanowires is shown in Figure 11.2. 1-2g of GaN powder is placed at the center of an alumina tube that is inserted in a horizontal tube furnace, where the temperature, pressure and reaction time are controlled. A long silicon wafer stripe (10×1 cm^2) is placed on the middle part of a wide alumina plate (10×3 cm^2), which is located 10 cm away from the GaN powders at the end of the alumina tube. After evacuation of the alumina tube to ~2.66×10^{-5} Pa the reaction is conducted at 1420 K for 5 h under a pressure of 5.3×10^4 Pa and argon gas flow rate of 50 sccm (standard cubic centimeters /minute). At the reaction temperature of 1420 K the GaN powder is decomposed into a dense, hot vapor of Ga and N_2 The hot Ga vapor rapidly condenses into small Ga clusters as the Ga species cool through

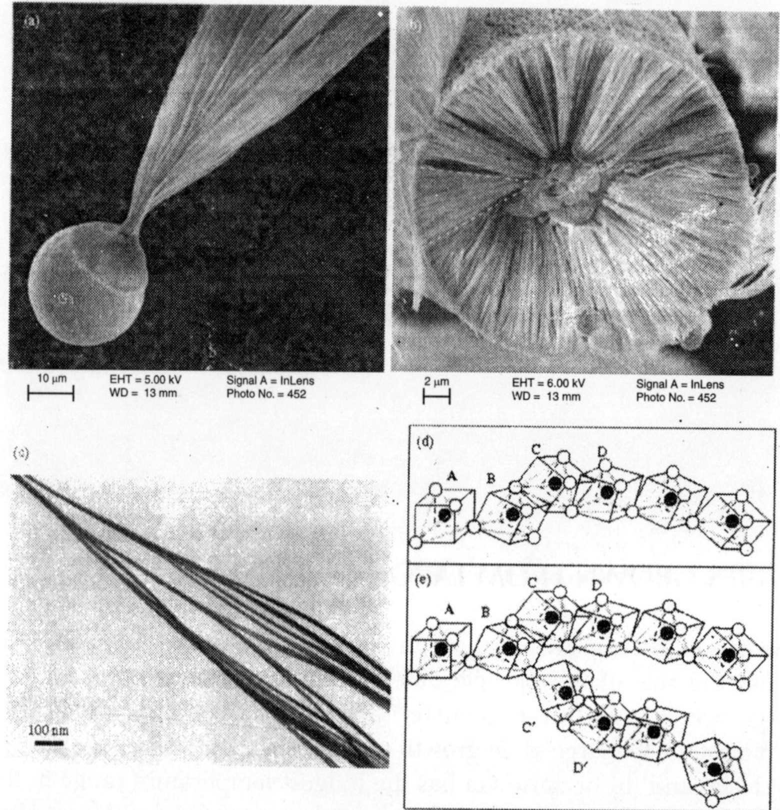

Figure 11.1 *(a–b) Aligned silica nanowires synthesized by evaporating GaN with the presence of a silicon substrate. The Ga droplet serves as the catalyst that leads to the nucleation and growth of uniform silica nanowires of diameters ~ 100 nm (c) The silica nanowiores have the characteristic of self-splitting, resulting in thicker boundles as the frowth proceeds (d) Interpretation about the self-splitting process of the silica glass structure.*

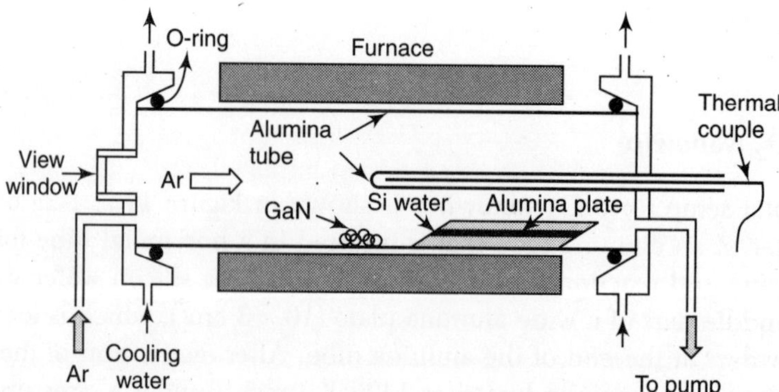

Figure 11.2 *The experimental setup for the growth of SiO_2 nanowires.*

collision with the buffer gas (with the presence of oxygen in the reaction chamber; there may also exist gallium oxide clusters, such as Ga_2O, in the vapor). The formed Ga clusters are transferred to the downstream end of the alumina tube by the carrier gas and then deposited on the surface of the silicon wafer and the naked part of the alumina plate in the regions with temperatures less than 1340K. Ga droplets with diameters < 1 μm are evenly distributed on the surfaces of the silicon wafer and alumina plate at the early reaction stage; however the diameters of the Ga droplets increase with the reaction time through continuously accepting the upcoming Ga clusters and are in the range of 5-50 μm after 5 h of reaction. The Ga droplets deposit on the silicon wafer. They etch silicon to form Ga-Si alloy and thus create a dense vapor of Si species around the silicon wafer and alumina plate which acts as Si source for the growth of SiO_2 nanowires.

During evaporation the temperature at any point between the tube center and the tube's downstream end is measured by a thermocouple, and the temperature range of the SiO_2 nanowire in their formation region is also measured.

(2) Morphology, Structure. and Composition Characterization:

The as synthesized products fabricated by the above experiment –(1) are characterized and analyzed by a scanning electron microscope (SEM), transmission electron microscope (TEM) 100 kV at 200 KV and energy-dispersive X-ray spectroscope (EDS) attached to the SEM and TEM. For SEM the products together with the growth substrates are directly transferred into the SEM chamber without destroying the location and orientation of the products on the substrates. For TEM some samples are taken from the growth substrates and are mounted on Cu folding TEM grids.

After the reaction, colorless and transparent products are formed on both the silicon wafer and alumina plate, covering approximately a 4-cm region and temperature range of 1120-1270. SEM observations show that the products formed on the silicon wafer and alumina plate have different morphology, size, size distribution and orientation. On the silicon wafer, carrot-shaped rods (CSRs) with diameters 10-50 μm and lengths of up to ~1 mm grow in groups in a high yield (Figure 11.3(a). For each group, several tens of CSRs radially grow upward and form a sisallike structure (Figure 11.3(b)). Each CSR terminates at its top end in a large spherical ball

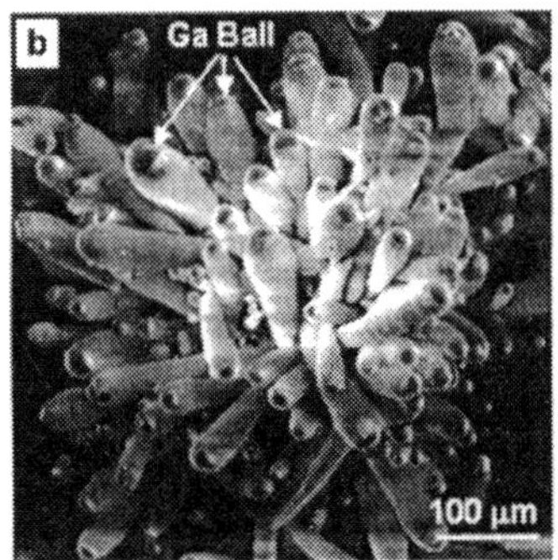

Figure 11.3 *SEM images of the SiO_2 nanowires grown on a silicon wafer: (a) low-magnification SEM image showing carrot shaped rods (CSRs) growing in-groups on the silicon wafer; (b) high magnification SEM image of the boxed area in (a), showing several tens of CSRs forming a sisallike structure. Note that each CSR has a liquid Ga ball at its tip.*

with diameter comparable to that of the connected rod. This corresponds to the morphology of VLS-grown nanowires and suggests that the growth of the CSRs is governed by the VLS mechanism. However, the large terminating balls are in the liquid state at room temperature The melting point of metal Ga is 300K, ~4.78K higher than room temperature. The Ga-Si phase diagram show a eutectic temperature of 300K at 0.006 mol % Si. The liquid state of the Ga balls at room temperature is due to the impurities in the volume. EDS analyses reveal that the balls are liquid Ga covered by a thin oxide layer composed of Ga, O, and a small amount of Si.

For the internal structures of the CSRs, several CSRs are dissected either perpendicular or parallel to their long axes as shown Figure 11.3(a). and several growth phenomena are observed.

First, the CSRs are not solid rods; instead each CSR is composed of numerous, highly aligned, and closely packed nanowires (Figure 11.4(b) to (g). These nanowires are of uniform diameters (10 – 30 nm) and lengths (10 – 40 μm). Quantitative EDS analyses of the nanowires confirm that the nanowires have a composition close to SiO_2.

Second, each Ga ball attaches to hundreds of thousands of SiO_2 nanowires that grow out perpendicularly from the surface of the ball's lower hemisphere (Figure 11.4(b)); that means, each Ga ball can catalyze growth of many SiO_2 nanowires, which is different from the VLS processes, in which one catalyst particle usually catalyzes growth of just one nanowire. Both high-magnification SEM image (Figure 11.4(c)) and EDS analyses show that the SiO_2 nanowires connect with the molten Ga ball through a thin oxide layer composed of Ga, Si, and O. The function of this oxide layer is to provide Si and O species for the growing SiO_2 nanowires.

Third, the CSRs have a tubular structure; that is, there is a central hole along the long axes of the CSRs (Figure 11.4(d) to (g). It is very interesting that the walls of the tubular structures are composed of a large quantity of highly aligned SiO_2 nanowires. The nanowires within the wall arrange at an angle of 30-60° to the axis of the tube, and the two ends of the wires respectively construct the tube's inner and outer walls. The outer wall of the tubular structure also contains many other wire's tips. These nanowires grow out from the existing nanowire forming a branch structure. Such growth manner is called "split growth". The outer wall of the tube is relative smooth, while the inner wall is rough, and it exhibits a stairlike structure with almost the same step height of ~ 10 μm (Figure 11.4). Most tubes have a continuous hole running through out their entire lengths (Figure 11.4(f)), but in some case (see Figure 11.4(g)) the hole is discontinuous and consists of a series of upside down bell-like cavities with almost the same interval between bells. The outer and inner diameters of the tube gradually increase at approximately the same rate from the tube's bottom end to the top end; for an individual tube, the nanowires in it have similar lengths and growth directions, resulting in the thickness of the wall being relatively constant. The regularity of the internal structures of the tubes suggests that the nanowires within a tube grow batch by batch. For each batch, the nanowires nucleate, grow at nearly the same rate and direction, and simultaneously stop growing. The hole or cavities inside the tube comprise the moving track of the Ga ball, which is periodically lift upward by the nanowires grown batch by batch.

The products formed on the alumina plate and shown in Figure 11.5(a) is a typical SEM image. Numerous cometlike structures grow out perpendicularly and separately from the surface of the alumina plate to form a comet array. Both high-magnification SEM image (Figure 11.5(b)) and EDS analyses reveal that the tail part of the cometlike structure is composed of a

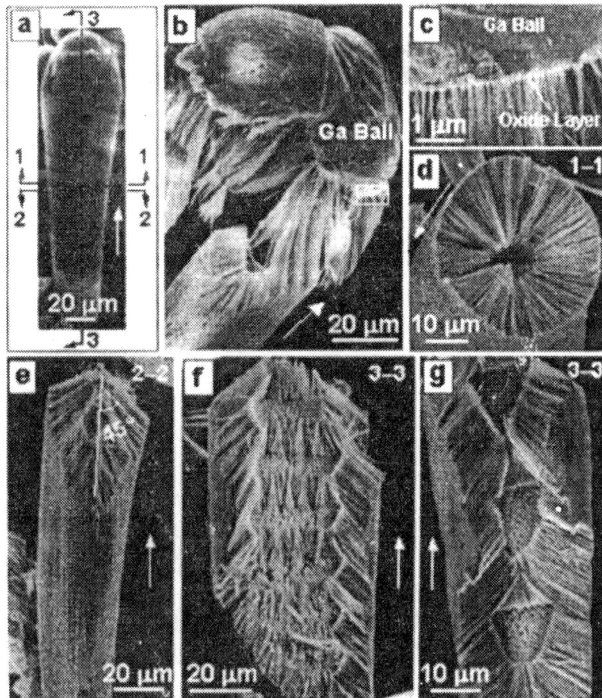

Figure 11.4 *SEM images of the inner structure of the CSRs: (a) an individual CSR shown for the dissection along direction either perpendicular (1-1, 2-2) or parallel (3-3) to its long axis; (b) SEM image of a dissected CSR at its tip region, showing a large quantity of nanowires growing out from the lower hemisphere surface. of a Ga ball; (c) high-magnification SEM image from boxed area in (b) with the oxide layer (composed of Ga, Si, and O); (d) cross section of a CSR viewed along the 1-1 direction, showing a tubular structure whose wall is composed of closely packed and highly aligned nanowires where the two ends of the nanowires respectively construct the tube's inner and outer walls; (e) cross section of a CSR viewed along the 2-2 direction, displaying an angle of ~ 45° between the growth direction of the nanowires and the axis of the tube; (f–g) two cross sections viewed along the 3-3 direction, displaying two kinds of inner structures of the CSRs. The image in (f) shows a continuous central hole with stairlike structure; the image in (g) shows discontinuous upside down bell-like cavities. The white arrows in (a), (b), and (d)-(g) show the growth direction of the CSRs.*

large quantity of oriented SiO_2 nanowires with diameter of 50-100 nm and length of 10- 50 μm while the tip is molten Ga covered with a thin oxide layer. Thus, the products formed on the alumina plate have the same composition as those formed on the silicon wafer. However some differences are found in morphology and inner structure between these two kinds of products from SEM. First, the cometlike structures are much thinner than the CSRs, with typical outer diameters in the range of 5-30 μm and lengths <500 μm; the diameters of the molten Ga balls are in the range of 5-20 μm. Second, the SiO_2 nanowires only grow from a small area of the ball's lower hemisphere, resulting in a low wire density in the cometlike structures. In some cases, only hundreds of nanowires connect to the ball, forming a cherrylike morphology, as that shown in Figure 11.5(c). In addition, the Ga ball in the cometlike structure can be easily removed, leaving a bowl-shaped cavity on the top of the comet (Figure 11.5(d)) Third, no apparent hole or cavity is found in the broken, cometlike structures (Figure 11.5(e)). Figure 11.5(e) also shows that the SiO_2 nanowires in the cometlike structure grow at a sharp angle to the axis, with a value usually less than 30°.

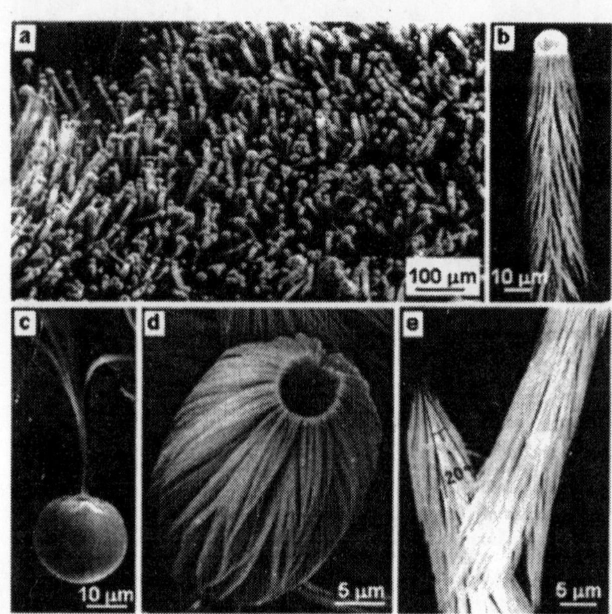

Figure 11.5 *SEM images of the SiO_2 nanowires grown on an alumina substrate: (a) low-magnification SEM image of the as-synthesized products, show a large quantity of cometlike structures, growing perpendicularly from the substrate to form an array; (b) high-magnification SEM image of a cometlike structure, showing that the tail part of the comet is composed of highly oriented SiO_2 nanowires; (c) cherrylike structure; (d) cometlike structure with the Ga ball being removed off, leaving a bowl-shaped cavity as its tip; (e) broken cometlike structure, displaying an angle of ~20° between the growth direction of the nanowires and the axis of the comet.*

The microstructures of the SiO_2, nanowires were characterized by TEM. Figure 5a, b shows the TEM Images of the SiO_2 nanowires grown on the silicon wafer and alumina plate, respectively. Both kinds of nanowires have a uniform diameter along their entire length and a very narrow diameter distribution. The average diameters of the nanowires shown are 20 nm for Figure 5a and ~60 nm for Figure 11.6(b) with the later ~ 3 times thicker than the former. Analysis of images taken from a number of nanowire samples grown on silicon wafer and alumina substrate constantly show nanowire diameters of 15 – 30 nm and 50 – 100 nm, respectively. The following similarities are observed for two kinds of nanowire samples. First, electron deflection patterns (insets in Figure 11.6(a) and (b) and EDS analyses show that the nanowires are pure amorphous SiO_2, which is contradictory to the conventional VLS-grown nanowires that are single crystal. Second, split growth occurs in SiO_2 one nanowires; i.e., one nanowire splits into two branches, and the newly formed branch also splits into two subbranches,

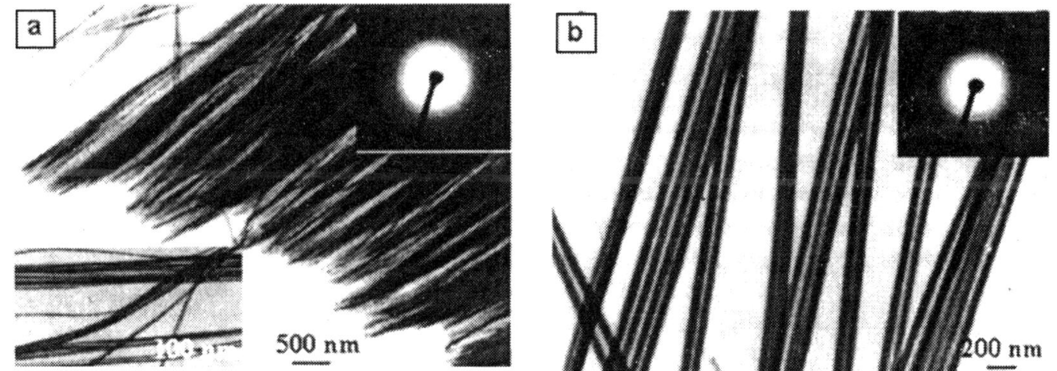

Figure 11.6 *TEM images of the SiO_2 nanowires: (a) bundle of SiO_2 nanowires grown on silicon wafer, showing amorphous (upper right inset) and very thin nanowires with average diameter of ~20 nm (lower left inset); (b) SiO_2 nanowires grown on alumina substrate, showing paired amorphous (inset) and straight nanowires with average diameter of ~60 nm.*

and so on (Figure 11.7(a)). This growth phenomenon is observed in cobalt-catalyzed SiO_2 nanofibers. The branches or the subbranches have diameter and growth direction similar to their parent one. This split growth occurs quickly and severely for the nanowires grown on the silicon wafers, resulting in the amount and volume of the nanowires increase dramatically within a short distance (Figure 11.7(b)). The split growth phenomenon is responsible for the formation of the compact tubular structures shown in Figure 11.3.

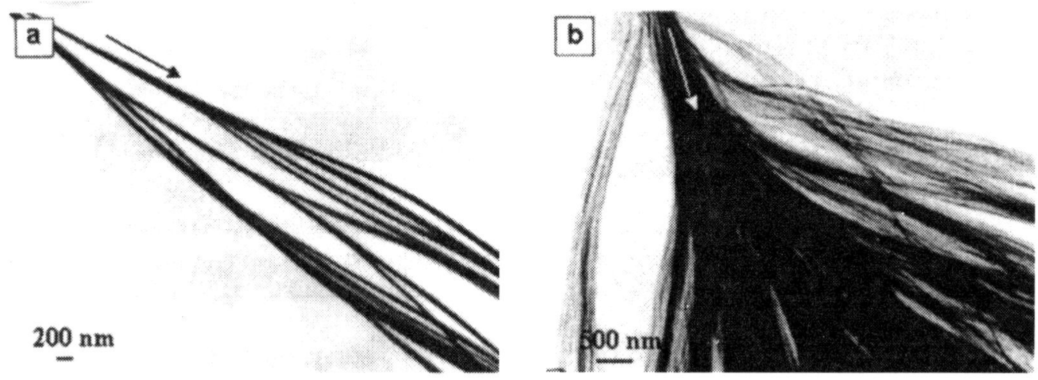

Figure 11.7 *TEM images showing the split growth in the SiO_2 nanowire samples: (a) nanowires grown on alumina substrate, displaying split growth phenomenon; (b) image of the nanowires grown on silicon wafer, showing the amount and volume of the nanowires increasing tens of times within a short distance (~5 μm) through the split growth. The arrows show the nanowires' growth direction.*

Thus, the low melting point metal Ga serves as an effective catalyst for the growth of amorphous SiO_2 nanowires via a VLS process. The above results indicate that molten Ga has different catalytic behaviors for nanowire growth from the commonly used metal catalysts such as Au and Fe. On the basis of the VLS growth mechanism and the SEM and TEM results described above, a growth model is for CSRs with stairlike inner structures (see Figure 11.4(f)) grown on the silicon wafer is the one depicted in Figure 11.8 is the and is described below.

First, nanowire growth of similar composition is observed on both the silicon wafer and the remote alumina plate from the model in Fig. 11.8 (see detail synthesis). This couples with the fact that the SiO_2 nanowires are catalyzed by a Ga ball that is not in contact with the growth substrates suggest that the nanowire growth is governed by a VLS process in which Si is fed from the vapor phase. The Si vapor is generated from etching of the silicon wafer by the hot liquid Ga (Figure 118(b)). Indeed,_at the region of SiO_2 nanowire formation, the silicon wafer is etched away to depths of ~50–100 μm after 5 h of reaction and most of the etched region is covered with many large molten Ga droplets. These large Ga droplets are formed in the early stage of the reaction. Since in the later stages this region is covered by a thick layer of high

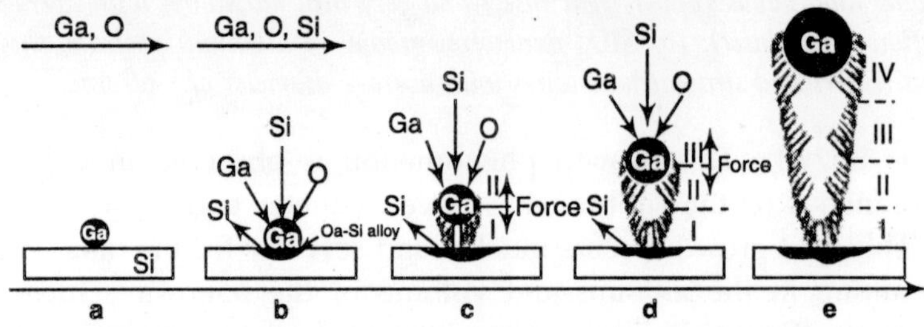

Figure 11.8 *Growth model for CSRs with stairlike inner structures (a) The decomposition of GaN powders produces a vapor of Ga that rapidly condenses into liquid Ga clusters. These Ga clusters then deposit onto the surface of the silicon wafer and grow into small Ga balls as the upcoming Ga clusters are absorbed from the vapor. (b) The hot liquid Ga ball etches the silicon wafer to form Ga-Si alloy. The Si in the Ga-Si alloy evaporates into the gas to create a dense vapor of Si species around the silicon wafer region. At this stage, the vapor consists of Ga, O, and Si, and thus, the Ga ball can also absorb Si species from the vapor. (c) When the concentrations of Si and O in the Ga ball are high enough, the Si and O reacts to form many SiO_2 nanoparticles on the surface of the lower hemisphere of the Ga ball. These particles act as the nucleation sites, initiating the growth of the first batch (batch I) of the SiO_2 nanowires. The Ga ball is then pushed away from the silicon wafer by the growing SiO_2 nanowires. From this stage, the Ga ball can only absorb Si species from the vapor. As this first batch of nanowires proceeds to grow, a second batch (batch II) of nanowires simultaneously nucleates and grows at nearly the same rate,and direction above the first. As growth continues, the newly formed nanowires begin to exert a force on the batch below. Note that split growth proceeds during the entire nanowire growth process. (d) When the force is great enough, the second batch of nanowires lifts the Ga ball upward, thereby detaching the first batch of nanowires from the Ga ball and halting their growth. Since the Ga ball is in the liquid state and the nanowires connect to it though a thin oxide layer, the second batch of nanowires will then sink to the position the first batch initially occupied. A third batch (batch III) of nanowires then nucleates and grows above the second. (e) The process of growth and detachment allows the formation of a tubular structure with regular stairlike inner wall. For this case, the nanowires only grow from a band around the lower hemisphere surface of the Ga ball. If the nanowires also grow from the region below the band CSRs with a series of upside down bell-like cavities (see Figure 11.4(g)) is formed.*

density CSRs, as shown in Figure 11.3 Throughout the entirety of the synthesis, the Ga droplets on the surface of the silicon wafer continuously etch the silicon wafer, allowing the Si species to be continuously fed into the vapor phase and sustain the SiO_2 nanowire growth. The differences in morphology and inner structure between the products grown on the different substrates can be correlated to concentration differences of the Si species above the two kinds of substrates. The Si concentration above the silicon wafer is much higher than that above the alumina plate, resulting in the products grown on silicon wafer having a larger size and higher density than those grown on alumina plate.

Second, it is apparent that considerable amounts of oxygen are involved in the growth of SiO_2 nanowires. The oxygen comes from (i) the oxide layer on the silicon wafer surface, (ii) the residue oxygen in the reaction chamber, (iii) the Ar gas, or (iv) the leakage of the vacuum system. Although the oxygen from the former three sources contributes to the formation of the SiO_2 nanowires, the amount of oxygen is not high enough to be responsible for the whole product of SiO_2 nanowires. Therefore the dominant source of oxygen can originate from the system leakage. To further understand the effect of oxygen, 1sccm of pure oxygen along with 50 sccm of Ar is introduced into the reaction chamber (the partial pressure of oxygen in the total reaction pressure is about 1.03×10^3 Pa). As a result, neither Ga ball nor SiO_2 nanowire is found on the silicon wafer or alumina plate because whole GaN powders are oxidized into stable Ga_2O_3 powders. This indicates that the partial pressure of oxygen in the reactor is a critical factor in the growth of SiO_2 nanowires.

Third, due to the presence of oxygen in the reaction chamber, the following reactant species are involved in the vapor gas: Si, O, Ga, SiO, SiO_2, Ga_2O, GaO, and Ga_2O_3 (in the growth model in Figure 11.8) These reactant species deposit and dissolve into the Ga balls, and the following main reactions may occur:

$$Si + 2O = SiO_2 \tag{1}$$

$$2Ga + 3O = Ga_2O_3 \tag{2}$$

$$2SiO = SiO_2 + Si \tag{3}$$

$$2Ga_2O + Si = SiO_2 + 4Ga \tag{4}$$

$$2GaO + Si = SiO_2 + 2Ga \tag{5}$$

For element Si and Ga, since the bond energy of Si-O bond , (185 $kJmol^{-1}$) is about three times higher than the Ga-O bond (59 $kJmol^{-1}$), the Si combines with O to form stable SiO_2. For reactions 4 and 5, the standard free energy changes are −198 and −1468 $kJmol^{-1}$ respectively; therefore, both reaction should proceed. These thermodynamic data that SiO_2 is the main products in the reaction. The experimental results are in good agreement with this prediction.

Fourth, the strong ability of liquid Ga for absorbing O from vapor is responsible for the formation of the SiO_2 nanowire tow experiments are conducted for this with varying catalysts and SiH_4 as a Si source in the reaction system shown in Figure 11.2. When Ga from the thermal decomposition of GaN is used as a catalyst, a large quantity of cometlike structures and a small amount of carrot-shaped rods. as those shown in Figures 11.5 and 11.3, respectively, are obtained. However, when Au nanoparticles are used as a catalyst, randomly distributed Si nanowires are observed.

Finally, by directly dispersing the liquid Ga onto the silicon wafer, SiO_2 nanowires grow perpendicularly from the Ga ball's entire surface, forming a nanowire layer around the Ga ball; however, the yield is very low and neither CSRs nor cometlike structure is formed.

SiC NANOWIRE

Experiment

The synthesis process of SiC NWs include the xerogel preparation and a subsequent carbothermal reduction. First, 6 g of phenolic resin and 0.5 g of lanthanum nitrate are dissolved in 18 mL of ethanol (AR) and then mixed with 25 mL of tetraethoxysilane. Then 0.5 mL of hydrochloric acid is added into the mixture and the mixture is then stirred for up to 30 h to enhance the hydrolysis of TEOS. Finally,5 mL of hexamethylenetetramine (HMTA, 35.8 %) aqueous solution is added into the above mixture for a rapid gelation. The xerogel is obtained by drying the gel product at 380 K for 12 h. The carbothermal reduction, is conducted in a horizontal alumina tubular furnace. The xerogel, is heated in Ar flow {60 cm^3min^{-1}) to 1290K at a rate of 10 K min^{-1}, then to 1590K at a rate of 2 $Kmin^{-1}$ and maintained at this termperature for 6 h. The raw product is heated in air at 790K for 4 h to remove the residual carbon, and subsequently treated by nitric acid (HNO_3) and then hydrofluoric acid (HF) to eliminate the residual silica and other impurities. SiC NWs are then synthesized. The crystalline structure of the as synthesized product is evaluated by the X-ray diffraction (XRD) analysis with CuKα radiation. The product is examined by field emission gun. A small piece of the as synthesized product is dispersed ultrasonically for 15 min in ethyl alcohol and placed on a Cu supporting carbon microgrid for TEM examination. TEM analysis and high resolution electron microscopy (HREM) are conducted at 200 kV. Morphology observation and selected-area electron diffraction patterns of the samples are performed. The chemical composition of the samples is detected using energy dispersive X-ray (EDX) spectroscopy and electron energy loss spectroscopy (EELS). This process is used for knowing super plasticity properties of Beta SIC nanowires at low temperatures.

SUPER-PLASTICITY OF BETA-SiC NANOWIRE

Super-plasticity is the ability of a material to exhibit an exceptionally large strain deformation rate during extensile deformation process. Ceramics and semiconductors, at room temperature or below a transition temperature, behave as brittle solids that fail due to fracture without evidence of plastic deformation. SiC as a reinforcing material behave with high hardness. High strength and high stiffness but brittleness. Bulk SiC ceramic material can transform to be ductile above 1270K Nanomaterials at the one dimensional (lD) form show some unusual properties such as the extreme high strength, good flexibility fracture toughness and inverse Hall-petch effect. As a wide bandgap semiconductor material, SiC nanowires (SiCNWs) are very attractive for applications in nanoelectronics that are operated at high temperatures, high powers, high frequencies, and in harsh environments. The electronic and mechanical properties of SiCNWs such as elastic-plastic response, brittle-ductile transition (BDT) are useful. The mechanical properties of semiconductor nanowire are rather different from those of bulk materials. Atomistic molecular dynamics (MD) simulations provide detailed information about dislocation motion, crack branching, fracture, and BDT of nanowires. However, brittle-fracture features with small strain are predicted by molecular dynamics (MD) simulations for SiC. In contrast, large strain elasticity and ductile features are observed .The nanoscale plastic deformation and fracture experiments on individual SiC NW obtained by the above experiment are observed in a scanning electron microscope (SEM). A super-plasticity of 200 % is observed for SiC nanowires at low temperature less than 350K. The results are critical for understanding the intrinsic and/ or universal mechanical properties of the materials with tetrahedral covalence bonded structure at small size.

The Structure and Substructures of Silicon Carbide Nanowires

Figure 11.9 a shows the general morphology of the SiC nanowires under SEM observation. The X-ray diffraction pattern indicates the well-developed cubic (3C) structure (with stacking faults) of the SiC NWs as shown in Figure 11.9(b). The average SiC NWs have a length of several tenths of microns and with a diameter up to 150 nm. The details of structure and substructures of the SiC NWs, the SiC specimens are obtained with TEM. Figure 11.10 is a morphology of the SiC NWs under TEM observation and Figure 11.10(b) is a magnified image of a SiC NW showing that the NW consists of two type intergrowth segments as indicated by 1 and 2. Figure 11.10(c) to (e) shows the corresponding electron diffraction patterns (EDPs) taken from area 1 along zone axes of [001], [-110], and [111], respectively, and Figure 11.10(f) to (h) are the [001], [-110], and [111] zone axis diffraction patterns taken from area 2. The diffraction patterns along [001] and [111] zone axis show no difference for segment 1 and 2 while there are long streaks along the [111]* reciprocal direction in the (-110)* reciprocal plane in Figure 11.10(g). As revealed in the high-resolution EM (HREM) image of Figure 11.11, the segment 1 possesses typical 3C structure with an atomic stacking sequence of ABCABC... and segment 2 has disordered structure along [111] direction, i.e., it has stacking-faulted sequence of (111) plane which is along the longitude growth axis of the NW. Here the structure in region 2

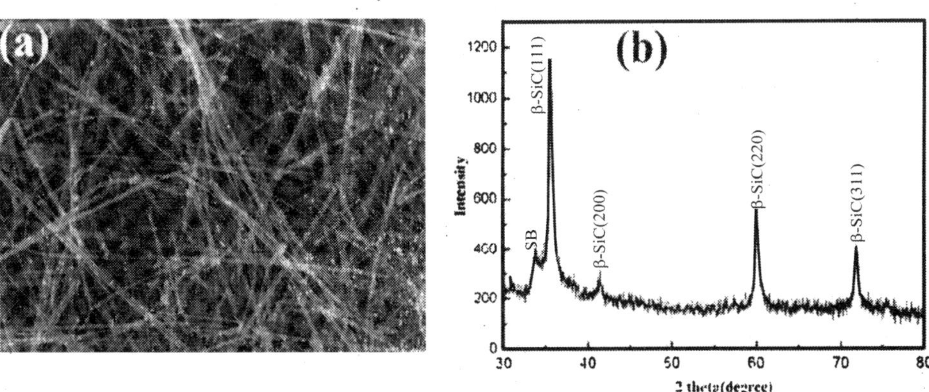

Figure 11.9 *(a) Morphologies of the SiC NWs from SEM observation. (b) XRD diffraction pattern indicaticating the SiC structure is cubic with stacking faults.*

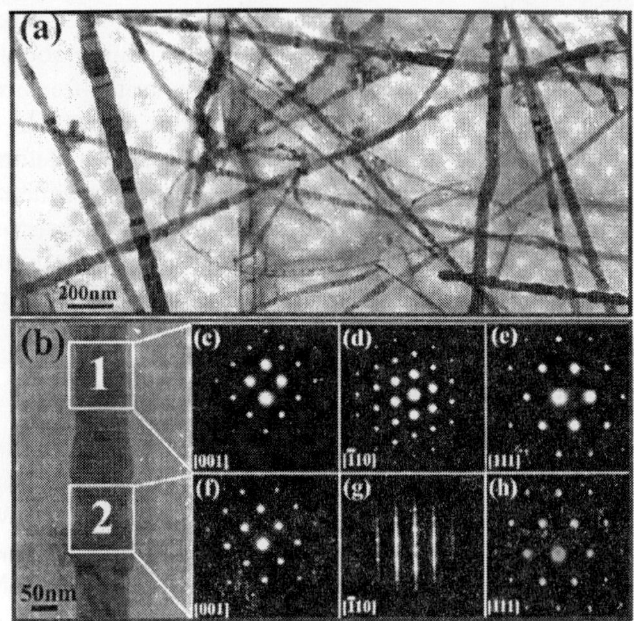

Figure 11.10 *(a) SiC NWs under TEM observation; b) is a selected piece of SiC NW showing the intergrowth feature with 3C and ODD/HD substructure segments. c)-e) are the [001], [-110] and [111] zone axis EDPs taken from area 1 of (b) and f-h) are the EDPs taken from area 2 of (b).*

of Figure 11.10(b) is donated as one dimensional disordered (ODD) or high defective (HD) structure. The structure difference of the two intergrowth segments of the SiC NWs is known from HREM analysis and electron energy loss spectroscopy (EELS) as well as energy dispersive X-ray (EDS).

Figure 11.11(a) is a HREM image taken along the [-110] zone axis of a SiC NW containing both 3C structural and ODD/HD segments. Distinctive atomic structure are revealed in the HREM images for regions 1 and 2 which correspond to 3C structure and HD structure, respectively. The fast Fourier transformed (FFT) diffraction patterns of region 1 and 2 are shown as insets of Figure 11.11(a) in the bottom left and bottom right corner, respectively. Figure 11.11(b) and (c) are enlarged HREM images Figure 11.11(a) The two dimensional projected atomic structural models are inserted in Figure 11.11(b) and (c) for the 3C and ODD/HD structures, respectively. Based on these structural models, the HREM images are overlaid as insets in Figure 11(b) and (c) for the regions of 1 and 2. The white dots in the HREM images represent the holes among atoms and the dark-gray regions are those of Si-C atomic pair projections. It is revealed that the 3C-SiC structure repeats the stacking sequence by ABCABCABC... while the HD structure disturbs the correct stacking sequence ABC but with the random order of ABC such as ACABABCB..... It is worthwhile to notice that the ABCABC stacking sequence of 3C structure creates straight atomic plane structures along all of the (111), (11-1), (1-11) and (-111) planes, i.e., there are translational symmetry and periodicity in these planes. In contrast, the stacking sequence of ACABABCB..., the translational symmetry and periodicity are broken in the three sets of (1-11), (11-1), and (-111) planes. Here A, B, C

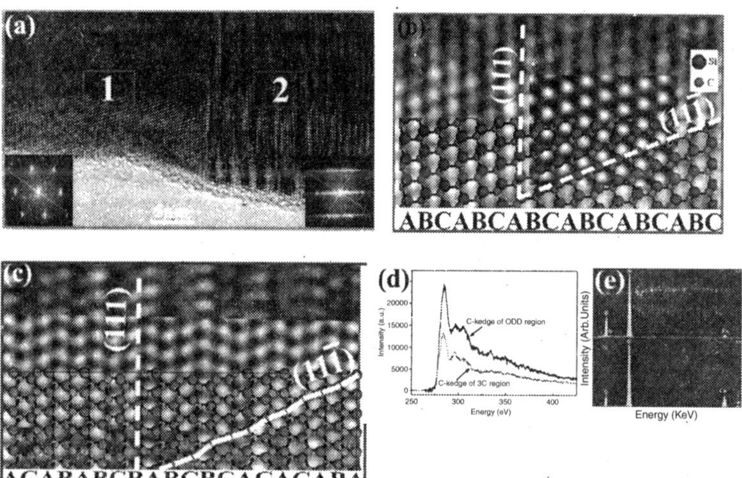

Figure 11.11 *(a) is the HREM image of a SiC NW showing the intergrowth segments of 3C and ODD/HD structures at the atomic level. The FFT diffraction patterns for the area 1 and 2 are shown as insets in the left bottom corner and right bottom corner, respectively. b) and c) Are the enlarged HREM images showing more clearly the atomic structure of 3C and ODD/HD structures. The atomic structural models and the simulated HREM images based on these models are overlaid in the corresponding figures. d) the EELS spectra for Carbon K edge for the 3C and ODD/HD structure segments and e) shows the EDS spectra taken from the regions 1 and 2 of (a).*

represents the three basic structure modules of tetrahedral bonding in SiC These two segments behave sharp different mechanical properties when external force is loaded from the axial direction.

The regions 1 and 2 show only structural difference other than compositional variation that is conducted for both EELS analysis and EDS on the regions 1 and 2 of Figure 11.11(a) Figure 11.11(d) shows the EELS spectrums taken from the areas 1 and 2, respectively, and Figure 11.11(e) shows EDS of the above two regions. The EDS indicates that both regions contain Si and C elements. The EELS spectrums show slight variance which is from the distinctive electronic structures of these two type structures.

Tensile Testing in the SEM

Following the structure and substructure determination of the SiC NWs, the tensile tests are conducted. For this two thermal bimetallic strips are mounted on the opposing position of a heating nanotensile testing stage. The bimetallic strip is made of two materials with different thermal expansion coefficients. A significant deflection at lower operational temperature is achieved by mismatch thermal expansion coefficients of the two materials. The bimetallic strips are induced to bend by heating from a filament. The bimetallic strip is made of $Mn_{72}Ni_{10}CU_{18}$ and Ni_{36} with the thermal expansion coefficients of 26×10^{-6} K^{-1} and 2.9×10^{-6} K^{-1} respectively. The movable part of the bimetallic strips (rectangular shape) is 20 mm long, 3 mm wide and 0.25 mm in thickness The bimetallic strip is operated below, 420K to avoid damage to sample.

A significant deflection, is visible under the gentle heating conditions. The detailed experimental setup is shown in Figure 11.12.

The as synthesized NWs are scattered between the two manipulators. Randomly distributed nanowires are bridged across the two manipulators. Nanowire that bridged on the manipulator are analyzed at an right angle. The SiC NWs are long and straight and are tangled and bridged across the two manipulators. Two methods are employed to "hold" the NWs on the manipulators tightly. One is the friction force among the NWs (this force is very big due to the tangling features of the long SiC NWs). The other is "soldering" using the electron beam to create "contamination" points on the contact locations of NWs/manipulators. With these two methods, the manipulators hold the NWs tightly without slipping. With heating the bimetallic strips, the manipulators are gradually moved away. A continuously increased temperature (up to 350K) drives the manipulator, so that a tensile load is applied to extend the clamped nanowire until fracture occurs. Image analysis of a series of SEM images taken during the tensile processes, for calculating the maximum fracture strain and image fracture surface is done. The manipulators are continuously driven to move away and the "soldered" SiC NWs and are gradually flattened (zero strain), then elastically extended with large strain and plastically deformed. The series image are checked and the calculations performed after the image with SiC NWs completely flattened.

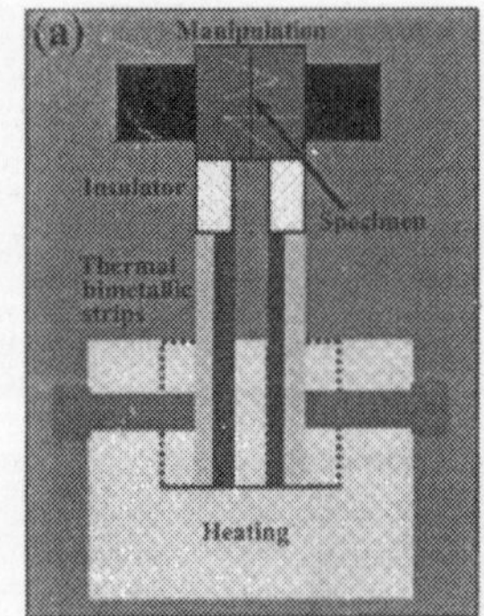

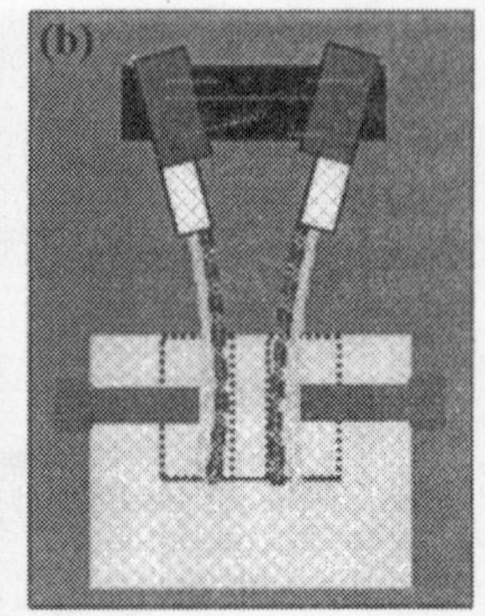

Figure 11.12 *(a) Illustration of the tensile tool prior to extensile experiment with the SiC NWs scattered on the manipulator. b) The conducting extensile experiment on the SiC NWs (top view).*

Here, eight single SiC NWs are pulled by the bimetallic extensor and the entire process is recorded by SEM imaging. All of the NWs show extremely large tensile strain (with an average fracture strain ≥25%) and plastic deformation characters. Figure 11.13(a) to (g) demonstrates SiC NW suspended and clamped between the two bimetallic actuating manipulators. The manipulators are used to pull the SiC NW to increase the strain with increasing temperature from 300–350K. The strain rate is about 5×10^{-4}, A 25 μ m long, 96 nm diameter SiC NW is axially extended and broken in the middle. The resulting fragment attached on the left has a length of at least 22 μm, whereas the fragment on the right side halt a length of at least 10 μ m Thus, the sum of the fragment lengths exceed the original section length, As revealed in Figure 11.13(g) of the fracture tip of the broken SiC NW, there is an abrupt diameter change along the tensile direction which indicates a plastic deformation features.

From the series images of Figure 11.13, the total length of the SiC NW prior to the extensile experiment is measured to be 25 μ m, an elongation of 7 μm is received after tensile testing. This gives an average elongation rate of 28%. However, the paired arrowheads indicate that local elongation rate of the SiC NW exceeds 200 %. As indicated by a pair of white arrowheads by tracking the reference nodes on the nanowire in the images shown in Figure 11.13 the starting segment length by L_0 is measured as indicated in Figure 11.13(b). The near end final

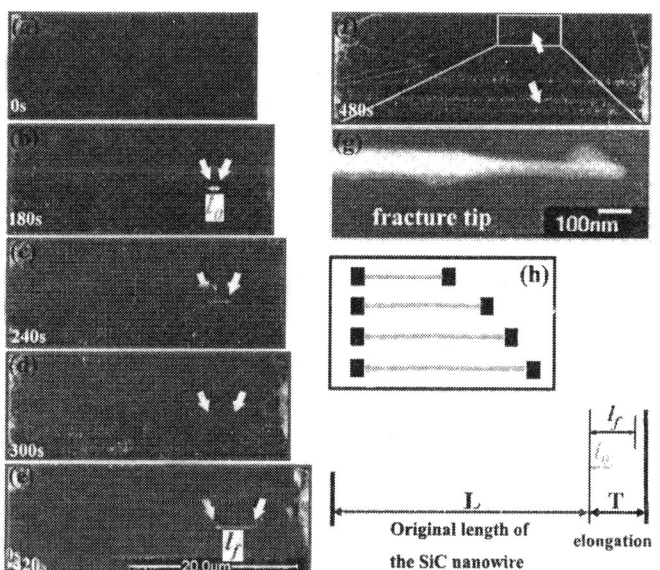

Figure 11.13 *(a-f) SEM images showing extensile experiments on a elongated SiC. NW. g) A high magnification back scattering electron (BSD) image showing a broken tip h) model of the tensile deformation process.*

length of the tracked super-extended segment is indicated by l_f in Figure 11.13(e). The local elongation rate of the SiC NW is calculated by $(l_f-L_o)/L_o$. With L_o= 1.5 × m and l_f=4.8 µ m, the local elongation rate of the SiC NW is close to 220 %, This shows a super-plastic deformation strain. Figure 11.13(g) is a magnified back scattering electron (BSD) image of the broken tip showing a clear feature (the top end is bent). The diameter is reduced 2-fold in average from 96 nm to 50 nm. Assuming the volume fraction remains constant prior to and after the plastic deformation, the 2-fold diameter reduction agrees well with the ~ 200 % elongation in length. The diameter of the plastically deformed region is identical which indicates a uniform deformation process. Without necking appearance, this deformation process shows a super-plastic deformation character. Thus the 200 % tensile strain and 2-fold reduction rate in diameter are unprecedented for a SiC NW at low temperature. A model of tensile deformation is illustrated in Figure 11.13(h)

Now Figure 11.14(a) shows typical TEM image of the 3C-structured SiC NW. Figure 11.14(b) shows the HREM images of the two structural modules of the SiC NWs. One is the defect-free segment with perfect cubic (3C) structure, while the other is high defective with a high density of stacking faults. When the SiC NW is pulled in tension: the deformation mechanism is a shearing action based on the resolved shear stress on active slip systems. The primary slip system is {111}/<110> for face-centered cubic (FCC) SiC NW is with the most favourable Schmidt factor. For the ODD/HD parts of SiC NW, Figure 11.14(c) the slip systems lie on the (111) plane in which the Schmidt factors are zero. In contrast, there are multiple slip systems such as (111)[-110] and (11-1)[011] in 3C structured segments in which the Schmidt factor on one (11-1) plane is 0.272 with respective to the loading axis along [111]. So the initial atomic

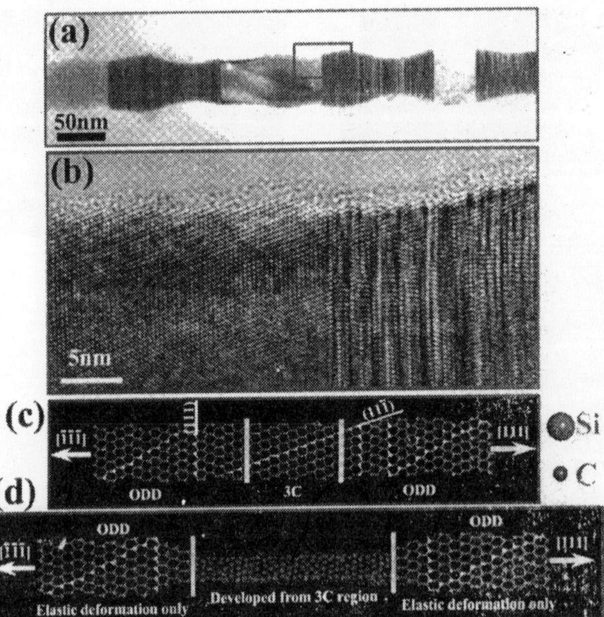

Figure 11.14 *(a) A typical TEM im age of a SiC NW and (b) is the HREM image of the 3C and ODD structured segments. (c) Is the two-dimensional atomic projection model of a SiC NW with two ODD/HD segments and one 3C segment in the middle. (d) Illustration of the atomic model of plastic deformation process in 3C strutured segment.*

slip is favorable to occur on the type of (11-1) plane of the 3C-structured region when the load, is applied along [111] or [−1−1−1] axial orientation. The subsequent extended tensile process along [111] orientation result dislocation nucleation and propagation continuously on the three sets of (11−1), (1−11) and (1−1−1) planes with favorable Schmidt factors. The dislocation activities happen within the region highlighted by the two blockers as indicated in Figure 11.14(c). It is therefore inferred, that a continuous plastic/super-plastic flow results in the cubic structural SiC NW segments to continuously transform to be amorphous until fracture failure. The super-plastic flow is confined in the 3C-structured region as illustrated in Figure 11.14(d) within the two blockers as indicated in Figure 11.14(c) and (d). The ODD/HD segments at the two-ends outside the 3C-structured segment only conduct elastic deformation. As revealed from the direct atomic observation the entire plastic deformation process is in three stages: dislocation initiation, dislocation propagation and amorphization. But here one more process of super-plastic flow is observed that follows amorphization process. Regarding the dislocation initiation and progagation as one integrated process of dislocation activity it also includes the super-plastic flow process so that the deformation and fracture process of SiC NWs is a quadra-step process, i.e.., dislocation activity → amorphization → super-plastic amorphous flow → plastic fracture. This is consistent with bending deformation by TEM and supports the prediction of the crystalline to amorphous phase transition procedure of SiC.

SILICON CARBIDE - SILICA BIXIAL NANOWIRES

The properties of 1D nanowires depend on their geometrical shape/configurations, which include tubular structures (multi/single-walled carbon nanotubes), solid cylindrical nanowires, or coaxial constructions. Devices made using nanowire heterojunctions can be critical for nanoelectronics. Silicon carbide is a wide band gap semiconducting material used for high-temperature, high-frequency, and high-power applications The growth of β-SiC whiskers is achieved using a variety of techniques, example vapor-liquid-solid technique. It is used to develop the laser ablation of metal containing silicon targets as a means of obtaining bulk quantities of silicon nanowires. The synthesis of bulk quantities of biaxially structured silicon carbide-silica nanowires, composed of side-by-side subnanowires is described below. The techniques of high temperature synthesis are applied to synthesize dislocation-free Si nanowires. Amorphous SiO is brought into intimate contact with carbon/graphite in an appropriate mix at elevated temperatures for an extended time period. To obtain the biaxial silicon carbide-silica nanowire configurations the system is operated at temperatures close to 1770K for 12 h.

Figure 11.15 (a) depicts a low-magnification transmission electron microscopy (TEM) image of the nanowires dispersed on a carbon film. The nanowires are uniform with diameter 50-80 nm, and a length which is as long as 100 μm. The as synthesized materials are grouped into three basic nanowire structures: pure SiO_x, nanowires, coaxiall SiO_x, sheathed βSiC nanowires [Fig 11.15(b)] (~50% of material and biaxel β SiC-SiO_x, nanowires [Fig. 11.15(c)] (~30% of material). The coaxial SiC-SiO_x, nanowires have a <111> growth direction with a high density of twins and stacking faults perpendicular to this growth direction, From the crosssectional TEM images of the as-synthesized nanowires, the geometrical shapes of the coaxial SiC-SiO_x, [Fig. 11.15(d)] and the biaxial SiC–SiO_x, nanowires [Figs. 11.15(e) and (f)] are revealed. The SiC nanowires have faceted shapes, and one type of the facet is {111}. The growth mechanism and the mechanical properties of the biaxial nanowire are discussed below.

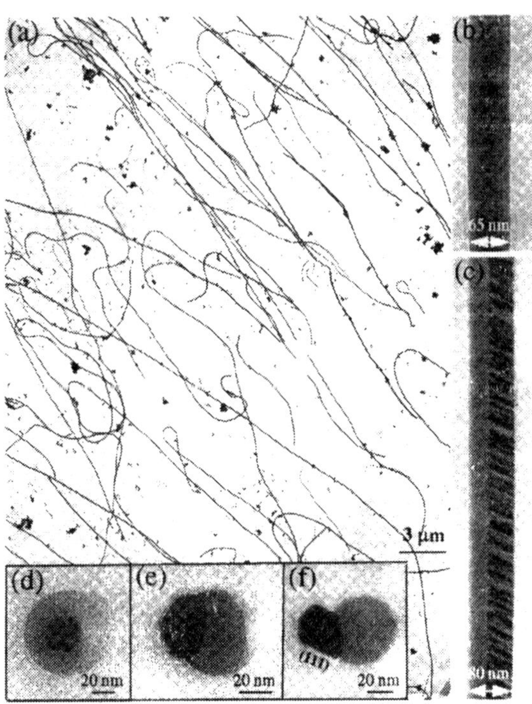

Figure 11.15 *(a) Low-magnification TEM image of thc as-synthesized nanowires supported by a carbon film, showing their, ultra long length and uniformity in diameter. Side views , of a (b) coaxially and (c) biaxially structured SiC-SiO_x, nanowire. (d), (e), and (f) Cross-sectional TEM images of coaxially structured SiC - SiO_x,. and biaxially structurcd SiC - SiO_x,. nanowires. respectively.*

The biaxial SiC-SiO_x nanowires consist of two side-by-side subnanowires of silica and β-SiC [Fig. 11.16.(a)], which are referred as a composite nanowire. If an image from the [011] direction perpendicular to the nanowire is obtained then the overlap between the amorphous SiO_x, side and the β-SiC is only ~4 nm. There is also a thin layer of silica passivated on the surface of the β-SiC. The β-SiC side has a high density of stacking faults and a few twins (so-called micro twins). This is shown in an enlargement of a local region [Figure 11.16(b)] where the outermost surface of the β-SiC is composed of nanosize {111} facets. At the interface between SiC and SiO_x, small segments of {111} faces are also identified [Fig. 11.16(c)]. The presence of a high density of planar defects results in a [311] axial direction for the biaxial nanowires, in contrast to [111] for the coaxial nanowires. Biaxial nanowires with a much lower density of stacking faults and a [211] growth direction are observed. Fig. 11.17 shows a group of TEM images that indicate the structural transformation in the SiC-SiO_x nanowire systems. The segment containing a high density of planar defects grows straight along the axis, while the presence of a defect free region produces bending [Figs 11.17 (a)-(c)] and sharp turns [Fig 11.17 (d)] along the nanowires. This is due to switching of the growth direction from [311] or [211] for the biaxial structure to [111] for the coaxial structure, while a common (111) plane is preserved. The SiC and SiO_x sides also switch across a bending area. [Figs. 11.17(b)-(d)] while the orientation of the planar defects does not change significantly. This indicates an exchange in position of the two sides rather than a twisting along the growth direction.

Figure 11.17 (c) shows a case where a short segment of coaxial nanowire (along [111]) links two biaxially structured nanowires as the side-by-side SiC and SiO_x subnanowire are exchanged. This is an interface junction between biaxial and coaxial nanowires in the same nanomaterial. Figure 11.17(e) displays structural transformation from coaxial growth to biaxial growth. The transition region is a thin neck and the twin results in the bending of the nanowire.

The unique structure of the nanowire is determined by growth kinetics. For the mechanism of the structural transformation from a coaxial to a biaxial nanowire, a structure block is introduced which is based on the high resolution TEM image shown in Fig. 11.16. It is enclosed by {111} facets as viewed along [011]. Although each block is a defect-free crystal slab of SiC, stacking faults and twins are introduced in stacking the blocks into a wire structure [Fig. 11.17(f)]. If the density of the twins is dominant, the nanowire grow along [111]. The growth direction switches to [211] if a larger defect-free block is stacked onto the (111) plane [Fig. 11.17(f)]. Alternatively, if the density of stacking faults is high so that there is a constant translation between the adjacent blocks toward one direction, the nanowire grows along [311] or an alternate direction depending on the relative shift between the blocks [Fig. 11.17(g)]. The atomic scale kinks and ledges created by the formation of the [311] nanowire increase the surface energy. Thus, to minimize the surface energy, a silica subnanowire should form adjacent to the SiC subnanowire on its rough surface. This process may operate simultaneously with kinetics to determine the development of silica passivation.

The synthesized nanowires are potentially useful for high-strength composites, in which mechanical properties are critical. For an TEM analysis one of the nanowire is fixed onto a gold ball using conductive glue and an oscillating voltage is applied across the ball and its counter electrode. Mechanical resonance occurs when the applied frequency matches the natural resonance frequency, however, care must be exercised in identifying the fundamental resonance frequency.

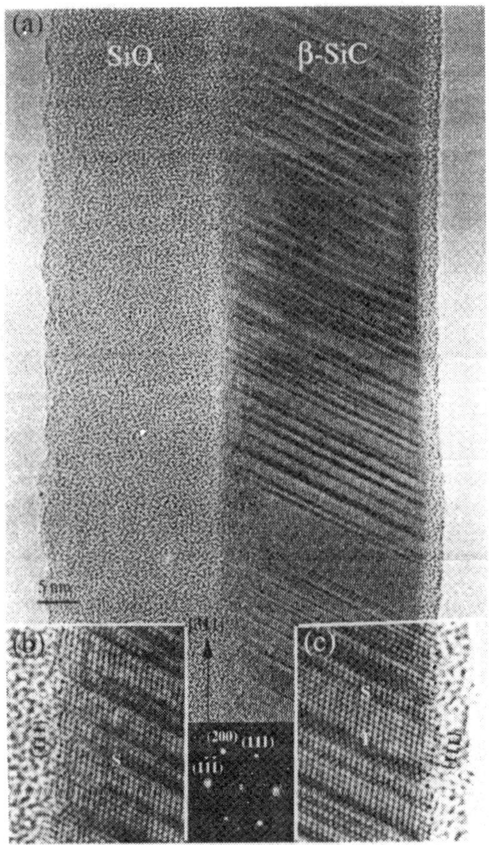

Figure 11.16 *High-resolution TEM image of a biaxially structured β-SiC–SiO_x, nanowire, showing the crystallographic and defect structures of the nanowire. The electron beam is along [01$\bar{1}$], perpendicular to the nanowire. A Fourier transform of the image from a defect-free area, given at the bottom. shows that the growth direction of the nanowire is [311]. (b) and (c) Enlargements of the areas between the interfaces of SiO_x–SiC at the contact area and near the surface of SiC-SiO,. respectively, showing the {111} faceted nanoareas at the interface. Sand T stand for stacking limit and twin. respectively.*

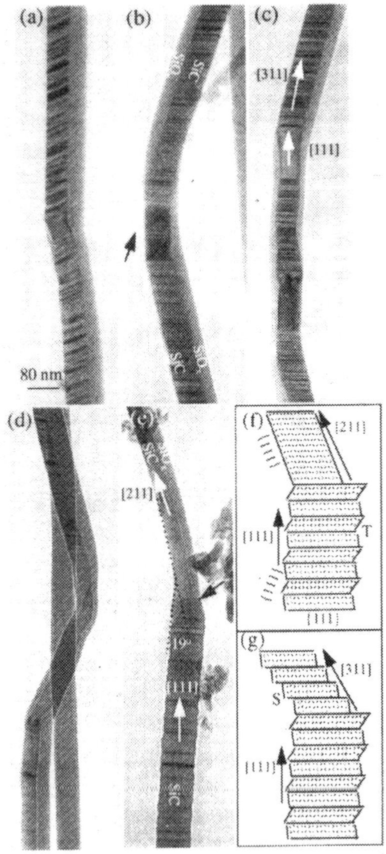

Figure 11.17 *(a)-(d) TEM images from the SiC-SiO_x, biaxially structured nanowires, displaying structural evolutions and interface junctions in the nanowires. Bending appears wherever a defect tree segment is created due to a change in growth direction from [311] or [211] to [111]. The electron beam is near [01$\bar{1}$] (e) Structural transformation from a coaxial to biaxial nanowire. where a neck-shape junction is formed at the interfacial region. (f), (g) The structural transformation from a coaxial to a biaxial nanowire growing along the [211] and [311] directions.*

MEASURING THE YOUNG'S MODULUS OF SIO_2/SIC

Synthesis of Nanowires

The nanowires used for measuring the young's modulus of SiO_2/SiC by TEM are SiO_2/SiC composite nanowires, which are synthesized by a solid-vapour phase reaction process. Amorphous SiO is brought into intimate contact with carbon/graphite in an appropriate mix at elevated temperatures for an extended time period. This is facilitated as a SiO and carbon mix placed into a low porosity carbon crucible at the centre of a temperature controlled tube furnace operated with a double concentric alumina tube combination. The appropriate mixture is heated to the desired temperature to form composites as the sample is bathed in an entraining argon flow at a flow rate of 100 sccm. The total pressure in the Inner alumina tube ranges from 2.6×10^4 Pa to 4.0×10^4 Pa, controlled by a mechanical pump attached to the inner alumina tube through a downstream water-cooled stainless steel end piece. This end piece is mechanically attached to a 'water cooled' cold plate, whose temperature is adjustable, at the fringes of the tube furnace hot zone. The entrance and exit regions of the alumina tube furnace are insulated by a matching set of zirconia blocks. To produce both coaxial and biaxial silicon carbide-silica nanowire configurations, the system is operated at a temperature of 1770K 12 h.

Nanomeasurements by TEM

To carry out property measurements on nanowires, a specimen holder for a TEM (100 kV) is built to facilitate the application of a voltage across a nanowire and its counter electrode (Fig. 11.18), The specimen holder requires the translation of the nanowire via either mechanical movement by a micrometer or by an axial directional piezo. The fibre is glued using silver paste onto a gold wire, through which an electric contact is made. The counter electrode is a gold ball directly facing the nanowire. The nanowire whose properties are to be measured is imaged directly under TEM, for recording electron diffraction patterns and images the nanowire. The information provided by TEM directly reveals both the surface and internal structure of the nanowire. An oscillating voltage with tuneable frequency is applied to the nanowire. Here, a mechanical resonance is induced in the nanowire if the applied frequency approaches the resonance frequency (Fig. 11.19). This technique works for either conductive or insulating nanowires.

Figure 11.18 *Diagram of the specimen holder used for measurement in TEM.*

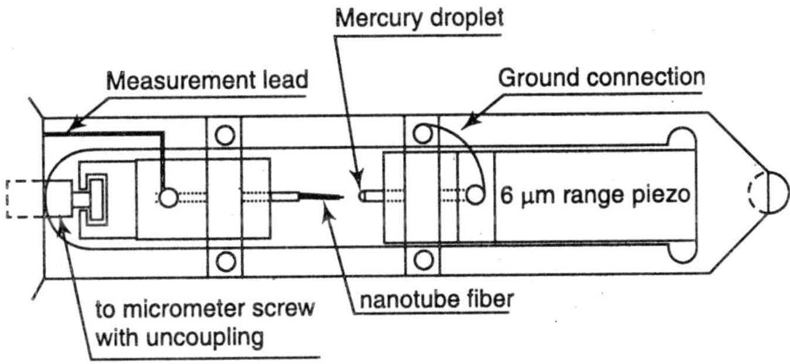

Figure 11.19 *A silica crystalline silicon or amorphous silica at (a) stationar and (b) th first harmonic resonance induced by an externally applied electrie field.*

Now as for a rod with one end hinged and the other free, the resonance frequency is given by;

$$f_0 = (\beta^2/2\pi)(EI/m)^{1/2}/L^2 \quad (1)$$

where f_0 is the fundamental resonance frequency, $\beta = 1.875$, EI is the flexural rigidity (or bending stiffness), E is the Young's modulus, I is the moment of inertia about a particular axis of the rod, L is the length of the rod, and m is its mass per unit length.

To apply equation (1) in a data analysis, it is vital to identify the true fundamental frequency. In practical experiments, the nanowires are positioned against a counter electrode. Because of the difference between the surface work functions of the nanowire and the counter electrode (e.g. Au), a static charge exists to balance the potential difference that exists even at zero applied voltage. Therefore, under an applied field the induced charge on the nanowire can be represented by $Q = Q_o + \alpha V_d \cos 2\pi ft$, where Q_o represents the charge on the tip that balances the difference in surface work functions, α is a geometrical factor, and V_d is the amplitude of the applied voltage. The force acting on the nanowire is

$$\begin{aligned} F &= \beta(Q_o + \alpha V_d.\cos 2\pi ft)^2 \\ &= (\beta Q_o^2 + \alpha^2 \beta V_d^2/2) \\ &\quad + 2\alpha\beta Q_o V_d \cos 2\pi ft \\ &\quad + \alpha^2 \beta V_d^2/\cos 4\pi ft. \end{aligned} \quad (2)$$

where β is a proportionality constant. Thus, resonance is induced at f and 2f with vibrational amplitudes proportional to V_d and V_d^2, respectively. The former is a linear term, as the resonance frequency equals the applied frequency, while the latter is a non-linear term for which the resonance frequency is twice that of the applied frequency. There are two ways to determine the fundamental frequency. Using the linear relationship, the vibration amplitude is linearly dependent on the magnitude of the voltage V_d (Fig. 11.20). Alternatively, the

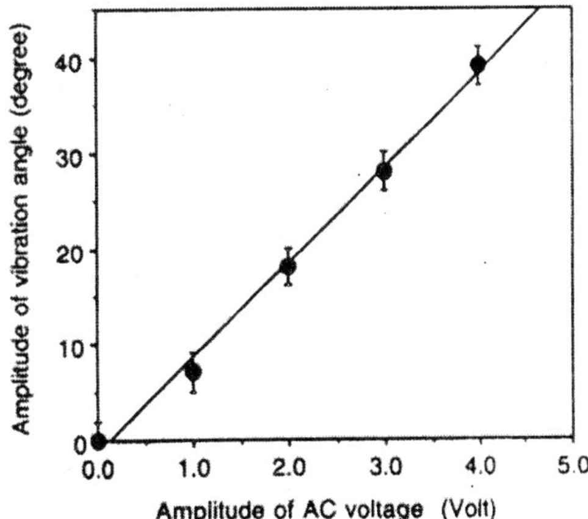

Figure 11.20 *Experimentally measured full vibration angle of a carbon nanotube as a function of the amplitude of the applied AC voltage.*

resonance is examined at a frequency that is half or close to half of the observed resonance frequency to ensure that no resonance occurs. The latter is the most convenient technique.

The diameters of nanowires are also determined directly from TEM images with high accuracy. The determination of length requires to consider the 2-D projection effect of the nanowire. It is essential to tilt the nanowire in order to determine its maximum length (which is the true length) in TEM. This requires that the TEM be operated at a tilting angle as large as ±60°. Further, the operation voltage of the TEM controlled carefully to minimize radiation damage. The 100 kV TEM produces almost no detectable damage to a nanowire.

To trace the sensitivity of the recorded resonance frequency to electron beam illumination and radiation damage at 100 kV, a carbon nanotube is resonated or more than 30 min. The resonance frequency shows an increase of ~ 1.4% over the entire period of the experiment (Fig. 11.21). The full width at half maximum (FWHM) for the resonance peak is measured to be $\Delta v / v = 0.6\%$ in a vacuum of (0.1 to 1) 10^{-4} Pa. The slight increase in the resonance frequency relates to a change in nanowire structure under the influence of the electron beam or to the loss of the root end. However, these slight changes cause a negligible effect on the measurement of Young's modulus.

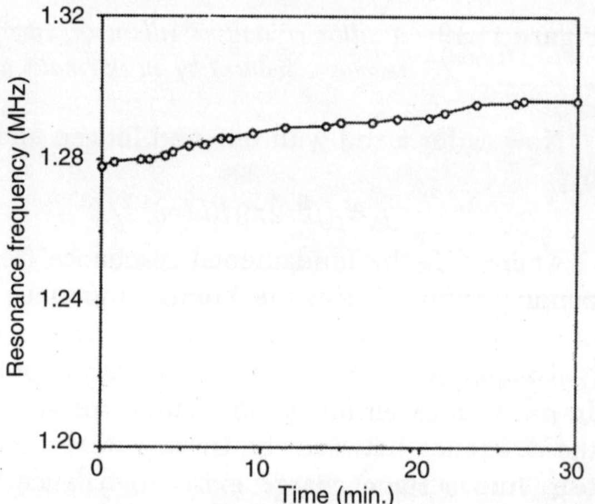

Figure 11..21 *Time dependence of the resonance frequency of a carbon nanotube being illuminated by 100 kV electrons.*

Young's Modulus of SiO_x Solid Nanowires

The as synthesized materials are grouped into three basic nanowire structures: pure SiO_x nanowires, coaxial SiO_x sheathed β-SiC nanowires and biaxial β-SiC/SiO_x nanowires. The Young's modulus is first measured for pure silica nanowires (Fig. 11.22) From eq. (1), for a uniform solid rod with diameter D, $I = \pi D^4/64$ the Young's modulus is given by

$$E = \rho(8\pi f_0 L^2/\beta^2 D)^2 \tag{3}$$

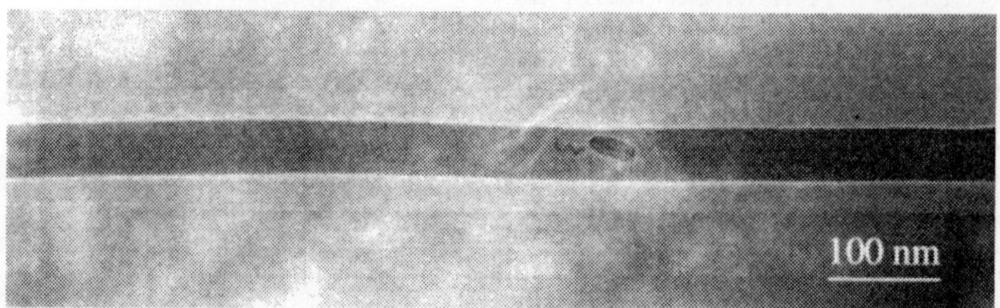

Figure 11.22 *TEM image of amorphous silica nanowires*

where ρ is the volume density of the rod. Table 11.1 shows the experimentally measured Young's moduli of silica nanowires. It is seen that the data are fairly consistent. The Young's modulus for the larger fused silica fibres for $D = 102$ µm, $E = 72.3$ GPa; $D = 20$ µm, $E = 71.9$ GPa; and $D = 4.1 - 6.0$ µm, $E = 56.3$ GPa. This gives the modulus, about half that of the values for the larger silica fibres. A density of point defects or chemical non stoichiometry in the wires is observed for this difference.

Table 11.1 *Measured Young's modulus, E. for solid silica nanowires. E is calculated using. the density of bulk amorphous SiO_2*

D (nm) (±2 nm)	L (m.tm) (±0.2 f.lm)	f_0 (MHz)	E (GPa)
42	11.7	0.160	31±5
53	6.3	0.562	20 ± 4.0
58	3.8	1.950	27 ± 7.5
70	13.0	0.200	26±3.1
95	14.8	0.232	32±3.1

f_o, fundamental resonance frequency; L, nanowire length; D, nanowire diameter.

Young's modulus of the coaxial composite nanowires

Coaxial SiO_2 and SiC nanowires are constituted of a cubic structured SiC (β-SiC) core and the SiO_2 sheathed layer (Fig. 11.23(a) High-resolution TEM demonstrates that the wire direction [111] (Fig. 11.23c). Cross-sectional TEM images indicate the coaxial structure of the nanowire (Fig. 11.23b). The core size and the thickness of the sheathed layer are found to vary from wire to wire, For such a coaxially structured nanowire whose core material density is ρ_c and diameter is D_c. and a sheath material density that is ρ_s with outer diameter D_s the average density of the nanowire is evaluated from D_s and D_c are the outer and inner diameters of the SiO_x sheath, respectively.

$$\rho_{eff} = \rho_c(D_c^2/D_s^2) + \rho_s(1 - D_c^2/D_s^2) \qquad (4)$$

If the volume density in equation (1) is replaced by the effective density given in equation (4). The effective Young's modulus of the composite nanowire. E_{eff} is given by

$$E_{eff} = \rho_{eff}(8\pi f_0 L^2/\beta^2 D_s)^2 \qquad (5)$$

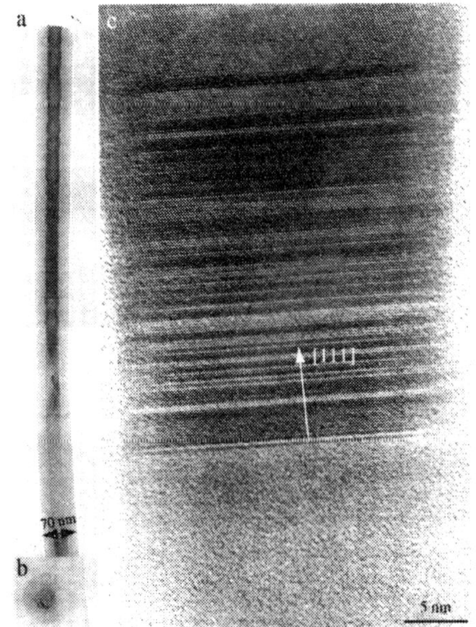

Figure 11.23 *(a) Low magnification and (b) high-resolution TEM image of a coaxial SiC/SiO_2 nanowire viewed from the side. (C) Cross-sectional TEM image of a coaxial SiO_2/SiC nanowire.*

The experimental results are shown in Table 11.2, where the Young's modulus is found to increase as the diameter of the nanowire increases. The Young's modulus for the coaxially structured SiC/SiO_x nanowires results from the combination of SiC and SiO_x where the contribution from the sheath layer of SiO_x exceeds that from the SiC core because of its larger flexural rigidity (or bending stiffness).

Table 11.2 *Measured Young's modulus for coaxially structured SiC/SiO_x, nanowires (SiCs the core and silica is the sheath). The densities of SiC and SiO_2 are taken from the bulk values ($P_{Silica} = 2.2 \times 10^3$ kg m^{-3}; $P_{Si} = 3.2 \times 10^3$ kg m^{-3})*

D_s (run) (± 2 nm)	D_c (mm) (± 1 nm)	L mm) (± 0.2 mm)	f_0 (MHz)	E_{eff}(Gpa) Exp.	E_{eff} (GPa) Them. (20)
51	12.5	6.8	0.693	46±9	73
74	26	7.3	0:.953	56±9.2	78
83	33	7.2	1.044	52± 8.2	82
132	48	13.5	.0.588	75±7 0	.79
190	105	19.0	0.419	81± 5.1	109

To compare the measured modulus with the theoretical modulus. A mode is assumed in which the coaxial nanowire is taken as a composite material. If the force that induces the mechanical vibration is closely perpendicular to the axis of the nanowire from the standard mechanical theory for composites. The Young's modulus of the composite is given by

$$1/E_{eff} = V_{sic}/E_{sic} + V_{silica}/E_{silica} \tag{6}$$

where $V_{SiC} = g$ and $V_{silica} = 1 - g$, with $g = D_c^2/D_s^2$, are the volume fractions of the SiC core and the silica sheathed layer, respectively and E_{sic} and E_{Silica} are the corresponding Young's moduli of the bulk materials (E_{SiC}= 466 Gpa; E_{Silica} = 73 Gpa).

Now, consider an effective medium theory for a composite rod. The effective flexural rigidity of a composite coaxial nanowire is $E_{eff}I_{eff} = E_{Silica}I_{silica} + E_{SiC}I_{SiC}$ Thus, the effective Young's modulus of a nanowire with cylindrical symmetry is given by

$$E_{eff} = g^2E_{sic} + (1 - g^2)E_{Silica} \tag{7}$$

The calculated results from equations (6) and (7) are in Table 11.2. The moduli are both consistent with the measured values. The tendency of the modulus to increase as the diameter increases is in agreement with observation. The data matches well with the values calculated for the larger diameter nanowires.

Young's modulus of the biaxial composite nanowires

Biaxial nanowires consisting of SiC and SiO_2 (Fig. 11.24(a) have also been synthesized The SiC side appear to have a high density of microtwins and stacking faults Fig 11.24(b), as the growth direction is [511] or [311] depending on the density of these stacking faults. Biaxial nanowires without stacking faults or twins are also found to grow along [211] (Figure 11.24(c) and (d)). The biaxial structure is demonstrated by the cross-sectional TEM image (Fig. 11.24(e) from which the outermost contour of the cross-section of the nanowire is approximated to be elliptical. For an elliptical cross-section of half long-axis *a* and half short-axis *b*, the moments of

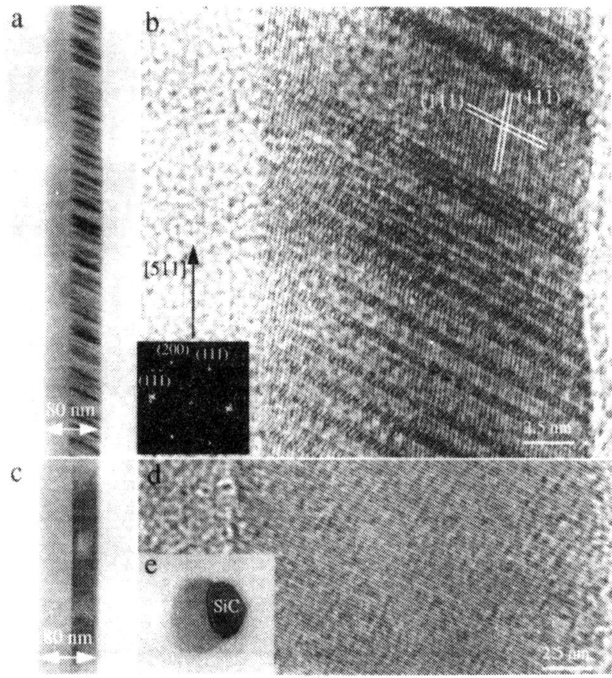

Figure 11.24 *(a) Low magnification and (b) high-resolution TEM image of a biaxial SiC/SiO_2 nanowire viewed from the side, where the wire has a high density of twins and stacking faults. (c) Low magnification and (d) high-resolution TEM image of a biaxial SiC/SiO_2 nanowire viewed from the side. where the wire is defect free. (e) Cross-sectional TEM image of a biaxial SiO_2/SiC nanowire.*

inertia are $I_x = \pi\ ab^3/4$ and $I_y = \pi ba^3/4$, .where a and b are calculated from the widths of the composite nanowire With consideration of the equal probability for resonance with respect to either the x or y axes, the effective moment of inertia introduced in calculation is taken to be approximately $I = (I_x + I_y)/2$. The density per unit length is $m_{eff} = A_{SiC}P_{SiC} + A_{Silica}P_{Silica}$ where A_{SiC} and A_{Silica} are the crosssectional areas of the SiC and SiO_x sides, respectively. Thus, the effective Young's modulus of the nanowire is calculated using the formula for a uniform rod with the introduction of an effective moment of inertia and density. The experimentally measured Young's modulus is given in Table 11.3.

Table 11.3 *Measured Young's modulus for biaxially structured SiC/SiO_x nanowires*

D_{wire} (nm) (±2 nm)	D_{SiC} (nm) (±1 nm)	L (mlm) (±0.2 m m)	f_o (MHz)	E_{eff} (GPa) Exp.
58	24	4.3	1.833	54±13.8
70	36	7.9	0.629	53±8.4
83	41	4.3	2.707	61±14.3
92	47	5.7	1.750	64±11.8

D_{wire} and D_{SiC} are The widths across the entire nanowire and across the SiC sub-nanowire respectively.

Tin Oxide Nanowires Nanoribbons and Nanobelts

Tin Oxide (SnO_2) is a typical n-type semiconductor with a wide direct band gap of 3.6 eV. It is applied as a good visible-blind ultraviolet (UV) detector. Due to its chemical stability and optoelectronic characteristics, SnO_2 nanowires (NWs) are broadly used in electrodes, solar cells, and many other photoelectronic devices. In addition, because of a large surface-to-volume ratio of 1D nanostructures, they are used in gas sensors for the detection of CO, NO, and H_2.

Tin Oxide is widely used as a solid-state gas sensor for detection combustible and toxic gases. The sensing mechanism relies on a change in the electrical conductivity due to the interaction with ambient reducing and oxidizing gases. Charge transfer between the substrate and the adsorbed species leads to a depletion or injection of charge carriers into the active region of the material. This results in band bending in the vicinity of charged adsorbates. The shift of the conduction band relative to the Fermi level changes the number of charge carriers and, as a consequence, results in a measurable change in conductivity.

Tin oxide nanobelts and nanoribbon are also used as materials for gas sensing applications. For small single crystals charge carrier depletion runs across the whole crystal, influencing nearly all electrons inside. This results in conductivity changes as a function of the adsorbing gas partial pressure. The large surface-to-volume ratio of the nanobelts, and lateral dimensions are comparable to their charge carrier screening length and thus make them might sensitive and efficient transducers of surface chemical process into electric signals.

The surface properties as well as molecular processes at the surface of tin oxide define its sensitivity and selectivity towards target gases. Well-defined facets of SnO_2 single crystals are considered model systems for SnO_2 based gas sensing materials. Their structure, composition, electronic properties, surface oxygen chemistry, Pd growth and adsorption of water is inverstigated.

As a continuation of SnO_2 surface investigation, nanobelt systems provide a more direct comparison with realistic gas sensing devices. In order to fully exploit the potential of these materials. It is important to characterize their surface morphology, modification with reactive metal clusters (which allows to functionalize the NBs), and molecular adsorption on the surface of NBs. Solid-state gas sensors are usually operated at high temperature (570 - 670 K) thus it is important to know the safe operation limits for NBs based gas sensors.

Nanowires sandwiched nanoribbons, and nanotubes of SnO_2 are synthesized using elevated temperature synthesis techniques, and their structures are characterized by scanning electrons microscopy (SEM) and transmission electron microscope TEM.

Synthesis

The-synthesis and structures of SnO_2 nanowire nanoribbons, and nanotubes produced using primarily layered mixtures of Sn foil and SnO powder and its growth directions and morphology is given below:

The technique of high temperature thermal oxide synthesis is adopted to synthesis procedures which, with the control of flow rates, reactant mixtures and annealing procedures, allows synthesis of silicon or silica based nanotubes and nanoarrays. With some modification these synthesis techniques allow the formation of tin oxide (SnO_2) nanowires, nanoribbons and nanotubes through the use of SnO/ Sn and SnO based mixtures in a double concentric alumina tube heated to 1320 – 1420K.

The inner tube of the furnace is Vacuum sealed by-two water cooled stainless steel end pieces, attached and tightly lock-press fit against custom viton O-rings. At one end of the furnace for the tin-based synthesis ultrahigh purity nitrogen gas enters through the upstream stainless steel end piece, passes through a matched set of zirconia insulators to the central region of the inner tube oven, and flows over an alumina inn crucible containing the source Sn/SnO or SnO mixtures, usually at a flow rate of 100 sccm. The total pressure in the inner tube can be changed from 2.6×10^4 Pa to 10.4×10^4 Pa, controlled by a mechanical pump attached to the inner alumina tube through a downstream cooled stainless steel and piece. This end piece is mechanically attached to a water cooled plate, whose temperature is adjustable, located at the fringes of the tube furnace hot zone. The entrance and exit regions of the alumina tube are insulated by a machined set of zirconia blocks. A light to dark gray fluffy product is collected on the alumina tube wall near the point where the double concentric alumina tube configuration exits the 720K oven and where the temperature is in the range between an 720K. and, 770K Additional products are collected on the cold plate which is positioned ~15cm into the tube furnace, fronting the upstream zirconia insulator. This cold plate is maintained at a temperature of 298 – 310K.

If the fluffy gray to gray-black, SNO*x*, product is heated slowly to 1370 K at $P_{total} \sim 2.6 \times 10^4$ Pa under oxidative conditions (1 % O_2 seeded argon) using a crucible directed concentric nozzle flow configuration, its complete conversion is observed to a white powder-like product which is more dense and consistent with bulk SnO_2. Further, if this fluffy gray product is heated slowly under a reducing atmosphere (Ar/4 % H_2, at 1370K.) it is converted completely to metallic tin. The conditions for the synthesis of all the product materials discusser below are presented in Table 12.1.

Table 12.1 *Synthesis Conditions*

Sample	*source material*	*T (K)*	*pressure (mm Hg.)*	*collecting place*
1; Figure 1(a)	Sn foil + SnO layered N_2 flow gas	1395	400	tube wall
2; Figure 1(b)	Sn foil + SnO layered N_2 flow gas	1395	200	tube wall
3; Figure 1(c)	Sn foil + SnO layered N_2 flow gas	1395	200	cold plate (308–310 K)
4; Figure 2	Sn foil + SnO layered Ar flow gas	1310	200	tube wall
5; Figure 3, Figure 7 Figure 8	SnO powder + N_2 entertainment	1320	250–700	cold plate (298–310 K)
6; Figure 5, Figure 6	Sn foil + SnO layered N_2 flow gas	1395	200	cold plate (298–310 K)
7; Figure 1(a)	SnO powder N_2 entertainment	1370	255–500	tube wall

The as-synthesized products are characterized by scanning electron microscopy (SEM) transmission electron microscopy (TEM), energy-dispersive X-ray spectroscopy (EDS) at 200 kV and a high-resolution TEM (HRTEM) at 400 kV.

1. Morphology. The as-synthesized products are inspected using SEM. Figure 12.1 (a) depicts a typical distribution in morphology for the fluffy gray products collected on the wall of the alumina tube furnace following an SnO/Sn based synthesis. It is observed that the products are dominated by wire like structures whose diameter varies over a broad range from several tens of nanometers to a micrometer. The typical length of the wires range from several tens to several hundred micrometers. The cross sectional shape of the wires vary from circular to the larger rectangle-like structure indicated by a white arrowhead in the enlarged SEM image (Figure 12.1(b)) The chemical composition of the wires is determined by EDS which is close to SnO_2. In addition to the wire like structures, Figure 12.1 (c) demonstrates that disklike structures are observed in the sample collected at lower temperatures on the cold plate. The diameter of these disks is $1 - 2\ \mu$ m with a thickness of several tens of nanometers. Some tiny nanodisks are also identified. The chemical composition of these disks, corresponds to SnO and not SnO_2.

2. Tetragonal (Rutile) Structured SnO_2 Nanowires. All of the wire-like products including those several tens of nanometers in diameter (Figure 12.2(a)) are found to be straight. The crystallography of these SnO_2 nanowires is determined using select area electron diffraction (SAD) (Figure 12.2(d)) and HRTEM imaging (Figure 12.2(b)) combined with a fast Fourier transform (FFT) analysis technique (Figure 12.2(c)). The results indicate that the SnO_2 nanowires have a normal rutile crystal structure (a = 0.470 nm, and c = 3.188 nm) which is similar to SnO_2 nanobelts. The SnO_2 nanowires also display a rectangular cross section enclosed by ± (010) and ± $(10\bar{1})$ facet planes, and a thickness-to-width aspect ratio ranging from 1:2 to 1 :5, which is smaller than that for the SnO_2 nanobelts (1:5 to 1:10). The growth direction of these SnO_2 nanowires is parallel to the [101] crystal direction.

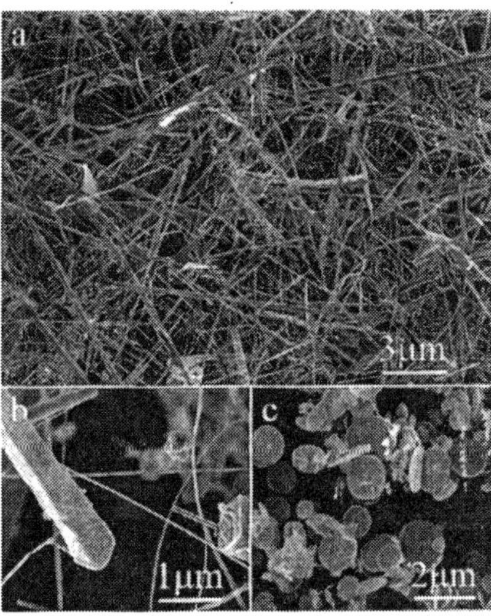

Figure 12.1 *(a) SEM image showing the morphology of the products for the Sn/SnO system. (b) Enlarged SEM image of a local area for image (a). The sample was made using a layered Sn foil/SnO mixture. The synthesis was conducted at a furnace temperature at the central region of 272 K and a chamber pressure of 5.2 × 10⁴ Pa at a flow rate of 100 sccm. (c) SEM image showing disklike products formed on a cold plate central to the upstream region of the flow system.*

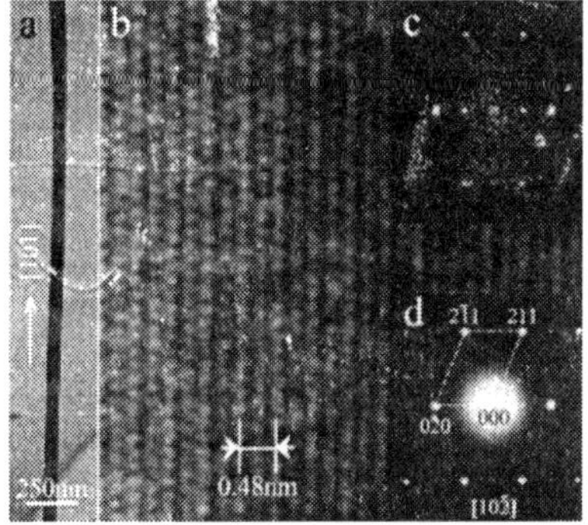

Figure 12.2 *(a) Low magnification TEM image of a rutile structured SnO_2 nanowire. (b) High-resolution TEM image of the nanowire. (c) Corresponding FFT of the image, and (d) SAD pattern from the nanowire.*

The diameter of the nanowires varies from wire to wire over a broad range. It is observed that structural defects are introduced in the nanowires of large diameter. Figure 12.3(a) depicts the TEM image of an SnO_2 nanowire with a single twin structure. The corresponding SAD pattern is shown in Figure 12.3(c). Here, the "twin" reflections are indexed with a subscript "T" and the remaining matrix reflection indices are marked without subscript. The zone axis of the diffraction pattern is [010] for the rutile structured SnO_2. The twinning plane is determined to be $(10\bar{1})$ and the twinning direction is [101], parallel to the growth direction of the SnO_2 nanowire. The twin is formed by one part of the crystal (twin) rotated 180^0 along the normal direction of the (10 1) crystal plane while the remaining sections of the crystal (matrix) maintain the original orientation. Shown in Figure 12.3(b) is a HRTEM image around the twin boundary that is completely coherent, with no relative displacement. The inset corresponds to the FFT of the HRT.EM image, which is consistent with the experimental electron diffraction pattern given in Figure 12.3(c).

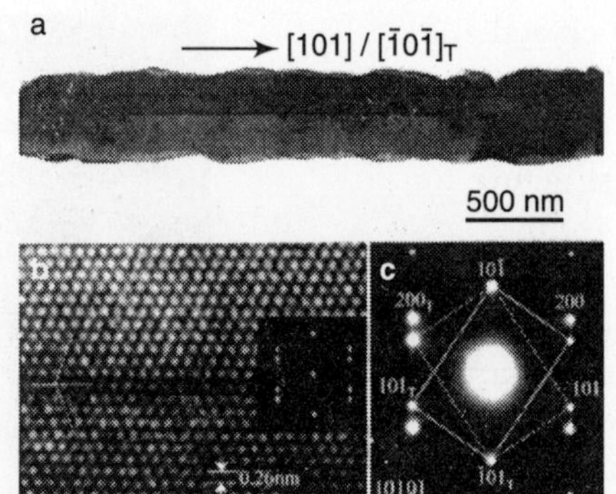

Figure 12.3 *(a) TEM image of an SnO_2 nanowire with twin structure, (b) HRTEM image of the twin, where the inset is the corresponding FFT of the image. (c) Electron diffraction pattern from the nanowire shown in (a). The sample is made by vaporizing an SnO powder sample at a furnace temperature 1320K and a chamber pressure (argon) of 3.3 $\times 10^4$ Pa The sample is collected from a cold plate placed in the flow channel mmHg.*

3. Orthorhombic Structured SnO_2 Naonowires It is found that some of the SnO_2 nanowires display a crystal structure different from rutile SnO_2. Depicted in Figure 12.4(a) is a low magnification TEM image of one of these SnO_2 nanowires. Figure 12.4(b) is the corresponding HRTEM image. The inset at the upper right-hand comer of Figure 12.4(b) is an experimental SAD pattern. A matched pattern is reproduced by the FFT of the HRTEM image, as inserted at the bottom right hand comer of Figure 12.4(b). The configuration of the reflections with strong intensities in the SAD pattern (inset) is the same as that for the

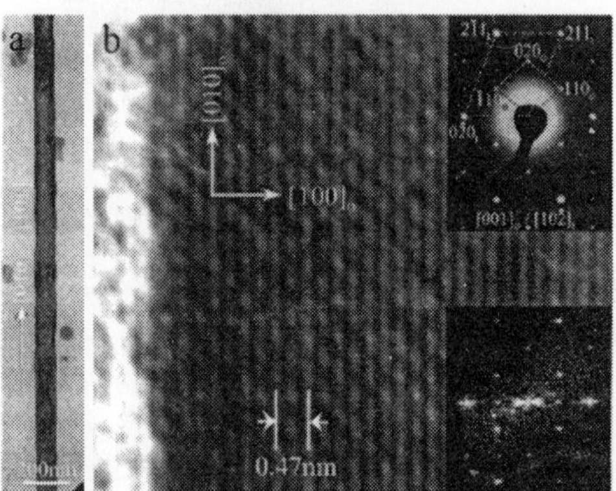

Figure 12.4 *(a) Low magnification TEM image of an individual orthorhombic SnO_2 nanowire, and (b) the corresponding HRTEM image. The inset at the upper right-hand corner is a select area electron diffraction pattern obtained for the nanowire shown in (a) and the inset at the bottom right-hand corner is a FFT of the HRTEM image. The sample is made using a layered Sn foil/SnO reactant mixture. The furnace temperature.is 1320 K and the chamber pressure 2.6×10^4 Pa. The sample is collected from the cold plate after being deposited at 298.*

$[10\bar{2}]$ crystal zone of rutile structured SnO_2 (Figure 12.2(d)). Some weak reflections, are seen in the SAD pattern inserted in Figure 12.4(b). Their geometrical configuration indicates the formation of a superlattice structure (or superstructure) associated with the normal rutile SnO_2. The superstructure matches that of orthorhombic structured SnO_2 having the lattice parameters: a = 0.4714 nm, b = 0.5727 nm, and c = 0.5214 nm, which is surprising as orthorhombic structure is formed only under high-pressure conditions. The SAD pattern corresponds to the $[001]_0$ crystal zone of the orthorhombic SnO_2 superstructure. The indexes with a subscript "o" correspond to the ortho-rhombic SnO_2 super structure, whereas the basic reflections indexed on the basis of the normal rutile structured SnO_2 are denoted with a subscript "t". The determined orientation relationship between the ortho rhombic SnO_2 and the tetragonal (rutile) SnO_2 corresponds to $[001]_0 \parallel [10\bar{1}]_t$ and $(100)_0 \parallel (010)_t$. Assuming the cross-sectional shape of the nanowire is rectangular, the orthorhombic SnO_2 nanowire is enclosed by $\pm (100)_o / \pm (010)$, and $\pm (001)_o / \pm (10\bar{1})_t$ its growth direction being along the $[010]_o$ crystal direction that is parallel to $[101]_t$ of the rutile structured SnO_2.

4. Sandwiched SnO_2 Nanoribbons Another morphology observed for the wire-like nanostructures is shown in Figure 12.5(a). Here, the nanostructure consists of two side layers whose thickness is about 20 nm and a core layer with a width ~ 120 nm. The nanostructure is found to have a rectangular cross section with a smaller aspect ratio than that observed for the nanobelts. It is, therefore known as a sandwiched nanoribbon (to distinguish it from the nanobelts). Although small parts of the side layers are missing, the nanoribbon appears to have a uniform width over its entire length. The composition of the sandwiched nanoribbon also approaches that of SnO_2 as determined by EDS. The SAD pattern (Figure 12.5(b) indicates that the nanoribbon is a rutile structured SnO_2 and significant orthorhombic structure (SnO_2) because the super-reflections of orthorhombic SnO_2 are observed. Further in the crystal structure of the nanoribbon, electron micro-diffraction patterns are recorded for the core layer (Figure 12.5c)) and side layer (Figure 12.5(d)), respectively. The reflections corresponding to the rutile structured SnO_2 are very strong in the micro-diffraction pattern of the center of the core layer (Figure 12.5(c))

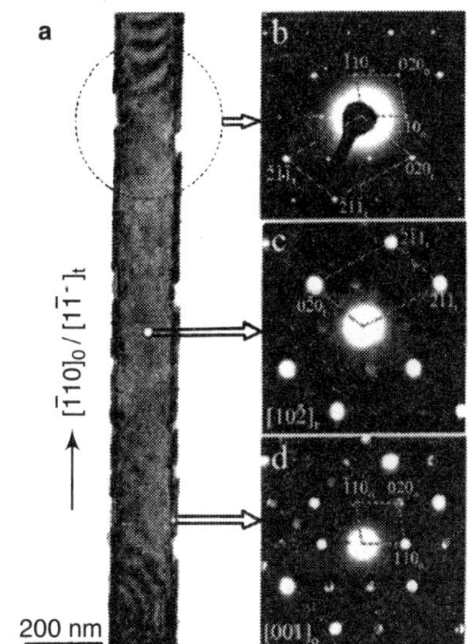

Figure 12.5 *(a) TEM image of an individual sandwiched nanoribbon. (b) Select area electron diffraction pattern from the ribbon, (c,d) Micro diffraction patterns taken from the center and edge of the nanoribbon, respectively. The sample is made using a layered Sn foil/SnO reactant mixture. The furnace temperature is as the chamber ranged from 3.3 × 10^4 Pa to 9.3 × 10^4 Pa. over the course of the run.*

although weak reflections correspond to orthorhombic SnO_2 The micro-diffraction pattern from the side layer (Figure 12.5(d)), however, displays a notably stronger diffraction feature for Orthorhombic SnO_2.

The results imply that the nanoribbon core layer adopts the rutile structure, whereas the side layers adopts the orthorhombic structure. The superlattice reflection observed in the pattern recorded from the center of the nanoribbon comes from the top and bottom surface layers that are dominated by the orthorhombic structure. The HRTEM image recorded near the edge layer in Figure 12.6, and the corresponding FFT given in the inset thus the rhombic structured SnO_2 is formed in the edge layer. The surface of the SnO_2 nanoribbon edge is not flat at the atomic level, as indicated by its HRTEM image (Figure 12.6). The orthorhombic SnO_2 and the normal rutile SnO_2 are coherent with orientation relationships: $[001]_o \parallel [10\bar{2}]$, and $(100)_o \parallel (010)_t$. This is same relationship as that deduced for the corresponding orthorhombic SnO_2 nanowires. The sandwiched SnO_2 nanoribbons, however, are enclosed by $\pm (110)_o / \pm (231)_t$, and $\pm (001)_o / \pm (10\bar{1})_t$ and their growth directions are parallel to $[\bar{1}10]_o / [1\bar{1}1]_t$.

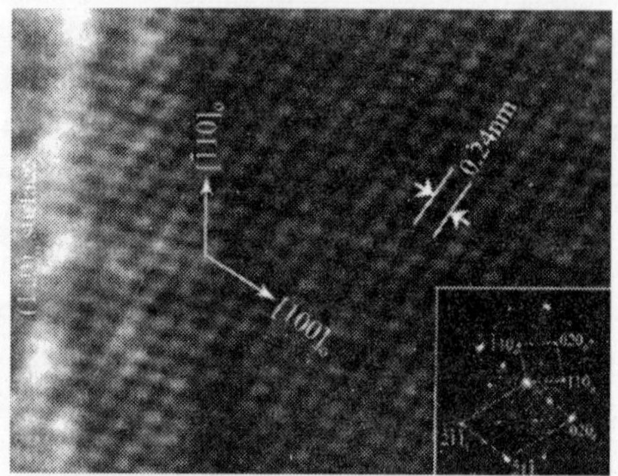

Figure 12.6 *(a) HRTEM image of the edge area of the sandwiched nanoribbon shown in Figure 12.5(a). Inset is the FFT of the corresponding image.*

The contrast over sandwiched SnO_2 nanoribbons with smaller widths, (similar to that shown in Figure 12.6(a), is uniform. With an increase in width, the contrast over the nanoribbons becomes complex, as indicated in Figure 12.7, parts (a) and (b), where two sandwiched nanobands with width of order 500 nm are displayed. One of the nanobands (Figure 12.7(b)) has only one side layer. The thickness of the surface layers does not increase as the width of the nanoribbon increases. Although the contrast displayed is complicated, the entire nanoribbon is single crystalline and the SAD pattern is identical to that given in Figure 12.5(b). Shown in Figure 12.7(c) and (d) are two enlarged images of local areas in the nanoribbon M (Figure 12.7(b)). Here, dislocations are identified, as indicated by the open arrowheads.

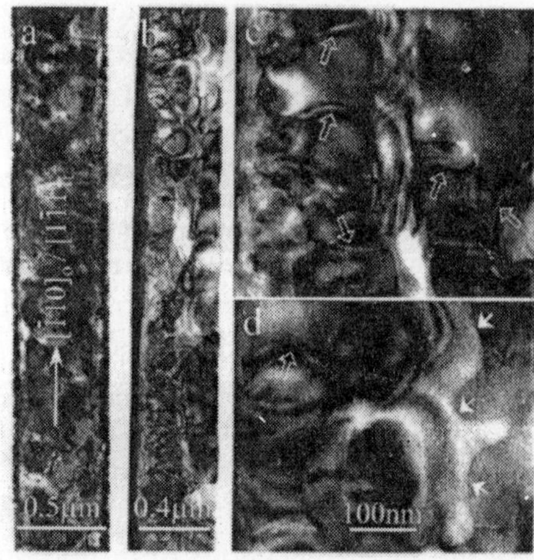

Figure 12.7 *(a) and (b), correspond to TEM images of two wide nanoribbons, and (c) and (d) are enlarged images showing the detailed microstmctures of the nanoribbons. The sample is made using SnO powder at a furnace temperature of 1320K a chamber pressure of 3.3 × 10^4 Pa. and the sample is collected from a cold finger placed in the channel.*

In addition, a domain-like boundary is identified in the nanoribbon, as marked by solid arrowheads in Figure 12.7(d). Figure 12.8 corresponds to an HRTEM image associated with the boundary region. This indicates that a domain boundary is formed by the rutile structured SnO_2 (t-SnO_2) and the ortho rhombic structured SnO_2 (o-SnO_2). The boundary is semicoherent and is not abrupt on the atomic scale. The dash line marked in Figure 12.8 represents this boundary. These observations thus suggest that the large width nanoribbons correspond to a mixed phase SnO_2, where dislocations and domain boundaries coexist.

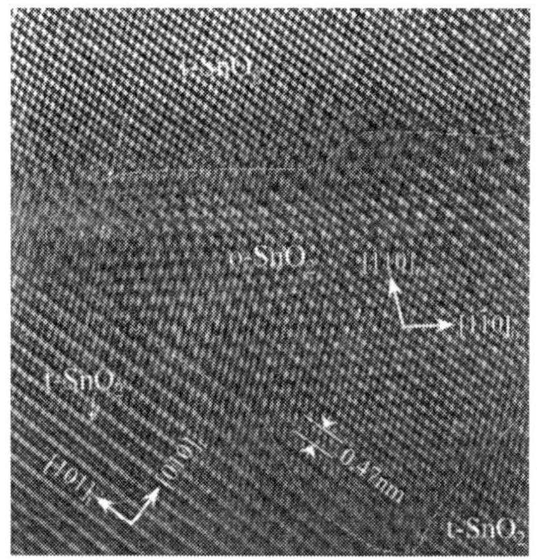

Figure 12.8 *HRTEM image showing the domains formed by rutile-SnO_2 and orthorhombic-SnO_2 in the nanoribbons displayed in Figure 12.7.*

Bulk orthorhombic SnO_2 is synthesized at a high pressure of 1.6×10^7 KPa. Although its lattice parameters are determined, the atomic positions in the crystal are not known. The structural relationship between the orthorhombic and normal rutile SnO_2 is known from a diagram of the crystal structures presented in Figure 12.9. This figure is constructed on the basis of the orientation relationships between the two structures, The rectangular parallelepiped $BOAC\text{-}B_1D_1,A_1C_1$, represents a unit cell for the rutile structured SnO_2 crystal. Here, the large spheres denote oxygen atoms and small spheres tin atoms. Six unit cells of rutile SnO_2 are drawn in the diagram, as indicated by the dotted lines. The rectangular parallelepiped $A_1OBC_1\text{-}E_1O_1G_1F_1$ is a unit cell for the orthorhombic SnO_2 This unit cell of the orthorhombic SnO_2 is formed the parallel-piped A_1OBC_1-EDGF in the rutile structured SnO_2 by shearing along the $[101]_t$ crystal direction, followed by a slight compression along the c-axis of orthorhombic SnO_2. The shearing angle is 20.2°, and the compression in length is 1.6%. The dimensions of the *a*-axis and *b*-axis of the orthorhombic SnO_2 are adjusted very slightly (0.5% compression in the *a*-axis and 0.3% expansion in the *b*-axis) relative to the corresponding spacing distances in the rutile SnO_2 structure.

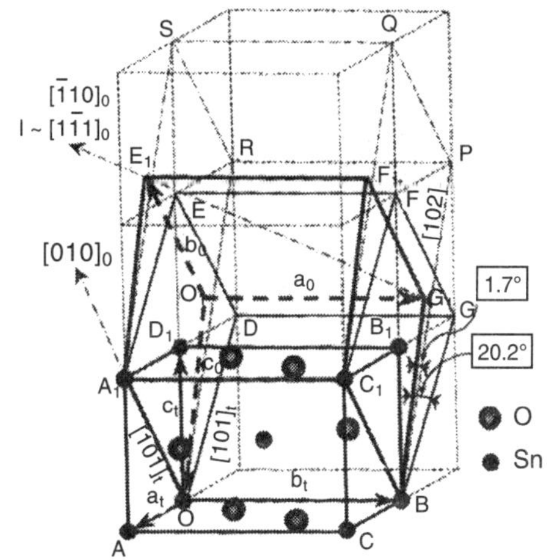

Figure 12.9 *Diagram of the crystal structures of rutile SnO_2 and orthorhombic SnO_2, where the a_t, b_t and c_t, represent the base vectors of the rutile structured SnO_2, and the a_o, b_o, and c_o denote the base vectors for the orthorhombic SnO_2.*

The volume decrease is 1.8%. The volume change results in the semi-coherent boundary between the orthorhombic and rutile structured SnO_2 (Figure 12.8) and is responsible for the missing parts of the side layers in the sandwiched SnO_2 nanoribbon shown in Figure 12.5. The $[001]_o$ (c-axis) crystal direction is not exactly parallel to $[10\bar{2}]$. Instead, there is an included angle of 1.7° between $[001]_o$ and $[10\bar{2}]_t$, which is detected from the select area electron diffraction patterns shown in Figures 12.4 and 12.5(b). The growth directions of the wire-like nanostructures are also indicated as $[010])_o/[101]_t$, and $[\bar{1}10]_o/[1\bar{1}1]_t$.

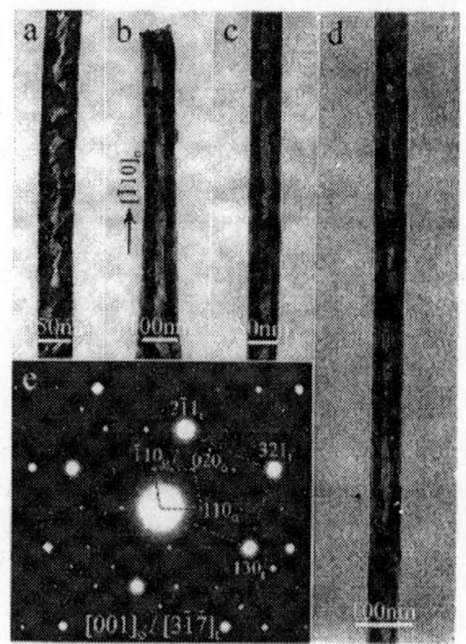

Figure 12.10 *(a), (b), and (c) are TEM images of SnO_2 nanotubes, and (d) a select area electron diffiaction pattern taken for the SnO_2 nanotube shown in (b). The sample is made using SnO powder at a furnace temperature of 1370K a chamber pressure of 3.3×10^4 Pa and the sample was collected from the walls of the tube furnace.*

5. SnO_2 Nanotubes. Figure 12.10 (a) – (c) correspond to TEM images. They show characteristic of the wire-like nanostructures as observed in the as-synthesized samples. The chemical composition of these wire-like nanostructures also corresponds to SnO_2. The bright contrast for the center part of the wire-like nanostructures is always accompanied by the dark sides even with tilting of the sample under TEM observation. This indicates that the wire-like nanostructures have hollow cores, i.e., they correspond to SnO_2 nanotubes. The diameter size of the SnO_2 nanotubes varies from 50 nm (Figure 12.10(c) – (d)) to 350 nm (Figure 12.10(a)) and their typical length is several micrometers. The hollow core is not continuous through the entire length of the sample, especially for those nanotubes of smaller (diameter) size (Figure 12.10(c) – (d)). For the SnO_2 nanotubes with large diameters, the hollow core displays a zigzag shape (Figure 12.10(a)).

Electron diffraction analysis indicate that the SnO_2 nanotubes correspond to a single crystal. Shown in Figure 12.10(c) is an SAD pattern taken for the SnO_2 nanotube depicted in Figure 12.10(b). Here, the reflections with strong intensities are indexed to belong to the $[31\bar{7}]_t$ zone axis of the rutile structured SnO_2, and the weak reflections correspond to an orthorhombic SnO_2 superstructure along the $[001]_o$ zone axis. The orientation relationship between the orthorhombic SnO_2 and the rutile SnO_2 structures is determined to be $[001]_o \parallel [3\bar{1}\bar{7}]_t$ and $\pm (110)_o \parallel (451)_t$ If only the orthorhombic SnO_2 is considered then the growth direction of the SnO_2 nanotubes is parallel to $[\bar{1}10]_o$ identical to that of the SnO_2 sandwiched nanoribbons.

Thermal Conductivities SnO_2

The metal-oxide nanobelts provide a unique system for phonon transport in low dimension materials. Thermal conductivities are observed in carbon nanotubes and semiconductor nanowires. Two opposite results are attributed to the unique crystalline structure of carbon nanotubes and an increased phonon boundary scattering rate in nanowires, while other phonon confinement effects in nanowires, are also obtained. These nanoscale thermal transport properties impact the performance and reliability of nanoelectronics, and find potential use for improving the thermoelectric figure of devices. The thermal conductivities of a 53 nm-thick, 204-nm-wide 64-nm-thick, 108-nm-wide SnO_2, nanobelt in the temperature range of 80 – 350 K are measured and it is found that the thermal conductivities of the nanobelts are strongly suppressed compared to the bulk values. So by using a full dispersion transmission function approach, it appears that the suppressed thermal conductivities can be attributed to an increased phonon boundary scattering rate although other confinement effects play a role.

A microfabricated device is used for the thermal measurement. The microdevice is a suspended membrane structure consisting of two suspended SiN_x, membranes, as shown in Fig. 12.11. A platinum (Pt) serpentine is patterned on each membrane, serving as a heater and resistance thermometer (RT). A separate Pt electrode is patterned on the edge of each membrane. The SnO_2 nanobelt is deposited on a Pt electrode pair using a wet deposition method. In this method, a collection of SnO_2 nanobelts, is dissolved in isopropanol or ethanol in an ultrasonic bath. A drop of the solution is then deposited on a diced wafer containing about twenty microdevices. The solution is subsequently spun off the wafer. This procedure yields SnO_2 nanobelts trapped on a Pt electrode pair and bridging the two suspended SiN_x membranes. The inset of Fig. 12.11 shows a 53-nm-thick SnO_2 nanobelt bridging the two membranes. The suspended segment of the nanobelt between the two membranes is 3.16 μ m long. Each of the two segments deposited on the two membranes is more than 6 μ m long.

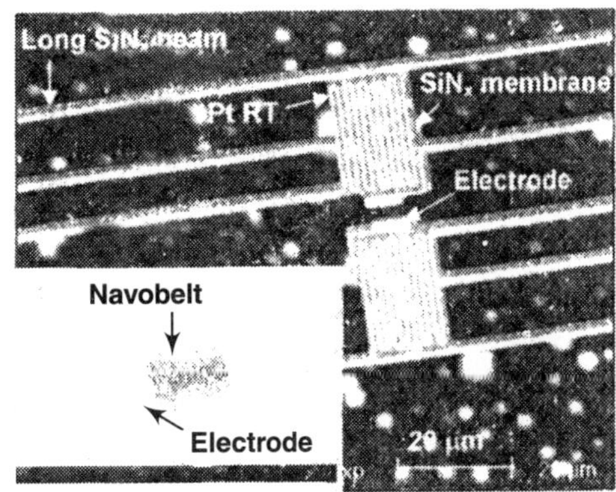

Figure 12.11 *A microfabricated device for measuring the thermal conductivity of a nanostructure. Inset: A 53-nm-thick SnO_2 nanobelt trapped on the two Pt electrodes of the device.*

During the measurement, the sample is kept in an evacuated continuous flow liquid-helium cryostat. A dc current (l) is supplied to one Pt serpentine to raise the temperature of the membrane supporting the serpentine. Part of the Joule heat generated in the heating membrane is conducted through the SnO_2 nanobelt to the other membrane (the sensing membrane). The temperature rise in the two membranes are measured using the two Pt RTs by measuring its differential electrical resistance (R_h). The thermal conductance (G_s,) of the nanobelt is obtained as

$$G_s = \frac{Q_h + Q_L}{\Delta T_h + \Delta T_s} \frac{\Delta T_s}{\Delta T_h - \Delta T_s} \tag{1}$$

where $Q_h = I^2 R_h$ is the Joule heating in the heating Pt serpentine, Q_L is the Joule heating in one of the two Pt leads supplying the decurrent *(I)* to the beating serpentine, and ΔT_h and ΔT_s are the temperature rise in the heating and sensing membranes, respectively. During the experiment ΔT_h is increased up to 2 K by ramping up the dc current (I). The small temperature excursion allows to assume that G_b and G_s are constant as ΔT_h is ramped up.

The measurement errors due to radiation and beat conduction through the residual gas in the evacuated cryostat are negligible. The major measurement error is the contact thermal conductance (G_c) between the nanobelt and the membrane, because the measurement result consists of a contribution from G_c as well as the intrinsic thermal conductance of the nanobelt (G_n), i.e.,

$$G_s = (G_n^{-1} + G_c^{-1})^{-1} \tag{2}$$

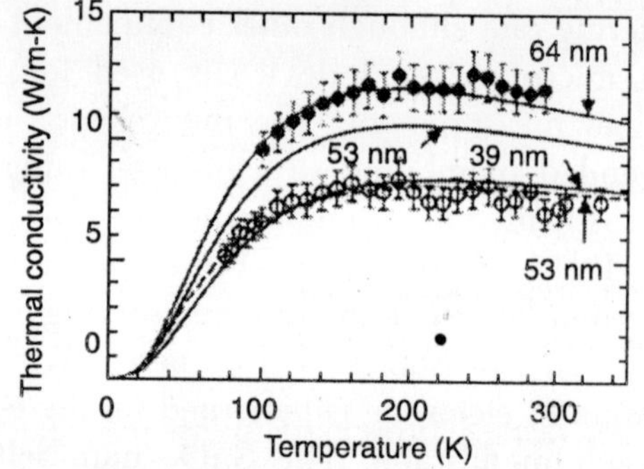

Figure 12.12 *Thermal conductivities of a 64-nm-thick (solid circles) and a 53 nm-thick (open circles) SnO_2 nanobelt as a function of temperature. Also shown are calculation results (lines) with the bulk parameters for the Umklapp process and diiferent FL values indicated for each line. An impurity scattering rate ten times of the hulk value is used for the dashed line; while the bulk value is used for the three solid lines.*

Assume $G_c \approx 2\pi(\kappa_n^{-1} + \kappa_m^{-1})^{-1}$ a, where κ_n and κ_m are the thermal conductivity of the nanobelt and membrane, respectively, and *a* is the contact spot size between the nanobelt and the membrane. For a very small contact spot a ~10 nm it is estimated that G_c is at least one order of magnitude larger than the measured G_s of the two nanobelts. Since the two nanobelts are more than 100 nm wide and each, of their two segments deposited on the two membranes is more than 3 μ m long, the contact spot size is much larger than 10 nm. To find $G_c >> G_s$, a 232-nm-thick and a 178 nm-thick nanobelt is measured for which G_n is closer to G_c than for the two thinner samples. The G_s of the two thick nanobelts are more than ten times larger than those of the two thinner samples. As all four samples have a similar contact area with the membranes, G_c does not vary significantly from the thick to the thin samples. Hence, as G_c is always larger than G_s for the thick nanobelts, G_c is at least ten times larger than the G_s of the thin ones. Therefore, for the two thin samples, $G_c >> G_s$ and $G_s = G_n$.

The thickness (*t*) of the nanobelt is measured using a tapping mode atomic force microscope to map the segment of the nanobelt laid on one membrane; while the length (L) and width *(w)* of the nanobelt are obtained using a high resolution scanning electron microscope. The thermal conductivities of the nanobelts are calculated according to $\kappa_n = G_s L/wt$, and plotted in Fig. 12.12. The thermal conductivities are found to be substantially lower than the bulk values.

For the origin of the reduced thermal conductivities, a full dispersion transmission function approach is used for thermal calculation. In calculation, the nanobelt dispersion relations are obtained for the rutile structure. Matthiessen's rule is used to obtain the frequency-dependent relaxation time of phonons as $\tau^{-1} = \tau_U^{-1} + \tau_b^{-1} + \tau_i^{-1}$. Here, $\tau_U^{-1} + \tau_b^{-1}$ and τ_u^{-1} are the Umklapp, boundary, and impurity scattering rates, respectively. A phenomenological expression for the Umklapp scattering rate is used: $\tau_u^{-1} = Be^{-b/T}.\omega^2T$, where B and b are two fitting parameters. The boundary and impurity scattering rates and written as $\tau_b^{-1} = v/FL$ and $\tau_i^{-1} = A\omega^4$. Here, L is the characteristic length of the system (thickness for the nanobelt samples), F is a parameter representing specularity of phonon refection at the boundaries, the FL product is referred as the effective thickness of the sample, and A is a parameter arising from Rayleigh scattering of phonons by atomic scale impurities.

In determining the parameters for $\tau(\omega)$, measurements of the bulk SnO_2 thermal conductivity in the (100) or $C_\perp$. direction and the (001) or $C_\parallel$ direction are used. Because bulk measurement data for the (101) direction is not found so, the bulk values for this direction are estimated by considering the (100) and (001) values as the components of the diagonal conductivity tensor.

After obtaining the parameters from the bulk measurement data, the thermal conductivities of nanowires are calculated by varying the effective thickness FL, while keeping other parameters same as the bulk values. The solid lines of Fig. 12.12 are three sets of calculated thermal conductivities in the (101) direction as a function of temperatures. The measurement data of the 64-nm thick and 53-nm. thick SnO_2 nanobelt is observed which agrees well with the calculation results, with FL = 64 nm and 39 nm, respectively. This shows that an increased phonon-boundary rate significantly suppresses thermal conductivities of the nanobelts (see note below) although other effects cannot be ruled out. For example, the measurement results of the 53 nm sample can be fitted with the bulk Umklapp parameters, a FL value of 53 nm, and an impurity scattering rate ten times higher than the bulk value, as shown as the dashed line in Fig. 12.12. The difference in surface specularity or impurity scattering rate between the two samples is due to different surface roughness or impurity concentrations.

Note: For an isotropic material a rectangular cross section an the F value in the range of 1.4 – 2 is expected for totally diffusive boundary scattering. Nevertheless, the fact that the rutile structure of SnO_2 is very anisotropic with higher speeds of sound in the directions perpendicular to the growth axis, results in lower values of F close to 1 in the present case.

Mechanical behavior-and magnetic separation of SnO_2 nanostructures

The as-synthesized nanowires and nanobelts usually have a large size distribution. A ball milling technique for narrowing the size distribution of oxide nanobelts and nanowires is described below. The mechanical behavior of SnO_2 nanobelts before and after low-energy all milling is also compared.

Experiment

The as-received growth product is milled for 6 h at room temperature in a mixer. The milling system contains several 3 mm-diameter stainless steel balls, and the mixture ratio of the steel balls and growth product, is kept of ~ 50:1 in weight percent. Hexane is used as surfactant to get uniform milling effect. Subsequent magnetic separation is employed using a magnet while the nanomaterial is ultrasonically vibrated in the hexane medium to achieve a better separation effect.

The as-deposited and milled samples are then characterized by scanning electron microscopy (SEM) and transmission electron microscopy (TEM) with energy-dispersive x-ray spectroscopy (EDS) at 200 kV and a high-resolution TEM, (HRTEM) at 400 kV.

Results

After evaporation and deposition, a large amount of long nanobelts is observed, Figures 12.13 (a) and (b) are SEM images displaying the morphologies of the nanobelts before and after ball milling, respectively. EDS analysis and electron diffraction show that the nanobelts are SnO_2. The as-synthesized SnO_2 nanobelts have a large size distribution with the width w varying from several tens of nanometers to several hundred nanometers and lengths of dozens of micrometers. All the nanobelts with $w > 150$ nm are straight. The irregular surface configuration originates from the imperfect grain growth during the deposition process. It is observed that nanobelts with $w < 100$ nm have curly morphology thus they demonstrate high mechanical flexibility.

After ball milling, most of the large size nanobelts are fractured into small particles, but the small size nanobelts survive. This is due to the mechanical alloying process. The thin nanobelts bent or twist easily to various degrees without rupture, while large nanobelts either fracture or are grounded into agglomerated nanoparticles. No contamination of Fe from milling vessel and / or balls is found in the milled nanobelts though the nanoparticles contain a significant amount of Fe as revealed by EDS patterns.

The graphs in Fig. 12.14 shows the size distributions of the nanobelt width before and after ball milling. The dominant size distribution of the original nanobelts is wide in a range from 50

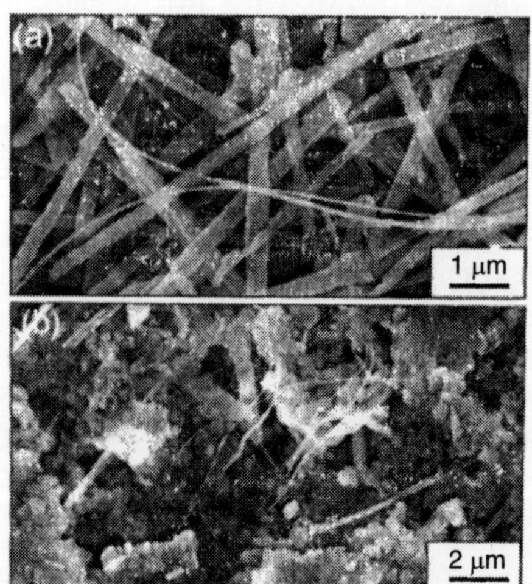

Figure 12.13 *SEM images of the SnO_2 nanobelts before (a) and after (b) ball milling process.*

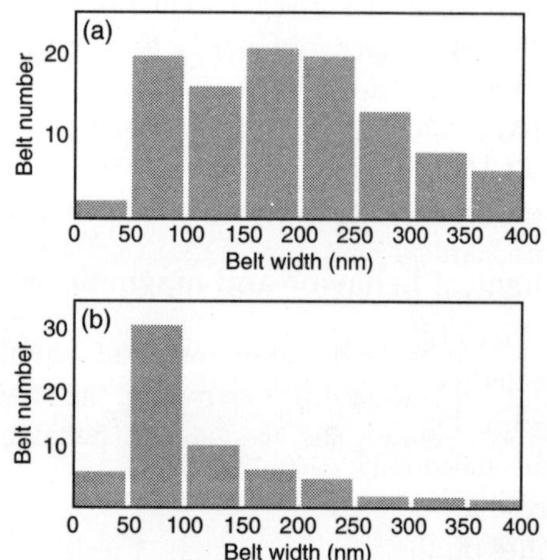

Figure 12.14 *Plots of the nanobelt size distributions before (3) and after (b) ball miliing process.*

to 300 nm, however a much narrow size distribution with peak between 50 and 100 nm is evident after ball milling. The milling process effectively narrows the size distribution of the nanobelts and preserves the small size nanobelts. The grinding and collision of balls result in the fracture of large-size nanobelts and the increase of volume fraction of nanoparticles.

In order to separate those nanoparticles which are embedded with magnetic iron from the nanobelts, a permanent magnet is used for the magnetic separation. The separation principle is based on the balance among the magnetic force, buoyant force and gravitational force, as;

$$F(B) = \frac{\chi}{\mu_0} B \frac{dB}{dz} + \rho_\ell g - \rho_n g,$$

where χ is the magnetic susceptibility of milled materials, μ_o is the vacuum permeability, B is the magnetic field strength, z depicts the field direction, g is the gravity acceleration, ρ_ℓ and ρ_n are the density of liquid and nanomaterials, respectively. When $F(B) > 0$, the materials are attracted to the magnet pole, and consequently, the materials move downward due to the gravity, thus leading to the materials separation. Because ρ_ℓ and ρ_n remain same so assume that the force $F(B)$ is predominately associated with the spatial distribution of magnetic field and the measurement of volume susceptibility χ- The χ value is closely associated with the content of impurity of Fe in the nanomaterials since tin oxide has an extremely low susceptibility compared with Fe. The materials , with high component of Fe are subjected to the external magnetic field. Figure 12.15 shows the SEM images and EDS patterns of milled samples subjected to a magnetic field (around 0.3 – 0.5 T). The nanomaterials with different morphology differ in how they interact with a magnetic field. The particles are very sensitive to the external field as shown in Fig. 12.15(a), verifying that soft magnetic Fe has embedded in these agglomerates due to the mechanical milling, which is consistent with the composition analysis by EDS [Fig. 12.15(b)]. The large-size nanobelts are also subjected to the attraction of magnetic field as shown in Fig. 12.15(c), while the fraction of Fe on the nanobelts as shown in Fig. 12.15(d) is lower than those of particles. The Fe mostly attaches on the surface of nanobelts. As evident from Fig. 12.15(e) these unstrained nanobelts of small size are free from Fe contamination (as revealed in Fig. 12.15(f)) and not subjected to the attraction of magnetic field, verifying

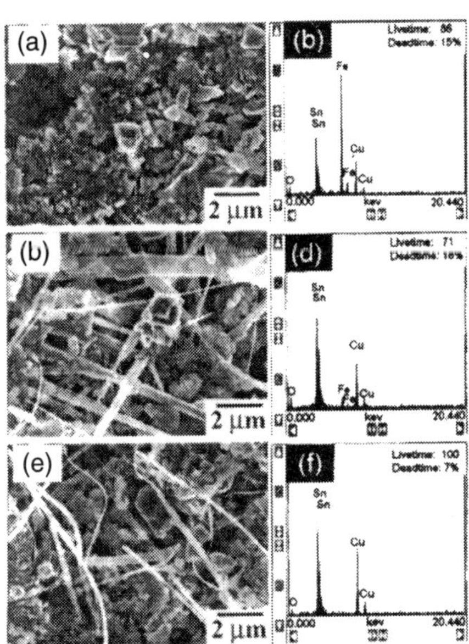

Figure 12.15 *SEM images and EDS pattems of the milled samples subjected to the magnetic separation. (a) Magnetic attracted particles. (c) Nanobelts with magnetic particles attached on the surface. (e) Nanobelts not subjected to the magnetic attractive force. (b), (d), (f) are the correlated EDS pattems to (a) particles, large-size nanobelts, and (e) smail-size nanobelts.*

the better anticontamination ability for these small nanobelts during the milling process. However the magnetic separation still has one difficulty. It is impossible to completely eliminate magnetic impurities due to the existing agglomeration under milling. In order to characterize the structures of the nanobelts before and after ball milling, TEM images and selected-area electron diffraction (SAED) patterns are recorded from the samples. Low magnification TEM images recorded from large and small belts before ball milling are shown in Fig 12.16. The SAED patterns of the large belt in Fig. 12.16 (a) and the, thin belt in Fig. 16 (b) are shown in Figure 12.16(d) and 12.16(e) respectively. The diffraction spots for the large-scale nanobelts are indexed using the SnO_2 tetragonal structure. The extra weak diffraction spots in Fig. 12.16 (d) are from the SnO_2 orthorhombic structure. The large nanobelts are composed of orthorhombic and tetragonal phases distributed as domains. Combining the images and the SAED patterns, the main growth front planes of the nanobelts are $(201)_{Tetragonal}$ and $(110)_{orthorhombic}$. The structural hybrid nanobelts are the dominant products before ball milling. However, the superlattice structure does not exist in the thin nanobelts which show a much more uniform and perfect tetragonal structure with $(201)_{Tetragonal}$ as the growth plane front [Fig. 12.16(e)] These single crystalline thin nanobelts are dozens of micrometers in length. The contrast observed is due to the bending contour.

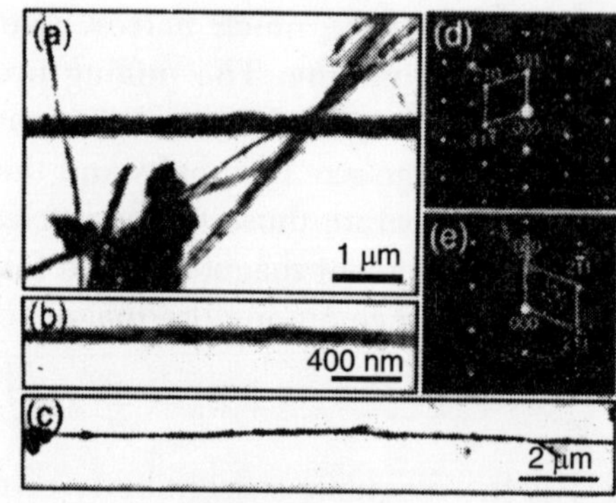

Figure 12.16 *A Typical SnO_2 belts of different sizes before ball milling. The SAED patterns in (d) and (e) coming from the large belts in (a) and thin belt in (b), respectively. The superlattices in (d) indicate the existence of orthorhombic structtlred SnO_2.*

After ball milling, two typical nanobelts with width of 80 and 50 nm are shown in Fig. 12.17(a). The inset SAED pattern from the large nanobelt indicates that the nanobelt retains its single crystalline tetragonal structure with the growth front plane of $(201)_{tetragonal}$, which is the same as that of the belt in Fig. 12.16(b). Its uniform contrast indicates that the belt is dislocation free. The thickness of the thin nanobelts is still quite uniform even after ball milling.

Figure 12.17 *(a) Nanobelt survived after the ball milling process. The SAED pattern (inset) shows the growth front plane is (201). Its uniform contrast in the enlarged image (b) indicates it is dislocation-free.*

The percentage of the orthorhombic structured nanobelts decreases after ball milling. The existence of the orthorhombic phase usually results in wide nanobelts due to the coexistence of the orthorhombic and tetragonal phases. The wider nanobelts are easier to be fractured, thus, they vanish the belt shape after ball milling, demonstrating that ball milling is useful in purifying the phase of the nanobelts.

Besides the nanobelts described above, many belts take the $(2\bar{2}\bar{1})$ plane as their growth front before and after the ball milling process. Figure 12.18(a) shows a nanobelt that survived ball milling. The inset is an enlarged image of the belt. The contrast in the image is attributed to the strain induced bending effect. Figures 12.18(b) and (c) the SAED pattern and HRTEM image of the belt in Fig. 12.18(a). The HRTEM image of the nanobelt displays a dislocation free volume in the milled samples. Although several types of nanobelts are observed it seems that the mechanical behavior of the SnO_2 nanobelts is sensitive to their size rather than their growth directions.

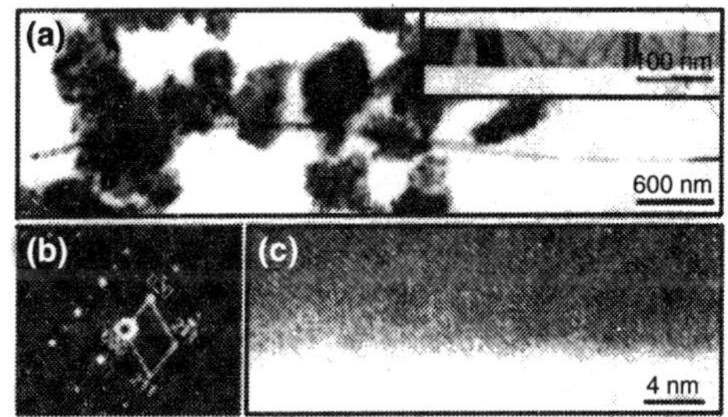

Figure 12.18 *Low magnification (a) and enlarged (inset) TEM images and the corresponding SAED pattern (b) from a nanobelt after ball milling, showing the growth front plane of $(2\bar{2}\bar{1})$, (c) is a HRTEM image of the belt.*

Characterization of Individual SnO_2 Nanobelts with STM

The as-grown NBs, deposited on a Si wafer, are characterized with SEM (see Fig. 12.19). They have a rectangular cross section and are uniform in width and thickness. The typical widths/thickness of the NBs (determined from the STM line profile analysis) are in the range of 100-700 nm/80-300 nm, although there is some variation between batches. The NBs are tens to hundreds of microns long.

X-ray photoelectron Spectra (XPS) are acquired at normal emission using Mg Kα (radiation and it shows carbon contamination. On macroscopic single crystals such carbon deposits are usually removed by Ar^+ sputtering; a subsequent anneal to 1020K is necessary to render large and flat terraces in STM. Attempts to use the same cleaning recipe for the NB's results in structural damage of e NBs. The thermal stability of the NBs in different environments by SEM is shown in Fig. 12.19 NBs deposited on the Si substrate are heated to different temperatures (in increments of 20-30 K) in air, UHV, and an O_2 atmosphere (~10^4 Pa in the load lock of the UHV chamber). Decomposition of the NBs material in UHV and in air is observed at 1100K and 1370K respectively. The NBs heated in O_2 to 1070 K do not decompose. The SEM images in Figs. 12.19(a) and (b) show the features characteristic of the onset of thermal decomposition.

In order to avoid damaging the samples and altering their surface morphology from the as-grown state, sample cleaning for the STM experiments is performed by exposing the dry-deposited NBs to oxygen plasma for 20 min at 10^{-3} Pa in the UHV chamber. This results in a significant reduction of the carbon deposits below the detection limit of XPS. Stable tunneling in STM is only possible after the plasma treated the NBs are briefly heated to 620K in UHV.

STM measurements are performed in the constant-current mode at room temperature with the sample at positive bias voltages in the range of ~0.5 – 5 V. One challenge in obtaining atomically-resolved images of the relatively high, large nanostructures is to determine the correct loop gain and Z-input gain, i.e. the signal amplification value of the z-piezo of STM. It is crucial to set up these parameters properly in order to accommodate rapid changes in the vertical position of the STM tip.

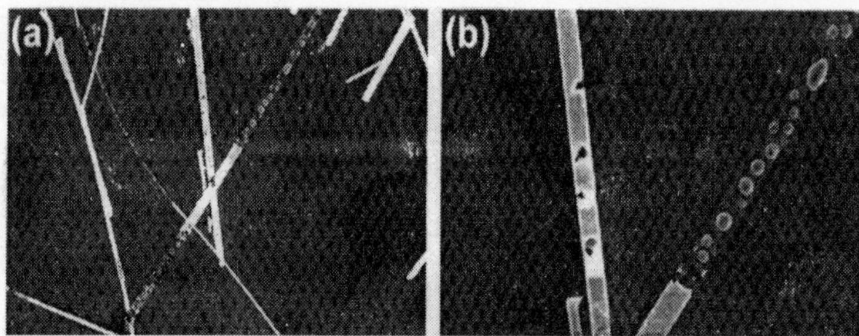

Figure 12.19 *SEM images of SnO_2 dry deposited on a Si wafer and annealed to 1100K in UHV .(a) Fragments of partially decomposed nanowires, (b) initial decomposition process (arrows) and further decomposition of SnO_2 nanobelts.*

Parts of three NBs dry-deposited on a slightly-reduced $TiO_2(011)$ substrate, are shown in Fig. 12.20(a). There is no apparent registry between the orientation of the NBs and the atomic-scale structure of the $TiO_2(011)$ An STM image obtained on the surface of an individual NB is shown in Fig. 12.20(b). Previous HRTEM results show that SnO_2 NB exhibit a rectangular cross section terminated by (101)/(100) facets on the wide/narrow sides. The image in Fig. 12.20(b) are taken on the wide side of a NB, and the morphology is similar to one obtained on a macroscopic $SnO_2(101)$ single crystal. Large, flat terraces are bordered by straight step edges oriented along the [101] direction and short, jagged steps that are mainly aligned along [010] (Fig. 12.20(b)). The terrace size appears slightly larger than that for the reduced surface of a macroscopic $SnO_2(101)$ single crystal. Bright features observed on the terraces are due to contaminants that still remain on the surface after the plasma cleaning.

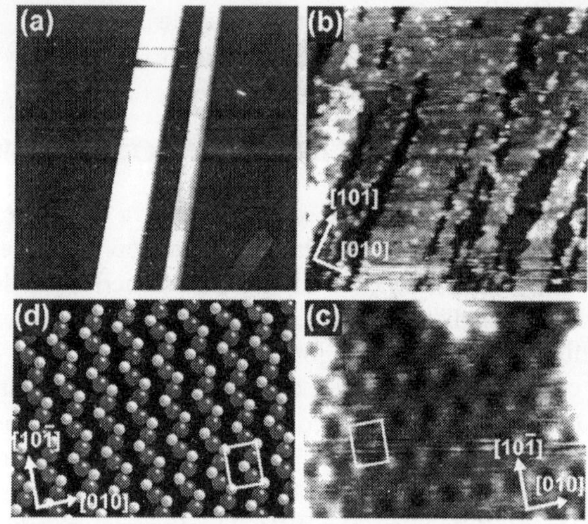

Figure 12.20 *Constant current STM images of SnO_2 nanobelts (NBs). (a) Fragments of three NBs dry-deposited on a slightly-reduced $TiO_2(011)$ substrate (3×3 μ m^2, $V_{sample} = +4.5$ V,$I_{tunnel} = 0.2$ nA). (b) Surface of the (101) side of a NB prepared by oxygen plasma treatment (80×80 nm^2, +2.0 V, 4.3 nA). (c) Atomically resolved STM image (4.6×3.3 nm^2, +0.8 V, 3.3 nA) of the reduced SnO_2 (101) surface. (d) Model of the reduced $SnO_2(101)$ surface light grey balls are tin atoms.*

An atomically-resolved STM image (Fig. 12.20(c) of a NB (large facet) reveals the 1×1 structure of $SnO_2(101)$. The $SnO_2(101)$ surface (and the $SnO_2(100)$ surface, present on the narrower sides of the NB's) is prepared reproducibly in either fully-oxidized or reduced form. In both cases, the surface maintains the (1×1) unit cell, as indicated in the STM image (Fig. 12.20(c)). Fully-oxidized surfaces exhibit a bulk-termination (with some relaxations), and contain two-fold coordinated O atoms and. 5-fold coordinated Sn atoms. Upon heating to 570K in UHV, all two-fold coordinated oxygen atoms desorb from the surface, leaving behind 3-fold Sn atoms with a (formal) oxidation state that is reduced from 4^+ (as in the bulk) to 2^+. Thus, annealing of the NBs to 620 K followed after plasma treatment results in the reduced surface formation. Such a reduced surface has a stoichiometry of SnO, and is characterized by a pronounced, $5s^2$ – derived surface state located at the upper edge of the valence band. The reversible surface oxidation/reduction is accompanied by large changes in work functions (on the order of ~ 1 eV), which influences the adsorption of water and benzene. A top-view (ball model) of the reduced SnO_2 (101) surface (unit cell 3.2 A × 4.7 A), is shown in Fig. 12.20(c). Under empty-states tunneling conditions, cations are often imaged on metal oxide surfaces, because these have a higher density of empty states than the surface oxygen atoms. Sn surface atoms are imaged on top of the NB which is consistent with STM obtained for the reduced SnO_2(101) surface. Morphologically, the top surface of the SnO_2 nanobelts closely resembles the reduced SnO_2(101) surface of a macroscopic single crystals.

Photocurrent Enhancement of SnO_2 Nanowires

In order to improve the sensitivity of gas sensors, the catalytically active metals, such as Pd and Ag, on semiconductor NWs are incorporated. The metal nanoparticles on the surface results in the formation of a localized Schottky junction, which creates a charge depletion region in the NWs in the vicinity of the metal particles. The metal-particles then become an intermediate in chemical combustion reactions which reduce the current through the NWs and than enhance the sensitivity. The photoresponse of SnO_2 NWs with Au-nanoparticles is considered below.

Experiment

The growth of SnO_2 NWs is based on vapor-liquid-solid (VLS) method. In the synthesizing process, the Au layer with 10 nm in thickness is deposited on M-plane (100) sapphire (Al_2O_3) to serve as the catalyst. Then Sn powder is placed on the ceramic boat and put into a furnace with Argon flow as gas carriers at a flow rate of 200 sccm. The temperature is increased from room temperature up to 1270K by a rate of about 100 min^{-1}. After the vaporized Sn is mixed with oxygen in air and turns into SnO_2, SnO_2 is then blown onto the Au layer. When SnO_2 dissolves into Au, the supersaturation of the alloy SnO_2 -Au droplet results in the nucleation, and SnO_2 NWs are fabricated. The as-grown SnO_2 NWs have diameters of 100 ~ 300 nm and length above 40 μm. After the fabrication, the electric contacts of SnO_2 NWs are made by shadow-mask technique. Sapphire is chosen to be the nonconducting substrate, and metallic Ni (15 nm)/Au (150 nm) electrodes are vapor deposited. A post treatment of annealing at 770 K for 60 s, is done to achieve the ohmic contact.

A scanning electron microscopy (SEM) image of the sample is shown in the inset of Fig. 12.21. The distance of the channel between two terminals is 10 μ m, and the diameters of the three NWs are respectively 137, 295, and 479 nm. The dark current vs bias (*I-V*) measurement is checked to ensure ohmic contact between nanowires and metal electrodes as shown in Fig. 12.21. The x-ray diffraction (XRD) shown in the inset of Fig. 12.21 identifies the structure of

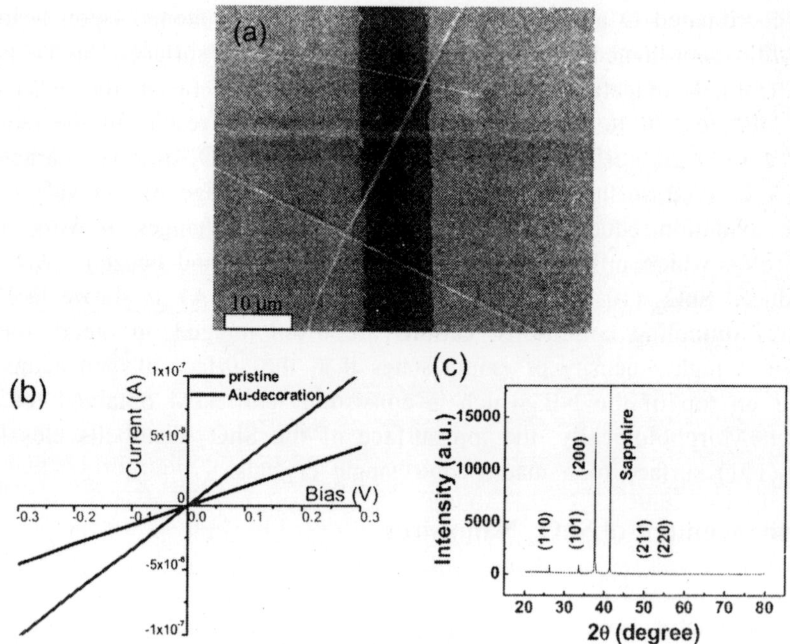

Figure 12.21 *I-V characteristics of* SnO_2, *nanowires with and without Au-nanoparticles in ambient air.*

synthesized products. Most of the peaks are indexed are according to the tetragonal rutile structure of SnO_2.

For the photocurrent (PC) measurement, a He-Cd laser working at 325 nm is used as the excitation light source whose energy is larger than the band gap energy of SnO_2 (~345 nm). This generates free electron hole pairs. A measurement system is used to supply the dc voltage (0.1 V) and to record the photocurrent. Au-nanoparticles are sputtered on the NWs in low-pressure ambient environment (below 5 Pa). By applying a high voltage gas ions are generated which bombard Au target, and generate Au nanoparticles from the target to be ejected with enough energy to travel to sample.

Now as shown in Fig. 12.22.(a), the diameter of the Au-nanoparticles is about 10 nm, and the random distribution of the Au-nanoparticles does not cover the whole surface of the NWs. Thus the decorated Au-nanoparticles do not screen the incident light, an are not able to produce a new conductive path on the surface and the substrate. As shown in Fig. 12.21, the *I-V* curves of the NWs show ohmic characteristic. Figure 12.22(b) shows the results of photoresponse under different excitation intensity performed on SnO_2 NWs in ambient air. It is clear that the PC increases with increasing light intensity from 2.5 Wm^{-2} to 774 Wm^{-2}. When the NWs are decorated with Au-nanoparticles the measured PC is enhanced by up to about 110%. The underlying mechanism that causes the PC enhancement after Au-nanoparticles decoration is given below.

The photoresponse is quantitatively described by current gain (Γ), a factor to determine the efficiency of electron transport and carrier collection during illuminating process. The current gain (Γ) indicates the number of electrons induced by photon, which be obtained by.

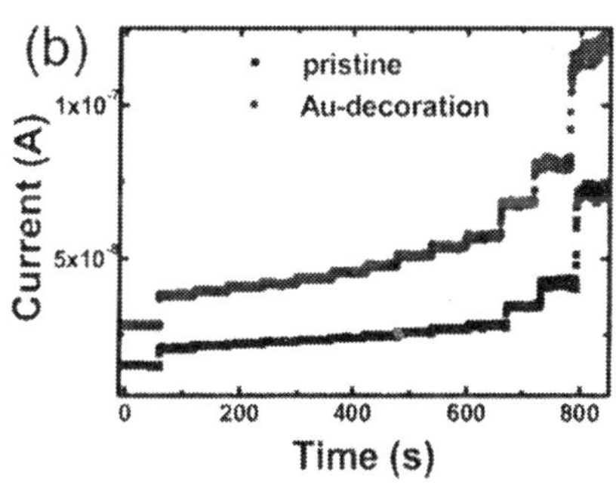

Figure 12.22 *(a) Scanning electron microscope (SEM) image of the Au-decorated SnO_2 nanowires. (b) Photoresponse of SnO_2 nanowires by UV illumination under different excitation intensity. The corresponding intensity of each step is 3.2, 4.2, 6.5, 9.6, 13.1, 18.9, 25, 33.1. 56.9. 96.7, 150,224, 774 Wm^{-2}, respectively.*

$$\Gamma = \frac{\Delta i/q}{P/hv} \times \frac{1}{\eta}, \tag{1}$$

where Δi is the current difference between photocurrent and dark current, q the electron charge, hv the photon energy of incident light of 3.81 eV. P is the power of incident photon that has been absorbed by the NWs, which means $P = I \times l \times d$. I is the excitation light intensity on the sample, I and d are respectively the length and the width of the NWs. η is the quantum efficiency which is set to be 1 for simplicity. For the SnO_2 NWs the measured PC has an extremely high PC gain with a value up to about 900.

The high current gain in SnO_2 is attributed to the presence of oxygen vacancies. Due to the existence of oxygen vacancies on the surface of SnO_2 NWs, free electrons are accumulated on the surface. Consequently, the existence of upward band-bending forms a low-conductivity depletion region near the surface referred to space charge regions (SCRs). After the electron-hole pairs are photogenerated, the photoinduced holes migrate to the surface by the built-in electric field. As a result, a conductive volume increment is produced. The spatial separation of electrons and holes also reduces the electron-hole recombination rate, and therefore, the electron lifetime increases and the photoresponse is enhanced. The PC induced by the modulation of surface SCRs causes Γ following an inverse power law with excitation intensity, i.e., $\Gamma \propto \Gamma^K$, and the exponent K is between 0.5 and 0.9. As shown in Fig. 12.23, the gain logarithmic plot in the intensity ranging from 2.5 Wm^{-2} to 774 Wm^{-2} shows a well defined power law with K~ 0.601± 0.009 which is in good agreement with the theoretical prediction.

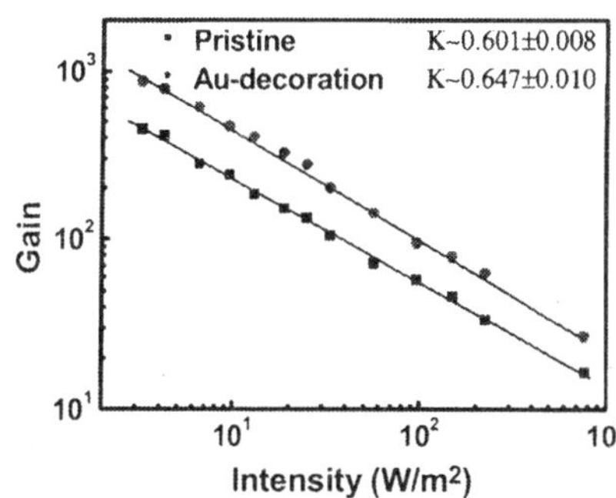

Figure 12.23 *Power dependence of pristine and Au-decorated SnO_2 nanowires.*

Consider now the origin of the PC enhancement after the Au decoration. It is known that depositing the metallic particles on the surface of a semiconductor, such as SnO_2 results in a localized Schottky barrier in the vicinity of the metallic particles. The formation of the Schottky barrier on the. surface enhances the surface electric field and increase the width and height of space charge region as shown in Fig. 12.24.

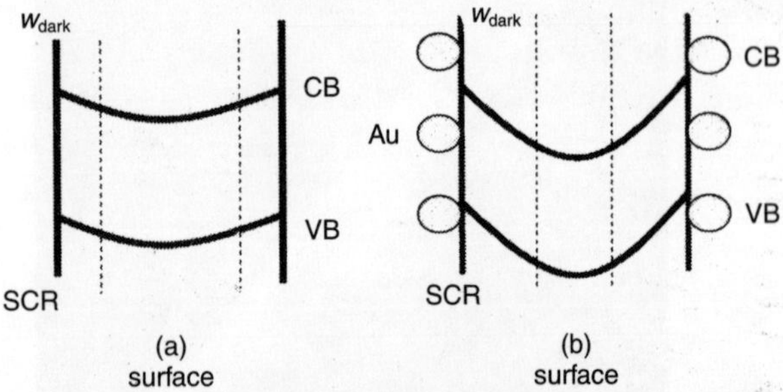

Figure 12.24 *(a) Accumulation of free electrons and upward band-bending on the surface. (b). Aunanopartieles result in a localized Schottky barrier in the vicinity of Au-nanoparticles.*

This is due to the fact that the work function of SnO_2 NWs has a value of 4.7 eV, which is smaller than that of Au cluster with 5.1 eV. The increase in the barrier height of SCRs on the surface will enhance the spatial separation of photoexcited electrons and holes, and the electron lifetime is enhanced. In turn, the measured PC is enlarged. In order to provide a further evidence to support the above proposed mechanism, the dependence of the photoresponse on excitation intensity for NWs without and with Au-nanoparticles decoration is compared. As shown in Fig. 12.23, the exponent K of inverse power law changes from 0.601± 0.009 to 0.647 ± 0.010 after the decoration of SnO_2 NWs.

This behavior is understood according to the following equation. SCRs inside a semiconductor produce a variation of the conductive volume when carriers are photogenerated, and the Δi is expressed as

$$\frac{i_{dark}}{d - w_{dark}} \times \left\{ \left[\frac{2\varepsilon\Delta\Psi_o}{qN_d} \right]^{1/2} - \left[\frac{2e(\Delta\Psi_o - \Psi_{ph})^{1/2}}{qN_d} \right] \right\}, \tag{2}$$

Where,

$$V_{ph} = V_T \ln\left[1 + e^{\Delta\Psi_o/V_{VT}} \left(q\eta \frac{P}{h\nu A^* T^2}\right)\right], w_{dark} \approx \left(\frac{2\varepsilon\Delta\Psi_o}{qN_d}\right)^{1/2}. \tag{3}$$

ε is the permittivity, and $\Delta\Psi_o$ is the barrier height. N_d is the dropping level, V_T= kT/q $\mathbf{A}^*$ is Richardson constant =1.2 × 10^6 A/m^{2}K^2, Figure 12.25 shows the simulated result according to equation (2), which indicate that both of the gain Γ and the exponent K increase with increasing barrier height. This result is again consistent with the proposed Schottky junction model as described above

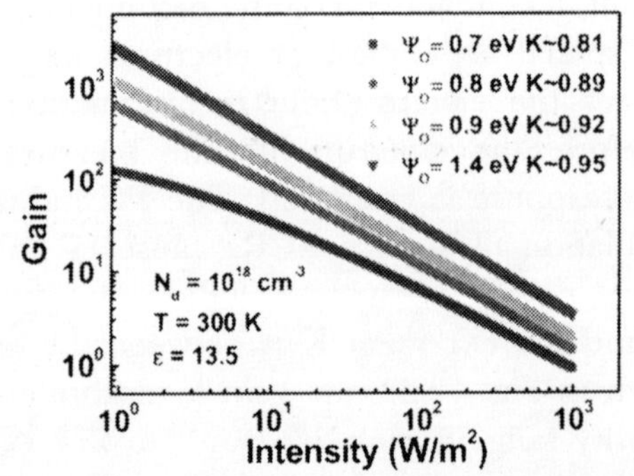

Figure 12.25 *Computer simulation of gain versus intensity and barrier height.*

However, Fig. 12.21(b) shows that the current of Au-decorated SnO_2 NWs before illumination is higher than that of pristine SnO_2 NWs. The dark current does not decrease by the expansion of SCRs after the Au-decoration. This behavior is understood in terms of the fact that a large number of electrons in Au-nanoparticles have been transferred into the conduction channel of SnO_2 NWs. In addition, because of the enhanced band-bending, the conduction carriers are more concentrated on the center of the wires, which reduces the surface scattering and therefore the dark current conductivity is enhanced:

13

α-Fe_2O_3 Nanowires and Nanobelts

α-Fe_2O_3 is an n-type semiconductor with a small bandgap of 2.1 eV. The crystal structure of α-Fe_2O_3 is identical to that of corundum (space group=R3c) and is described is a hexagonal close-packed system with oxygen ions stacked along the [001] direction i.e., Planes of anions are parallel to the (001) planes (it can also be an index in a rhombohedral system), with lattice constants of α-Fe_2O_3 of $a_o = 0.5034$ nm and $C_o = 1.375$ nm. Systematic investigation as done after experimental process for fabrication of α-Fe_2O_3 (see experiments I and II below.)

NWs of α-Fe_2O_3 are synthesized using solid-phase reactions by oxidizing bulk iron in a mixed gas ambient containing CO_2, O_2, N_2, SO_2, H_2O, etc. in temperature ranges of 870 – 970 K. Fe_3O_4 on the other hand, has an inverse spinel structure with face-centered cubic unit cell of 32 O^{2-} ions with a lattice constant of 0.839 nm, in which one third of the iron ions occupy the tetrahedral sites (all Fe^{3+}), While two thirds of the iron ions occupy both the octahedral, and tetrahedral sites (one half is Fe^{2+} and the other half is $=Fe^{3+}$). The synthesis of Fe_3O_4 nanomaterials occurs via a chemical route, based on the hydrothermal process. The synthesis of Fe_3O_4 NWs using a vapor-phase method is rare.

For the other oxide nanowires, such as ZnO, MgO, and In_2O_3 the growth phenomena is investigated using various growth parameter.

The experimental conditions, such as type of substrate, local growth and geometrical environment, gas-flow rate and growth temperature, under which high-density α- Fe_2O_3 NW arrays can be grown by a solid-phase route via tip-growth mechanism are examined.

An n-p switch is demonstrated in an a-Fe_2O_3 thin-film sample after annealing with oxidation and reduction processes, for which the main principle is based on the surface adsorption of oxygen to increase band bending near the surface. It is known that the surface of a nanowire (NW) is very unstable, due to the large surface-to-volume ratio, and it easily adsorbs foreign molecules for stabilization. By using NWs with unstable surface states, the n-p transition through surface adsorption is easily controlled and achieved with thick films.

The electronic properties of α -Fe_2O_3 NWs (with oxygen vacancy orders) are described in which the p-type is found in as-grown NWs without any additional annealing. After a process of annealing in a reductive ambient, the p-type to n-type transition is observed. The detailed electronic structure of the α-Fe_2O_3 NWs before and after annealing is measured by electron energy loss spectrometry (EELS). The finding of a p-n transition suggests the potential application of the NWs in future nanodevices. Growth of aligned and uniform α-Fe_2O_3 nanowire (NW) arrays is achieved by a vapor-solid process. The experimental conditions, such as type of substrate, local growth and geometrical environment, gas-flow rate, and growth tempoerature, under which the high density α-Fe_2O_3 NW arrays are grown by a vapor-solid method via the tip-growth mechanism as in experiment I and II below.

I. Vapor-solid Process

Four types of iron-based substrates are used, including Fe-Ni alloy plates with a composition of 1:1 and 64:36 (atomic percentage), bulk iron (purity > 99.995 %), and iron film deposited on a Si wafer using a non-ultrahigh-vacuum electron-beam system with a thickness of 1 μ m. All of the substrates are cut into small pieces of 1.5 cm^2 and ultrasonically cleaned using alcohol (99.95 %). The furnace used is a three-zone tube furnace, including a vacuum pumping system an alumina tube, a quartz tube, and a gas-flow system. The vacuum level for this configuration is 4Pa and the maximum output temperature is 1770K. A diagram of the experiment is shown in Figure 13.1. The quartz tube is set into the alumina tube to avoid contamination of the external alumina tube during synthesis of the NWs. The system is pumped to a base pressure of 4Pa and then maintained at the pressure of 5×10^4 to 10×10^4 Pa by introducing Ar carrier gas (99.995 %) at a flow rate of 70-150 sccm, which, is controlled by the maximum flow controlled troller (MFC) All the sample are tilted by 90° on an Al_2O_3 stage and put into the lower-temperature region of 620K to 970K for 10 hours. 5 g of α-Fe_2O_3 powder is placed in the high-temperature region of 1470K. A constant rate of 5 $Kmin^{-1}$ is used to raise the temperature to a desired value and the system is allowed to cool to room temperature.

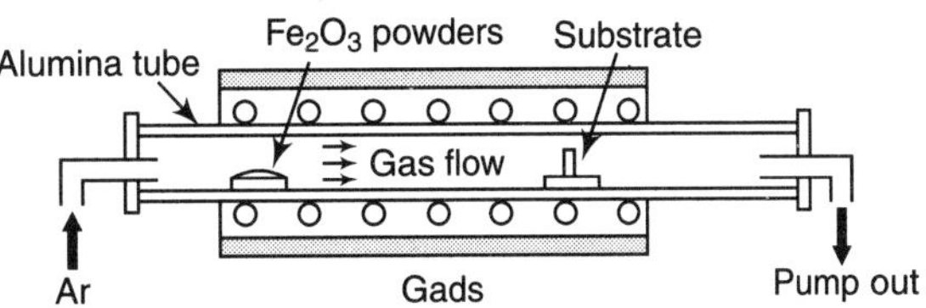

Figure 13.1 *The configuration of furnace for the synthesis of Fe_2O_3 NWs.*

During the conversion of the nanowires, the reduction procedure is carried out in a mixed reductive ambient of H_2(5 %) + Ar(95 %) using the as-prepared a-Fe_2O_3 NWs annealed at a temperature of 670 - 870K for 20 min in the furnace.

The surface morphology is examined by a field-emission scanning electron microscope operated at 15 kV.

All samples are sonicated in ethanol and then dispersed on a copper grid supported by a whole carbon film. A field-emission transmission electron microscope operated at 300 kV with a point-to-point resolution of 0.17 nm is equipped with an energy dispersion spectrometer (EDS) and an electron energy loss spectrometer which are used to characterize the microstructures and chemical compositions. The magnetic properties are characterized by vibrating sample magnetometry. The electron field-emission behaviour is measured in a vacuum of 133×10^{-7} Pa using a spherical stainless steel probe (1 mm in diameter) as the anode. The lowest emission current is recorded on the level of nanoamps. The measurement distance between the anode and the emitting surface is fixed at 100 μm.

II. Thermal Oxidation Method

Thermal oxidation method is used to synthesize the α -Fe_2O_3 nanobelt and nanowires arrays. The experimental setup consists of a horizontal tube furnace of length 120 cm and diameter 10 cm, a quartz tube 100 cm in length and 5 cm in diameter, and a gas flow/control system. Fresh iron foils (10 mm× 5 mm × 0.25 mm) with a purity of 99.9 % are used as both a reagent and a substrate for the growth of α-Fe_2O_3 nanowires. The iron foil are cleaned with absolute ethanol in an ultrasound bath before being loaded into a quartz boat, which is positioned at the end of the quartz tube. The quartz tube is then mounted in the middle of the tube furnace. A flow of high-purity nitrogen (>99.995%) is introduced into the quartz tube at a fast rate (~200 sccm) for 20 min to remove air in the system, and then adjusted to 20 sccm accompanied by a flow of oxygen at a rate of 2 – 5 sccm. At this time, the tube furnace is heated at a rate of 20 K min^{-1}. After being held at the designated temperature for ~10 hours. the O_2 flow is stopped, with only the N_2 gas being kept flowing, and the system is allowed to cool naturally to room temperature. The product samples are then collected, which considered of a scarlet layer homogeneously coated on the substrate.

(A) GROWTH OF ALIGNED ARRAYS OF α - Fe_2O_3 AND Fe_3O_4 NANOWIRES

Effect of Alloyed Substrates on the Synthesis of α-Fe_2O_3 NWs

The first condition to be changed is the substrate used for the growth. Figure 13.2(a) to (d) shows the morphologies of NWs taken from the central areas of substrates consisting of different alloys, including $Fe_{0.5}Ni_{0.5}$, $Fe_{0.64}Ni_{0.36}$, Fe foil, and Fe film. The substrates are placed standing vertically on an alumina stage so that the vapor flow directly against the surface of the substrate. The growth is conducted at 670K for 10 h in an Ar ambient at a vapor-flow rate of 100 sccm. It is found that the density of NWs increases on increasing the Ni concentration in the alloy substrates. The growth mechanism of the α-Fe_2O_3 NWs is quite different from that of other 1D nanomaterials. It is characterized by several important characteristics: the growth of micro-/ nanostructures from the alloy substrate with a taperlike feature, the lengthening of the diameter at high temperatures, and the long growth time. This growth mechanism is called the tip-growth mechanism. Here, the surface defects inside the iron substrate play an important role in lowering the energy barrier, to serve as the nucleation sites for NW growth. In addition, the surface defects display various forms, such as linear and planar defects, resulting in different morphologies of Fe_2O_3 nanostructures under different growth conditions. During the growth of one-dimensional iron oxide nanostructures, several iron oxide layers with thicknesses of several micrometers are subsequently formed before the growth of the NWs. These iron oxide layers, including those of FeO, Fe_xO ($x > 1$), Fe_3O_4 and Fe_2O_3, are formed below the surface of the alloy substrate and., serve as the reactants for interdiffusion after growth at high temperature, so that the iron atoms constantly diffuse to the surface and form nucleation sites for growth of Fe_2O_3 NWs. Thus, growth of the NWs is mainly dominated by surface diffusion and internal diffusion.

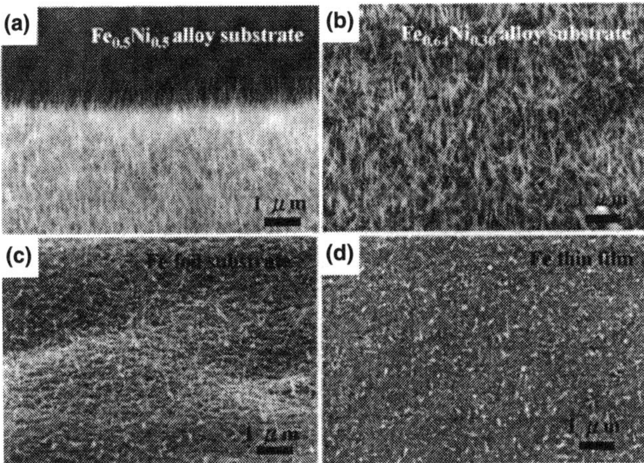

Figure 13.2 *Scanning electron microscopy (SEM) images of a-Fe_2O_3 NWs synthesized on a) an $Fe_{0.5}Ni_{0.5}$ alloy substrate; b) an $Fe_{0.64}Ni_{0.36}$ alloy substrate; c) Fe foil; d) an Fe film evaporated by electron-beam deposition at 670K.*

The thickness of the iron film deposited on the Si substrate is 1-2 μ m, and most of the iron film is oxidized during the growth process, resulting in insufficient iron atoms from the substrate for NW growth. This is why the density of NWs on iron films deposited on Si substrates decreases (Fig. 13.2(d)). An increase in the concentration of Ni atoms in the Fe-based alloy enhance the self-diffusion of Fe atoms inside the alloy substrate and thus result- in an increased density of NWs The carrier gas used is Ar. However, oxygen vapor is formed from the decomposition of a-Fe_2O_3 powder at 1470K.

Dependence of α-Fe_2O_3 NW Growth Morphology on the Local Substrate Environment

The local environment at the substrate affects the growth morphology of the nanostructures. In the tube furnace, a vertically positioned substrate is chosen so that its temperature is constant. The local "turbulence" or flow speed and pressure of the vapor phase and carrier gas is investigated, and is shown in Figure 13.3(a). Although the gas flow is stable and uniform in the entire tube furnace, the variation is significant. Four regions from bottom to top (labeled A to D) are choosen and the structures grown on the $Fe_{64}Ni_{36}$ substrate at 670K for 10 hours in an Ar ambient at a flow rate of 100 sccm as shown in Figure 13.3(b) to (e). Different morphologies are formed in the bottom and top regions. In region A, the morphology, is sheetlike, and becomes a combination of sheetlike and NW-like in region B (Fig. 13.3(b) and (c)). The inset in Figure 13.3(c) shows a magnified scanning electron microscopy (SEM) image of the sheetlike feature. A wirelike morphology of uniform high density is found in region C (Fig. 13.3(d)), while the density of NWs is significantly decreased in region D. This is due to different concentrations of oxygen vapor in the local regions. The high flux provides a high concentration of oxygen vapor, resulting in an increased growth rate of the 2D structures.

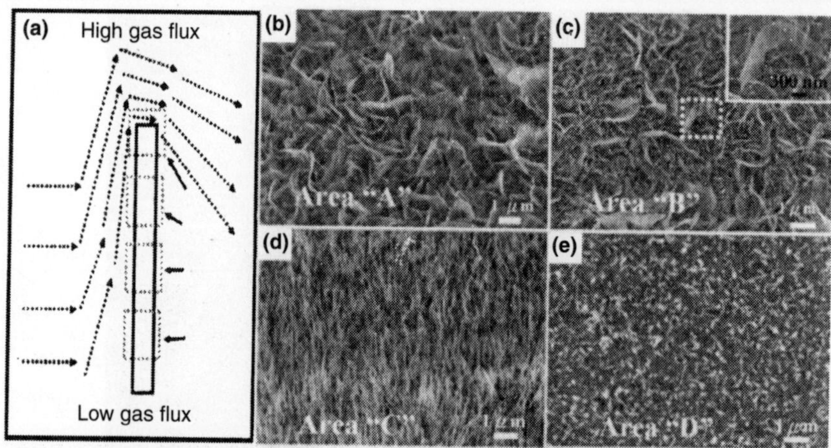

Figure 13.3 *(a) A gas flux around the substrate that is placed vertically on an alumina stage. The different areas are marked as A-D. The corresponding SEM images are shown, respectively in (b - e) for* $Fe_{0.64}Ni_{0.36}$.

Microstructure and Electron Energy-Loss Spectroscopy Analysis of the Samples

To understand the detailed structures of thee nanomaterials, transmission electron microscopy (TEM) and electron energy-loss spectroscopy (EELS) are used. Figure 13.4(a) shows a TEM image taken from region B in Figure 13.3(a). The corresponding diffraction pattern is shown in the inset of Figure 13.3(a), where the individual planes are indexed. It reveals that the NW phase is α-Fe_2O_3. Figure 13.4(b) shows a low-magnification TEM image of the α-Fe_2O_3 NW with diameters of 10-20 nm taken from region C. The corresponding diffraction pattern, as shown in the inset of Figure 13.3(c) also confirms the α-Fe_2O_3 phase with a [001] zone axis. Extra spots are found in the diffraction pattern with d = $1/5(3\bar{3}0)$, which has five times the distance of the (330 plane, as shown by arrow heads shown in Figure 13.4(d). Figure 13.4(e) shows a corresponding high-resolution TEM image the highlights the ordering feature. This superstructure is caused by oxygen vacancies inside the α-Fe_2O_3 NWs during growth. Similar results are found in other materials. The magnified high-resolution TEM image in Figure 13.4(f) show the superstructure caused by the oxygen deficiencies. The two almost identical d-spacings of 0.25 nm are consistent with the d-values of $(2\bar{1}0)$ and (110). It indicates the single-crystal feature with a growth direction along [110]. The long-range ordering phenomena with a period of 0.72 nm are located and marked by white arrow heads. The inset in the bottom of Figure 13.4(f) shows Fast Fourier transform (FFT) image. It is consistent with the results of the diffraction pattern. So to clarify the ordering phenomenon the diffraction pattern and FFT pattern from the high-resolution TEM image, respectively, the ideal high-resolution TEM image is processed by inverse FFT via selecting only the matrix reflections from the ideal α-Fe_2O_3 structure, and is shown in Figure 13.4(g). On comparing the two high-resolution TEM images (Fig. 13.4(f) and (g), the lattice image along the white arrow heads between the two nearest bright dots is caused

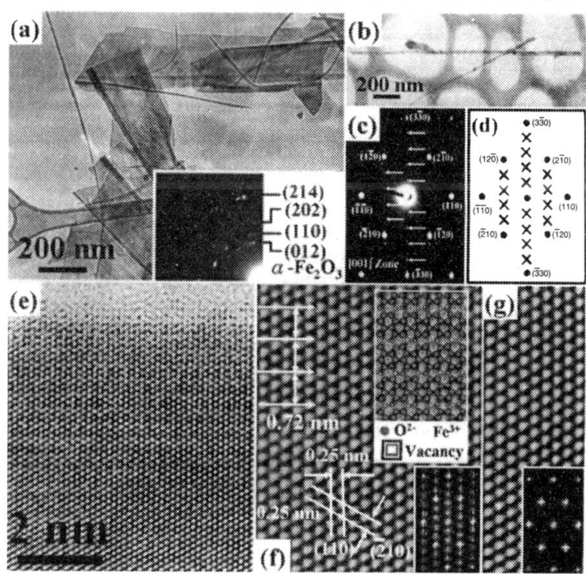

Figure 13.4 *(a.b) TEM images taken from the regions B and C shown in Figure 2a. Inset in (a) shows the corresponding diffraction pattern. c) The diffraction pattern with [001] zone axis extracted from (b). The extra spots with five times ($3\bar{3}$ 0) can be clearly found and are illustrated in (d). e) The high-resolution TEM image recorded from (b). It is clear that regular lines can be observed. f) The corresponding high-resolution TEM image. The hard-ball model fits the relative atomic position. The inset shows the fast-Fourier transform (FFT). which is consistent with the diffraction pattern in (c). g) The Fourier-filtered image formed by removing the extra reflection spots. The inset shows the corresponding FFT image without the extra spots.*

by ordered oxygen vacancies. Further the hard-ball model, formed by removing the oxygen atoms along the five period distances of the Fe-O lattice plane, fits coherently with the individual sites of the corresponding high-resolution TEM image, as shown in the middle inset of Figure 13.3(f), where the dark-grey and light-grey balls and the double rectangle represent the oxygen anion, iron cation, and oxygen vacancies caused by the oxygen deficiency, respectively.

No Ni atoms is found inside the α-Fe_2O_3 NWs when FeNi alloy is used as the substrate.

Three layers, Fe_2O_3, $(Fe,Ni)_3O_4$, and $(Fe, Ni)O_x$ are subsequently formed on top of the Fe-Ni alloy substrate The Ni atoms in the Fe-Ni alloy are bounded inside the $(Fe,Ni)_3O_4$ layer during the oxidation process and the top Fe_2O_3 layer is free of Ni. As a result α-Fe_2O_3 NWs grow via the tip-growth mechanism. In addition, the tip-growth mechanism of NWs occurs via surface and internal diffusion. So The surface diffusion has a lower energy barrier than internal diffusion, whereas the oxygen vacancies found at the α-Fe_2O_3 NWs reduce the barrier height of internal diffusion, resulting in a decreased synthesis temperature for the growth of the NWs. In addition, it is found that the density of oxygen deficiencies is decreased as the annealing temperature is increased.

Influence of Gas-Flow Rate

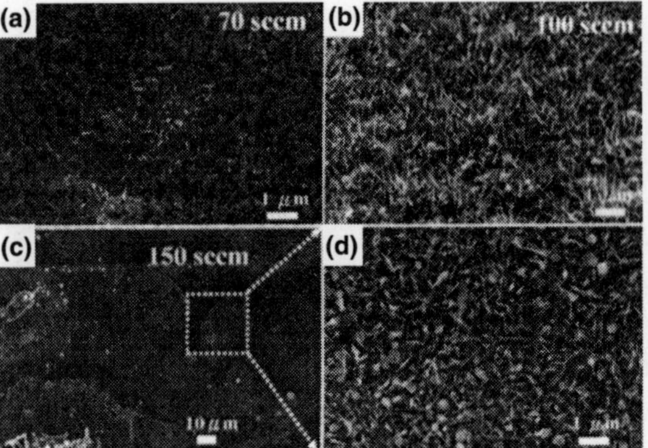

Figure 13.5 *SEM images of α-Fe_2O_3 NWs grown on $Fe_{0.5}Ni_{0.5}$ 670K 10 h in Ar ambient at a flow rate of a) 70. b) 100, and c) 150 sccm. d) SEM image taken from the rectangular area in (c).*

Firstly for the influence of gas-flow rate on the growth morphology, the gas-flow rate is changed from 70 to 100, to 150 sccm. It is found that the density of NWs significantly increases and the NWs are uniformly distributed on the Fe-based alloy substrate on increasing the Ar gas flow rate. The NW density remains fairly uniform until the Ar gas flow rate is increased to 150 sccm; for example a series of SEM images is shown in Figure 13.5(a) to (c) for the α-Fe_2O_3 NWs grown on an Fe film at 670K for 10 h with Ar flow rates of 70,100, and 150 sccm, respectively. Figure 13.5(d) shows a magnified SEM image recorded from the rectangular area. In Figure 13.5(c). There are two possible factors that influence the NW growth: the oxygen concentration and resident time. If the concentration of oxygen vapor in the carrier gas is no more than the minimum value, the possibility for NW growth via the tip-growth mechanism is limited, resulting in few nanowires. Second, the different gas-flow rates also influence the resident time, T, of the carrier gas during the growth of the NWs, which is a quantity defined to specify the interaction time of the gases with the growth substrates/products; thus, a simple equation is given

$$\tau = \frac{P \times V}{Q} \tag{1}$$

where P, V, and Q represent the pressure and volume of the reaction chamber, and the flow rate, respectively. At a given pressure, the low flow rate results in a long resident time that all causes different growth rates at different reaction areas, thus leading to the growth of non-uniformly distributed NWs on the alloy substrate. By reducing the resident time via increasing the flow rate, growth of uniformly distributed NWs is enhanced. However, the density is nonuniform when the flow rate is more than 150 sccm because the oxygen vapor from decomposed α-Fe_2O_3 in the Ar carrier gas is oversaturated, resulting in different NW growth rates in different areas (Fig. 13.5(c). Accordingly, the flow rate of 100 sccm is a moderate value used to strike a balance between the oxygen-vapor concentration and resident time.

Influence of Growth Temperature

Figure 13.6(a) to (d) shows SEM images of α-Fe_2O_3 NWs grown on the $Fe_{0.5}Ni_{0.5}$ substrate with an Ar gas flux of 100 sccm at growth temperatures 670,770,870 and 970K respectively. Uniform, high-density α-Fe_2O_3 NWs with diameters of 10 - 40 nm and lengths up to several micrometers can be synthesized at a temperature of 670K. The diameter of these α-Fe_2O_3 NWs increases when the annealing temperature is increased. The increase of the annealing temperature results in the enhancement of the diffusivity and reaction rate to elongate the NWs so that the density is decreased owing to Ostwald ripening. At high annealing temperatures, the deviation in chemical potential between large and small materials becomes significantly important. If the dimension (size) of the nuclei in the initial growth of NWs is larger than the minimum size, they tend to coarsen, or they disappear, resulting in a decreased density of NWs during the growth period.

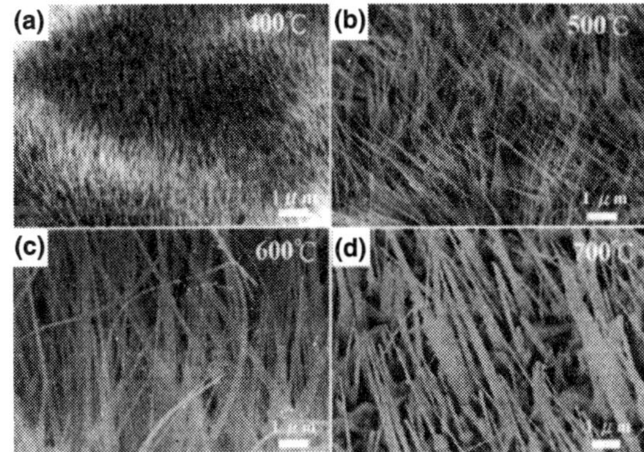

Figure 13.6 *(a-d) SEM images of α-Fe_2O_3 NWs grown on an $Fe_{0.5}Ni_{0.5}$ substrate at temperatures of 670,770,870, and 970K respectively, using an Ar gas flux of 100 sccm.*

Reducing α-Fe_2O_3 NWs to Fe_3O_4 NWs
(a) Reduction Process

α-Fe_2O_3 NWs are converted into Fe_3O_4 NWs by reduction. Starting from the α-Fe_2O_3 NWs synthesized on an $Fe_{0.64}Ni_{0.36}$ substrate, the reduction process is carried out in a mixed reductive ambient of H_2(5 %) + Ar(95 %) at different annealing temperatures. Figure 13.7(a) to (c) show the SEM images of the NWs after being annealed at 720,770 and 870K respectively, in a reductive ambient. The sample reduced at an annealing temperature of NWs is destroyed during the formation of the bulge like shape, but most of the NWs don't change in morphology. When the reductive temperature is increased to branched shapes these are found at the roots caused by fusing of the NWs during the reductive process, and the bulge-like morphology is also observed in many regions. When the reductive temperature is increased to 870K most of the NWs form branch and bugellike features and the density is significantly decreased as a result of fusing and bunching.

Figure 13.7(d) shows the TEM image of an α-Fe_2O_3 NW sample annealed at 720K in a reductive ambient. The diameters of these NWs are about 10-80 nm and the corresponding diffraction pattern in the inset of Figure 13.7(d) obviously confirms that the phase of these NWs is Fe_3O_4 The upper inset in Figure 13.7(e) shows a low magnification TEM image of an Fe_3O_4 NW with a taperlike feature, indicating that the NW keeps the original shape after reduction. The diffraction pattern, as shown in the bottom inset in Figure 13.7(e), confirms and reveals the single-crystal nature with the [111] zone axis. Two spacings of 0.29 nm are consistent with the planes of $(02\bar{2})$ and $(20\bar{2})$, and the NW growth direction is $[1\bar{1}0]$.

Figure 13.7(f) shows the EELS spectrum of oxygen K-edge and Fe L-edge ELNES for the α-Fe_2O_3 NW before and after the reduction process. No chemical shift in the L_3 line or change in the separation between L_3 and L_2 is found but the intensity ratio I(L3)/I(L2) decreases from 4.2 for α-Fe_2O_3 NWs to 3.9 for Fe_3O_4 NWs. In addition, the intensity of peak c in the oxygen K-edge ELNES spectrum of the α-Fe_2O_3 NW, after the reduction process is increased, also confirms the Fe_3O_4 phase. Though a similar spectrum is found in γ-Fe_2O_3 NWs, the analysis of the diffraction pattern and EELS spectrum show that the phase of the NWs after the reduction process is Fe_3O_4.

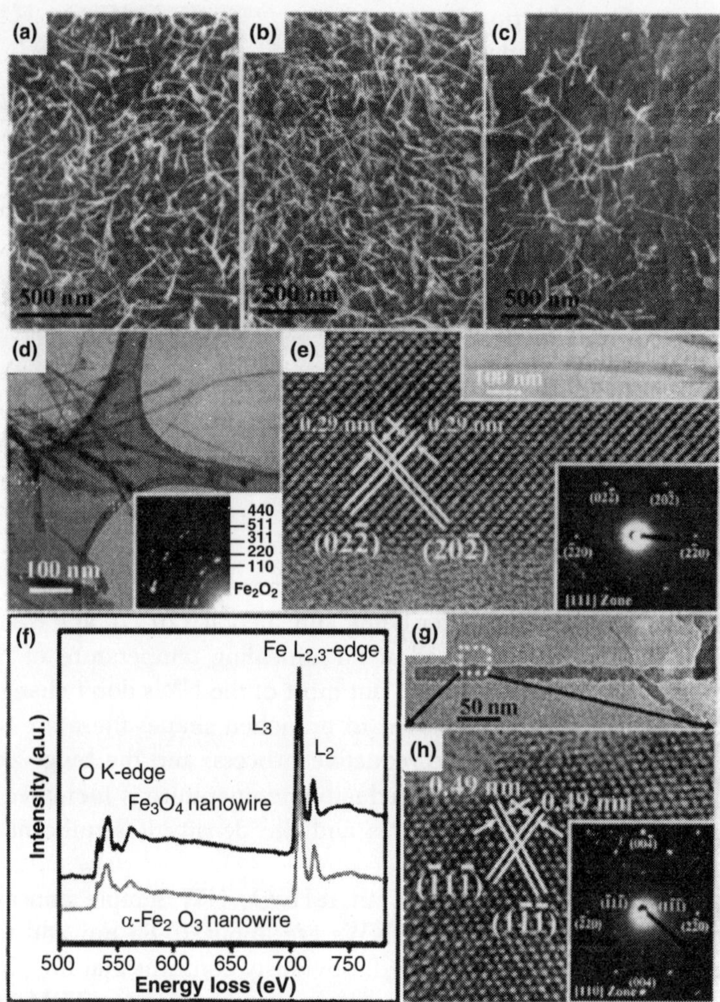

Figure 13.7 *SEM images of Fe_3O_4 NWs synthesized by converting α-Fe_2O_3 NWs in a reductive ambient for 20 min at a) 720K b),770, and c) 870K. d) TEM image of the converted NWs. The inset shows the corresponding diffraction pattern. e) High-resolution TEM image of the Fe_3O_4 NWs. The upper inset shows the low magnification TEM image. The bottom inset shows the corresponding diffraction pattern with [111] zone axis. f) The EELS spectrum of the oxygen' K-edge ELNES and Fe L-edge for both the α-Fe_2O_3 and Fe_3O_4 NWs, respectively. g) TEM image of Fe_3O_4 NWs at converted from α-Fe_2O_3 NWs 870K in a reductive ambient. h) The corresponding high-resolution TEM image recorded from the rectangular area shown in (g). The inset shows the diffraction pattern with[110] zone axis.*

Figure 13.7(g) shows the TEM image of an Fe_3O_4 nanowire taken from the α-Fe_2O_3 NW sample after annealing at 870K. It is seen that the surface is very rough. Also the high-resolution TEM image obtained from the rectangular area in Figure 13.7(g) is shown in Figure 13.7(h), confirms the single-crystal nature with a growth direction of [1 $\bar{1}$0], where the corresponding diffraction pattern with the [110] zone axis is shown.

(b) Phase Transformation

The synthesis of Fe_3O_4 NWs by annealing the α-Fe_2O_3 NWs in a reductive ambient, is also made by direct observation of phase any structure transformation from α-Fe_2O_3 to Fe_3O_4 conducted by TEM. Though it is hard to maintain the reductive ambient in the TEM chamber, the annealing of α-Fe_2O_3 NWs in the vacuum chamber is another kind of reductive reaction. Here, an α-Fe_2O_3 NW is dispersed on a Mo grid with one end of the NW fastened and the other suspended, as shown in Figure 13.8(a). The dark line in the α-Fe_2O_3 NW is a defect and is used as a reference point during annealing. The morphology of the NWs do not change when the annealing temperature is increased to 770K but starts to change when the annealing temperature is increased to 970K .The surface becomes much rougher but maintains the original shape when the annealing temperature is further increased to 1170K. The corresponding high-resolution TEM image taken from the rectangular area in Figure 13.8(e) is shown in Figure 13.8(f). The diffraction pattern, as shown in the inset of Figure 13.8(f), confirms the Fe_3O_4 phase with a [110] zone axis. Two spacings of 0.48 nm are observed to be in consistent with the $(\bar{1}1\bar{1})$ and $(1\bar{1}\bar{1})$ planes. From the diffraction pattern and high-resolution TEM image,

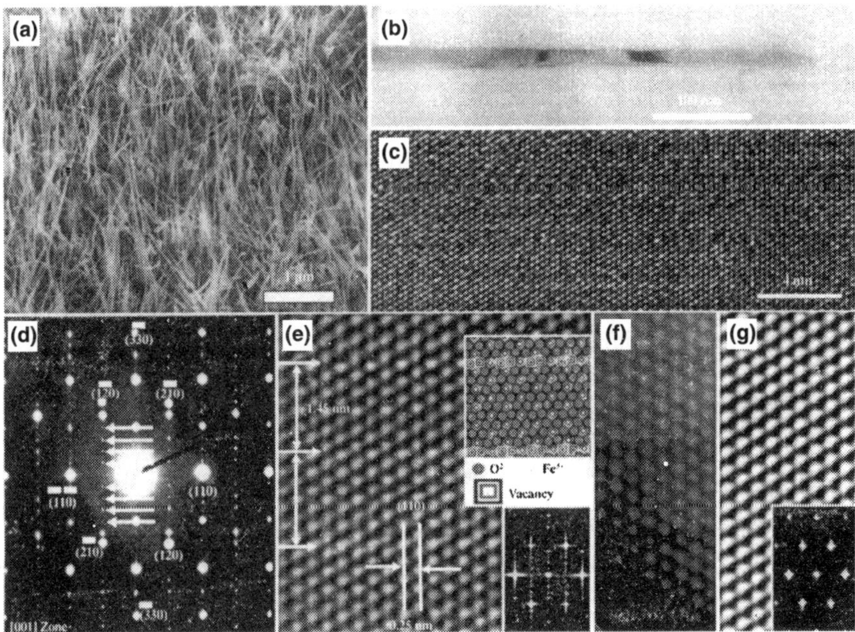

Figure 13.8 *α-Fe_2O_3 NWs: a) A tilted SEM image. b) A low-magnification TEM image. c) A medium-magnification TEM image. d) The diffraction pattern with [001] zone axis as extracted from (c). The extra spots with ten times the distance of $(3\bar{3}0)$. e) A high-resolution TEM image recorded from (c). The lower inset shows the fast Fourier transform (FFT) f) A high. resolution low-angle annular dark-field (LMDF) image. g) The Fourier filtered image made by removing the extra reflection spots. The inset shows the corresponding FFT image without the extra spots.*

the NW growth direction is found to be $[1\bar{1}0]$, which is perpendicular to the (111) close-packed plane. O atoms inside the α-Fe_2O_3 and Fe_3O_4 are found to have ABAB... and ABCABC... stackings, respectively. The corresponding hard-ball models are illustrated in Figure 13.8(g) and (h). The oxygen vacancies found in α-Fe_2O_3 NWs play an important role in the mediation of the structure transformation, resulting in the remaining original morphology where the growth direction of the NWs converted from [110] for α-Fe_2O_3 NWs to $[1\bar{1}0]$ for Fe_3O_4 NWs.

p-Type α-Fe_2O_3 Nanowires and their *n*-Type Transition In a Reductive Ambient

Figure 13.8(a) shows an SEM image of α-Fe_2O_3 NWs synthesized on an $Fe_{64}Ni_{36}$ substrate at 720K for 10 hours in an Ar ambient of 100 sccm. It is obvious that α-Fe_2O_3 NWs are uniformly formed on a large scale. A TEM image of all α-Fe_2O_3 NW, with a diameter of $\approx$ 18 nm; is shown in Figure 13.8(b). The sequential periodic structure with a regular spacing of 1.45 nm is observed in the image shown an in Figure 13.8(c). The corresponding diffraction pattern, shown in Figure 13.8(d). confirms the phase of α-Fe_2O_3 with a [001] zone axis. The extra spots are found in the diffraction pattern. They have ten times the distance of the $(3\bar{3}0)$ plane, as marked by arrows. Figure 13.8(e) shows the high-resolution TEM image highlighting the ordering structure. The two almost-identical *d* spacings of 0.26 nm are consistent with the *d* values of the $(2\bar{1}0)$ and (110) planes thereby indicating the single-crystal nature of the wires with growth along the [110] direction. The long range ordering phenomenon with a periodicity of 1.45 nm is located and marked by white arrows. The inset in the bottom of Figure 13.8(e) shows the corresponding Fast Fourier transfer (FFT) of the image, which is consistent with the diffraction pattern. To clarify the ordering phenomenon found in both the diffraction pattern and the FFT pattern directly transformed from the high-resolution TEM image, an high-resolution TEM image is processed by using an inverse FFT method through selection of only the matrix reflections from the stoichiometric α-Fe_2O_3 structure (see Figure 13.8(g)) In Figure 8 (e) to (g), the ambiguous lattice image (dark image) between the two nearest bright dots are seen (along the white arrows in Figure 13.8(e)); this is caused by ordered oxygen vacancies. The atomic resolution image acquired with scanning tunneling electron microscopy (STEM) by using an annular dark-field (ADF) detector, as shown in Figure 13.8(f), confirms the vacancy-ordering phenomenon owing to the presence of a strain field caused by oxygen deficiencies due to both the diffraction effect and weakened local scattering. Furthermore, the hard-ball model, with removal of the oxygen atoms along the tenperiod distance of the Fe-O lattice plane fits with the individual sites in the corresponding high-resolution TEM image, as shown in the upper inset of Figure 13.8(e) where the big/small circles and the double rectangles represent the oxygen anions, iron cations, and oxygen vacancies caused by oxygen deficiency, respectively.

The influence on electronic characteristics by high-order oxygen inside α-Fe_2O_3 example Figure 13.9(a) shows the current as a function of the bias voltage under different gate voltages. The inset shows the two-terminal probe measurements for an α-Fe_2O_3 NW with a length and diameter of 0.12 μm and 37 nm, respectively. A resistance of 3.17 MΩ is found from the linear region, as the bias voltage goes through zero voltage. However, this linear resistance does not

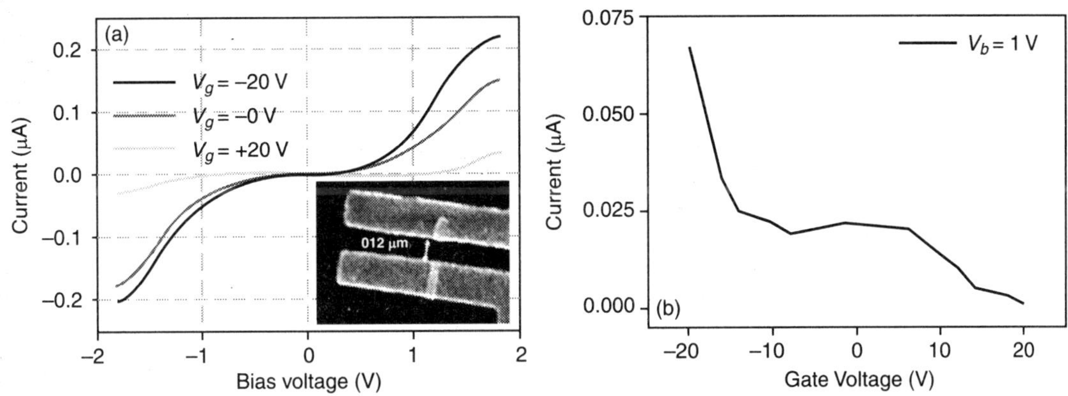

Figure 13.9 *α Fe_2O_3 NWs: a) Measurement of current versus bias voltage by applying different gate voltages (V_g= + 20. 0. and −20 V). The inset shows the corresponding SEM image. b) Measurement of current versus different gate voltages at a bias voltage of V_b = 1 V.*

represent the entire resistance of the α-Fe_2O_3 NW due to the presence of contact resistance between the NW and electrodes. With the increase of the gate voltage (− 20 V to + 20 V), an decrease of channel conductance is found, which indicates that α-Fe_2O_3 NW with high-order oxygen vacancies has a p-type semiconductor feature. In addition, a graph of the current as a function of the applied gate voltage (*I/Vg* curve show in Figure 13.9(b) gives evidence of the p-type nature, since the current increases as the gate voltage decreases when a fixed voltage of 1 V is applied across the NW. For a typical cylindrical NW, the relationship of carrier density $n(cm^{-1})$ and carrier mobility $u(cm^2V^{-1}s^{-1})$ is given by

$$n = \frac{V_{gt}}{e} \times \frac{2\pi\varepsilon_0\varepsilon_r}{\ln(2h/r)} \tag{1}$$

$$u = \frac{dI}{dV_g} \times \frac{\ln(2h/r)}{2\pi\varepsilon_0\varepsilon_r} \times \frac{L}{V_{ds}} \tag{2}$$

where, V_{gt}, e, ε_r h, and L, V_{ds} and r represent the threshold gate voltage, electron charge, relative dielectric constant (ε_r= 3.9 for SiO_2) thickness of the gate oxide layer, channel length, bias voltage, and radius of the nanowire, respectively. The hole concentration and carrier mobility are 1.05×10^7 cm^{-1} and $4.22 \times 10^{-2} cm^2V^{-1}s^{-1}$ respectively, by extracting V_{gt} = 2.3 V and $dI/dV_g = 2.2 \times 10^{-9}$ AV^{-1} from Figure 13.9(b). Another device of the same α-Fe_2O_3 NW is prepared, and the I/V curve from this device, with different applied gate voltages, shows the same trend: the p-type semiconductor with a hole concentration and carrier mobility of 4.9×10^6 cm^{-1} and 3.09×10^{-2} $cm^2V^{-1}s^{-1}$ respectively.

In order to determine the specific p-type nature of the α-Fe_2O_3 NW, a chemical analysis on the electronic structure EELS spectrum is done which is sensitive to the electronic structure of the material. All raw EELS spectra are calibrated in the relative-energy position by a zero-loss peak and then subtracted by a power law to remove the background signal generated by plural scatting events. After recalibration of the energy position and removal of the multiscattering effect, the EELS measurements are deconvoluted by plasmons and a low-loss spectrum through a Fourier log to obtain the true single-scatting spectrum. Figure 13.10 a shows the EELS

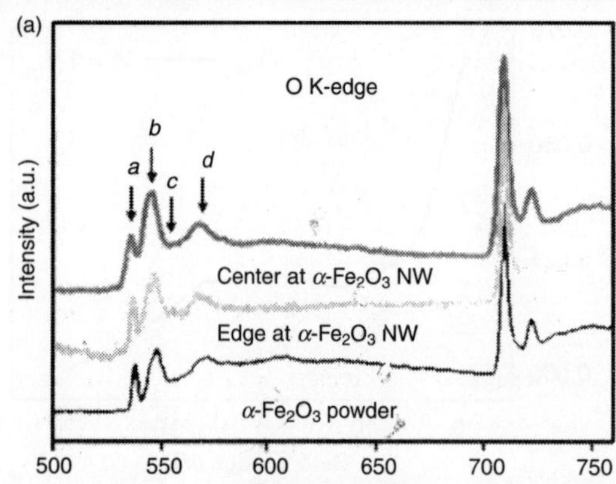

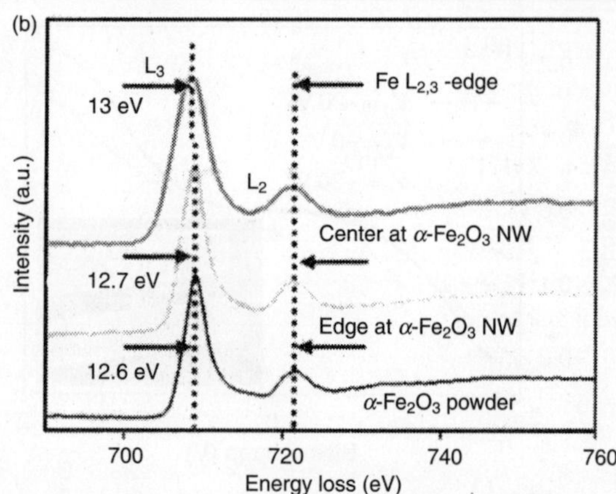

Figure 13.10 *(a) The EELS spectra of oxygen ELNES for α-Fe_2O_3 NWs at the center and edge regions and for the α-Fe_2O_3 powder. b) The corresponding EELS spectra of the $L_{2,3}$ edge for α-Fe_2O_3 NWs at the center and edge regions and for the α-Fe_2O_3 powder.*

spectrum of the oxygen k-edge energy loss near the edge fine structure (ELNES), taken from the center and edge regions of an α-Fe_2O_3 NW. The oxygen k-edge ELNES spectrum from α-Fe_2O_3 powder is also shown at the bottom for comparison. Four peaks, labeled *a-d* are found in all of the spectra. In general, peak a is derived from the oxygen 1s to 2p core level hybridized with the iron 3d orbital, while peak *b* is derived from the oxygen 2p states hybridized with the transition-metal 4s and 4p states Peaks *c* and *d* result from the scattering of the third and first oxygen coordination shells by outgoing or backscattering electrons.

In a comparison of the oxygen k-edge ELNES spectra acquired from the center and edge regions of the α-Fe_2O_3 NW, the relative intensity between the *a* and *b* peaks gradually becomes smaller. The decrease in intensity in peaks *a* and *b* is caused by oxygen vacancies inside the α-Fe_2O_3 NWs, which result in a diminishing hybridization of the metal 3d orbitals with the oxygen 2p orbitals. The relative intensity between the *a* and *b* peaks at the edge regions of the α-Fe_2O_3 NWs is similar to that of the α-Fe_2O_3 powder, a result indicating that the oxygen deficiencies are compensated for by foreign molecules (oxygen and water molecules) through adsorption. However, the oxygen vacancy inside the α-Fe_2O_3 NWs accompany a change in the oxidation states of the nearest neighbor Fe cations, which is observed by EELS, as shown in Figure 10 b, for which the EELS spectra of the iron $L_{2,3}$ edge are taken from the center and edge regions of the α-Fe_2O_3 NWs and from α-Fe_2O_3 powder, respectively. The L_3 and L_2 lines are the transitions of $2p^{3/2} \rightarrow 3d^{3/2}3d^{5/2}$ and $2p^{1/2} \rightarrow 3d^{3/2}$, respectively. For the α-Fe_2O_3 powder, the difference between the relative positions of L_3 and L_2 is about 12.6 eV, while for α-Fe_2O_3 NWs the differences between the relative positions of L_3 and L_2 at the edge and center regions are $\approx$ 12.7 and 13 eV respectively. The energy difference of 0.4 eV obtained from the relative positions of the L_3 and L_2 lines in the center region of the α-Fe_2O_3 NWs results from the oxygen vacancies, which partially decrease the oxidation states of the Fe cations from Fe^{3+} to Fe^{2+}. However, for the edge region of the α-Fe_2O_3 NWs no energy shift is observed a result indicating that the oxidation states of the Fe cations are the as the bulk counterpart.

After an annealing process at 720K in a reductive ambient the morphology do not change, as shown in Figure 13.11(a). In addition, the high-resolution TEM image and the corresponding diffraction shown in Figure 13.11(b) in which the individual planes in the diffraction pattern with a [001] zone axis and two d spacings in the high-resolution TEM image are indexed and illustrated, confirm no oxygen vacancy ordering. Therefore, annealing in a reductive ambient not only limits the oxygen molecules adsorbing on the surface but also creates more oxygen vacancies inside the α-Fe_2O_3 through removal of oxygen anions from the α-Fe_2O_3 lattice,. During the reduction process, more oxygen anions diffuser to the surface, thereby resulting in the creation of oxygen vacancies, Thus, this diffusion behavior destroys the oxygen vacancy ordering inside the α-Fe_2O_3 NW. From the EELS spectra of the oxygen k-edge ELNES for the α-Fe_2O_3 NWs before and after the annealing in a reductive ambient, as shown in Figure 13.11(c), the significant change in the a:b ratio after the annealing process in a reductive ambient proves that the density of oxygen vacancies increases. Figure 13.11(d) describes the *I-V* characteristics as a function of different gate voltages (from – 20 to +20 V) for a α Fe_2O_3 NW with a length and diameter of 1.7 μm and 32 nm, respectively, after annealing in a reductive ambient; the corresponding device is shown in the inset image, The conductance increase the forward gate voltage is increased, a result indicating the n-type nature, Accordingly, equations (1) and (2) give the electron concentration and mobility to be $1.452 \times 10^7 cm^{-1}$ and 1.574×10^{-3} cm^2-$V^{-1}s^{-1}$ by taking $V_{gt}= -3.45V$ and $dI/dV{=}25{\times}10^{12}AV^{-1}$, as extrapolated from the linear region (+ 4 to + 20 V) of the I/V_g curve.

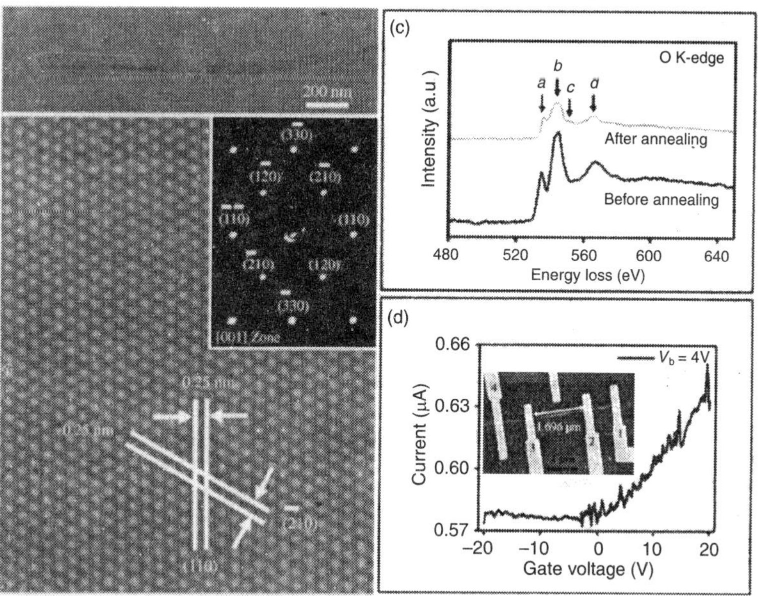

Figure 13.11 *(a) A low-magnification TEM image of an α-Fe_2O_3 NW after annealing in a reductive ambient. (b) The corresponding high-resolution TEM image and diffraction pattern (inset), (c) The EELS spectra of oxygen ELNES for α-Fe_2O_3 NWs before and after annealing in a reductive ambient.d) Measurement of current versus various gate voltages at V_b = 4, The inset shows the SEM image of the corresponding α Fe_2O_3 nanodevices.*

In general, the oxygen deficiencies inside the metal oxide nanomaterials result in an n-type nature, and the corresponding band diagram is drawn in Figure 13.12(a), where the Fermi level (E_F) is above the intrinsic Fermi level (E_i). Due to the larger density of the unstable surface state, the adsorption of oxygen and water molecules on the surface of metal oxide NWs cause bending and a change in work function, which results in the formation of a depletion layer and a decrease in the conductance near the surface. However, NWs with wide-bandgap features retain an n-type conductivity, for example, ZnO NWs (3.4 eV), SnO_2 NWs (3.6 eV), and TiO_2, NWs (3.2 eV). Now a narrow bandgap of 2.2 eV makes α-Fe_2O_3, typical n-type semiconductor at low synthesis temperatures with pure oxide stoichiometry, but it then becomes a p-type semiconductor at high synthesis temperatures with impure oxide stoichiometry. The n-type nature of α-Fe_2O_3 NWs is due to surface adsorption, which causes a depletion layer on the surface, but the major carriers are electrons, which indicates the n-type feature. By contrast the opposite tendency namely, a p-type nature is observed. Based on the EELS results for the center and edge regions of α Fe_2O_3, the p-type tendency results, from the formation of an inversion layer, which originates from the strong surface absorption of oxygen and water molecules near the surface due to the large density of unstable surface states because of the oxygen vacancy ordering, as shown in Figure 13.12(b). After annealing in a reductive ambient, the switch from p- to n-type is seen in the I/V_g characteristics. During the reductive process, the removal of oxygen anions from the α-Fe_2O_3 lattice result in an increase of oxygen vacancies to destroy the balance of the oxygen valency order. Although the density of oxygen vacancies is high, the surface adsorption causes a depletion layer instead of an inversion layer, as shown in Figure 13.12(c). In addition, the increase in oxygen vacancies results in deviation from the stoichiometry, that is, α-Fe_2O_{3-x} where x is proportional to the density of oxygen vacancies, and thus leads to a lower oxide compound magnetite (Fe_3O_4) after consecutive annealing processes at high temperatures and long annealing times.

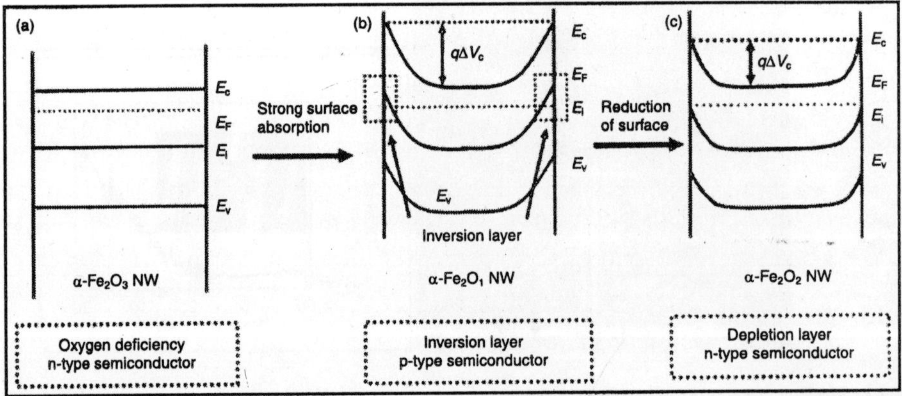

Figure 13.12 *Energy-level diagrams of a) an n-type semiconductor caused by oxygen vacancies. b) a p-type semiconductor through formation of an inversion layer caused by strong surface absorption owing to the oxygen vacancy order. and c) the depletion layer caused by surface absorption.*

Characterization

1. Magnetic Properties

Consider the magnetic properties of α-Fe_2O_3 NWs. Figures 13.13(a) to (d) provide magnetic measurements on the α-Fe_2O_3 and Fe_3O_4 NWs. For the measurements, all of the NWs are dispersed on a Si substrate in order to avoid the magnetic signal from the Fe-based alloy substrate interfering with the magnetic measurements. Figure 13.13(a) shows the hysteresis loops of the NWs measured by vibrating sample magnetometry measurements by applying the magnetic field parallel and perpendicular to the Si substrate. The hysteresis loop indicates very weak ferromagnetism as the applied magnetic field is applied parallel to the Si substrate (parallel to the α-Fe_2O_3 NWs), while it indicates diamagnetism as the magnetic field is applied perpendicular to the Si substrate (perpendicular to the α-Fe_2O_3 NWs). The corresponding SEM image is taken after applying the magnetic field along the direction of the arrow that is parallel to the Si substrate as shown in Figure 13.13(b). The inset in Figure 13.13(b) shows the magnified SEM image taken from the rectangular area shown in Figure 13.13. All of the α-Fe_2O_3 NWs are randomly distributed after applying the magnetic field. α-Fe_2O_3 is an antiferromagnetic material and the two antiferromagnetic spins are along the [001] direction below temperature 260 K. When the temperature is between 260 and 955 K, α-Fe_2O_3 exhibit

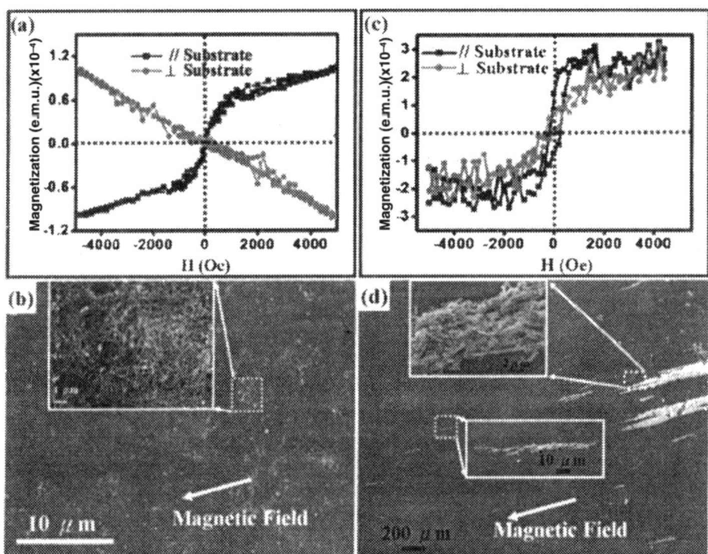

Figure 13.13 *(a) Hysteresis loops of α-Fe_2O_3 NWs measured by vibrating sample magnetometry with the magnetic field applied perpendicular and parallel to the substrate. (b) SEM image of α-Fe_2O_3 NWs dispersed on Si substrate after applying the magnetic field parallel to the substrate. (c) Hysteresis loop of the α-Fe_2O_4 NWs measured by vibrating sample magnetometry with the magnetic field applied perpendicular and parallel to the substrate. (d) SEM image of Fe_3O_4 NWs dispersed on Si substrate after applying the magnetic field. Insets in (b) and (d) show the magnified SEM images recorded from the corresponding rectangular areas.*

weak ferromagnetic behavior, which is called parasitic ferromagnetism. However, oxygen vacancies inside the α-Fe_2O_3 NWs destroy the balance of antiferromagnetic behavior, resulting in the extremely weak ferromagnetism. Accordingly, the force generated by the external field is insufficient to rearrange the NW parallel to the applied field due to the extremely weak interaction between the external magnetic field and the α-Fe_2O_3 NWs. Now, the magnetic signal from the α-Fe_2O_3 NWs is much smaller than that from the Si substrate when the applied magnetic field is perpendicular to the Si substrate (perpendicular to the NWs) so it results in the diamagnetic feature in the hysteresis loop originating from the Si substrate (the magnetic susceptibility $X_{m'} = -5.2 \times 10^{-6}$)

Figure 13.13(c) shows the hysteresis loop of Fe_3O_4 NWs dispersed on the Si substrate obtained by applying the magnetic field parallel and perpendicular to the Si substrate. Ferromagnetic behavior is exhibited by the hysteresis loops for both applied directions (perpendicular and parallel to the Si substrate). A high remanent (residual magnetization) with a coercive field of about 220 Oe (1 Oe = $1000/4\pi$ A m^{-1}) is found, which is consistent with the applied magnetic field when it is parallel to the substrate. The remanent decreases when the applied magnetic field is perpendicular to the Si substrate, with a coercive field of 140 Oe. The high remanent and coercive fields result from the shape anisotropy, forcing the magnetic moments to mostly align along the axis of the NWs. The larger data variation is found at both the hysteresis loops (magnetization-magnetic field (M-H) curves), which is caused by the movement of the NWs during measurement. Figure 13.13d shows the SEM image of Fe_3O_4 NWs after applying the magnetic field parallel to the Si substrate. All of the NWs are aligned along one direction after applying the magnetic field parallel to the substrate.

2. Field-Emission Properties of α-Fe_2O_3 NWs

The high density of α-Fe_2O_3 NWs has the potential for use in field-emission devices. The corresponding current density (J) as a function of the applied electric field (E) with a separation of 100 μ m between the anode and emitting surface is shown in Figure 13.10. The turn-on and threshold fields are defined as the fields required for achieving current densities of, 0.01 and 10 $mAcm^{-2}$, respectively. The turn-on and threshold fields are measured to be about 6.3 and 10 V μ m^{-1}, respectively. The Fowler-Nordheim (F-N) lot of In (J/E^2) against $1/E$ is shown in the inset of Figure 13.14. The straight line indicates that the field-emission behavior follows the F-N mechanism where the electrons tunnel through the potential barrier from conduction band to a vacuum state. The F-N relationship is expressed by the equation.

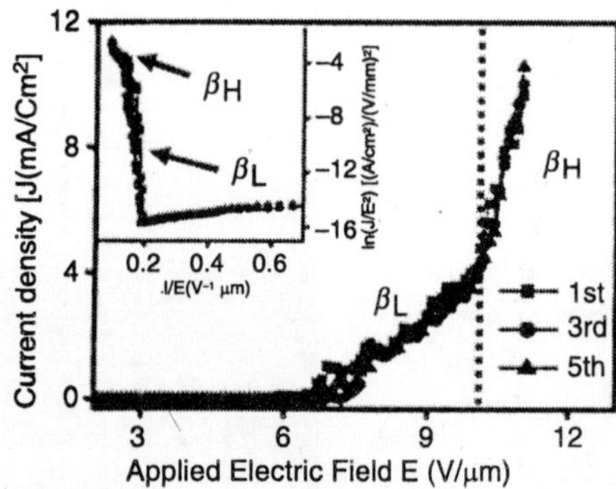

Figure 13.14 *J as a function of E. Two slopes are found. The inset shows a plot of ln (J/E2) against 1/E.*

$$J = (A\beta^2 E^2/\Phi) \exp(-B\Phi^{3/2}/bE)$$

where ϕ is the work function; A and B are constants corresponding to 1.56×10^{-10} AV^{-2} e V and 6.83×10^3 Ve $V^{-3/2}$ μ m^{-1},respectively; β is (a) field-enhancement factor (which indicate the degree of the field-emission enhancement of any tip shape on a planar surface) and (b) a parameter depending on the geometry of the NW, crystal structure, and the density of the emitting points. It is worth pointing out that two slopes are found at the low- and high-field regions in the J-E and F-N plots. Accordingly, the field enhancement at low and high electric fields, β_L and β_H, determined to be 560 and 1500; from the slopes of In(J/E^2)-$(1/E)$ with the work function of 5.6 eV. These two slopes are attributed to the semiconductor feature of NWs, where the electrons emitted from the NWs originate from two successive processes: the excitation of electrons from the valance to the conduction band, and the emission from the conduction band to the vacuum. At low applied voltages, a few electrons excited from the valance band to the conduction band and emitted from the NWs. When the applied voltage is increased to a critical value, all of the electrons totally excited from the valence band to the conduction band, exhibiting the quasimetallic behavior that results in the increased β value in the high-field region. Compared to the P values of other materials, such as Si nanowires ($\beta = 1000$),. $NiSi_2$ nanorods (β-630), TiSi2 nanowires ($\beta = 501$), SnO_2.($\beta = 1402.9$), and AlN ($\beta = 950$), the low turn-on field and high β value of α-Fe_2O_3 NWs exhibit a promising potential in field emission.

Characterization of Vertically Aligned Arrays of α-Fe_2O_3 Nanobelts and Nanowires

Vertically aligned iron oxide nanobelt and nanowire arrays are synthesized on a large-area surface by direct thermal oxidation (see Experiment II above) of iron substrates under the flow of O_2. The as prepared products on the substrate by thermal oxidation are directly subjected to chareterizations by scanning electron microscopy (SEM) and powder X-ray diffraction (XRD), For transmission electron microscopic (TEM) observations, the nanowire products are transferred onto carbon-coated copper grids by carefully drawing the surfaces together and gently sliding them over each other. The XRD analyses are performed on a X-ray diffractometer with Cu Kα irradiation ($\lambda = 1.54$ Å) at a scanning speed of 0.025 deg/s over the 2θ range of 25-75°. The morphologies and elemental compositions of the nanowires are characterized using SEM at an accelerating voltage of 15 kV. TEM observations are carried out on different microscopes operating at 200 and 400 kV, respectively.

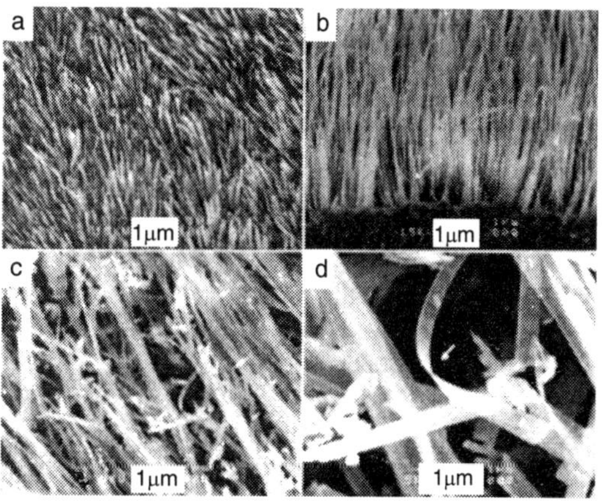

Figure 13.15 *Typical SEM images of the α-Fe_2O_3 nanobelt array (a) top view; (b) side view; (c), (d) low- and high-magnification images of the nanobelts lying down on the substrate. The nanobelt growth conditions are 970K, 5 sccm of O_2 and 20 sccm of N_2. The reactor pressure is kept at ~ 1 atm.*

Figure 13.15 shows SEM images at 970K (20 sccm of N_2 and 5 sccm of O_2) Only wire-like features are produced, which are aligned in a dense array, approximately perpendicular to the substrate surface (Figure 13.15 (a) and (b)). The coverage of the array on the substrate is uniform. The SEM images after rubbing with a forceps reveal that the wires that the wire-like features are actually nanobelts (see Figure 13.15(c) and (d)). Figure 13.15(d) shows a belt twist with a thickness of several nanometers. The belt bending is observed on the top tips of all of the as-grown arrays as is seen in Figure 13.15(a) and (b). The nanobelts are about 5 – 10 nm in thickness, 30– 300 nm in width, and 5–50 μ m in length. The width and thickness of a nanobelt is uniform along the whole length except at the top tip, where thinning and bending usually occur. The temperature dependence of the nanowire K is shown in Figure 13.16 in the form of a series of SEM images. At 670K only very few wire-like structures are observed Figure 13.16(a), which are meandering and sporadic. Instead, the surface is covered with well crystallized flake-like structures. XRD data confirms the flake-like structures are pure α-Fe_2O_3. Increasing the temperature to 870K is accompanied by an increase of the density of the wire-like features and the extent of the alignment Figure 13.16(b). However, the flake-like structures are still all pervading. It is worth noting that, without exception, the wirelike features are all grown out of the cleft between the protruding flakes. When temperature is increased to 970K, the surface of the substrate is all covered by a dense nanobelt array and no flake-like structure is found (Figure 13.16(c)). With a further increase of the temperature to 1070K, the nanobelt array is an even more uniform array consisting of cylindrical nanowires (Figure 13.16(d)). The nanobelt and nanowire morphologies are also shown in the insets of Figure 13.16(c) and (d) respectively. Another distinction lies In the winding top tips of the nanobelts (Figure 13.16(c) and the straightness of the nanowires (Figure 13.16(d)). The cylindrical nanowires average ~30 – 60 nm thick and several tens of micrometers long.

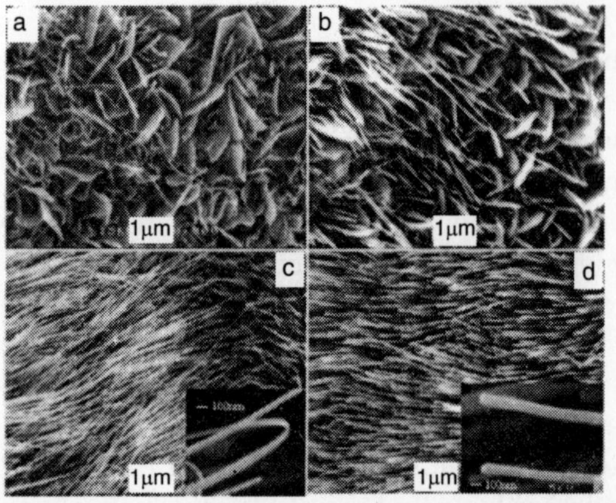

Figure 13.16 *SEM images of the α-Fe_2O_3 970K, synthesized at different temperatures: (a) 670K (b) 870K (c) 970K (d) 1070K Gas flows: 5 sccm of O_2 and 20 sccm of N_2.*

Shown in Figure 13.17 are XRD patterns of the as-prepared nanobelt (a) and nanowire (b) arrays. It is seen that both XRD patterns are in conformity with rhombohedral α-Fe_2O_3 (a = 5.038 Å, c = 13.772 Å). in the XRD pattern of the nanobelt array, the relative peak intensities of the diffraction planes (110) and (300) are much higher as compared with those of the standard powder diffraction pattern of bulk α-Fe_2O_3 (Figure 13.17(c)). This is due to the preferential orientation and alignment of the nanobelt crystals on the substrate. The α-Fe_2O_3 nanobelts grow along the <110> direction, which gives rise to a relatively intense diffraction peak of the (110) plane. The nanobelt top tips are generally bowed; the bending is not severe but probably to the extent that the (300) plane becomes roughly parallel to the substrate surface (<100> is at an angle of 30° to <110>). This is why the (300) diffraction peak

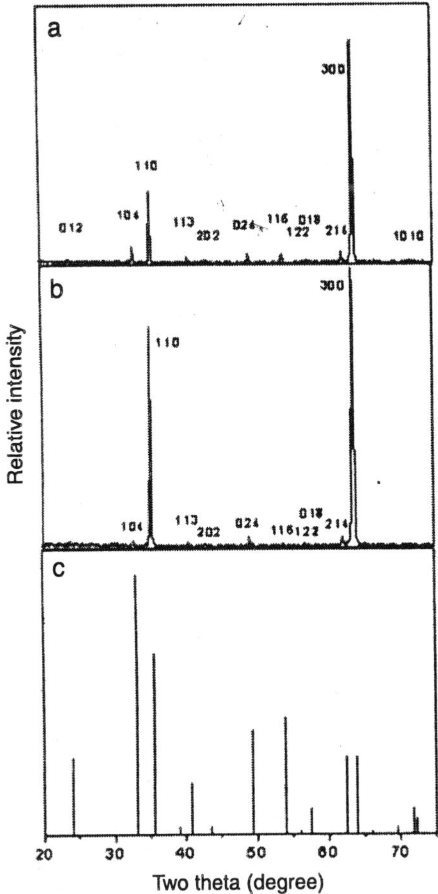

Figure 13.17 *XRD patterns of the vertically aligned α–Fe_2O_3 nanobelt array (a) 970K and nanowire array (D, at 1070K) (c) Standard XRD pattern of α–Fe_2O_3 powder.*

is even much stronger than the (110) diffraction peak. As for the nanowires, the intensity of the (300) diffraction peak becomes comparable to (110), the tilting of the nanowires away from the surface is also seen from the SEM image in Figure 13.16(d).

TEM data and analyses of the α-Fe_2O_3 nanobelt grown at 970K are given Figures 13.18 and 13.19. After being transferred to the TEM grid, the nanobelts are in a lying-down configuration and still aligned with a regular width of 150 – 250 nm (Figure 13.18(a)). The nanobelts get broken at both ends during the transfer but the nanobelt ends are always cleaved at the same angle (~60^0) and the cleaved plane is probably (100). Most of these nanobelts have a uniform width (30–300 nm) on the whole length (several to several tens of micrometers), and the thickness of the nanobelts is only about several nanometers (estimated from SEM observations). The high resolution TEM image of a single nanobelt (Figure 18b) exhibits clear fringes perpendicular to the nanobelt axis. The fringe spacing measures 0.251 nm. which goes well with the interplanar spacing of (110) and alludes to the nanobelt growth direction along [110].

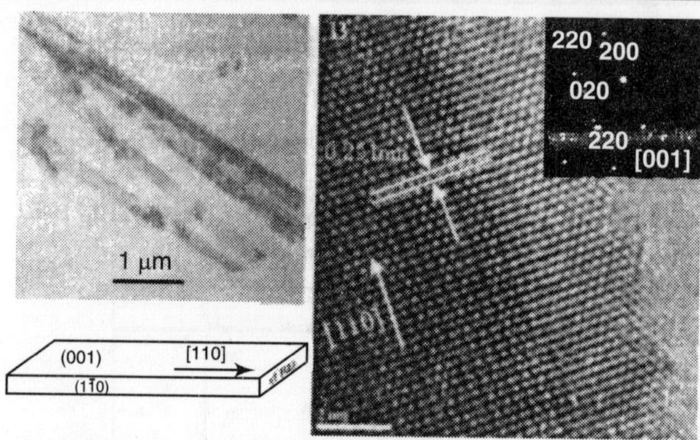

Figure 13.18 *Structure of the α-Fe_2O_3 nanobelts prepared at 970K (a) low-magnification TEM image of the nanobelts; (b) HRTEM image of a single nanobelt along with an ED pattern (inset); (C) a structural model of the α-Fe_2O_3 nanobelts.*

This is consistent with the selected area electron diffraction (SAED) pattern in the inset of Figure which is indexed to the [001] zone axis of rhombohedral α-Fe_2O_3. On the basis of the XRD, TEM, and SAED results above, a structure model of the α-Fe_2O_3 nanobelts in Figure 13.18(c) which extends along [110] and is enclosed by the top/bottom surfaces (001)/(00$\bar{1}$) and the side surfaces (1$\bar{1}$0)/($\bar{1}$10).

Extra weak diffraction spot are observed in the inserted SAED pattern in Figure 13.18 Figure 13.19 is used to characterize the local structural details of these nanobelts. Figure 13.19(a) is a bright-field image of a single nanobelt obtained at 970K. The SAED pattern in Figure 13.19(b) is composed of two sets of diffraction patterns, which correspond to an electron beam parallel to the [001] and [$\bar{1}$11] directions. The image contrast suggests a boundary in the middle of the belts. The high resolution TEM (HRTEM) images from the rectangles C and D enclosed areas are displayed in Figure 19(c) and (d). The inset in each of them is the Fast Fourier transform (FFT) of the corresponding HRTEM image, which indicates that the D area, (bottom part of the belt), is composed of at least two overlapped grains to form a bicrystal structure. The two overlapped grains share the same [110] direction. One of the grains is comparably large and crosses the whole width of the belt as the FFTs indicate. Figure 13.20 shows TEM images of the α-Fe_2O_3 nanowires synthesized at 1070K. After being transferred to the copper grids, the nanowires look more twisted. The nanowires are much thinner (d = ~40 nm) than the nanobelts as found in Figure 13.18. The

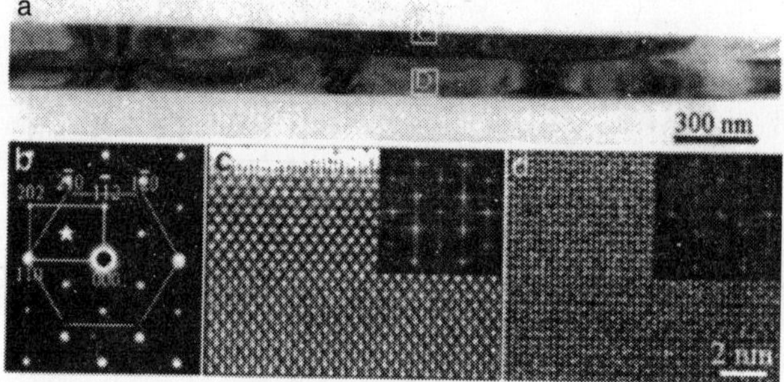

Figure 13.19 *(a) Bright-field TEM image of the Fe_2O_3 nanobelt in the temperature zone. (b) SA ED pattern of the nanobelt in (a). (c,d) HRTEM images recorded from the rectangles C and D enclosed areas in (a). Their FFTs are shown in the insets.*

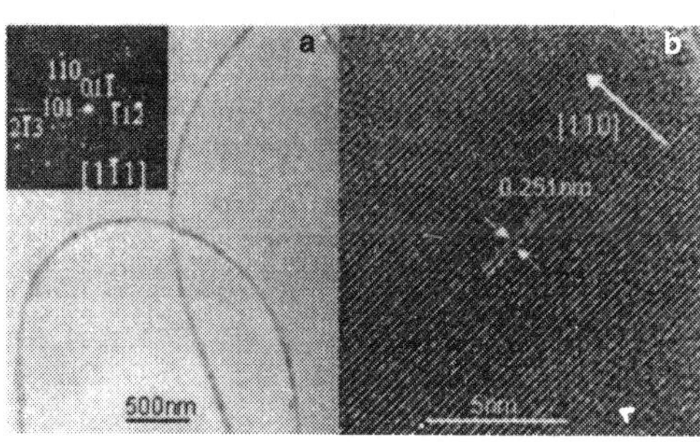

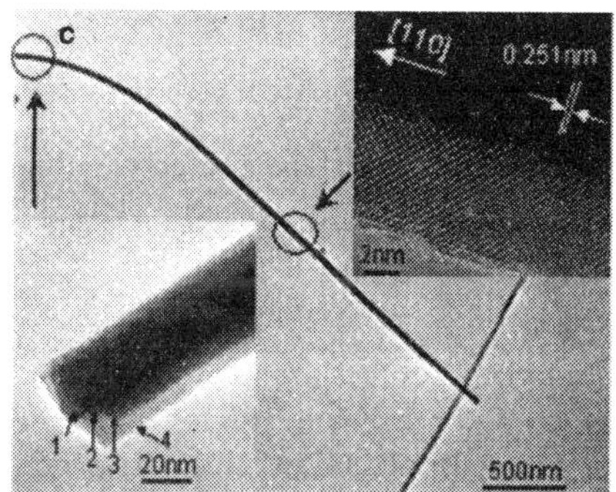

Figure 13.20 *Structure of the α-Fe_2O_3 nanowires and nanoscrolls prepared at 1070K (a) low-magnification TEM image of the nanowires along with an ED pattern of a single nanwire (inset); (b) HRTEM image of a single nanowire; (c.) low-magnification TEM image of a single nanoscroll. Bottom left inset: high-magnification TEM image of the nanoscroll tip. Top right inset: HRTEM image of the same nanoscroll in the shaft region.*

TEM images indicate that the products synthesized at high temperature are not belt-like but probably cable-like judging from the uniform thickness along the wires in the bending configurations. This is consistent with the inference above. The stronger set diffraction spots in the inset of Figure 13.20(a) are assigned with the $[1\bar{1}1]$ zone axis of α-Fe_2O_3. The HRTEM image of a single nanowire shows regular fringes (d = 0.251 nm) perpendicular to the axial direction thus suggesting the same growth direction as for the nanobelts, i.e., [110]. Surprisingly, some nanowires are found to possess a scroll-like structure as is seen in the bottom left inset of Figure 13.20(c). Here four rolling layers are observed at the tip, which form a cylindrical nanoscroll with a diameter of 40 nm. The top right inset of Figure 13.20 shows a bright-field HRTEM image of the same nanowire in the middle portion. The edge area is lighter compared to the inner region and exhibits a better contrast because it is only a single sheet extending out. Regular fringes are also observed perpendicular to the wire axis with a spacing of 0.251 nm. This indicates that the rolling direction of the sheet is perpendicular to [110].

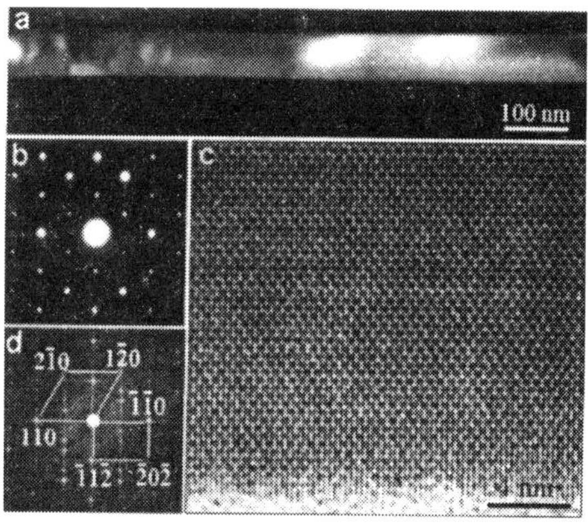

Figure 13.21 *(a) Dark-field TEM image of the Fe_2O_3 nanowire in the 1070K temperature zone. (b) SAED pattern of the nanowire in (a). (c) HRTEM image recorded from the nanowire in (a). The FFT of the HRTEM image is shown in (d), which is consistent with the SAED pattern (b).*

Shown in Figure 13.21 is a nanowire prepared at 1070K. The diffraction contrast in Figure 13.21(a) suggests its wire morphology. The SAED pattern in Figure 13.21(b) is the same as that in Figure 13.19(b), indicating the bicrystal structure. The FFT (Figure 13.21(d)) from the HRTEM image (Figure 13.21(c)) recorded from the wire conforms to its bicrystal structure. It is clear that most of the synthesized α-Fe_2O_3 nanobelts and nanowires have a bicrystal structure and the nanostructures grow uniquely along the [110] direction. The bicrystal structure not be indexed as a twin because the normal twin planes in bulk α-Fe_2O_3 are the basal plane (001) and the R-plane {012}.

Now, consider the α-Fe_2O_3 nanowire growth under the same flow conditions but with different compositions. The results for the 970K growth are shown in Figure 13.22. When the preparation is carried out in the static laboratory air, interconnected rod-like structures 30 – 60 nm in diameter are formed with random orientations (Figure 13.22(a)). In addition some flake like structures are also found. The random orientations of the nanorods is caused by the static gas condition testing mixtures of flowing O_2 and N_2; With O_2 (5 sccm) and N_2 (20 sccm), the surface of the substrate is covered with a well-aligned Fe_2O_3 nanobelt array (Figure 13.22(b)). However, in the atmosphere of pure O_2 (25 sccm), only a small amount wire-like features are obtained (Figure 13.22(c)), which are not uniform and poorly oriented. The nanowires synthesized under this condition are cylindrical, which are more brittle.

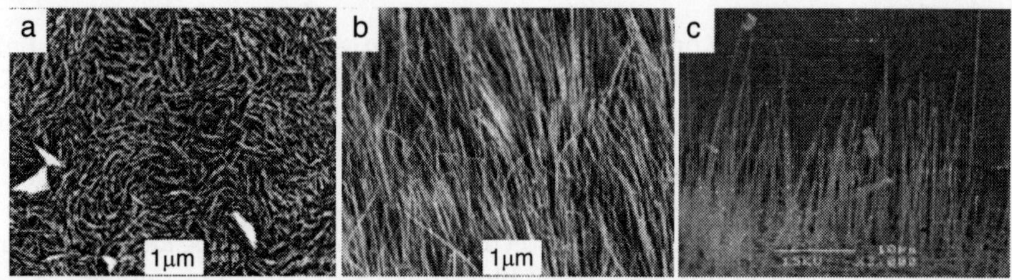

Figure 13.22 *SEM images of the α-Fe_2O_3 nanostructures synthesized under different gas flow conditions: (a) in air; (b) N_2(20 sccm) + O_2(5 sccm); (c) pure O_2(25 sccm).*

Thus due to the very low Fe and α-Fe_2O_3 vapor pressures as the melting points of Fe and α-Fe_2O_3 are 1800K and 1620K. 1D growth of α-Fe_2O_3 is inexplicable by the vapor-liquid-solid (VLS) and vapor-solid (VS) mechanisms, The former is rejected by the fact that no spherical particles at the tips of the nanowires, and the latter is ruled out because all of the 1D features observed are grown out of the substrates. Also low-temperature synthesis prefers the formation of the nanobelts and high temperature synthesis favors the growth of the nanowires. This nanobelt-to-nanowire morphology transition is probably related to surface energy and different growth rates along different crystal directions. The nanobelts are all grown out of the gaps between flakes; no nanobelts are nucleated on the well-crystallized flakes. This indicates that the nanobelts grow from defects such as planar defects of twins and grain boundaries. Finally, tip-growth mechanism is supported. With the lapse of reaction time, narrowing and thinning of the nanobelts occur. If a root-growth mechanism are operative, the nanobelt root appears to grow wider and wider with time. This is not expected because the diffusion of O_2 to the root region is more restricted as the nanobelts grow longer.

A structure model of hematite is sketched in Figure 13.23 to illustrate the growth of the nanobelts and nanowires, As is seen from Figure 13.23(a), the structure comprises the alternatelayers of O triangle nets (a and b) and Fe hexagonal nets (c_1, c_2, and c_3) in the sequence of $Ac_1Bc_2Ac_3Bc_1Ac_2Bc_3$... This layer structure accounts for the smallest dimension of

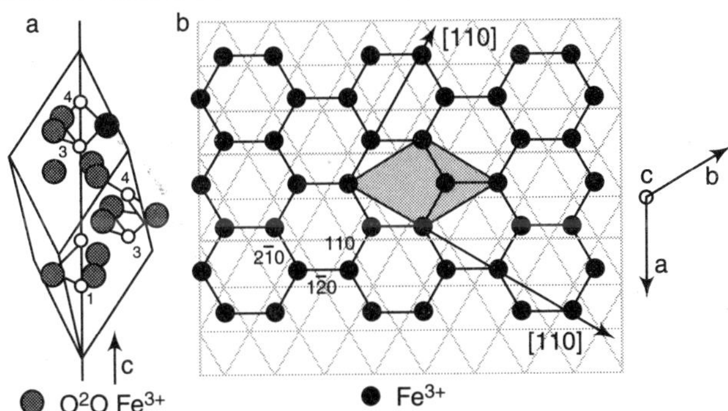

Figure 13.23 *(a) Unit cell (b) an O layer/Fe layer alternate packing structure model of α-Fe$_2$O$_3$.*

the nanobelts-the thickness-because the layers are close-packed and the laver-by-layer growth is expected to be slow, It is also seen in Figure 13.23 that the O atoms are close-packed in the (110) plane, whereas the packing of the Fe atoms in this plane is less close than that in, say, the ($\bar{1}$10) plane. The O-rich and Fe-deficient character of the (110) plane is the driving force for the preferential growth along the [110] direction. On the basis of the considerations above, surface diffusion of Fe atoms is a determining factor, which brings Fe atoms from the Fe foil base to the nanowire tip to sustain the nanowire growth. Because Fe atoms are more reactive and more deficient in the (110) plane and the Fe diffusion in the [110] direction is more facile, the nanobelt growth along the [110] direction is more favorable as observed.

Similarly, with the experimental observation the hematite nanowire growth is rationalized. At low temperatures (<670K), although surface oxidation does occur, very few hematite nanowires are formed because the Fe surface diffusion is too slow. The surface oxidation mainly produces, hematite flakes due to its layer structure. At 870K some fiber like structures start to grow from the clefts of the flakes due to the increased diffusion rate of Fe in the defect sites. When the temperature is increased to 970K the increased diffusion rate of Fe significantly facilitates the hematite nanobelt growth so that the aligned nanobelts uniformly cover the whole substrate surface. The thinning of the nanobelt tips result in the bending of the tips, which is due to the reduced supply of Fe owing to the increased diffusion length, Finally. when the temperature is 1070K the nanowires are much more straight and their cross section becomes cylindrical, at least some of which have a scroll-type structure. Here the temperature is sufficiently high that the Fe diffusion can maintain the nanowire growth without thinning. Moreover, this, temperature is also high enough to induce structure reorganization from a belt to a cylindrical structure so as to minimize the surface energy.

Nanowires (NWs) with different -physical properties and structure characteristics are used for a wide range of applications, Metallic NWs are applicable as interconnects for nanodevices semiconductor NWs are used to fabricate transistors and electro-optical devices, and magnetic 1D nanomaterials have applications in perpendicular data recording and spintronic devices. Among magnetic materials, Iron oxides, such as α-Fe$_2$O$_3$ and Fe$_3$O$_4$, are the most popular materials. In hexagonal or cubic close-packed structures, Fe ions partially fill in the, octahedral or, tetrahedra coordination. α-Fe_2O_3 is the most stable iron oxide compound material and is widely used in photoelectrodes gas sensing catalysts, magnetic recording, and medical fields. Determination of the p-type of α-Fe_2O_3 for as grown samples and then conversion to n-type nature by simply annealing at lower temperatures in a reductive ambient also facilitates promising and various applications in nanodevices.

(B) PULSED LASER GROWTH OF IRON OXIDE NW AND NB

The synthesis and characterization of aligned α-Fe_2O_3 (hematite), $\in Fe_2O_3$, and Fe_3O_4 (magnetic) nanorods, nanobelts, and nanowires on alumina using a pulsed lase deposition (PLD) method is described below;

Experiment III

The iron oxide nanorods nanowire and nanobelts are synthesized via pulsed laser deposition (PDL) using a pellet of pressed magnetic powder as target. Approximately 5 g of Fe_3O_4 powder is pressed in a 9.5 mm diameter steel diameter stéel die at ~ 680 kg pressure to from a cylindrical pellet. The Fe_3O_4 target is placed in a quartz tube in the center of a single-zone horizontal tube furnace. The basic apparatus is shown in Figure 13.24. Next to the target is placed a quartz boat containing polycrystalline alumina wafer substrates coated with 2 nm of Au film to form catalyst islands. The tube ends are closed and water-cooled to maintain a uniform thermal gradient, with one end connected to a rotary vacuum pump and the other to a gas supply. To minimize residual oxygen, the tube is evacuated to 1Pa and held for 1hour prior to running the experiment. Argon gas (99.99%) is intermittently added during this time at a flow rate of 50 sccm to facilitate reduction of adsorbed oxygen and then stopped to allow the system to return to a low pressure of around 2 Pa. The furnace is then allowed to heat up at a rate of 20 K min^{-1} to the desired maximum temperature under a well maintained ambient pressure prescribed according to the chart shown in Table 13.1. Argon flow gas is introduced into the system 50 sccm and is maintain through the duration of the synthesis.

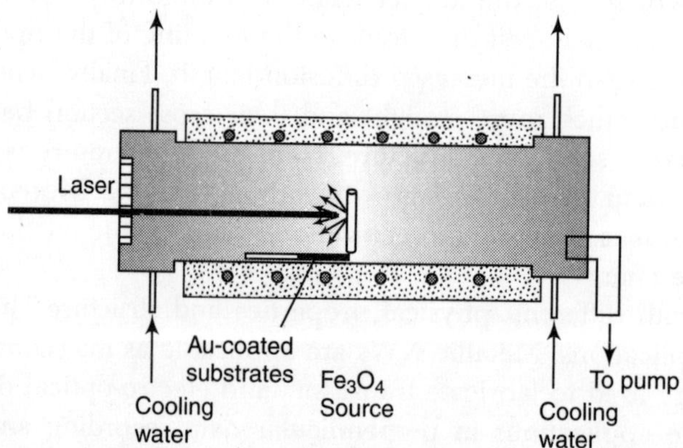

Figure 13.24 *PLD systhesis apparatus, Laser energy is directed through a treated glass window to a pressed magnetitie target vapor are released and recondensed as ID nanostructure nucleated on gold catalyst particles.*

(a)	P_{eq} (Pa)	P_{eq} (Pa)	P_{eq} (Pa)	P_{eq} (Pa)	P_{eq} (Pa)
T_{max}	1	10^3	10^4	3×10^4	5×10^4
1270 K					
1120 K					
1170 K					
1120 K					
1070 K					
220 K					
270 K					
920 K					
870 K					
770 K					
670 K					
570 K					
470 K					
370 K					

Table 13.1 showing tested pressure/temperature combinations and observed outcomes. Each cell represents a unique experiment. Green cells indicate combinations that produced high-density 1D nanowire growth. Red cells indicate no or very low growth. White cells indicate untested parameter combinations or those in which results are inconclusive.

Once the temperature and pressure are stabilized, pulsed laser energy generated (20 Hz, 30 kV, ~ 300 mJ) is directed through a transparent window on one end of the tube and focused toward the source material placed in the beam path. The laser spot on the target measures approximately 1 mm wide by 5 mm long. The laser is then allowed to ablate the target for 60 min, while pressure and temperature remain constant. After 60 min, the laser, furnace, and flow gas are turned off and the system is evacuated back down to 2 Pa. A fan is placed near the tube to ensure rapid cooling. After the system reaches room, temperature, nitrogen gas is pumped into the tube to equilibrate the pressure, the tube, is opened, and the substrate with the deposited sample is carefully collected and loaded onto an SEM stub for analysis. Most samples appear red/orange to gray in color with rainbow-like fringes on the substrate radiating away from the leading edge.

Note that the PLU method allows an additional level of control due to the ability to maintain stoichiometry defined by a premixed target material. The ability of PLD to retain the target stoichiometry in the deposited structure is the result of the extremely high heating rate of the target surface (~108 K/s) due to the pulsed laser irradiation. This leads to the congruent evaporation of the target irrespective of the evaporating point of the constituent elements or compounds of the target. Also, because of the high heating rate of the ablated materials, laser deposition demands a much lower substrate temperature than other evaporation growth techniques. The synthesis of $\in$-Fe_2O_3 long nanobelts is achieved by this method.

The structure of the nanowires is investigated by transmission electron microscopy (TEM) and X-ray diffraction (XRD). The magnetic properties of the synthesized nanostructures are investigated by superconducting quantum interference device (SQUID).

The as-synthesized sample obtained by the above experiment is directly imaged using a field emission scanning electron microscope (SEM) with a resolution of 1 nm. The accelerating beam voltage is limited to 3 kV to reduce sample charging and potential damage. Some nanobelts are observed to interact with the electron beam, occasionally vibrating and later breaking if under beam focus. Several images as well as EDS data on the sample are collected.

Grazing incidence X-ray diffraction measurements are obtained using a PANalytical X-Pert Pro MRD with CuKα radiation. A 1° grazing angle is used to minimize substrate signal noise, along with a parabolic mirror, 0.27° parallel plate collimator with a flat graphite crystal diffracted beam monochromator. The monochromator reduces iron fluorescence under Cu radiation. Measurements are taken with a tube power of 45 kV and 40 mA, from 28 to 60° 2θ, with a 0.04° 2θ step size and a 100 s count time. For structural analysis, a carbon film coated copper grid is swept over the sample and loaded an high-resolution transmission electron microscope (TEM) operating 400 kV. Electron diffraction patterns as well as high-resolution image of several nanostructure are recorded and developed on film. The conditions necessary for 1D nanostructure growth using the VLS-assisted PLD method are achieved by correlating systematic variations in experimental parameters. Table 13.1 show the parameters mostly explored at the maximum ambient temperature (T_{max}) inside the tube and the controlled equilibrium pressure (P_{eq}) during laser ablation. Each cell in the table is a possible unique experiment keeping all other variables, such as substrate type, catalyst density, target species, Ar flow rate, temperature ramp rate, ablation time, laser energy, and laser frequency constant. After each experiment is complete, samples are examined in the SEM to determine the ability of the chosen parameters to sustain growth of 1D ferrite nanostructures. In the figure, red cells correspond to experiments in which nanowire growth is unsuccessful, green cells correspond to parameters yielding successful growth, and white cells correspond to untested parameter combinations, or those in which results are inconclusive.

Since the ferrite nanostructures grow only on catalyst particles, successful growth is measured according to nanostructure length rather than density. So, a uniform high-density growth of nanostructures is experimented with length dimensions at least 3 times that of the catalyst particle diameter. The smallest nanorods in this experiment display greater length. The best 1D ferrite nanostructure growth occurs from 970 to 1220K at 10^3 Pa pressure. Experiments utilizing these parameters are repeated several times.

Figure 13.25(a) and (b) show; length variation of nanostructures according to chamber temperature (T_{max}) at time of synthesis. Nanorods grown at 970K as shown in Figure 13.25(a) are aligned, and single-crystalline ones, just like those shown in Figure 13.25(b), grown at 1170K with the only difference being nanostructure length, This illustrates the way to control ferrite nanostructure size by varying reaction chamber temperature. Although P_{eq} and T_{max} are the primary parameters investigated the effect of laser ablation time on 1D nanoferrite growth is also determined the experiment carried out at 1170K and 10^3 Pa show that increasing ablation time from 60 to 90 or 120 min do not increase mean nanostructure length, but increases branching and secondary growth Figure 13.25(c) demonstrates this with an SEM image taken from a sample ablated for 120 min at 1170K and 10^3Pa the same P_{eq} tind T_{max} combination shown in Figure 13.25(b) after the normal 60 min ablation time. As the laser

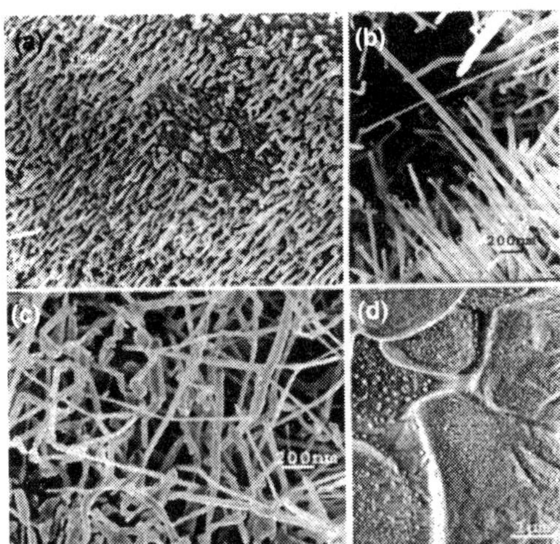

Figure 13.25 *(a) SEM image showing short rods synthesized a 970 K, 10^3Pa (b) SEM image showing long belts synthesized at 1170K, 10^3Pa. (c) SEM image displaying greater secondary growth after 120 min of laser energy exposure compared to usual 60 min growth time 1170K, 10^3 Pa. (d) SEM image showing no 1D growth at 1170K, 10^3 Pa, without laser energy.*

energy is the primary energy source for nanowire growth; So The experiment is conducted at 10^3 Pa and 1170K to produce very high density ferrite nanostructures when laser energy is applied. This results in Figure 13.25(d) Clearly, there is no growth at all on the substrate the laser radiation is the principle energy source responsible for target vaporization and results nanostructure growth. This result also indicates the ability for greater control over growth time in the PLD process relative to thermally driven vapor evaporation growth techniques in which critical growth initiation temperature.

SEM, TEM Characterization of Magnetite and Hematite Nanostructures

Figure 13.26 shows SEM images recorded from an as-synthesis N.W with the inset containing EDS data gathered from TEM. The EDS data taken from TEM analysis indicate the presence of Fe, Mg, and O, with Cu and C radiation originating from the sample grid, and Si peaks due to the signature of the detector. Both Mg-doped and pure ferrite nanostructure are grown via this method. Images in Figure 13.26(a) and (b) illustrates ferrite nanowires, including some nanobelts with approximately 6 atom % Mg present, while Figure 13.26(c) and (d) shows pure iron oxide nanostructures. From these images, both pure and Mg-doped 1D ferrites are of high-quality single-crystalline nanostructures. Mg incorporation occurs almost exclusively in the longest nanobelts. Nanowires and nanobelts in Figure 13.26 are grown at a maximum ambient temperature of 1170K. Although nanostructure dimensions vary greatly within samples, nanowires and nanobelts grown at 1170K are generally highly dense and on the order of μ m long × 30 nm wide, with the longest nanowires growing up to approximately 5 μ m. Most long structures appear to possess a rectangular cross section, giving them a belt-like morphology; however, those with more circular cross section are observed, as well, especially at the higher temperatures.

Figure 13.26 *SEM image of Mg-doped and pure iron oxide 1D nanostructures. (a) Low-magnification SEM image of aligned iron oxide nanowires doped with Mg showing local alignment along alumina crystallites. Inset shows EOS chemical signature. (b) Higher magnification SEM image showing aligned Mg-doped iron oxide nanorods. (c) SEM image showing long nanobelt structure most likely having* ∈ *-Fe_2O_3 microstructure, without presence of Mg. (d) SEM image showing group of aligned iron oxide nanowires with Mg absent.*

Under the 970K condition, nanorods have dimensions of 500 nm × 30 nm. Long nanobelts up to 7 microns in length and averaging around 100 nm in width are observed under both temperature conditions and appear to grow in random orientations in or near grain boundaries. Density of these structures likewise increases at elevated temperature.

The uniform length and alignment of the nanorods localized on different alumina grains is seen at low magnification in Figure 13.26(a). Each grain promotes aligned growth in a specific direction relative to the plane of the substrate. Some grain faces are more favorable and initiated growth earlier, as nanostructures are longer in some orientations relative to others. This effect is not due to simple vapor flow variation as evidenced by the observation that nanostructures growing on adjacent grains exhibit significant length differences. The nanostructures growth at specific orientations and according to substrate crystallography indicates that they are controlled by the angle of their growth relative to the substrate by carefully engineering surface orientation on single-crystalline substrates. In fact, synthesis on c-plane and α-plane sapphire (single-crystalline Al_2O_3) produce high density arrays with perpendicular and in-plane growth, respectively. Figure 13.26(b). shows each nanowire tip terminated with a spherical gold particle. The ability to determine nanowire location through precise placement of catalyst material provides further spatial control over the nanowire synthesis.

Figure 13.26(c) is a image of the observed long nanobelts Electron diffraction analysis as well as XRD data indicate that these long nanobelts exhibit an ∈-Fe_2O_3 phase of iron oxide. Some are shown in EDS to contain a small amount of Mg species.

Figure 13.27(a) to (e) shows TEM data of magnetite and hematite nanowires. Both hematite and magnetite nanostructures are found with no observable local phase domains. Although early N.Ws contain majority of hematite structures, predominantly magnetite samples are later synthesized by flowing Ar (50 sccm) for 10 – 20 min intervals during initial tube evacuation to better eliminate residual oxygen in the system. Adding a piece of iron foil to the chamber to act as an oxygen "getter' is - helpful in increasing the relative magnetite yield. This finding indicates that phase is controlled through careful variation of oxygen partial pressure within the tube. In Figure 13.27(a), a bright field image of a magnetite nanowire with the spherical gold catalyst particle is visible at the tip. The high-resolution TEM image of this structure seen in Figure 13.27(b) illustrates the near perfect single-crystalline arrangement of atoms in the nanowire. Very little amorphous material is present on the surface. The electron diffraction pattern in the inset shows the single-crystalline nature of this nanowire and that wires along the $[1\bar{1}1]$ direction. This is not the sole growth direction of magnetite nanowire as TEM they are found to grow

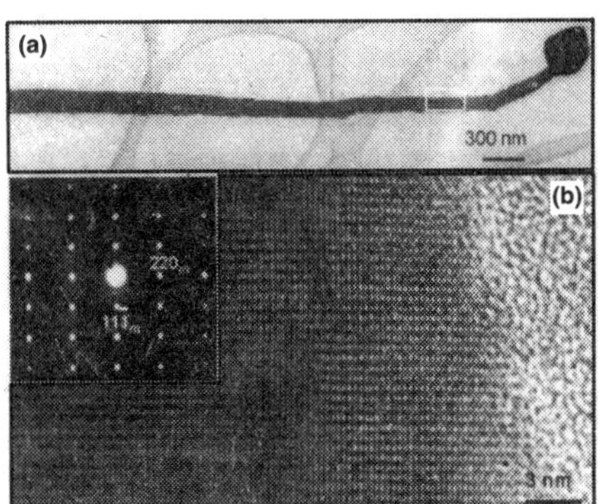

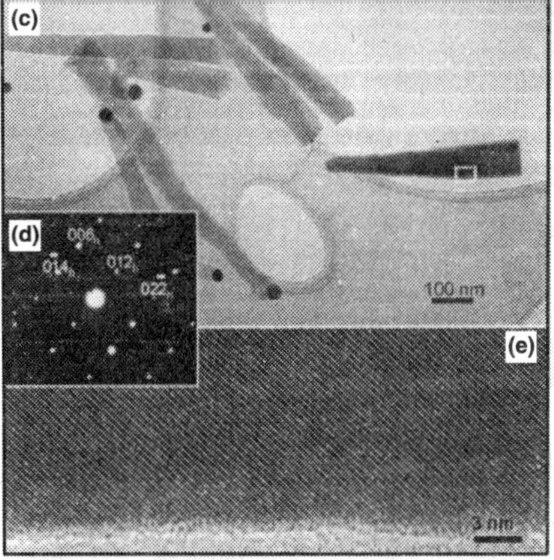

Figure 13.27 *TEM data on magnetite and hematite nanostructures. (a) Low-magnification bright field TEM image of magnetite nanowire. (b) High-resolution TEM image of nanowire taken from the side edge of the wire in (a), illustrating high-quality single-crystalline nature of the wire. (c) Low-magnification bright field TEM image of hematite nanorods with Au particle tips. (d) Electron diffraction image taken from dark rod in previous image shows single-crystalline nature of the rod. (e) High-magnification TEM image of the section outlined in (c), showing well ordered structure and very little amorphous material on the nanorod surface.*

along the <110> directions. Presence of magnetite rather than magnetic (γ-Fe_2O_3 defect spinel) in the samples is confirmed through determination of lattice parameters from TEM diffraction data as well as XRD phase analysis. Both electron diffraction and XRD distinguish the structures based on differences in the space groups exhibited by each structure. The *d* spacing of Fe_3O_4 is about 0.48 nm, while that of γ-Fe_2O_3 0.59 nm.

Figure 13.27(c) shows a bright field TEM image of several hematite nanorods. Again, Au catalyst particles are observed at the tips, indicating VLS as the dominant growth mechanism in the sample. The high-resolution image in Figure 13.27(e) clearly shows the ordered arrangement of atoms and the lack of a significant amorphous layer at the surface. The nanorod in this figure grew along the $[01\bar{1}]$ direction as seen in the electron diffraction pattern shown in Figure 13.27(d).

XRD Characterization of Ferrite Nanostructures

X-ray diffraction characterization of several N.W is given in Figure 13.28. The image in Figure 13.28(a) shows XRD data from a sample grown at 970K and 10^3Pa, on a polycrystalline alumina substrate. The diffraction pattern analyzed. Although the polycrystalline alumina peaks are the strongest, the pattern shows the presence of gold, magnetite, and hematite. Figure 13.28(b) shows a pattern performed on an iron oxide nanowire grown at 1020K 10^3Pa, on a *c*-plane sapphire substrate. Sapphire is used to eliminate the influence of the substrate on the XRD diffraction pattern. The structures grow perpendicular to the substrate in a high-density nanowire array tipped with gold particles. Phase identification of the diffraction pattern in Figure 13.28(b) yields five known phases: Au hematite (Fe_2O_3,) a magnetite-like phase, an orthorhombic iron oxide found to correspond to ϵ-Fe_2O_3, and a cubic iron oxide The cubic iron oxide phase is identified as a BCC phase.

A weight fraction-is ratio of 0.57 hematite to magnetite is obtained. ϵ-XRD analysis of alumina substrate indicates small amounts of a spinel aluminum oxide phase within the corundum Al_2O_3 wafer. This is a $MgAl_2O_4$ phase added to and in sintering and is the source of the Mg seen in some of long nanobelt structures.

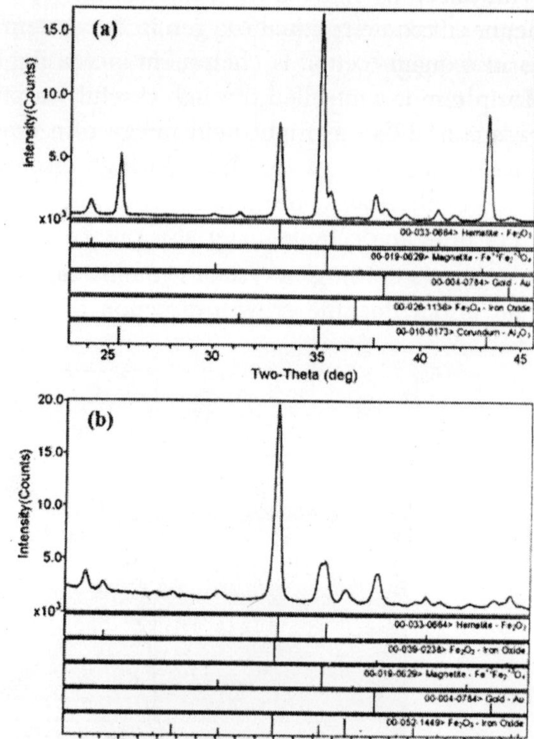

Figure 13.28 *XRD data taken from samples grown at 1070K 10^3Pa, and 10^3Pa, respectively. (a) XRD signal from grazing incidence measurements of iron oxide nanobelts and nanowires grown on polycrystalline alumina. (b) Similar measurements on iron oxide nanostructures grown on a c-plane sapphire substrate.*

SQUID Characterization of Ferrite Nanostructures

SQUID (superconducting quantum interference device) measurements on NW created at 970K and 10^3Pa; 1170K and 10^3Pa, are shown In Figures 13.29 and 13.30 respectively Figure 10.29 demonstrates magnetic hysteresis behaviors of these samples at 300 K (room temperature) and 5 K. Data for nanostructure created under the 970K condition are shown in Figure 13.29(a) and (b) with Figure 13.29(c) and (d) correlating to the 1170K synthesized sample. As shown in Figure 13.29(a), the nanostructures grown at 970K exhibit strong ferromagnetic behavior at room temperature. The kink, or "wasp waist", shown at approximately 4 kOe in the curve indicates presence of two compatible magnetic phases. The diamagnetic nature of the alumina substrate is the cause of the curve's decay at high field strengths. While ferromagnetic behavior is for the magnetite 1D nanostructures, above the temperature of hematite ~ 263 K bulk), magnetic moments are away from the anti ferromagnetic axis, resulting only in a very small net positive magnetization. This effect is not diminished by the small size of the nanowires. The Morin temperature is found to decrease as particle size decreases in hematite, depressing this transition further away from the temperature shown in this figure. The magnetic characterization

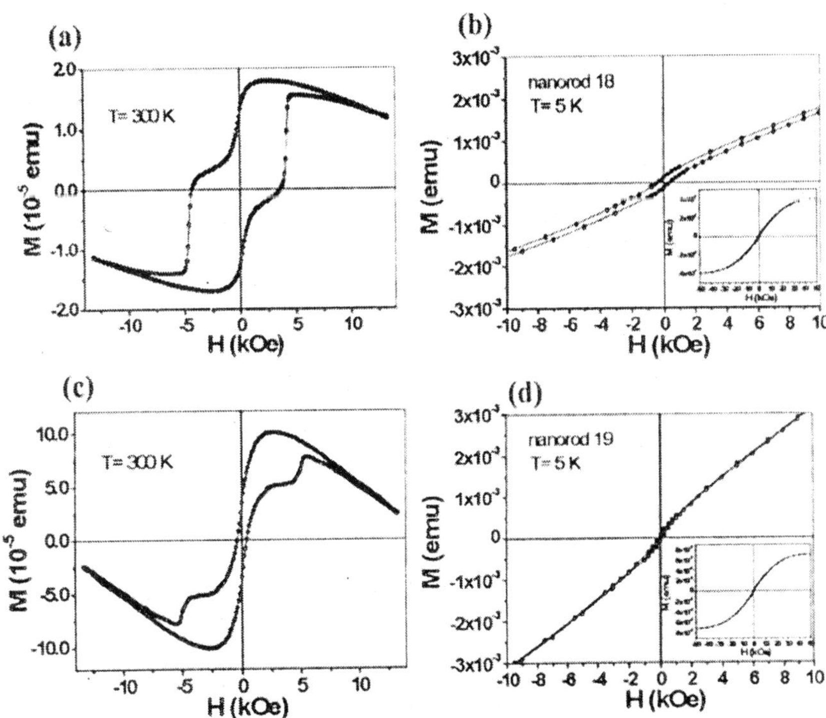

Figure 13.29 *Magnetic hysteresis behavior of iron oxide nanostructures grown at 970 K measured with SQUID (a) M – H lop at 300K showing ferromagnetic behavior and wasp-waist shape most likely derived from multiphase nature of the sample. (b) M H loop at 5 K with low field region expanded showing soft magnetic behavior Right inset: M -H loop from –5 to 5T (c and d) similar results in a sample prepared at 1170K in (c).*

of hematite nanowires and nanoparticles show only very weak ferromagnetic or paramagnetic behavior at room temperature. Therefore, it is possible that the two compatible magnetic phases displayed in the graph shown in Figure 13.29(a) are magnetite and the nanobelts of the ferrimagnetic $\in$-Fe_2O_3 phase. The figure shows an H_c value of about 4 kOe. This value is much less than the maximum room temperature coercivity of 20 kOe for $\in$-Fe_2O_3 nanoparticles. The decrease in coercivity is due to interactions with the magnetite and hematite phases. Since 20 kOe value is obtained in samples containing IIA metal Ions with much lower values for pure phase samples the absence of Mg can be responsible for the lower coercivity.

Figure 13.29(b) shows the M-H behavior of this NW at 5 K. At this low temperature, the hysteresis behavior disappears, producing a curve that looks like a soft magnetic material. At 5 K, the magnetic response from the hard phase seen at room temperature disappears.

This indicates the presence of two phases with different temperature-dependent magnetic behaviors. This figure provides further evidence for the presence of magnetite and $\in$-Fe_2O_3 as magnetic phases in the sample. $\in$-Fe_2O_3 observes-an unusual temperature dependence in which the coercivity decreases significantly below 100 K. This effect is due the Arrhenius nature of domain wall motion. The critical magnetic domain size reduces to a value below the size of the nanostructure as T and therefore k (the reaction rate constant) decreases, causing an inhomogeneous state and loss of single domain character. The inset shows the expanded M-H loop, indicating magnetic saturation around 50 kOe.

Figure 13.29(c) and (d) show similar data for the 1170K sample. In this sample, the magnetic response from the softer phase has a greater influence on the overall hysteresis curve. Quantitative XRD measurements show that the hematite concentration in this sample is greater, thus indicating that hematite nanostructures contribute to the change in magnetic response. Long nanobelts are observed in large density on this sample. This indicates that the influence of the $\in$-Fe_2O_3 phase cannot be ignored, The stronger soft magnetic response is evident in Figure 13.29(d), in which the area contained within the loop is smaller than that in the other sample.

Figure 13.29 also demonstrates hysteresis behavior, such as that of a soft magnetic material with small remnant magnetizations. These small portions contain a local phase domain such that the magnetic behavior appears like that of a homogeneous group of ferrite nanowires. In these samples, SQUID measurements reveal saturation around 15 kOe when exposed to temperatures from 5 to 300 K.

The temperature dependence of the magnetization under ZFC-FC (zero field cooling-field cooling) conditions at 1000 Oe is shown in Figure 13.30(a) and (b). The same sample shown in Figure 13.30(a) and (b) is measured again at 100 Oe, and the result is shown in Figure 13.30(c). These figures show significant irreversibility from room temperature, with FC curves demonstrating Curie like increasing. In Figure 13.30(c) it is observed that the irreversibility occurs at 250 K. The magnetic field application direction has a little effect on these, since the nanowires are aligned only locally according to each grain.

Nanowire Fe_3O_4 to ϵ-Fe_2O_3

Among the polymorphs of ferric oxides, maghemite (γ-Fe_2O_3) and hematite (α-Fe_2O_3) are well known, how ever, knowledge of $\in$-Fe_2O_3 considered an intermedite phase between magnemite and hematite is relatively limited. The disparity is most likely due to difficulties in

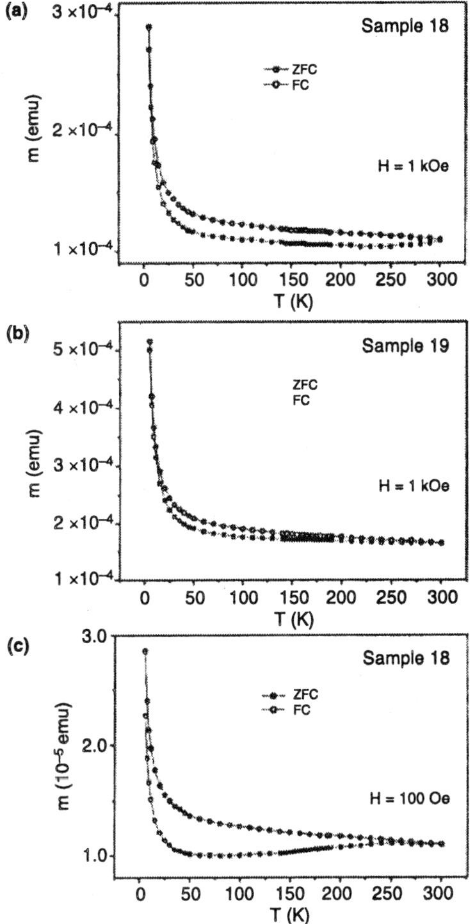

Figure 13.30 *Temperature-dependent magnetization under ZFC-FC (zero field cooling-field cooling) conditions at applied field of 100 – 1000 Oe. (a) Effect of temperature on magnetization for iron oxide nanowire on alumina substrate prepared at 970K at 1000 Oe field strength. b) Effect of temperature on magnetization for iron oxide nanowire sample on alumina prepared at under 1000 Oe. (c) ZFC-FC curve for under 100 Oe field strength.*

stabilizing the ϵ-Fe_2O_3 now the transformation of γ-$Fe_2O_3 \rightarrow \epsilon$-$Fe_2O_3 \rightarrow \alpha$-$Fe_2O_3$ in small particles .through using spectroscopy noting that, most likely, the formation of ϵ-Fe_2O_3 an intermediate is the path way the only known to the synthesis of ϵ-Fe_2O_3, however a completely novel synthesis path, one involving magnetite (Fe_3O_4) , rather than maghemite as the precursor phase. It is not surprising that if maghemite can give rise to the epsilon phase, magnetite can also do so. Both magnetite and maghemite phases share the same basic structure, with close-packed oxygen atoms resting in an alternating ABCABC stacking sequence along the [111] direction, and Fe occupying octahedral or tetragonal lattice sites. The distinguishing feature of the gamma phase is that some Fe positions are left occupied as random vacancies, reducing the ordering in crystal structure. In transmission electron microscopy (TEM), this loss of symmetry corresponds to the appearance of diffraction spots that are forbidden in a more perfect lattice.

The observations of $\in$-Fe_2O_3 describe the phase as an intermediate of γ-Fe_2O_3-and not Fe_3O_4 The conversion from Fe_3O_4 to $\in$-Fe_2O_3 implies, equivalently, an overabundance of iron or an oxygen deficiency in the starting phase. Therefore, the forming material ($\in$-Fe_2O_3) must either get rid of the excess iron or adsorb additional oxygen atoms. For high-surface-area nanowires growth via solid state physical evaporation, oxygen adsorption is not problematic At low pressures, irons desorbed from the lattice. If oxygen is adsorbed into the lattice (or iron desorbed from the lattice) such that Fe site occupation is no longer periodic, the maghemite phase is observed. A direct transition from Fe_3O_4 to $\in$-Fe_2O_3 without other intermediate phases indicates that the formation of the epsilon phase is accompanied by perfectly ordered species evolution.

A standard synthesis method for $\in$-Fe_2O_3 nanocrystals is the heat treatment of γ Fe_2O_3 nanoparticles dispersed in silica xerogels. Drawbacks of this method include the existence of silica impurities and small grain size, which normally fans under 200 nm. For many potential applications, pure crystal $\in$-Fe_2O_3 are required. Impurities alter the performance of devices based on $\in$-Fe_2O_3 materials, and create obstacles in further refining or treatment. Small particle sizes limit the applicability of these materials due to maneuverability and processing, as well as integration into larger systems. Other synthesis methods such as thermal decomposition, and high-energy deposition syntheses produce products of mixed phase or other impurities, and often with low yield; for these reasons $\in$-Fe_2O_3 is not fully exploited. Hence a pulsed laser deposition (PLD) method and Fe_3O_4 powder as a precursor is used to synthesize one-dimensional F_3O_4 nanowires and high-yield $\in$-Fe_2O_3 nanowires.

The structural evolution of three iron oxide phases are found to exist in nanowires grown by using this method is given below.

Structure and Structural Evolution

The synthesized Fe_3O_4 nanowires are classified according to growth direction into three categories, as those growing along the $[111]_M$ the $[211]_M$ and the $[110]_M$ directions, which are displayed separately in Figures 13.31 to 13.33 respectively, (Note notation convection of using M, $\in$, and H subscripts to denote the plane and orientation indices in Fe_3O_4 (magnetite), $\in$-Fe_2O_3 and α-Fe_2O_3, (hematite) phases, respectively.

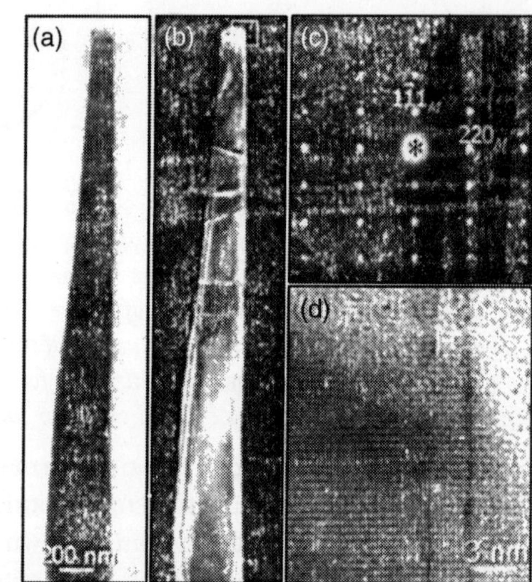

Figure 13.31 *General morphology of the nanowires, TEM (a) bright, field and b) dark-field images of the [111] growth magnetite nanowire, c) SAED pattern of the wire, d;) HRTEM image recorded from the white rectangular area in (b).*

Figure 13.31(a) and (b) shows the bright field and dark-field TEM images of a Fe_3O_4 nanowire grown along the $[111]_M$ direction. The select-area electron diffraction (SAED) pattern in Figure 13.31(c) is uniquely indexed using cubic-phase magnetite, Although some stacking faults are identified in the dark-field image in Figure 13.3l(b), the entire wire is single crystalline and is described as a pure magnetite structure. The high-resolution TEM (HRTEM) image in Figure 13.31(d) confirms the $[111]_M$ growth direction of the nanowire.

The nanowire in Figure 13.32(a) takes the <211>$_M$ growth direction. Twin boundaries and stacking faults are seen in the image. Further, the existence of twin structures also is seen in the SAED pattern in Figure 13.32(b). Without expectation, all diffraction spots in Figure 13.32(b) are indexed using the twinned magnetite structure, taking $(111)_M$ as the twin plane. Although the planar defects exist within the wire, it is best described as a phase of pure magnetite.

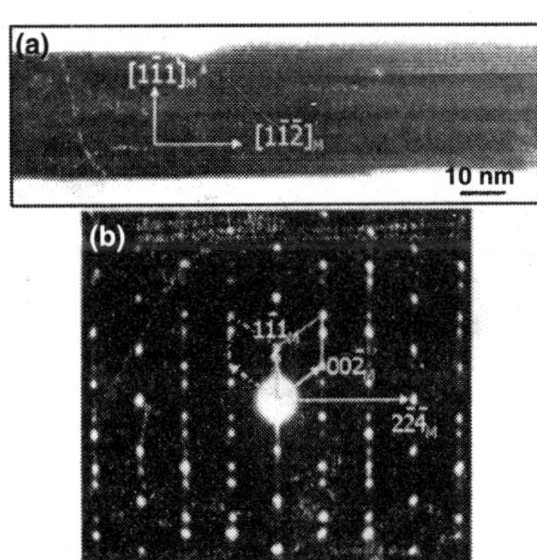

Figure 13.32 *(a) HRTEM image and b) SAED pattern of a magnetite nanowire grown along the <211> direction. Stacking faults and (111) twins observed in the HRTEM image.*

While the nanowire shown in Figure 13.33(a) grows along the $[110]_M$ direction, there are extra diffraction spots in the SAED pattern recorded from the whole nanowire, as shown in Figure 13.33b, which is not indexed by the magnetite phase. Changing to a small select-area aperture, the SAED patterns are created as shown in Figure 13.33(c) and 13.33(d) recorded from the circled areas "C" and "D" in Figure 13.33(a). The pattern in Figure 13.33(e) is the lattice projection when the electron beam is parallel to $[\bar{1}\,\bar{1}2]_M$, while the SAED pattern in Figure 13.33(d) is the projected $\in$-Fe_2O_3 structure with the electron beam along $[100]_\in$. The presence of the $\in$-Fe_2O_3 phase is firmed from the varying geometric projections observed in electron-diffraction patterns. The series of electron-diffraction patterns (See Figure 13.34) in which the sample is rotated along a fixed axis so that the electron beam project atomic positions at known angles. The reciprocal space lattic shown in Figure 13.33(f) is constructed. Each diffraction pattern in Figure 13.34(a) to (e) corresponds to a slice through this reciprocal lattice at a unique orientation. This reciprocal space lattice is found to correspond to a real-space orthorhombic unit cell, with α ~ 5.11 Å, b ~ 8.72 Å, and c ~ 9.42 Å, parameters which match well to the $\in$-Fe_2O_3 phase. Looking back to the SAED pattern in Figure 13.33(b), the $(333)_M$ and $(444)_M$ diffraction spots are not single points, but are separated into pairs, with the additional spots corresponding to $(006)\in$. and $(008)\in$ as shown in

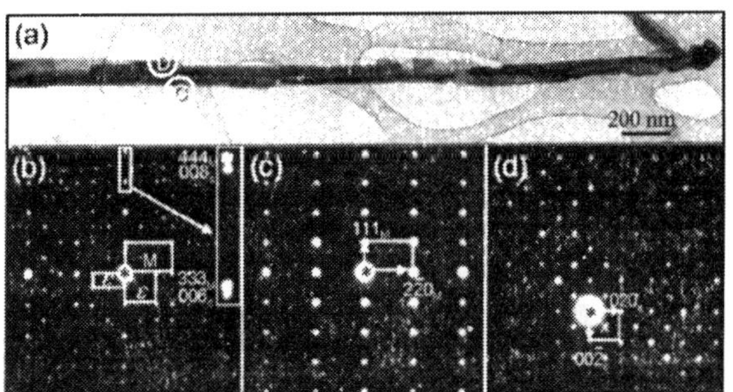

Figure 13.33 *(a) Low-magnification TEM image of a [110] growth magnetite nanowire. The SAED patterns from the whole wire, and circled areas "C" and "D" are displayed (b), (c), and (d), respectively.*

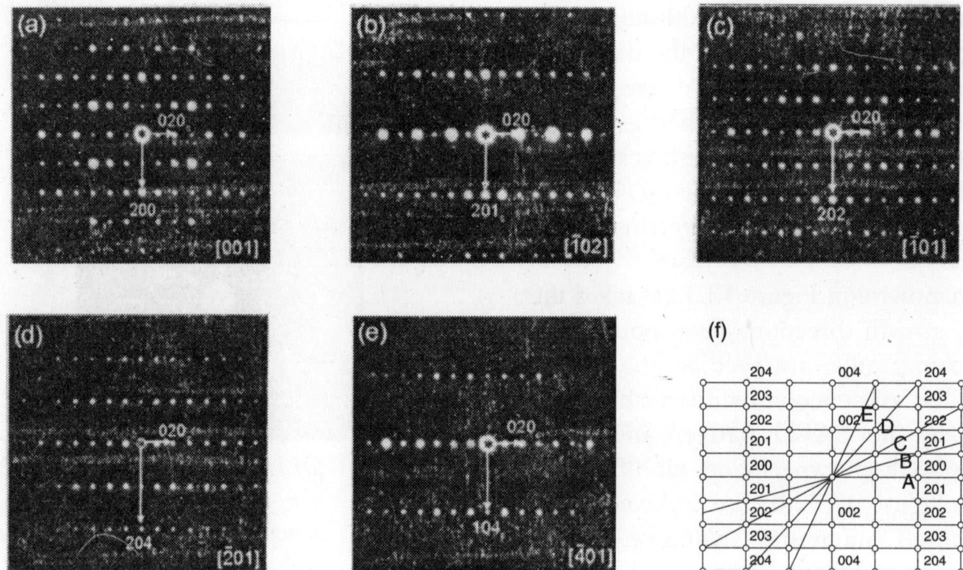

Figure 13.34 *Multiple directional examination of the diffraction behavior of the ε-Fe_2O_3 nanowires. a-e) A series of zone axis diffraction patterns of the ε-Fe_2O_3 phased wire with beam directions along the [001], [$\bar{1}$ 02] [$\bar{1}$ 01] [$\bar{2}$ 01], and [-401] directions. respectively. These diffraction patterns provide the 3D structure of the ε-Fe_2O_3 phase. f) The reciprocal space lattice built up from these patterns, where the reciprocal planes correspond to the patterns displayed in (A-E) are marked.*

the inset. This effect is due to the small difference between the planar distance of magnetite (111) and $\in$-Fe_2O_3 (001). There is no splitting of the $(333)_M$ and $(444)_M$ diffraction spots in Figures 13.31c and 32b, indicating the absence of the $\in$-Fe_2O_3 phase in these nanowires. Besides the $[\bar{1}\,\bar{1}2]_M$ pattern, two sets of $\in$-Fe_2O_3 patterns distinguished in Figure 13.33(b). The structural relationship between magnetite and the $\in$-Fe_2O_3 phase is described as, $(001)\in||.(11\underline{1})_M$, $[010]\in.||$ $<110>_M$, and $(001).||$ $(111)_M$, $[1\,\bar{1}0]||$ $<110>_M$ The two different structural relationships are due to the existence of 120° rotation domains in the $\in$-Fe_2O_3 section.

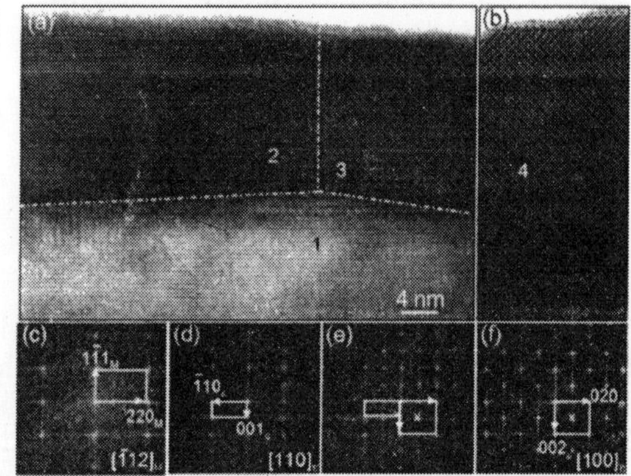

Figure 13.35 *Domain structure in a single nanowire. a,b) HRTEM images recorded from the wire shown in Figure 13.33(a). c-f) FFT patterns from the 1 – 4 areas shown in (a) and (b).*

The HRTEM images in Figure 13.35(a) and (b) are recorded from the nanowire shown in Figure 13.33(a). The fast Fourier transforms (FFT) from areas 1, 2, 3, and 4 are displayed in Figure 35(c) to (f), respectively. After indexing these FFT patterns, the area that corresponds to the magnetite phase is determined (area 1), while areas two and four correspond to the ∈-Fe_2O_3 phase, and area three is best described by two ∈-Fe_2O_3 phase domains overlapped. The two individual structural relationships between magnetite and the ∈-Fe_2O_3 phase is confirmed in the HRTEM images.

From a crystallographic point of view, both magnetite and ∈-Fe_2O_3 are composed of close-packed oxygen layers with Fe ions occupying the interstitial oxygen octahedral or tetragonal sites. The close-packed oxygen layers in magnetite stack in a ...ABCABC... sequence along the $[111]_M$ direction, whereas the stacking sequence in ∈-Fe_2O_3 changes to ...ABACABAC... along the [001]∈. direction. Figure 13.36a shows a ball-and-stick model of the magnetite structure, and Figure 13.36(b) and 13.36(c) shows models for the ∈-Fe_2O_3 phase. In the two ∈-Fe_2O_3 models, I and II in Figure 13.36, are based on the two structural characterizations model I is the ideal structure whereas II is the refined structure by tilting the nanowires, HRTEM images with the electron beam along the [$\bar{1}$01]∈ [001]∈ direction are recorded in Figure 13.36(d) and (e), respectively. Taking into account a sample thickness of 3.6 nm and defocus effects, simulated images based on models I and II are created. The simulated pattern are seen in the small squares superimposed on the actual high-resolution images shown in Figure 13.36(d) and (e)

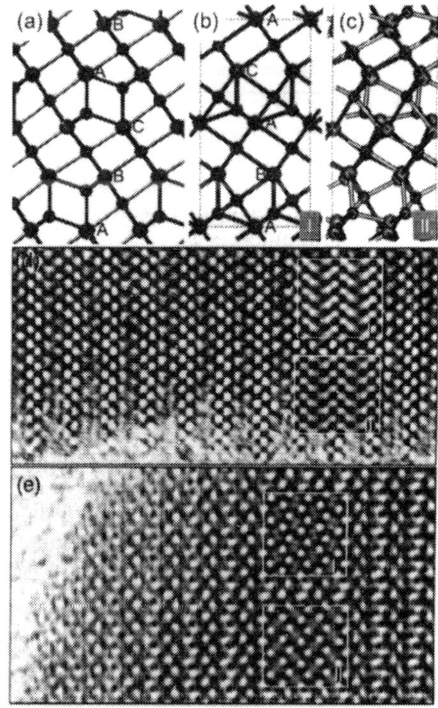

Figure 13.36 *Atomic models of a) magnetite , b) ∈-Fe_2O_3 and c) ∈-Fe_2O_3 d,e) Experimental and simulad (inset) HRTEM images of the ∈-Fe_2O_3 phase with the beam along the [$\bar{1}$ 01] and [001] directions, respectively The insets are the simulated images based on model I and II as displayed in (b) and (c) tespectively for the ∈-Fe_2O_3 phase.*

∈-Fe_2O_3 an intermediate phase between γ-Fe_2O_3 and α-Fe_2O_3. Magnetite and ∈-Fe_2O_3 phases coexist in nanowires without the presence of γ-Fe_2O_3. It seems that the images in Figure 13.37 of a long wire and its SAED pattern seen in Figure 13.38 provide phase transition process during the synthesis. The existence of the Au catalyst at the tip of the wire show that the synthesis process follows the vapor-liquid –solid growth mechanism. Figures 13.37(a) and (b) show low-magnification TEM images from two connected parts of the same nanowire. The SAED patterns in Figure 13.38(a) to (e) are from the circled areas "A"–"E" in Figure 13.37. The SAED pattern in Figure 38(a) shows that area "A" close to the Au catalyst is composed of the pure magnetite phase. Area "B" is composed of two phases, magnetite and ∈-Fe_2O_3 with a

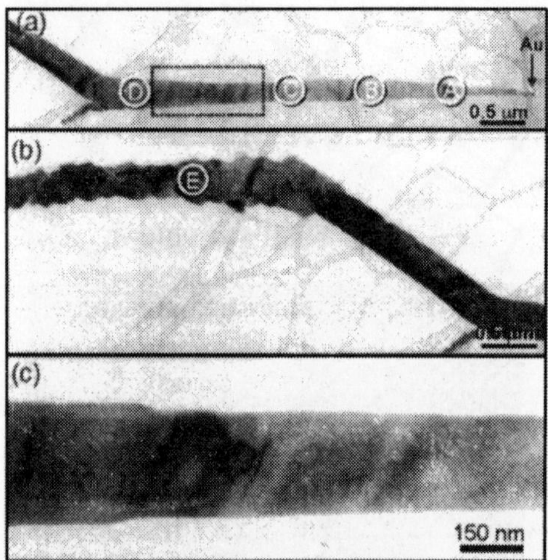

Figure 13.37 *Structural evolution among the three phases along a single nanowire. a,b) Low-magnification TEM images showing different parts of the same nanowire. c) Enlarged image of the enclosed rectangular area in (a).*

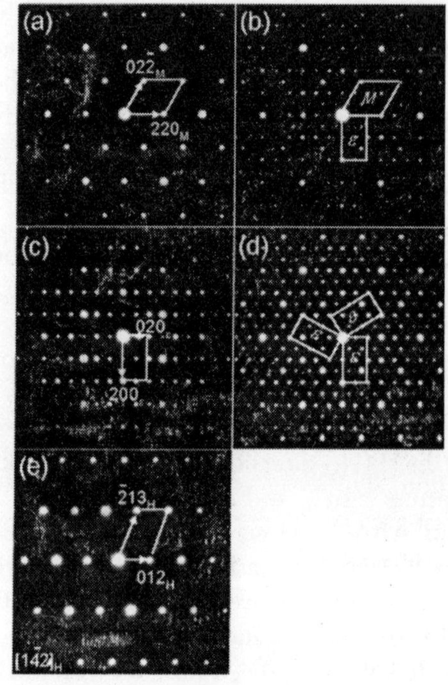

Figure 13.38 *(a-e) SAED patterns recorded from the circled areas A-E for the same nanowire presented in Figure 13.37, showing the different phases along the length of the same nanowire.*

fixed orientation relationship between them. Based on the SAED pattern shown in Figure 38(b), this orientation relationship described as $(001)_{\in} \parallel (111)_M$, $[010]_{\in} \parallel <110>_M$, which is the same as that shown in Figure 13.33. Progressing to area "C", the nanowire changes to the $\in$-Fe_2O_3 phase. At area "D", the diffraction pattern in Figure 13.38(d) is composed of three $[001]_{\in}$ patterns. By rotating 120^0 clockwise or anticlockwise around the $[001]_{\in}$ axis, each of these $[001]_{\in}$ patterns are superimposed onto any of the other two $[001]_{\in}$ patterns. The diffraction pattern in Figure 13.38(d) signifies the formation of rotation domains. There is some misalignment between the structure in areas "E" and in "A"-"D". The wire the right zone-axis SAED pattern, as shown in Figure 13.38(e). Neither magnetite nor the $\in$-Fe_2O_3 phase can be used to index the pattern displayed in Figure 13.38(e). Indeed, it belongs to the hematite phase, with the electron beam along the $[1\bar{4}2]_H$ direction. Although the orientation relationship between the connected $\in$-Fe_2O_3 and α-Fe_2O_3 regions cannot be found owing to , the wire bending. The phase structure from the tip to the end of the nanowire changes continuously, from magnetite to $\in$-Fe_2O_3, then to α-Fe_2O_3. There is no evidence of the γ-Fe_2O_3 phase at any point. If it exist the diffraction pattern unambiguously separates it from the above three phases. So, the wire should be the away from the catalyst, with freshest growth occurring close to the catalyst. The as-deposited nanowires form the magnetite phase, with time, oxidize into $\in$-Fe_2O_3, and later into α-Fe_2O_3. Figure 13.37(c) provides an enlarged image of the rectangular-enclosed area in Figure 13.37(a). The bubblelike contrast in the TEM image is regarded as evidence of oxidization after the wire's initial formation.

The TEM analysis show that while some magnetite nanowires grow along the $[111]_M$ and $<211>_M$ directions, the $\in$-Fe_2O_3 phase in these structure is not observed. However, Fe_2O_3 nanowires grown along the $[110]_M$ direction; and the presence of the $\in$-Fe_2O_3 phase is there. It is also known that nanowires grow along the <110>M direction and have the longest length (normally > 2μm) compared to those that grow along the $[111]_M$ and $<211>_M$ directions. Therefore, the fast-growth direction of Fe_3O_4 nanowires is the $<110>_M$ direction. The magnetite nanowires grow along the fast-growth $<110>_M$ direction. This provides sufficiently exposed surface for oxidation to the $\in$-Fe_2O_3 phase, while the nanowires growing along the $[111]_M$ and $<211>_M$ directions provide small surface area and thus make difficult for this transition to proceed. The SAED patterns in Figures 13.33(b) and 13.38(d) reveal the existence of 120^0 rotation domains. These are in the bright-field and dark-field images as shown in Figure 13.39(a) and (b). Based on the contrast in Figure 13.39(b), besides the 120^0 rotation domain boundaries (the interfaces between bright and dark domains), another type of planar defect, antiphase boundaries (APBs), also exist as the white arrowhead indicates. The HRTEM image in Figure 13.40(a) shows three separated 120^0 domains. The FFTs of domains "D1" and "D2" are displayed in Figure 13.40(b) and (c). After indexing the FFT patterns, the identified *b* axes of each domain are marked in Figure 13.40(a). The flat 120^0 rotation domain boundaries lie in (010) or (110) planes, irregular domain boundaries are also observed, as shown in Figure 40(d). The formation of these rotation domain boundaries is related simply to the redistribution of Fe ions in the oxygen interstitial sites from one domain to another. The stacking of the close-packed oxygen ions, which serve as the framework of the structure, remains intact Therefore the formation of the irregular-shaped domain boundaries is made.

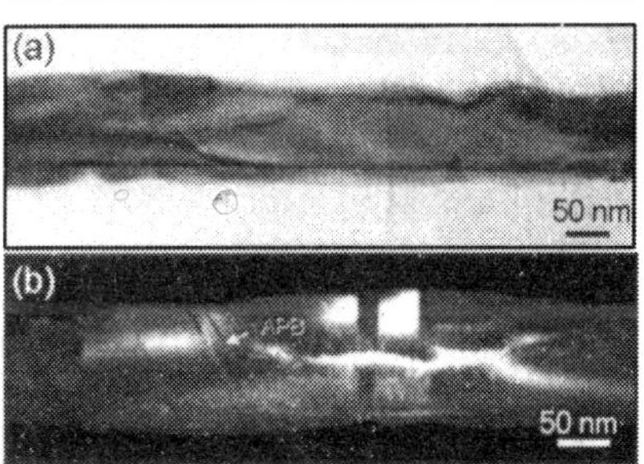

Figure 13.39 *(a) Bright field and b) dark field image to show the antiphase do main boundaries and 120^0. rotation domain walls in the $\in$-Fe_2O_3 phased nanowire.*

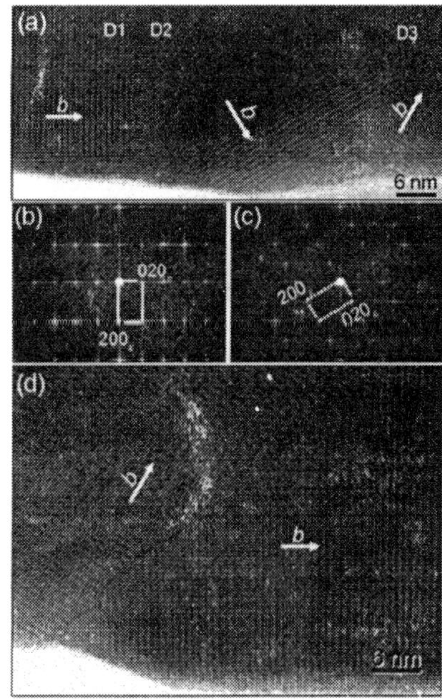

Figure 13.40 *(a, d) HRTEM images of the 120^0 rotation domains as shown in Figure 9. b,c) FFT of the domain D1 and D2 in (a).*

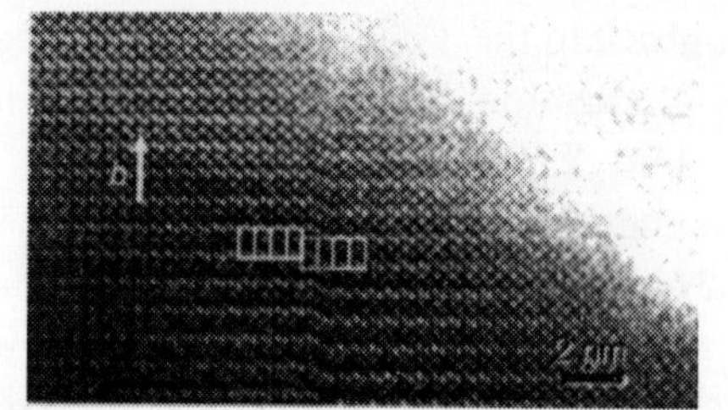

Figure 13.41 *HRTEM image of a ε-Fe_2O_3 phase nanowire, showing the atomic scale structure of an antiphase domain boundary.*

The HRTEM image of an APB is shown in Figure 13.41. The boundary plane normally takes the (100) plane. The displacement across the defect is 1/3 b ~ 2.9 Å. Similar to the case of the 120^0 rotation domain walls, the framework of close-packed oxygen remains intact; only the Fe ions readjust positions across the defect. In such κ-alumina crystal structures, the displacement of the elementary APB corresponds to a cation shift of half the O - O distance along the *b* axis, which is around 1.45 Å in the ∈-Fe_2O_3 crystal. The observed APBs in Figure 13.41 is considered as two elementary APBs combined together.

Considerable pures ∈-Fe_2O_3 phase nanowire characterized by using TEM such as the wire shown in Figures 13.39 to 13.41. More than 50 % of the ∈-Fe_2O_3 nanowires do not show the diffraction pattern of Fe_3O_4. It indicates that wires are fully oxidized into the ∈-Fe_3O_4 phase so the yield of the ∈-Fe_2O_3 phase nanowire is fairly high.

Pure Iron oxides, including maghemite and magnetite are widely used in many fields, from magnetic recording storage, and magnetizable printing to medical applications, such as cell separation and magnetic resonance imaging. While many applications require soft magnetic properties, large coercivities are required in applications such as high-density magnetic recording, where the magnetic domain needs to be stable, and desire for ever-smaller pixel sizes has pushed some materials to the superparamagnetic limit. Ferric oxides like magnetite and maghemite are attractive choices, for several reasons, including their low toxicity, low cost, availability, stability, high corrosion resistance, and high resistivity which correlates to low eddy current energy loss. However, their face-centered cubic crystal structure leads to a low magnetocrystalline anisotropy constant therefore augmentation of the coercive field (H_c) is difficult. The Fe_3O_4 phase takes the orthorhombic crystal structure, producing a room-temperature He as large as 20 kOe. The highly anisotropic magneto plumbite barium hexaferrite ($BaFe_{12}O_{19}$) produce a coercive field as about 7.5 kOe 1 Oe -= $(1000/(4\pi))$ A m^{-1} In addition, this magnetic behavior of $BaFe_{12}O_{19}$ is stable at elevated temperatures in the epsilon phase, which has a Curie temperature or 510 K. Crystals of Fe_3O_4 larger in size retain the favorable characteristics observed at the nanoscale and are deal candidates for almost any technology requiring hard magnetic materials, and provide the additional benefits of low toxicity, high corrosion resistance, anti so forth, inherent to ferric oxide materials. Further long nanowires of this material are ideal for current and future electronic, storage, and printing technologies. With further understanding of the phase stability and exploring new synthesis methods, Fe_3O_4 is a replacement for maghemite and magnetite in many of their current applications, as well as those of more complex and expensive hard magnetic and materials.

Magnetic and Electrical Characterization of Fe_3O_4 Nanowire

Half - metallic materials, such as CrO_2, $La_{0.7}$ $Sr_{0.3}$ MnO_3 (LSMO), and Fe_3O_4 are highly attractive for spintronics applications because of their high spin polarization. Among these materials, magnetite (Fe_3O_4) is superior to others because of its high Curie temperature (T_c) of

858 K, which is crucial for thermal stability in device applications. In addition, magnetite has proven to be a ferromagnetic material with a high spin polarization (100 %) at the Fermi level, which results in a metallic minority spin channel and a semiconductor majority spin channel.

Experiment IV

A typical Zimberger-made three-zone tube furnace system, including a vacuum pump, a gas flow system, and a quartz/alumina tube combination is used to synthesize the NWs. The vacuum level of this configuration is higher than 4 Pa. The annealing temperature can reach a maximum of 1770K. After the α-Fe_2O_3 NWs are synthesized on a FeNi substrate, magnetite NWs are fabricated by converting the α-Fe_2O_3 template NWs in a mixed reductive atmosphere of H_2 (5 %) + Ar (95 %), while the temperature is maintained at 720K for 20 min. The surface morphology is èxamined by using a field-emission scanning electron microscope operated at 15 kV. Afield-emission transmission electron microscope operated at 300 kV with a point-to-point resolution of 0.17 nm. Electrical measurements are performed by electron-beam lithography, metal evaporation, and device evaluation. The sample with the NWs is grown on the surface and is brought into tight contact with the designed pattern substrate with a common clip. This is followed by transfer of the NWs from the sample surface to the designed pattern substrate by ultrasonic vibration. The field-emission scanning electron microscope (FESEM) is used to locate the positions of the randomly dispersed nanowires on the chips. Ni (35 nm) and Au (65 nm) are used as contact electrode materials. The width of the electrodes on the nanowires is kept 0.2 μ m. A cold field emission scanning electron microscope (SEM) with a nanopattern generation system (NPGS) is used. A program is used to control the *I-V*. Electron holography is performed by using another FETEM dedicated to magnetic domain observations, that is, the microscope has a magnetic-shielded objective lens in which the magnetic field is reduced to 0.6 mT. A superconducting quantum interference device (SQUID) with a biprism setup is used to measure the magnetic properties of the NWs.

Figure 13.42(a) and (b) shows a top view morphology image ofα-Fe_2O_3 and Fe_3O_4 NWs, respectively. After the reduction process, the morphology of the Fe_3O_4 NWs is very similar to that of the α-Fe_2O_3 template. Figure 13.42(c) shows a transmission electron microscopy (TEM) image of the magnetite NWs; the high resolution TEM (HRTEM) image of the modulated α-Fe_2O_3 NW due to the oxygen deficiency and the magnetite diffraction pattern (DP) are shown in the insets of the image. The HRTEM image in Figure 13.42(d) reveals the single crystalline structure of the NWs, without linear or planar defects: The two *d*-spacings of 0.29 nm are identified as Fe_3O_4 $\{02\bar{2}\}$ planes. The diffraction pattern, shown in the inset in Figure 13.42(d), also illustrates the single-crystal nature of the NWs at the [111] zone axis.

Figure 13.43(a) shows a scanning electron (microscopy (SEM) image of the nanodevices; an enlarged image of one of the devices in Figure 13.43(a) is shown in Figure 13.43(b). The two-point *I-V* measurements are performed at room temperature under, ambient conditions. The linear *I-V* curves (shown in Fig. 13.43(c)) indicate that the characteristics fit well to Ohm's law. The zero-field resistivities of the nanodevices are estimated by the following equation: $R = \rho L/A$ (R: resistance, ρ: resistivity, A: cross-section area, L: NW length). The diameter and length of the measured NWs are 25 nm and 0.7526 μ m, respectively. Assuming that the Fe_3O_4 NWs are

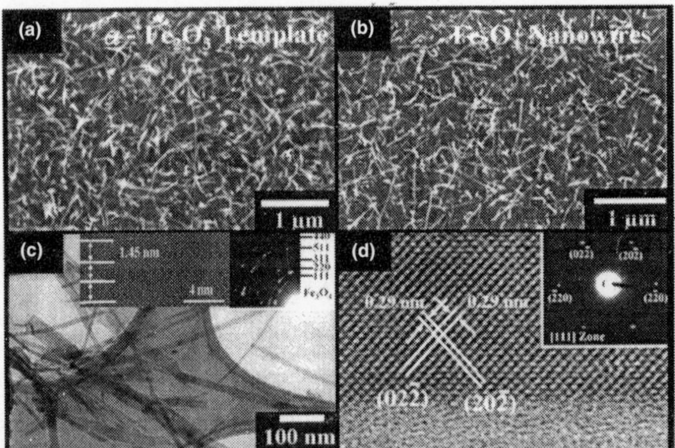

Figure 13.42 *Scanning electron microscopy (SEM) oimage of α -Fe_2O_3.NWS synthesized on an FeNi alloy substrate. b) SEM image of the magnetite nanowires after the reduction process. c) Transmission electon microscopy (TEM image of magnetite NWs; the high resoulution TEM image of the modulated α -Fe_2O_3 NWs from oxygen dificiency, and the magnetite diffraction pattern (DP) are shown in the upper corner inset. d) High resolution TEM image of a magentie NW.*

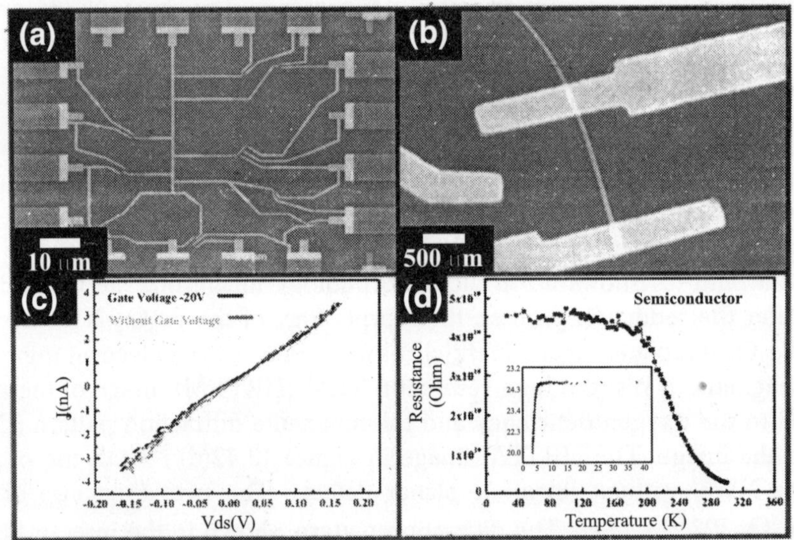

Figure 13.43 *(a) (SEM) image of the nanodevices; b) enlarged;image of one of the devices shown in (a). (c) I-V charterristics measurements in varouis gate biases, and d) the resistances measured at different temperatures.*

of a circular cross-section, the obtained resistivity, is 10.30 Ω cm; approximately three orders-of magnitude larger than that of bulk magnetite crystal (19000 μ Ω cm) The large measured difference between the NWs and the single crystal due to contact resistance and surface scattering, resulting from the high surface ratio. However the surface-scattering mechanism, based on the

FS theory, indicates that the aspect ratio is an important factor to the total resistance of nanostructures. According to FS theory when surface scattering is the dominant mechanism the resistivity of a nanowire decreases with increasing wire width. The positive slopes of the temperature versus resistance curves indicate that the larger the surface scattering, the higher the temperature. Here, the curves are as shown in Figure 13.43(d). The extraordinary high resistivity is not be the result of the surface-scattering mechanism

The transition describes the charge transportation mechanism for Fe_3O_4. Theoretically, magnetite is a mixed-valency 3d transition metal compound, in which one third of the iron ions occupy tetrahedral A sites (all Fe^{3+}), and two thirds of the iron ions occupy octahedral B sites (one half of which as Fe^{2+}. and the other half as Fe^{+3}, both in B2 and B3 sites). The transition is related to the ordering of the Fe irons, although the mechanism is controversial. The de conductivity increases abruptly by two orders of magnitude (from 10^{-1} to 10^{-1} a Ω^{-1} cm^{-1}) as the measured temperature increases above the transition temperature (Tv). The electrical properties of Fe_3O_4 change during the transition due to a transformation of its electronic state which strongly depends on the equilibrium positions of Fe atoms in the lattice. The magnetite NWs are obtained from hematite NWs, which exhibit oxygen vacancy modulation properties, while the size confinement of the 1D nanowires and the oxygen vacancy modulations result in the formation of charge ordering of iron ions at the B sites of the magnetite NWs. The charge ordering effect hinders carrier movement, which seems to be the main reason for the discrepancy between single-crystal NWs and bulk Fe_3O_4 crystal.

Figure 13.43(d), indicates that the resistance is a function of the temperature. As the temperature increases above 120 K, the NWs show semiconductor behavior. Below 120 K, the NWs show insulator characteristics. This suggests that carrier hopping is frozen at a temperature close to 120 K. In addition, the Arrhenius plot, (In(R) ~ 1000/T, shown in the inset of Fig 13.43(d) suggests that the transport mechanism is dominated by thermal activation at high temperatures (200-300 K). The thermal-activation model predicts that the conductivity follows the equation

$$\sigma(T) = \sigma_0 \exp(-\Delta E/k_B T) \qquad (1)$$

where σ is the conductivity, k_B is the Boltzmann constant, and ΔE is the activation energy. From the slope of the In(R) ~ 1000/ T curve an activation energy of $\Delta E \approx 0.154$ e V is estimated.

For the magnetic measurements, the magnetite NWs are dispersed on a Si substrate to avoid mixed signals from the FeNi alloy substrate. Figure 13.44 shows the magnetization of magnetite NWs as a function of temperature in an applied field of 500 Oe (1 Oe=1000/(4π) Am^{-1}) between 2 and 300 K using field cooling (FC) and zero-field cooling (ZFC) procedures. When the specimen is treated by ZFC, the individuals magnetic moments are randomly oriented at 2 K.Upon

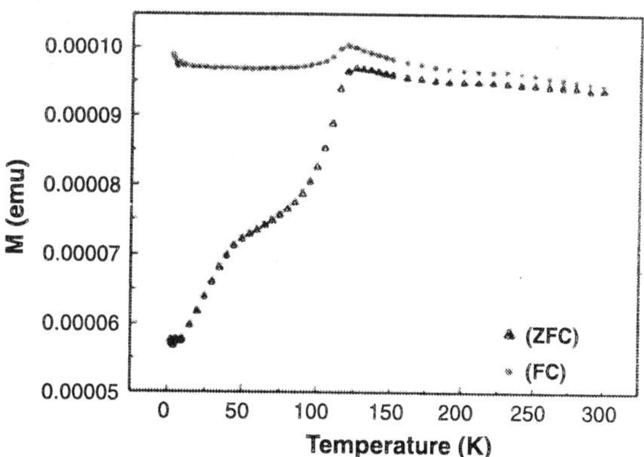

Figure 13.44 *Magnetization of magnetite NWs as a function of temperature in the applied field of 500 Oe between 2 to 300 K using field cooling (FC) and zero-field cooting (ZFC) procedures.*

applying the external magnetic field, the original random distributed moments are forced to align along the applied field. The magnetization M increases with the temperature, and reaches a maximum value at 125 K, which is defined as the blocking temperature (TB). At this temperature, the energy of the aligned magnetic moments is balanced with the thermal energy. Above this temperature thermal perturbation destroys the alignment of the moments. Both ZFC and FC magnetization curves exhibit a maximum moment at a T_B value of 125 K. Below T_B the ZFC curve shows a gradual increase as the moments progressively reorient along the applied field line at a low temperatures (2-125 K), while the FC curve shows a flat line at a low temperature because of saturation of the magnetic moment. Similar results are observed in magnetite nanoparticles and ultrathin magnetite films.

In addition, both ZFC and FC magnetization show an abrupt increase at the transition temperature of about 120 K. The structure transition results in a change of the electrical and magnetic properties at 120 K. The large divergence of ZFC and FC (below 300 K) result from the large amounts of magnetic isotropy energy contributed by the external magnetic field during the cooling process. Hence, it is also the reason why the ZFC and FC curves decrease slightly above 125 K.

The magnetization distribution in the Fe_3O_4 nanowires is examined by electron holography, which deduces the magnetic information from the phase shift of electrons. Figure 13.45(a) shows a bright-field image of the nanowires supported on an amorphous microgrid carbon film. The nanowire marked as "X" makes contact with another nanowire, "Y". Figure 13.45(b) shows a reconstructed phase image, acquired in the rectangular area of Figure 13.45(a). A transparent bright-field image is superimposed on the contour map in Figure 13.45(b). The phase shift of electrons is due not only to the Figure 13.45 Electron hologram, indicating the magnetization distribution in the magnetite nanowires magnetic field but also to the electric field i.e., mean inner potential of the specimen). The unwanted signal originating from the mean inner potential y the "time reversal operation of an electron beam" method. Therefore, Figure 13.45(b) displays only the magnetic information. The contour lines and arrows in Figure 13.45(b) represent the lines of magnetic flux and their directions, respectively. An essential point is that the magnetic flux is parallel to the longitudinal axis of the nanowires. Several specimens having small diameters (on the order of 10 nm) are observed by the holography technique but no magnetic domain walls are found The figure also demonstrates that some amount of magnetic flux leaks out of the nanowires as is seen from the leakage near the terminal end (Fig 13.45(b)), wherein the flux lines are no longer parallel to the nanowire. Apparently, the shape anisotropy dominates the magnetization distribution in these nanowires, and the result is consistent with the small magnetocrystalline anisotropy in cubic Fe_3O_4 at room

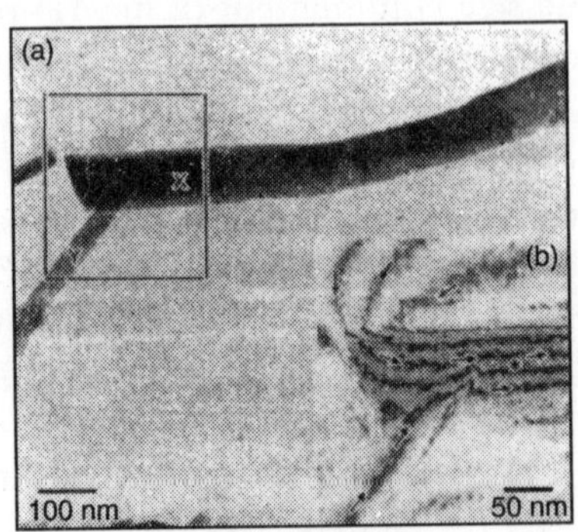

Figure 13.45 *Electron hologram, indicating the magnetization distribution in the magnetic nanowires.*

temperature. The magnetic flux density B of the nanowire X is estimated by following the equation B = h/(etl), where h stands for Planck's constant, e is the elementary electric charge, t the specimen thickness, and L the spacing of the contour lines in the reconstructed phase image. The magnetic flux density B is evaluated at 0.46 T at the position of the asterisk in Figure 13.45(b) The parameter t is estimated at 122 nm assuming a mean inner potential of 12.0 V, which is deduced from the phase shift in the rod shaped nanowire.

The flux density is lower than that of bulk Fe_3O_4 (0.60 T). The discrepancy is from surface spin disorder, which is significant in nanowires with a large surface-to-volume ratio. Further the flux density is sensitive to the crystallinity of nanowires. The observations in Figure 13.45 implies the possibility of regulating the spin current with the half-metallic nanowires due to the controlled magnetization distribution in the 1D form.

14 Nanowires and Nanobelts of ZnS

ZnS is a direct wide band gap (3.91 eV) compound semiconductor that has a high index of refraction and a high transmittance in the visible range and is one of the most important material in photonics. As a one dimensional nanostructure, ZnS is synthesized as nanowires nanobelts and nanocombs but these nanostructures are randomly distributed on the surface of the substrate.

In the electronics industry it has a wide range of applications including light-emitting diodes and efficient phosphors in flat panel displays. An assortment of luminescent properties excited by ultraviolet, X-ray, cathode ray or electrical currents observed by doping ZnS with various metals.. Excellent light transmission with a high index of refraction (2.27 at 1 μm) also makes ZnS useful in photonic crystal devices that operate in the region from visible to near-infrared.

The most stable bulk form of ZnS at room temperature is the zinc blende structure which can transfer to wurtzite structure after heating to 1300K in ambient pressure. Phase control in the growth of ZnS crystal is important for instance, the different phases show different lattice vibration properties and nonlinear optical coefficient.

Zinc sulfide has two types of crystal structures hexagonal wurtzite ZnS (referred to as the hexagonal phase) at high temperature and cubic zinc blende ZnS (referred to cubic phase) at low temperature. Typically, the stable structure at room temperature is zinc blende with few observances of stable wurtzite ZnS. Using a reverse micelle solution based technique, zinc blende structured ZnS nanowire are fabricated. Based on ZnO nanorods and nanobelts made by a physical process ZnO is converted into zinc blende-structured ZnS nanorods and nanotubes that have porous structures made of nanocrystallites with gram sizes of ~ 7 nm. Wurtzite ZnS is much more desirable for its optical properties than the sphalerite phase. For instance, phosphors of ZnS are synthesized in the wurtzite phase to optimize luminescence at a temperature near 1270K, because bulk wurtzite crystallizes at temperatures above l300K. However, the quenched wurtzite easily transforms to the sphalerite form at ambient conditions. Decreasing the particle size lowers the temperature boundary and nanosized wurtzite can be synthesized at temperatures 670K lower than the bulk. Thus, it is possible to synthesize the ZnS wurtzite phase and keep it stable at room conditions by adjusting the surface energy through particle size tuning. Synthesis of nanoscale semiconductors demonstrates that the quantum-confinement effect varies the optical and electrical properties, mostly resulting in their improvement relative to those of their bulk counterparts. These enhanced properties are also observed in three-dimensional wurtzite ZnS nanocrystals obtained in a variety of ways, but the structural instability remains analogous to its bulk counterpart.

Synthesis of ZnS NW and NB by VLS for Mass Diffusion Growth

The formation of zinc blende structured ZnS nanobelts and NWs is related not only to the lower deposition temperature (950 – 1020K), but also to the size of the Au catalyst particles (smaller than 50 nm). Pure wurtzite structured ZnS nanobelts are synthesized in a relatively high deposition temperature (above 1020K) regardless the size of the Au catalyst particles.

Vapor – solid (VS) and vapor – liquid-solid (VLS) growth mechanisms are widely applied to steer and interpret the growth of various 1D nanostructures via vapor routes.

The synthesis of two types of ZnS nanostructures an is done via Au-catalyzed VLS process: (1) periodically twinned CZB nanowires (PTNWs) and (2) asymmetrical polytypic nanobelts (APNBs) in which CZB ZnS straight strips parallel to the belt axis are embedded in HWZ nanobelts throughout the entire length.

Experiment

The synthesis of ZnS nanostructures is performed in a horizontal tube furnace under atmospheric pressure ZnS powder (1 g, 99.99%) loaded in an alumina boat is placed in the central heating zone of the alumina tube in the furnace. Several slices of single-crystal Si (100) with a thin gold coating film of ~5 nm are placed downstream as deposition substrates. Initially, the alumina tube is purged by high purity Ar for 2 h and then the Ar gas is switched off. Second, the system is heated to 1270K at 25 $Kmin^{-1}$ and maintained at this temperature for 5 min. Third, the system is heated to 1470K in 1 min and maintained at the peak temperature for another 30 min. During the peak temperature heating, high-purity Ar is introduced into the reaction system as carrier gas at 50 sccm. After the system is cooled naturally, white wool-like or powder-like products grow on the substrates. The morphological and structural features of these products are investigated by field emission scanning electron microscopy (FESEM, at 10 kV), transmission electron microscopy (TEM, at 200 kV), and energy dispersive X-ray spectroscopy (EDX, equipped with TEM).

Mass Diffusion growth by VLS

There are two mass diffusion processes regarding the vapor-liquid-solid (VLS) growth of nanostructures: one is inside the catalyst droplet toward the liquid solid interface; the other is along the side surface planes of the growing nanostructures. In this mechanism, the liquid metallic nanodroplets act as catalysts to adsorb and dissolve reactant species and guide the growth of nanowires with controlled diameters. Four consecutive steps are involved as shown in Figure 14.1: (1) the adsorption of reactant species on both the catalyst droplet surface and the side surface of growing nanostructures; (2) the dissolution of the species at the droplet surface; (3) the diffusion of the species inside the droplets; and (4) precipitation, incorporation and crystal growth at the liquid solid interface. There are also two types of mass diffusion processes during the VLS growth. One is inside the catalyst droplet and then to the liquid-solid interface; the other is from the side surface of the growing nanostructures toward the catalyst droplet for dissolution. Therefore, the nature of the catalyst droplets and the side surface planes of the growing nanostructures influences the mass diffusion process. But modulating experimental conditions can modify the mass diffusion procedure, and thus desired nanostructures can be achieved.

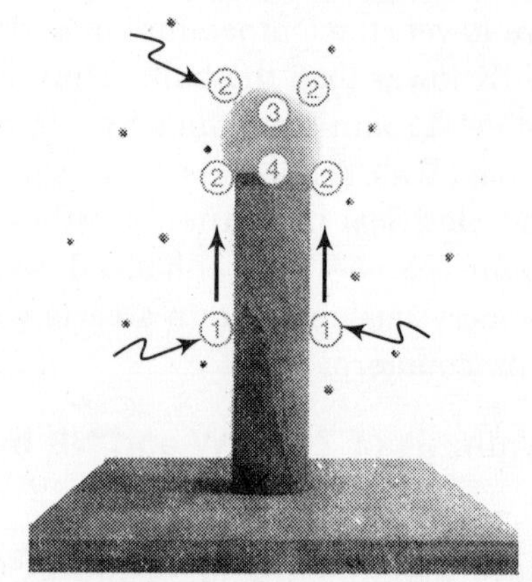

Figure 14.1 *Illustration of detailed, continuous VLS growth steps for 1D nanostructures.*

Figures 14.2(a) and (b) reveal low and high-magnification FESEM images of the two kind of ZnS nanostructures. Detailed observations reveal that a small amount (~20- 30%) of the products is long ZnS nanowires with uniform diameters around 60 – 100 nm. One of these nanowire observed in the upper-right part of Figure 14.2(c). The nanobelts, at a significant percentage (~70 – 80%) of the yield, have a typical width of 1 – 3 μm, a thickness of 40 – 80 nm (Figure 14.2(d)), and lengths of hundreds of micrometers. All of the ZnS nanobelts show tapered morphology toward their ends, where an Au nanoparticle is located at the tip of the nanobelt (lower-left part in Figure 14.2(c)).

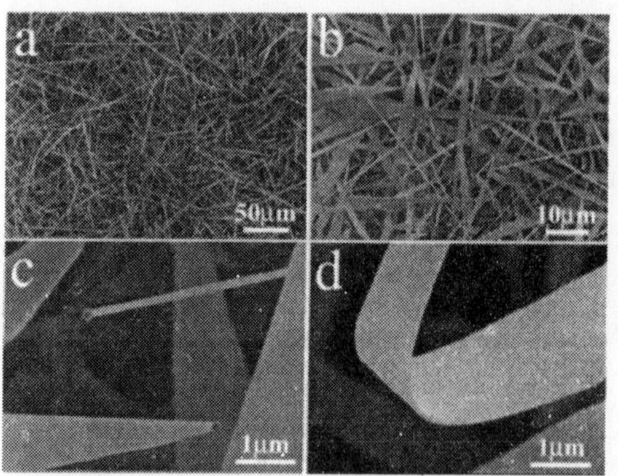

Figure 14.2 *SEM images of two type of ZnS nanosturctures a and b are low and high magnification images, respectively. (c) Both type of nanostructures are terminated by catalytic particles, and the nanobelts show tapering features toward the fold particles. (d) Typical bending contour of ZnS nanobelt.*

The NWs are characterized by using TEM, Most of the ZnS nanowires (~90%) reveal sequentially bright/dark contrast stripes throughout the entire length of the wires (Figure 14.3(a) and (b). Different from hexagonal nanowires, a selected area electron diffraction (SAED, Figure 14.3(c) pattern taken from the wire in Figure 14.3(b) shows a CZB structure from the [110] zone axis, indicating the existence of a twin defect along the [111] growth direction. A high-resolution TEM image (Figure 14.3(d)) reveals periodically alternating twins along the [111] axis of the wire: atomic sharp twin boundaries appear every 7 – 9 Zn – S layers. The zigzag angles are 141°(70.5° + 70.5°) in accordance with the relative rotational angle of (111) twin crystals in face-centered cubic (fcc) structures. Twin crystals along {111} facets fcc

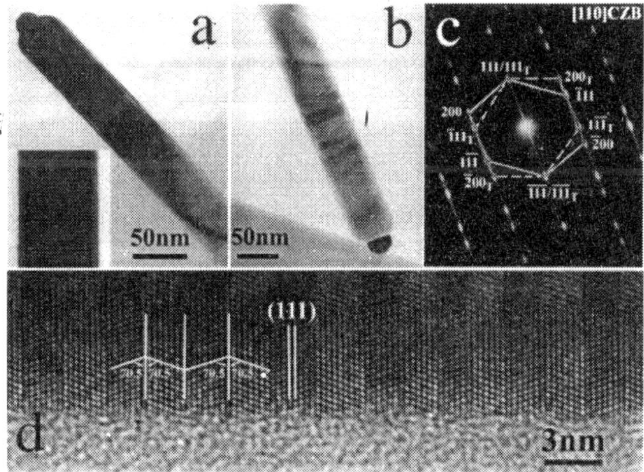

Figure 14.3 *TEM images of PTNWs. a and b are typical alternating bright/dark stripes along the wires. Gold particles are terminated at their ends; the inset in a is a close-up view of the bright/dark stripes in a ZnS wire. (c) SAED pattern recorded from the nanowire in b, showing twinning features labeled by normal and dashed lines. (d) HRTEM image of one nanowire showing periodical twin structures. The amorphous coating on the wire in d is a surface-oxidized sheath.*

structures are observed due to their crystallographic characteristics; however, the occurrence of such ordered twin defects in PTNWs via VLS growth is considered to originate from intrinsic successive self-oscillations.

For the ZnS nanobelts, TEM observation reveal a high width-to-thickness ratio which gradually narrow towards the catalytic nanoparticles at the ends. It is worth noting that straight strips parallel to the belt axis are always embedded inside the belt throughout the entire length, as shown in Figure 14.4(a) and (b). The typical widths of the strips are 30 – 50 nm. An SAED pattern (Figure 14.4(c)) indicates that the nanobelt contains both HWZ and CZB phases. HRTEM images (Figure 14.4(d) and (e)) show that the strip (as interlayer) is the CZB phase, whereas the left- and right-hand parts of the strip are HWZ structure. The detailed orientation relationships are HWZ(0001) || CZB(111) and HWZ[$2\bar{2}\bar{1}0$] || CZB[$0\bar{1}\bar{1}$]. Moreover, the interfaces of the two phases are atomic smooth (Figure 14.4(d)). Such a tapered nanobelt looks like an Au-guided nanowire with another segment including HWZ and CZB phases, which grow only on one side of its lateral surface planes of the wire. As the solubility of an Au droplet is identical during growth so, the wires and the tapered segments attached are not initiated only by the gold particle. Taken together, the nanowire grows along the [$01\bar{1}0$] direction and is bounded by orthogonal {$2\bar{2}\bar{1}0$} and ± (0001) planes as side surfaces; a belt-like segment is attached by its {111} facet of CZB strip onto only one basal surface of the HWZ phase, but with the opposite side surface left flat and without growth. As a whole, a well-faceted ZnS APNB is built up, where CZB and HWZ phases coexist in a single ZnS nanobelt. Generally speaking from the structure point of view, a stacking of ABABAB layers form a hexagonal structure, whereas a stacking of ABCABC constitutes a cubic phase Thus the two phases are transformed by introducing so-called "stacking faults". Although stacking faults sometimes occur in ZnS nanostructures, and CZB{111} and HWZ {0001} facets are atomically identical, it is seen that

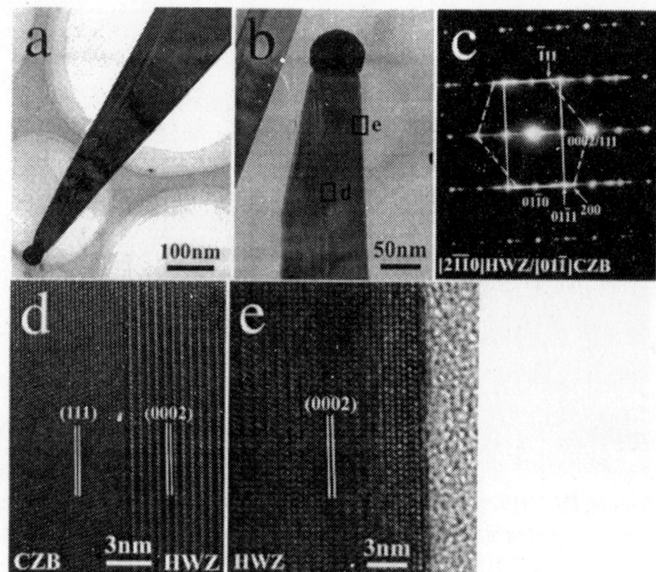

Figure 12.4 *TEM images of ZnS APNBs. (a) and (b) are typical polytypic nanobelts at different magnifications. (c) The SAED pattern clearly indicates the coexistence of HWZ and CZB phases, which are denoted by normal and dashed lines, respectively. (c) and (e) are HRTEM images from the boxed areas in c, revealing that the gold-assisted HWZ ZnS nanowire is atlached by CZB part on only one side surface.*

planar defects with a high density within a width of nearly 50 nm appear at certain locations, which is related to some particular growth mode. In addition, EDX spectroscopy analyses indicate that both of the PTNWs and APNBs are composed of Zn and S, whereas Au is present only in the catalytic nanoparticles.

The liquid droplet is a unique characteristic of VLS growth. The above-mentioned detailed steps related to the droplets are the key to the growth procedures and the formation of nanostructures. Under steady-state supply of vapor reactants, either of the last two steps may become rate-determining under different conditions: the crystal growth at the liquid – solid interface at low temperature, or the diffusion inside the catalyst droplets at high temperature. Here the deposition temperature is high and the crystal growth rate is directly related to the availability of chemical species in the droplet and its diffusion rate through the droplet; thus, a fast growth of the nanowire leads to the depletion of ZnS species in the liquid droplets, which then depresses the crystal growth rates and increases the contact angle between the solution droplet and solid ZnS wire due to reduced wettability at the liquid – solid interface (as shown in Figure 14.5(a) and Stage 1 in the model of Figure 14.5(b). Accordingly, under constant supply of vapor reactants, the concentration of chemical species in the droplet increases slightly after some time so the wettability and growth rate are increased (stage 2). Hence the growth is initiated and leads depletion in chemical species in the droplet (stage 3). During the entire process, nanowires grow continuously with a periodically modulated growth rate due to the periodic change/modulation of available chemical species in the droplet. This process assumes

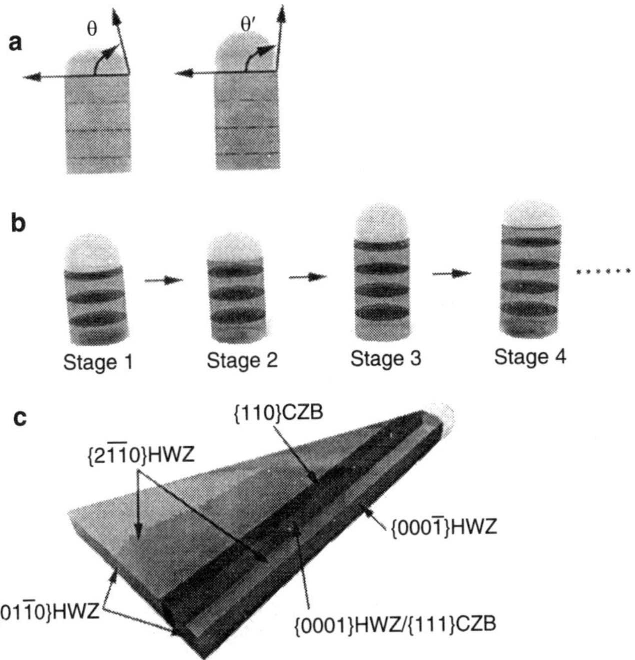

Figure 14.5 *(a) The variation of contact angle (θ) with regard to liquid-solid interface wettability. (b) and (c) are model for the formation of ZnS PTNWs and APNBs. respectively. Twin boundaries are indicated by dark green slices in (b).*

that the rate at which the atomic species forms a crystalline structure is faster than the diffusion rate of the chemical species through the catalyst droplet. In the meantime, because of the change in wettability of the liquid-solid interface, the droplet reshapes periodically between nearly spherical and nearly ellipsoidal, followed by successive release of surface and interface energy. Therefore, planar twin defects are formed periodically to release the stored energy. The periodical reshaping catalyst particles is found by imaging the catalyst-induced growth of carbon whiskers that is the shape change occurs during the catalyst-assisted growth of crystal filaments.

Actually, the reactant adatoms on the side surfaces undergo a "competitive diffusion" process between the liquid droplets and other reactive sites on the side surface. Due to the ideal roughness of liquid droplets, most adatoms are dissolved into the droplet through side-surface diffusion. Nanostructures grown via the VLS mechanism are therefore characterized by wire-like morphologies because their side surfaces are bounded by low-energy facets. However, HWZ ZnS is described as alternating stacking of positive (0001)-Zn and negative (000$\bar{1}$) -S layers along the polarized C axis, much similar to the hexagonal ZnO, in which the (0001)-Zn plane is chemically more active than the (000$\bar{1}$)-O plane and the surface energy of the (0001)-Zn plane is the highest among those of all of the low-index crystal planes. Thus, for HWZ ZnS nanowires growing along the [01$\bar{1}$0] direction (Figure 14.5(c)), both the droplets and the chemically active (0001)-Zn side surface planes simultaneously become two preferential diffusion directions to exhibit the competitive process and initiate the tapered nanobelt growth under a relatively high supersaturation ratio. Compared with the growth of the high temperature stable HWZ wires through the intermediate dissolution and precipitation in droplets (this process normally reduces barriers for chemical reaction and reduces the activation energy of nucleation), direct sink and fast incorporation of adatoms on the (0001)-Zn side surface planes usually favor the high-temperature metastable CZB phase, which is kinetically controlled. The segments growing without the assistance of droplets, form a thermodynamics-favored HWZ phase at high deposition temperatures (See Figure 14.9) Thereby, the CZB dominates at confined locations of APNBs. However, the occurrence of the CZB phase is due to surface diffusion.

Following are the SEM and TEM images for the products that are deposited at 1270 – 1320K. There are also other types of ZnS nanostructures at different substrate temperature zones, which are valuable comparisons for understanding the controlled synthesis of desired structures.

When the same experimental conditions peak heating temperature, carrier gas flow rate, etc are applied the qualitative supersaturation profile corresponds to the solid line in Fig. 14.6.

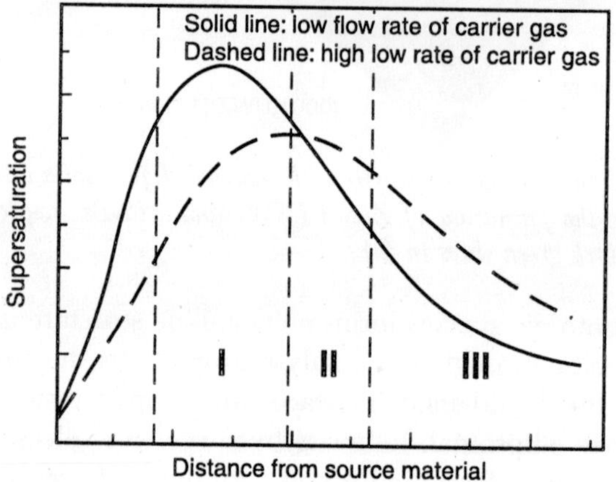

Figure 14.6 *Under the steady state advection by the carrier gas the supersaturation profile is qualitatively shown above which is used to choose appropriate substrate temperatures.*

Region I: The products deposited on the substrate above 1320K are mainly irregular shaped structures, although minor Au-directed structures are observed. This corresponds to extremely high supersaturation all the surface planes obtain large quantities of reactant atoms, and the effect of preferential diffusion can be neglected.

Region II: Desired nanostructures, PTNWs and APNBs, are formed in relatively high supersaturation and appropriate substrate temperature (1270 – 1320K)

Region III: At the substrate temperature of 1170 – 1220K ordinary nanowires are obtained. Nanowires with [0001] orientation are frequently observed. This is usual growth direction for wurtzite nanowires (the right in Fig 14.7(b) and its SAED pattern in Fig 14.7(d). Under such moderately low supersaturation, weak surface diffusion effect is also revealed (the left in Fig 14.7(b) and its SAED pattern in Fig 14.7(c).

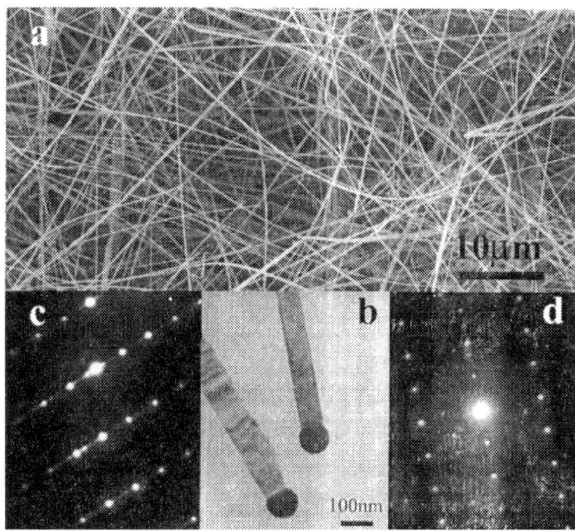

Figure 14.7 *The substrate temperature around 1270K when Au catalyst is not used, wide and single-crystal wurtzite ZnS nanobelts dominated by low energy facets of* $\{2\bar{1}\,\bar{1}\,0\}$ *are main products. Here VS mechanism determines the product morphologies.*

Synthesis of Aligned and Ordered ZnS Nanowires

The orientation-ordered ZnS nanowire bundles are synthesized through a two-step thermal evaporation process in a horizontal tube furnace. The CdSe powder is placed in the center of a single-zone tube furnace. A vacuum is pulled into the tube for several hours to purge oxygen from the chamber. After evacuation to less than 2.66 Pa, the temperature in the center of the tube is elevated to 1020 K, at a rate of 30 K min^{-1}. A nitrogen (N_2) gas flow is introduced into the system at a rate of 50 sccm. The gas act as a carrier for transporting the sublimated vapor to cooler regions within the tube furnace for deposition. The pressure is maintained at 3×10^4 Pa. The Silicon substrates reaches a temperature of about 850K. After 1hour, the furnace is allowed to cool. Then, the source material of CdSe powder is replaced by ZnS powder The temperature in the center of the tube is elevated 1320K. Again a nitrogen gas flow is introduced at 50 sccm and the pressure maintained at 3×10^4 Pa. The silicon substrates reach a temperature of about 1020K The system is held in this condition for a period of 1 hour. Single-crystal silicon substrates are used for synthesizing the nanostructures. These substrates are placed downstream from the source material to collect the deposited nanostructures.

SEM Analysis

The sample is analyzed on a scanning electron microscope (SEM). The deposited material has an extremely high yield on the substrate and shows high reproducibility. The deposited material in the synthesized sample has a dominant morphology consisting of "bundles" of wire shaped ZnS nanostructures (Figure 14.8(a) and (b)). All of nanowires arc orientationally aligned, and they grow uniformly along the bundle. With energy dispersive X-ray microanalysis (ESD) it is found that the ZnS nanowires grow on top of a layer of CdSe film, which is a buffer layer

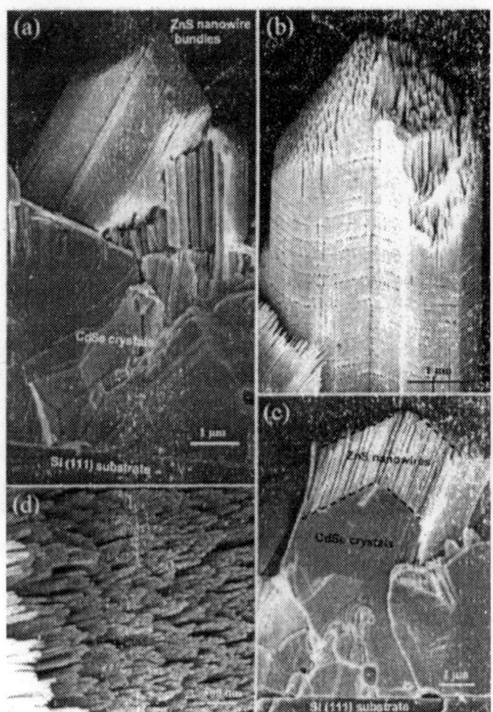

Figure 14.8 *(a) SEM image of ZnS nanowire bundles grown on a CdSe Substrate. The CdSe is a solid film, but the ZnS is a bundle of aligned nanowires. (b) Enlarged SEM image of a ZnS bundle showing traces created due to fluctuation in growth condition, presenting the equal growth rate of all of the nanowires.(c) Fractured surface of the sample showing the direct growth of ZnS nanowires on the CdSe crytals and the preservation in the ZnS nanowire bundles of the surface morphology of the CdSe crystal. (d) Enlarged top view of the aligned ZnS nanowires.*

between the Si(111) substrate and the nanowires. Each bundle preserves a fairly good face structure (Figure 14.8(a) and (b)), as if they are a single entity. SEM of a factured cross-section indicates that top morphology of the ZnS nanowire bundle preserve the shape of the top of the CdSe crystal (Figure 14.8(c)), indicating the equal growth rate of all the nanowires. This also shows that the ZnS nanowires are initiated at the top surface of CdSe crystal, and that they are distributed following the surface of CdSe. Each individual nanowire is clearly identified, A top view of the as-grown nanowire bundles, shows their vertical alignment as well as a high degree of size uniformity and a high density (Figure 14.8(d)).

Transmission electron microscopy shows that the individual nanowires are single-crystal wurtzite ZnS, with trace amount of CdSe interspersed within the lattice. From the diffraction pattern recorded from a bundle ((Figure 14.9(a) and (b)). the single-crystal pattern indicates the orientation alignment among all of tile nanowires in the bundle along [0001]. and they share the same side surface of (2–1–10) and a top surface of (01–10). The nanowires are aligned not only in length direction but also in crystallography orientation. This is a key characteristic.

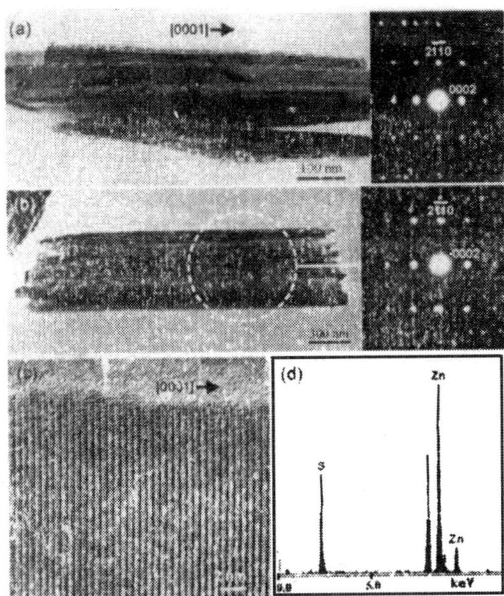

Figure 14.9 *(a and b) Low – magnification TEM images of bundles of aligned ZnS nanowires and the corresponding electron diffraction patterns from the bunches showing orientation ordering among the [0001] nanowires. (e) High- resolution TEM image of a nanowire showing uniform lattice structure (d) EDS of ZnS nanowires showing the chemical composition.*

High-resolution images of an individual wire confirm that each nanowire is single crystal and that the *c* direction is the fastest growth direction (Figure 14.9(c). There are no dislocations observed in the volume of the nanowire. EDS spectrum taken in the TEM shows that the wires are ZnS (Figure 14.9(d) with negligible signals from CdSe.

The nanowire bundles are formed by a two-step growth process as a result of the two-step synthesis procedure. In the first step, the CdSe thin film grows on the silicon (111) substrate. The [0001] oriented CdSe has a 6-fold-symmetric a-plane, {10–10}, with an interplanar distance 0.3724 nm, which matches well to the 6-fold symmetric Si(111) substrate, with an interplanar distance 0.3839 nm, resulting in a c-axis-oriented growth of CdSe on Si(111). Due to a larger lattice mismatch single-crystal thin films are not grown but multiple nucleation of CdSe results in the grown of a polycrystalline CdSe film. In the second step, the CdSe film serves as a substrate for epitaxial growth of the ZnS nanowires. Because both CdSe and ZnS have wurtzite structure an epitaxial growth along the c-axis is preferred to minimize the interface lattice mismatch. All of the nanowires have the same growth rate so that the morphology of the CdSe crystal is preserved (see Figure 14.8(c) All of the aligned nanowires in one bundle are synchronized in response to the growth condition; the topographical shape of the CdSe nanocrystal is maintained as the wire-bundle grows. The bundles also maintain the width and depth profile of the Cdse crystal upon which the wires are originated. The nanowire bundles exhibit the same morphology characteristics that larger wurtzite nanocrystals exhibit, including the 6 fold symmetry along the [0001] direction (see below fig 14.10)

Figure 14.10 *Low magnification SEM image showing the ZnO nanowire bundles grown on CdSe.*

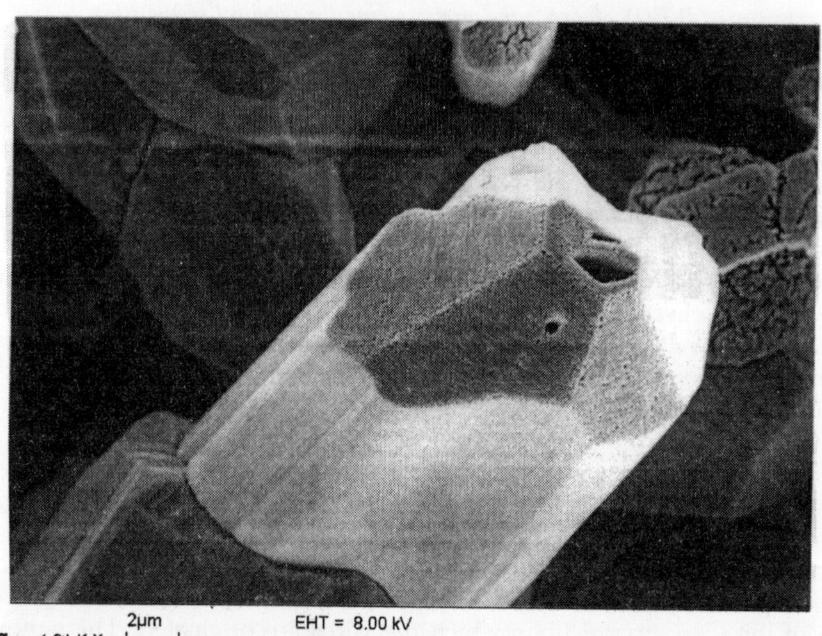

Figure 14.11 *Low magnification TEM image showing the faceted morphology of ZnS nanowire bundles.*

Phase Controlled Synthesis of ZnS nanobelts

Experiment

ZnS nanobelts are synthesized through a vapor liquid-solid (VLS) process High-purity ZnS powders are put in an alumina boat located at the center of the alumina tube. The entire length of the tube furnace is 75 cm. Silicon substrates covered with Au film are placed in the tube at a distance from 5 to 25 cm from the center of the alumina boat. After evacuation of the tube to 2.66 Pa a carrier gas of high purity argon premixed with 5% hydrogen is kept flowing throughout the tube. The flow rate and pressure inside the tube are kept, 50 sccm and 4×10^4 Pa throughout the experiment respectively. The furnace is then heated at a rate of 25 $Kmin^{-1}$ to 1070 K and held at the temperature for 20 min. The furnace temperature is then further raised to 1270 K and held at that temperature for 50 min. The local temperature distribution and the position of the substrate is calibrated and the result is shown Fig. 14.12(a). After the deposition, the furnace is cooled to room temperature. Au thin films with an average thickness around 5 and 10 nm are deposited onto silicon substrate, which produces gold particles of sizes, ~35 and ~100 nm, respectively. The deposited products are characterized by scanning electron microscopy (SEM) and transmission electron microscopy (TEM) operated at 200 and 400 kV.

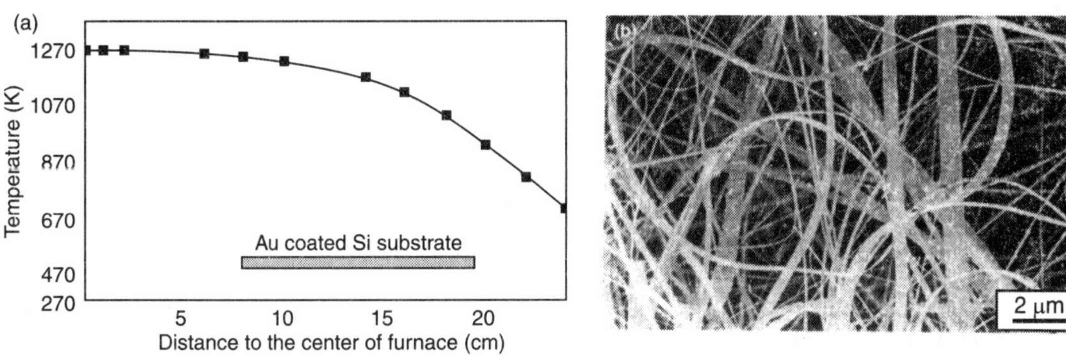

Figure 14.12 *(a) A plot of the relationship between the local temperature and the distance to the center of the furnace for the experimental Set up. (h) SEM image recorded from a sample collected in the temperature region of ~ 1000K grown using thin film ~5 nm in thickness.*

1. Nanostructures formed at ~ 1000K using 35 nm size Au catalyst particles

The SEM image in Fig. 14.12(b) displays the morphologies of the ZnS nanostructures grown in the temperature grown with Au catalyst with an average deposition thickness of ~5 nm. Uniform nanobelts and asymmetric nanosaws are the two dominant morphologies of the ZnS nanostructures. The sizes of the nanobelts and nanosaws obtained from the TEM image as shown in Fig 14.13(a) are: The width of the uniform nanobelts is under 50 nm, while the width of the nanosaws is over 100 nm. The gold catalyst particles are at the tips of the nanobelts and their sizes are ~35 nm.

The narrower nanobelts and the wider nanosaws are of the same crystal structure. This analysis is achieved from the selected area electron diffraction (SAED) and high resolution

TEM (HRTEM). For the narrower nanobelts, Figure 14.15(b) is the SAED pattern from a fine nanobelt shown at the center of Fig 14.13(a). The diffraction pattern is indexed as the zinc blend structure with the incident electron beam parallel to [011]. The corresponding HRTEM image recorded from the nanobelt is presented in Fig 14.13(c) revealing that the nanobelt grows along [100] and takes $\pm$ $(0\bar{1}1)$ as the side surfaces. The inset atomic structure model in Fig. 14.13(c) reveals that the nanobelt growth front is a polar surface terminated either with Zn or S. The Zn-terminated surface is catalytically more active than the S terminated surface, so the growth front is terminated with Zn, leading to the fastest growth. The HRTEM image shows that the $\pm(0\bar{1}1)$ side surfaces are flat and without reconstruction as viewed along [011].

Now, considered the nanosaws. Although the nanosaws are deposited in the same temperature zone, they do not have the same structure as the narrow uniform nanobelts (see Fig. 14.13). The structure of the nanosaws is displayed in Fig. 14.14 The SAED pattern shows (Fig 14.14(b)) zinc blende (as indicated by arrowheads) and wurtzite (rectangle marked diffraction spots) structures co-exist in the nanosaw. The HRTEM images in Fig. 14.14(c) to (e) from different areas of the

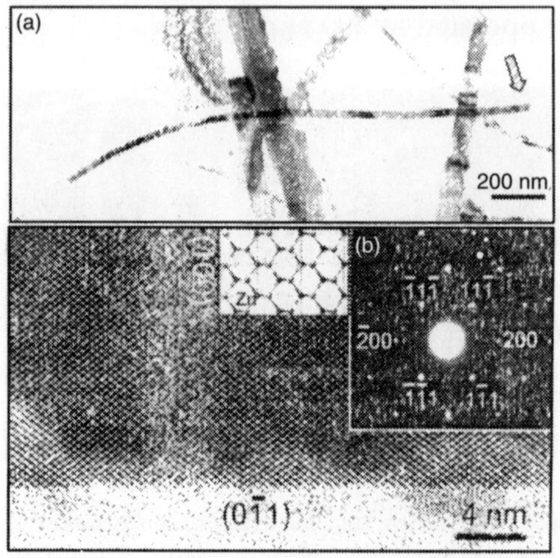

Figure 14.13 *Structure of narrow nanobelts grown at ~1000K and using ~35 nm Au particles (a) TEM image showing the uniform nanobelts and asymmetric nanosaws (b) SAED pattern and (c) HRTEM image recorded from a narrow nanobelt as pointed in (a) which has the zinc blend structure. The inset in (c) is the atomic model of zinc blende ZnS.*

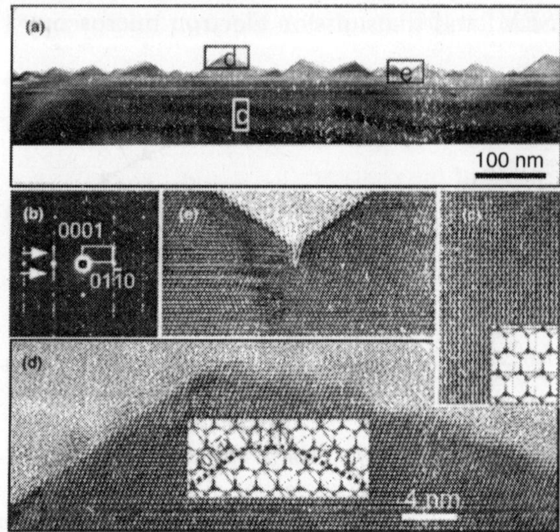

Figure 14.14 *Structure of nanosaws grown at ~1000K and using ~35 nm Au particles. (a) Low-magnification TEM image of a nanosaw. (b) SAED pattern of the whole nanosaw, showing the co-existence of the zinc blend and wurtzite pbases. (c-e) HRTEM images recorded from the rectangle enclosed areas labeled in (a). The inset in (c) gives the atomic model of wurtzile phased ZnS. and the inset in (d) is for the zinc blend.*

nanosaw, as marked in Fig. 14.14(a), reveal the local structures. The main body of the saw is wurtzite structure (see Fig. 14.14(c) but the teeth are zinc blende structure (Fig. 14.14(d)) The transformation from wurtzite to zinc blend occurs at root of the saw teeth (Fig. 14.14(e)). Comparing the atomic model inserted in Fig. 14.13(c) with the image shown in Fig. 14.14(c) the two structures are identified.

For the zinc blende structure, the (111) plane is the close packing plane; the (0001) plane of wurtzite is the close packing plane. The relationship between zinc blende and wurtzite is a change in stacking sequence of the atomic planes. Zinc blende is a stacking of ABCABC , parallel to the close packing plane, and wurtzite is a stacking of ABAB.

The formation of the nanosaw morphology is explained using a combination of vapor-liquid-solid (VLS) and the self-catalyzed growth process in two steps. First, the Au catalyst guides the fastest growth along $[01\bar{1}0]$ to form the wurtzite structured nanobelt, with $\pm(2\bar{1}\bar{1}0)$ and $\pm$ (0001) as its top/bottom and side surfaces; the sulfur terminated $(000\bar{1})$ side surface is inert and is stable, while the zinc terminate (0001) surface is catalytically active and can initiate growth in the direction perpendicular to the nanobelt. In the second growth process, small islands nucleate on the flat (0001) surface, and each island epitaxially grows up to form a saw tooth, but strain is produced when the two adjacent teeth meet. This is proved by the existence of mismatch dislocations in the region between the two teeth (islands), as displayed in Fig. 14.14(e). Each island has the zinc blende structure. If the growth front of the teeth is $(11\bar{1})$ plane, the two side surfaces of the teeth are $(01\bar{1})$ and $(31\bar{1})$ planes as marked in the atomic model in Fig. 14.14(d). The high index $(31\bar{1})$ plane has higher surface energy, thus, it is preferred to reconstruct into a series $(11\bar{1})$ and (100) facets.

2. Nanostructure formed at temperatures >1020K using Au catalyst particles of 35 nm in size

The nanostructures formed at temperatures higher than 1020K using the 35 nm size Au catalyst particles. The higher deposit temperature results in the formation of pure wurtzite nanobelts as shown in Fig. 14.15. No matter the size of the nanobelts is, the SAED patterns recorded from them indicate that they are all of wurtzite structure (see in Fig. 14.15(c) and (d). The fine nanobelts always take [0001] as the growth direction and $\pm$ $(2\bar{1}\bar{1}0)$ and $\pm$ $(01\bar{1}0)$ as top/bottom and side surfaces. The nanobelts growing along $[01\bar{1}0]$ directions are normally with larger width.

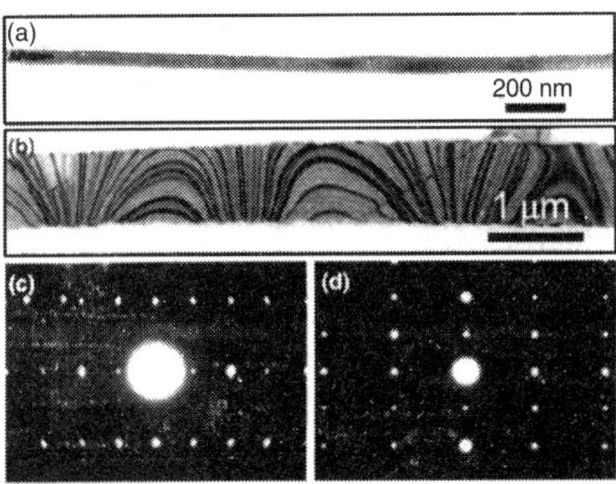

Figure 14.15 *Structure of nanobelts grown at > 1020K and using ~35 nm Au particles (a,b) TEM images a narrow and wide nanoblelts and (c,d) are corresponding SAED patterns respectively.*

3. Nanostructure formed using Au catalyst particles of 100 nm in size

The relationship between the phase formation and the size of the catalyst particles using a thicker Au film of an average thickness ~10 nm; which produces Au particles of ~ 100 nm in sizes. The SEM image from the sample collected in a deposition temperature zone of ~1000K is shown in Fig. 14.16(a). Similar to those shown in Figs. 14.13 and 14.14 uniform nanobelts and asymmetric nanosaws are the two main morphologies (Fig. 14.16(b) and (c)). The Au catalysts have an average size around 100 nm, which guide the growth of larger size ZnS nanostructures. In order to identify the phase of each nanostructures, SAED patterns from many nanostructures are recorded. The diffraction pattern presented in Fig. 14.16(e) is from the nanobelt shown in Fig. 14.16(c) which is a pure wurtzite phase. The pattern in Fig. 14.16(d) recorded from the nanosaw in Fig. 14.16(b), showing that it is a mixture of wurtzite with zinc blend, similar to the nanosaw presented in Fig. 14.14 After examining over 50 nanobelts, no pure zinc blende phased nanobelts are found even for the small width nanobelts. This is a distinct difference from the nanostructure grown using ~35 nm size Au catalyst particles:

After analyzing the as-received product, nanostructure formed in temperature zones of higher than 1020K all have wurtzite structure. If the wurtzite phased ZnS is needed, the high deposit temperature is the key, no matter about the size of the Au catalyst particles.

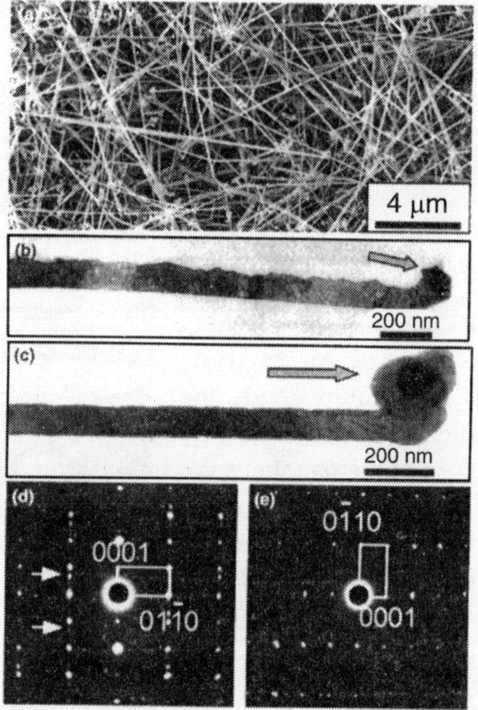

Figure 14.16 *Structure of nanobelts and nanosaws grown at 1000K and using ~100 nm Au particles. (a) SEM image of the sample. (b, c) TEM images of a nanosaw and nanobelt, and (d, e) are corresponding SAED patterns, respectively. The nanosaw is a mixture of zinc blend and wurtzite, and the nanobelt is wurtzite structure.*

WURTZITE-TYPE ZNS NANOBELTS NANOCOMBS, AND NANOWINDMILLS

Experiment

Wurtzite structure ZnS nanobelts are synthesized by heating bulk ZnS in flowing argon with the assistance of Au particles acting as catalyst. The resulting ZnS nanobelts are deposited on an alumina substrate. The as-deposited materials are characterized and analysed by SEM, TEM (at 200 kV at 400 kV) and energy-dispersive X-ray spectroscopy.

The wurtzite ZnS nanobelt bundles are removed from the substrate and loaded into a diamond anvil cell. A stainless-steel gasket is pre-indented to 50 μm in thickness by two opposing diamond anvils with 450 μm culets. A 250 μm-diameter hole is made as a chamber. Organic oil and a small ruby chip serves as the hydrostatic pressure medium and pressure marker respectively. The mass of ZnS nanobelt is estimated to be approximately 0.01 mg from the volume of the chamber and the density of ZnS nanobelts with consideration of the existence of the pressure medium. High-pressure X-ray-diffraction measurements are performed at room temperature with energy-dispersive synchrotron radiation Energy calibrations are made using well-known radiation sources (^{55}Fe and ^{133}Ba). and angle calibrations are made at a fixed angle of 15° using the six Bragg peaks of Au powder standard. X-ray-diffraction patterns, collected up to pressures of ~11.4 GPa are used for structure analysis and cell-parameter refinement. The recovered samples are also examined by SEM and TEM. Particle sizes are estimated by TEM and the FWHM of the observed X-ray-diffraction peaks. The standard equation for energy dispersive X-ray diffraction is modified as

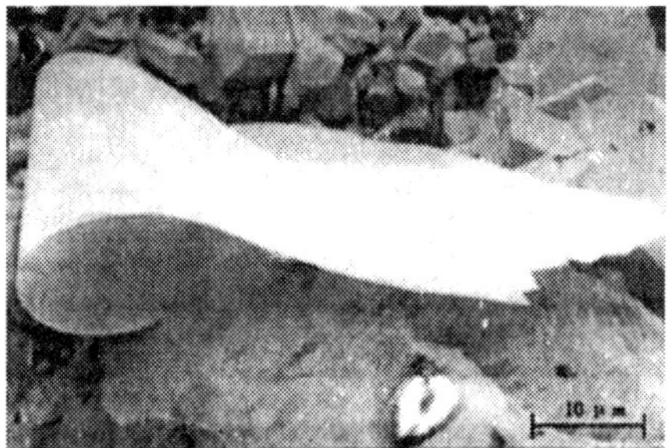

Figure 14.17 *SEM image of ZnS comb- like and sheet – like structures.*

$$\tau = (0.94 \ll Ed)/\Delta E,$$

where τ is the particle size in ångströms, E and ΔE are the energy and FWHM of the observed Bragg peak in keV and d is the inter-planar spacing in ångströms.

The calculations are based on the total Gibbs energy difference between the two ZnS polymorphs of wurtzite and sphalerite in combination with the size-induced surface-energy contribution. At 1 atm. the Gibbs thermodynamic equation is modified as $\Delta G = \Delta H - T\Delta S + \Delta \gamma s$. where ΔG, ΔH, ΔS and $\Delta \gamma$ represent the difference of the Gibbs free energy, enthalpy entropy and the unit surface energy between wurtzite and sphalerite respectively and s is the surface area.

In wurtzite ZnS nanobelts the surface energy is dominated by the ± $(2\bar{1}0)$ facets and in the newly formed sphalerite nanocrystals, it is dominated by (111) and (100) facets. The corresponding cubic and tetragonal morphologies are characterized in the recovered samples (Fig 14.18 inset: HRTEM image). At a certain thickness of nanobelt, the Gibbs free-energy difference ($\Delta G_{298\,K}^{1\,atm}$) at 298 K and 1 atm pressure is calculated. On compression, although the energy gap of $\Delta G_{298\,K}^{1\,atm}$ is overcome by a newly added energy term of P* Δ V. wurtzite transforms to the sphalerite phase. Here. P* and Δ V represent the transition pressure of the wurtzite-to-sphalerite phase transformation and the volumetric difference between the two ZnS polymorphs at P* respectively.

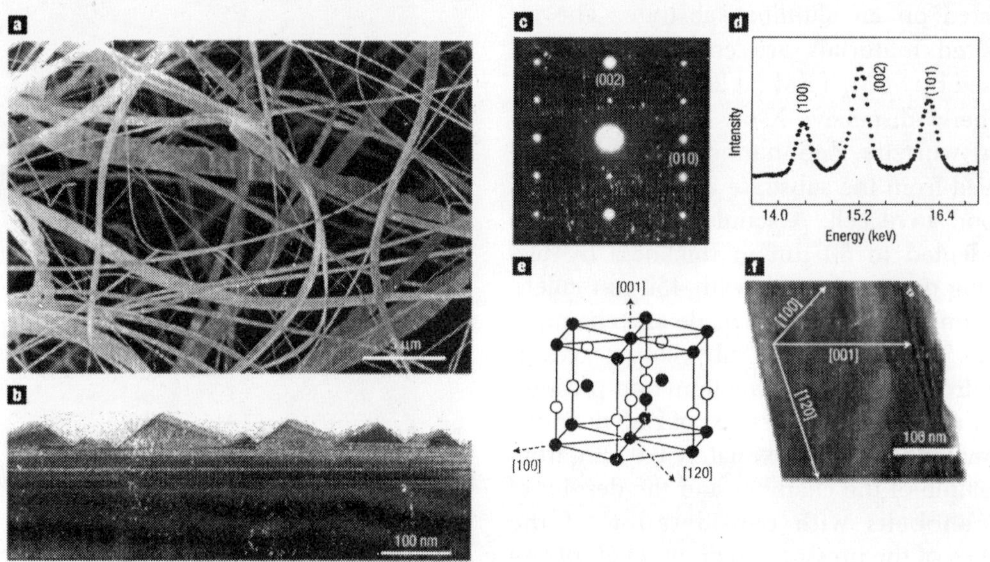

Figure 14.18 *The synthesized wurtzite-structure ZnS nanobelts. (a) Low-resolution SEM image; (b) HRTEM image showing saw-like edge; (c) electron-diffraction image; (d) X-ray-diffraction pattern (e) the wurtzite sturcture: and (f) the crystalographic dimensions of nanobelts.*

Below is discussed the ultrastable one-dimensional ZnS nanobelts the morphology-tuned explosive transformation mechanism to the sphalerite form, and a model for the kinetic behaviour. High-pressure synchrotron X-ray diffraction demonstrates that wurtzite ZnS nanobelts have a wide field of structural stability up to 6.8 GPa, which is different from the bulk and three-dimensional monodisperse spherical nanoparticles that transform to the sphalerite structure slowly at ambient conditions and easily with even slightly applied compression (Table 14.1). The calculations show that the nanobelt morphology produces a particular low-energy state that is the main reason for the improved stability of this metastable structure, in which the ± $\{2\bar{1}0\}$ surface facets dominate the surface structure of wurtzite ZnS nanobelts. High-resolution transmission electron microscopy (HRTEM) shows that twinning and stacking faults provide further stabilization constraints. This provides the stability of the metastable wurtzite ZnS nanobelts, and controlled synthesis of stable nanobelts is optimized to produce belts, wires or other functional units for biological labelling or the fabrication of nano-optoelectronic devices.

Table 1 *Comparison of the transition pressures observed in different wurtzite ZnS forms.*

Wurtzite form	*Transition pressure (GPa)*
Bulk	0
Nanocrystal	<0.5
Nanobelt	6.8

Wurtzite-structure ZnS nanobels are synthesized by a solid-vapour phase thermal-sublimation technique. A scanning electron microscopy (SEM) image indicates that the as-synthesized nanobelts have a uniform cross-section along their length, with a typical width in the sub - micrometre range, extending to over 100 μm in length (Fig. 14.18(a)) and comblike and sheet like structure (see Figure 14.17). HRTEM shows that partial nanobelts have a saw like edge Fig. 14.18(b). X-ray and electron diffraction demonstrate that the ZnS nanobelts adopted the hexagonal wurtzite structure (Fig. 14.18(c) to (e). Nanobelts are ~10 nm thick on average and have a specific growth direction along with side surfaces (001) and top surfaces (010) (Fig. 14.18(f).

The mechanical properties and structural stability of wurtzite type ZnS nanobelts in a diamond anvil cell are obtained by high-pressure synchrotron X-ray diffraction. High-pressure X-ray-diffraction patterns collected under hydrostatic (or quasi hydrostatic at higher pressure) conditions are shown in Fig. 14.20. At 1 atm pressure, both X-ray-diffraction (Fig. 14.20) and electron diffraction patterns (Fig. 14.18(c)) indicate that the ZnS nanobelts have hexagonal symmetry, and unit cell parameters computed from the X-ray-diffraction pattern are $a_0 = 3.7943(2)$ Å, $c_0 = 6.2679(5)$ Å and $V_0 = 78.15(8)$ Å^3, where a_0, c_0 and V_0 are the unit cell parameters.

The comb-like saw-toothed belts, and the regular belt structures are found in the same growth temperature range. Bright-field transmission electron microscopy (TEM) images of these structures are shown in Figure 14.19. No particle is observed at the ends of any of the structures. The ripple-like contrast observed in the TEM images is due to the strain resulting from the bending of the nanobelts. The growth characteristics of these structure show some consistency. Shown in Figure 14.19(a) is a TEM image of a ZnS nanobelt with one side having

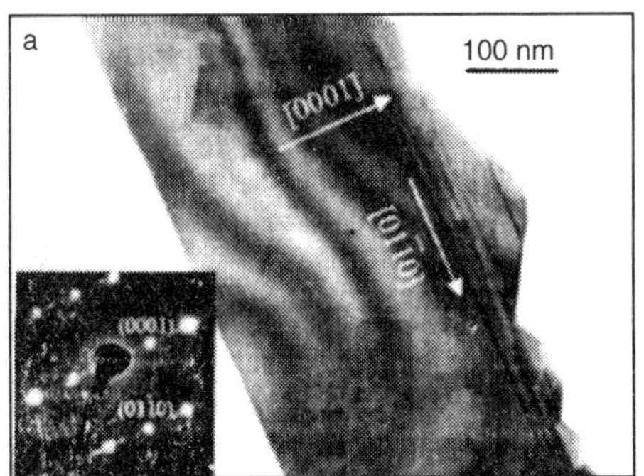

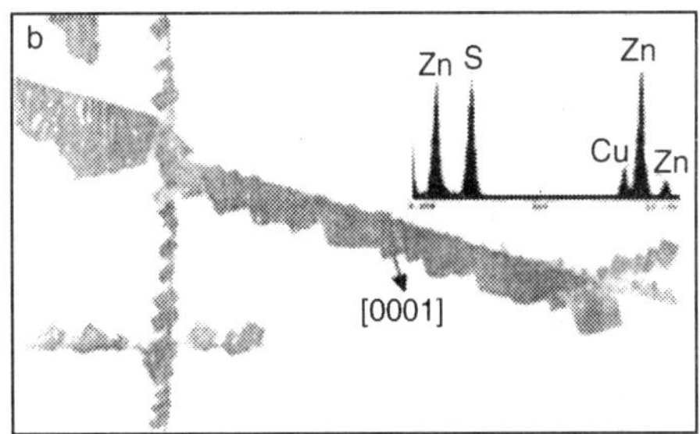

Figure 14.19 *(a) SEM image and the corresponding electron diffraction pattern from a belt-and saw-like ZnS structure. (b) TEM image of the comb-like ZnS nano-structure. The inset is an EDS (energy-dispersive X-ray spectroscopy) spectrum showing the existence of sulfur and zinc atoms in the specimen, while the Cu signal came from the TEM grid used for supporting the sample.*

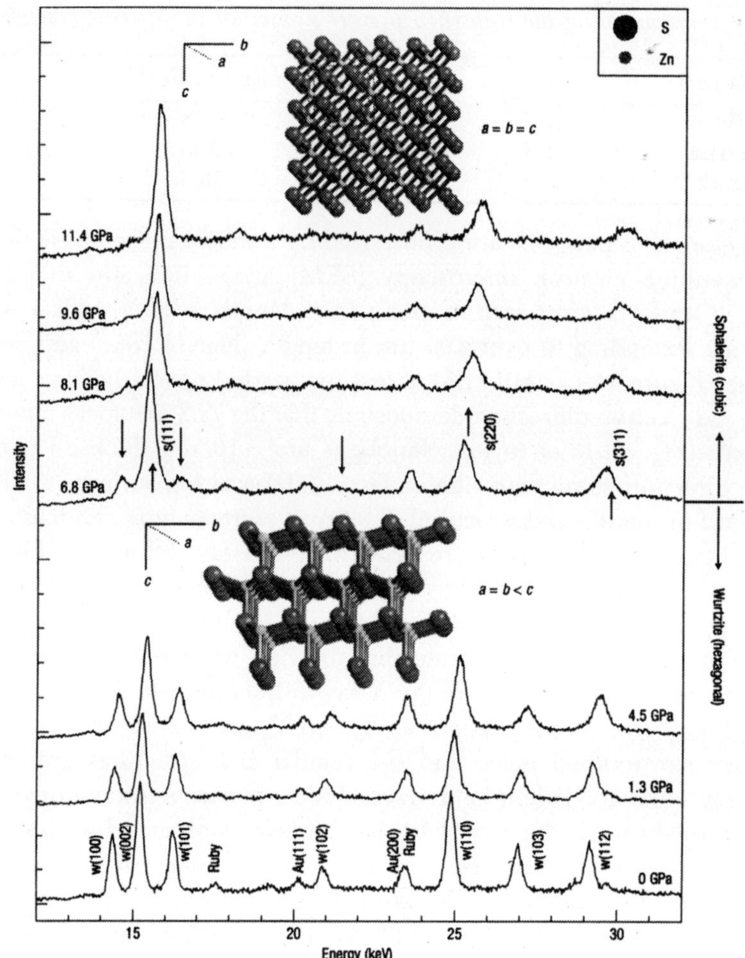

Figure 14.20 *High-pressure X-ray-diffraction patterns of the wurtzite ZnS nanobelts. The upward and downward arrows (↑ and ↓) represent the occurrence and disappearance of X-ray-diffraction peaks of the new cubic phase and hexagonal phase. Their crystallographic characteristics are shown in the two inset figures: bottom for hexagonal wurtzite and top for cubic sphalerite. Weaker X-ray-diffraction peaks are due to the small amount of gold nanoparticles used as a catalyst during synthesis and by the ruby used as a pressure calibrant.*

the saw-tooth structure. The nanobelt grows along [0001] with side surfaces $(01\bar{1}0)$ and the top surfaces $(2\bar{1}\bar{1}0)$. The extrusive sawteeth point along $[01\bar{1}0]$, and their top surfaces are $(2\bar{1}\bar{1}0)$. The sawteeth are defined by facets close to $(01\bar{1}3)$ and $(0\bar{1}13)$. A common feature observed is that (0001) stacking faults are usually present between the teeth and the nanobelt. The comb-like fingertip structure is formed by the formation of low-energy facets of (0001) and $(01\bar{1}0)$, while the top surfaces are still $(2\bar{1}\bar{1}0)$. The fingers for the comb-like structures are pointing toward [0001]. Two fastest growth directions for ZnS are [0001] and $[0\bar{1}10]$, and the largest facets are $(2\bar{1}\bar{1}0)$.

Analysis of these data does not support a size-induced contraction as observed for sphalerite with a slight tetragonal distortion from cubic symmetry. Several weak X-ray-diffraction peaks occur. As pressure approaches 6.8 GPa, several new peaks of the high-pressure phase appear, whereas the characteristic X-ray-diffraction peaks of the starting hexagonal ZnS phase drastically weaken. The newly observed peaks are consistent with the sphalerite structure. Wurtzite (w) and sphalerite (s) show a significant overlapping of w(002) and s(111), and of w(110) and s(220). On the basis of the constant intensity ratios between the three characteristic peaks of w(100), w(002) and w(101), it is obvious that the wurtzite structure remains stable to 6.8 GPa. The full-width at half-maximum (FWHM) of wurtzite peaks also does not change, so the nanobelt maintains a constant shape below the transition pressure of 6.8 GPa. Above this pressure, the two peaks of w(100) and w(101) reduce significantly in intensity and become extremely weak; in contrast, the intensity of s(002) increases. Simultaneously, the new peaks show an abrupt and very noticeable broadening. These changes are interpreted as arising from a collapse of the nanobelt occurring on the wurtzit-to-sphalerite phase transformation. Such a phenomenon is observed in the nanowire form of the semiconductor CdSe, where the pressure-induced wurtzite-to rocksalt phase transformation takes place. Although the phase boundary is not determined, it is reasonable to assume that the hexagonal phase is stable to ~6.8 GPa, and thus quickly transforms to the sphalerite structure. Organic oil pressure medium, enables a hydrostatic state to be maintained at pressures below 5 GPa; above 5 GPa, a quasi-hydrostatic condition generates the deviatoric stress and slight pressure gradient across the sample chamber. This explain the existence of a very small ratio of wurtzite phase and its extremely weak X-ray-diffraction peaks. A quick accomplishment of the phase transformation in nanobelts differs from the sluggish kinetics observed in bulk and nanoparticle ZnS forms that show a wide transition-pressure range of 5-8 GPa. This implies that the transformation mechanism is explosive. The resulting sphalerite phase remains stable to the peak pressure of 11.4 GPa, and is quenchable on release of pressure.

With decreasing particle size, surface energy plays a significant role in the optical and electronic properties and structural stability, in particular with extension to ambient conditions for technological applications. It is found that ZnS nanocrystals show enhanced optical and electronic properties, but the structural stability in wurtzite does not improve similarly to bulk wurtzite. However, because the nanocrystal is tuned to belt like morphology, wurtzite ZnS nanobelts retain their morphology with hexagonal crystallographic symmetry to pressures as high as 6.8 GPa. It is difficult to understand this remarkable observation from simple considerations of either the thermodynamics applied in bulk materials or the nanosize-induced enhancement of surface energy. The morphology of the wurtzite ZnS nanobelts is responsible for the exceptional structural stability of the materials. Specifically, the enhanced stability is a consequence of the fact that nanobelts are regular extended solids along the length and width of the structure, but have nanometre-scale thicknesses. A direct consequence is the very different surface-energy density on the top and bottom surfaces of a nanobelt compared with the surface-energy density for the faces (the 'sides' of the nanobelt), which are nanometresized in one dimension. To explore this effect of the nanobelt morphology, the structural stability field of a wurtzite ZnS nanobelt as a function of the thickness of the nanobelt is calculated by minimizing the total free energy of the wurtzite and sphalerite., taking into consideration the extra contributions associated with surface size, morphology, shape, twinning and stacking faults, and volumetric contraction.

At 298 K and 1 atm. bulk ZnS polymorphs of wurtzite and sphalerite have a Gibbs free-energy difference of 10.25 kJ mol^{-1}, which reflects the greater stability of sphalerite compared with wurtzite. With decreasing particle size surface energy starts to play an increasingly dominant role in determining structural stability Simulations indicate that each facet associated with a well-defined diffraction plane possesses a different surface energy In wurtzite, (001) has the highest surface energy of 0.91-1.52 J m^{-2}, whereas the (110) face has the lowest surface energy of 0.28-0.49 J m^{-2}, and the (100) face has an intermediate energy of 0.52-1.0 J m^{-2}. In sphalerite, both (111) and (100) faces have the two highest surface energies of 1.84 and 2.56 J m^{-2}, respectively, whereas the other facets have lower surface energy (<1.0 J m^{-2}). The total surface energy of a three dimensional spherical nanoparticle of wurtzite or sphalerite is then the mean surface energy, computed from the surface energy of all of the existing crystal facets that are averagely and randomly exposed on the particle surface. Accordingly, three-dimensional spherical nanoparticles of sphalerite have a mean surface energy of 0.79 J m^{-2}, greater than the surface energy of 0.57 J m^{-2} in wurtzite. The observation that the transformation temperature of sphalerite to wurtzite reduces with decreasing particle size strongly supports this surface-energy estimation.

The wurtzite ZnS nanobelts have a specific growth direction along [120], with ± (210) planes defining the two dominant surface (see Fig. 14.21 inset top and bottom faces and Fig. 14.18(d). Such a ± $\{2\bar{1}0\}$-dominant surface structure is different from the common surface structure of nanocrystals that have a high percentage of cubic, tetrahedral and octahedral-like shapes, dominantly bounded by the combined facets of {111}, {110], {001} and {100} The surface-energy difference associated with different crystallographic planes holds a general sequence as $\gamma_{\{111\}} < \gamma_{100} = \gamma_{001} < \gamma_{\{110\}}$ in the cubic symmetry nanocrystals. The high-energy {110} surface is mostly observed in nanorods, but its instability is often observed by the formation of spherical clusters in terms of the atom sublimation, such as Au nanorods. However, the surface energies in ZnS with hexagonal symmetry have a reverse sequence and the (hk0) facets, including six equal planes of ± (110), ± $(2\bar{1}0)$ and ± $(1\bar{2}0)$, have the lowest surface energy and so favour the formation of this type of low-energy nanobelt growing along the [120] direction with the lowest energy. As a result, the front cross-section ± (010) and side ± (001) surface faces are neglected in the estimation of the total surface area and surface energy. A slight volumetric reduction of 1.25% takes place on the wurtzite-to-sphalerite phase transformation at 6.8 GPa. The nucleating sphalerite crystal shows the (111) plane that results directly from the (001) plane of the wurtzite phase. In addition, (111) and (100) represent the two largest *d*-spacing planes, so both of them dominate the surface area of the sphalerite when no specific crystal direction growth exists.. This is demonstrated by the observed cubic and tetrahedral morphologies of recovered nanoparticles (Fig. 14.21 inset: HRTEM image). Therefore, it is reasonable to assume that the newly formed sphalerite represents a structure with a higher surface energy than wurtzite. In

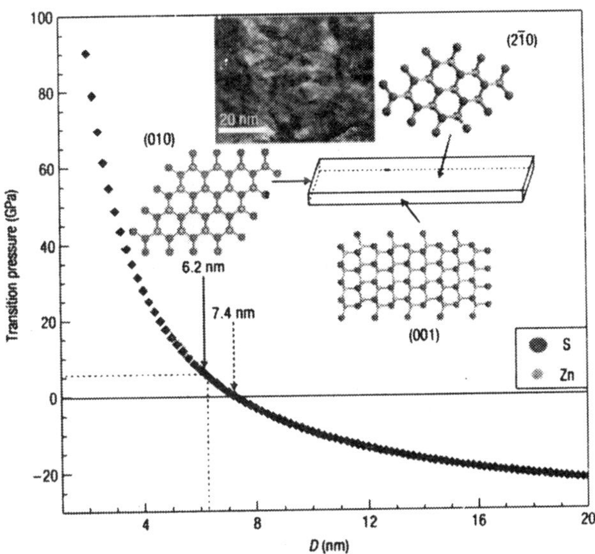

Figure 14.21 *Correlation between the nano thickness (D) and transition pressure of one dimensional wurtzite nanobelts. Inset figure show the three characteristic surface and the corresponding atomic arrangement in reciprocal lattice demonstrations. Inset HRTEM figure shows that the recovered samples have the cubic and tetragonal shapes with an average size of 10 nm*

combination with the bulk Gibbs energy ($\Delta G_{298K.1atm}$), surface-energy difference ($\Delta\ \gamma$) and the volumetric collapse the structural stability field of wurtzite ZnS nanobelts (the wurtzite-to-sphalerite transition pressure) as a function of nanobelt thickness is calculated. Below the critical thickness of 7.4, nm a wurtzite nanobelt is more stable than sphalerite (Fig. 14.21) on reducing the thickness, the stability seems to be enhanced. In wurtzite ZnS nanobelt the observed transition pressure is found to be 6.8 GPa for a thickness of 6.2 nm. Transmission electron microscopy (TEM) observation indicate that the recovered N.Ws have an particle size of ~10 nm (Fig. 14.21, inset HRTEM image) However, the particle size of ~ 12 nm is calculated from the broad x-ray diffraction peaks (Figs 14.22). These values are greater that the calculated nanobelt thickness of 6.2 nm which corresponds to the observed transformation pressure of 6.8 GPa, or the critical thickness of 7.4 nm Thus a reasonable explanation requires the incorporation of further effects induced by crystallographic defects, twins, stacking faults and volumetric contraction generated in the newly formed sphalerite on phase transformation.

Both the Zn and S atoms in wurtzite and sphalerite are four-coordinated so the phase transformation only requires a partial atomic rearrangement. Wurtzite has the simple hexagonal close-packed stacking order of ABABABAB along the [001] direction with a resultant base of (001) and sphalerite has the cubic-close-packed stacking order of ABCABC along the [111]: direction with the characteristic crystallographic facet of (111). On transformation from wurtzite to sphalerite, the (001) plane of wurtzite converts directly to the (111) plane of sphalerite. On the basis of such a fundamental genetic relation and combined with the particular crystallographic characteristics of the two Zns polymorphs, two types of transformation mechanisms, are suggested (Fig. 14.23). The first involves the rearrangement of three {ZnS} layers (Fig. 14.23(a)) leading to

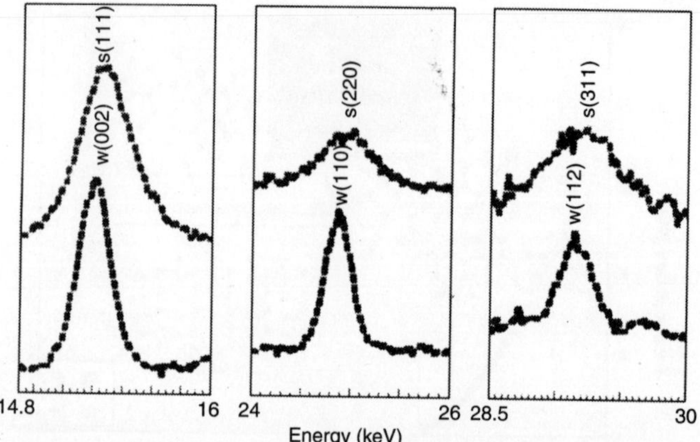

Figure 14.22 *Comparison of the three characteristic X-ray-diffraction peaks between wurtzite and sphalerite polymorphs at 1 atm pressure. Here w and s represent the wurtzite and sphalerite, respectively.*

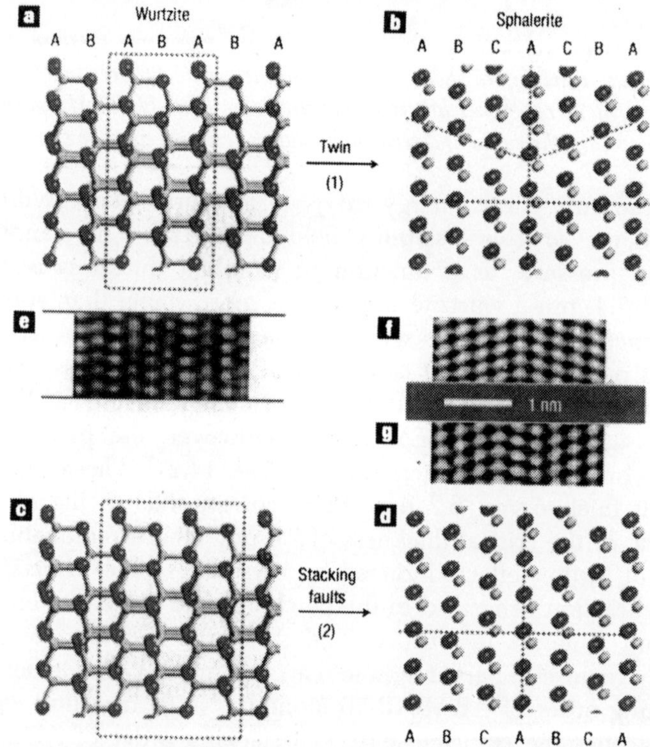

Figure 14.23 *Representation of the wurlzite-to-sphalerite phase transformation. (a,b,) The first (1) involves the rearrangement of three {ZnS} layers (a), leading to (111) twinning in sphalerite related by a* 180^0 *rotation (b). c,d, The second (2) involves the rearrangement of four (ZnS) layers (c), resulting in the development of stacking faults (d). e-g, Filtered HRTEM images of the samples: starting wurtzite phase (e); recovered sphalerite phase (f,g).*

(111) twinning in sphalerite related by a 180^0 rotation (Fig. 14.23(b)). The second involves the rearrangement of four {ZnS} layers (Fig. 14.23(c)), resulting in the development of stacking faults (Fig. 14.23(d)). It seems that the transformation initiated by the first type of atomic rearrangement requires a higher energy than the second type of transition mechanism, because the rotation in the first type of mechanism is produced as a consequence of a faulted stacking of perfect crystals, and the bonding configuration at the stacking faults and the twin boundaries remain close to that of the perfect structure.

The two types of mechanism are observed in the HRTEM images taken from the recovered sphalerite (Fig. 14.23(e) and (f) that show the formation of the boundary twins (ABCACBA) (Fig. 14.23(e)) and stacking faults (for example, the double stacking faults shown in Figure 14.23(f). The resultant stacking faults and twins imply a high-energy sphalerite structure. Although numerous types of energy contribution, are added to the total energy of sphalerite, the nanobelt thickness (also including the critical thickness) is greater for the wurtzite nanobelts that transform to sphalerite at modelled pressure with thinner thickness. It is observed that sphalerite nanoparticles show a size-induced contraction by undergoing a tetragonal distortion from the cubic structure, which leads to a volumetric decrease of the order of ~2% compared with bulk sphalerite with cubic symmetry. At the transition pressure of 6.8 GPa, this results in an energy change as large as ~12.5 kJ mol^{-1} between sphalerite and wurtzite, responsible for a transition pressure jump of ~2.4 GPa. Combining the above contributions provides an understanding of the high transition pressure of 6.8 GPa in the wurtzite ZnS nanobelt synthesized here.

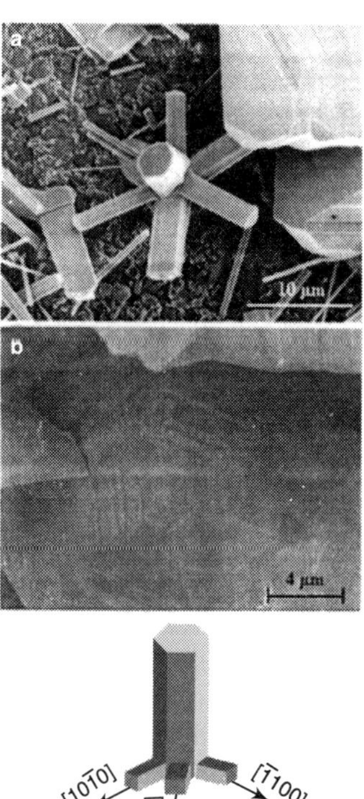

Figure 12.24 *(a) SEM image of windmill structure b) SEM image of the polyhedral structure of ZnS. c) Structural model of the windmill structure.*

Another structure observed is the six-fold windmill structure (Fig. 14.24(a)) in which the axis is [0001] and the six wings are along ± [10$\bar{1}$0], ±[0$\bar{1}$10], and ± [$\bar{1}$100] as shown in Figure 14.22(c). This type of structure is observed for hexagonal ZnO. The six wings grow laterally and interconnect to form a faceted structure like the one shown in Figure 14.22(b). where the side facets are likely to be (10$\bar{1}$1), ($\bar{1}$011), (01$\bar{1}$1), (0$\bar{1}$11), ($\bar{1}$101). and (1$\bar{1}$01). This type of faceted three-dimensional polyhedral structure is common in the synthesized material.

The wurtzite-structured ZnS is unstable and it transforms zinc blended structure shown in Figures 14.25(a) and (b) are two images recorded from the same area before and after the sample is illuminated for about 10 min under 200 kV electrons, showing an increase in the density of planar defects. The electron diffraction pattern recorded from the area shows the coexistence of the hexagonal wurtzite structure and the cubic zinc blende

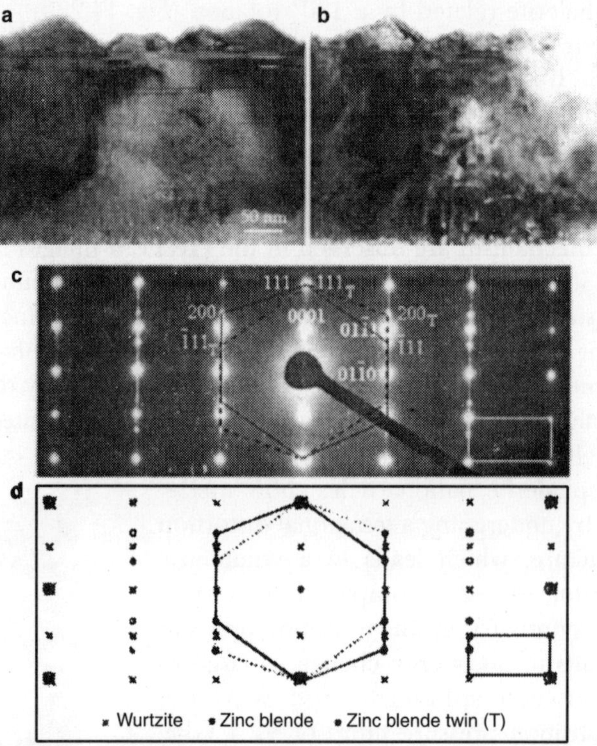

Figure 14.25 *(a.b) TEM images of a saw-like ZnS nanostructure prior and after illumination by an electron beam for ca. 10 min. showing the formation of planar defects. c) Electron diffraction pattern recorded from the area which can be indexed as the coexistence of the hexagonal and cubic phases with the presence of twins in the cubic phase. d) The systematic reflections in corresponding to the experimental pattern shown in (c).*

structure. (Fig. 14.25(c)). The orientation relationships between the two phases are: $[2\bar{1}\bar{1}0] \parallel [01\bar{1}]$, and $(0001) \parallel (111)$. The two phases coexist by sharing the same (0001) or (111) plane. The cubic phase of ZnS has the {111} twins. The existence of the twins is indicated by the electron diffraction pattern. The diffraction spots and the corresponding indexes from the hexagonal phase, the cubic phase and its twin are illustrated in Figure 14.25(d)

Unit-cell models for the hexagonal and cubic phases of ZnS are shown in Figure 14.26(a). The presence of the two phases in the nanobelts are directly identified by high-resolution TEM. Figure 14.26(b) shows a TEM image recorded from the hexagonal phase that is oriented along $[2\bar{1}\bar{1}0]$, which matches fairly well to the projected position of the Zn atoms in the unit cell as shown in the inset. The sulfur atoms are too light to be resolved by the TEM Figure 14.26(c) is a $[01\bar{1}]$ projected image of the cubic phase and the projection of the unit cell is marked by a rectangular frame. The projected positions of the Zn atoms in the cubic phase, as shown in the inset, match well to the bright contrast observed in the image.

The cubic phase of ZnS is viewed as being composed of a face-centered cubic (fcc) Zn sublattice with a fcc S sublatice that is displaced by 1/4 in the [111] axis direction with respect to the Zn sublattice. From the structural point of view, a face-centered cubic phase is related to a hexagonal close-packed phase by changing the stacking sequence in parallel to the (111) or (0001) plane. For close-packed configurations, there are three distinct stacking layers, named A, B, and C, each of which is composed of one layer of Zn atoms and one layer of S atoms. A stacking of ABAB forms the hexagonal phase, while a stacking of ABCABC forms the cubic phase. The two structures can be transformed simply by changing the stacking sequence. Figure 14.27 shows a high-resolution TEM image from an area in which the hexagonal and cubic phases coexist; the corresponding structure model for the image is illustrated at the right-hand side in the frame. The incident beam direction is $[2\bar{1}\bar{1}0]$ or $[01\bar{1}]$ for the hexagonal or cubic phase, respectively. The (0001) or (111) plane is the stacking plane. Although the positions of the S atoms cannot be determined from the image, the projected positions of the Zn atoms can be directly provided. The region is composed of two phases and a twin of the cubic phase. The orientation relationship provided by this image matches exactly to that derived from the diffraction pattern given in Figure 14.25(c).

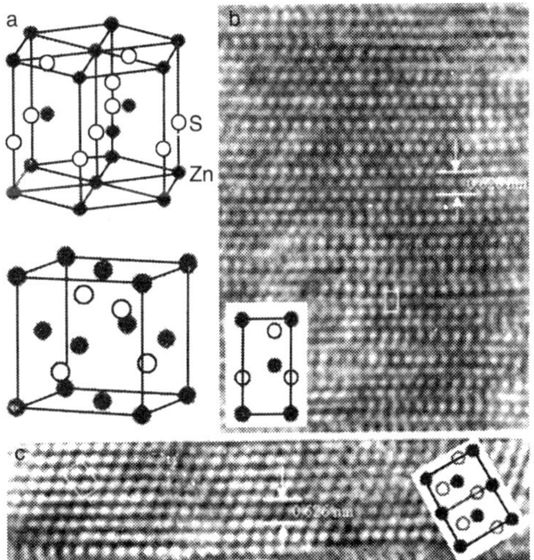

Figure 14.26 *(a) Unit-cell models for the hexagonal and cubic phases. b) $[2\bar{1}\bar{1}0]$ High resolution TEM image of the wurlzite ZnS. c) $[01\bar{1}\bar{1}]$ High-resolution TEM image of the zinc blende ZnS. The insets arc thc projections of the corresponding unit cells.*

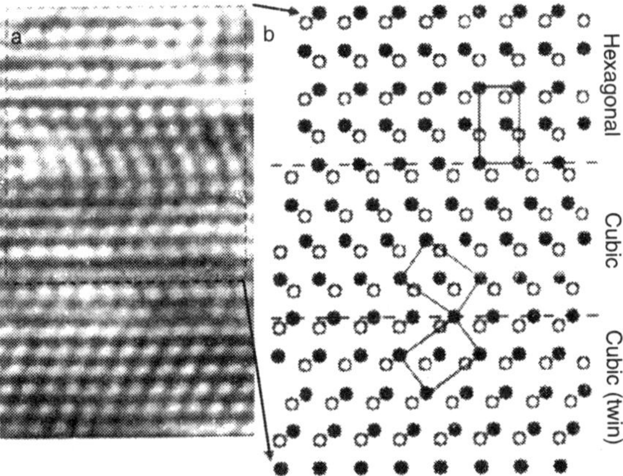

Figure 14.27 *High-resolution TEM image showing the coexistence the wurtzite. zinc blende, and the twinned zinc-blend structures; the corresponding structure model is illustrated at the right-hand side showing the transformation among the structures by adjusting the stacking sequence of the atomic layers.*

15

Core-shell Nanowire

GROWTH OF ULTERLONG ZnS/SIO$_2$ CORE – SHELL NANOWIRE

A large scale synthesis of laterally aligned ultralong crystalline ZnS nanowires covered by an amorphous silica (SiO_2) shell is described below A novel synthesis of such nanowires exhibiting both dense and well aligned organization while maintaining high aspect ratios. These nanowires, with lengths on the order of 1 mm to 1 cm. are long enough for simple manipulation into devices, The single-crystalline nature of the ZnS core ensures high-quality semiconductor material characteristics, while the amorphous silica shell helps to prevent mechanical or radiation damage and suppress chemical reactivity, which leads to oxidation and contamination in bare ZnS nanomaterials. For biorelated applications, the low toxicity of both ZnS and silica is highly desired. In fact it is found that amorphous SiO_2 has low toxicity even in systems that do not respond well to its crystalline form.

A series of systematic experiments coupled with careful analysis, including transmission .electron microscopy (TEM), X-ray diffraction (XRD), and photoluminescence (PL), allow to examine how variations in synthesis time affect the nanowire growth. These experimental measurements are used to formulate a growth mechanism. A dual diffusion process is experimented in which ZnS atoms migrate to the growing nanowire core via bulk diffusion through the gold catalyst particle, while the SiO_2 species migrate around the particle surface before becoming incorporated into the amorphous nanowire shell. Although several other routes leading to the growth of such core-shell nanowire can be envisioned, each of these is eliminated through careful examination of the final growth product. Finally, the theoretical feasibility of this model through diffusion calculation is tested.

Experiment

Synthesis of the ZnS/SiO_2 nanowires is carried out via a simple vapor deposition process. ZnS powder is placed in the center of a single-zone horizontal tube furnace where the atmosphere, evaporation time, pressure, and temperature are controlled. Single-crystal silicon substrates with 20 nm of gold deposited by thermal evaporation are placed "downstream" at a lower temperature region in the furnace. Using a rotary vacuum pump, a vacuum is pulled in the tube for several hours to purge oxygen from the chamber After evacuation to a pressure of about 0.266 Pa the temperature in the center of the tube is elevated to 1270K at a rate of 50 K min^{-1}. A N_2 gas flow is introduced into the system at a rate of 50 standard cm^{-3} min^{-1} (sccm). The pressure in the system is a allowed to increase to a value of 0.266 Pa and is held at that level through the duration of the synthesis. The silicon substrates reach a temperature of about 1020K. After these conditions are maintained for between 1 and 2 hours, the N_2 gas flow is turned off and a vacuum with a pressure around 0.266 Pa is again achieved. At this time the furnace is cooled to room temperature.

Optically each sample appears a thick deposition of white powder or fluffy white fibrous material covering the silicon substrate, Samples grown longer time show increased deposition density on the substrates with some deposition masking the silicon wafer completely. Some samples exhibit deposition product several millimeters past the silicon substrate. Care is taken while preparing samples for analysis not to destroy or contaminate the as-grown nanowires. Samples are analyzed initially using XRD and scanning electron microscopy (SEM). Initial XRD scans are fast scans (~10 min/sample) performed to determine basic phase information. Longer subsequent scans are then taken to gather more in-depth information from the samples. The SEM used is operated at 5 kV. The samples are examined using TEM at 200 kV

X-ray diffraction measurements are performed on each sample using a X-ray machine with Cu Kα radiation. As-grown samples are affixed to a aluminum SEM stubs with carbon tape loaded into the sample holder (for solid samples affixed to the cradle) Alignment is performed via manual scans to check the sample position with respect to the 2 θ, Z, and ω axes. Sample offsets are input into the s stem to calibrate zero positions for each axis. Thus the sample is aligned parallel to the beam without the necessity to perform otherwise potentially destructive sample preparation.

For the measurements, the radiation beam emitted from an X-ray tube set to line focus, coupled with a parabolic mirror and 1/80 fixed divergence slits, is directed toward the unmodified nanowire coated substrate. After interaction with the sample, a 0.27 deg parallel plate collimator with a flat graphite crystal diffracted beam monochromator helps to refine the signal before reaching the Miniprop large window point detector. The $2\theta - \omega$ measurements are taken with a tube power of 45 kV and 40 mA, from 5 to 120^0 in 2θ with a 0.01^0 step size and a 2 hours and 23 min total scan time.

Results

SEM images like those illustrated in Figures 15.1 and 15.3 show that the aligned nanowires on the top layer of the substrate are fine and very long. Figure 15.1(a) to (d) reveals several images of a single representative nanowire sample. Figure 15.1 shows a composite SEM image taken across cross the ~ 1 cm deposition substrate. It shows that the nanowires are aligned along the flowing direction of the carrier gas. It appears that some nanowires begin at the leading edge of the substrate and continue well past substrate making these nanowires approximately 1.5 cm in length considered an individual nanowire followed by a microscope along the entire length, of the substrate. The nanowire is lost about two-thirds of the way due to SEM image shift. Figure 15.1(b) contains an optical image of this sample, illustrating the length and density of the nanowires at low magnification. Note that the silicon substrate is completely masked by the nanowire sample. Figure 15.1(c) illustrates an SEM image of the substrate's leading edge. Figure 15.1(b), (d) confirms that the nanowires grow beyond the substrate. Note that the boundary observed in Figure 15.1(d) corresponds to the edge of the sample stub seen at the far right of Figure 15.1(b), rather than the edge of the silicon substrate. The EDS profile taken in the SEM, (seen in the inset) ,reveals the presence of Zn, S. Si, and O in the sample.

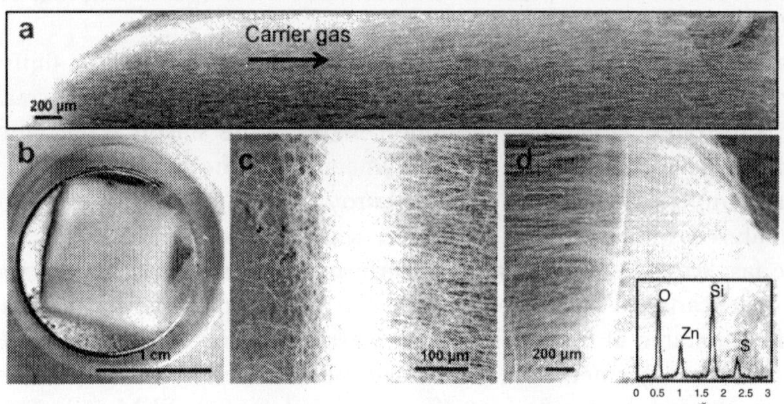

Figure 15.1 *SEM and optical images showing a representative sample of ultralong ZnS-SiO$_2$ nanowires (a) composite of several SEM images demonstrating nanowires length; (b) optical image taken from above sample, fixed on metal stub; (c) SEM image of the sample's leading edge; (d) SEM image of opposite edge, showing long nanowires reaching past the metal stub. The inset shows corresponding EDS data.*

TEM images and electron diffraction patterns are presented in Figure 15.2. Several nanowires are imaged using low-resolution TEM, as shown in Figure 15.2(a) to (c). The images shown in Figure 15.2(a) and (b) show the tip of the nanowire, including the terminating metal catalyst particle. EDS scans across the area of the particle (not shown) indicated the presence of Zn, S, Si, O and Au. Note the contrast change seen in the amorphous shell layer at the tips of these nanowires. This contrast change in the 2D TEM projection indicates the presence of a sphere or halfsphere of amorphous material surrounding the metal particle. The "neck" region of the nanowire just below the particle in Figure 15.2(a) is also of interest. Figure 15.2(c) shows a slightly higher magnification image of a nanowire center. Electron diffraction patterns are taken of the core and shell areas of this nanowire and are shown in Figure 15.2(d) and (e), respectively. The electron diffraction pattern shown in Figure 15.2(d), together with the EDS data taken in the TEM, reveals that the nanowire core consists of a wurtzite-structured ZnS. The presence of Cu seen in the EDS spectrum is due to the Cu TEM grid (affix the sample). Using analysis of the electron diffraction pattern it is found that the ZnS core of the nanowire grows long the [0001] direction, the fast growth direction of wurtzite ZnS. Figure15.2(e) shows TEM and EDS information corresponding to the shell of the nanowire sample. The electron diffraction pattern

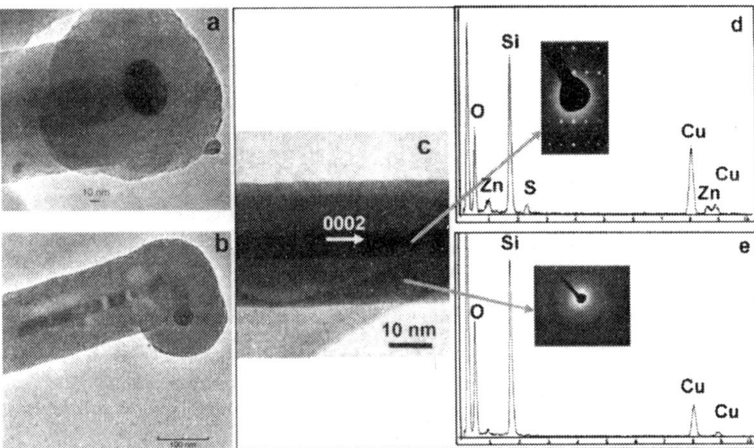

Figure 15.2 *(a, b) Low-magnification TEM images showing the tip of two ZnS- SiO_2 core–shell nanowires, including the terminating metal particle. (c) Low-magnification TEM image of a nanowire center. (d) Electron diffraction pattern and corresponding EDS data from the nanowire illustrating the single-crystalline nature of the core. (e) Electron diffraction pattern and corresponding EDS data from the nanowire shell, showing the shell to be amorphous in nature and consisting of only Si and O.*

indicates that the shell is made up of an amorphous material, while the EDS graph shows presence of Si and O but not Zn or S. These results indicate that the nanowire is composed of a crystalline ZnS core and an amorphous SiO_2 shell.

For, the growth mechanism of these core–shell nanowires synthesis at different time periods is conducted in an effort to observe the growth evolution of the nanowire structures. Synthesis runs are performed in which the maximum reaction chamber temperature is maintained for 15,30,45,60,90, and 110 min. To ensure that results are consistent and reproducible, at least three experiments are run for each set of parameters. The results observed in these experiments suggests a high degree of repeatability in the growth process. SEM images corresponding to each of these synthesis times are shown in Figure 15.3. Although ZnS/SiO_2 nanowires are synthesized with growth times as short as 15 min, nanowire alignment is not observed in growth times shorter than 45 min. Longer synthesis times corresponded in general to longer and denser growth product. Although Figure 15.3(f) appears to show a lower density of nanowires but this image is taken at higher magnification to show greater sample details Note that the images in Figure 15.3 are rotated 90° relative to those in Figure 15.1. such that the flow gas here is indicated to be in the vertical direction.

Several analysis techniques are performed to compare samples of varying synthesis times. First, a TEM of the various widths of the ZnS core and the SiO_2 shell is conducted. This consist of measuring, via TEM images, the widths of both the core and the shell and comparing them according to synthesis time. With measurement of over 30 nanowires from each of the six time intervals, the average (mean) widths of core and shell for nanowires in each section are determined. While core width are measured directly shell widths are determined by subtracting the core size form the total width of the nanowire, a graph of these data is shown in Figure

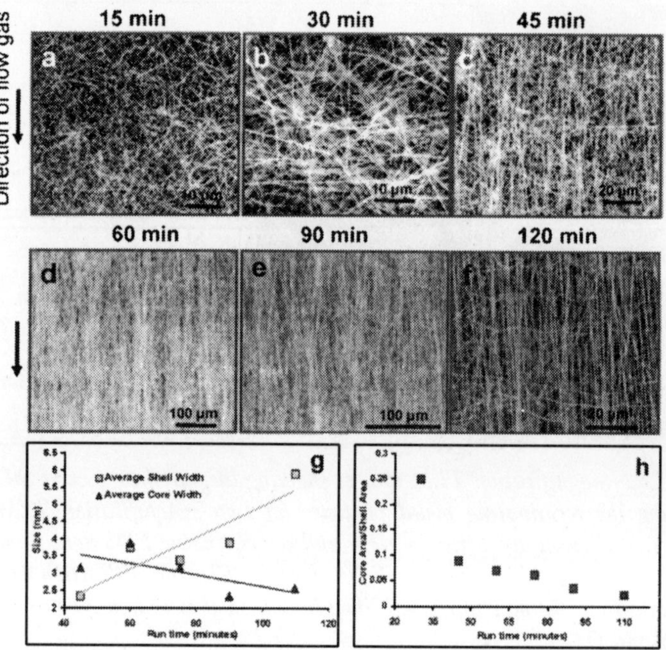

Figure 15.3 *Ultralong nanowires as a function of growth time. SEM images show sample characteristics after (a) 15 min, (b) 30 min, (c) 45 min. (d) 60 min. (e) 90 min, and (f) 120 min, at maximum chamber temperature. (g) Plot showing variation in average nanowire core and shell widths as a function of growth time as determined from TEM analysis. (h) Plot illustrating variation in the core/shell area ratio as a function of growth time.*

15.3(g) Not included on this graph are measurements for the 30 min synthesis timine interval. These values are omitted as outliers, determined initially by distortion of the graphs , and then confirmed via a standard statistical analysis technique in which the values are found to lay more that one standard deviation away from the next closest data point. For completeness note that measurements of sample run in the 30-min interval confirms the overall trend observed in other samples, showing an average ZnS core size of 13.84 nm and a silica shell width average of 1.59 nm. From graph, the ZnS core tends to become smaller over time (although only by a maximum of a few nanometers) while the SiO_2 shell grows larger with longer synthesis times.

While core and shell width values differ between samples according to time interval, each nanowire maintains a constant width along its entire length without the presence of dips or tapering. Therefore the nanowires relative volumes are determined from their respective widths according to the following equation:

$$\frac{\text{core volume}}{\text{shell volume}} = \frac{r_c^2}{[(r_c + w_s)^2 - r_c^2]} \tag{1}$$

Here $r_c = w_c/2$ is radius of the ZnS core and w_s is the thickness of the SiO_2 shell. This analysis for nanowires in each time interval is shown in Figure 15.3(h). As seen in this figure it is found that the core volume relative to the shell volume decreases over time.

While TEM analysis give detailed information regarding the nature of a small subset of nanowires in each sample, X-ray diffraction (XRD) techniques are utilized to provide a better understanding of the global nature of the nanowires. In particular, the overall changes in relative core and shell thickness whit respect to synthesis time for an area of nanowires measuring a few square millimeters. Six samples are examined, each differing only according to experimental reaction time, as in the quantitative TEM analysis. These times are 30, 45, 60, 75, 90, and 110 min. The results of scans from each sample are shown in Figure 15.4(a).

Note that while intensities vary, scans show similar peak profiles for each sample. This result confirms that the experimental design and sample preparation are such that phases remain consistent and no contamination occurs. Variations in intensity occur for many reasons including instrument effects, sample alignment, nanowires density, relative phase concentration, crystallite size and preferred sample orientation, etc. Phase identification of the diffraction pattern in Figure 15.4(a), yields three known phases: Au, the wurtzite phase of ZnS and SiO_2. To determine relative ZnS and SiO_2 core and shell thickness from the XRD data, a semiquantitative phase analysis is used for the ZnS phase. Software is employed to perform the XRD analysis.

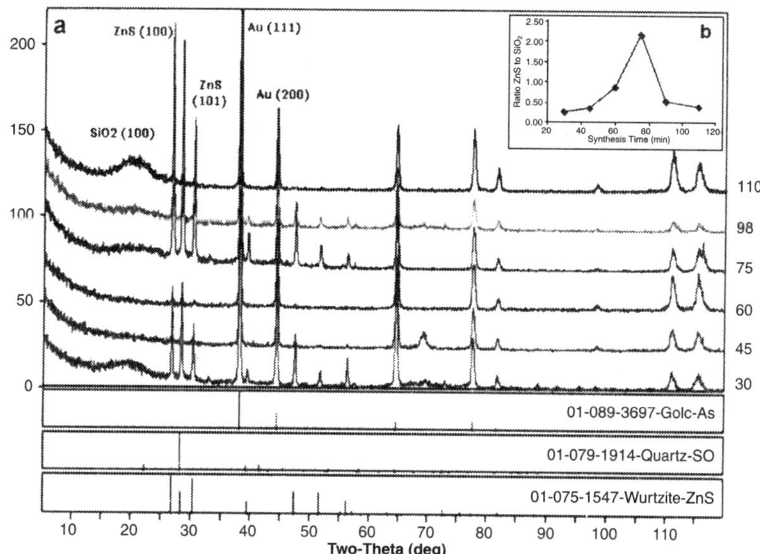

Figure 15.4 *(a) Series of XRD patterns from samples with reaction times varying from 30 to 110 min. (b) The ZnS:SiO_2 ratio in nanowire as a function of reaction time. derived from XRD pattern semiquantitative phase analysis,*

The semiquantitative analysis on the basis of reference intensity ratio (RIR) values, divide the intensity of the pure phase to be determined by the intensity of the pure phase to be determined by the intensity of a standard material, often corundum (Al_2O_3). To perform analysis on a simple containing multiple phases, normalized RIR method is used The normalization used in this method assumes that the sum of all identified phases is 100%. This means that unidentified or amorphous phases reduce the meaning and accuracy of this analysis.

TEM shows that the SiO_2 shell is mostly amorphous in nature, which is validated through broad peak signatures in the XRD pattern. Figure 15.4(b) shows a graph illustrating the relative concentration of ZnS versus SiO_2 for each of the samples, as reaction time increases. Note that, due to the presence of some amorphous SiO_2 in the sample and for an oxidation layer on the silica substrate surface, these values are indicative of a trend rather than as stand - alone empirical values.

Through the use of semiquantitative phase analysis on XRD diffraction patterns, a clear trend appears in the variation of core size in the ZnS/SiO_2 core–shell ultralong nanowires as reaction time progresses as illustrated in Figure 15.4(b). At short reaction times, both the relative percentage of ZnS in the nanowires and the average size of the ZnS wurtzite crystal begin at a low base value. As the relation progresses the ZnS core grows large and a large proportion of ZnS is present in the nanowires, relative to SiO_2 amorphous shell. As the reaction continues the size of the core appears to reach a maximum value and then the ZnS again decrease both in size and concentration.

Comparing, the photoluminescence spectra for the samples grown at different time periods, it results in graphs as shown overlaid in Figure 5. All of the spectra shown at least two distinct peaks, one peak appears at 340 nm and the other, around 532 nm. The lower wavelength peak is a pure Zns PL peak. This peak is associated with a material band gap of about 3.65eV, a good match with the theoretical wurtize value of 3.91 eV. Because ZnS with very few effects, vacancies, or intersitials, it is rare to see the 340 nm ZnS peak. The 532 nm peak is more

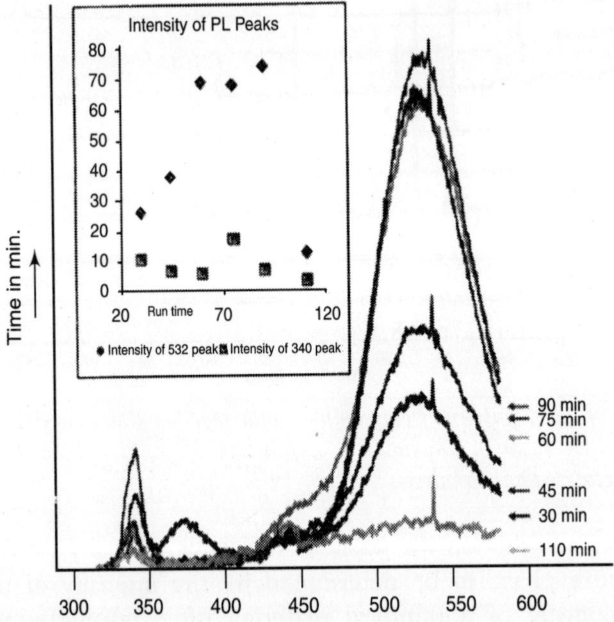

Figure 15.5 *Composite of photoluminescence data from several ultralong ZnS-SiO_2 core-shell nanowire samples. Each pattern corresponds to a unique sample exposed to synthesis conditions for a time interval between 30 and 110 min. The inset illustrates a plot of this data, correlating peak intensities with reaction time.*

common. It is associated with Zn defects, vacancies, and interstitials in the ZnS crystal. This peak translates into a band gap of 2.33 eV. The sharp peak that appears to be convoluted with the much broader signature around 540 nm is due to an instrumental effect and is present in all profiles originating from the PL equipment used for this analysis.

In the lowest growth time measured, that of 30 min, a broad –peak is observed at 375 nm, which corresponds to a band gap of 3.30 eV. This indicates the presence of a silica layer on the substrate surface. It exhibits observable luminescence only in the sample with the lowest density of nanowires. Comparing the intensities of the peaks and how they change with longer synthesis times, is shown in the inset of Figure 15.5. Focusing on the 532 nm peak it is observed that the intensity of the peak increases with increasing synthesis times from 30 up to 90 min. However, for nanowires grown for 110 min, the peak disappears. Also, the ratio between the two major peaks increases with longer synthesis times from 30 to 90 min. Again, at the 110 min measurement, the ratio drops dramatically.

Since the first observations of catalyst assisted PVD (physical it vapor deposition) nanowire synthesis ,the underlying growth mechanisms governing the production of these materials are identified. Observation of a metal catalyst particle at the nanowire tip is no longer considered sufficient evidence to support the classic VLS formation process in whole or in part, especially with GaAs and InAs structures and other metal catalysts, which indicate that nanowires are grown by particles remaining in the solid state throughout the synthesis process. Sn-catalyzed ZnO,1D nanostructures indicate that the crystalline and electronic structures of the substrate, nanostructure, c and catalyst particle all playa role in determining the nature of the final growth product.

Silicon forms nanostructures via bulk diffusion through the Au catalyst particle. It is not observed as a tube or shell structure. Synthesis of ZnS-Si core-shell nanowires is achieved through templating of the silicon onto a previously grown ZnS nanowire and not direct PVD growth. ZnS nanostructures are formed by bulk transport means, as 1D nanostructure of this material tends toward formation of solid products and core, rather than shell, components. Further evidence is from has EDS analysis, as a TEM has revealed traces of Zn and S as well as Au when focused on the terminating Au particle of a ZnS nanowire.

Conversely, SiO_2 appears as a sheet or as a tubular structure, frequently forming coating layers over other growth products and in some cases even capping the metal particle. Such behavior is not surprising when considered in light of silica's reduced solubility in gold compared to elemental Si. It is not uncommon to observe areas of increased gold mobility on silicon substrates with localized oxidation, as low Si nanowire growth and high Au mobility are correlated to locally oxygen-rich regions on a Si substrate. This phenomenon is demonstrated as an effective technique for patterning of gold particles on selectively etched Si-SiO_2, wafers. Here, the Si substrates are the only source of silicon while oxygen species originate from residues left in the reaction chamber or even from slight oxidation of the substrate surface. Additionally, SiO acted as an intermediate in the process.

These processes occur by mass diffusion through the metal particle, surface diffusion around the metal particle, selective incorporation entirely at the particle–nanowire interface, transport up the sides of the growing nanowire, and direct adsorption onto the nanowire and are all identified as mechanisms of formation in ZnS and SiO_2 1D nanostructures. The mechanism responsible for the growth of nanostructures, is examined with regard to observations of the final ZnS-SiO_2 nanowire growth product. It is determined that any of four distinct growth processes, is the growth of ZnS-SiO_2 nanowire structures. Each of these is examined in Figure 6.

Figure 15.6(a) illustrates a growth process in which ZnS molecules diffuse through or around the gold catalyst particle, with silica species originating on the substrate surface and moving up the sides of the growing nanowire. A similar process is illustrated in Figure 15.6(b); however, here the shell is formed via direct adsorption of silica species from the gas phase. For very long nanostructures containing an amorphous silica shell, synthesis via these processes produces structures that vary in similar shell thickness along the length. In short, variations in silica shell thickness are found along a single nanowire. Firstly, this phenomenon occurs due to the

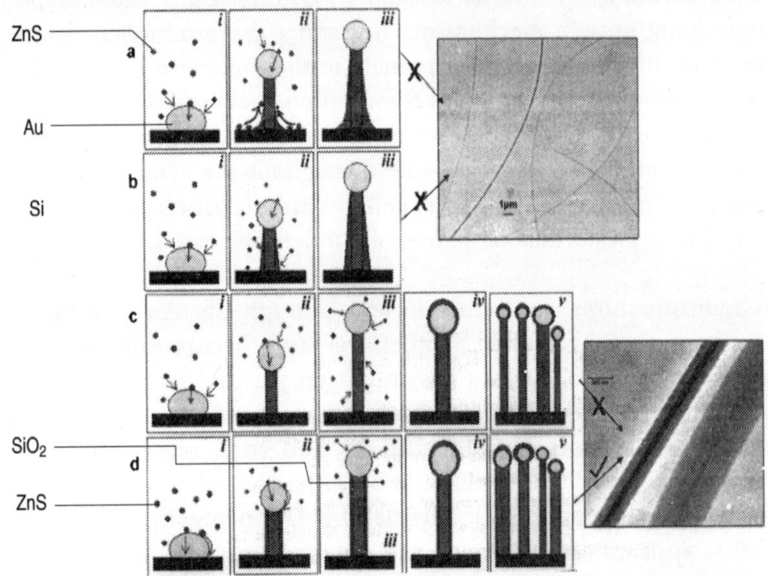

Figure 15.6 *Sample formation mechanisms and corresponding TEM images. Here yellow signifies Au. dark blue represents the Si substrate. light blue represents ZnS, and red coloration is used for SiO_2. (a) In this model. ZnS species are preferentially attracted to the Au catalyst particle and migrate to form the nanowire core.SiO_2 species form and migrate up the sides of the ZnS nanowire. (b) This model is similar to (a) except here silica species adsorb as gaseous species rather than migrating up the ZnS nanowire from its base. Final TEM image shows that these mechanisms are unlikely, since the samples demonstrate uniform shell thickness along their entire length. (c) In this model. the ZnS nanowire forms first and then is completely encased in an amorphous silica coating. If this were the case. the silica coating should be the same thickness for all nanowires in a sample. The included TEM image demonstrates that this condition is not observed. (d) The proposed formation mechanism: ZnS diffuses through the bulk of the Au catalyst droplet and forms the crystalline nanowire core while SiO_2 species travel around the droplet surface to create the amorphous shell.*

increasing distance that the silica species travel up the nanowire to reach close to the tip. Then, the portion of the nanowire forming earliest, or that at the bottom is exposed to the silica vapor for a longer period and, therefore, forms a thicker shell. An amorphous silica layer does not experience a strong driving force to promote the SiO_2 species up the nanowire such that newly formed ZnS core material at the top maintains the same shell thickness as earlier areas near the substrate surface. Its is seen in the last image of these figures, nanowires in the samples demonstrate uniform thickness along their entire length, (sometimes of several hundred microns) suggesting that neither of these mechanisms is applicable ,for their formation.

Secondly, a synthesis route is presented in Figure 15.6(c). The process is similar to an atomic layer deposition (ALD) process in two steps in which the ZnS nanowire forms first and then is uniformly coated with an amorphous silica shell. This process as a possible formation mechanism for solica nanotubes in which the metal catalyst particle becomes completely encased. However, as shown in part v of this figure, such a mechanism yields nanowires with completely uniform shell thickness, despite any variations in core size of structure. As in a conventional ALD process, similar shell thickness is observed among different nanowires in a single sample since at the time of silica deposition the ZnS core does not exhibit dynamic growth. The image shown in conjunction with Figure 15.6(c) illustrates significant variation in shell thickness between nanowires from the same sample indicates that, for synthesis, such a result is not observed. Therefore the most probable nanowire formation mechanism in samples is that illustrated in Figure 15.6(d). This series of images describes a process in which ZnS diffuse through the Au particle to form the nanowire core and SiO_2 travels on the surface of the Au particle to the nanowire interface in a simultaneous and dynamic growth process. Variations in particle size and surface contact angle among other factors lead to observed thickness variations in the core and shell portions of ultralong nanowires.

On the basis of this growth model, the results of the systematic analysis growth model, the results of the systematic analysis are examined. TEM indicates that the size of the silica shell increases over time, while the ZnS core remains relatively the same with a slight decrease in size. XRD analysis suggest an initial increase in the ZnS core material with increase in reaction time; however, this increase is peaked near 60–75 min and then declined. The XRD data differs somewhat from the trend revealed through TEM analysis because each technique contains inherent sources of error. Since the XRD techniques plot ZnS core ratio, if after the peak synthesis time, the silica shell thickness increases the ZnS core incorporates more defects, the XRD signatures, show a ZnS size decrease while the actual ZnS size remains constant.

Photoluminescence data is considered to explore the development of ultralong nanowires at each time interval. SiO_2 passivation layers on Si have the effect of increasing the intensity of measured photoluminescence peaks. The intensity of the emission correlates to the degree of surface passivation on the films. The SiO_2 shell has the same effect on the ZnS core's luminescence. This, along with the XRD suggests that the ZnS core initially increases, and explains the increase in intensity of the 532 nm peak. But this peak disappears at very long synthesis run times. As the ZnS core is reduced that the intensity of the 532 nm peak greatly diminishes. Because this peak is associated with defects in the crystal structure, so the defects are too energetic to remain in the ZnS core with the annealing time (growth time) being long. It is also possible that the SiO_2 shell is so large that it eclipses the effective luminescence of the much smaller ZnS core.

The XRD analysis also reveal that the nanowire growth process depends on time. As synthesis time increase the silica shell thickness also increases. The shell thickness increase is due to the increased number of available species in a very fast and efficient surface diffusion growth process. The size variation of the ZnS core however for both the PL and XRD data support an initial increase and then decrease of the ZnS core size relative to the silica shell. Because the shell thickness increases as time evolves, the indication of a decrease in these analyses is due to the role of a thicker SiO_2 layer, with the ZnS core size remaining the same. Whether the ZnS grows to a maximum size and then decreases or whether the average core size remains mostly unchanged after this periods, such a peak in core thickness shows that at 60-75 min there exists a set of conditions producing ideal growth of these structures, i.e., high-quality aligned nanowires of greatest length and density.

Now, when experimental conditions are kept consistent for producing the ultralong core-shell nanowires; the thickness of the gold catalyst layer is reduced (below 20 nm), the synthesized product changes, producing short, unaligned, low-density structures. Therefore, the gold catalyst layer is a component in the formation of these ultralong nanowires. The critical influence of the Au-Si alloy in formation of Si nanowires, and the Au-Si phase diagram show that such an alloy allows for a dramatically reduced temperature at which the material exhibits a liquid phase. The presence of a thick gold layer coupled with the elevated chamber temperature facilitates a formation of a Au-Si eutectic allow on the substrate surface. Analyses have shown that, in the case of Au on Si, an alloy layer is formed on the substrate surface which then quickly nucleates spherical Au particles. These continue to agglomerate until large enough to sustain nanowire growth. The Gibbs - Thomson effect places a lower limit on the width of wires which grows under a given set of conditions, such that very small catalyst particles cannot sustain growth due to an effective wire chemical potential greater that of the vapor phase. Whether through formation of a eutectic or simply gold migration, thinner gold layers most likely cannot sustain high-density production of these small yet growth sustaining Au catalysts needed for the formation of fine, dense, and ultralong nanowires.

Thus the role of thick gold layer but not the observed maximum in ZnS core thickness occurring at intermediate synthesis times is explained. SEM observation of a 20 nm gold coated substrate, imaged prior to and after exposure to the same experimental conditions but without the presence of ZnS powder indicate that while the surface initially appears like a rough film, after just 15 min at elevated temperature before cooling, the gold on the surface migrated to form regions of large islands and small particles. It is found that similarly to very small gold particles, those above a certain size also do not promote as efficient nanowire growth as those with higher surface contact angles. As short synthesis times, only the smaller more ideal catalyst sites initiate nanowire growth, while longer synthesis times allow the larger less efficient sites to nucleate ZnS nanowires. Therefore, the growth of ultralong ZnS-SiO_2 core-shell nanowires depends upon the presence of gold catalyst particles of a size within an ideal range. Above or below a critical size, gold islands do not support nanowire growth, leading to a peak in ZnS core size at around 60-75 min growth time and only with a sufficiently thick substrate surface gold layer.

The ZnS nanowire core forming via bulk diffusion through the gold catalyst is critically dependent on the size and condition of the catalyst particle. The Si species however, combine with residual oxygen in the chamber before diffusing around the catalyst particle to form an amorphous silica coating and is much less critically dependent on Au particle size.

Finally, calculations are performed to determine the feasibility of this growth. To find if dual diffusion can be predicted according to general diffusion relations, case of bulk diffusion for ZnS and then find the resulting diffusion constant for SiO_2. Absorption operations involve contact of a gas mixture with a liquid and preferential dissolution of a component in the contacting liquid. Depending on the chemical nature of the involved molecules, the absorption mayor may not involve chemical reaction. This calculation is based on the expression for adsorption of a gas in a liquid with a chemical reaction at the interface:

$$N_A|_{Z=0} = \frac{D_{AB}C_{AO}}{\delta}\left[\frac{\sqrt{k/D_{AB}}\,\delta}{\tan h(\sqrt{k/D_{AB}}\,\delta)}\right] \tag{2}$$

Here N_A is the flux of species A through the liquid, D_{AB} is the diffusivity of species A in B, C_{AO} is the maximum concentration of arriving species at the liquid surface, and δ is (the distance that the diffusing species must travel) the diameter of the catalyst droplet. This equation assumes a constant diffusivity and a low concentration of A in the liquid such that the liquid interface contains plenty of adsorption sites.

The constants of this equation are evaluated from the boundary conditions:

$$\begin{aligned} Z &= 0: C_A = C_{A0} \\ Z &= \delta: C_A = 0 \end{aligned} \tag{3}$$

With careful evaluation of equation 2 it is apparent that the term.

$$[\sqrt{k/D_{AB}}\,\delta)/(\tan h(\sqrt{k/D_{AB}}\,\delta))] \tag{4}$$

describes the influence of the chemical reactions at the liquid interface. This term is a dimensionless quantity, often termed the Hatta number. If the chemical reaction at the interface is neglected, the simplified expression correlating flux and diffusivity is

$$N_A \approx (D_{AB})(C_{AO})/\delta \tag{5}$$

which is the equation used for calculations.

Here, the diameter of the catalyst particle, is assumed to be δ is equal to 40 nm. To find the concentration of each gaseous species, assume an ideallike behavior and note that the evaporation temperature of Si (and SiO_2) is twice that of ZnS. Here, the total pressure is taken to be a simple sum of the partial pressure contributed by each species. Since the relative thicknesses of the core and shell in nanowires do not fluctuate or taper, the flux of ZnS molecule is equivalent to that of SiO_2 species at the nanowire-gold interface. The diffusivity of ZnS is found by observing the average experimental nanowire growth rate and applying this to the molecular system. For ZnS, the diffusivity, D_{AB} is found to be ~8×10^{-6} cm^2s^{-1}. This value is well within the range of bulk diffusion through a liquid. Equation (5) is then used to find the flux for ZnS. This value is then fixed for the case of SiO_2, and the diffusivity is calculated. This value is found to be 0.16 cm^2s^{-1} a very high diffusivity. Therefore, these calculations attest the theoretical feasibility of proposed growth model, in which the ZnS core is formed by volume diffusion across the catalyst particle while the silica shell is formed by surface diffusion on the catalyst particle.

SiO_2 /Ta_2O_5 CORE SHELL NANOWIRES

Ta_2O_5 is a fascinating functional material that is used in applications such as dynamic random access memory (DRAM) devices, antireflection coating layers, gas sensors, photocatalysts, and capacitors owing to its high dielectric constant, high refractive index, chemical stability, and high temperature piezoelectric properties. However, the synthesis of Ta_2O_5, nanostructures (e.g. nanowires) have a little success as a result of its high melting point. On the other hand, silica (SiO_2) nanowires are well developed for electronic and optoelectronic applications. Various methods of synthesizing SiO_2 nanowires include pulsed laser ablation, directed growth from a silica substrate or silica nanoparticles in a reductive atmosphere, direct growth from a Si substrate with catalyst, carbon-assisted and carbothermal reduction of silicon dioxide or metal oxides, and the sol-gel method. All of these approaches are direct, simple, and high-yielding, which are important factors for applications.

Experiment

Single-crystal Si(001) wafers (resistivity: 1-30 Ω cm) are cleaned using standard cleaning procedures and then dipped in diluted HF solution (1:50 HF/H_2O) for 30 s before being loaded into the deposition system (P> 6.65×10^{-4} Pa). A 2-nm-thick layer of Au is deposited on the Si substrate at a pressure of 6.65×10^{-4} Pa with a deposition rate of 0.01 nms^{-1}. Subsequently, as-deposited samples are annealed in a horizontal furnace at 1420K for 2 hours under an atmosphere of N_2 to row the SiO_2 nanowires. The high-density SiO_2 nanowire samples are transferred into a Ta-filament heating chamber for annealing (P>6.65×10^{-4} Pa) at 1220K for 12-32 hours to produce SiO_2/Ta_2O_5 core-shell nanowire structures; Ta atoms are constantly vaporized from the supplementary source in this chamber. Ta_2O_5 nanotubes are formed by dipping the core-shell structures into dilute HF solution (1:50 HF/ H_2O) to remove the inner SiO_2 nanowires.

Grazing incidence X-ray diffractometry (GIXRD) with a fixed incident angle at 0.5° is carried out to identify the phases of the nanostructures. The surface morphology is examined by a field emission scanning electron microscope operated at 15 kV. To prepare the TEM specimen, all samples are sonicated in ethanol and then dispersed on a copper grid supported by a holey carbon film. A field-emission transmission electron microscope operated at 300 kV, with a point-to-point resolution of 0.17 nm and equipped with an energy-dispersion spectrometer, an electron energy loss spectrometer, as well as a high-angle annular dark field detector, is used to characterize the microstructures and chemical compositions. Electron field emission behavior is measured in a vacuum of 133×10^{-7} Pa by using a spherical stainless-steel probe (1-mm diameter) as the anode. The lowest emission current is recorded on the level of nA. The measurement distance between the anode and the emitting surface is fixed at 100 μm. The CL spectrum is measured in the scanning electron microscope with an electron probe microanalyzer. CL spectra are accumulated in single-shot mode within a short time of 1s. In general, the CL excitation is performed with a beam current of about 100 nA in television scanning mode of 2.9 $\times 10^{-5}$ cm^2.

Fabrication

Consider, a technique for the fabrication of SiO_2/Ta_2O_5 core-shell nanotubes, Ta_2O_5 nanotubes and Ta_2O_5 nanowires by using SiO_2 nanowires as template. The structures are prepared by annealing SiO_2 nanowires in an atmosphere of Ta at 1220K a pressure of 133 $\times 10^{-6}$ Pa. The diameter of the SiO_2 core in the SiO_2/Ta_2O_5 core-shell structure is readily controlled and the silica template is removed from the core shell structures by using dilute HF solution to leave Ta_2O_5 nanotubes. Further. Ta_2O_5 nanowires are synthesized by increasing the reduction (annealing) time so that all of the SiO_2 in the coreshell structure is reduced by Ta vapors. The subsequent characterization of the core-shell structures as well as the Ta_2O_5 nanotubes and nanowires is carried out by cathodo luminescence (CL) and field-emission measurements.

The scanning electron microscopy (SEM) image of SiO_2 nanowires, prepared by annealing a 2nm-thick layer of Au on a Si wafer under N_2 atmosphere at 1420K for 2 hours, is shown in Figure 15.7(a). The diameter of these SiO_2 nanowires is almost uniform and the length is up to several hundred micrometers. The corresponding transmission electron microscopy (TEM) image indicates that most of the SiO_2 nanowires have a smooth morphology, with a diameter of 100-150 nm (Figure 15.7(b)). The diffraction pattern (upper inset in Figure 15.7(b)) displays a highly diffusive ring, indicating that the silica nanowires are amorphous. The atomic concentration of Si and O for these synthesized nanowires is about 34 % and 66 %, respectively, and a ratio of 1:2 Si/O is inferred from quantitative TEM/EDS (energy-dispersive spectrometry) measurements (lower inset in Figure 15.7(b) After annealing the nanowires at 1220K for 12 hours and subjecting them to a reductive Ta atmosphere, the morphology of the resultant SiO_2/Ta_2O_5 structures is similar (Figure 15.7(c)). The corresponding magnified SEM image (upper inset in Figure 15.7(c)) shows the wirelike features. The phase and structure of these nanowires is characterized by X-ray diffraction (XRD; Figure 15.7(d)) and reveals the Ta_2O_5 phase to have an orthorhombic structure (C2mm space group) and lattice constants of a = 0.618, b = 0.366, and c =0.388 nm, respectively. Note that the $TaSi_2$ phase found in the XRD spectrum originates from a silicide reaction between the Si substrate and Ta vapors during the reduction procedure.

TEM analysis is essential to examine the detailed microstructures of these nanowires (Figure 15.7(e)). The different contrast between the inside and the outside of these nanowires provide significant evidence of a core-shell structure. The lower inset in Figure 15.7(e) shows the diffraction pattern and plane indices recorded from a SiO_2/Ta_2O_5 core-shell structure. The pattern is consistent with the results of XRD, indicating the polycrystalline characteristic of the SiO_2/Ta_2O_5 core-shell structure. Diffusive white contras is also observed confirming that SiO_2 nanowires are surrounded by a Ta_2O_5 shell. Note that SiO_2 or Si is incorporated inside the Ta_2O_5 shell during the reduction process, but the main phase remains that of Ta_2O_5 as determined by both the XRD and diffraction patterns (Figure 15.7(d) and (e), respectively). TEM/EDS measurements of the shell of a SiO_2/Ta_2O_5 structure reveal that it consists of 20% Ta, 71 % O, and 9% Si. Upon examining in detail many SiO_2/Ta_2O_5 core-shell structures, the maximum concentration of Si in these Ta_2O_5 shells is found to be no more than 15 %.

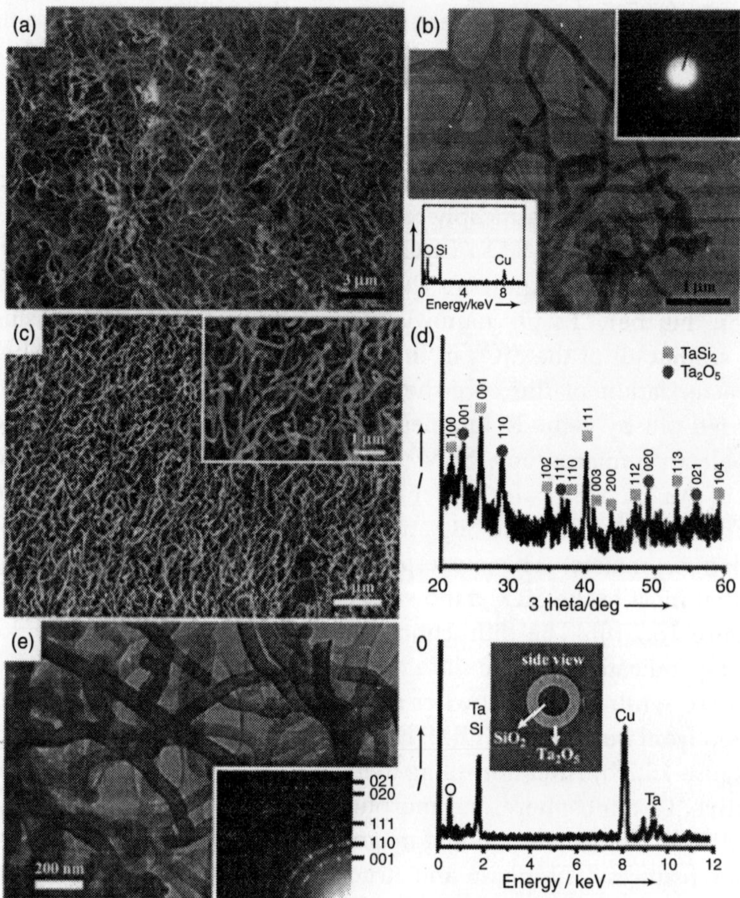

Figure 15.7 *(a) Top-view SEM image and b) corresponding TEM image of SiO_2 nanowires. The upper inset in (b) shows an electron diffraction pattern from a nanowire, while the lower inset shows the EDS spectrum. c) Top-view SEM image of SiO_2/Ta_2O_5 core-shell nanostructures formed by annealing SiO_2 nanowires under a Ta atmosphere at 1220 K for 12 h. The inset in (c) shows the magnified SEM image. d) XRD spectrum corresponding to the sample in (c). e) TEM image of SiO_2/Ta_2O_5 core-shell . nanostructures. The inset in (e) shows the diffraction pattern. f) EDS spectrum recorded on the Ta_2O_5 shell of a SiO_2/Ta_2O_5 nanostructure and the configuration of the core-shell nanostructure.*

The morphology of the core-shell structure is tunable, depending on the original shape of the SiO_2 nanowire. If the SiO_2 nanowire had a spiral morphology, the SiO_2/Ta_2O_5 core-shell structure formed after the reduction process also reveales a spiral morphology. An example of the spiral morphology of a SiO_2/Ta_2O_5 core-shell structure is high lighted in Figure 15.8 while the corresponding TEM image is shown in Figure 15.8(b). The phase is confirmed to be that of Ta_zO_5 by the diffraction pattern (see inset in Figure 15.8(b)). The thickness of the Ta_2O_5 layer outside the SiO_2 nanowires can tuned by controlling the annealing (reduction) time The thickness of the Ta_2O_5 shell increases as the reduction time is increased at a constant temperature of 1220K or the temperature is increased.

The Ta atoms reduce the SiO_2 to form the SiO_2/ Ta_2O_5 core-shell structures. Figure 15.8(c) shows the heat of formation (– Δ H_f [kcal $atom^{-1}$]) per oxygen atom for various transition-metal oxides and SiO_2. Note that the heats of formation for transition-metal oxides such as HfO_2 TiO_2, and Ta_2O_5 are lower than that of SiO_2, which result in the reduction of SiO_2 at high annealing temperatures. On the other hand, the reduction of SiO_2 is prohibited if the heat of formation of transition-metal oxides such as MnO_3 is higher than that of SiO_2. Figure 15.8(d) shows plots of the thickness of the Ta_2O_5 shell of the SiO_2/Ta_2O_5 core-shell structure as a function of the reduction time at 1220K. The relationship between the thickness of the Ta_2O_5 shell and the reduction time is nonlinear, revealing that the reduction mechanism is under diffusion control, that is $X^2 = Dt$ (X D, and t denote the oxide thickness, parabolic reduction rate constant, and reduction time, respectively). The reduction rate constant D is evaluated as about 2×10^{-16} cm^2s^{-1} by plotting the thickness of the oxide shell as a function of the square root of the reduction time (see inset in Figure 15.8(d)). The thermal kinetic motion of Ta ions during the reduction process involves their diffusion through SiO_2 and reduction of the SiO_2 layer. The diffusion-limited mechanism indicates that the diffusion through the SiO_2 layer is rather slow, resulting in easier control over the reduction process. The measured rate constant represents the rate constant of diffusion. Also, the Si signal detected in EDS in various Ta_2O_5 shells indicate that the Si atoms from SiO_2 reduced by Ta vapors are involved in the Ta_2O_5 sublattice or excluded from the grain boundary.

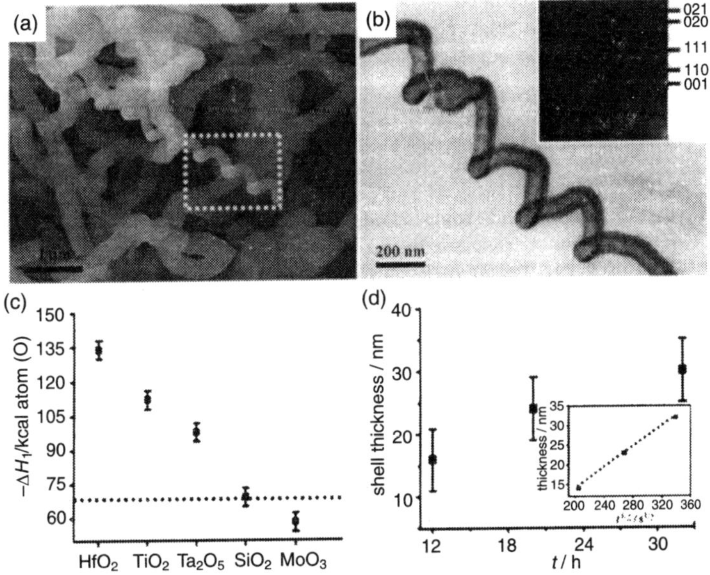

Figure 15.8 *(a) SEM image of a spiral SiO_2/Ta_2O_5 core-shell nanostructure obtained by annealing SiO_2 nanowires in a Ta atmosphere at 1220 K for 12 h. b) TEM image corresponding to the dashed rectangular area in (a). The inset shows the corresponding electron diffraction pattern. c) Heats of formation ($-\Delta H_f$) for different metal oxides (the dashed line is shown to compare other metal oxides with SiO_2). d) Variation in the thickness of the Ta_2O_5 shell as a function of reduction (annealing) time. The inset shows the linear relationship between the square root of reduction time and the shell thickness.*

After dipping the core-shell structures in a diluted solution of HF to remove the SiO_2 inside the SiO_2/Ta_2O_5 core-shell structure, the morphology is unchanged. The inner SiO_2 nanowires are removed to leave Ta_2O_5 nanotubes intact (see Figure 15.9(a)) The residual $TaSi_2$ formed by the reduction procedure could be completely etched by HF solution. The XRD results for the nanotubes show no peaks for $TaSi_2$ and indicate that the amorphous SiO_2 region is eliminated

Ta_2O_5 nanowires are also prepared by increasing the reduction time of the silica nanowires. For example, after annealing SiO_2 nanowires with a diameter of less than 60 nm in a reductive Ta atmosphere at 1220K for 32 hours, then Ta_2O_5 nanowires with lengths of over several hundred micrometers are formed instead of the SiO_2/Ta_2O_5 core-shell structure (Figure 15.9(b)). The TEM image of a Ta_2O_5 nanowire with a diameter of 100 nm is shown in Figure 15.9(c) and indicates the polycrystalline feature of the structure. From the quantitative EDS measurements, it is seen that the concentration of Si is about 3-14 %. Although the concentration of Si inside the Ta_2O_5 nanowires is fairly high, the main phase remains that of Ta_2O_5, as confirmed by the diffraction pattern shown in the lower inset in Figure 15.9(c). The morphology of the Ta_2O_5 nanowires are modified according to the original morphology of the SiO_2 nanowire template (lower right inset of Figure 15.9(b) shows the spiral morphology of Ta_2O_5 nanowires). The magnified SEM image of the spiral morphology of the Ta_2O_5 nanowires as well as a magnification of this spiral structure are clearly seen in Figure 15.9(d). The polycrystalline feature is due to two possible factors: 1) the segregation of Si during the reduction process through the grain boundary of Ta_2O_5 nanowires and 2) anisotropic reduction along the SiO_2

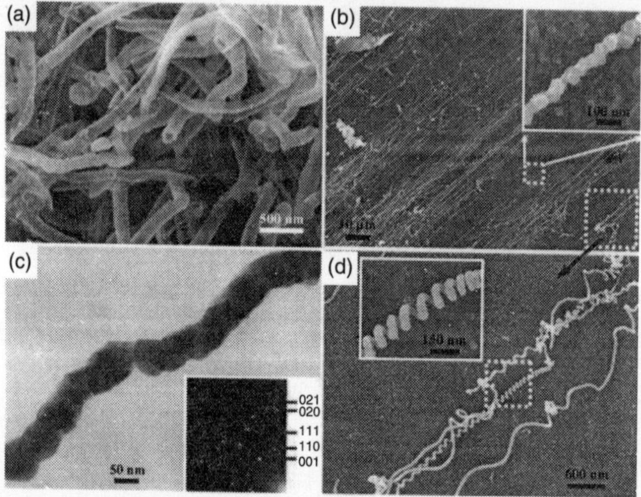

Figure 15.9 *(a) Top-view SEM image of Ta_2O_5 nanotubes prepared by dipping a sample of SiO_2/Ta_2O_5 core-shell nanostructures in diluted HF solution to remove the SiO_2 nanowire core. b) Top-view SEM image of Ta_2O_5 nanowires. The inset shows the magnified SEM image recorded from the dashed rectangle area indicated by arrows. c) Typical TEM image of a Ta_2O_5 nanowire. The inset shows the corresponding electron diffraction pattern. d) SEM image showing the spiral morphology of Ta_2O_5 nanowires taken from the larger dashed rectangular area in (b). The inset shows the magnified SEM image.*

nanowires. The dislocations are found inside the grain of Ta_2O_5 which is caused by the location of Si atoms in the substitutional or interstitial sites in the sublattice of Ta_2O_5.

Figure 15.10(a) shows the SEM image of a Ta_2O_5 nanotube after dipping it into a dilute HF solution of HF for 5 hours. The CL spectra recorded from two areas of the sample labeled as A and B are shown in the insets of Figure 15.10(a). Two peaks from the area labeled. as A are detected, whereas no peak is found from the area B. The CL image under excitation at 15 kV is also measured (Figure 15.10(b)). Figure 15.10(c) shows the CL spectrum from the Ta_2O_5 nanotube sample (area A) after Gaussian fitting and reveals two clear peaks at 563 nm (2.2 eV) and 301 nm (4.1 eV), as well as a broad peak in the range of 2-5 eV. The band gap of Ta_2O_5 is about 4.1-4.2 eV, which results in some radiative recombination emission caused by oxygen deficiency. The peak at 563 nm (2.2 eV) is attributed to the oxygen vacancies inside the Ta_2O_5 shell. The broad band at 2-5 eV is derived from a combination of two peaks at 460 and 355 nm which originate from the residual SiO_2 inside the Ta_2O_5 nanotube or from the initial Si substrate. The peak at 301 nm (4.1 eV, violet region) in the CL spectrum originates from the d-band transition between the e_g and t_{2g} states, induced by ligand-field splitting. A similar phenomenon is found in other materials, such as α-Fe_2O_3 nanoparticles. The violet peak derived from the oxygen

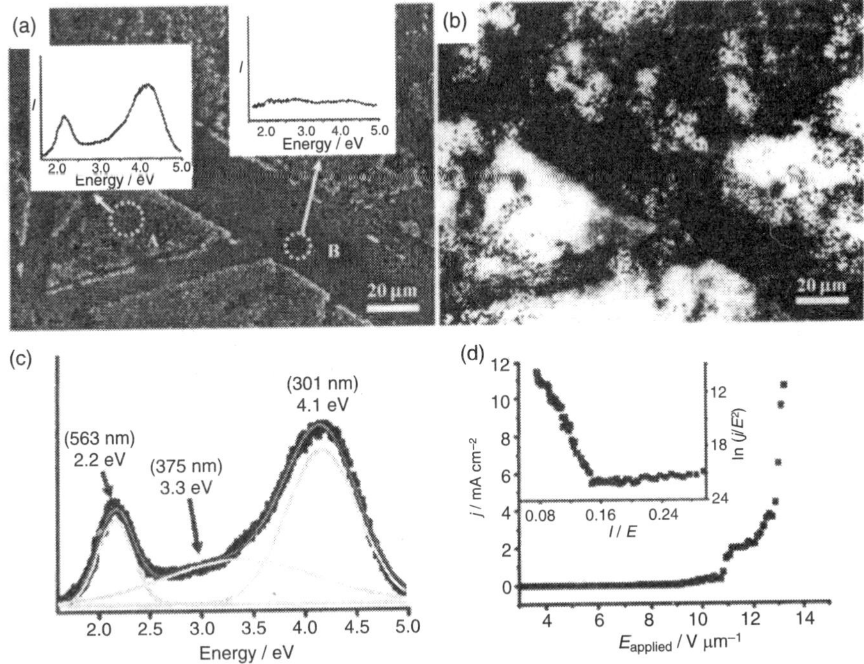

Figure 15.10 *(a) SEM image of a Ta_2O_5 nanotube after dipping it into diluted HF solution; the insets show CL spectra recorded from two areas labeled A and B. b) CL image of the sample in (a) excited at 15 kV. c) CL spectrum recorded from a Ta_2O_5 nanotube (black line) after Gaussian curve fitting (pale gray line; deconvoluted spectra are shown in dark gray). d) Field-emission properties of a Ta_2O_5 nanotube obtained from dipping a SiO_2/Ta_2O_5 core-shell structure into diluted HF solution (j = current density). The inset shows the corresponding In(j/E^2) versus 1/E plot.*

deficiency transition inside the Ta_2O_5 is also the factor for the CL peak at 301 nm. Violet light is of interest for applications in full-color displays. Here, by controlling the thickness of the SiO_2 core inside the SiO_2/Ta_2O_5 core-shell structure, which in turn defines the CL wavelength, the nanotubes are used in optical transportation phenomena, such as light propagation.

The current density as a function of the applied electric field for the Ta_2O_5 nanotube sample at a fixed distance of 100 μm between the anode and the surface of the nanotube is shown in Figure 15.10(d). Two parameters, the turn-on field and the threshold field, are defined as the applied voltage (E) needed to produce a current density of 0.01 and 10 $mAcm^{-2}$, respectively, for which respective values of 8 and 13 V μm^{-1} are determined. The inset in Figure 15.10(d) shows a plot of In (j / E^2) versus 1/ E. The linear relationship is consistent with the so-called Fowler-Nordheim plot (F-N plot; Figure 15.10(d)) and indicates that the field-emission behavior obeys the F-N rule, that is, electrons tunnel through the potential barrier from the conduction band to the vacuum state. Although these values are somewhat higher than for other oxide materials, such as ZnO,- SnO_2 and $W_{18}O_{49}$ the core-shell structures show great promise in display applications owing to their straightforward and high-yielding production and the ease with which they can be integrated in silicon based industries.

16

Molecular Assembly of Nanowire

Fabrication of one-dimensional (1-D) fundamental nanoscale structures from molecular systems through bottom up approach is one of the steps to realize nanoscale electronic devices. Because 1-D systems are the structures with the lowest dimension that permit efficient electron transport, the nanowires functionalize and integrate with the nanoscale electronic devices. Nanowires are important units in constructing electronic circuits, particularly in electrical conducting; thus a variety of nanowires are used in nanoscale electronic systems. A wide range of compounds from inorganic metals, semiconductors, and carbon nanotubes are employed as nanowires (Fig. 16.1). The diameter, length, and electronic structure of these nanowires vary significantly. The electronic properties of nanowires range from insulating, semiconducting, metallic, to superconducting. Electrical conduction within nanowires is dominated by the carriers at around Fermi level of the band structure, just as in bulk metals and semiconductors (Fig. 16.1). The electronic density of states along the short axis of nanowire has a discrete character because of the restricted lattice translation, it is represented as the band structure along the wire direction. The doped polymers, single-walled carbon nanotubes (SWCNTs), semiconductor, and metal nanowires have Fermi surfaces. On the other hand, DNAs are the insulator with large band gap. Semiconductor nanowires are more useful than metal nanowires for device application.

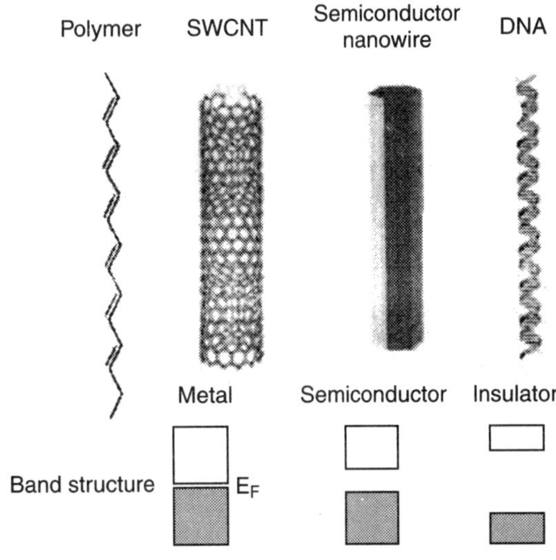

Figure 16.1 *Nanowires constructed from doped-polymer, SWCNT, semiconductor, and DNA. The electronic band structures of these are illustrated below.*

Only a few examples are there for the preparation of nanowires of molecular assembly though these molecular nanowires play an important role in the complete bottom-up manufacture of the molecular electronics. Such nanowires can be assembled from π-molecules through molecule-by-molecule π-stacking. The anisotropic charge-transfer (CT) interaction in molecular conductors is advantageous to form the 1-D π -π stacking nanowire structure. In addition, "supramolecular chemistry" offers powerful tools to fabricate molecular nanowires through the self-assembly process. The design of molecule to orient and integrate the molecular-assembly nanowires on the substrate surface is effective to realize molecular assembly electronic devices. Also, the techniques of "Langmuir-Blodgett (LB) Films" are useful methods to fabricate nanoscale molecular-assembly structures on a variety of substrate surfaces.

Molecular Conductors

A large number of molecular conductors, ranging from semiconductors to metals and superconductors, are prepared. A stable organic molecule has a closed-shell electronic structure without conduction carriers, thereby making molecular solids highly insulating. Conduction carriers are generated via the intermolecular CT interaction between the highest occupied molecular orbital (HOMO) of the electron donor (D) and the lowest occupied molecular orbital (LUMO) of the electron acceptor (A) molecules. Fig. 16.2 shows the molecular structures of typical D and A molecules utilized in the field of molecular conductors. Among them, tetrathiafulvalene (TTF) and 7,7 ,8,8-tetracyano-*p*-quino-dimethane (TCNQ) are the well-known D and A molecules, respectively. This the first molecular metal of $(TTF^{+0.59})(TCNQ^{-0.59})$. The degree of CT ($\delta$) in the binary $(D^{+\delta})(A^{-\delta})_x$ CT complex depends on the ionization potential of D and electron affinity of A molecules. The planar π π-conjugated D and A molecules have a tendency to form the 1-D columnar structure through the π - π stacking interaction, which also forms 1-D π-band structure (Fig. 16.2(b)). Because the π-orbitals exist orthogonal to the molecular plane, the direction of π - π interaction is highly anisotropic. Therefore, the 1-D π - π interaction in the molecular conductors is suitable for obtaining molecular nanowires.

Langmuir-Blodgett Technique

Langmuir-Blodgett (LB) method is one of the conventional fabrication techniques of nanoscale thin-film structures utilizing the air-water interface. Fig. 16.3 illustrates typical procedures to obtain Langmuir mono layer at the air-water interface and the film-forming processes of LB multilayers on substrate. Amphiphilic molecules having hydrophilic and hydrophic moieties are dissolved into conventional organic solvents, which are spread at the air-water interface. The monolayer with a thickness of molecular scale can form a variety of molecular-assembly structures such as gas-, liquid-, and solid-like short- and long-range orders through the control of surface pressure (F, mN m^{-1}). Increase in the F enhances the magnitude of intermolecular interactions between the molecules. The chemical designs of hydrophilicity and hydrophobicity of the component molecules are important to obtain stable monolayer at the air.water interface.

Stable monolayers at the air-water interface can be transferred onto hydrophilic or hydrophobic substrates via dipping of the substrate. Glass, quartz, CaF_2, Si, and mica are utilized as typical hydrophilic substrates. The surface wetting of these substrates largely influences the surface

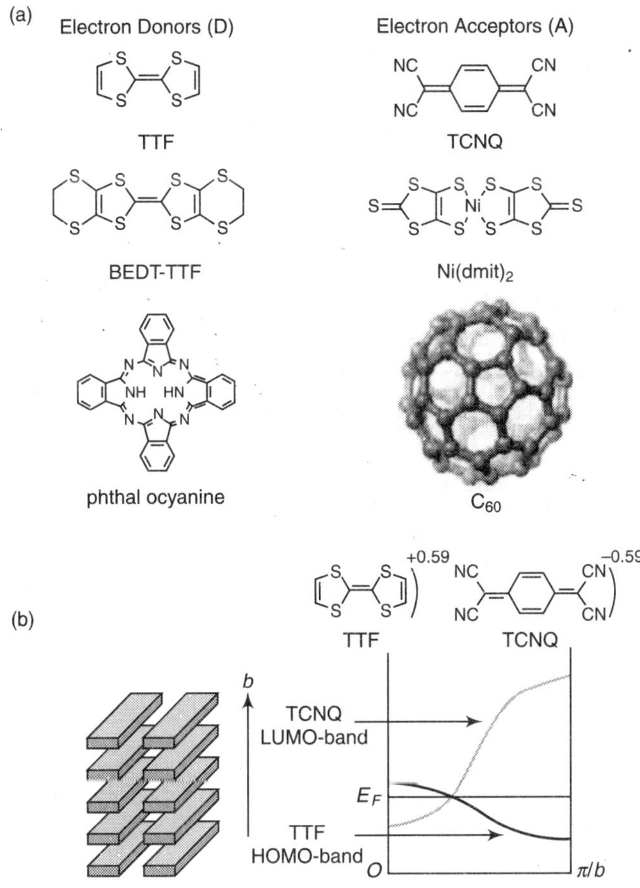

Figure 16.2 *Molecular structures of typical a) electron donor (D) and acceptor (A) molecules employed in molecular conductor b) the 1-D columnar structure through the π -π stacking interaction in crystals (left) and metallic 1-D π band structure of (TTF) (TCNQ) (right)*

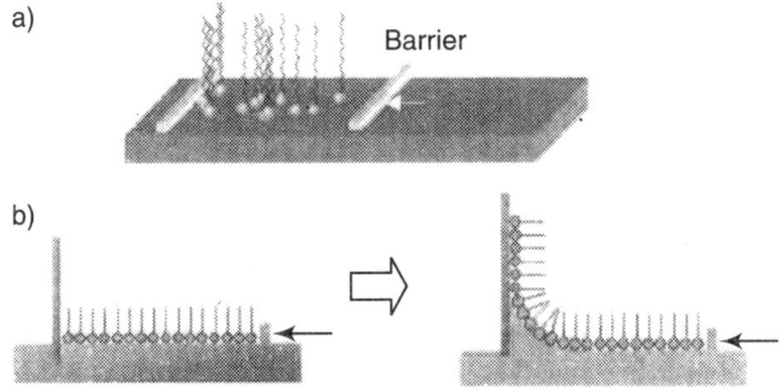

Figure 16.3 *Langmuir-Blodgett technique to fabricate nanoscale thin-film sturcture. a) Langumuir monolayer at thir water interface. Amphiphilic molecule has hydrophilic and hdrophobic tail. Surface pressure (F, mNm^{-1}) is controlled by moving barriers. b) Transfer process of Langmuir monolayer onto substrate surface*

morphologies of the transferred LB films. Another coneventional deposition technique is the horizontal-lifting method, which directly transfers the monolayer at the air water interface onto the hydrophobic substrate.

Electrically active thin films are fabricated by the LB techniques. For example, metallic and semiconducting LB films have been obtained from amphiphilic molecular conductors based on TTF and TCNQ derivatives. Although the LB films possess a periodicity along the film-forming direction, random distribution of two-dimensional crystalline domains on the substrate surface dispels bulk periodicity within the substrate surface.

Supramolecular Chemistry

The term of supramolecular chemistry is a last key science to realize the bottom-up self-assembly approach. Supramolecule is defined as an entity composed of several or large numbers of molecules, which are connected to each other through the non covalent weak intermolecular interactions such as van der Waals (dipole-dipole, dipole induced dipole and dispersion interactions), charge transfer, hydrogen-bonding interactions, etc. The design of these intermolecular interactions is essential to obtain self-assembly molecular nanowires. Self-assembled supra-molecules bearing the desired structure can be achieved by using programmed molecules that are appropriately designed. From the flexibility in supramolecular designs together with molecular conductors, electrical active molecular nanowires are used for constructing nanoscale devices.

A large number of complex supramolecular assemblies already constructed are: A simple complex between cation and crown ether is one of the typical model systems of supramolecules (Fig. 16.4). Crown ethers such as 12-crown-4, 15-crown-5, and 18-crown-6 have hydrophilic cavity to bind cation through metal oxygen interatomic interactions. According to the size of hydrophilic cavity, cations are selectively included into the cavity. For example, 15-crown-5 and 18-crown-6 molecules show high Na^+ and K^+ affinity, respectively. Two design concepts of molecular conductors and LB technique are used to fabricate molecular-assembly nanowires. For an effective design of molecular nanowires, the supramolecular approach is made to obtain nanowire orientation on the substrate surface.

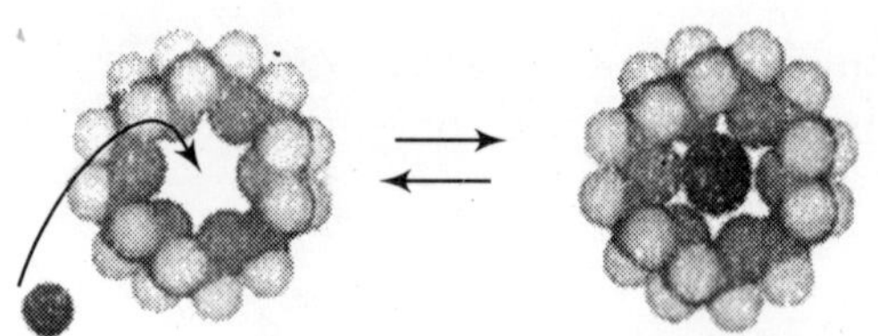

Figure 16.4 *Supramolecular complex between the cation and crown ether. Cations are complexed into the crown ether cavity through cation-oxygen interactions.*

DESIGN OF MOLECULES FOR FABRICATING MOLECULAR-ASSEMBLY NANOWIRES

Amphiphilic bis(tetrathiafulvalene) [bis(TIF)] macrocycel 1 is designed from the concepts of molecular conductors, Langmuir-Blodgett films, and supramolecular chemistry. The molecule has two redox-active TTF units that are linked via a [24]crown-8 macrocycle and two long hydrophobic decylthio-chains (Fig. 16.5). Two TTF units within the molecule 1 can act as

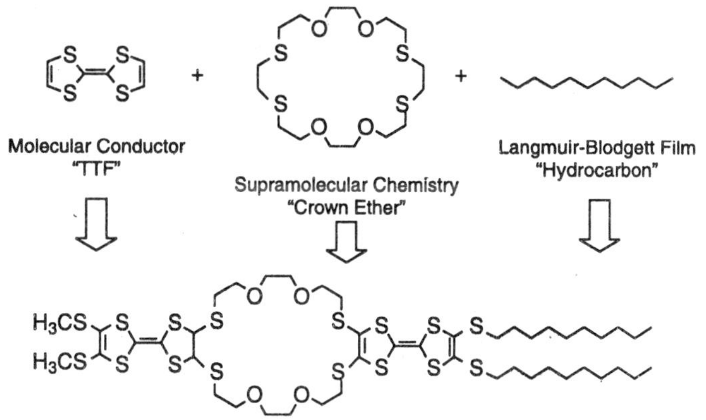

Molecular Conductor "TTF"

Supramolecular Chemistry "Crown Ether"

Langmuir-Blodgett Film "Hydrocarbon"

Figure 16.5 *Molecular design of amphiphilic bis (tetrathiafulvalene) [bis (TTF)] macrocycle 1 from the viewpoints of molecular conductor, (TTF) supramolecular chemistry (crown ether), and Langmuir-Blodgett films (hydocarbons).*

electron donor for realizing intermolecular CT interaction with electron acceptors, which forms electrically conducting 1-D π-π stacks. The second structural point is the introduction of two hydrophobic chains ($-SC_{10}H_{21}$), which are introduced into one side of the TTF unit. By introducing these hydrophobic chains, the molecule has amphiphilic character to apply the LB technique. The last designing point is carried out via the supramolecular approach-the introduction of ion-recognizing crown ether moiety into the molecule. The ion-recognizing property is employed to fabricate oriented molecular-assembly nanowires on the substrate surface.

Compound 1 is synthesized by the cyanoethyl protected TTF building block. Stepwise deprotection/alkylation procedure is performed by CsOH. H_2O/ 2,6-bis(2-iodoethoxy)ethane. Because a TTF molecule possesses a two-step redox process (TTF → TTF^+ and TTF^+ → TTF^{2+}), donor 1 possesses a four-step redox process. The cyclic voltammetry (CV) diagram of donor 1 in 1, 2-dichloroethane ($C_2H_4CI_2$) vs. SCE shows the two-step, two-electron oxidation waves at 0.56 and 0.90 V, respectively. The two TTF units in donor 1, linked via crown-8 unit, independently exhibit redox reaction.

The CT complex between one molecule of donor 1 and two molecules of 2,3,5,6-tetrafluoro-7,7,8,8-tetracyano-*p*-quinodimethane (F_4-TCNQ), (1)(F_4-TCNQ)$_2$ is prepared to fabricate oriented molecular-assembly nanowires. The electronic ground state of the CT complexes in terms of the differences between the redox potentials is $\Delta = E_{1/2}$(donor)$-E_{1/2}$(F4-TCNQ), for electron donor and F_4-TCNQ. Partial CT states, $(D^{+\delta})(A^{-\delta})$ with $0.5 < \delta < 1$, are necessary for the formation of metallic CT complexes having a segregated-stack structure, which is achieved when ΔE has a value between –0.02 and 0.34 V. Fully ionic $(D^+)(A^-)$ and neutral $(D^0)(A^0)$ CT complexes are observed when $\Delta E < -0.02$ V and $\Delta E > 0.34$ V respectively. In the case of donor 1 and F_4-TCNQ ($E_{1/2}(1) = 0.73$ V), the ΔE values of CT complex of (1)(F_4-TCNQ)$_2$ is –0.17 V. These values are far from the conditions of partial CT complexes, therefore fully ionic ground state $(1^{2+})(F_4\text{-TCNQ}^-)_2$ is expected for the molecular nanowires.

Monolayer of Charge-Transfer Complexes at the Air-Water Interface

The CT complex (1) i.e. $(F_4\text{-TCNQ})_2$ is prepared by mixing donor 1 and two equivalent of F_4-TCNQ in $CHCl_3/CH_3CN$ (9:1, v/v). Concentration of the spreading solution is fixed at 1 mM with respect to $(1)(F_4\ TCNQ)_2$. Surface pressure (F)-area per molecule) (A) isotherms are recorded at 291.5 K with a barrier speed of 50 $cm^2\ min^{-1}$. The LB monolayers are transferred onto freshly cleaved mica surfaces by a single up-stroke drawing. Because the macrocyclic moiety recognizes ions introduced into the subphase upon the formation and deposition of the monolayers, the K^+ ions are introduced into the subphase for realizing ion recognition at first.

The F-A isotherms of the floating monolayers of the CT complex of $(1)(F_4\text{- TCNQ})_2$ on pure water and aqueous 0.01 M KCl subhase are shown in Fig 16.6. The surface areas of CT complexes extrapolated at 0 mNm^{-1} on pure water and 0.01 MKCl subphase are almost consistent with each other (A_0~ 1.2 nm^2). Because these values are larger than that of neutral 1 ($A_0 = 0.6\ nm^2$), the formation of CT complex causes much expansion of the surface area. The films are deposited under a controlled surface pressure of $10 mNm^{-1}$ at which point the surface areas of (1) $(F_4\text{-TCNQ})_2$ on pure water and 0.01 M KCl (A_{10}~ 1.1 nm^2) are almost identical to each other.

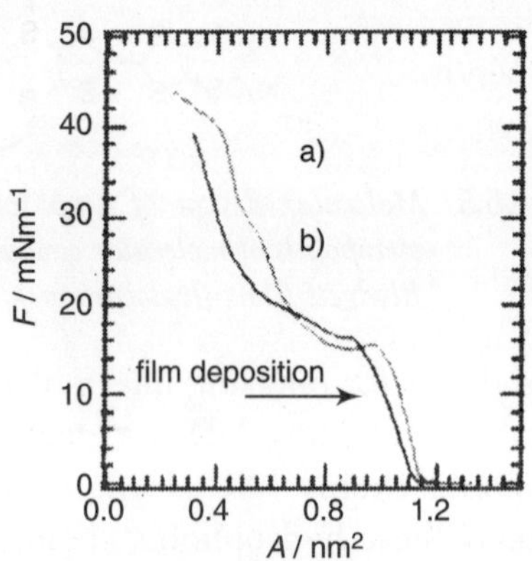

Figure 16.6 *F-A isotherms of CT complex of $(1)(F_4\text{-TCNQ})_2$ on (a) pure water and (b) 0.01 M KCl subphase. The films are transferred at 10 mN m^{-1}*

Molecular-Assembly Nanowires on Mica Surface

Fig. 16.7 shows the surface morphologies of the CT complex of $(1)(F_4\text{-TCNQ})_2$ transferred onto freshly cleaved mica by a single withdrawal from monolayer on a) pure water and b) 0.01 M KCl subphase. Although both films show the formation of molecular-assembly nanowires, significant difference about the nanowire orientation is confirmed between them. Both nanowires had typical width of 50 nm and height of ~2.5 nm. By introducing

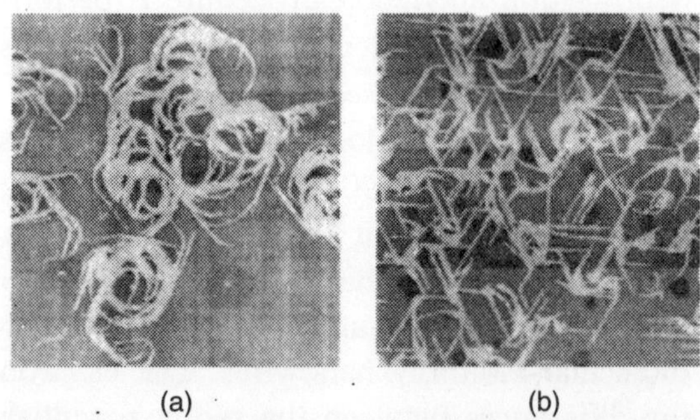

Figure 16.7 *Surface morphology of transferred nanowires of $(1)(F_4TCNQ)_2$ from (a) pure water and (b) 0.01 M KCl. The scales of AFM images are $10 \times 10\ \mu m^2$. 0.01 M KCl into the subphase, the nanowire orientation appears as network structure connecting the well developed nanowires to each other.*

Because the F-A isotherm of $(1)(F_4\text{-TCNQ})_2$ between pure water and 0.01 M KCl subphase is almost consistent with each other, the CT complex does not recognize K^+ ion at the air-water interface. Therefore the orientation of nanowire occurs during the film deposition processes because the orientation direction of the nanowires is consistent with the crystal lattice of mica. The morphology of nanowires on mica surface strongly depended on the cation species that are introduced into the subphase (Fig. 16.8). When Li^+ (ion radius r_i=0.6Å) or Na^+ (r_i=0.95 Å) are introduced into the subphase, the nanowire orientation completely disappears. Much larger cation of Cs^+ (r_i = 1.69 Å) yield the ill-developed molecular nanowires with maximum length of less than ~1 μ m. In the case of Rb^+(r_i = 1.48 Å), well-developed oriented nanowires are observed at a typical length of above 2 μ m. The most densely developed oriented nanowires are confirmed by the introduction of K^+ (r_i = 1.33 Å) into the subphase, which results in a typical nanowire dimension of 2.5 × 50 × ~2000 nm^3. Although Ba^{2+} (r_i= 1.35 Å) has almost similar ionic radius to that of K^+, the nanowire orientation completely disappears on the mica surface.

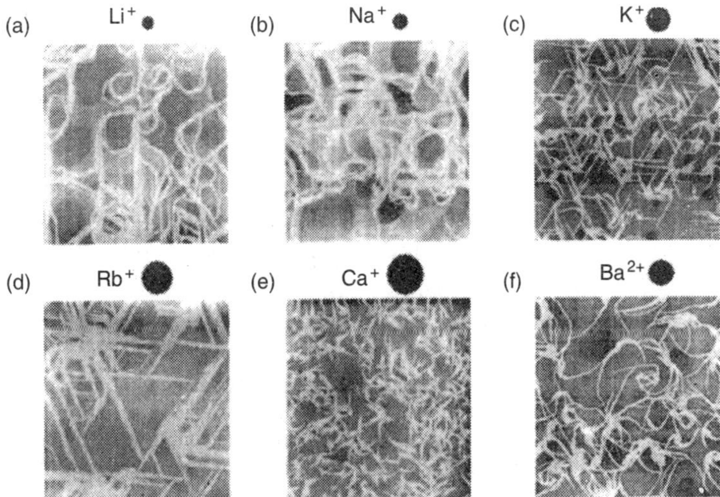

Figure 16.8 *Surface morphology of transferred nanowires of (1)(F_4-TCNQ)$_2$ from (a) LiCI (b) NaCI, (c) KCL (d) RbCi, (e) Cs CI and (f) $BaCI_2$ containing subphase. The concentration of ions in the subphase was fixed at 0.01 M.*

The difference of the nanowire orientation according to the ions is related with the surface property of mica rather than the ion radius. Mica has a typical composition of $KAl_3Si_3O_{10}(OH)_2$, and vacant potassium sites of sixfold symmetry appear by the cleavage (Fig. 16.9) These K^+ sites are negatively charged and attract cations when the mica substrates are immersed in the K^+- containing subphase. The nanowire orientation occurs by matching of the surface potential of mica surface and that of the nanowires, rather than by ion recognition of individual molecules. The anionic species of F_4-TCNQ anion radical are embedded on K^+, and nanowires orient in an epitaxial way through the lattice matching between mica and nanowires. Such recognition mechanism is responsible for the nanowire orientation. Although the dimension of nanowires is larger than the unit cell of hexagonal array of K^+ on mica surface, the nanowire orientation is affected by the K^+ array on mica surface.

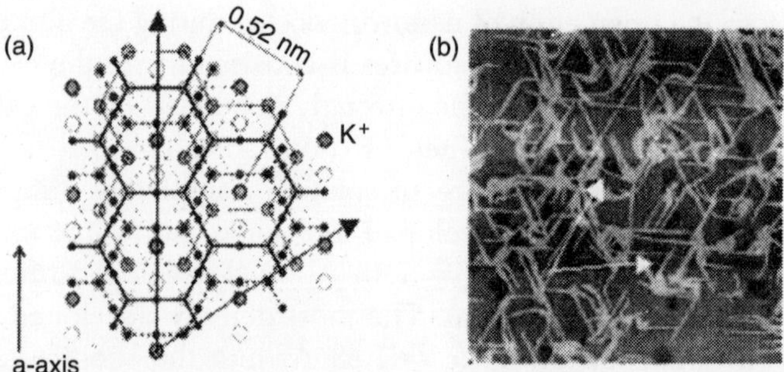

Figure 16.9 *(a) Crystal structure of mica surface. Hexagonal arrays of K^+ sites are indicated by closed circles. b) Atomic force microscopy (AFM) image of oriented nanowires on mica surface deposited from 0.01 M KCI subphase. The white circles allows correspond to hexagonal lattice of K^+ site on mica, showing the relation between the nanowire orientation and K^+ array.*

Fig. 16.10 shows AFM images ($1 \times 1\mu m^2$) of oriented molecular-assembly nanowires on mica. The nanowires oriented 60° to each other, provides nanoscale structures such as nanowire tree and triangles. The six-folded connecting modes of each nanowire form nanoscale junction structures, which are used for constructing the integrated nanoscale electronic circuits.

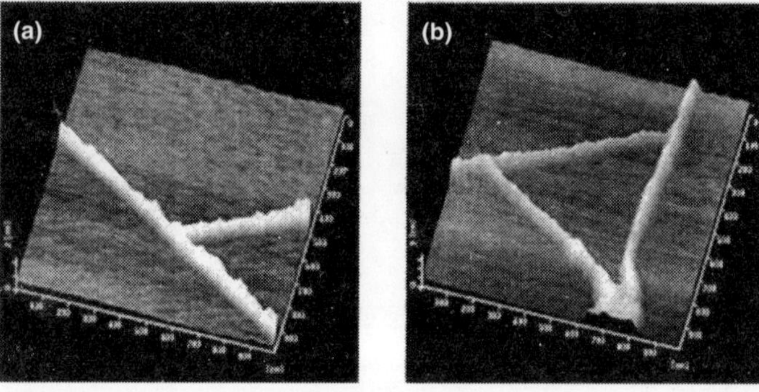

Figure 16.10 *Selected AFM images of oriented molecular-assembly nanwires. a) Nanowire tree and b) nanowire trinagles.*

Molecular-Assembly Structures Within Nanowire.

The single crystal of $(2)(F_4\text{-TCNQ})_2$ is obtained, donor 2 is the four-alkyl chain derivative of compound 1. Assuming a molecular packing structure of CT complex between $(2)(F_4\text{-TCNQ})_2$ and $(1)(F\ _4\text{-TCNQ})_2$ the molecular-assembly structure in the nanowires can be based on the crystal structure of $(2)(F_2\text{-TCNQ})_2$. Fig. 16.11 shows the unit cell of $(2)(F_4\text{-TCNQ})_2$ viewed along the b axis. Donor 2 forms intramolecular π-π dimer through folding of the flexible 24-crown-8 moiety, in which the π-planes of the two TTF units overlap at their inner C_3S_5 rings. Along the c axis, the D-A layers are separated by the hydrophobic decylthio chains; moreover, the hydrophilic and hydrophobic layers are alternately arranged. The direction normal to the substrate surface in the LB film is consistent with the c axis (c= 1.72 nm). In Fig. 16.11 the π-π

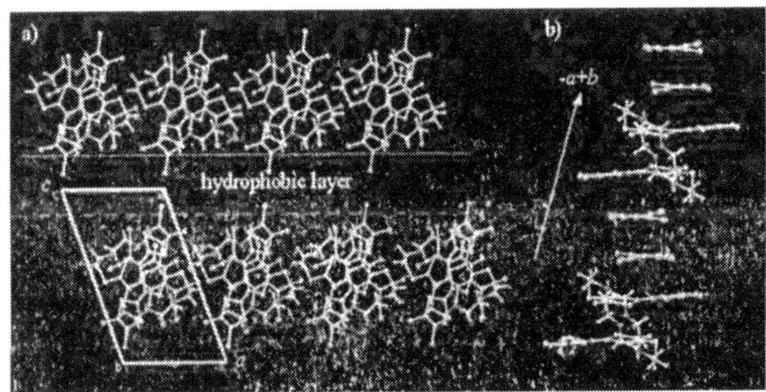

Figure 16.11 *Crystal structure of CT complex (2)(F_4-TCNQ)$_2$ a) Unit cell viewed along the b axis. (b) Alternate π- π stacking sturcture between donaor 2 and F_4-TCNQ dimer along the −a+b axis.*

stacking structure within the ab plane, the π - π.stacking sequence between donor 2 and F_4-TCNQ is –D-A-A-D-, forming a 1-D mixed-stack structure.

The molecular orientation within the LB films is evaluated by using the polarized ultraviolet-visible-near infrared (UV-vis-NIR) spectra. The polarized UV-vis-NIR spectra of the LB film of (1)(F_4-TCNQ)$_2$ do not indicate any in-plane anisotropy with the normal incidence, indicating the isotropic distribution of molecular orientation within the substrate surface. Two sharp absorptions (B Band) are assigned to the intramolecular transition of the F_4-$TCNQ^-$ anion radical, while the C and D bands are assigned to the intramolecular transitions of TTF^+ and F_4-$TCNQ^-$, respectively. The broad A bands are assigned to the intramolecular transition of bis(TTF)-macrocycles. which overlap with the B band with 45° incidence (Fig 16.12(a)), the absorption intensity for the s-polarization is larger than that for transition moments of donor 1 and F_4-TCNQ lie along the long axes of TTF and F_4-TCNQ molecules, the long axes of donor 1 and F_4-TCNQ are parallel to the substrate surface.

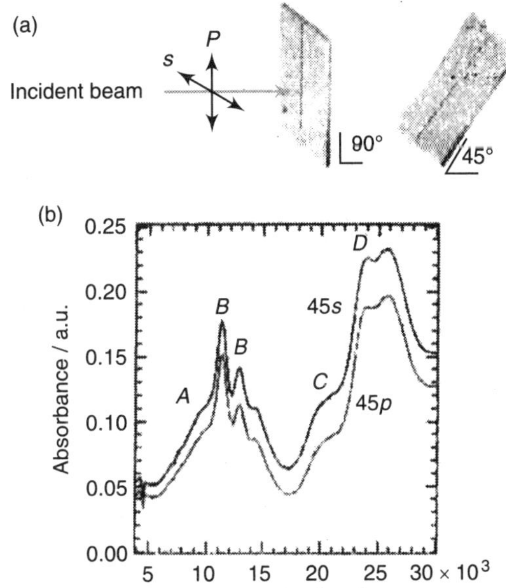

Figure 16.12 *Polarized UV-vis-NIR spectra of the LB film of (1)(F_4-TCNQ)$_3$ (a) Optical arrangements of the substrate to measure the polarized UV-vis-NIR spectra at 90° and 45° incidence. (b) Polarized UV-vis-NIR spectra of 45° incidence with p- and s-polarization.*

From the X-ray structural analysis and polarized UV-vis-NIR spectra a molecular packing model is constructed with in the nanowire, (Fig. 16.13). The nanowire transferred from K^+ containing subphase have dimensions of 2.5 × 50 × ~1000 nm^3. Considering the molecular size of donor 1, the nanowire is composed of 1-2 molecular layers in the height direction and 20-30 molecules in the width direction. Assuming that the

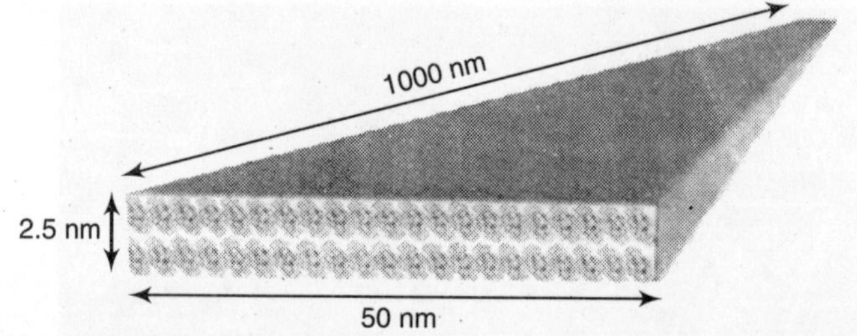

Figure 16.13 *Possible molecular arrangements within a nanowire.*

nanowire has a local structure similar to that of crystal structure of (2)(F_4 TCNQ)$_2$, the longitudinal direction of the nanowire corresponds to the π -π stacking direction of TTF unit and/or F_4-TCNQ molecules. Therefore, a few thousand molecules are necessary to form over μm-long nanowire, taking into account that the standard π - π stacking distance is around 0.35 nm.

Electrical Conductivity and Magnetic Properties

The electrical conductivity of nanowire the temperature-dependent electrical conductivity and paramagnetic susceptibility are all measured in the form of the LB fiim. Fig. 16.14 shows the temperature-dependent electrical conductivity (right scale) and paramagnetic susceptibility (left scale) of LB film of (1)(F_4- TCNQ)$_2$. The room-temperature electrical conductivity (σ_{RT}) is 1.5×10^{-3} S cm^{-1} with an activation energy of 0.17 eV. The semiconducting temperature dependence is consistent with that of the mixed-stack arrangement of donor and F_4-TCNQ dimer. Although the accumulated film is composed of piles of nanowires and monolayers , of (1)(F_4-TCNQ)$_2$, each nanowire has the same order of electrical conductivity.

The temperature-dependent paramagnetic susceptibility (χ) of the LB films, of (1)(F_4-TCNQ)$_2$ is measured via electrons in resonance (ESR). As shown in Fig. 16.14, χ increases monotonically with decreasing temperature which follows the Curie-Weiss law. The line-shape of the ESR signals are fitted using Lorentzian. Line width around 0.5 mT do not show anomaly within the measured temperature range. Because g-values for BEDT-TTF$^+$ cation and F_4-TCNQ$^-$ anion radical are observed at 2.0074 and 2.0025, respectively, the g-values (~2.004) of the LB films for (1)(F_4-TCNQ)$_2$ are observed in the intermediate range between BEDT–TTF and F_4-TCNQ, which indicates that both the cation and the anion radical species contribute to the paramagnetic susceptibility of the LB films.

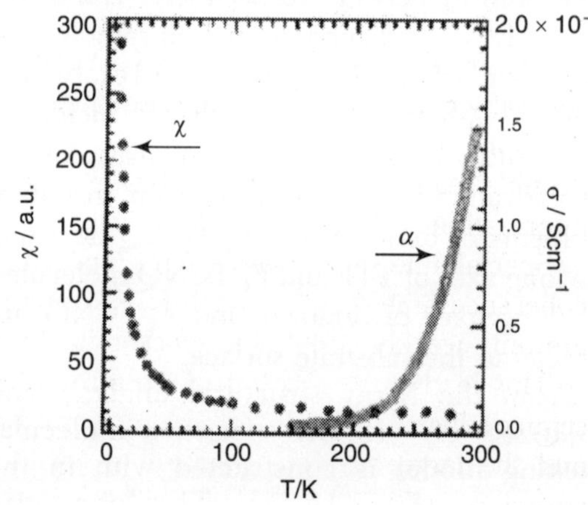

Figure 16.14 *Temperature-dependent electrical conductivity (right) and paramagnetic susceptibility (left) of the LB film of (1)(F_4-TCNQ)$_2$.*

Semiconductor Nanowires and Nanoribbons

One dimensional nanostructures not only exhibit interesting electronic and optical properties associated with their low dimensionality and the quantum confinement effect, but they also represent critical components in potential nanoscale devices.

FABRICATION OF SEMICONDUCTOR NANOWIRES FOR-ELECTRONIC TRANSPORT MEASUREMENTS

Nanometer-size structures weakly coupled to electric contacts, so called quantum dots, are generally considered as the fundamental functional unit in nanoelectronic devices single charges and spins are controlled with high precision in quantum dots. Therefore quantum dots are a possible realization of quantum bits (qubits) in solid-state based quantum computers and quantum information processing schemes.

Special interest is paid to the spin degree of freedom. An important motivation is that coherence times can be much larger for spin than for charge in solid-state systems since spins are only weakly coupled to electric fluctuations of the environment.

This leads to the vision of 'spintronics' instead of pure electronics. InAs is an appropriate semiconductor material for building spintronic devices because of its strong spin-orbit interaction and the large effective g-factor. In addition, the very small effective electron mass leads to strong quantum confinement effects and hence to large energy level spacings, making it experimentally feasible to observe quantum mechanical effects.

Nanowires grown from InAs show the combined properties mentioned above: narrow confinement in lateral directions without the need for lithographic steps and promising spin-related properties. They are an interesting alternative to carbon nanotubes (CNT), where the difficulty to control the tube, even if it is semiconducting or metallic before contacting makes it a specific device fabrication.

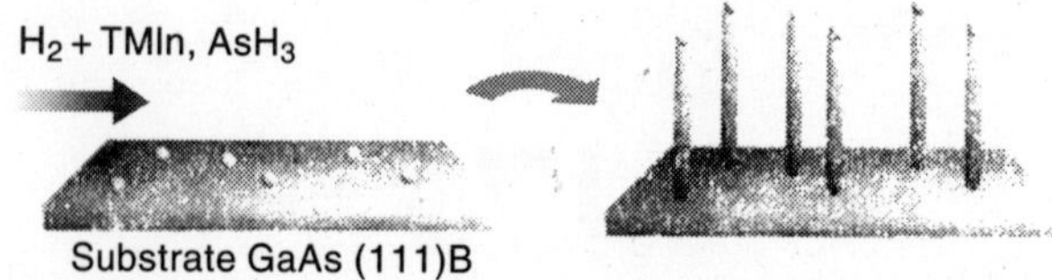

Figure 17.1 *The principle of nanowire growth from catalytic Au nanoparticles.*

Growth of InAs Nanowires

Single-crystal line InAs nanowires of several micrometers length are fabricated to make processing for electrical contacts feasible. To observe quantum effects at temperatures accessible with standard low temperature equipment (T = 30 mK up to 4.2 K), the wires must have a width of less than 200 nm.

1. Catalytic Nanowire Growth

Growth of semiconductor nanowires is accomplished with several different growth techniques. A very controllable approach is the catalytic growth from metallic nanoparticles which are distributed on the growth substrate (Fig. 17.1). This is explained and demonstrated for Si nano whiskers and for III-V semiconductors.

The catalytic growth process is based on the vapor-liquid-solid (VLS) mechanism. Here, the growth is fed by precursors in the gas phase surrounding the catalytic nanoparticle and the substrate. The growth species and the metal catalysts form a eutectic alloy at a temperature below the melting point of the semiconductor. The metal catalyst itself has to be insoluble in the materials to be grown. For most III- V-semiconductors this condition is met by Au, but also growth from Pt, Pd, Ni, Fe is demonstrated. The driving force of the growth is a supersaturation of source materials in the alloy, which reduced by nucleation at the liquid-solid interface. So only longitudinal growth takes place, as long as the temperature does not rise above a critical value, where bulk growth in lateral direction becomes thermodynamically, favorable. Properly sized Au particles allow straight-forward control of the average diameter of the nanowires. The position of the nanowire to be grown is well defined by the catalytic particle. This high degree of control is crucial for an up-scaling of the growth process which is necessary for nanowire applications.

2. Growth with Metal Organic Vapor Phase Epitaxy

Several different epitaxy schemes are applied to grow nanowires such as Metal-Organic Vapor Phase Epitaxy (MOVPE), Chemical Beam Epitaxy (CBE), Molecular Beam Epitaxy (MBE) and laser ablation.

For, MOVPE starting from mono-disperse, colloidal Au nanoparticles ,which are randomly distributed on the growth substrate eg; InAs nanowires growth, trimethyl-indium (TMI) and arsine (AsH_3) are used as precursors in a constant carrier flow of hydrogen (H_2). In contrast to CBE and MBE, which are ultra-high vacuum approaches where the precursors are brought to the substrate in a focused beam, a gas phase with a total pressure of around 10^4 Pa, surrounds the substrate. This leads to higher growth rates of typically 10 $nm^{-1}s^{-1}$ in longitudinal direction compared to about one atomic monolayer per second in CBE. An disadvantage of this is stronger gas diffusion making it more difficult to grow heterostructures with sharp interfaces, since this requires instantaneous changes of the source materials at the growth interface. The surface of the growth substrates should be free of native oxide. Colloidal Au particles with different diameters (20 nm, 40 nm, 60 nm, 100 nm) are dispersed on the substrate.

Crucial parameters for the nanowire growth are the growth temperature, the crystallographic orientation of the substrate surface and the partial pressures of the group-III and group- V precursors. But consider the growth of InAs nanowires on GaAs <111>B and InP <001> surfaces. The process is performed under low-pressure MOVPE conditions with 10^4 Pa. The total pressure the partial pressures for AsH_3 (8.2 Pa) and TMIn (0.15 Pa) is fixed. Fig. 16.2 shows the results of InAs nanowires grown for 15 minutes on GaAs <111>B surfaces.

All three growth runs are catalyzed by Au particles with 40 nm diameter. The length distribution of the wires depend strongly on the reactor temperature. For the lower temperatures wires as long as 30 μm but with strong variations in the length. At T = 735 K the length distribution is quite narrow around 7 μm.

Fig. 16.2 shows that, for higher temperatures, the nanowires are tapered from base to top. Typical widths are up to 150 nm at the bottom and less than a few tens nanometer at the tip.

These observations, together with an reduced growth rate at the higher temperature (Fig. 16.2) are obtained from kinetically activated competition between radial shell growth and the axial VLS growth for temperatures above the eutectic regime. These are in agreement with of the growth kinetics for different III - V-semiconductor nanowhiskers.

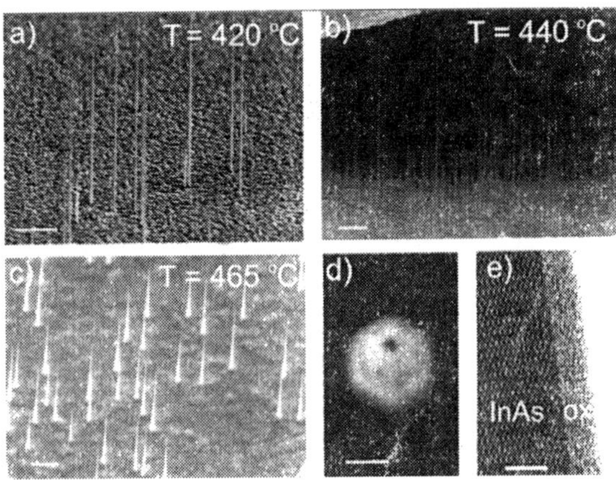

Figure 17.2 *(a)-(c) Scanning electron microscope (SEM) images of InAs nanowires grown on GaAs <111> B surface from 40 nm Au particles for three different growth temperatures. The longitudinal growth rate is smallest for T = 735 K. d) Top view SEM image showing the hexagonal cross section for growth in <111> orientation; e) TEM image showing the crystal lattice and the surface oxide of a wire from (b).*

InAs nanowire growth is been demonstrated at temperatures below the eutectic point of Au/In which casts doubt on the validity of the VLS mechanism. This and other ambiguities are attributed to a significant change of the phase diagram and the details about the eutectic regime for structures on the nanometer scale.

3. Stacking Faults

That for growth on GaAs <111> B surfaces, the InAs nanowires are not single-crystalline. InAs whiskers show perfect hexagonal cross sections, as show in Fig. 16.2(d). The two maximally close-packings lead to this structure. They are the zinc-blende and the wurzite lattice. Because of the energetic equivalence, there is the possibility for stacking faults between (111) layers of the zinc-blende lattices and their rotational twins. These planes are turned by 180° around the [111] growth direction and a pair of a plane and its rotational twin constitute a single (001) wurzite plane at the boundary. This way, segments of zinc-blende and wurzite are generated along the nanowire by twining planes.

Although the preferred crystal structure is determined by the other growth conditions (especially temperature and ratio of the group III and V precursors), the stacking faults are frequently observed for growth on GaAs (111)B surfaces. Their separation ranges from a few nanometers up to several hundred nanometers. This reduces the electronic mean free path to values shorter than expected for epitaxially grown crystalline structures.

Different from (111) substrates, nanowires emerge with square cross section for growth in <001> direction. Here, stacking faults are very rare due to the lack of energetically equivalent twinning planes. Perfect defect-free structures are found on (001) oriented substrates for InP nanowires. As can seen in Fig. 17.3, most of the nanowires grow along the two equivalent <-1-1-1> directions. Nevertheless there are some whiskers growing perpendicular to the (001) surface, which have a square cross section. The growth rate for the [001] orientation is much smaller than for <-1-1-1>. Compared to typically 10 11μm for <-1-1-1>, the <001> wires only reach lengths of about 3 μ m after 15 min of growth.

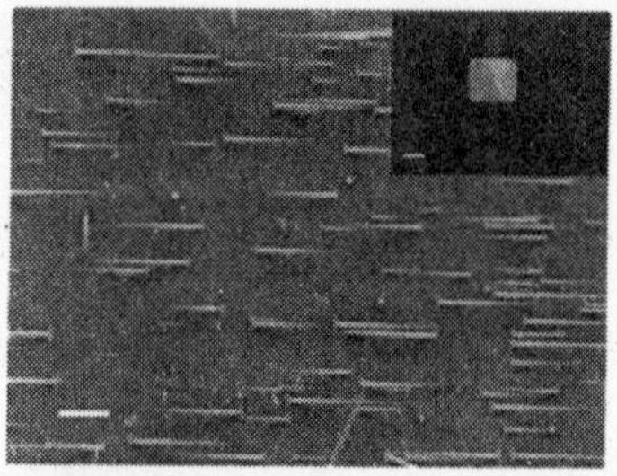

Figure 17.3 *Top view SEM image of InAs nanowires grown on (001) InP substrate at 710K from 40 nm Au particle. Most of the wires grow in <-1-1-1> directions (left and right). Wires growing perpendicular to the surface in [001] orientation have a quadratic cross section (inset).*

Electrical Transport Measurements

1. Contacts to Individual Nanowires

From the technological point of view, a major advantage of the InAs semiconductor is its tendency to form highly transparent electrical contacts to many metals A narrow band gap of 350 meV (compared to 1.4 eV in GaAs) avoids the formation of high Schottky barriers. For the bulk material it is known that the Fermi level is pinned in the conduction band at the surface which avoids the formation of a Schottky contact if a metal is deposited. This is in sharp contrast to other commonly used semiconductors like silicon or GaAs. In those materials, electronic surface depletion requires intricate alloying procedures to obtain ohmic contacts of the electron gas in the semiconductor to metals.

After growth, the nanowires are transferred to a conductive, degenerately doped Si substrate with a 300 nm insulating SiO_2 over-layer. A conventional method for this transfer is to dissolve the nanowires in a small amount of ethanol using ultrasonic pulses to break the wires off the substrate and subsequently putting a microliter droplet of the solution on the silicon chip. By mechanically brushing the nanowires from the growth sample and tipping onto the Si substrate with a piece of clean paper good results are obtained. For thin nanowires with diameters smaller than 100 nm, the Si substrate should contain markers with specified positions. The nanowires are then located with a scanning electron microscope (SEM) and contact electrodes can be fabricated relative to the markers using electron beam lithography (EBL).

Fig. 17.4(a) shows an SEM image of a sample, where a long InAs nanowire with <111> growth direction and a short <001> wire with square cross section are contacted simultaneously.

Thicker wires with diameters larger than 100 nm are visible in an optical microscope with magnification of 200. For these nanowires a simpler procedure based on a single step optical lithography is used. After transfer on a Si/SiO_2 chip without the need for predefined markers, the nanowires are optically located in a standard mask aligner and contacts are defined with an optical mask containing four electrode fingers with a fixed separation of 1.5 micrometers. An example is shown in Fig. 17.4(b). The native semiconductor oxide (see Fig. 17.2(e)) is removed from the contact surfaces before metallization to achieve good ohmic contacts. Good results are achieved with the commonly used etch treatment in a buffered HF solution. One disadvantage of this etchant is that it also reacts with the insulating SiO_2 layer leading to current leakage in later electronic measurements. Another issue is the rapid reoxidation of the surface. A reliable contacting with about 90% yield is obtained by using a simultaneous etching and surface passivation scheme: A highly diluted (1: 1000) ammonium polysulfate ($(NH_4)_2S_X$) solution removes oxide efficiently and provides protection against reoxidation during the transfer of .the samples to the metal evaporation equipment. The InAs nanowires themselves and the SiO_2 substrate are almost unaffected by this procedure. After the lithographic steps and etching, a titanium 'wetting' layer of 20 nm thickness followed by about 180 nm Au are deposited as electrode material.

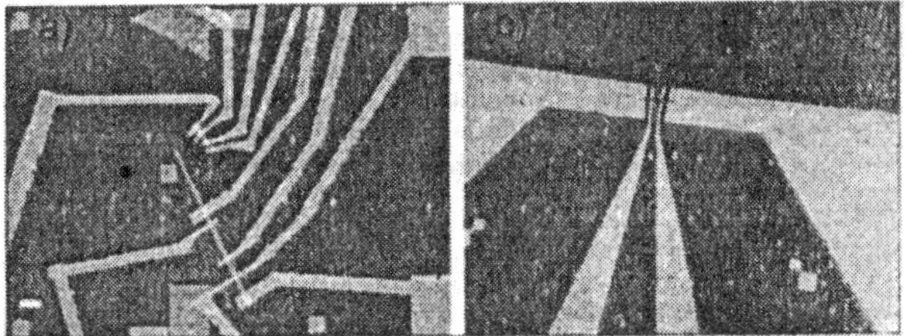

Figure 17.4 *(a) SEM image of a long <111> InAs nanowire and a short <001> nanowire contacted by EBL on a Si/SiO_2 substrate. b) Optical microscope image of a thicker nanowire contacted with optical lithography. The separation between the contact fingers is 1.5nm.*

Electrical Characterization

The accumulation of charge carriers at the surface of bulk InAs does not straight forwardly imply the same behavior for nanometer-sized wires. Nevertheless, other groups have already reported on ohmic contacts to InAs nanowires with excellent quality, even better than to interfaces between metal and bulk InAs. Two-terminal resistances of several k Ωμm wire length are measured at room temperature for <111> InAs wires as shown in Fig. 17.5. From four terminal measurements on devices like those in Fig. 17.4, where a current is driven between the outer pair of electrodes and the voltage drop is measured over the two inner fingers. A contact resistance of less than 100 Ω is obtained. At low temperatures, quantum mechanics influence the transport so these measurements are interpreted with care. For ballistic transport every electrode disturbs the system by scattering a considerable amount of electrons into the nanowire. To obtain the correct resistance, the voltage probes must have smaller contact transparency than the source and drain electrodes. The gate-dependent measurements described below indicate diffusive rather than ballistic transport for the nanowires even at low temperatures. This is in agreement with the mean free path and scattering in InAs nanowires. Further the conductive doped Si substrate is used as a global back-gate to perform field effect measurements. In Fig. 17.6(a) the source-drain current is plotted as a function of bias voltage for different backgate voltages between +20V and –20V. The conduction is n-type, i.e. a negative gate voltage reduces the density of mobile electrons and it is possible to completely pinch off the current through the nanowire.

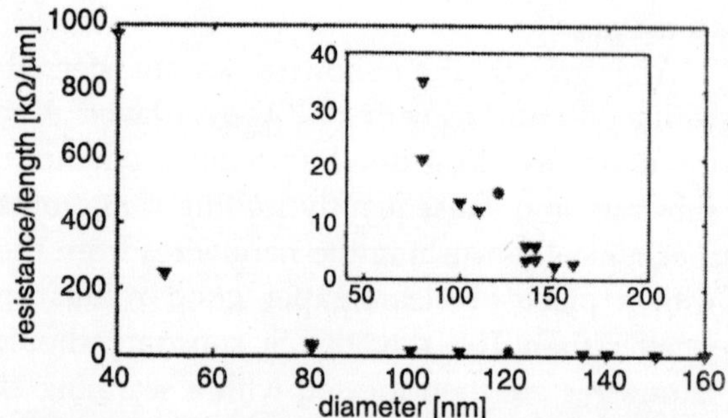

Figure 16.5 *Diameter dependence of the two terminal resistance per length for nanowires grown in <111> direction measured at room temperature. Inset: Zoom for d >80 nm.*

Measurement of the differential conductance $G = dI_{SD}/dV_{SD}$ is dependent on the gate voltage is shown in Fig. 16.6(b). It allows the estimation of electron mobility and density with a field-effect transistor (FET) model for a wire-shaped channel on top of a back-gate.

For zero back-gate voltage, intrinsic carrier densities of 2–5 × 10^{18} cm^{-3}. This high concentration is due to doping with carbon, which is contained in the metal-organizing precursors used in MOVPE. It is however not easy to of extract the donor density, since the above mentioned pinning of the Fermi level in the conduction band at the surface leads to a considerable amount of mobile electrons. The carrier mobilities in <111> of InAs nanowires are found to be 200-1500 $cm^2V^{-1}s^{-1}$. This value is much smaller than theoretical expectations of more than 10^6 nm, Cm^2 $V^{-1}s^{-1}$ for one-dimensional electron channels in perfect crystalline heterostructures but it is about two orders of magnitude larger than for the intensively studied silicon nanowhiskers. In a three dimensional diffusive system this mobility corresponds to a mean free path of less than 100 nm. A reason for these findings is scattering at stacking faults along the growth direction for <111> InAs nanowires. Since this kind of structural changes is not expected for

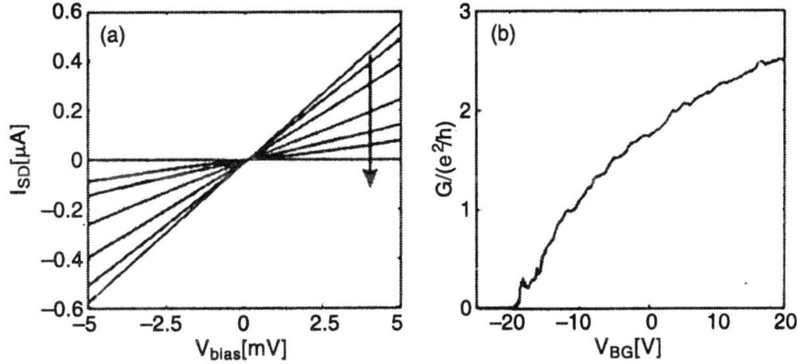

Figure 16.6 *Electrical measurements on devices as shown in Fig. 17.12 at T =-4.2 K. (a) Source-Drain current as a function of the applied voltage. The voltage on the back-gate is decreased from +20 V to -20 V along the arrow. b) Field effect measurement of the zero bias conductances as a function of backgate voltage.*

growth in <001> orientation, the resistivity of both types of nanowires is compared. As shown in Fig. 17.5 no improvement is found for <001> wires. The main scattering mechanism therefore is of a different nature. Because of the electron accumulation at the surface, the short mean free path is due to scattering at surface states. This is compatible with the diameter dependence of the resistance shown in Fig. 17.5, which indicates an abrupt transition to high resistivities for nanowire diameters smaller than 80 nm. Further reasons for this are an increasing contact resistance for smaller diameters or the effect of the narrow lateral confinement pushing the lowest transverse modes above the Fermi level.

Despite the diffusive nature of transport in long nanowires, coherent electronic effects are experimentally probed in regions confined to sizes shorter than the mean free path.

Quantum Dots in Nanowires

In semiconductor quantum dots, only relatively small number of mobile carriers are confined to a sub-micrometer sized region. Electrons (or holes respectively) occupy discrete quantum levels corresponding to the situation known from basic atomic physics. Quantum dots are therefore often considered as 'artificial atoms' with properties that can be tailored at will during fabrication and tuned with gate electrodes as described below.

If the island is only weakly coupled to electric contacts by tunnel barriers, charges can be added or removed to the dot. An important characteristic is the classical charging energy needed to add another electron from the contact at the cost of the Coulomb repulsion energy. For low enough temperatures, the charging energy is larger than the thermal fluctuations. The quantum dot then acts as a single-electron transistor (SET), if the energy levels inside the island are tuned with a gate electrode. Current through the SET is switched on and off when the energy level corresponding to a certain number N of electrons on the island are in or off resonance with the electrochemical potentials in the source and drain contacts. The off-state of the SET is referred to as the Coulomb blockade. Since no carriers can enter or leave the dot in this situation, the number of charges is then completely fixed. On resonance, a finite current can flow and the charge on the island fluctuates between N and N+1.

Under suitable conditions, also the level spacing between the quantum states of the individual electrons become larger than the thermal energy. In this case not only classical Coulomb blockade, but also the quantum mechanical spell structure of the addition spectrum is probed. Single charges and individual spins can be controlled in quantum dots in a coherent way, making them promising for the realization for quantum bits (qubits) in solid-state based quantum computation schemes.

Quantum dots are realized with various methods in semiconductor nanowires.

Earlier nanowire were used between weakly transparent contacts to form an island which could be charged and discharged, similar to devices demonstrated in CNTs. Alternatively, quantum dots were defined along the nanowire by incorporation of barriers made of different material during growth. The result in dots show highly resolved single-level features and could be emptied up to the very last electron. The coulomb blockade was tuned with a global back-gate and the transparency of the tunnel barriers could not be adjusted independently.

Local gating is an important step for precise control over quantum states and the integration of nanowires into larger nanoelectronic circuits. Therefore consider the improvements on attempts to form quantum dots within the nanowire, where both the barriers and the energy levels are tuned with local electrostatic gates. Fig. 16.7(a) shows an SEM image of one of the measured devices. Source (S) and drain (D) ohmic contacts to the nanowire are fabricated with optical lithography and then top-gates are defined with electron beam lithography (EBL). The gate-fingers have a width and separation of about 70 nm. In contrast to the contacts, no etching and surface passivation is done before deposition of 10 nm Cr and 100 nm Au. The native oxide layer on the surface of the nanowire, is shown in Fig 16.2(e), electrically insulates the gates from the wire. Typical breakthrough voltages are above ±1.5 V.

A single quantum dot is formed by applying a negative voltage to two of the top gates (GC and GR). This builds up potential barriers for the electrons in the nanowire. The gate GL is kept at a positive voltage is during this experiment. A bias voltage is applied between source (S) and drain (D) and the current through the quantum dot is measured at low temperatures in a dilution refrigerator with 30 mK base temperature. In Fig. 16.7b, the differential conductance $G = dI_{SD}/dV_{SD}$ is plotted with a color scale as a function of the voltage on the top-gate GR and the bias V_{SD}. The voltage V_{GC} is varied proportional to V_{GR} to ensure symmetric barriers. Inside the typical 'diamonds' the current is suppressed due to Coulomb blockade. The pronounced influence of quantum mechanics on electronic transport through the dot is demonstrated The size of the diamonds vary corresponding to the shell filling of this artificial atom. At certain electron numbers, especially stable configurations are formed leading to an increased addition energy. Lines running parallel to the edges of the diamonds are related to transport through excited single particle states in the dot.

Beyond local gating, the possibility to create quantum dots by locally changing the shape and hence the potential landscape of the nanowire is also investigated. One method to create nanometer-sized constrictions is local anodic oxidation with a scanning force microscope (SFM). Here, a negative voltage is applied to the tip of an SFM, which can be located at a specified position. In controlled temperature and humidity, the semiconductor is locally oxidized in the vicinity of the tip due to an activation of the electrochemical reaction by the strong local

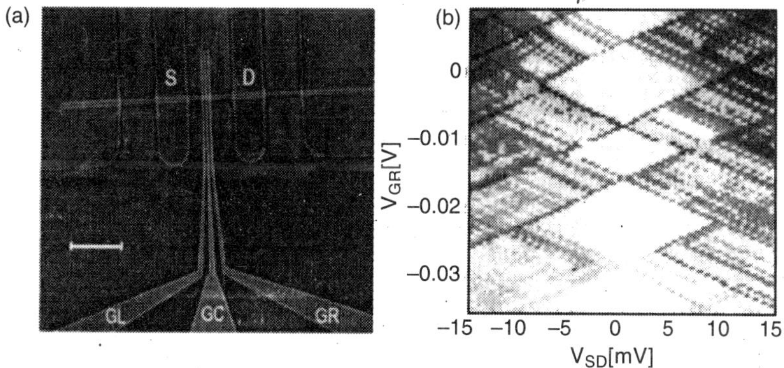

Figure 17.7 *(a) InAs nanowire with ohmic contacts (S, D) and three top-gates (GL. GC, GR). The scale bar is 2μ m. b) $G = dI/_{SD}/dV_{SD}$ for a dot formed between GR and GC.*

electric field. Oxidation effectively removes parts of the semiconductor and hence a local constriction is formed. Fig. 17.8(a) shows a topographical image of a piece of a nanowire with an oxide spot created by the described method. The image is recorded with the same SFM tip right after the oxidation.

Quantum dots are realized in InGaAs quantum wells by local etching of geometrical constrictions into lithographically defined channels. In line with that, EBL is used to write 30 nm wide lines into standard EBL-resist (PMMA) over the nanowire. After exposure and development, the sample is etched in a dilution of HBr and HNO_3 (conc.) in water (1:2000). An SEM image of the resulting trenches is shown in Fig. 17.8(b).

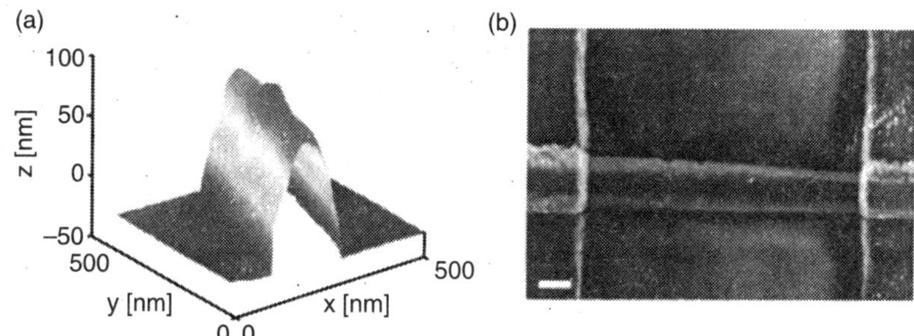

Figure 17.8 *(a) Scanning Force Microscope topography of a piece of InAs nanowire. An oxide spot is created by local anodic oxidation in humid ambience. b) SEM image of a locally etched nanowire. The trenches have been predefined by EBL and etched in a dilution of HBr and HNO_3.*

CONTROLLED BUCKLING OF SEMICONDUCTOR NANORIBBONS FOR STRETCHABLE ELECTRONICS

Although various methods have been developed for defining the material composition, diameter, length and position of nanowires and nanoribbons, there are relatively few approaches for controlling their two and three-dimensional (2D and 3D) configurations. Certain configurations are controlled during growth to yield geometries such as coils, rings and branched layouts or after growth to produce, for example, sinusoidal wave-like structures by coupling these elements to strained elastomeric supports or tube-like (or helical) structures by using built-in residual stresses in layered systems. Semiconductor nanoribbons with wavy geometries are of interest because they enable the production of high-performance stretchable electronic systems for potential applications such as spherically curved focal plane arrays, intelligent rubber surgical gloves and conformable structural health monitors. The approach for forming wavy geometries in semiconductor nanoribbons is different from an alternative route to these same applications that use rigid device islands with stretchable metal interconnects. The wavy nanoribbons have two main disadvantages: First, they form spontaneously, with fixed periods and amplitudes defined by the moduli of the materials and the thicknesses of the ribbons, in a way that offers little control over the geometries or the phases of the waves. Second, the maximum strains are in the range of 20-30%, limited by the non-optimal wavy geometries that result from this process. The procedures introduced below, uses lithographically defined surface adhesion sites together with elastic deformations of a supporting substrate to achieve buckling configurations with deterministic control over their geometries, Periodic or aperiodic designs are possible, for any selected set of individual nanoribbons, in large-scale organized arrays of such structures. Specialized geometries designed for stretchable electronics enable strain ranges of up to nearly 150%, even in brittle materials such as GaAs, which is consistent with analytical modelling of the mechanics.

Experiment

Fabrication of GaAs Ribbons

Photolithography and wet chemical etching is used to generate GaAs ribbons from GaAs wafers with epitaxial layers. An photoresist is spin-cast on the GaAs wafers (5,000 r.p.m., 30 s) and then backed at 373K for 1 min. Exposure through a photomask with patterned lines oriented along the $(0\,\bar{1}\,\bar{1})$ crystallographic direction of GaAs, followed by development, generates line patterns in the photoresist. Mild O_2 plasma removes the residual photoresist. The GaAs wafers are then etched for 1 min in the etchant:4 ml H_3PO_4 (85 wt%), 52 ml H_2O_2, (30 wt%) and 48 ml deionized water which is cooled in the ice water bath. The AlAs layer are dissolved with an HF solution diluted in ethanol (1:2 in volume). The samples with released ribbons on mother wafers are dried in a fume hood. The dried samples are coated with 30-nm SiO_2 deposited by electron beam evaporation.

Fabrication oF Si Ribbons

Si ribbons are fabricated from SOI wafers (Soitec top silicon 290 nm, buried oxide 400 nm). The wafers are patterned by conventional photolithoraphy using photoresist and etched with SF_6 plasma. After the photoresist is washed away with acetone, the buried oxide layer is then etched in HF (49%).

Fabrication of UVO Masks

Fused quartz slides are cleaned in piranha solution (at 330K) for 15 min and thoroughly rinsed with plenty of water. The dry, cleaned slides are coated with sequential layers of 5-nm Ti (adhesive layers) and 100-nm Au (mask layers for UV light) by electron-beam evaporation. A negative photoresist is spin-cast on the slides (3,000 r.p.m., 30.s) to yield ~5-μm-thick films. Patterns are generated in the photoresist by soft baking, exposure to UV light, post baking and developing. Mild O_2 plasma removed the residual photoresist. The photoresist serves as a mask to etch Au and Ti using gold etchant (aqueous solution of I_2, and KI) and titanium etchant (diluted solution of HCI), respectively.

Preparation of PDMS Stamps

PDMS substrates of thickness ~4 mm are prepared by pouring the prepolymer 1:10, Sylgard 184, Dow Corning into a Petri dish and baking at 335K for 4 hours. Rectangular Slabs of suitable sizes are cut from the resulting cured piece, then rinsed with isopropyl alcohol and dried using nitrogen blowing. A specially designed stage is used to mechanically stretch the PDMS to the desired levels of strain. Illuminating these substrates with short-wavelength UV light (low-pressure mercury lamp, 173 μW cm^{-2} from 240 to 260 nm, BHK) for 5 min through a UVO mask placed in contact with the PDMS generates the patterned surface chemistries.

Formation and Embedding of Buckled GaAs(Or Si) Ribbons

GaAs wafers with released ribbons coated with SiO_2 (or Si ribbons with native oxide layer) are laminated against the stretched PDMS with patterned surface chemistry, Banking in an oven at 360K for 5 min, cooling to room temperature in air, and then slowly relaxing the strain 111 the PDMS generates buckles along each ribbon. Embedding the buckled ribbons involve, flood exposure to UV light for 5 min followed by the casting of liquid PDMS prepolymers to a thickness of ~4 mm. Leaving the sample either in an oven at 335K for 4hours or at room temperature for 36 hours cures the prepolymer, to leave the buckled ribbons embedded in a solid matrix of PDMS.

Characterization of Buckled Ribbons

The ribbons are imaged with an optical microscope by tilting the sample by ~90^0 (for non-embedded samples) or ~30^0 (for embedded samples). The SEM images are recorded on a field-emission scanning electron microscope after the sample is coated with a thin layer of gold (~5 nm in thickness). The same stage used for pre-stretching the PDMS stamp is used to stretch and compress the resulting sample.

Fabrication and Characterization of MSM PDs

Fabrication of PDs starts with samples in the configuration shown in the bottom frame of Fig. 17.9(b). An ~0.8-mm-wide strip of polyethylene terephthalate (PET) sheet is gently placed on the PDMS with its longitudinal axis perpendicular to the longitudinal axes of the ribbons.

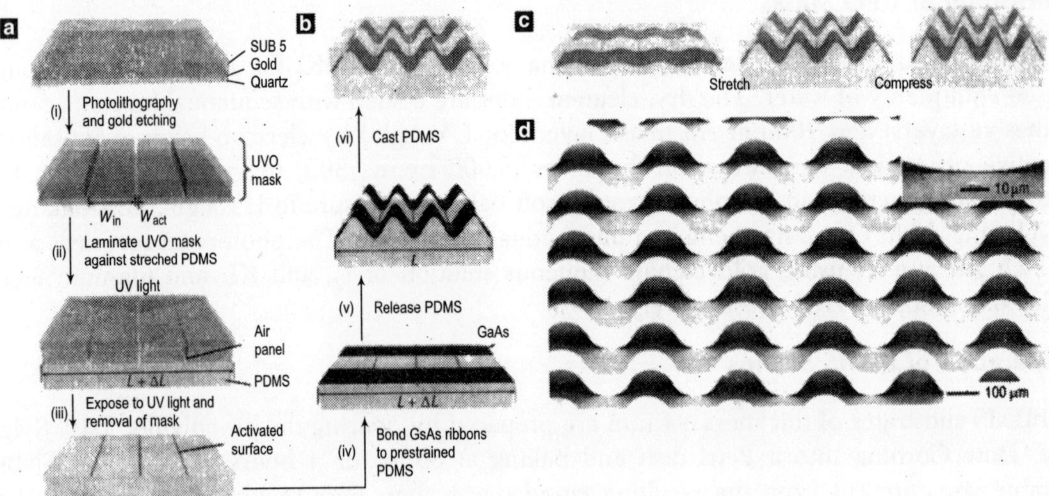

Figure 17.9 *Processing steps for 3D buckled shapes in semiconductor nanoribbons. (a) Fabrication of a UVO mask and its use in patterning the surface chemistry on a PDMS substrate. (b) Formation of buckled GaAs ribbons and the process of embedding them in PDMS. (c) Response of buckled GaAs ribbons to stretching and compressing. (d) SEM image of a sample formed using the procedures in a and b.*

This strip serves as a shadow mask for electron-beam evaporation of a 30-nm-thick gold film (to form Schottky electrodes). Removing the PET strip, and relaxing the prestretched PDMS stamp results in MSM PDs, built with buckled GaAs ribbons. Liquid PDMS prepolymer is cast onto the regions of the ribbons without electrodes and then cured in an oven. The gold electrode are extended beyond the top PDMS to enable probing with a semiconductor parameter analyser. In measurements of the photoresponse, the PDs are manipulated using a mechanical stage for stretching and compressing. An infrared LED source (with wavelength 850 nm) provides the illumination.

Fabrication of Buckles In Nanoribbons

Figure 17.9 shows the steps in the fabrication of buckles in nanoribbons. The fabrication starts with the preparation of a mask for patterning surface chemical adhesion sites on an elastomeric substrate of poly (dimethylsiloxane) (PDMS). This process involves passing deep ultraviolet (UV) light (240-260 nm) through an unusual type of amplitude photomask (fabricated in step (i)) referred to as a UVO mask., while it is in conformal contact with the PDMS. The UVO mask possesses recessed features of relief in the transparent regions, such that exposure to UV light creates patterned areas of ozone proximal to the surface of the PDMS., The ozone converts the unmodified hydrophobic surface, dominated by –OSi $(CH_3)_2O$- groups, to a highly polar and reactive surface (activated surface), terminated with $-O_nSi(OH)_{4-n}$ functionalities, which allow reactions with various inorganic surfaces to form strong, chemical bonds. The unexposed areas retain the unmodified surface chemistry (inactivated surface) and generate only weak van der Waals interactions with other surface. The procedures involve exposures on

PDMS substrates (thickness ~4 mm) under large, uniaxial prestrains ($\varepsilon_{pre} = \Delta L/L$ for lengths changed from L to L + ΔL) (step(ii)).For masks with simple, periodic line patterns denote the widths of the activated (in the bottom frame of Fig. 17.9(a)) and inactivated stripes as W_{act} and W_{in} (step iii). The activated areas bond strongly and irreversibly to other materials (for example, GaAs ribbons covered with SiO_2 films and Si ribbons with native oxide layers) that have exposed –Si–OH groups on their surfaces due to the formation of strong siloxane linkages –O–Si–O–) through condensation reactions. These patterned adhesion sites are exploited to create well-defined 3D geometries in nanoribbons,

The nanoribbons consist of both single-crystal Si and GaAs. The silicon ribbons are prepared from silicon-on-insulator (SOI) wafers using procedures described above. The GaAs ribbons involve multilayers of Si-doped n-type GaAs (120 nm; carrier concentration of 4×10^{17} cm^{-3}), semi-insulating GaAs (SI-GaAs; 150 nm) and AlAs (200 nm) formed on a (100) SI-GaAs wafer by molecular-beam epitaxy (MBE). Chemically etching the epilayers in an aqueous etchant of H_3PO_4 and H_2O_2 using lines of photoresist patterned along the $(0\bar{1}\bar{1})$ crystalline orientation as etch masks, defined the ribbons. Removing the photoresist and then soaking the wafer in an ethanol solution of HF (2:1 volume ratio of ethanol and 49% aqueous HF) removes the AlAs layer, thereby releasing ribbons of GaAs (n-GaAs/SI-GaAs) with widths determined by the photoresist (~100 μm for the examples shown here). The addition of ethanol to the HF solution reduced the probability of cracking of the fragile ribbons due to the action of capillary forces during drying. The low surface tension (compared to water) also minimizes drying-induced disorder in the spatial layout of the GaAs ribbons. In the final step, a thin layer of SiO_2 (~30 nm) is deposited to provide the necessary –Si–OH surface chemistry for bonding to the activated regions of the PDMS.

By laminating the processed SOI or GaAs wafers against a UVO-treated prestretched PDMS substrate (ribbons oriented parallel to the direction of pre-strain), baking in an oven at 360K for 5 min, and removing the wafer, all of the ribbons are transferred to the surface of the PDMS (step (vi)) Heating facilitates contact and the formation of strong siloxane linkages between the activated areas of the PDMS and the native SiO_2 layer on the GaAs ribbons. Relatively weak van der Waals interactions bond the ribbons to the inactivated regions of the PDMS. Relaxing the strain the PDMS a to the formation of buckles through the physical separation of the ribbons from the inactivated regions of the PDMS (step (v)). The ribbons remain tethered to the PDMS in the activated regions owing to the strong chemical bonding. The resulting 3D ribbon geometries (that is, the spatially varying pattern of buckles depend on the pre-strain and the patterns of surface activation. (Similar results are achieved through patterned bonding sites on the ribbons). For the case of a simple line pattern, W_{in} and the pre-strain determine the width and amplitude of the buckles. Sinusoidal waves with wavelengths and amplitudes much smaller than the buckles also formed in the same ribbons when W_{act} is > 10 μ m, owing to mechanical instabilites of the type that generate wavy silicon (see Fig. 17.10 for images of samples formed with different values of W_{act}) As a final step in the fabrication, the 3D ribbon structures are encapsu-lated in PDMS by casting and curing a liquid prepolymer (step vi). Because of its low viscosity and low surface energy, the liquid flows and fills the gaps formed between the ribbons and the substrate (see Fig. 17.11)

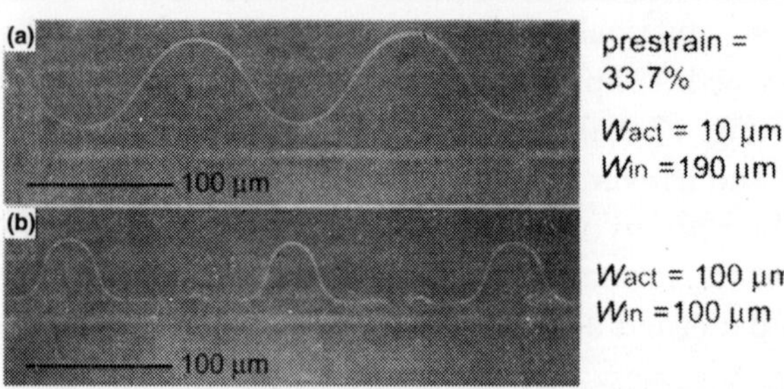

Figure 17.10 *Side-view profiles of buckles formed on PDMS substrates using prestrains of 33.7% and with: (a) W_{act} = 10 μm and Win = 190 mm; and (b) W_{act} = 100 μm and W_{in} = 10 mm. Both samples exhibit buckles in the inactivated regions due to detachment opf ribbons from the PDMS. Sinusoidal waves with small peaks formed only in the activated regions with W_{act} = 100 μm. A comparison of these two samples indicates that selecting W_{act} smaller than a critical value is important to avoid the formation of small wavy structures.*

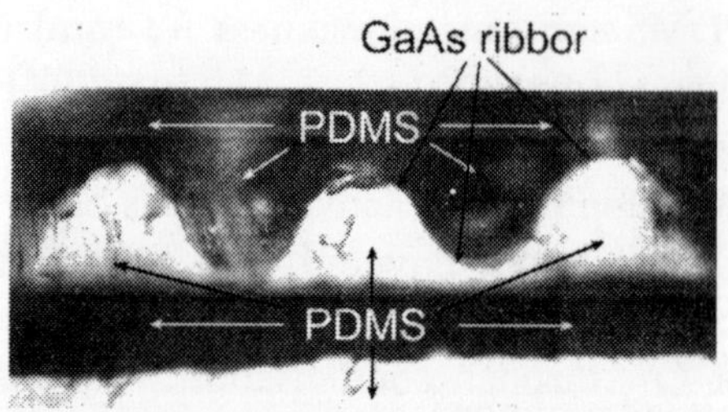

Figure 17.11 *Side-view image of a buckled GaAs ribbon embedded in PDMS after microtoming. The image shows that the PDMS fully fills the gaps between the ribbons and the underlying substrate.*

Figure 17.9(d) shows a tilted- view scanning electron microscope (SEM) image of buckled GaAs ribbons on PDMS, in which ε_{pre} =60%, W_{act} = 10 μ m and W_{in} 400 μm. The image reveals uniform, periodic buckles with common geometries and spatially coherent phases for all ribbons in the array. The anchoring points are well registered to the lithographically defined adhesion sites. The inset shows an SEM image of a bonded region: the width is ~10 μ m, consistent with W_{act}. The images also reveal that the surface of the PDMS is flat, even at the bonding sites. This behaviour is very different from the strongly coupled wavy structures .It suggests that the PDMS induces the displacements but is not involved in the buckling process (that is, its modulus does not affect the geometries of the ribbons) Thus the PDMS represents a soft, non-destructive tool for manipulating the ribbons through forces applied at the adhesion sites.

Stretchability and Bendabiliy of Buckled Nanoribbons.

Figure 17.12(a) shows side-view optical micrographs of buckled ribbons formed on PDMS with different ε_{pre} (W_{act} = 10 μm and W_{in} = 190 μ m). The heights of the buckles increase with ε_{pre}. The ribbons in the inactivated regions do not fully separate at low ε_{pre} (see the samples formed with ε_{pre} = 11.3% and 25.5%), At higher ε_{pre} the ribbons (thickness h) separate from the PDMS to form buckles with vertical displacement profiles characterized by $y=(1/2)A^0_1(1 + \cos(\pi/L_1)\, x)$. where $A^0_1 = 4/\pi\sqrt{L_1L_2(\varepsilon_{pre} - h^2\pi^2/12L_1^2)}$ and $L_1 = W_{in}/(2 \times (1 + \varepsilon_{pre}))$, $L_2 = L_1 + (W_{act}/2)$, as determined by nonlinear analysis of buckles formed in a uniform thin layer. The maximum tensile strain in the ribbons is, approximately, $\varepsilon_{peak} = \kappa|_{max}\cdot(h/2) = y''|_{max}(h/2) = (h/4)A_1^0(\pi/L_1)^2$. The width of the buckles is $2L_1$ and the periodicity is $2L_2$. Because $h^2\pi^2/12L_1^2$ is much smaller than ε_{pre}, the amplitude is independent of the mechanical properties or thickness of ribbons and is determined by the layout of adhesion sites and the prestrain. This suggests that– ribbons made of any material will form into similar buckled geometries. This prediction is consistent with the results obtained for Si and GaAs ribbons. The calculated profiles, plotted as dotted lines in Fig. 17.12(a) for prestrains of 33.7% and 56.0%, agree well with the observations in GaAs ribbons. Further, the parameters (including periodicity, width and amplitude) of the buckles shown in Fig. 17.12(a) are consistsnt with analytical calculations, except at low ε_{pre} (see Table 1). So, the maximum tensile strains in the ribbons are small (for example, ~1.2%), even for large ε_{pre} (for example, 56.0%). This scaling enables stretchability, even with brittle materials such as GaAs.

The lithography defined adhesion sites enable more complex geometries than the simple grating patterns associated with the structures in Fig. 17.9. For example, buckles with different widths and amplitudes are formed in individual ribbons. Figure 17.12(b) shows, an SEM image of buckled Si ribbons (width 50 μm and thickness 290 nm), formed with a pre-strain of 50% and adhesion sites characterized by W_{act} = 15 μm and W_{in} = 350, 300, 250, 250, 300 and 350 μm

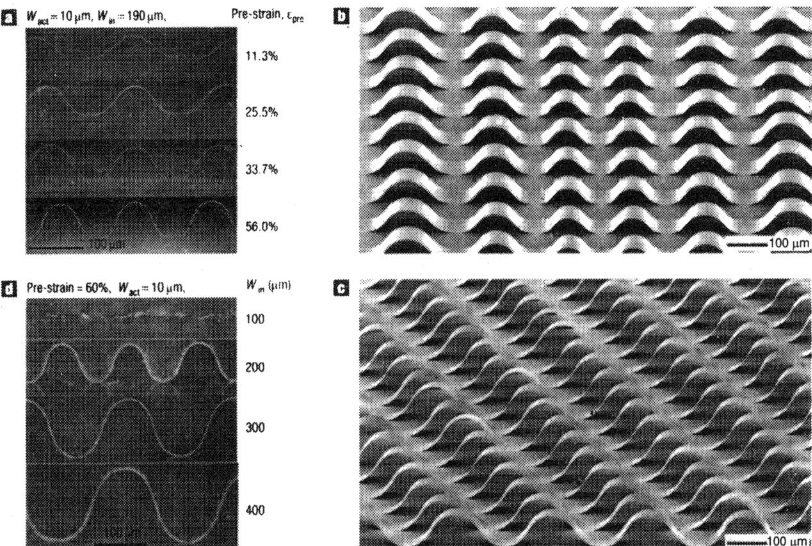

Figure 17.12 *Optical and SEM micrographs of the side-view profiles of buckled GaAs and Si ribbons a, GaAs ribbon structures formed on PDMS substrates with different pre-strains, ε_{pre} b, Si ribbon structures formed on a PDMS substrate pre-strained to 50% and patterned with W_{act} = 15 μ m and W_{in} = 350, 300, 250, 250, 300 and 350 μ m (from left to right), The image was taken by tilting the sample at an angle of 45°. c, Si ribbon structures formed on a PDMS substrate pre-strained to 50% and patterned with parallel lines of adhesion sites (W_{act} = 15 μm and Win = 250 μ m) oriented at angles of 30° with respect to the lengths of the ribbons. The image was taken by tilting the sample at an angle of 75°. d, GaAs ribbon structures formed on PDMS substrates patterned with different values of W_{in}.*

along the lengths of the ribbons. The image shows the variation of widths and amplitudes of adjacent buckles in each of the ribbons. Buckled ribbons are also formed with different phases for different ribbons. Figure 17.12(c) presents an example of a Si system designed with phases in the buckles that vary linearly with distance perpendicular to the lengths of the ribbons. The UVO mask used for this sample has W_{act} and W_{in} of 15 and 250 μ m, respectively. The angle between the activated stripes on PDMS stamp and Si ribbons is 30°.

Buckled GaAs ribbons on PDMS with $\varepsilon_{pre} = 60\%$, $W_{act} = 10$ μm and different values of W_{in} as shown in Fig. 17.12(d), illustrate an aspect that is important for applications in stretchable electronics. The profiles, show failure due to cracking in the GaAs when $W_{in} = 100$ μ m (and smaller). The failure results from tensile strains (~2.5% in this case) that exceed the yield point of the GaAs (~2.%). An optimized configuration for -robustness to stretching and compressing is achieved by selecting $W_{in} >> W_{act}$. In this situation, pre-strains up to and greater than 100% can be accommodated. The only limit is in the resolution with which the adhesion sites on the PDMS are patterned, which is ~2 μ.m. This stretchability is demonstrated directly by applying forces to the PDMS supports, Changes in the end-to-end distances ($L_{projected}$) of segments of the ribbons provide a means to quantify the stretchability and compressibility, according to $|(L^{*}p_{rojected} - L^{0}_{projected})/L^{0}_{projected}| \times 100\%$, where $L^{*}_{projected}$ represents the maximum or minimum length before fracture and $L^{0}_{projected}$ is the length in the relaxed state. Stretching and compressing correspond to $L^{*}_{projected}$ greater and less than $L^{0}_{projected}$ respectively. Buckled ribbons on PDMS with $W_{act} = 10$ μm and $W_{in} = 400$ μ.m and $\varepsilon_{pre} = 60\%$ exhibit stretchability of 60% (that is, ε_{pre}) and compressibility of up to 30%. Embedding the ribbons in PDMS, mechanically protects the structures, and also produces a continuous, reversible response, but with slightly different mechanics compared to the unembedded case. In particular, the stretchability and compressibility decreases to ~51.4% (Fig. 17.13(a)) and ~I8.7% (Fig. 17.13(b)), respectively. The PDMS matrix on top of the ribbons flattens the peaks of the buckles slightly, owing partly to curing-induced shrinkage of the overlying PDMS. Small period waves form in these regions under large compressive strains, due to the spontaneous mechanics of the type that generate the wavy ribbon structures. Mechanical failure tends to initiate in these areas, as illustrated in Fig. 17.13(b), thereby reducing the compressibility. Buckled structures with $W_{act} = 10$ μ.m and $W_{in} = 300$ μm avoid this type of behaviour. Although these samples exhibit slightly lower stretchability than the one shown in Fig, 17.13(a), the absence of the short period waves increases the compressibility to ~26%. Overall, single-crystalline GaAs nanoribbons with buckles that are formed on the prestretched PDMS substrates with patterned surface chemical adhesion sites exhibit stretchability higher than, 50%, and compressibility larger than 25%, corresponding to full-scale strain ranges approaching 100%. These numbers are further improved by increasing ε_{pre} and W_{in} and by using a substrate material capable of higher elongation than PDMS. For even more elaborate systems, these fabrication procedures repeat to generate 3D multilayered architectures (see Fig. 17.14)

A direct consequence of this large stretchability and compressibility is extreme levels of mechanical bendability. Figure 17.15(a) to (c) presents optical micrographs of bent configurations.

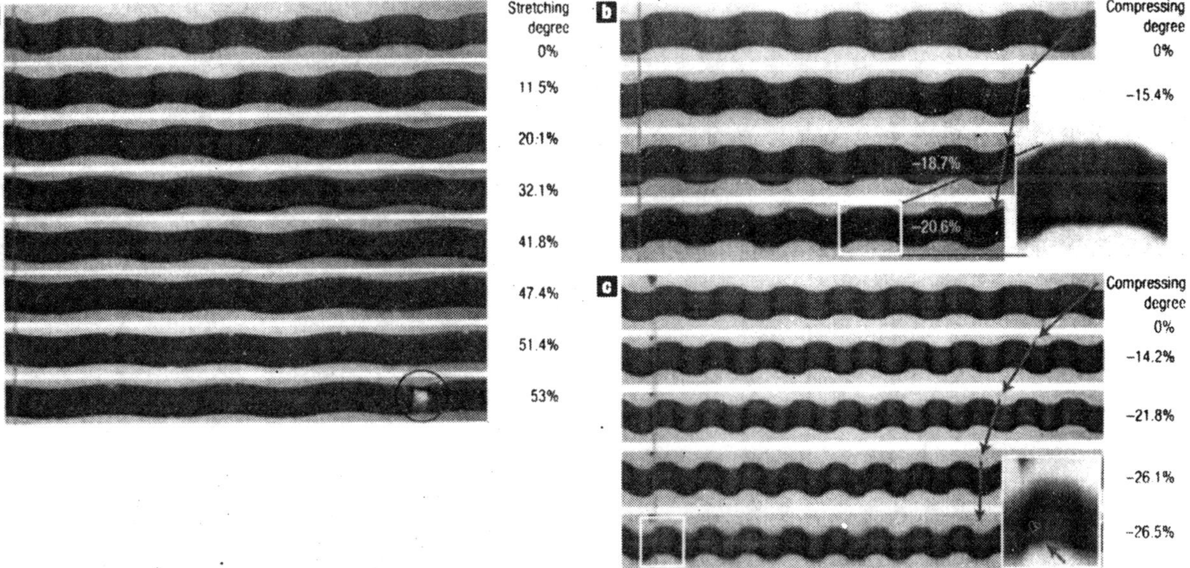

Figure 17.13 *Stretching and compressing of buckled GaAs ribbons embedded in PDMS (a) Images of a single buckled ribbon stretched to different levels of tensile strain (positive %).(b) Images of a single buckled ribbon compressed to different levels of compressive strain (negative %).*

Figure 17.14 *Photograph of a sample with two layer of buckled GaAs ribbons arrays.*

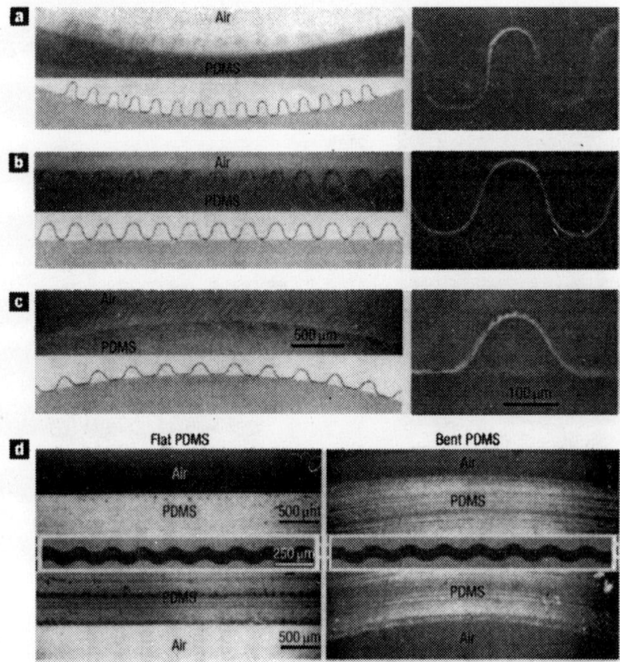

Figure 17.15 *Bending of buckled ribbons on surfaces and in matrixes of PFMS. (a–c) Optical microscope images with low magnification (top left frames) and high magnification (right frames) and schematic illustrations (bottom left frames) of buckled GaAs ribbons on PDMS with concave (a), flat (b) and convex (c) surfaces. (d) Images of buckled ribbons rmbedded in PDMS before (left) and after (right) bending. A single embedded GaAs ribbon can be seen in the enlarged area enclosed in th ewhite box. The top and bottom frames show the curvatures of the top and bottom surfaces, respectively. The bucled ribbons are the same as those shown in Fig. 17.9(d).*

The PDMS substrate (thickness ~4 mm) is bent into concave (radius ~5.7 mm), flat and convex (radius ~6.1 mm) curvatures, respectively, The images show the profile changes to accommodate the bending induced surface strains (~ 20-25% for these cases). The shapes are similar, in fact, to those obtained in compression (by strains of ~20%) and tension (by strains of ~20%), The embedded systems exhibite even higher levels of bendability due to neutral mechanical plane effects-. When the top and bottom layers of PDMS have similar thicknesses, there is no change in the buckling shapes during bending (Fig. 17.15(d)).

Stretchable Electronics With Buckled Nanoribbons

To demonstrate these mechanical properties in functional electronic devices, metal-semiconductor-metal photodetectors (MSM PDs) are built using buckled GaAs ribbons with profiles similar to those shown in Fig. 17.15 by depositing thin gold electrodes onto the SI-GaAs sides of the ribbons to form Schottky contacts. Figure 17.16(a) shows the geometry and equivalent circuit, and top-view optical micrographs of an MSM PD before and after stretching by ~50%. In the absence of light, little current flows through the PD; the current increases with an increase of illumination with an infrared beam (wavelength ~850 nm) (Fig. 17.16(b)). The asymmetry in the current/voltage (I–V) characteristics is due to differences in the electrical properties of the contacts. Figure 17.16c (stretching) and 8d (compressing) show I–V characteristics measured at different degrees of stretching and compressing. The current increases when the PD is stretched by up to 44.4% and then decreases with further stretching. Because the intensity per unit area of the light source is constant, the increase in current with stretching is due to increases in the projected area (referred to as effective area, S_{eff}) of the buckled GaAs ribbon as it flattens. Further stretching of the PD induces the formation of defects on the surface and/or in the lattice of the GaAs ribbon, resulting in a decrease of current and, eventually, at fracture, in

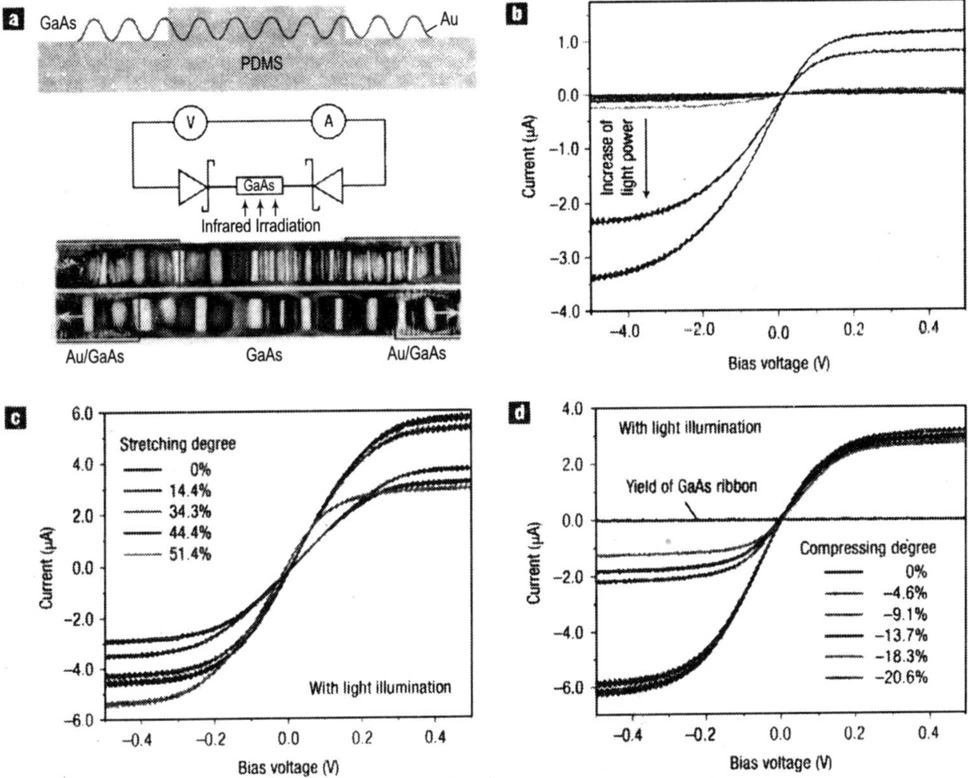

Figure 17.16 *Characterization of stretchable metal-semiconductor-metal photodetectors (MSM PDs). (a) Illustrations of the geometry (top), an equivalent circuit (middle) and optical images of a buckled PD before and during stretching (bottom, the white arrows indicate the direction of force applied during stretching), (b) Current/voltage (1 – V) curves recorded from a buckled PD that was irradiated by an Infrared lamp with different output intensities. (c,d) I – V characteristics of PDs illuminated with constant luminance and stretched (c) or compressed (d) to different degrees.*

an open circuit. Similarly, compression leads to a decrease in S_{eff} and also a decrease in the current (Fig. 17.16d), These results indicate that buckled GaAs ribbons embedded in a PDMS matrix provide a fully stretchable/compressible type of photosensor that is useful in various applications, such as wearable monitors, curved imaging arrays and other devices.

Table. 17.1 *Parameters extracted (from experiments and calculations) from the buckles as shown in Figure 17.13(a). The calculations assume that the widths (i.e., 10 m m for the samples shown in the figure) of the activated regions are the same before and after stretching.*

pre-strain	*Measured width(μm)*	*calculated width(μm)*	*measured amplitude Am (μm)*	*calculated amplitude Acal (μm)*	*calculated peak strain ε peak (%)*
11.3%	*136.6*	*170.7*	*37.5*	*37.6*	*0.38*
25.5%	139.6	151.4	51.5	50.3	0.65
33.7%	140.1	142.1	56.4	54.3	0.80
56.0	124.3	121.8	63.6	6004	1.2

Thus soft elastomers with lithographically defined adhesion sites can be used as tools for creating certain classes of 3D configurations in semiconductor nanoribbons. Stretchable electronics provide one example of the many possible application areas for these types of structures. Simple PD devices demonstrate some of these capabilities. The high level of control over the structures and the ability to separate high-temperature processing steps (such as formation of ohmic contacts) from the buckling process and the PDMS suggest that more complex devices (for example, transistors and small circuit sheets) are possible. The well controlled phases of buckles in adjacent ribbons provide an opportunity for electrically interconnecting multiple elements.

Schottky Diodes and FETs of NWs and NBs

The ZnO nanobelts used for fabricating the FET devices are synthesized through a solid-vapor process in a high-temperature horizontal furnace system. The Au electron patterns are defined with photolithography on a SiO_2 substrate. The electrodes consist of two 3-μm-wide fingers pointing head to head at a distance of 4 μm. These two fingers are connected to two 500 × 500 μm^2-contacting pads for probe contacts. The as-synthesized nanobelt samples are placed in ethanol and ultrasonicated for 15 min to disperse the bundles into individual nanobelts.

By aligning single ZnO nanobelts/nanowires across paired Au electrodes using dielectrophoresis, rectifying diodes of single nanobelt/nanowire-based devices, are fabricated to reveal their electronic properties.

After applying a droplet of the nanobelt suspension onto the electrodes, the electrodes are connected to a 5 V and 1 MHz AC signal which is chosen for optimizing the alignment of a single nanobelt. This signal generates, an alternating electrostatic force on the nanobelts in the solution. Under the electrical polarization force, the nanobelts are deposited on the electrodes. By precisely controlling the concentration of the nanobelt in the solution, a circuit with only a single nanobelt across the two electrodes is made (inset of Figure 18.1). The most striking feature is that the *IV* characteristics of the devices formed by this process display a rectifying behavior as shown

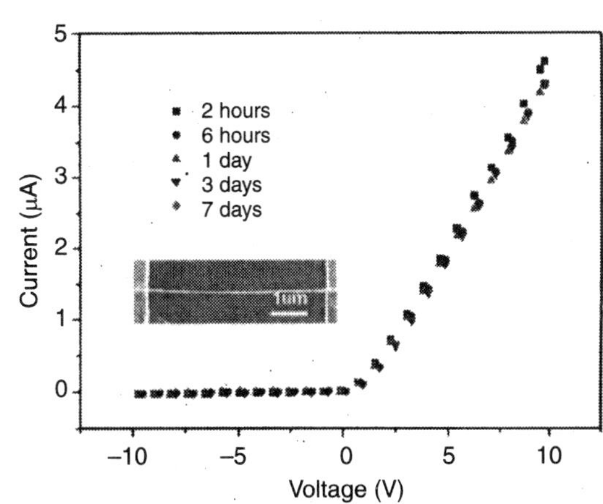

Figure 18.1 *Rectifying IV characteristics of a single ZnO nanobelt lying on Au electrodes at different times after the fabrication, showing the stability of the device. The current ratio at "on" and "off" state is 2000. Measurements are done at room temperature.*

in Figure 18.1. The inset is an SEM image of the electrodes. Data corresponds to different measurements of the same devices at different time intervals after the device is made. It shows that the rectifying behavior of the nanodevice is very stable.

Figure 18.2(a) is the inversed current characteristic. At a 10 V reverse bias voltage, the reverse current is only 0.8 nA, corresponding to a current density of 10.5 A cm^{-2}. The ideality of the diode is determined from the forward bias characteristic as shown in Figure 18.2(b), The diode current is

$$I = I_0\left[\exp\left[\frac{V - V_{th}}{nkT}\right] - 1\right] \qquad (1)$$

where n is the ideality factor that is a quantity for describing the deviation of the diode from an ideal Schottky barrier, for which $n = 1$, I_0 is the saturation current, V_{th} is the threshold voltage, k is Bolzmann's constant, and T is the absolute temperature. The ideality factor of the device is determined to be 2.87, larger than one. This higher value of n is due to a barrier at the ZnO^- Au contact formed during the electrophoresis alignment.

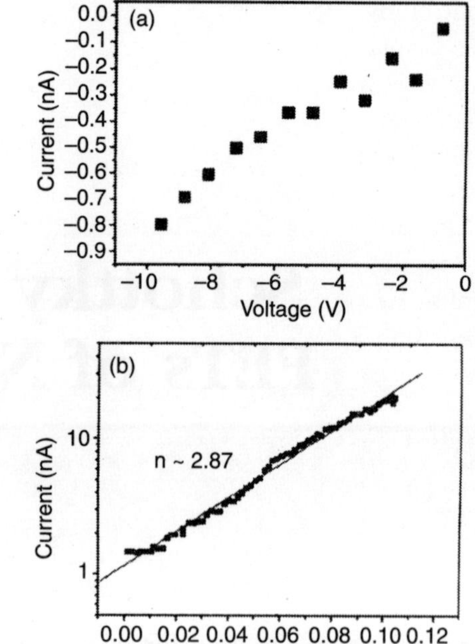

Figure 18.2 *(a) Detailed IV characteristic of the Schottky diode under reverse bias. (b) The IV characteristics of the device at forward bias.*

The temperature dependence of the diode is measured (Figure 18.3). The current under the forward bias decreases with the lowering of the temperature. The resistivity of the nanobelt increases by a factor of 2 when the temperature decreases from 215 to 104 K, showing a typical semiconductor *IV*, characteristic, When the temperature is below 50 K, the carrier transportation is depressed greatly, The device acts like an open circuit with a resistivity greater than 10^5 Ω cm.

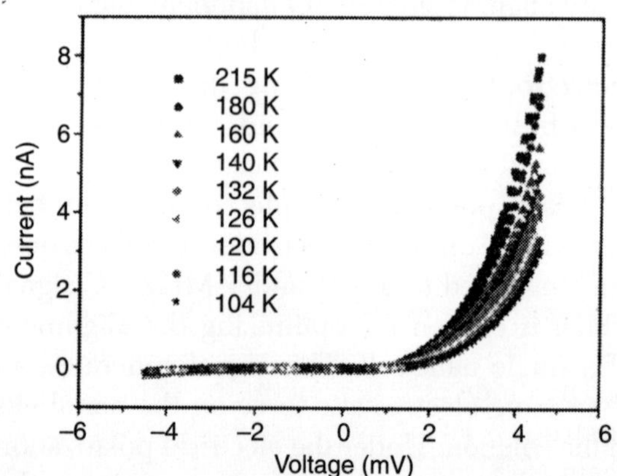

Figure 18.3 *IV characteristics of the Schottky diode at different temperatures showing the semiconducting behavior.*

Table 1 shows *IV* characteristics of all of the devices made using ZnO nanobelts. The data shows a high reproducibility of the diodes. Devices fabricated on two different substrates, SiO_2 and Si_3N_4, are tested. The data shows that there is no difference in the *IV* characteristics of the devices; so, the influence of the substrate on the rectifying behavior of the devices is eliminated.

By controlling the size of the nanobelts at around 200 nm in width, most of the fabricated devices display the rectifying behavior. For nanobelts with widths greater than 500 nm, the devices show an infinity resistance under both forward and inverse bias, which is due to the poor contacts on both electrodes. Because of the nonflatness of the gold electrodes, the wider nanobelts do not have a stronger contact with the substrate and so result in large contact resistance; The smaller one have a good contact due to stronger van der Waals interaction. Thus, the size of the nanobelt plays an important role in determining the contact property of the diode; the smaller nanobelt has a better contact with the electrodes. For the nanobelts of ~ 200 nm in width, the forward current is as high as 0.5 μA at 1.5 V forward bias, corresponding to a resistivity of 8.7×10^{-2} Ωcm; the reverse current is around 0.2 nA. The on-to-off current ratio is as high as 2000.

Table 18.1 *I-V Characteristics of Dielectrophoresis Aligned Nanobelts with Different Sizes and on Different Substrate.*

sample	*number of devices with rectifying behavior*	*number of devices with infinity resistance*	*number of devices with linear IV*
~200-nm-wide belts (SiO_2 substrate)	27	7	1
> 500-nm-wide belts (SiO_2 substrate)	2	17	0
~200-nm-wide belts (Si_3N_4 substrate)	7	2	1
total no. of devices	36	26	2

Chemically synthesized ZnO nanorods have lower crystallinity and higher density of defects. By characterizing the *IV* characteristics of a single ZnO nanobelt nanowire/produced by a solid-vapor process at high temperature, it is observed that current flowing through the diode is ~0.5 μA at 1.5 V forward bias. The ideality factor of the devices is ~3, indicating a much better performance of devices. Therefore, the factor that produces this rectifying behavior is the alternating zinc and oxygen layers parallel to the basal plane: They produce a dipole moment that leads to a potential gradient and then introduce the asymmetry of current flow along the c axis; by assuming that the noncentral symmetric and layered distribution of cations and anions in the wurtzite structure, and the contact, is at the top and bottom (0001) surfaces (for nanowires), as the contacts are at the same side-surface of the nanowire lying on the gold electrodes (see Figure 18.4(a) and the cation- or anion-terminated surfaces have no effect unless the contacts are at the bottom ends of the nanowire.

The diode effect is due to the asymmetrical contacts between the ZnO nanobelts and the Au electrodes. To identify the origin of the rectifying behavior Pt is deposited using a focused ion beam (FIB) microscope at the contacts between the nanobelt and Au electrodes, as shown in Figure 18.4(a). This is done carefully to avoid any ion contamination on the nanobelts so that the change in the *IV* characteristic is determined solely by the change in contact properties. The devices show a linear *IV* curve after the Pt deposition (Figure 18.4(b) indicating the disappearance of the rectifying effect. For a total of over 20 devices deposited with Pt, the current flowing through the nanobelts is increased greatly, 5-20 times. This means that the contact resistance is reduced significantly and the diode effect disappears after Pt deposition at the contacts or that the Pt deposition results in symmetric contacts at both sides.

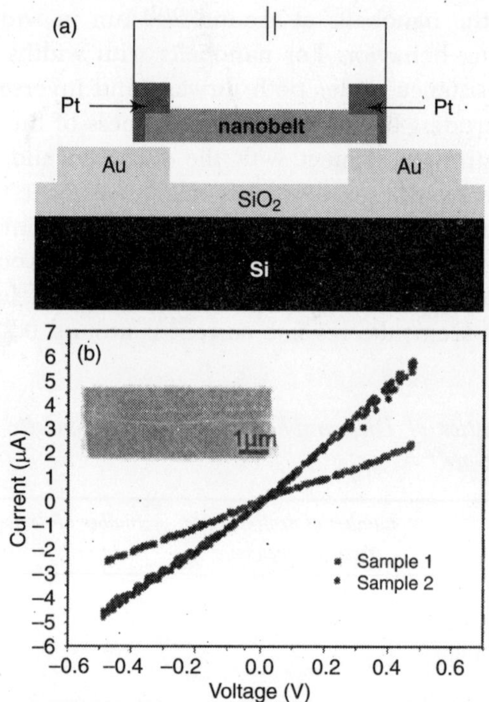

Figure 18.4 *(a) View of the electrode structure after Pt deposition using an FIB. (b) Linear IV characteristics of the device after Pt deposition. The inset is the SEM image of the device after Pt deposition at the two ends.*

The linear *IV* of the nanodevice after Pt deposition proves that the rectifying behavior comes from the contact between the ZnO nanobelt and the Au electrodes. The ZnO-Au ,contact is usually a Schottky contact. So, an as-fabricated device is composed of two inversely connected Schottky contacts and there is no rectifying behavior if the contacts are symmetric. However, in the dielectrophoresis deposition process, the AC voltage introduces an electrostatic force on the nanobelt; then, the nanobelt moves toward the electrodes until it finally deposits on them. The contacts of both ends of the nanobelt onto the electrodes occur in a consecutive order. The side of the nanobelt that touches the electrode first has a firm contact with the electrode, thus forms a better contact with lower barrier height, whereas the other end that contacts later has a higher barrier, thus leads to the formation of the Schottky diode for device.

In the dielectrophoresis alignment, the heat generated by the AC signal during dielectrophoresis is significant. It produced an asymmetric heating effect in the drain and source areas because of the different barrier height of the two contacts. This effect is examined. As shown in the Figure 18.5 the electrode on the left is molten, whereas the electrode on the right is left unattained with a 5 V and 1 MHz AC signal applied. The difference in the heat generated at the two electrodes introduces a different degree of annealing/sintering at the two sides. For the contact on the lower temperature side, the annealing effects is different compared to the other side, indicating a difference in contacts at the two ends. For the side with moderate annealing temperature, a local oxygen vacancy is created; thus, the local conductivity is enhanced. The contact at this end with the gold electrode is similar to a metal-metal Ohmic contact. For the side with higher annealing temperature, the contact is degraded, forming a metal-semiconductor Schottky contact. Then the entire structure is equivalent to a Schottky diode.

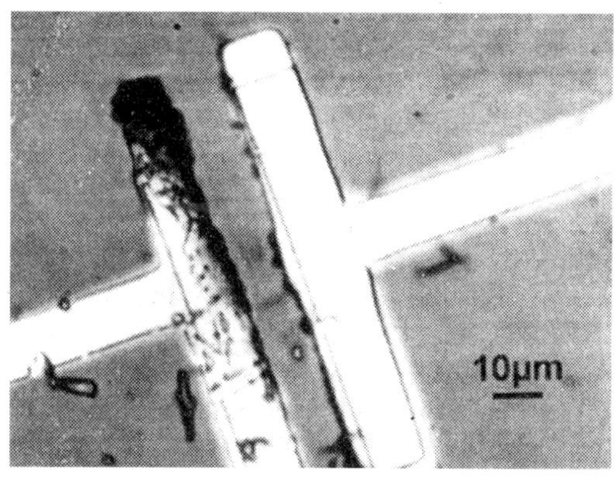

Figure 18.5 *Optical image of a pair of Au electrodes after dielectrophoresis alignment of the nanobelt at a 5 V, 1 MHz AC signal.*

VERTICAL NANOWIRE ARRAY – BASED LIGHT EMITTING DIODES

Electoluminescence from a nanowire array based light emitting diode is a nanowire based light emitting diode (LED) consisting of a p-type GaN thin film and an n-type ZnO vertical nanowire array. The vertical nanowire array offers, a, number of potential advantages over the conventional thin film architecture. Firstly, the ZnO nanowire array is grown directly on the p-GaN thin film using a low temperature solution procedure, This is a simple process towards making functional p-n junctions. Secondly, the vertical nanowire array creates natural waveguiding cavities where part or all of the emission can be manipulated to travel to the top of the device, an improvement towards better extraction efficiency.

1. Experiment

A p-GaN thin film is grown on a c-plane sapphire substrate (Al_2O_3), through metal-organic chemical vapor deposition (MOCVD). The dopant source is magnesium and following deposition, a post-anneal of 20 min at 1070K is utilized to ionize the dopants. The resulting 1 mm thin film has a hole concentration of $4.5 \times 10^{17} cm^{-3}$ and hole mobility of 12 $cm^2V^{-1}s^{-1}$ as determined by a four-point Hall measurement. A low temperature solution growth method is then used to form a vertical ZnO nanowire array on top of the thin film. The growth method is the hydrolysis of a zinc salt in water. The substrate is suspended in solution for 2 hours at 365K affording 2 μm tall ZnO nanowires. The procedure is repeated two more times, resulting in nanowire lengths of around 5-6 μ m

The resulting nanowire array is entirely vertical with nominal diameters of 100-600 nm (Fig. 18.6). The structure is wurtzite hexagonal as shown by hexagonal cross sections in Fig. 18.1(a), The nanowires are grown epitaxially from the p-GaN thin film because there is a good lattice match between the two materials. After nanowire growth, the substrate is spin coated with poly(methyl methacrylate) (PMMA) to protect the nanowires and to form a buffer layer between the p-GaN thin film and eventual nanowire metal contacts (Fig. 18.7). An oxygen plasma etch is used to remove the PMMA from the tips of the nanowires in order to allow

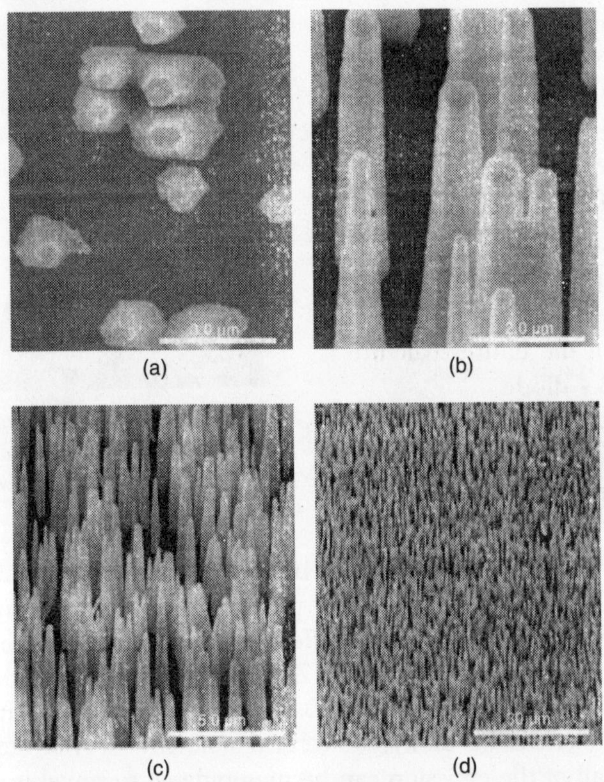

Figure 18.6 *(a) Top down and (b)–(c) 45° tilted SEM images of solution-grown ZnO nanowire array*

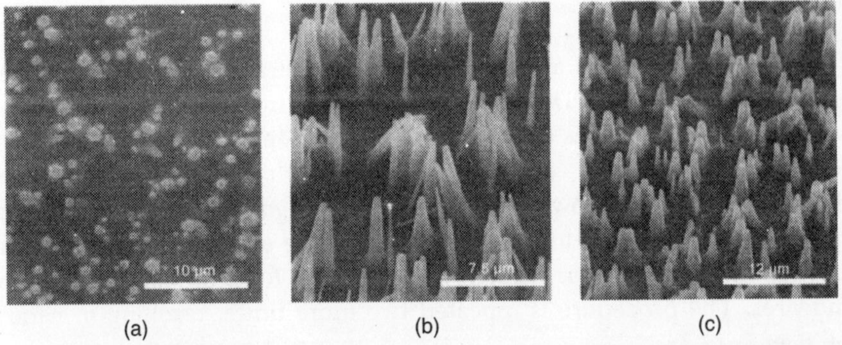

Figure 18.7 *(a) Top and (b), (c) 45° tilted SEM images of solution grown ZnO nanowires after spin coating of a PMMA insulating layer*

them to form effective ohmic contacts. Nil Au contacts are thermally evaporated onto the p-GaN thin film and Ti/ Au contacts onto the exposed ZnO nanowire tips.

The resulting devices display electroluminescence visible to the human eye when subjected to a forward bias over 10 V. The emitted light appears to originate completely or predominantly from the p-type GaN thin film at the junction between the p-type thin film and n-type nanowire array (Fig. 18.8(a)). The localization of the electroluminescence at the p-n junction proves that the phenomenon is injection electroluminescence in which luminescence occurs from the recombination of minority carriers that are injected across the junction.

Almost all the devices show diode like current voltage behavior which demonstrates proper p-n junction formation with minimal contribution from contact resistances. The rectifying behavior is shown in Fig. 18.8(b). A linear current-voltage behavior is observed when electrodes a probed simply across the p-GaN thin mm layer (the two left hand metal contacts in Fig. 18.8(a). Further back to-back reverse rectifying current-vòltage behavior is observed when electrodes are probed across the nanowire array (the two right-hand metal contacts in Fig. 18.8(a).

Room temperature photoluminescence (PL) spectra of the device are collected by means of excitation with a 325-nm HeCd continuous-wave laser. Emission is collected by a microscope coupled by multimode fiber to a spectrometer with a liquid nitrogen cooled silicon charge coupled device (CCD). As the spot size of the beam at the sample is around 100μm in diameter it is possible to localize the excitation area. PL is observed from the bare p-GaN thin film region of the device as well as from the n-ZnO nanowire array region of the device (Fig. 18.8(c)). The emission from the p-GaN thin film is red-shifted from the bandgap to around 430 nm. This shift is well known and is commonly observed in Mg-doped GaN due to dominant band to acceptor transitions.

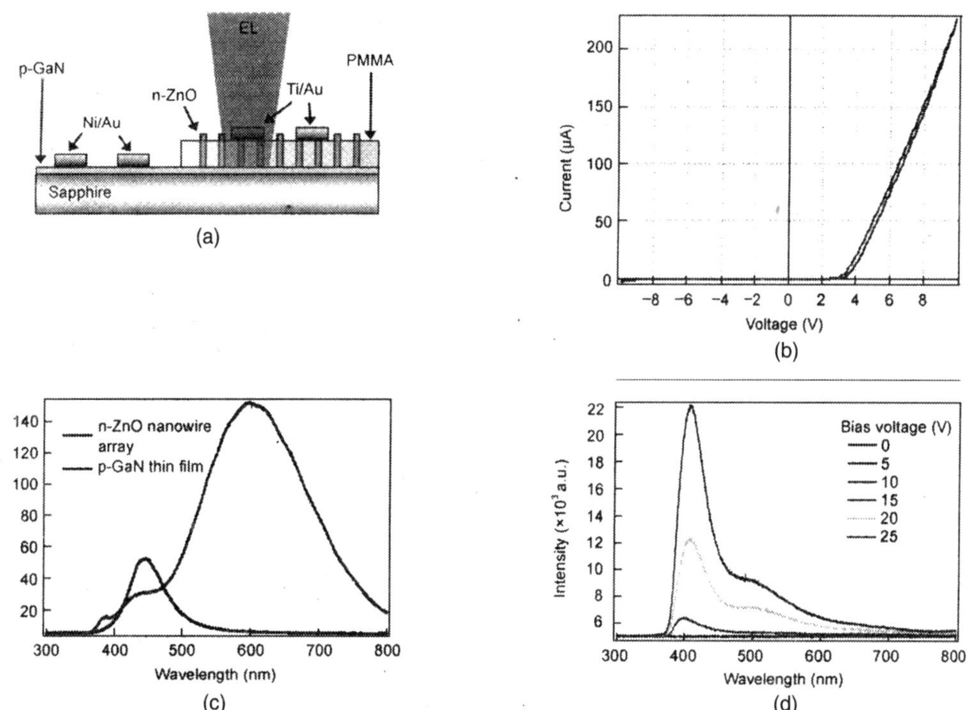

Figure 18.8 *Solution-grown-ZnO nanowire LED device: (a) view of the device and location of electroluminescence; (b) I–V curves; (c) photoluminescence spectra; (d) electroluminescence spectra*

PL spectra collected from the nanowire array region reveals three peaks. The first peak at around 380 nm corresponds to interband transitions in the ZnO material. The second peak at around 430 nm results from the p-GaN thin film directly below the nanowire array. The third broad peak centered around 600 nm is a "yellow peak" commonly seen in ZnO materials and is attributed to defects, perhaps due to interstitial oxygen ions.

Room temperature electroluminescence (EL) is collected by electrically stimulating the device using a voltage source and emission is detected in the same fashion as the PL measurements. Figure 18.8(d)). shows spectra generated at 5 V increments between 0 and 25 V. Electroluminescence is detectable by the naked eye at 10 V. The emission generated is believed to be a result of a combination of the various transitions observed in the PL spectra. The dominant peak is centered at ~410 nm and thus the electroluminescence is believed to be dominated by acceptor to band transitions in the p-GaN thin film and is blue shifted by contributions of luminescence from the interband transitions in the nanowires and perhaps interband transitions in the thin film as well. The low contribution from the defect peak shows that deep level traps saturate and thus allow interband transitions to dominate.

The change in EL intensity per unit voltage step increases as the voltage increases. This is consistent with the band bending of a p-n junction. As voltage increases, the band bending between the p and n materials is reduced; therefore, the number of carriers able to traverse the junction increases. In contrast, the increase in the defect peak per voltage step is not as substantial. As applied voltage increases band-to-band transition; increases above defect-related transitions.

An advantage of using a vertical nanowire array over a thin film is the potential for waveguided emission, and therefore the potential for improved extraction efficiency. The natural unseeded epitaxial growth of the ZnO nanowires from the p-GaN thin film yield fairly large diameters around 100-600 nm. An ideal single mode waveguide cavity is around 200 nm for a ZnO material surrounded by a medium of PMMA. For the waveguiding, the collecting optical fiber is placed directly above the luminescence site at angles ranging from 0° to 90° from the normal (Fig. 18.9(a)) At every measurement, the radial position of the end of the optical fiber is kept exactly 1 cm from the surface of the device. Figure 18.9(c) shows electroluminescence spectra at various angles from the normal and Fig. 18.9(d) shows the correlation between normalized integrated EL intensity for the various positions.

The strongest emission intensity is observed at 0° from normal (directly on top of the device), and the intensity rapidly decays as detection angle increases. This observation shows that part of the junction emission is directly waveguided to the tip of the nanowires. A similar setup employing a vertical ZnO nanorod array as a waveguiding mechanism for emission from a traditional GaN-based multiple quantum well (MQW) thin film LED shows that increases of 50% and 100% in light output at applied currents of 20 and 50 mA. 38.4% light output increase by a current injection of 200 mA is also achieved.

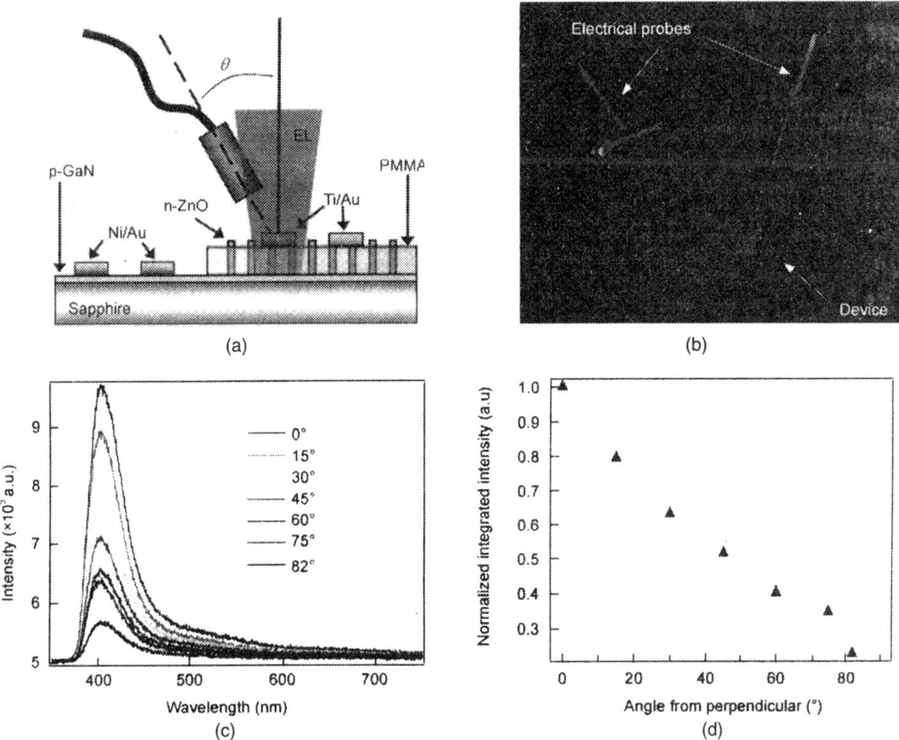

Figure 18.9 *(a) A view of waveguiding experimental setup. (b) Electroluminescence intensity captured by CCD camera at 25 V forward bias. (c) Electroluminescence spectra at various angles from vertical orientation of the nanowire array. (d) Normalized integrated intensity of EL emission as a function of angle from the perpendicular.*

FIELD EFFECT TRANSISTOR BASED ON SINGLE SEMICONDUCTOR OXIDE NANOBELTS

Experiment

Single crystalline SnO_2 and ZnO nanobelts of thicknesses between 10 and 30 nm are synthesized by thermal evaporation of oxide powders in an alumina tube without the presence of a catalyst. Large bundles of either SnO_2 or ZnO nanobelts are dispersed in ethanol by ultrasonication until mostly individual nanobelts are isolated. These ethanol dispersions are dried onto a SiO_2/Si substrate for imaging by non contact mode AFM.SnO_2 field-effect transistors are fabricated by depositing SnO_2 nanobelt dispersions onto SiO_2 / Si (p^+) substrates, followed by treatment m an oxygen atmosphere at 1070K for 2 hours. The SiO_2 substrate are then spin-coated with PMMA, baked, exposed to electron-beam lithography for the definition of electrode arrays, and developed. A 30-nm-thick layer of titanium is deposited by electron-beam evaporation to serve as the source and drain electrodes, and the remaining PMMA is lifted off in hot acetone.

A alternative way of contacting the nanostructures is applied to ZnO nanobelts. The field-effect transistors are fabricated by deposition dispersed ZnO nanobelts on predefined gold electrode arrays. The SiO_2 gate dielectric thickness is 120 nm, and the back gate electrode is fabricated by evaporation of gold on the Si (p^+) side of the substrate. Also the electrode arrays, are variable spaced. They include electrode gaps as small as 100 nm and as large as 6 μm.

Field Effect Transistors. Both SnO_2 and ZnO nanobelts from field effect transistors. The structurally and morphologically controlled semiconducting oxide nanobelts are ideal entities for fabricating functional devices. Figure 18.10(a) shows a field effect transistor (FET) fabricated using a single ZnO nanobelt, and its design principle is given in Figure 18.10(b) A typical SnO_2 field effect transistor pretreated in a 1 atm oxygen atmosphere at 1070 (Figure 18.11(a)) demonstrate a gate-threshold voltage of –2.5V, a current switching ratio *I(ON)/I(OFF)* of nearly 10000, and (ignoring voltage drops at the Ti contacts) a peak conductivity of 8.14 $(\Omega cm)^{-1}$. Measured conductivities range from 4 to 15 $(\Omega\ cm)^{-1}$ By controlling the backgate voltage, switching ratios as large as six order of magnitude, a conductivity as high as 10 (Ωcm^{-1}), and a mobility as large as 35 $cm^2V^{-1}\ s^{-1}$ are observed. The typical leakage conductance between electrodes is measured between 20 and 80 pS and is gate-bias independent. The switching ratios of FETs are limited by this conductivity. From an analysis of the transconductance dI_d/dV_g, for gate biases the electron mobility of an n-type field effect transistor is estimated if the geometry of the device is known. Without subtracting voltage drops at the contacts, from analysis of Figure 18.11(a) the electron mobility in the SnO_2, nanobelt is ~35 $cm^2V^{-1}s^{-1}$. Other measured SnO_2 nanobelt FETs exhibit electron mobility ranging from 10 to 125 $cm^2V^{-1}s^{-1}$. From four-probe measurements on other devices, the contact resistance between the Ti electrodes and the SnO_2 nanobelt is on the same order of magnitude as the four-probe "ON" channel resistance.

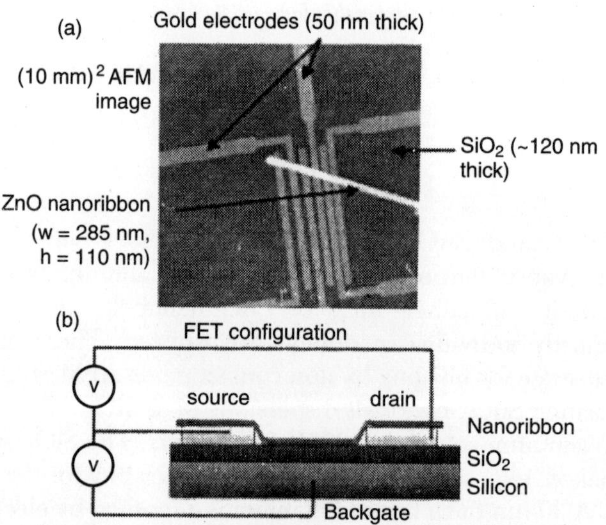

Figure 18.10 *(a) E-beam lithography fabricated field-effect transistor (FET) using a single ZnO nanbelt. b) The working principle of the FET device.*

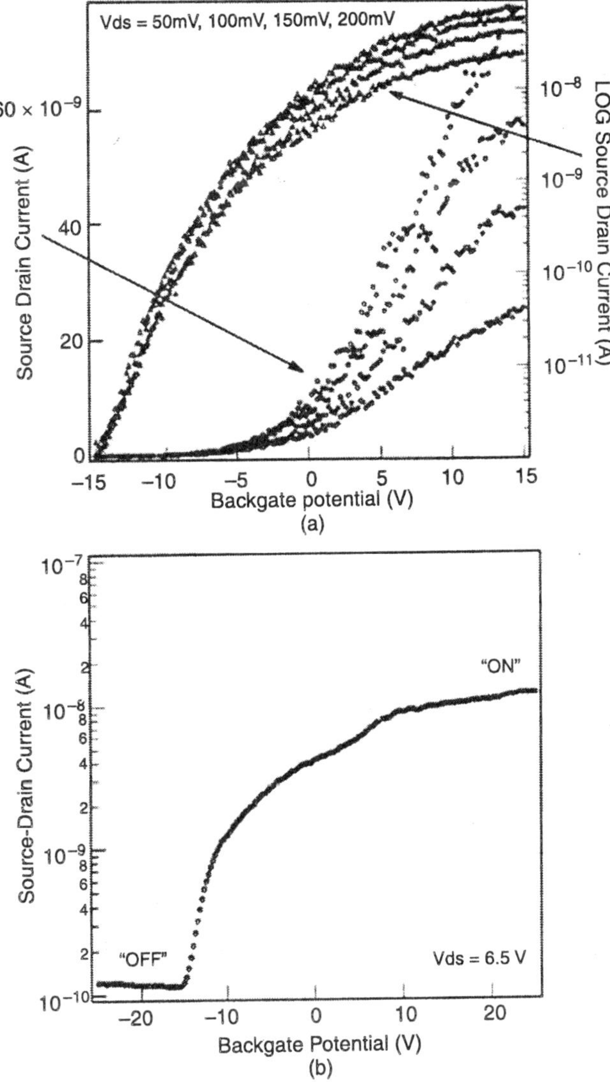

Figure 18.11 *Nanobelt FET, (a) Source-drain current versus gate bias for a SnO_2, FET in ambient. (b) Source-drain current versus gate bias for a ZnO FET in ambient.*

The alternative way of contacting the nanobelts is by depositing them on top of prefabricated gold electrodes led to very resistive contacts. SnO_2 nanobelts are also doped by annealing in reduced oxygen environment, thus increasing conductivity and decreasing the gate threshold voltage, indicating the feasibility of tuning device characteristics by controlling oxygen vacancies. A typical ZnO field effect transistor (Figure 18.10(b)) shows a gate threshold voltage of −15 V, a switching ratio of nearly 100, and a peak conductivity of 1.25×10^{-3} $(\Omega cm)^{-1}$ (Figure 18.11(c)). A completely analogous behavior is observed in the case of carbon nanotubes deposited on top of Au electrodes or covered by Ti electrodes.

Sensitivity of the Electrical Properties of Nanobelts to-Environment

Before electrical measurement, SnO_2 nanobelts are annealed in a 1 atm oxygen environment at 1070K for 2 hours. Without annealing, the as-produced nanobelts exhibit no measurable conductivity for source-drain biases from −10 to + 10 V and for gate biases from −20 to +20 V, whereas after annealing the SnO_2 nanobelts exhibit considerable conductivity. The SnO_2 devices, with their conductivity activated, demonstrate electrical properties. By further annealing of the devices at lower temperatures in a vacuum, oxygen, or ambient, the electrical properties of the nanobelts can be tuned.

After annealing of the SnO_2 devices in a vacuum at 470K nanobelt conductivity is observed to increase along with an associated negative-shift in gate-threshold voltage. Smaller, additional increases in conductivity are observed after additional vacuum anneals. Eventually, the nanobelt behaves like a metal with the gate field being unable to affect the current flowing through the device. In contrast, annealing nanobelt devices in ambient, at 470K leads to a decrease of the conductivity, along with a shift in the gate-threshold voltage in the opposite, positive direction (Figure 18.12(a)). The source-drain conductivity at zero gate bias spans 3 orders of magnitude from 0.09 $(\Omega\ cm)^{-1}$ after annealing at 470K in ambient, to 75.3 $(\Omega\ cm)^{-1}$ after annealing at 520K in a vacuum. The changes in conductivity with low temperature annealing most likely result from variations in the number of oxygen species adsorbed on the SnO_2 surfaces or in the number of oxygen vacancies in the SnO_2 bulk with the amount of oxygen in the environment. The number of equilibrium surface and bulk oxygen defects are a function of the environmental oxygen partial pressure an temperature. Annealing in a vacuum decreases the number of adsorbed oxygen species and increases the number bulk oxygen vacancies, whereas annealing in oxygen or ambient do the opposite. It is found that bulk and surface oxygen vacancies in SnO_2 act as electron donors, which increases SnO_2 conductivity and decreases the gate threshold voltage.

The SnO_2 nanobelt conductivity is observed to increase and the gate threshold voltage is observed to decrease by taking a nanobelt device from ambient into vacuum without annealing (Figure 18.12(a)). Because the, diffusion of bulk oxygen vacancies at room temperature are limited. This indicates that surface oxygen desorption takes place at this temperature. Because of their small dimensions, semiconducting oxide nanobelts are on the order of 10^{20} surface oxygen sites per cubic centimeter of material. Thus, for partial changes in the concentration of adsorbed oxygen species, large changes in nanobelt conductivity is observed.

Annealing at 470K and 520K induces further changes in conductivity (Figure 18.12(a)). Therefore either the surface or bulk nanobelt composition must change as a result of these annealing steps. For temperatures below 1170K the equilibrium bulk nonstoichiometry of SnO_2 is insignificant in comparison to surface nonstoichiometry in samples of similar surface/volume ratios. So surface oxygen vacancies control the conductivity after 470K and 520K anneals. Annealing also desorbe species than oxygen from the nanobelt surfaces. Water acts as a mask on the nanobelts surface, decreasing sensitivity to environment oxygen until removed.

The sensitivity of the FET to gas exposure is demonstrated by measuring the SnO_2 nanobelt conductivity as a function of time after introduction of oxygen. The conductivity of a hot 470K SnO_2 nanobelt in a vacuum is observed to decay to about $^1/_3$ its initial value after exposure to

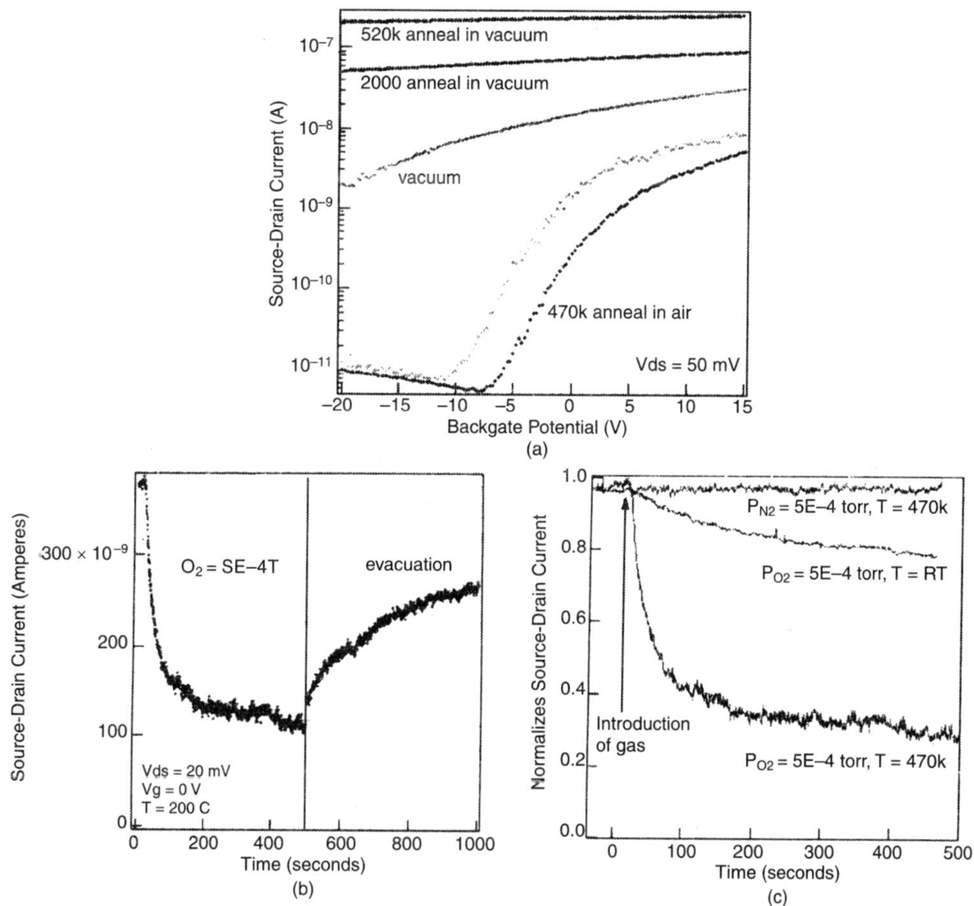

Figure 18.12 *SnO_2, Nanobelt FET oxygen sensitivity, (a) Source-drain current versus gate bias for a SnO_2, FET after various treatments measured in this order: air vacuum, 470K vacuum anneal, 520K vacuum anneal,470K air anneal. All measurements taken at room temperature with vacuum anneal measurements taken in a vacuum and air anneal measurements in air. (b) Source-drain current at zero gate bias of a hot 470K SnO_2, FET as a function of time with exposure to air and subsequent recovery, (c) Comparison of source-drain current at zero gate bias of a SnO_2 FET in nitrogen and oxygen,*

oxygen at a pressure of 6.65×10^{-2} Pa (Figure 18.12(b)). The exponential time constant for this decay is 37.2 seconds. After removal of the oxygen, the conductivity recovers about twice as slow, with time constant of 76.5s. In contrast the conductivity of the same SnO_2 nanobelt at the same temperature is invariant upon exposure to nitrogen at the same pressure (Figure 18.12(c)). Exposure of the nanobelt to 6.65×10^{-2} Pa oxygen at room temperature leads to the decay of the conductivity at a much slower rate with a time constant of 156.9s, reflecting the activated nature of the oxygen desorption process. At room temperature a smaller percentage of surface vacancy sites are affected; thus a smaller current change is produced.

Channel Length Dependence of the Transistor Operation

SnO_2 nanobelts are contacted with electrodes spaced at 127, 254,1080 and 1840 nm and the dependence of the conductance and electrode gaps than between large electrode gap. Between electrodes spaced 1080 and 1840 nm and the dependence of the conductance of the nanobelts on gate bias as a function of electrode spacing is observed. A nonlinear relationship between the conductance and electrode spacing is found. The conductance of the SnO_2 nanobelt is modulated less effectively between short electrode gaps than between large electrode gaps. Between electrodes spaced 1080 and 1840 nm apart sweeping the gate voltage from –15 to +15V (source drain bias 0.05 V) the conductance is modulated by more than 5 order of magnitude (from less than 10^{-11} S to greater than 10^{-6} S, Figure 18.13(a)). Between electrodes spaced 127 and 254 nm apart, sweeping the gate over the same range of voltages only the conductance the same nanobelt is modulated by less than 1 order of magnitude. Considering the SnO_2 nanobelts as a resistor the conductance is expected to scale inversely with length regardless of gate bias. By normalizing the conductance by electrode spacing, the nonlinear effects are observed (Figure 18.13(b)). At large positive gate biases near + 15V, the conductivity of the nanobelt is same for each segment. Small variations are observed (less than 1 order of magnitude) in conductivity and are due to variations in the contact resistances to each segment. The length dependence is evident at negative gate biases near –15V, where the conductivity of the nanobelt varies by more than 5 orders of magnitude depending on electrode spacing. It is observed that the current in short segment cannot be turned "OFF" by the gate, whereas in long segments, the current is modulated by the gate field. The effects observed are not the "short channel effects" well known in metal-oxide semiconductor field effect transistors (MOSFETs). Other SnO_2 nanobelt contacted with variably spaced electrodes exhibit the same

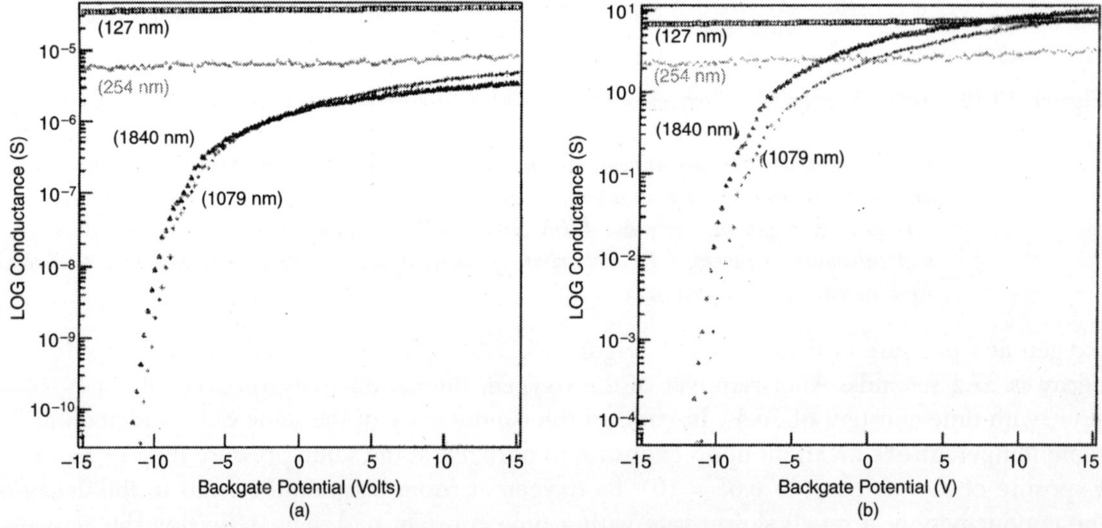

Figure 18.13 *SnO_2 Nanobelt FET I- V_g curves as a function of back gate bias and electrod spacing. (a) Source-drain conductance versus back gate voltage for electrode spacing of 127. 254. 1079, and 1840 nm. (b) Conductance normalized by area and electrode length versus back gate voltage for the same electrode spacings.*

short channel effects. It is observed that conductivity of devices between electrodes spaced far apart is effectively modulated by a gate bias (by as much as 6 orders of magnitude), whereas the conductivity of devices between closely spaced electrodes is not effectively modulated. Qualitatively, the transition between these two regimes is broad and centered near 500 nm.

Photoconductivity of Nanobelts

It is shown that the FET is very sensitive to UV illumination, and, in turn, the device is controlled by UV illumination. In ZnO nanobelts, which have a direct bandgap in the UV, two processes are observed to contribute to photoconductivity. The first process is due to the photogeneration of electron-hole pairs. The second process is most likely due to the desorption of oxygen from the ZnO surface under UV exposure.

Ultraviolet light irradiation of ZnO in air is observed to result in a significant increase of the conductivity. Light with a wavelength of 350 nm (E_λ = 3.54 eV) is used, exceeding the direct band-gap of ZnO, which is of 3.2 eV. The increase in the conductivity results from both photogeneration of electron-hole pairs as well as doping by UV light-induced surface desorption. These processes are observed by introducing a shutter between the light source and the ZnO nanobelt so that the flux of UV photons is turned "ON" and "OFF". In the insert to Figure 18.14 the conductivity is observed to rise and fall in response to opening and then

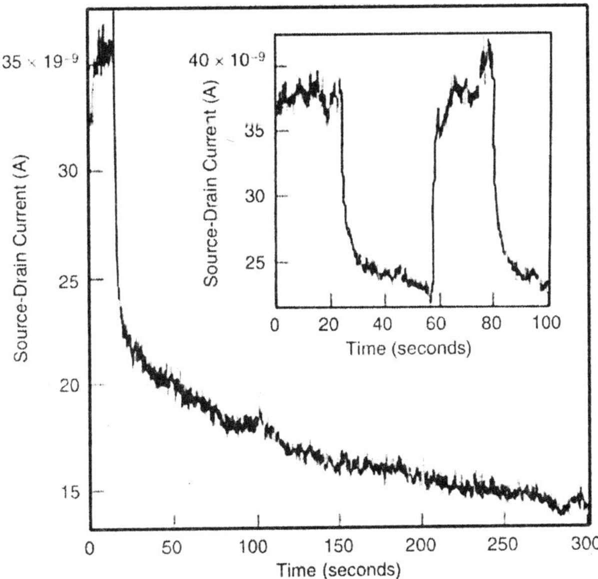

Figure 18.14 *ZnO nanobelt response to UV light. Source-drain current versus time at zero-gate bias and a source drain potential of 6.5 V. The initial state of the nanowire in these responses is one of steady state with a 350 nm light source ON. Insert: Source-drain current versus time as a 350 nm source is cycled OFF and ON. Figure: Source-drain current versus time plot depicting two regimes of current decay after UV stimulation is turned OFF. The first regime is the sharp drop in current which most likely represents the cessation of photo generation of carriers. The second regime is the slow decay of current.*

closing the shutter. After the establishment of a steady state with the UV light "ON", the decay of conductivity after closing the shutter reveals two regimes (Figure 18.14). The first regime is characterized by a sharp drop in conductivity immediately after the shutter is closed. This sharp drop is due to ,the recombination of photogenerated electron-hole pairs, The second regime is characterized by the slow decay of current, This slow gradual change suggests a surface chemical reaction.

Similar effects are observed in ZnO thin films. It is observed that during UV exposure, photogenerated holes react with negatively charged adsorbed oxygen species liberating O_2, example, the photogenerated holes react with negatively charged carbon dioxide species and liberate CO_2. The UV light is turned "OFF" the surface conductivity slowly decays as oxygen is readsorbed thus increasing the number of electron acceptors in the n-type material.

Gas sensors are fabricated using the single-crystalline SnO_2 nanobelts based on two electrodes contact. The conductance of the nanobelt is affected by the surface adsorbed molecules, resulting in a change in the electric current. Electrical characterization show that the contacts are ohmic and that the nanobelts are sensitive to environmental pollutants such as CO and NO_2, as well as to ethanol-useful for breath analyzers and food control applications. The sensor response defined as the relative variation in conductance due to the introduction of the gas, is $\Delta G/G$ = 200 % for 250 ppm of ethanol and $\Delta G/G$ = 3000 % for 0.5 ppm nitrogen dioxide at 470K (Fig. 18.15) The sensor displays selectivity between ethanol and NO_2. These results demonstrate the potential of fabricating nanosize sensors using the integrity of a single nanobelt with a sensitivity at the level of a few parts per billion.

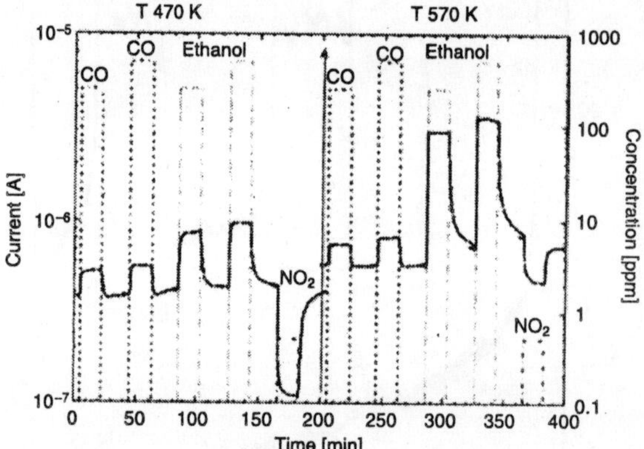

Figure 18.15 *Response of SnO_2 sensors to CO, ethanol, and NO_2 gases at 470K and 570K at different gas concentrations. The left y-axis shows for the change in electronic current and the right y-axis shown million the gas concentration in parts per million.*

CHARACTERISTIC OF SnO_2 AND ZnO NANOBELTS

Effects of Schottky Contacts vs Low Resistance Ohmic Contacts

Field-effect transistors (FETs) based on individual semiconducting oxide (SnO_2 and ZnO) nanobelts with multiterminal electrical contacts are fabricated and characterized. Below are compared the effects of-Schottky contacts and low-resistance Ohmic contacts on the characteristics of SnO_2 and ZnO nanobelt FETs.

While a large modulation of the source-drain current by gating are obtained in both cases, the FET source drain current *(I)*-voltage *(V)* characteristics, transfer characteristics, and magnitude of the "on" current are distinctly different. A direct correlation, appears between the different FET behaviors and the nature of the source-drain electrical contacts with the nanobelt channel using multiterminal devices and simultaneous two- and four-terminal measurements. Low-resistance Ohmic contacts on the SnO_2 and ZnO nanobelts lead to high-quality n-channel depletion mode FETs with well-defined linear and saturation regimes, large on current, and, on/off ratio as high as 10^7.

The SnO_2 and ZnO nanobelts are synthesized by thermal evaporation of pure oxide powders under controlled conditions without any catalyst. The as synthesized nanobelts are pure single crystalline, and structurally uniform. They have a rectangular like cross section with a typical thickness of tens of nanometers and width-to-thickness ratio of 5-10. Bundles of nanobelts retrieved from the alumina substrate are separated into individual nanobelts in an isopropanol solution by ultrasonic agitation, and then dispersed onto a substrate. An individual nanobelt is located and multiple electrodes are defined via photolithography, pulsed laser ablation and/or thermal evaporation, and lift-off. A n-type degenerately doped (100) Si substrate (p ~1 m Ωcm, n - 10^{19} cm^{-3}) is used as a back gate, and a 250 nm thick thermally grown SiO_2 serves as the insulating gate dielectric. All measurement are carried out in air at room temperature.

Low-resistance Ohmic contacts are realized on both SnO_2 and ZnO nanobelts and verified via direct comparison of two-probe and four-probe *I* - *V* measurements. In order to achieve low-resistance Ohmic contacts consistently on SnO_2 nanobelts, pulsed laser deposition to deposit a coating of metallic RuO_2 (~200 nm) at room temperature, followed by a layer of Au (~100 nm) by thermal evaporation in order to ease wire bonding. Figure 18.16(a) shows the four-probe and two-probe (two center contacts) *I-V* characteristics of a SnO_2 nanobelt device with four RuO_2/ Au contacts, as shown in the scanning electron microscopy (SEM) image in the inset. Both *I* - *V* curves are linear and the contact resistance is less

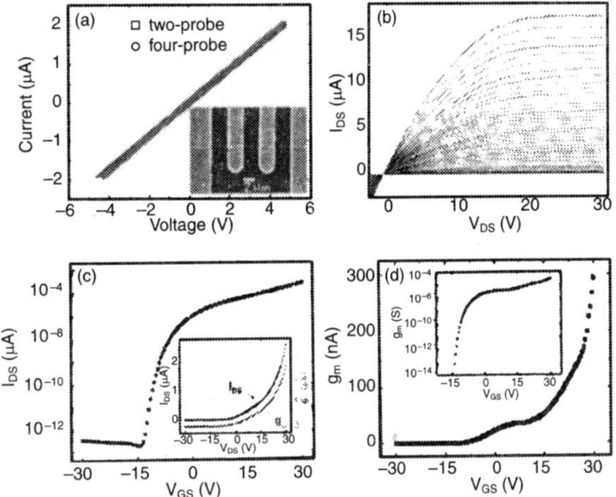

Figure 18.16 *(a) Four-probe (red circle) and two-probe (black square) I- V curves of a SnO_2 nanobelt device with RuO_2/ Au contacts. Inset: SEM micrograph of the device. (b) Source-drain 1- V characteristics of the SnO_2 nanobelt FET at increasing gate voltage (-30 to +30 V) in steps of I V from bottom to top. (c) Transfer characteristics of the SnO_2 nanobelt FET at V_{DS}= 1 V. Inset: Linear plots of I_{SD}- V_{GS} and g_D- V_{GS} at V_{DS}=I V. (d) Transconductance of the device at V_{DS}=I V. Inset: Semilog plot of g_m -V_{GS} at V_{DS}=I V.*

than 1% of the channel resistance. The consistent formation of Ohmic contacts is due to the limitation of interfacial diffusion and reaction at the SnO_2/RuO_2 interface. In contrast, direct metallization with Cr/ Au and especially Ti/ Au on SnO_2 nanobelts results in poor electrical contacts. For ZnO nanobelts, Ti/ Au metallization shows a high degree of reliability in producing low resistance Ohmic contacts, although in some rare cases Schottky contacts are observed. Cr/ Au, en the other hand, lead to non-Ohmic contacts on ZnO nanobelts.

The nature of the contacts has a significant impact on the FET characteristics of the nanobelt devices. Figures 18.16 (b)-(d) display the room temperature FET behavior of the device shown in Fig. 18.16(a). The two middle leads are used as the source and drain electrodes. Since the contact resistance is determined to be at least two orders of magnitude smaller than the channel resistance, the measured FET characteristics reflect the intrinsic response of the SnO_2 channel. Figure 18.16(b) shows the source-drain *I-V* curves at various gate voltages from −30 to +30 V in steps of *I-V*. In the 'ON' state, the *I-V* curves have a well-defined linear regime at low biases, and the drain current saturates upon further increase in the bias. The saturation behavior is similar to the OFF effect in conventional silicon FETs due to a local depletion of carriers around the drain electrode. The channel can be turned OFF with a negative gate voltage of −14 V, which is shown in the transfer characteristic in Fig. 18.16(c) The I_{DS} vs V_{GS} plot in semilogarithmic scale at a constant V_{DS}= 1V (in the linear regime of the source-drain *I-V*) shows a threshold voltage of −14V and ON-OFF of ratio as large as 10^7. The inset of Fig. 18.16(c) shows that I_{DS} and conductance $g_D=I_{DS}/\ V/_{DS|VG\,=\,const}$ increase with increasing gate bias and reach their maximum values of 2.5 μA and 2.5 μS, respectively. The FET has an "OFF" state current of ~10^{-13} A, which remains constant below V_{th}. The typical gate leakage current is measured to be at least two orders of magnitude smaller than I_{DS} in the off state and is gate bias independent. The subthreshold swing, $S = V_{GS}/\partial\text{-}\log(I_{DS})$, is 1.02 V/decade(for this device). Although for conventional Si metal-oxide-semiconductor field effect transistor, the typical S value is 70–100 m V / decade, the subthreshold swing is improved by reducing the thickness of the gate oxide later 250 nm SiO_2 is used). Figure 18.16(d) and the inset show the transconductance, $g_m = I_{DS}/\partial V_{GS}$, at V_{DS} = 1 V in the same range of gate voltages. The effective field-effect carrier mobility μ_{eff} of the FET is estimated- by $g_m=Z\mu_{eff}\ C_O\ V_{DS}/\ L$, where Z is the channel width (corresponding to the nanobelt width of 150 nm), L the channel length (corresponding to the electrode separation of 4 μm), and C_O the oxide capacitance per unit area (for the 250 nm thick SiO_2 substrate, C_O= 1.4 × $10^{-4}Fm^{-2}$) μ_{eff} shows a maximum value of around 334 $cm^2\ V^{-1}s^{-1}$ at V_{GS} = 30 V and a value of 26.1 $cm^2V^{-1}\ s^{-1}$ at V_{GS}= 0V. Since the FET action in the device is channel limited, the mobility is intrinsic to the SnO_2 nanobelt. The device shows an excellent stability. Measurements before and after six month storage in ambient air yield the same transistor behavior with identical transfer characteristics and consistent transconductance values.

Figure 18.17 shows the FET characteristics of a ZnO nanobelt device with Ti/ Au Ohmic contacts. Compared with the SnO_2 nanobelts, the ZnO nanobelts are wider (see the inset of Fig. 18.17(a) and less resistive. The source drain *I-V* curves show well-defined linear and saturation regimes [Fig. 18.17(a)]; I_{DS}-V_{GS} and g_D-V_{GS} at a drain bias of 0.3 V [Fig. 18.17(b) and the inset] show a threshold voltage of −14V, maximum conductance of 18 μS, and ON/OFF ratio

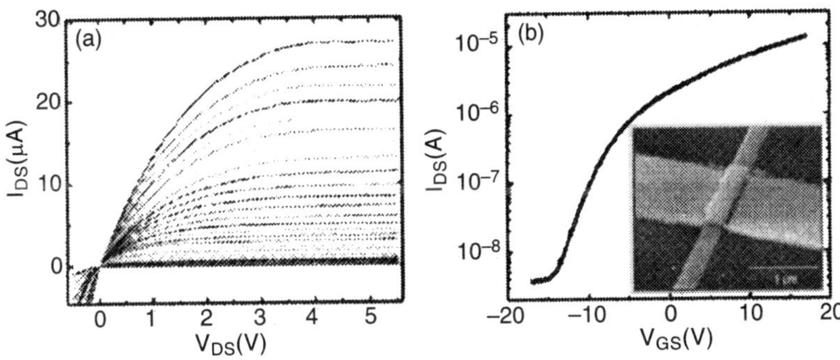

Figure 18.17 *n-channel depletion mode FET based on a ZnO nanobelt: (a) I_{DS} vs V_{DS} at different gate voltages (Gas) in steps of 1 V from -17 V (bottom) to +17 V (top). (b) Transfer characteristics of the FET at V_{DS}=0.3 V. Inset: A SEM image of the ZnO nanobelt contacted by a Ti/ Au electrode.*

approaching 10^4. The calculated effective mobility peaks at ~1.35 ×10^3 cm^{2}V^{-1} s^{-1} at V_{GS}= 17 V and is 4.40 × 10^2 cm^{2}V^{-1} s^{-1} at V_{GS} = 0 V, which is larger than that in the ZnO thin film FETs.

The electrical characteristics and performance of the FET devices are dependent upon the nature of the electrical contacts to the nanobelts. The source-drain *I-V* characteristics of a ZnO nanobelt device with non-Ohmic Cr/ Au contacts are shown in Fig. 18.18. The source-drain *I-V* at V_{GS}=0 shows a strong rectifying behavior. It indicates Schottky source-drain contacts of different barrier heights. I_{DS}– V_{SD} at different V_{GS} (from +22 to −22V in steps of 1 V) shows a large gate modulation typical of a directionally dependent Schottky barrier FET. The inset shows the transfer characteristics at drain biases of V_{SD}= +0.5 and −0.5 V. In the forward direction, the maximum conductance go of the on state is about 7 × 10^{-7} S and the off state is ~10^{-12} S, with an ON/OFF ratio greater than 10^4. The maximum transconductance g_m is about 10^{-8} S. In the reverse direction, these parameters are suppressed by more than two orders of magnitude. The directionally dependent behavior of the ZnO nanobelt FET resembles that in an intentionally fabricated asymmetrical carbon nanotube FET with one Ohmic contact and one Schottky contact. The FET behavior is determined predominantly by one contact. A similar rectifying behavior is found in ZnO nanobelt devices with structurally symmetrical contacts. The characteristics of nanobelt Schottky barrier FETs show large variations in the

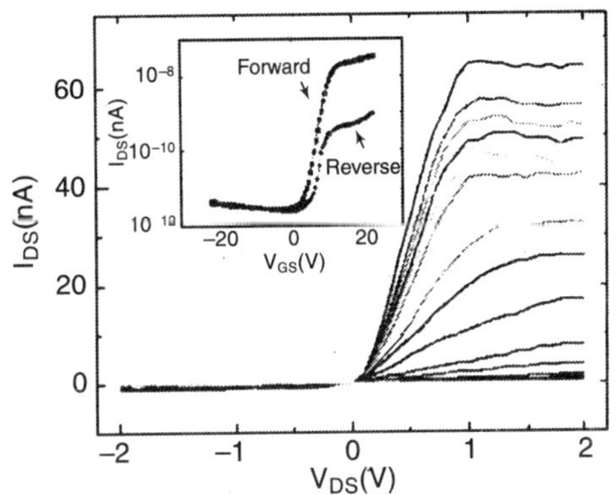

Figure 18.18 *ZnO nanobelt asymmetric Schottky barrier PET: (a)I_{DS}-V_{DS} plot with Vas from -22 to 22 V in steps of 1 V.lnset: n-channel enhancement mode transfer characteristics of the PET in forward (V_{DS} = +0.5 V) and reverse (V_{DS} = -0.5 V) directions.*

degree of asymmetry and gate modulation from device to device. This is in contrast to the channel-limited nanobelt FETs with Ohmic contacts in which the FET characteristics are highly systematic and reproducible. These observations are due to the fact that the Schottky barrier formation is due to surface states. Although the Schottky barrier FETs have some advantages over normal FETs in certain electronics applications, precise control and reproducibility of the Schottky barrier contacts must be obtained. For quantitative and scalable sensing applications with the nano-FETs, it is necessary to establish the FET action from the channel rather than form the contacts.

With identification of the channel-limited nanobelt FETs, the potential of the device can be evaluated for sensor applications by examining the intrinsic response of the nanobelt channels. The response of a channel-limited SnO_2 nanobelt FET (with RuO_2/Au contacts) is 0.2% H_2 balanced with N_2 at room temperature. Figures 18.19(a) and (b) show two senes of I_{SD}- V_{SD} plots in ambient air and in 0.2% H_2, respectively, at different gate voltages (–30 to +30 V). The FET characteristics show a significant modification upon exposure to 0.2% H_2. The channel conductance (in the linear regime) increases by around 17% at all gate voltages. The hydrogen reacts with and removes the oxygen adsorbed on the metal oxide surface and thus increases the electron concentration and the conductance of the nanobelt channel. This also makes the pinchoff at the drain contact more difficult, which is evidenced by the fact that the linear region in the I_{DS}- V_{DS} now extends to much higher drain biases. The response of the SnO_2 nanobelt FET to hydrogen at room temperature without any catalyst demonstrates great promise of such devices in chemical gas sensing under ambient conditions.

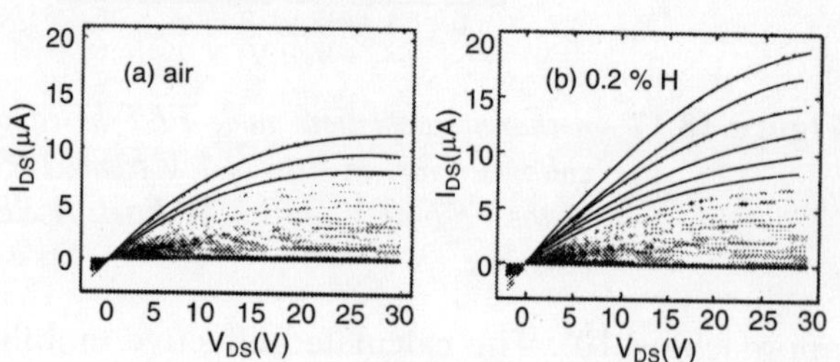

Figure 18.19 *I_{DS}–I_{DS} at various gate voltage (-30 to +30 V) in steps of 2 V for a SnO_2 nanobelt FET with Ohmic RuO2/Au contacts (a) in air and (b) in 0.2% H_2 balanced with N_2 at room temperature.*

Nanogenerators

The operation of nanodevices fabricated with one-dimensional nanostructures such as nanowires, nanotubes, and nanobelts usually requires very low power which is provided by an external source, such as a battery that may have to be replaced or recharged regularly. The reliance on an external power source may present a limitation for these systems. Various approaches have been developed for energy scavenging with applications in wireless electrronics, such as thermoelectric, piezoelectric thin-film, and vibrational cantilevers. Developing wireless nanodevices and nanosystems are of critical importance for sensing, medical science, defense technology, and even personal electronics. It is highly desirable for wireless devices and even required for implanted biomedical devices that they be self powered without use of a battery.

A approach for converting mechanical energy into electrical energy by piezoelectric zinc oxide nanowire (NW) arrays is described below. The operation mechanism of the electric generator relies on the unique coupling of the piezoelectric and semiconducting properties of ZnO as well as the gating effect of the Schottky barrier formed between the metal tip and NW. The piezoelectric effect converts mechanical strain into ionic polarization charges that generate a piezoelectric potential; the Schottky barrier at the interface between the electrode and the nanowires gates and directs the flow of electrons under the driving of the piezoelectric potential.

(A) THE CONCEPT OF PIEZOELECTRIC NANOGENERATORS (NG)

The concept of the NG was first introduced by examining the piezoelectric properties of ZnO nanowires (NWs) with an atomic force microscope. ZnO has a wurtzite structure, in which the Zn cations and O anions form a tetrahedral coordinator. ZnO has two important characteristics: One is the presence of polar surfaces, such as Zn^{2+} - terminated (0001) and O^{2} - terminated $(000\bar{1})$. The interaction of the polar charges at the surface results in the growth of a wide range of unique nanostructures such as; nanobelts, nanosprings, nanorings, nanohelices. Another characteristic is the lack of center symmetry, which results in a piezoelectric effect, by which a mechanical stress/strain can be converted into electrical voltage, and vice versa, owing to the relative displacement of the cations and anions in the crystal. It is important to point out that the polar surfaces are a surface effect, while piezoelectricity is an integrated volume effect. ZnO NWs and nanobelts exhibit piezoelectricity, and the piezoelectric coefficient is even larger than that of the bulk, so use is made of ZnO NWs to convert mechanical energy into electricity and then transport it through an external load.

Therefore aligned ZnO NWs are grown on a conductive solid substrate (Fig. 19.1(a)). The measurements are performed by atomic force microscopy (AFM) using Si tip coated with a Pt film which has a tetrahedral shape with an apex angle of 70° and height of 14 μm and a spring constant of 1.42 Nm^{-1}. The measurements are done in AFM contact mode under a constant normal force of 5 nN between the tip and sample surface and the scan area is 70 × 70 μm^2 (Fig. 19.1(b)). The tip is scanned over the top of the ZnO NW, and the tip's height is adjusted according to the surface morphology and local contacting force. In the corresponding voltage-output image for each contact position, many sharp output peaks are observed (Fig. 19.1(c)). The voltage signal is inverted (for display purposes); it is actually negative in reference to the grounded end. The topological profile of an NW and its corresponding output potential show a delay in the voltage signal (Fig. 19.1(d)), which means that there is no electric power output when the tip is in contact with the NW but a sharp voltage peak is generated at the moment when the tip is about to break contact with the NW. This delay is due to the power-output process. Also, the voltage V_L presented is converted from the current flowing through the external load of resistance R_L. The NG based on the ZnO NWs has the following characteristics:

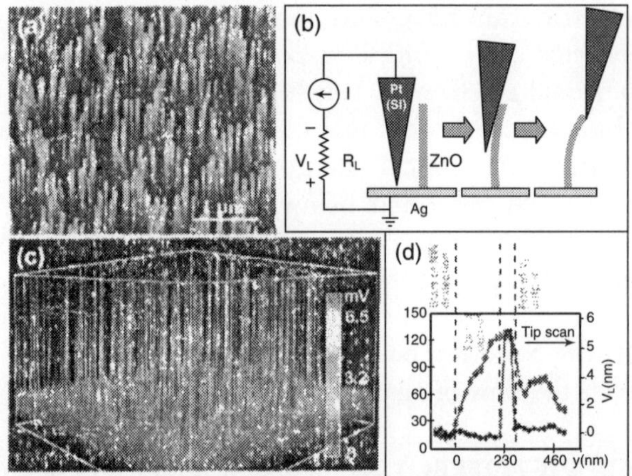

Figure 19.1 *(a) Scanning electron microscopy (SEM) image of aligned ZnO NWs grown on a GaN/sapphire substrate. b) Experimental setup and procedure for generating electricity by deforming a piezoelectric NW using a conductive AFM tip. The AFM scans across the NW arrays in contact mode. c) Output-voltage image of the NW arrays when the AFM tip scans across the NW arrays. d) An overlap plot of the AFM topological image and the corresponding generated voltage for a single scan of the tip across an NW.*

(i) The output potential is a sharp peak that is negative in reference to the grounded end of the NW.

(ii) No output current is received when the tip touches the NW and pushes the NW; electrical output is observed when the tip is about to leave the NW at the second half of the contact.
(iii) The power output occurs only when the tip touches the compressive side of the NW.
(iv) An output signal is observed only for piezoelectric NWs. No electrical output is received if the NWs are tungsten oxide, carbon nanotubes, silicon, or metal. Friction or contact potential play no role in the observed output power.
(v) The magnitude of the output signal is highly sensitive to the size of the NWs.
(vi) To effectively output electricity, the contact between the tip and the NW is required to be Schottky in nature and the contact between the NW and ground is ohmic.

Positive output voltage peaks are observed for CdS NWs under visible light illumination and ZnO NWs coated with a p-type polymer This is due to a change in contact transport characteristic from Schottky to quasi-ohmic.

The piezoelectric discharging; process of a single ZnO wire/belt is based on mechanical manipulation of a single ZnO wire/belt by an atomic force microscope. By use of a long ZnO wire/belt that is large enough to be seen under an optical microscope, one end of the ZnO wire is affixed on a silicon substrate by silver paste, while the, other end is left free. The substrate is an intrinsic silicon, so its conductivity is rather poor. The wire is laid on the substrate but kept a small distance from the substrate to, eliminate the friction with the substrate except at the affixed side (Figure 19.2(a)). Both the topography (feedback-signal from the scanner) and the corresponding output voltage (V) images across a load are recorded simultaneously when the AFM tip is scanned across a wire/belt. The topography image reflects the change in normal force perpendicular to the substrate, e, which shows a bump only when the tip scans over the wire. The output voltage between the conductive tip and the ground is continuously monitored as the tip scans over the wire/ belt. No external voltage is applied. The entire process and the output images are captured by video recording so that the electric generation process is visulazed directly. The output data are obtained from snapshots of the characteristic events observed during the experiments.

The AFM tip scanned line-by-line at a speed of 105.57 μm s^{-1} 'perpendicular to the wire either from above the top end to the lower part of the wire or from the lower part toward the top end. For a wire with hexagonal cross section (Figure 19.2(a)), three characteristics features are observed.

When the tip scans above the top end of the wire without touching the wire but the silicon substrate, the output voltage signal is nothing but noise (Figure 19.2(b)). When the tip scans, until it touches the top end of the wire, a spark output voltage signal is observed (Figure 19.2(c)). The output voltage is negative for the load R_L for all the cases indicates the tip has a lower potential than the ground silver paste. When the tip scans down along the wire, it deflects the wire but does not go over it, and. the output voltage shows no peak but noise (Figure 19.2(d)).

The observation for a large belt is given in Figure 19.3. The SEM images clearly display the rectangular cross section of the belt (Figure 19.3(a)). Hence a strong belt is used. Also an AFM tip pushes perpendicular to the belt at its middle section. When subjected to a displacement force, one side of the belt is stretched (left-hand side) and the other side (the right-hand side) is compressed. The objective is to observe if the piezoelectric discharge occurs when the tip

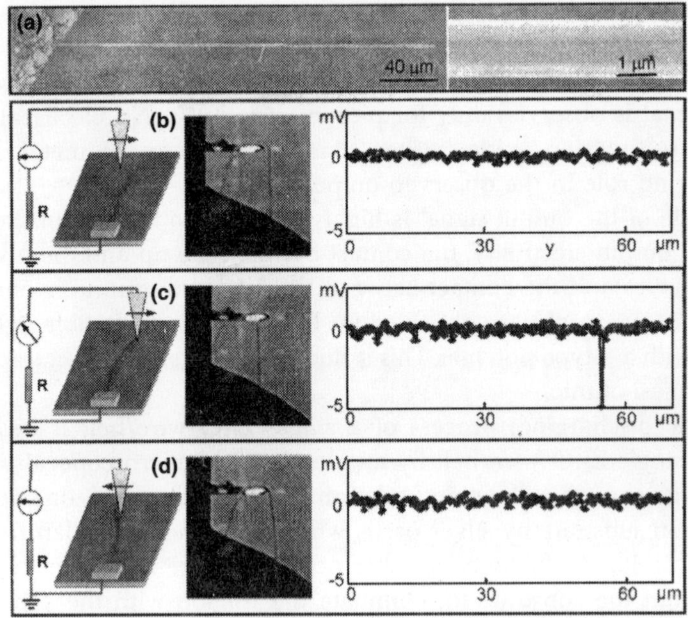

Figure 19.2 *The process for converting mechanical energy into electrical energy by a piezoelectric ZnO wire, (a) SEM images of a ZnO wire with one end affixed by silver paste onto a silicon substrate into the other end free. The wire has a hexagonal cross section. (b, c, and d) Three characteristic snapshots and the corresponding output voltage images when the tip scanned across the wire. The illustration of the experimental condition is shown at the left-hand side, with the scanning direction of the tip indicated by an arrowhead. The scanning speed of the AFM tip was 105.57 µm/s, and the sampling density is 256 per line. (b) The AFM tip scanned over the top of the wire on the silicon substrate, and the voltage output is just in the noise level. (c) The AFM tip touched the top end of the wire, and the output voltage image showed a sharp negative peak. (d) Right after the AFM tip passed the top end of the wire and scanned toward its lower part, but without across the wire. No output voltage is detected.*

touches either the stretched side or the, compressed side or both sides, if the topography image is directly captured (if the tip passes over the belt) or due to a representation of the normal height received by the cantilever. When the tip pushes the wire but does not go over and across it, as shown by the flat output signal in the topography image (Figure 19.3(b)), no voltage output is produced, indicating the stretched side produces no piezoelectric discharge event. Once the tip goes over the belt and is in touch with the compressed side, as indicated by a peak in the topography image, a sharp voltage output peak is observed (Figure 19.3(c)). By analyzing the positions of the peaks observed in the topography image and the output voltage image it is observed that the discharge occurs, after the tip nearly finishing across the wire. This indicates that the compressed side is responsible for producing the negative piezoelectric discharge voltage. For another belt, when the tip retracts from the right hand side to the left hand side (Figure 19.3(d)), no output voltage is detected because the tip just touches the stretched side of the belt without crossing it.

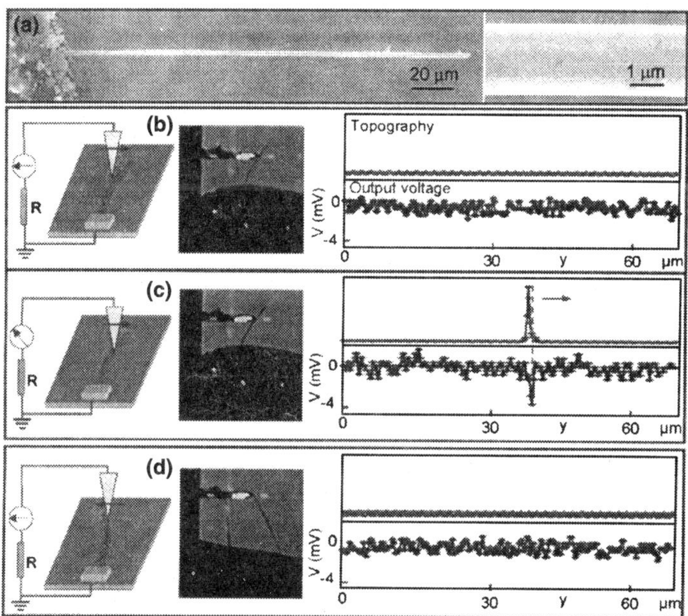

Figure 19.3 *In situ observation of the process for converting mechanical energy into electric energy into electric energy by a piezoelectric ZnO belt. (a) SEM images of a ZnO belt with one end affixed by silver paste onto a silicon substrate and the other end free. The belt has a rectangular cross section. (b, c, and d) Three characteristic snapshots and the corresponding topography (red curve) and output voltage (blue curve) images when the tip scanned across the middle section of the belt. The schematic illustration of the experimental condition is shown at the left - hand side, with the scanning direction of the tip indicated by an arrowhead. The scanning speed of the AFM tip was 105.57 μm/s, and the sampling density is 256 per line. (b) The AFM tip pushed the belt toward the right-hand side but did not go above and across its width, as judged from the topography image. No output voltage was detected. (c) The AFM tip pushed the belt toward the right-hand side and went above and across its width, as judged from the peak in the topography image. The output voltage image showed a sharp negative peak. There is a delay in the output voltage peak in reference to the normal force image (the peak in the topography image). (d) For an alternative belt, the AFM tip pushes the belt toward the left but did not go above and across its width, as judged from the topography image. No output voltage was detected beyond noise level although the deflection was large.*

Thus there are three key results. First, piezoelectric discharge is observed for both wire and belt, and it occurs only when the AFM tip touches the end of the bent wire/belt. Second, the piezoelectric discharge occurs only when the AFM tip touches the compressed side of the wire/belt, and there is no voltage output if the tip touches the stretched side of the wire/belt. Third, the piezoelectric discharge gives negative output voltage as measured from the load R_L. Finally, in reference to the topography image, the voltage output event occurs when the AFM tip is about finished crossing the width of the wire/belt, which means that the discharge event is delayed to the last half of the contact between the tip and the wire/belt.

The Piezoelectric Potential in ZnO NWs and its Experimental Detection

When a ZnO NW is elastically deformed, a piezoelectric potential field is created in the NW. The piezoelectric potential is created by the polarization of ions in the crystal rather than the free-mobile charges. Since the charges associated with the ions are rigid and affixed to the atoms, they cannot freely move. Free carriers in the semiconductor NW, screen the piezoelectric charges, but they cannot completely deplete the charges. This is a distinct difference from the p - n junction in semiconductor physics. Therefore, the piezoelectric potential is still preserved, although a possible reduction in magnitude is possible owing to the finite conductivity of the NW.

For a NW of 50 nm in diameter and 500 nm in length the piezoelectric potential drop across the NW is ~0.6 V if the NW has no conductivity. With consideration of the finite electric conductivity at a moderate carrier density of ~ 10^{16} – $10^{17}cm^{-3}$ which is determined by the synthesis conditions, the free carriers screen the piezoelectric charge but they donot cancel out the piezoelectric charges, which means that magnitude of the piezoelectric potential is reduced by the charge carriers but is large enough to drive the flow of external electrons. Considering the conductivity of ZnO, for a carrier density of 10^{16} – 10^{17} cm^{-3} the magnitude of the piezoelectric potential drop remains at ~ 0.3 V. This is sufficient to operate a Schottky diode. Using a micrometer probe, the piezoelectric. potential at the tensile and compressive side surfaces of a ZnO wire is measured experimentally.

By using a two-end bonded ZnO piezoelectric fine-wire (PFW) (nanowire, microwire) on a flexible polymer substrate, the strain-induced, change in *I-V* transport characteristics from symmetric to diode-type are observed. This phenomenon is due to the asymmetric change in Schottky-barrier heights at both source and drain electrodes, as caused by the strain-induced piezoelectric potential drop along the PFW, which is quantified using the thermionic emission-diffusion theory. The existence of a piezoelectric potential in the ZnO wire is achieved although it has a moderate conductivity. This means that the free carriers partially screen the piezoelectric potential/charges, but they cannot neutralize all of the charge. The existence of the piezoelectric potential not only supports the mechanism proposed for NGs, but are also used to fabricate piezoelectric diode and switch. How long can the piezoelectric potential last. The life time of the piezoelelctric potential is measured by an experiment. Using a long ZnO wire that is bonded at the two ends to metal contacts, the current transported through the wire at a fixed applied external voltage once it is bent, is monitored continuously which means that the wire is under a constant strain. After the wire is bent and held stationary, a trend is seen in the recovery curve of the conductance in the current-time (*I-T*) curve. Since the piezoelectric potential effectively modify the height of the Schottky barrier at the contacts, the change in conductance directly reflects the change in barrier height. Therefore, the conductance of the device is tuned by the piezoelectric potential, which, in reverse, proves the existence of the piezoelectric effect and its life time.

Schottky Barrier at the Electrode - Nanowire Interface

TheSchottky contact between the metal contact and ZnO NW is another key factor affecting the current generation and output process. To examine the role played by the Schottky Barrier in the NG an AFM-based manipulation and measurement system is used for demonstrating the piezoelectric NG (see Fig. 19.4(a)). When a 100 nm thick Pt coated Si tip is used for scanning the NWs in contact mode, voltage peaks are observed (Fig. 19.4b),. and the output voltage is in the order of −11 m V (the negative sign means that the current flows from the grounded end through the external load). By changing the tip to an Al - In (30 nm/30 nm) alloy- coated Si tip, the ZnO NWs show no piezoelectric output (Fig. 19.4(c)). To understand these two distinct performances, corresponding *I-V* characteristics are measured with the ZnO NWs. The stability of the contact is achieved with a large electrode that is in contact with a group of NWS, as shown in Figure 19.4(d), The Pt – ZnO contact presents a Schottky diode (Fig. 19.4(e), while the AlIn – ZnO is an ohmic contact. Figure 19.4(b) and (c), shows that a Schottky contact between the metal electrode and ZnO is necessary for a working NG.

By tuning the height of the Schottky barrier and the density of charge carriers in ZnO using UV illumination, the output of an NG gets changed. No voltage/current is generated if there is no Schottky barrier at the tip/electrode ZnO interface.

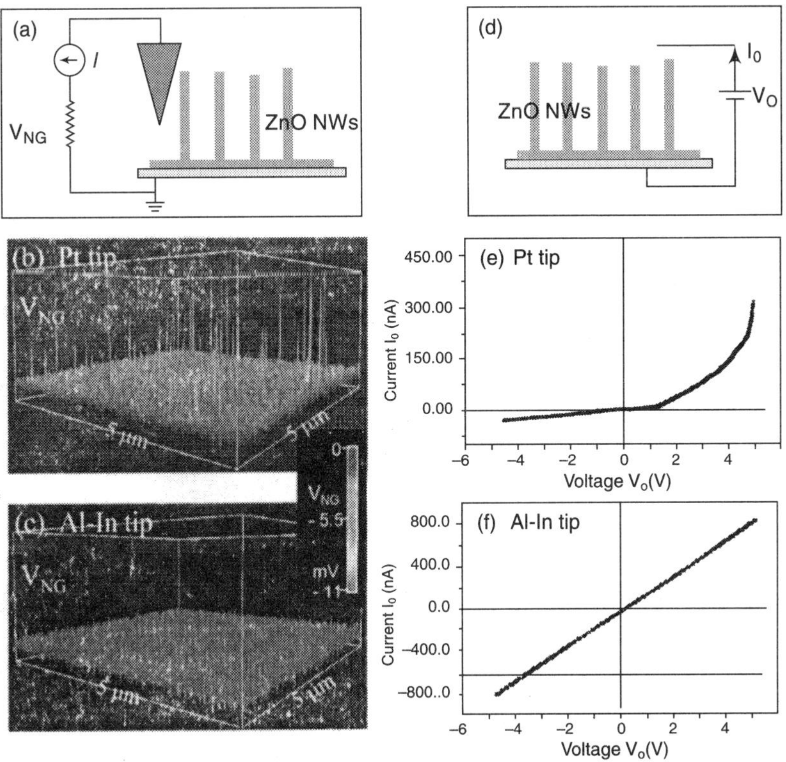

Figure 19.4 *(a) AFM-based measurement setup for correlating the relationship between the metal ZnO contact and the NG output. b) Output potential generated by ZnO NW array when scanned by a Pt-coated Si tip. c) No output potential is generated by ZnO NW array when scanned by an Al-In-coated Si tip. d) Experimental setup for characterizing the I-V characteristics of a metal-. ZnO NW contact. e) I-V curve of a Pt-ZnO NW contact, showing the Schottky diode effect. f) I-V curve of an alloyed AlIn-ZnO NW contact, showing ohmic behavior.*

PIEZOELECTRIC POTENTIAL OUTPUT FORM ZNO NANOWIRE

(a) The Charge Generation and Output Process of a Piezoelectric Nanowire

The physical principle for creating, separating, accumulating, and outputting the piezoelectric potential in the NW is a coupling of the piezoelectric and semiconducting properties the material. For a vertically straight ZnO NW, the deflection of the NW by the AFM tip creates a strain field, with the outer surface being tensile (positive strain ε) and the inner surface compressive (negative strain ε). The asymmetric strain introduces an asymmetric potential at the two surfaces, with the compressed surface having negative V $(=\varphi^C)$ and the tensile surface positive $V^+(=\varphi^T)$.

Now consider the power-output process. Firstly, the AFM conductive tip that induces the deformation is in contact with the tensile surface of positive potential V^+ (Fig. 19.5(a)). The Pt metal tip has a potential of nearly zero, $V_m = 0$, so the metal tip-ZnO interface is negatively biased for : $\Delta V = (V_m - V^+) < 0$. By consideration of the n-type semi- conductor characteristic of the as-synthesized ZnO NWs, the Pt metal-ZnO semiconductor (M – S) interface, is a reverse-biased Schottky diode, and little current flows across the interface. This is the process of creating, separating, and accumulating the piezoelectric charges. Secondly, when the AFM tip is in contact with the compressed side of the NW (Fig. 19.5(b)), the metal tip-ZnO interface is positively biased for $\Delta V = V_L = (V_m - V^-) > 0$. The M - S interface is a positvely biased Schottky diode, and if produces a Sudden increase in the output electric current. The current is the result of ΔV driven flow of electrons from the semiconducting ZnO NW to the metal tip. This is the power-output process:

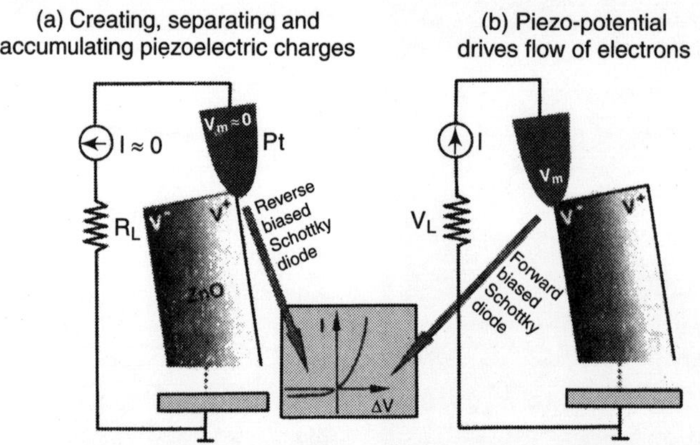

Figure 19.5 *Principle of the observed power-generation process of a piezoelectric ZnO NW, showing a unique coupling of piezoelectric and semiconducting properties in this metal semiconductor Schottky-barrier-governed transport process. (a,b) Metal and semiconductor contacts between the AFM tip and the semiconductor ZnO NW at two reversed local contact potentials (positive and negative), showing reverse- and forward-biased Schottky rectifying behavior, respectively. The Schottky barrier is responsible for separating, accumulating, and later releasing the charges.*

Also, the piezoelectric is created by the polarization of ions in the crystal rather than the free, mobile charges. Since the charges associated with the ions are rigid and affixed to the atoms, they cannot freely move. Free carriers in the semiconductor NW screen the piezoelectric charges, but they do not completely deplete the charges. This is a distinct difference from the p-n junction in semiconductors. Therefore, the piezoelectric potential is preserved, although a possible reduction in magnitude is possible owing to the finite conductivity of the NW.

An as-synthesized ZnO NW is n-type. The presence of oxygen vacancies and impurities and a large portion of surface atoms (surface states) introduce a moderate conductivity to the NW. These free carrier screen the piezoelectric charges, but they do not neutralize the charges. Therefore, the piezoelectric potential is preserved, at a reduced magnitude, even when the moderate conductivity of ZnO is considered.

An explanation of the charge-releasing process for an NG based on the band structure model is given below. The AFM tip (T) has a Schottky contact (barrier height Φ_{SB}) with the NW, while the NW has an Ohmic contact with the grounded side (G) (Fig. 19.6(a)). When the tip slowly pushes the NW, a positive piezoelectric potential V^{+} is created at its tensile surface. As the tip continues to push the NW, electrons slowly flow from the grounded electrode through the external load to reach the tip, but the electrons cannot pass across the tip-NW interface because of the presence of a reverse-biased Schottky barrier at the contact (Fig. 19.6(b)). In such a case, the accumulated free charges at the tip affects the piezoelectric potential distribution in the NW because of the screening effect of the charge carriers. The piezoelectric potential is generated as a result of the rigid and non - mobile ionic chares in the NW; the potential completely depleted by the free carriers . The local newly established potential V'^{+} lowers the conduction band (CB) slightly.

When the tip scans in contact mode across the NW and reaches the middle point of the. NW (see Fig. 19.6(c)), the local piezoelectric potential is zero. In such a case, with sudden drop in local potential, the originally accumulated electrons in the tip flow back through the load to the ground. This process is faster than the charge-accumulation process presented in Fig. 19.6b. An alternative but, similar result is that the tip temporarily lifts off the NW. which also leads to the back flow of the accumulated electrons to the ground. When the tip reaches the compressive side of the NW (Fig. 19.6(d)), the local potential drops to V''^{-} (negative), which results in the raising of the conduction band near the tip. If the raise in local potential energy is large enough to: overcome the Schottky barrier threshold at forward bias, as determined by the degree of NW bending, the local accumulated n-type carriers in the NW can quickly flow through the contact to the tip, which creates a circular flow of the electrons in the external circuit, thus producing a current. This process is a lot faster than the charge-accumulation process; therefore the created transient potential at the external load is large enough to be detected beyond the noise level. The presence of a Schottky contact at the tip-NW interface is mandatory for an NG, which acts like a "gate" for, separating and slowly accumulating the charges and then quickly releasing them. The absence of a "gate" at an Ohmic contact results in no charge accumulation or release; thus, no detectable signal is received.

Now how large is the output voltage? This question is answered by the energy-band diagram shown in Figure 19.6(e) for the NG. The role played by the piezoelectric potential is to drive the electrons from thc ZnO NW, overcoming the threshold energy at the metal –ZnO interface,

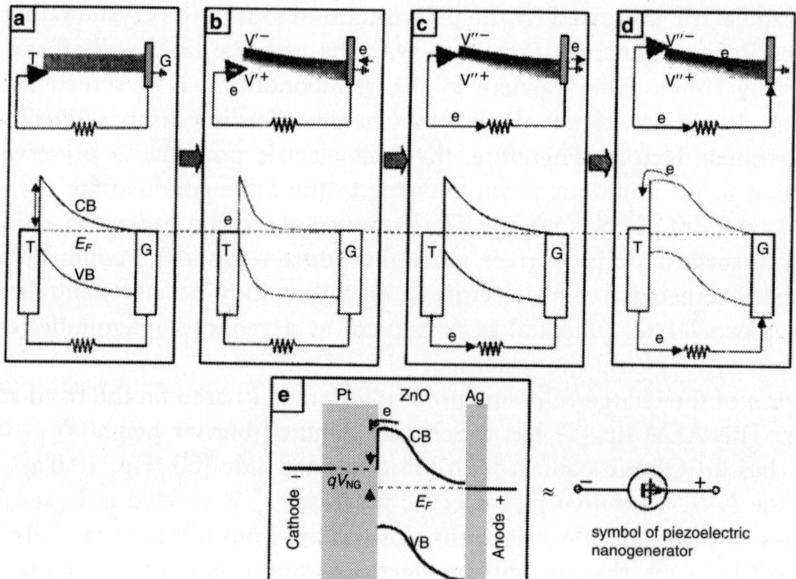

Figure 19.6 *Band diagram for charge output and flow processes in a nanogenerator.*

into the Pt electrode but it does not directly determine the, magnitude of the output voltage. As more electrons are "pumped' into the Pt electrode, the local Fermi surface is suddenly raised. Therefore, the output voltage is the difference between the Fermi energies of Pt and the bottom Ag electrode (5 - 30 mV depending on NW size and degree of bending). The measured voltage at load is considered for contact resistance.

Although the total amount of charge Q generated by a single NW is small (1000 - 10000 electrons), quick release of these electrons produces a significantly large electric current/voltage pulse, because $V_L \approx R_L\, Q / \Delta\, t$, where $\Delta.t$ is the time interval for the charge releasing process. A calculation of the output voltage by the capacitance C of the NWs and system ($V = Q/C$) less than the magnitude of the pulse.

Thus the charge – flowing processes shown in Figure 19.6, the first process is a forward and back flow of the electrons through the load from the ground to the tip; the second process is a piezoelectric – potential - driven circular flow of the electrons through the NW. The two processes generate constructive currents flowing in the same direction. resulting in a negative voltage at the load in reference to the grounded electrode. This is an important characteristic: The- output current is significant large if the charge-releasing process is fast. In the case of the AFM tip contact, the second process is likely the dominant process. The entire process of energy generation in one sentence is the piezoelectric - potential-driven flow of external electrons is the power-output process of the nanogenerator.

Piezoelectric Potential at Surface of a Bent Nanowire (Quantitative Analysis)

To have a better understanding about piezoelectric voltage output from a single NW, the magnitude of the potential created in an NW when it is subject to mechanical deformation is calculated.

The theoretical background for the nanogenerator and nanopiezotronics is based on a voltage drop created across the cross section of the NW (as described above) when it is laterally deflected, with the tensile side surface in positive voltage and compressive side in negative voltage. It is essential to quantitatively calculate and even develop analytical equations that gives a direct calculation of the voltage at the two side surfaces, of the NW, which is important for calculating the efficiency, of the nanogenerator and the operation voltage of the nanopiezotronics. Numerous theories for one - dimensional (1 D) nanostructure piezoelectricity are proposed, including first-principles calculations, molecular dynamics (MD) simulations, and continuum models. However, first - principle theory and MD simulatior are difficult to be implemented to the nanopiezotronics system. (the typical dimension of which is ~50 nm in diameter and ~2 μm in length) due to the huge Number of atoms. The continuum model is used as it gives a criterion to distinguish between a mechanically dominated regime and an electrostatically dominated regime. As piezotronics systems involve lateral bending, the strain across the NW cross section is observed. So, a continuum model for the electrostatic potential in a laterally bent NW is considered. A technique is introduced to solve the coupled differential equations, and the derived analytical equation gives a result within 6% of full numerical calculation. The theory establishes the physical basis of nanopiezotronics and nanogenerator.

(i) **System Setup and Governing Equations** The typical A continuum model is used for the electrostatic potential in laterally bent NW. The typical setup of a vertical piezoelectric nanowire nanogenerator is shown in Figure 19.7. The NW is pushed laterally by an AFM tip. Piezoelectric potential/field is created by the NW. The objective is to derive a relationship between the potential distribution in the NW and the dimensionality of the NW and magnitude of the applied force at the tip.

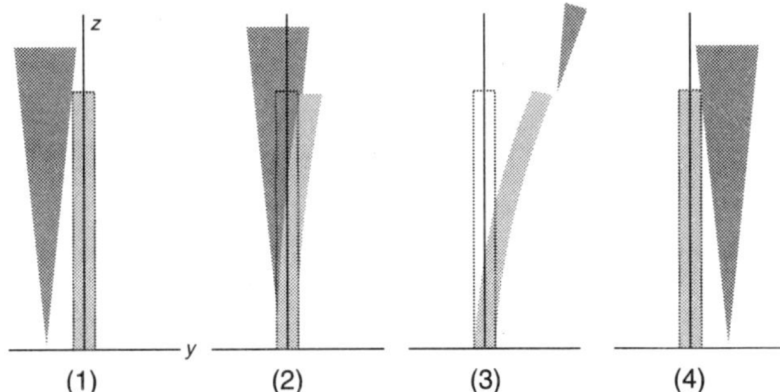

Figure 19.7 *The typical configuration of a ZnO nanogenerator. When pushed by an AFM tip, mechanical det1ection gives rise to an electrical field, the power of which can be released. The pushing force increases from (1) to (3), until (4) the wire is released.*

The governing equations for a static piezoelectric material, are three sets: mechanical equilibrium equation (equation 1), constitutive equation , (equation.2), geometrical compatibility equation (equation.3), and Gauss equation of electric field (equation.4). The mechanical equilibrium condition, when there is no body force $(\vec{f}_e^{(b)}=0)$ acting on the nanowire, is

$$\nabla.\sigma= \vec{f}_e^{(b)}=0 \tag{1}$$

where σ is the stress tensor, which is related to strain $\in$, electric field $\vec{E}$, and electric displacement $\vec{D}$ by constitutive equations

$$\begin{cases} \sigma_p = c_{pq}\in_q - e_{kp}E_k \\ D_i = e_{iq}\in_q + \kappa_{ik}E_k \end{cases} \tag{2}$$

where C_{pq} is the linear elastic constant, e_{kp} is the linear piezoelectric coefficient, and κ_{ik} is the dielectric constant. Here, equation 2 does not contain the contribution from the spontaneous polarization introduced by the polar charge on the ± (0001) polar surfaces, which are the top and bottom ends of the NW, respectively. By considering the C_{6v} symmetry of a ZnO crystal (with wurtzite structure), C_{pq}, e_{kp} and κ_{ik} are written as

$$c_{pq} = \begin{pmatrix} c_{11} & c_{12} & c_{13} & 0 & 0 & 0 \\ c_{12} & c_{11} & c_{13} & 0 & 0 & 0 \\ c_{13} & c_{13} & c_{33} & 0 & 0 & 0 \\ 0 & 0 & 0 & c_{44} & 0 & 0 \\ 0 & 0 & 0 & 0 & c_{44} & 0 \\ 0 & 0 & 0 & 0 & 0 & \frac{(c_{11}-c_{12})}{2} \end{pmatrix} \tag{2.1}$$

$$e_{kp} = \begin{pmatrix} 0 & 0 & 0 & 0 & e_{15} & 0 \\ 0 & 0 & 0 & e_{15} & 0 & 0 \\ e_{31} & e_{31} & e_{33} & 0 & 0 & 0 \end{pmatrix} \tag{2.2}$$

$$\kappa_{ik} = \begin{pmatrix} \kappa_{11} & 0 & 0 \\ 0 & \kappa_{11} & 0 \\ 0 & 0 & \kappa_{33} \end{pmatrix} \tag{2.3}$$

The compatibility equation is a geometrical constraint that must be satisfied by strain $\in_{ij}$

$$e_{ilm}e_{jpq}\frac{\partial^2 \in_{mp}}{\partial x_l \partial x_q} \tag{3}$$

In equation 3 the indices are in the normal definition e_{ilm} and e_{jpq} are antisymmetric tensors. For simplicity of the derivation, assum that the NW bending is small.
Finally, by assuming no free charge $\rho_e^{(b)}$ in the nanowire, the Gauss equation must be satisfied

$$\nabla \cdot \vec{D} = \rho_e^{(b)} = 0 \tag{4}$$

(ii) Theory for the First Three Orders of Electromechanical Coupling.

Equaltion 1 - 4 along with appropriate boundary conditions give a complete description of a static piezoelectric system. However, the solution of these equations is complex, and analytical solution does not exist in many cases. Even for a two-dimensional (2D) system the solution has a partial differential equation of order 6. In order to derive an approximate solution of the equations, is done to simplify the analytical solution. Then, the accuracy of the perturbation theory is examined in reference to the exact results calculated by the finite element method.

In order to derive the piezoelectric potential distributed in the NW for the different orders of electromechanical coupled, effect, a perturbation parameter λ is introduced in the constitutive equations by defining $\tilde{e}_{kp} = \lambda e_{kp}$, which is introduced to trace the magnitudes of contributions made by different orders of effects in building the total potential. Consider a virtual material with linear elastic constant C_{pq}, dielectric constant k_{ik}, and piezoelectric coefficient $\tilde{e}_{kp}$. When $\lambda = 1$, this virtual material becomes the realistic ZnO. When $\lambda = 0$, it corresponds to a situation of no coupling between mechanical field and electric field. For virtual materials with λ between 0 and 1, the mechanical field and electric field are both functions of parameter λ, are written in an expansion form

$$\sigma_p(\lambda) = \sum_{n=0}^{\infty} \lambda^n \sigma_P^{(n)} \; ; \; \epsilon_q(\lambda) = \sum_{n=0}^{\infty} \lambda^n \epsilon_q^{(n)} \; ; \; E_k(\lambda) = \sum_{n=0}^{\infty} \lambda^n E_k^{(n)} \; ; \; D_i(\lambda) = \sum_{n=0}^{\infty} \lambda^n D_i^{(n)} \tag{5}$$

where the superscript (n) represents the orders of perturbation results. By substituting equation 5 into equation 2 for a virtue material with piezoelectric coefficient $\tilde{e}_{kp}$, and comparing the terms in the equations that have the same order of λ, the first three orders of perturbation equations are given as follows:

zeroth order:

$$\begin{cases} \sigma_p^{(0)} = c_{pq}\epsilon_q^{(0)} \\ D_i^{(0)} = \kappa_{ik} E_k^{(0)} \end{cases} \tag{6}$$

first order:

$$\begin{cases} \sigma_p^{(1)} = c_{pq}\epsilon_q^{(1)} - e_{kp} E_k^{(0)} \\ D_i^{(1)} = e_{pq}\varepsilon_q^{(0)} + \kappa_{ik} E_k^{(1)} \end{cases} \tag{7}$$

second order:

$$\begin{cases} \sigma_p^{(2)} = c_{pq}\epsilon_q^{(2)} - e_{kp} E_k^{(1)} \\ D_i^{(2)} = e_{kq}\varepsilon_q^{(1)} + \kappa_{ik} E_k^{(2)} \end{cases} \tag{8}$$

For equation 1, 3, and 4, since there is no explicit coupling, no decoupling process is needed while seeking perturbation solution.

Now consider the solutions of the first three orders. For the zeroth order equation (6), the solution is for a bent nanowire without piezoelectric effect, which means there is no electric field even with the presence of elastic strain. For the case of ZnO NW, it normally grows with its *c*-axis parallel to the growth direction. The ± (0001) surfaces at the top and bottom end of the NW are terminated by Zn^{2+} and O^{2-} ions, respectively. The electric field due to spontaneous polarization arising from polar charges on the ± (0001) surface is ignored for the following two reasons. First, since the NW has a large aspect ratio, the polar charges on the ±(0001) polar surfaces, which are the top and bottom ends of the NW in most of the cases, are viewed as two point charges. They do not introduce an appreciable intrinsic field inside of the NW. Second, the polar charges at the bottom end of the NW are neutralized by the conductive electrode, while the ones at the top of the NW are neutralized by surface adsorbed foreign molecules while exposed to air. Even if the polar charges at the top end introduce a static potential, it does not contribute to the power generated but shifts the potential baseline by a constant value, which goes to the background signal, because the polar charges are present and remain constant regardless the degree of NW bending. Therefore, take $E_k^{(0)} = 0$, $D_i^{(0)} = 0$, So from equation.7 and 8, $\sigma_p^{(1)} = 0$, $\epsilon_p^{(1)} = 0$, $D_p^{(2)} = 0$, $E_p^{(2)} = 0$. Equations 6-7 thus become zeroth order:

$$\sigma_p^{(0)} = c_{pq}\epsilon_q^{(0)} \tag{9}$$

first order:

$$D_i^{(1)} = e_{kq}\epsilon_q^{(0)} + \kappa_{ik}E_k^{(1)} \tag{10}$$

second order:

$$\sigma_p^{(2)} = c_{pq}\epsilon_q^{(2)} - e_{kp}E_k^{(1)} \tag{11}$$

The physical meaning of these equations is explained as follows, Under the different orders of approximation, these equations correspond to the decoupling and coupling between the electric field and mechanical deformation: the zeroth order solution is purely mechanical deformation without piezoelectricity; the first- order is the result of direct piezoelectric effect that strain-stress generates an electric field in the NW; and the second order shows up the first feedback (or coupling) of the piezoelectric field to the strain in the material.

As the nanowires are bent by AFM tip, the mechanical deformation behavior of the material is unaffected by piezoelectric field in the NW. Therefore, for the calculation of piezoelectric potential in the nanowire, the first-order approximation is sufficient. The accuracy of this approximation is examined in reference to full numerical solutions of the coupled equation 1 - 4.

(iii) Analytical Solution (under the Saint-Venant Approximation). To simplify the analytical solution, assume that the nanowire has a cylindrical shape with a uniform cross section of diameter $2a$ and length l. Let the Young's modulus be E and Poisson ratio ν and $a_{pq}^{isotropic}$ to be the inverse of matrix $c_{pq}^{isotropic}$. Therefore, the stress and strain relation is given by

$$\begin{vmatrix} \epsilon_{xx}^{(0)} \\ \epsilon_{yy}^{(0)} \\ \epsilon_{zz}^{(0)} \\ 2\epsilon_{yz}^{(0)} \\ 2\epsilon_{zx}^{(0)} \\ 2\epsilon_{xy}^{(0)} \end{vmatrix} = \sum_q a_{pq}^{isotropic} \sigma_q^{(0)} = \frac{1}{E} \begin{pmatrix} 1 & -\nu & -\nu & 0 & 0 & 0 \\ -\nu & 1 & -\nu & 0 & 0 & 0 \\ -\nu & -\nu & 1 & 0 & 0 & 0 \\ 0 & 0 & 0 & 2(1+\nu) & 0 & 0 \\ 0 & 0 & 0 & 0 & 2(1+\nu) & 0 \\ 0 & 0 & 0 & 0 & 0 & 2(1+\nu) \end{pmatrix} \begin{vmatrix} \sigma_{xx}^{(0)} \\ \sigma_{yy}^{(0)} \\ \sigma_{zz}^{(0)} \\ \sigma_{yz}^{(0)} \\ \sigma_{zx}^{(0)} \\ \sigma_{xy}^{(0)} \end{vmatrix} \quad (12)$$

In the configuration of the nanogenerator, the root end of the nanowire is affixed to a conductive substrate. while the top end is pushed by a lateral force f_y. Now, assume that the force, f_y is applied uniformly on the top surface so that there is no effective torque that twists the nanowire Thus, by Saint – Venant theory of bending, the stress induced in the nanowire is given by

$$\sigma_{xz}^{(0)} = -\frac{f_y}{4I_{xx}} \frac{1+2\nu}{1+\nu} xy \quad (13.1)$$

$$\sigma_{yz}^{(0)} = \frac{f_y}{I_{xx}} \frac{3+2\nu}{8(1+\nu)} \left[a^2 - y^2 - \frac{1-2\nu}{3+2\nu} x^2 \right] \quad (13.2)$$

$$\sigma_{zz}^{(0)} = -\frac{f_y}{I_{xx}} y(l-z) \quad (13.3)$$

$$\sigma_{xx}^{(0)} = \sigma_{xy}^{(0)} = \sigma_{yy}^{(0)} = 0 \quad (13.4)$$

where

$$I_{xx} = I_{yy} = \frac{\pi}{4} a^4$$

Equation 13 is the zeroth order mechanical solution to equation 1, 3, and 12. Because the Saint-Venant principle is used to simplify the boundary condition, solution equation 13 is valid only for regions far away (i.e >> NW diameter) from the affixed end of the nanowire.Full numerical calculation show that it is safe to use equation 13 when the distance from the substrate is larger than twice the NW diameter.

Equations 4 and 10 give the direct piezoelectric behavior. By defining a remnant displacement $\vec{D}^R$ as

$$\vec{D}^R = e_{kp} \varepsilon_q^{(0)} \hat{i}_k \quad (14)$$

$$\therefore \quad \frac{\partial}{\partial x_i} (D_i^R + \kappa_{ik} E_k^{(1)}) = 0 \quad (15)$$

From eqs 14, 13, 12, and 2.2, the remnant displacement is

$$\vec{D}^R = \begin{pmatrix} -\dfrac{f_y}{I_{xx}E}\left(\dfrac{1}{2}+\nu\right)e_{15}xy \\ \dfrac{f_y}{I_{xx}E}\left(\dfrac{3}{4}+\dfrac{\nu}{2}\right)e_{15}\left(a^2-y^2-\dfrac{1-2\nu}{3+2\nu}x^2\right) \\ \dfrac{f_y}{I_{xx}E}(2\nu e_{31}-e_{33})y(l-z) \end{pmatrix} \tag{16}$$

Note that, it is the divergence of $\vec{D}^R$ rather than $\vec{D}^R$ itself that induces $E_k^{(1)}$. By assuming $E_k^{(1)}$ $= (\kappa_{ik})^{-1} D_i^R$ one arrives at an absurd electric field with nonzero curl. Instead, by defining a remnant body charge

$$\rho^R = -\nabla \cdot \vec{D}^R \tag{17}$$

and remnant surface charge

$$\Sigma^R = \vec{n}\cdot(0-\vec{D}^R) = \vec{n}\cdot\vec{D}^R \tag{18}$$

Equation 15 is transformed into an elementary electrostatic problem with the Poisson equation

$$\nabla \cdot (\kappa_{ik}E_k^{(1)}\hat{i}_i) = \rho^R \tag{19}$$

with charge give by equation 18 on the cylindrical surface of the nanowire. Equation 17 and 16, gives;

$$\rho^R = \frac{f_y}{I_{xx}E}[2(1+\nu)e_{15} + 2\nu e_{31} - e_{33}]y \tag{20}$$

$$\sigma^R = 0 \tag{21}$$

It is very important to note that, in eqs 20 and 21, the remnant charge is independent of vertical height z. Therefore, electric potential $\varphi = \varphi(x,y) = \varphi(r,\theta)$ (in cylindrical coordinate) is also independent of z (the superscript $^{(1)}$ is dropped here for the first order approximation for simplicity). Physically, it suggests that the potential is uniform along the z direction except for regions very close to the ends of the nanowire. This means that the nanowire is approximately like a "parallel plate capacitor" Noting that $\kappa_{11} = \kappa_{22} = \kappa_\perp$ the solution of equations 19, 20, and 21 is

$$\varphi = \begin{cases} \dfrac{1}{8\kappa_\perp}\dfrac{f_y}{I_{xx}E}[2(1+\nu)e_{15}+2\nu e_{31}-e_{33}]\left[\dfrac{\kappa_0+3\kappa_\perp}{\kappa_0+\kappa_\perp}\dfrac{r}{a}-\dfrac{r^3}{a^3}\right]a^3\sin\theta, & r<a \\ \dfrac{1}{8\kappa_\perp}\dfrac{f_y}{I_{xx}E}[2(1+\nu)e_{15}+2\nu e_{31}-e_{33}]\left[\dfrac{2\kappa_\perp}{\kappa_0+\kappa_\perp}\dfrac{a}{r}\right]a^3\sin\theta, & r\ge a \end{cases} \tag{22}$$

where κ_0 is the permittivity in vacuum $\kappa_\perp$. Equation 22 is the potential inside and outside the NW.

From equation 22, the maximum potential at the surface ($r = a$) of the NW at the tensile (T) side ($\theta = -90°$) and the compressive (C) side ($\theta = 90°$), respectively, being

$$\varphi_{\max}^{(T,C)} = \pm \frac{1}{\pi} \frac{1}{\kappa_0 + \kappa_\perp} \frac{f_y}{E} [e_{33} - 2(1+\nu)e_{15} - 2\nu e_{31}] \frac{1}{a} \quad (23)$$

By elementary elastic theory, under small detlection, the lateral force f_y is related to the maximum deflection of the NW tip $\nu_{\max} = (z = l)$ by.

$$\nu_{\max} = \frac{f_y l^3}{3EI_{xx}} \quad (24)$$

Thus the maximum potential at the surface of the NW is

$$\varphi_{\max}^{(T,C)} = \pm \frac{3}{4(\kappa_0 + \kappa_\perp)} [e_{33} - 2(1+\nu)e_{15} - 2\nu e_{31}] \frac{a^3}{l^3} \nu_{\max} \quad (25)$$

where $\nu_{\max}$ is the maximum deflection of the NW at the tip. The maximum potential at the surface of the NW is directly proportional to the lateral displacement of the NW and inversely proportional to the cubic square of its length-to-diameter aspect ratio.

This means that the electrostatic potential is directly related to the aspect ratio of the NW instead of its dimensionality. For a NW with a fixed aspect ratio, the piezoelectric potential is proportional to the maximum deflection at the tip.

(iv) Numerical Results and Discussion. the anisotropic elastic constant C_{pq} by an isotropic one $c_{pq}^{isotropic} = (a_{pq}^{isotropic})^{-1}$. Representing E as elastic modulus, the validity of this approximation, is examined by error introduced in this approximation as;

$$\left|\frac{\Delta c}{c}\right| = \frac{\sum_{p,q} (c_{pq} - c_{pq}^{isotropic})^2}{\sum_{p,q} c_{pq}^2}$$

For bulk ZnO $c_{11} = 207$ GPa, $c_{12} = 117.7$ GPa, $c_{13} = 10.6.1$ GPa, $c_{33} = 209.5$ Gpa, $c_{44} = 44.8$ GPa, and $c_{55} = 44.6$ GPa, the approximating constants are $E = 129.0$ GPa and $\nu = 0.349$, with minimized error of $\sqrt{|\Delta c/c|^2} = 3.27\%$ .Therefore, the approximation is excellent. the relative dielectric constants are $\kappa_\perp^r = 7.77$ and $\kappa_\parallel^r = 8.19$ for bulk ZnO, and the piezoelectric constants are $e_{31} = -0.51\,C.m^{-2}$, $e_{33} = 1.22\,Cm^{-2}$ measured for film. The calculation based on equation 22 is calculated for two configurations: the first one is a *c*-axis vertically aligned ZnO nanowire with smaller diameter grown by vapor –liquid-solid process; the second is a large size ZnO nanowire typically grown by chemical approach. In both cases, the wires are well fixed on the substrate and pushed laterally by an AFM tip at the top.

In the first case, the wire diameter is $d = 50$ nm, length is $l = 600$ nrn. and lateral force applied by AFM tip is 80 nN.

Therefore the magnitude of piezoelectric potential is 0.3V at the stretched surface and is – 0.3 at the compressive surface (Fig. 8).

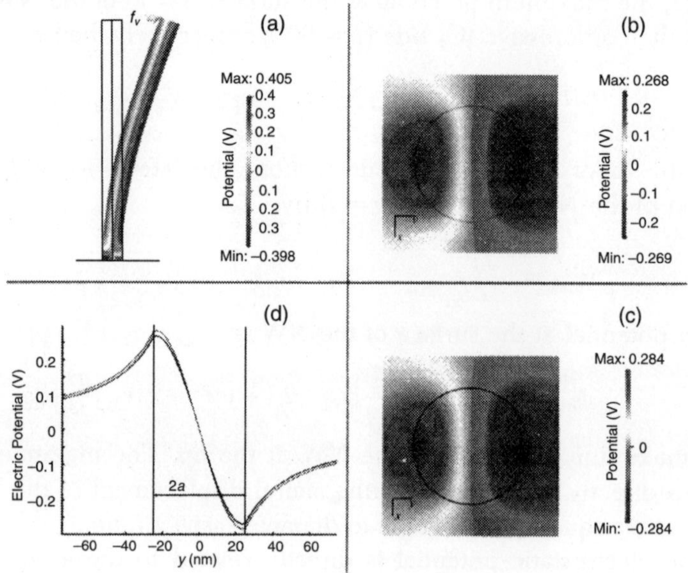

Figure 19.8 *Potential distribution for a ZnO nanowire with d = 50 nm and l = 600 nm at a lateral bending force of 80 nN. (a) and (b) are side and top cross-sectional (at z_0 = 300 nm) output of the piezoelectric potential in the NW (c) is the cross-sectional output of the piezoelectric potential. The maximum potential in (b) is smaller than that in (a), because here the potential in the bottom reverse region is larger than that in upper "parallel-plate capacitor" regions. (d) gives a comparison of the line scan profiles from both (b) and (c).*

To further confirm the validity of omitting the higher order terms in analytical derivations. a finite element method (FEM) calculation is done for a fully coupled electromechanical system using equation 1 - 4 with for a simplified, medium of isotropic elastic modulus tensor and with cylindrical geometry. The boundary condition assumes that the bottom end of the NW is affixed. ZnO is considered as a dielectric medium, Figure 19.8(a) and (b) shows the potential distributions calculated by full FEM in the bent NW as viewed from side and in cross section, respectively, clearly presenting the "parallel plate capacitor" model of the piezoelectric potential except at the bottom i.e. it does not depend on the z-coordinate along the NW unless very close to either end. As for the nanogenerator and nanopiezotronics, only the potential distribution in the upper body of the NW matters. By use of the analytical. equation 22, the calculated potential distribution across the NW cross section for a lateral defection of 145 nm (produced by 80 nN deflection force) is shown in Figure 19.8(c), with the two side surface having ± 0.28 V piezoelectric potential, respectively .

To compare the identity of the FEM result with the analytical result; a line scan is made across the output of the cross-section potential along the symmetry line following the lateral defl ection direction, and the results are directly compared in Figure 19.8(d) The difference is smaller than 6%,thus proving the accuracy of analytical solution.

As equation 22 is based on the Saint-Venant approximation and is therefore not valid in the bottom region of the NW. However, from the FEM solution the bottom reverse region is small and extends only for less than twice the nanowire diameter. Therefore, equation 22 is useful even for nanowires with rather small aspect ratio.

A similar calculation is done for a large size NW with $d = 300$ nm, length $l = 2$ μm, and a lateral force of 1000 nN (see Fig. 19.9). The value of pushing force is estimated based on the lateral deflection observed. This large NW gives a potential distribution of ± 0.59 V across its cross section. The analytical solution is within 6% of the FEM full numerical calculation, thereby proving the validity of the analytical solution. Therefore, the perturbation theory presented above is an excellent representation for calculating the piezoelectric potential across a NW.

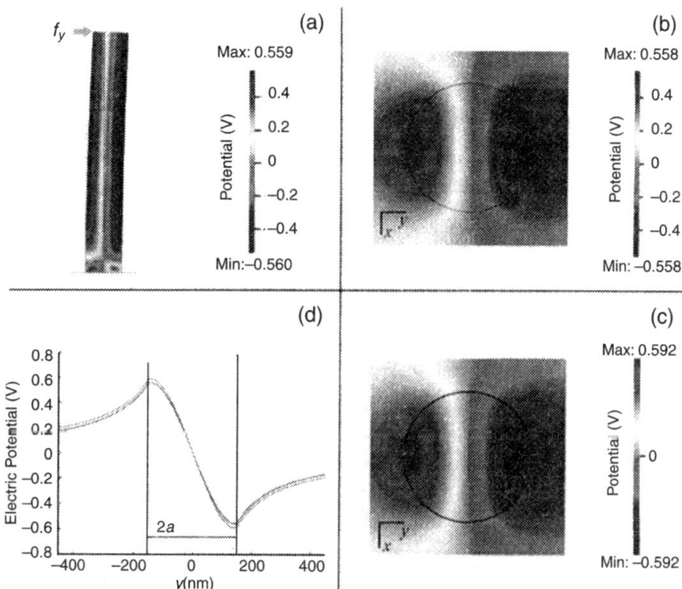

Figure 12.9 *Potential distribution for a nanowire with d = 300 nm and , = 2 m m at a lateral bending force of 1000 nN. (a) and (b) are side and top cross-sectional (at $z_0 = 1$ μm) output of the piezoelectric potential in the NW c) is the cross-sectional output of the piezoelectric potential. The maximum potential in b is almost the same as that in (a), (d) gives a comparison of the line scan profiles from both (b) and c).*

Note that the elastic modulus and piezoelectric coefficient used for the calculations are adopted from the values measured from bulk and film. The measurements show a reduction in elastic modulus for ZnO nanowire and an increased piezoelectric coefficient (by 200%). If these values are used in above calculations, the potential on the NW surface increases by a factor of 3 - 4.

The voltages at the stretched and compressed sides of the, NW are measured by using a metal tip that is placed either at the tensile side or the compressive side of the NW, while the NW is deflected by air blowing. When a periodic gas flow pulse is applied to a ZnO win:, the wire is bends and a corresponding periodic negative voltage output (Fig. 19.10(a)) is detected by connecting the surface of the compressed side of the NW to an external measurement

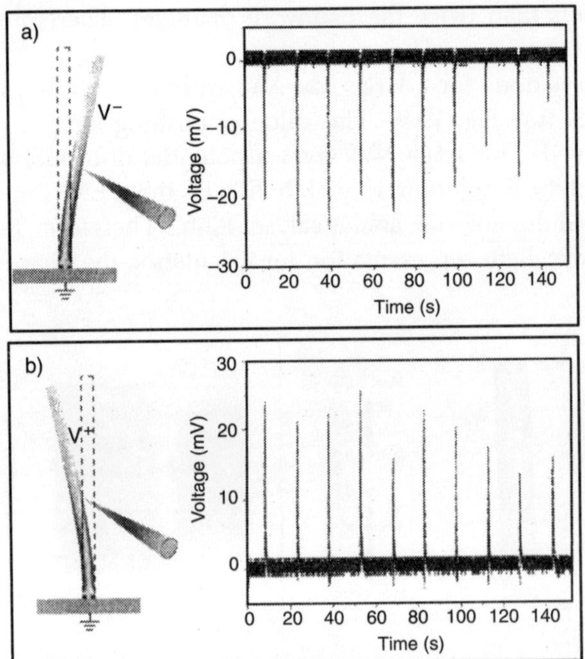

Figure 19.10 *Direct measurements of the asymmetric voltage distribution on the tensile and compressive side surfaces of a ZnO wire. a) By placing a metal tip at the right-hand side and blowing Ar pules at the left-hand side, negative, voltage peaks of approximately 25 mV are observed once the pulse is on. b) By quickly pushing and releasing the wire at the right-hand side by a metal tip, a positive voltage peak of approximately 25 mV is observed for each cycle of the deflection. The frequency of the deflection is once every 15 s.*

circuit. The voltage output detected is –25 mV. Correspondingly, a periodic positive voltage output (Fig 19.10(b)) is detected at the stretched side of the wire, by using a Au-coated needle, when the ZnO, wire is periodically pushed by the Au-coated needle. Such, measurements are done with the use of a voltage amplifier.

The measured voltage signal is much lower than that predicted by the static model calculation (see Fig. 19.8). The following reasons account for this difference. First of all, the contact resistance is large as a result of the small contact area between the ZnO wire and the tip. Therefore, the voltage created by the piezoelectric effect of the ZnO wire is largely consumed at the contact due to the contact resistance, and only a small portion is received as the output. Secondly, the capacitance of the measurement system is much larger than the capacitance of the ZnO wire. The large system capacitance consumes most of the charge produced by the ZnO wire and results in a lower voltage output. Finally, the finite conductivity of ZnO greatly reduces the magnitude of the piezoelectric potential. As the voltage measured across the load is negative in reference to the grounded electrode, the current flows from the tip into the NW.

From the calculation 0.3 V is created between the NW and the AFM tip during mechanical bending. This voltage is much higher than the thermal voltage of $k_B T/e$ ~ 25 mV, and it is applied to the Schottky barrier between the NW and the tip and is responsible for driving the rectifying behavior of the Schottky barrier. The received voltage is in the order of 10m V, much lower than that calculated theoretically for the following reasons. First, a single NW based nanogenerator (see Figure 19.7) simplified as a voltage source V_{nw}, which is created by the piezoelectric effect of the NW when subjected to mechanical deformation, a contact resistance R_s between the tip and the NW, and an inner resistance R_{nw} of the NW. Since the ZnO NW is a bare NW without a catalyst particle at the tip, the contact resistance (R_c) is very large due to the small contact area between the metal tip and the edge of the NW tip. Therefore, the voltage that drove the Schottky diode (V_{nw}) is largely consumed at the contacted resistance, and only a small portion is received as the output. Second, the capacitance of the nanogenerator system (~1 pF) is much larger than the capacitance of a single NW (~fF). The larger system capacitance consumes most of the charges produced by the NW, and the large system capacitance results in low voltage output, but the output current is not affected.

(B) DIRECT–CURRENT NANOGENERATOR

By an Ultrasonic Wave

For an innovative design one should improve the performance of NG in following aspects. First the use of AFM should be for making the mechanical deformation of the NWs so that the power generation is achieved by an adaptable, mobile, and cost effective approach over a larger scale. Secondly, all of the NWs should generate electricity simultaneously and continuously, and all of this electricity should be collected and used. Finally, the energy converted into electricity should in the form of a wave or vibration so the NG can operate "independently" and wirelessly.

To replace the role played by AFM tip in deflecting the NW, first an "inverted V-shaped" (*I-V*) electrode is used. Once the *I-V* electrode moves downward as driven by an external excitation, an NW deflects towards the left (say). At the contact point, a positive potential is created by the local tensile strain (Fig. 19.11(a)). In such a case, the piezoelectric potential is preserved as the local contact is a reverse - biased Schottky barrier that does not permit the flow of charge. When the electrode further pushes the NW until it bends enough to reach the other side of the *I-V* electrode (Fig. 19.11(b)), the local contact is a forward-biased Schottky barrier, thus, the local potential drop drives the flow of electrons. Therefore, the AFM tip can be replaced by an *I-V* electrode. The *I-V* electrode can be extrapolated into a zigzag electrode (Fig. 19.11(c)) made of Si coated with Pt. The Pt coating is not only for enhancing the conductivity of the electrode but also for creating a Schottky contact at the interface with ZnO. The coating metal is any conductive alloy as long as it can form a Schottky barrier with ZnO. The zigzag electrode acts as an array of parallel AFM tips. The electrode is placed and manipulated above the NW arrays at a controlled distance. The output electricity produced by the relative deflection/displacement of the NWs and the electrode via bending or vibration is continuous.

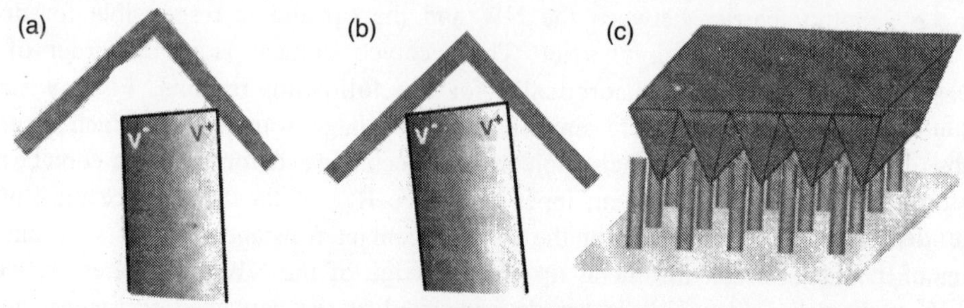

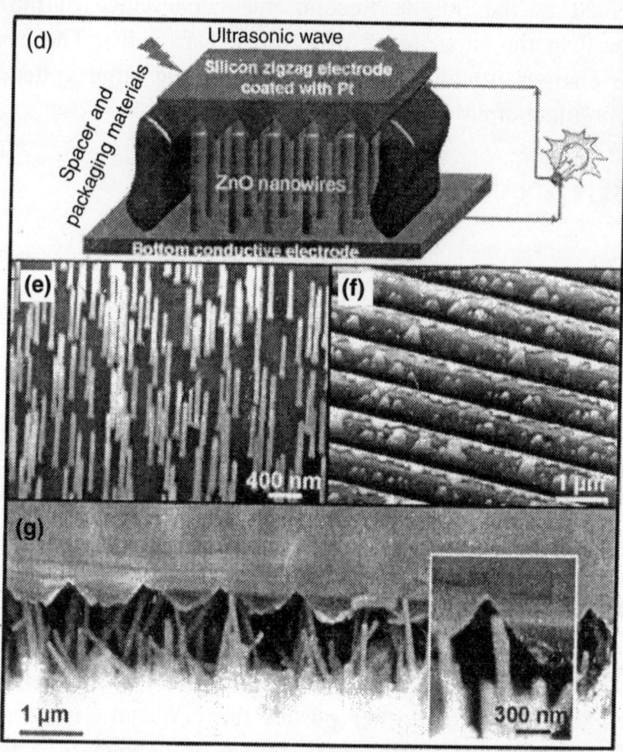

Figure 19.11 *(a)-(b) idea for replacing the role played by an ATM tip by an "inverted-V shaped" electrode. (c) The idea of introduction of a zigzag electrode to form hundreds of parallel tips, each acting like an AFM tip (d) diagram showing the design and structure of the nano-generator. Aligned ZnO NWs grown on a solid/polymer substrate are covered by a zigzag electrode. The substrate and the electrode are directly connected to an external load. (e) Aligned ZnO NWs grown on a GaN substrate. The gold catalyst particles used for the growth are mostly vaporized; thus, the final NWs are purely ZnO with flat top ends. (f) Zigzag trenched electrode fabricated by the standard etching technique after being coated with 200 nm of Pt. The surface freatures are due to nonuniform etching. (g) Cross-sectional SEM image of the nano-generator, which is composed of aligned NWs and the zigzag electrode. (inset). A typical NW that is forced by the electrode to bend.*

Experiment

The experiment setup is shown in Fig. 11(d). An array of aligned ZnO NWs are covered by a zigzag Si electrode coated with Pt. The Pt coating not only frequency of the ultrasonic wave is ~41 kHz. The output current and voltage are measured by an external circuit at room temperature.

The experimental design relies on coupling between piezoelectric and semiconducting properties of the aligned ZnO NWs. The asymmetric piezoelectric potential across the width of a ZnO NW and the Schottky contact between the metal electrode and the NW are the two key process for creating, separating, preserving, accumulating, and outputting the charges.

A top electrode is designed to achieve the coupling process and to replace the role played by the AFM tip, and its zigzag trenches act as an array of aligned AFM tips (fig. 19.11(c) and 19.12(a). When subject to the excitation of an ultrasonic wave, the zigzag electrode can move down and push the NW which leads to lateral deflection of NW I. This, in turn, creates a strain field across the width of NW I, with the NW's outer surface being in tensile strain and its inner surface in compressive strain. The inversion of strain across the NW results in an inversion of piezoelectric field E_z along the NW (Fig. 19.11(a) and (b), which produces a piezoelectric-potential inversion from V^- (negative) to V^+ (positive) across the NW (Fig. 19.22(b)). When the electrode makes contact with the stretched surface of the NW, which has a positive piezoelectric potential, the Pt metal-ZnO semiconductor interface is a reversely biased Schottky barrier, resulting in little current flowing across the interface. This the process of creating, separating, preserving, and accumulating charges. With further pushing by the electrode, the bent NW I will reach the other side of the adjacent tooth of the zigzag electrode (Fig. 19.12(c)). In such a case, the electrode is also in contact with the compressed side of the NW, where the metal/ semiconductor interface is a forward-biased Schottky barrier, resulting in a sudden increase in the output electric current flowing from the top electrode into the NW. This is the discharge process.

Figure 19.12(a) to (c) shows four possible configurations of contact between an NW and the zigzag electrode. NWs I and II are deflected towards the left and right sides, respectively, by the electrode. Regardless of their deflection directions, the currents produced by NWs I and II constructively add up. NW III is chosen to elaborate the vibration induced by an ultrasonic wave. As shown in Figure 19.12(c), when the compressive side of NW III is in contact with the electrode, the same discharge process as for NW I occurs, resulting in the flow of current from the electrode into the NW. NW IV, (short in height) is under, compressive strain by the electrode without bending. So the piezoelectric voltage created at the top of the NW is negative. Thus, across the electrode – ZnO interface, a positively biased. Schottky barrier is formed; hence, the electrons freely flow across the interface. As a result, electrons flow from the grounded substrate electrode into the NW into the top zigzag electrode as the deformation occurs. This discharging process, contributes to the measured current. The output current is a sum of those NWs that actively contribute to the output power, but the voltage of the NG is determined by a single NW because all of the NWs are "in parallel".

An equivalent electric circuit is shown in Fig. 2D to illustrate the measurement and outputs of the nanogenerator. The NWs producing current in the nanogenerator are equivalent to a voltage source V_s plus an inner resistance R_i that also contains the contact resistance between the active NWs and the electrode. There are a lot of NWs that are in contact with the electrode but cannot be bent or move freely; thus, they do not actively participate in the current generation, but they do provide a path for conducting current. These NWs are simply represented by a resistance R_w that is parallel to the portion that generates power. A resistance Rc is introduced to represent the contact resistance between the electrode and the external measurement circuit. The capacitance in the system is ignored in the circuit in order to simplify the discussion about dc measurement.

The current and voltage outputs of the nanogenerator are shown in Fig. 19.12(e) and (f) respectively, with the ultrasonic wave being turned on and off regularly. A jump of ~0.15 nA is

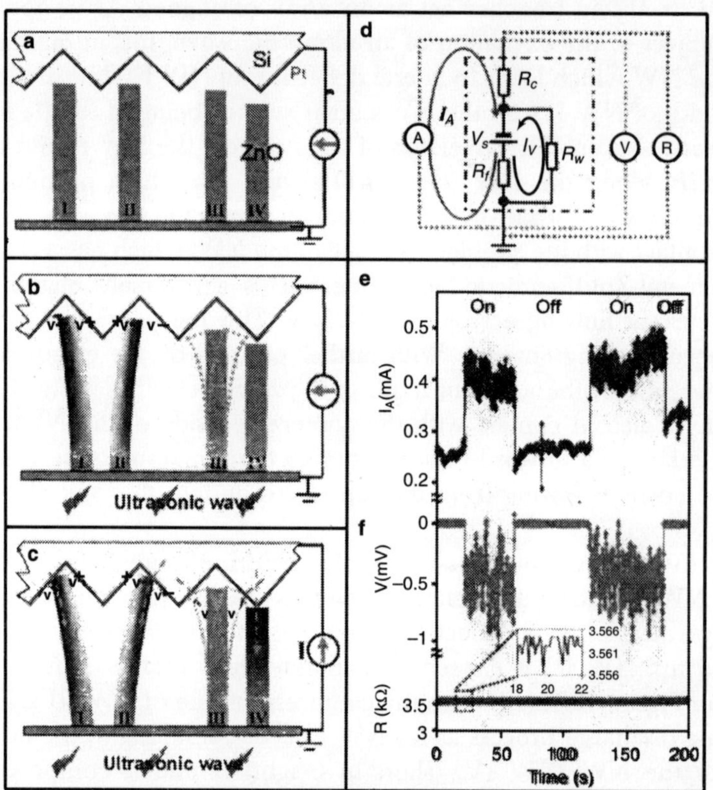

Figure 19.12 *The mechanism of the nanogenerator driven by an ultrasonic wave (a) illustration of the zigzag electrode and the four types of NW configurations. (b) The piezoelectric potential created across NWs I and II under the push/deflection of the electrode, as driven by the ultrasonic wave. NW III is in vibration under the stimulation of the ultrasonic wave. NW IV is in compressive strain without bending. (c) Once the NWs touch the surface of the adjacent teeth, the Schottky barrier at the electrode/NW interface is forward biased, piezoelzctric discharge occurs, resulting in the observation of current flow in the external circuit. (d) Equivalent circuit of the nanogenerator and the setup for measuring the output current I, output voltage V, and resistance R. (e to g) I, V, and R measured with the connections shown in (d), respectively.*

observed when the ultrasonic wave is turned on, and the current immediately falls back to the baseline once the ultrasonic wave is turned off. Correspondingly, the voltage signal exhibits a similar on and off trend but with a negative output of ~–0.7 mV. The negative sign of the voltage is consistent with the mechanism presented in Fig. 19.12(c). The signal-to-noise ratio of the current is substantially better than that of the output voltage for the following reasons: Because the resistance of the current meter (ideally, zero) plus R_c (20 to 30 ohms) is only ~1/1000 of R_w when the current is measured, the current generated by the nanogenerator can be assumed to be bypassing R_w; the current path is indicated by a solid curve in Fig. 19.12(d), so the measure current is $I_A < V_s/(R_c + R_i)$. However, because R_w is very much smaller than the inner resistance of a voltage meter (ideally, infinity) when the voltage is measured, a loop is formed between the power-generating portion of the system and R_w, as shown by another solid curve in Fig. 19.12(d). In this case, the current is I_V and the measured voltage V is that across the power-generating portion, V » –VsRw/(Ri + Rw). During the ultrasonic vibration, for the unstable contacts between the NWs and the electrodes, I_A is affected by the instability of Vs and Ri (but mainly by Vs because Rc is a constant), and V is affected by the instability of Vs, Ri, and Rw. As a result, V has about two times the noise level of I_A, consistent with the observations displayed in Fig. 19.12(e) and (f). On the other hand, because the voltages created by all of the NWs are in parallel, the output voltage is effectively the voltage created by one NW; thus, it appears relatively unstable and with a large noise level than that of I_A (Fig. 19.12(f)). Furthermore, the output voltage is naturally smaller than that created by deflecting a NW by AFM, because it is limited by the smaller degree of deflection amplitude of the NW, as induced by an ultrasonic wave in comparison to that induced by the AFM tip. Finally, the output current is a sum of the currents produced by many NWs. Therefore, the current signal is more stable and continuous, and it is used to characterize the performance of the dc nanogenerator. The resistance of the entire nanogenerator is also measured with and without turning on the ultrasonic wave (Fig. 19.12(g)). The resistance remains stable at R = 3.560 ± 0.005 kilohms. This measurement indicates that the jump in current is not due to the variation in resistance, as caused by the vibration of the NWs suggesting that the current signal in Fig. 19.12(e) is created by the nanogenerator.

The output electricity of the nanogenerator is continuous and reasonably stable. A continuous output current is generated when the ultrasonic wave is turned on, and the current disappears when the wave is turned off (Fig. 19.13(a)). The output current is in the nanoampere range. The current signal shows no direct coupling with the frequency of the ultrasonic wave, because the wave frequency is ~80 times smaller than the resonance frequency of the NWs (~3MHz). The size of the nanogenerator is ~2 mm^2 in effective substrate surface area. The number of NWs that are actively contributing to the observed output current is 250 to 1000. The nanogenerator works continuously for an extended period of time of beyond 1 hour (Fig. 19.13(b)).

The experimental design (as presented in (Fig. 19.11)] is tested in comparison to the experiments conducted using different materials or configurations. Using the design shown in Fig. 19.11(d), simply by replacing a ZnO NW array with an array of carbon nanotubes (CNTs),

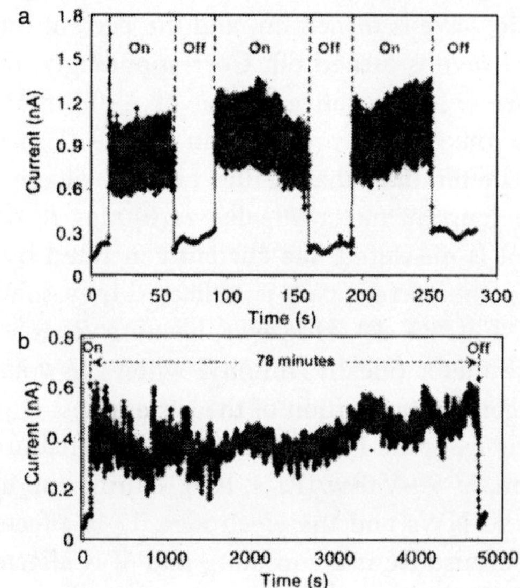

Figure 19.13 *Continuous dc output of the nanogenerator, as characterized by the current signal. (a) Reproducible and highly repeatable current output of the nanogenrator when the ultrasonic wave is turned on and off. (b) Continuous current output of the nanogenerator for an extended period of time. The baseline of the current signal is produced by the current signal produced by the electronic measurement system and the interference from teh ultrasonic-wave source. The size of the nanogenrator is ~2 mm² in effective substrate surface area.*

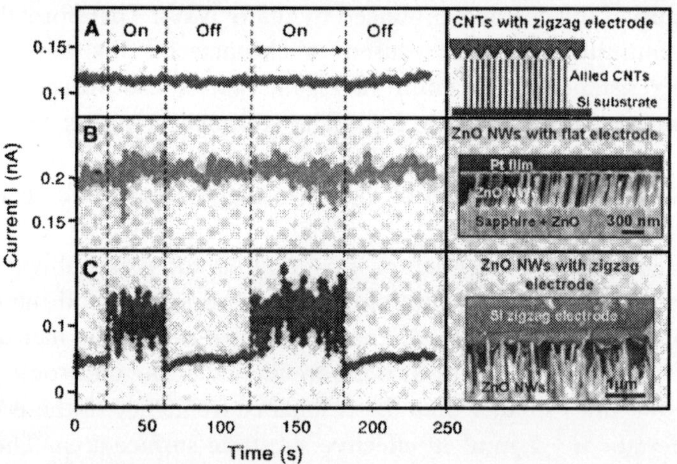

Figure 19.14 *Output of the nanogenerator with different materials and design configurations. The designs and true-device SEM images are shown on the right-hand side, and the corresponding current curve is on the left-hand side. (a) Nanogenerator based on array of CNTs with a zigzag top electrode. (b) Nanogeneraotr based on array of ZnO NWs but with a flat top electrode, showing no jump in current while stimulated by an ultrasonic wave. (c) Nanogenerator based on arrays of ZnO NWs with a zigzag top electrode, showing a large jump in current when the ultrasonic wave is on.*

no jump in current is observed when the ultrasonic wave is turned on (Fig. 19.14(a)). This is because the CNTs are not piezoelectric. In a system with a ZnO NW array but in which the top electrode is replaced with a flat, thin Pt film that totally covered the tips of the NWs (Fig. 19.14(b), no jump in output current is observed. A clear jump is observed only when the top electrode is in a zigzag shape and when ZnO NWs are present (Fig. 19.14(c)).

Three major objectives achieved by this method are: (i) the expensive and sophisticated AFM tip is replaced with ultrasonic waves/vibrations, induce the elastic deformation and vibration of the NWs, and a cost-effective prototype technology is achieved for fabricating the nanogenerator. (ii) an array of tips are integrated into a zigzag electrode for the simultaneous creation, collection, and output of electricity generated by many NWs, establishing the principle for raising the output power. (iii) A continuous and stable dc output is achieved with this system. The principle demonstrated here has set a platform for harvesting energy from the environment to power in vivo biosensors, wireless and remote sensors, and nanorobots, and has also established the basis for building zero-power force/pressure sensors.

The number of NWs that is effective for producing output electricity is estimated from the output power of the nanogenerator. The NW deformed by AFM produce an electric energy of ΔW_{AFM} ~ 0.01 fJ for each cycle of discharge, which lasts for 0.1 ms; thus, the power generated by one NW is ΔW_{AFM} ~0.1 pW. In this experiment the vibrational amplitude of the NW is much smaller than that of the NW directly deflected by an AFM tip; thus, the output voltage is ~1 mV, which is about 5 to 10 times smaller than that received when AFM is used as the deformation tool. In this case, the output power of a NW, as driven by ultrasonic wave, is ΔW_{wave} – 1 to 4 fW. The output-power volume density per NW is ~1 to 4 Wcm^{-3}, which is more than two orders of magnitude higher than that produced by a vibrational microgenerator. The output power of the nanogenerator fabricated with a substrate of area = 2 mm^2 is $W_{wave} = I_A V \approx 1$ pW (Fig. 19.12(c) and (d)). Therefore, the number of NWs that is active for producing electricity in Fig. 19.12(e) is $N = W_{wave}/\Delta W_{wave} \approx 250$ to 1000 NWs. As limited by the multiple contacts between the NWs and the electrode in the present design (Fig. 19.11g), the large majority of the NWs do not produce electricity because of their nonuniformity in height and distribution on the substrate surface; thus the output current is rather small. In addition, some of the NWs directly push the electrode at the top edges/apexes of the zigzag trenches, preventing the other NWs from reaching and contacting the electrode to produce electricity. These technical difficulties can be overcome by an optimized design to improve nanogenerator efficiency. For example, nanogenerator efficiency can be improved with the use of patterned-tip arrays as electrode, the designed and patterned growth of high-quality uniform NW arrays matching the design of the electrode and an improved packaging technology to keep a precise control on the spacing and alignment between the electrode and the NW arrays. If the area taken by each metal tip is 0.5 by 0.5 μm the grown density of NWs is ~10^9 cm^{-2}. If one NW produces 10 fW of power of optimizing its size and shape, the output power per unit of area can be 10 μW cm^{-2}. The power used to operate a device fabricated with one NW or nanotube is ~10 nW; thus, the nanogenerator built with the NWs grown on an area of 1 cm^2 can operate upto 1000 of such nanodevices.

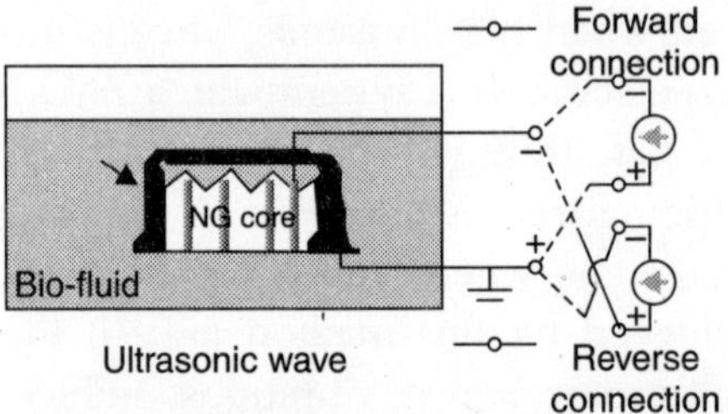

Figure 19.15 *An NG that operates in a biofluid and the top two types of connections used to characterize the performance of the NG. The black and red curves represent signals -from forward and reversely connected current/voltage (I/V) meters, respectively.*

Criteria for Identifying the True Output Signal from a Nanogenerator

Measurement of the current and voltage signal generated by an NG is rather challenging, especially at the level of nanoamperes and millivolts, because of system and/or t environmental interference, such as system capacitance, thermal/instability drifting, and the bias of thc amplifier. The electrical measurements of NGs is even more complex than conventional conductivity measurements, because for the current/voltage that is generated by the NGs is required rather than the bias current/voltage introduced by the electrical measurement that detects the signals. Another challenge is from the mechanical movement of the components during energy conversion, which introduces electronic interference and capacitance change. Actually no electrical output (or electrical signal) is observed Extreme caution must be exercised in each step to ensure the signal is generated by the NG instead of the measurement circuits. In many cases simulating circuits are built to identify the sources. The key for each measurement is to repeat it for many devices and under different experimental conditions. Two testing criteria are to identify the true signal from an NG. Also carefully check on the specification of the measurement system is needed to ensure it has an extremely small capacitance (~pf), low noise (< 0.5 mV), and very low bias current (< 2 pA).

The first criterion is the "switching-polarity" test. As illustrated in Figure 19.15, the current/voltage meter is first forward connected, for example, positive and negative probes are connected to the positive and negative electrodes, respectively. Then the connection polarity to the two electrodes of the NG is reversed. The corresponding *I*sc land V_{oc} signals are shown in Figure 19.15 and 19.12(e), respectively. A switch in sign in both current and voltage after switching the connecting polarity is a key test to eliminate the effects from measurement system error and confirm that the power is generated by the NG. Resistors and capacitors are symmetric devices and they cannot produce a reversal in output signal if it is initiated from the measurement system. A diode produces only a one-way current signal but zero in the other direction, which not result in a reversed current if It originates from the measurement system.

The second criterion is a linear superposition of the output current/voltage when two or more NGs are connected in parallel/series. As shown in Figure 19.16(a) and (b), two NGs I and II are measured under the same experimental conditions. NG I produced an average I_{sc} of about 0.7 nA and NG II generated a signal of 1 nA. These two NGs are then connected in parallel (inset of Fig. 19.16(c) and tested under the same conditions again. The resultant output

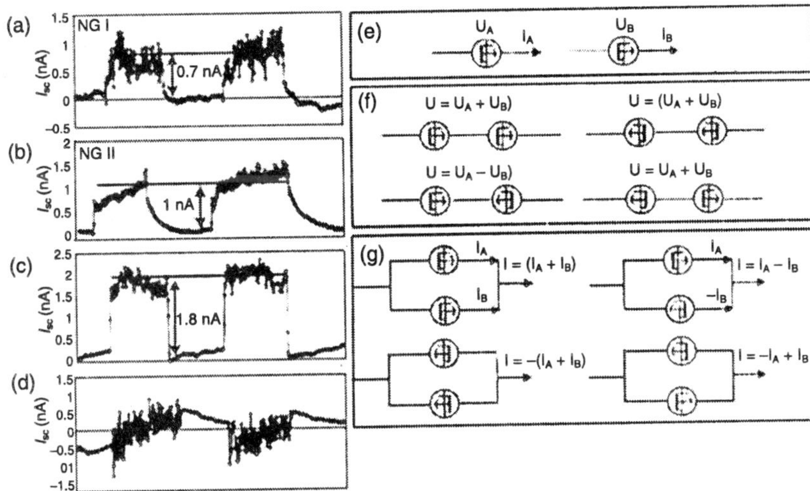

Figure 19.16 *Criteria adopted for differentiating between the true output signal versus that from artifacts. (a,b) Current signal measured from two individual NGs, I and II. (c,d) Current signal measured from parallel and anti parallel connected NG I and NG II, respectively. (e) Two NGs with expected output. The symbol presented here represents the piezoelectric NG. (f) Four testing configurations for serial connection of two NGs to identify true voltage output signals. (g) Four testing configurations for serial connection of two NGs to identify true current output signals.*

current reaches an average of about 1.8 nA, which is the sum of the two individual outputs (Fig. 19.16(c)). This concept is also proved by anti parallel connection of NG I and NG II (inset of Fig. 19.16(d)). Since the magnitude of I_{sc} for the two NGs is very close, the total current is cancelled out by the "head-to-tail" connection in parallel, and the received signal, is around the baseline (Fig. 19.16(d))

Capacitors, resistors, and inductors are symmetric devices and their output is equivalent regardless of which direction the current flows. However, diodes are nonsymetric devices. Once a diode is involved, one must demonstrate the "add-up" and "subtract-off" tests as illustrated above for identifying the true signals generated by the NGs. Figure. 19.16(f) and (g) summarize the eight connection configurations and the expected results for ruling out artifacts. These tests are essential for identifying the true output from NGs.

For technological applications, raising the voltage and current outputs of the NG is essential for raising the output power. If each NG is considered as a "battery", the most straightforward approach for increasing the current/voltage is to put them in parallel/series, respectively.

A Schottky barrier is required between the zigzag electrode and the NWs. By simply measuring the *I-V* characteristics of a packaged NG, the presence of a Schottky barrier is used as a criterion for identifying a working NG verses a defected one. This is useful for, quality inspection and control.

Functionalized with p-Type Oligomer

Based on mechanical manipulation of a single ZnO wire/belt using AFM: The selected ZnO wire/belt is long enough to be directly visualized under optical microscope built in AFM. Once end of the ZnO wire belt is fixed on to a flat intrinsic silicon substrate using silver paste that is electrically grounded; the other end is left free (see the lower left comer inset in Figure 19.17;). The silicon substrate is an insulator. The ZnO wire/belt is laid parallel to and a little bit above the substrate in free suspension to eliminate friction. A resistor $R_L = 500\ M\Omega$ is connected between the fixed end of the ZnO wire/belt and the AFM tip, and the output signal is characterized by a voltage drop across the resistor in reference to the grounded end. The measurement is performed using an AFM with a Pt-coated silicon tip, which has a tetrahedral tip of 14 μ m in height. The cantilever has a spring constant of 1.6 Nm^{-1}. All the measurements are performed in contact mode with a set point of 5 nN and a scanning speed between 60 and 160 $\mu\ m^{-1}s$. No external voltage is applied. Both the topography and potential output (V) images on the resistor are monitored simultaneously during the tip scanning.

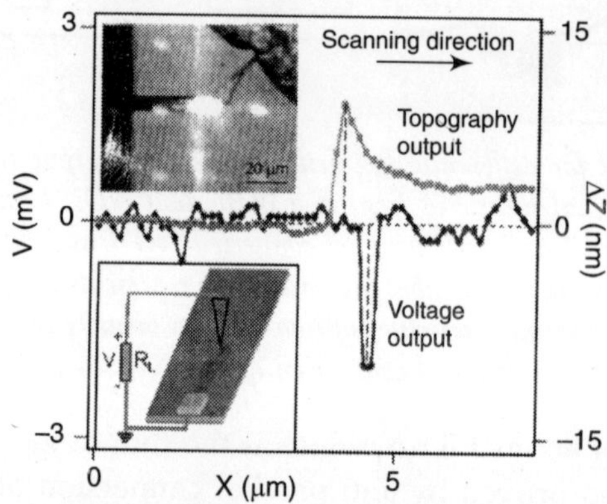

Figure 19.17 *The process of outputting piezoelectric potential by scanning an one-end-free ZnO wire/belt using an AFM tip. The ZnO is clean without oligomer coating. Top inset is a snapshot optical image of the AFM tip that is deflecting the wire/belt. Bottom inset is a experimental setup. The curves are the aligned plot of the output voltage over the external load and the topography profile.*

The piezoelectric output is observed from a clean ZnO wire/belt that has no surface coating. During the scan, when the tip only touches, the stretched side of the wire/belt and does not lift up, to go beyond the central line of the wire/belt to reach the compressed side, there is no voltage output signal. The output voltage peak is observed when the tip is beyond the wire/belt (to reach its compressed side as characterized by a peak in the topography image) in Figure 19.17. The corresponding output voltage is represented by the voltage output curve in Figure 19.17 which is always a negative peak in reference to the grounded end of the wire/belt. The center of the voltage peak is delayed with respect to the peak in the topography image. These are the characteristics for a ZnO wire/belt without surface functionalization, and they have a “switch effect” created by the Schottky barrier between the AFM tip and the ZnO wire/belt.

When the ZnO wire/belt is coated with a thin layer of OPV (oligomer (p-phenylene vinylene)) derivatives, which are selected because they are extensively used for p-type oligomers for light-emitting, field-effect transistors, and supramolecular assembly and are amphiphilic due to their high absorption coefficient, strong donating ability, and good thermal, chemical, and photochemical stability. The as-synthesized ZnO wire/belt is immersed in 1% Mol.L^{-1} (alpha,omega-bis(methylthioacetate) oligo(phenylene vinylene)-COOH) (OPV2) menthol solution for 48 hours. Then, the wire/belt is taken out and rinsed by deionized water and dried in air. A 2–3 nm thick OPV2 is coated on the surface of the ZnO wire/belt (see the inset TEM) image in Figure 19.18. Using this functionalized wire/belt, the above experiment as described in Figure 19.17 and a voltage output is presented in Figure 19.18(a). In contrast to the ZnO wire/belt without surface functionalization, a pair of positive and negative output voltage peaks are observed. The positive peak appears when the tip is in contact with the stretched side of the wire/belt, and the negative peaks show up when the tip, is in contact with the compressed side, which is determined by examining the peak positions in reference to the signal in the topography image. The uneven background in the topography image is caused by the downward bending and/or slight vibration of the fee-standing wire/belt during the scan. Some of the scans produce only positive voltage pulses, some generate only negative pulses, and some generated both. So, information provided by the topography image, the negative pulse is generated only when the AFM tip scans over the central line and reaches the compressed side of the wire/belt.

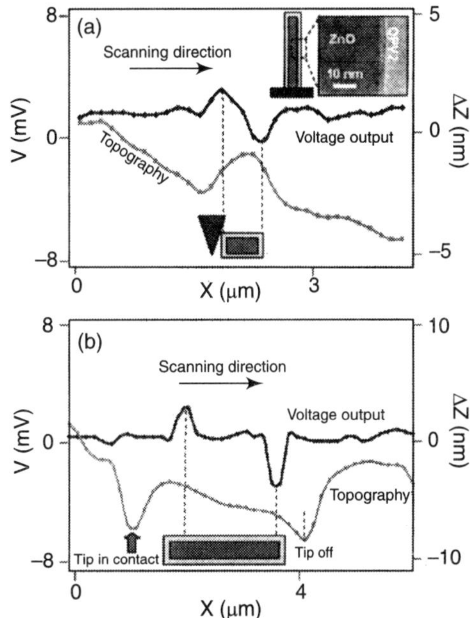

Figure 19.18 *The process of outputting piezoelectric potential of a one-end-free ZnO wire/belt that is coated with a thin layer of OPV2. (a,b) Aligned plot of the output voltage over the external load and the topography profile measured from two different wires/belts. Top inset is a TEM image showing the thickness of the oligomer coating. Cross-section of the wire/belt.*

Figure 19.18(b) shows a tip scan profile across a ZnO belt, in which both the positive and negative pulses are generated when the tip is scanned from the stretched side to the, compressed side of the belt. In the contact mode, the tip is lifted up by the belt to keep a constant normal force. Because the belt is freely suspended, it moves slightly downward or even vibrates when it is pushed by the tip, thus, uneven background is obtained. Moreover, the width of the topography profile is wider in comparison to the actual width of the wire/belt because the wire/belt is continuously pushed and bent when the image is obtained. The vertical distance, ΔZ, is not be the actual thickness of the wire/belt owing to its free-standing configuration and the shape of the tip. The positions at which the tip touches the belt and starts to hit up is identified, as indicated by an arrowhead in Figure 19.18(b). By correlating the relative position of the voltage peak to the topography image of the belt, no voltage, is released when the tip contacts the belt. The positive voltage peak appears after the tip pushes the belt for ~ 0.5 μ m along the scanning direction. This indicates that the belt has to be bent to a certain angle/distance before the voltage pulse is released. The delayed discharging peak shows that a simple physical contact cannot produce the voltage pulse, which rules out the contributions from surface/volume electrostatic charge, friction charge, or even measurement circuit. For the negative voltage peak, it appears that the voltage pulse is released before the tip loses contact with the belt, as shown by the dip in the topography curve.

Figure 19.18 shows the transport properties of the ZnO wire/belt, OPV2, and OPV2-functionalized ZnO wires are characterized. Two pieces of ZnO NW arrays are cut from the same sample grown on indium tin oxide (ITO) glass substrate with only one of them being functionalized with OPV2. To maintain the reliability and stability of the *I-V* measurements, two ~2 mm^2 size Pt electrodes are employed for the contact measurements. One Pt electrode is connected to the bottom ITO layer, and the other Pt electrode is put in contact to the tips of the nanowire arrays, as shown in the inset of Figure 19.19. There externally applied voltage is swapped from –5 to 5 V. The *I -V* curve shows that the contact between Pt and ZnO nanowire is a typical Schottky contact (Figure 19.16). To characterize the contact between Pt electrode and OPV2, an *I-V* measurement is performed on an OPV2-covered ITO glass substrate, where one electrode contacted ITO glass and the other one is in contact with the OPV2 layer. The contact between Pt and OPV2 is ohmic (black curve in Figure 19.19). The contact between OPV2-coated ZnO NWs and the Pt electrode is a nonsymmetric curve. When $0 < V_e < 1.5$ V, the *I–V* characteristic is the Schottky type, but when $V_e > 1.5$ V, the. *I–V* curve starts .to show the ohmic behavior. This indicates that the p–n junction is broken through when the local potential is larger than ~ 1.5 V. For a large belt a few hundred nanometers in width, the local piezoelectric potential is a to a few tens of volts. Thus, it is possible to have a local break through at large deflections. It is apparent that the OPV2 coating has changed the transport properties of the contacts, which is important for understanding the phenomenon presented in Figure 19.18.

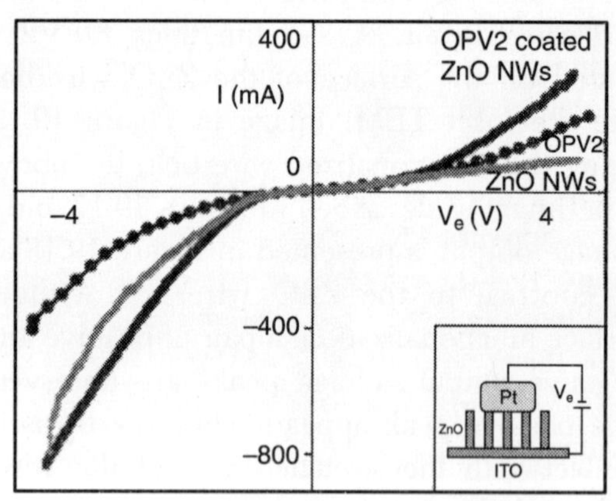

Figure 19.19 *I - V characteristics of the contacts between Pt electrode and ZnO nanowires without coating Pt electrode and OPV2 and Pt electrode and OPV2-functionalized ZnO nanowires. Inset is a schematic diagram of the experiment set up. A flat Pt electrode was placed on the top of the aligned ZnO nanowire with or without coating a layer of p-type oligomer.*

After knowing the contact characteristics, a model is made for understanding the voltage signal as shown in Figure 19.5. The model is based on the mechanism shown in Figure 19.5 for understanding the charge creation–accumulation and charge-releasing processes proposed for a clean ZnO wire/belt. By bending a ZnO wire/belt, an asymmetric strain distribution across its width results in an asymmetric potential distribution with a positive potential ($V_e^+ > 0$) at its stretched side and negative potential ($V_s^- < 0$) at the compressed side in reference to the grounded-fixed-end. The as-grown ZnO NW is n-type. For a clean ZnO wire/belt without oligomer coating, a Schottky diode at the interface between the Pt tip and the n-type ZnO wire/belt is shown, which controls the charge accumulating an releasing process. Because the work function of Pt is 6.1eV, and the electron affinity of ZnO is 4.5 eV, the Pt-ZnO contact is Schottky with a barrier height of 1.6 eV for bulk materials. The scanning tunneling microscopy. Measurement show that the barrier for nano-contact is 0.45 eV and is very sensitive to local strain, which is lower than that of bulk contact. The piezoelectric potential V_s^+ is proportional to the degree of bending of the wire/belt, and it is simply described by a formula

$$V_s^{\pm} \approx \pm 27\ (a/L)^3\ y_m \text{ (in V)} \tag{1}$$

where a and L are the radius and length of the NW, respectively, and y_m (in nm) is the lateral displacement at tip of the NW. Fora typical wire of $a = 1\ \mu$m, $L = 40\ \mu$m, and $y_m = 10\ \mu$m (see Figure 19.17), the maximum potential at the surface is $(V_s^{\pm})_{max} \approx \pm 4.2$ V.

When the ZnO wire/belt is functionalized with a thin layer of OPV2, the Pt tip is separated from the n-type ZnO wire/belt by this p-type oligomer. Because the contact between Pt and ZnO is ohmic, the carrier transporation from Pt to ZnO is controlled by the p-n junction formed between OPV2 and ZnO. From the *I-V* characteristic presented in Figure 19.19, it is likely that the p–n junction barrier height Δf between OPV2 adn ZnO is lower than Schottky barrier height between Pt and ZnO. When the tip first touches the OPV2 coated wire/belt (Figure 10.20(a), it deflects the wire/belt and the piezoelectric charges are continuously produced, but the local potential V_s^+ is smaller than the barrier height because V_s^+ is proportional to the lateral deflection of the nanowire (Equation 1). Thus, the p–n junction at the OPV2–ZnO interface is reversely biased (Schottky type). In this case, the charges are accumulated without releasing due to the rectifying effect of the p–n junction. With the further increase of bending, the local piezoelectric potential V_s^+ becomes larger than the barrier height $\Delta\phi$ (Figure 19.20c)

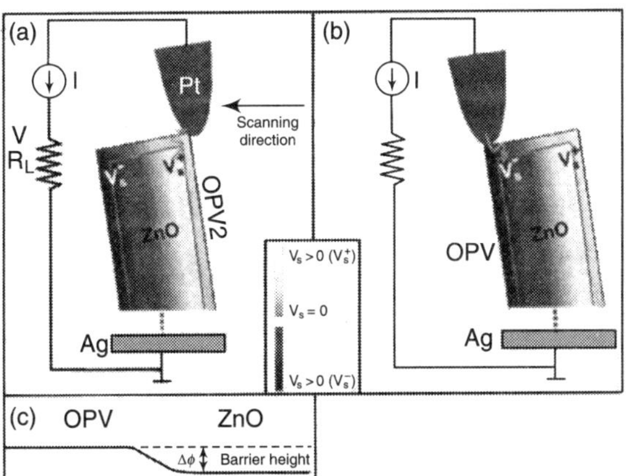

Figure 19.20 *The piezoelectric potential generated from an OPV2-coated ZNO wire/belt when deflected by an AFM tip. (a, b) Metal and semiconductgor contacts between the AFM tip and the semiconductor ZnO belt coated with OPV2 at positive adn negative local contact potential (see text). (c) Diagram showing the barrier height at teh OPV2-ZnO p–n junction.*

and the p–n junction is broken through (ohmic type). Thus, the current flows from the ZnO wire/belt into Pt tip through the OPV2 layer resulting in the positive potential peaks as presented in Figure 19.18.

If the Pt tip scans fast across the wire/belt, two possible consequences result. First, the contacting time is so short that the piezoelectric charges are not completely neutralized/screened by the external electrons if the carrier density and carrier mobility is low in OPV2. As a result, a small potential is preserved even after neutralizing/screening part of the charges. The other possibility is that the wire/belt bends when the tip is scanned from the stretched side to the compressed side, which generates additional potential with the increase of deformation. When the conductive tip reaches the compressed side of the ZnO wire/belt, the negative piezoelectric potential V_s^- at thee ZnO side sets the p–n junction to forward bias, resulting in a flow of current from the Pt tip into the ZnO wire/belt, thus negative voltage pulses are observed (Figure 19.18). In another, case, if all of the charges are neutralized/screened when the tip contacts the stretched side of the wire and no more deflection is introduced when the tip slides to the compressed side, there is no negative pulse output. This is the reason that only a positive pulse is observed in some of the scans.

From the conductivity of the OPV observed it is found that if its conductivity is too high, the created piezoelectric potential in the NW is screened by the charges in OPV within a time period shorter than the contacting time of the scanning tip with the NW. If its conductivity is too low, the flow of current through OPV is liimited, thus showing very low output voltage.

(C) THE SINGLE-WIRE GENERATOR (SWG)

The SWG consists of a single ZnO microwire/nanowire lying on a flexible substrate, with its two ends firmly fixed by metal contacts. Bending of the substrate results in stretching or compression of the microwire and develops a piezoelectric-potential drop along the microwire. A Schottky contact at one end is required to prevent the flow of electrons through the microwire and enables the deformed wire to be a "charge pump" and a "capacitor." The charging and discharging process when the ZnO microwire is stretched and released creates an oscillating electric current in the external load.

The output of the SWG is affected by the measurement system, change in capacitance of the microwire. and electric circuit during mechanical deformation, and the coupling of the SWG with the measurement system; it is thus easy to observe false signals. To differentiate the electric power that is generated by the SWG from possible artifacts three criteria consisting of 11 tests are developed to rule out artifacts. (1) The generator must satisfy the Schottky behavior test and (2) switching polarity tests, which are also a linear superposition of current and voltage for eight configurations. A true signal (current and voltage) generated from a generator must pass each and all of the tests. These criteria and configurations are applicable to all types of NGs and serve as standard test.

A piezoelectric fine wire (PFW) is placed laterally on a polyimede film [Fig. 19.21(a)]. Because the film thickness is greater than the diameter of the ZnO PFW, the ZnO PFW experiences a pure tensile strain when the substrate is bent inward as shown in Fig. 19.21(b).

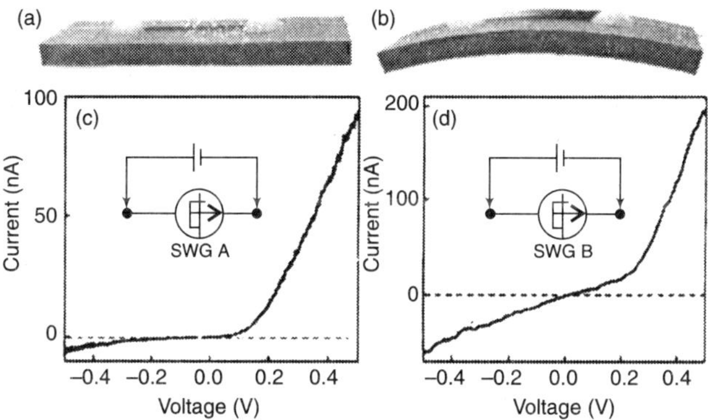

Figure 19.21 *A PFW lying on a polymer film substrate with two ends tightly bonded to the substrate. The leading wires connect to the measuring instrument (b) Bending of the substrate results in the axial tensile strain and corresponding piezoelectric potential drop along the PFW. (c) and (d) I-V characteristic of SWG A and SWG B.*

The transport property of a working device which generates good electricity output, always shows asymmetric behavior, as shown in Figs. 19.21(c) and (d) for two different SWGs. The nonlinear Schottky like transport behavior is required for a working NG. The asymmetric characteristic of Schottky barrier divides each cycle of driving action into two steps, charge accumulation and charge release, resulting in an electric out-put pulse. In the case of SWG, the Schottky contact serves as a one-way gate (to prevent the flow of electrons through the microwire such that the electric power can be effectively outputted. Without the presence of the Schottky barrier, there, is no charge accumulation, thus, no charge release. This criterion eliminates those defective devices which do not harvest energy from the environment. A symbol is created as shown in the insets of Figs. 19.21(c) and 19.21(d) to represent the SWG. For easy notation, the side that has a Schottky contact is the positive side.

For the short-circuit current measurement, first the forward connection in which the positive probe and negative robe of the measurement system are connected to the positive and negative ends of the generator, respective. Figures 19.22(a) and 19.22(b) show the results for SWG A and SWG B. The insets give the connection configurations. The positive peaks correspond to the stretching states of the PFW when the underneath substrate is bent inward. When the substrate is released, the PFW returns to free state, resulting in a negative electric peak.

In order to identify if the signal is the true electricity output due to piezoelectric property of the ZnO fine wire, the switching-polarity test is done in which the positive and negative probes of the current meter are connected to the negative and positive ends of the SWG respectively. The result is shown in Figs. 19.22(c) and 19.22(d). The output signal is the reversal of that presented in Figs. 19.22(a) and 19.22(b). Stretching of the PFW produces a negative pulse, and a positive pulse is generated when released. The satisfaction of the switching polarity test rules out the possible error from the system capacitor change. The change in contact resistance produces a signal when the SWG is deformed, but such signal does not change its sign from positive to negative when the connection is reversed. Additionally, the magnitude difference between the signals with forward connection and with the reverse connection is observed. The

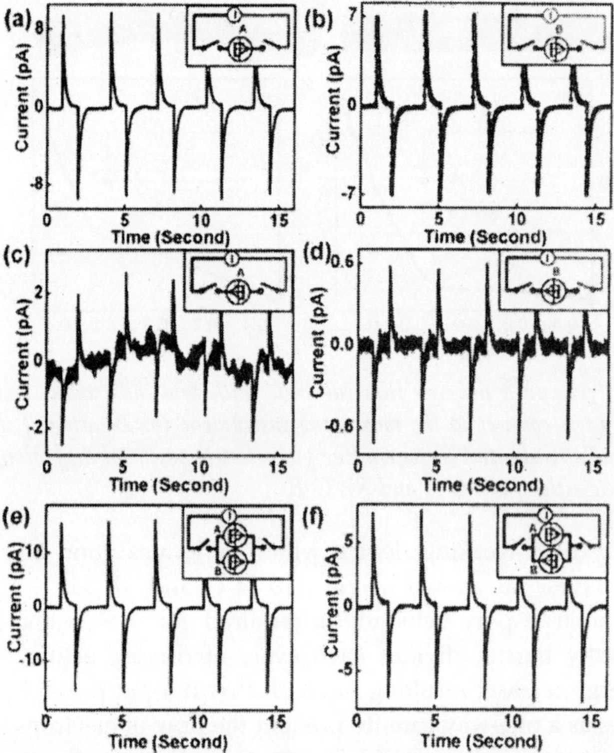

Figure 19.22 *(a and b) Short-circuit current power output from SWG A and SWG B, respectively, under forward connection. c) and d). Short-circuit current power output from SWG A and SWG B, respectively, under reverse connection. e) and f) Parallel connection of SWG A and SWO B demonstrates SWG's ability to "add up" or "cancel out" All insets illustrate the connection configuration of the two SWGs in reference to the measurement system.*

nonsymmetric output of the SWG prior and after switching the polarity is caused by the measurement system that has a bias current. If this bias current adds up to the current generated by the SWG at forward connection configuration, it is subtracted from the generated current at reversely connected configuration. The true signal generated by the SWG is an average of the magnitudes observed under forward and reverse connection configurations.

Due to the presence of Schottky at one end of the SWG, the linear superposition test is applied. The current is measured when two SWGs are connected in parallel (to examine the liner superposition of currents). Since there is a Schottky contact at one end of the SWG, attention is paid to the. connection direction. Figure 19.22(e) shows the result when two SWGs are in the same direction, in which output current is enhanced and approximately equal to the sum of signal frrom Figs. 19.22(a) and 19.22(b). When two SWG are connected in reversed directions (See Fig. 19.22(f), the output current is decreased and is approximately equal to the sum of the signal from Figs. 19.21 (a) and (d). As a result, a linear superposition of current is obtained for the two SWGs. In addition, the parallel connection of the two SWGs also gives the switching-polarity test.

The voltage output of the SWGs is. presented in Fig. 19.23 SWG A and SWG B generate a positive voltage signal when the PFWs are stretched and a negative signal when the PFWs are released as shown in Figs. 19.23 (a) and (b). The switching polarity is also obtained for output voltage, as shown in Figs. 19.23 (c) and (d). The signal and further characterization the generator, is measured by the voltage, when the two SWGs are in serial, as shown in Figs. 19.23 (e) and (f). When two SWGs are connected in the same direction, the final output is increased. When two SWGs are connected in opposite directions, the final output is reduced. As a result, the superposition of voltage is obtained for the two SWGs. The SWGs connected in serial also satisfy switching-polarity test.

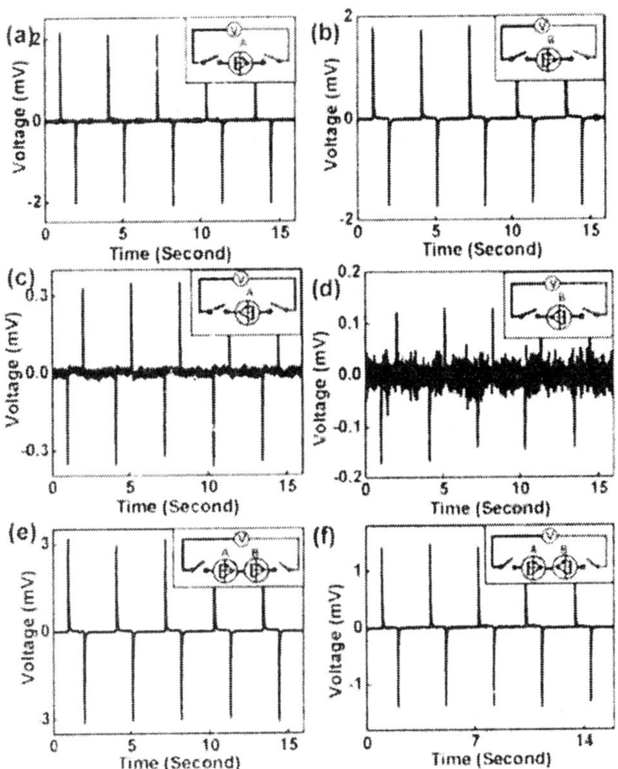

Figure 19.23 *a) and b) Open-circuit voltage output from SWG A and SWG B, respectively, under forward connection. c) and d) Open-circuit voltage power output from SWG A and SWG B, respectively under reversal connection. e) and f) Serial connection of SWG A and SWG B demonstrates the add up or cancel out effect. All insets illustrate the connection configurations of the SWGs in reference to the measurement system.*

For the superposition phenomena, two SWGs are deformed and released simultaneously. A slight delay in the deformation, the outputs of the two SWGs are revealed. Figure 19.24 (a) shows a case where two SWGs are being oppositely connected. A pair of positive-negative peaks is observed, one following immediately the other when PFWs are stretched or released.

The double peaks that are opposite in sign and one closely following the other correspond to the signals from two SWGs, respectively [Fig. 19.24 (b)]. Before the bending of SWG A is completed, bending of SWG B starts, which results in a negative peak in the first double peak owing to its reversal connection. Similarly, before the releasing of SWG A is completed, the release of SWG B starts, which results in a positive peak due to its reversal connection. The two positive-negative peaks in Fig. 19.24 further confirm that the generated voltages satisfy the linear superposition rule, and any discharge artifacts, if any, cannot generate signal as presented in Fig. 19.24 (b).

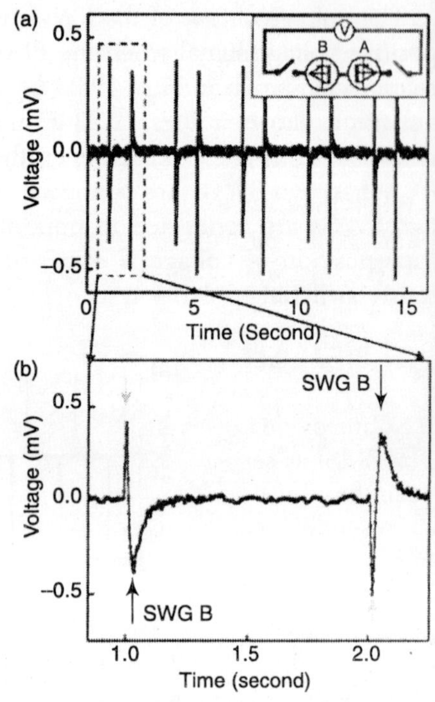

Figure 19.24 *(a) Voltage output of two reversely connected SWGs in serial when the deformation of the two were slightly off synchronization. The inset illustrates the connection configuration of the SWGs. (b) An expanded output of two double peaks in (a) to illustrate the sharp switch in output voltage.*

By taking the average value from the output signals when the SWGs are bent, the output current and voltage for Fig. 19.19 and 19.20 are tabulated in Table 19.1 Now consider; V^+_A and I^+_A as the voltage and current measured from SWG A under forward connection, and V^-_A and I^-_A when reversely connected. The same is applied to SWG B. In an ideal case $V^-_A = -V^-_A$;. and $I^-_A = -I^+_A$. However, with consideration of the contribution from the bias current from the measurement system, the measured voltage/current does not have the same magnitude. In any case, a true electricity output from an SWG changes its sign when the, SWG is reversely connected. The first two rows in Table 19.1 show that the SWG A and SWG B satisfy the first criteria of switching-polarity test for both output current and voltage.

Table 19.1 *The current and voltage output when the SWG A and SWG B are connected under various configurations illustrated in Figs. 19.22 and 19.23*

	Current (pA)		*Voltage (mV)*	
	Forward connection	*Reverse connection*	*Forward connection*	*Reverse connection*
SWGA	9.54	–2.42	2.13	–0.36
SWG B	7.31	–0.7	1.81	–0.15
SWG A + SWG B	16.2	–3.76	3.18	0.37
SWG A – SWG B	7.33	4.65	1.46	1.5

The last two rows in Table 19.1 demonstrate the second criteria of linear superposition of current and voltage under eight connection configurations. When the SWG A and SWG B are connected in parallel, as shown in the insets in Figs. 19.19 (e) and 19.19 (f), the measured currents (last two rows in Current columns) obey the following requirements:

$$I^+_{A+D} = I^+_A + I^+_D,\quad I^-_{A+B} = I^+_A + I^+_B,\quad I^+_{A-B} = I^+_A + I^+_B,\quad I^-_{A-B} = I^-_A + I^-_B$$

When the SWG A and SWG B are connected in series, as shown in the insets in Figs. 19.23 (e) and (f), the measured voltage (last rows in voltage columns) obey the following requirements:

$$V^+_{A+B} = V^+_A + V^+_B,\quad V^-_{A+B} = V^+_A + V^+_B,\quad V^+_{A-B} = V^+_A + V^+_B,\quad V^-_{A-B} = V^-_A + V^-_B$$

(E) PIEZOELECTRI CHARACTERIZATION OF ZINC OXIDE NANOBELT

Zinc oxide is a semiconducting piezoelectric material used in microelectromechanical systems (MEMS) as sensors and actuators and in communications as surface acoustic wave (SAW) and thin-film bulk acoustic wave resonator (FBAR) devices. Quasi-one-dimensional ZnO nanobelts are synthesized, and used for the fabrication of field effect transistors, gas sensors, resonators, and nanocantilevers. These applications mainly utilize the semiconducting properties and geometrical shape and size offered by nanobelts ZnO nanobelts dominated by different crystal surfaces have been synthesized. The most common growth direction of the nanobelt is along the c-axis [0001] or [01$\bar{1}$0], with the top surface being (2$\bar{1}\bar{1}$0) and side surface being (01$\bar{1}$0) or (0001), respectively (Figures 19.25(a) and (b). $\bar{1}$ Growth of (0001) surface dominated nanobelts overcomes an energy barrier due to surface polarization. Recently, free-standing piezoelectric ZnO nanobelts dominated by large (0001) top and bottom surfaces (Figure 19.25(c) are synthesized. Individual zinc oxide nanobelt is a promising piezoelectric material for nanosensor and nanoactuator applications due to its perfect single crystalline structure and that it is free of dislocation.

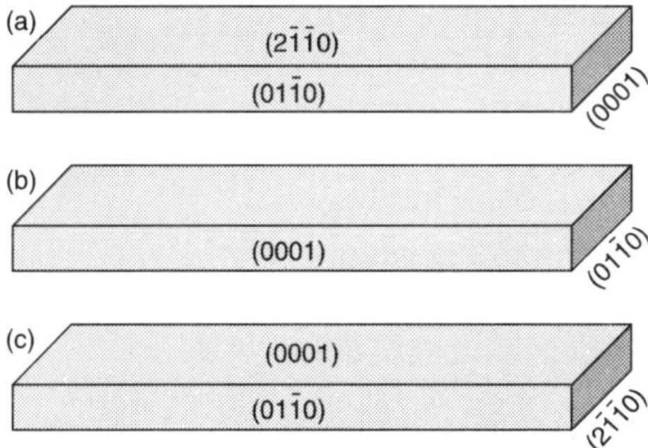

Figure 19.25 *Surface facets of ZnO nanobelts: (a) growing along [0001] (c axis), top surfaces (2$\bar{1}$ $\bar{1}$ 0), and side surfaces (01$\bar{1}$ 0), showing no piezoelectric property across thickness; (b) growing along [01$\bar{1}$ 0] (c axis), top surfaces (2$\bar{1}$ $\bar{1}$ 0), and side surfaces (0001). showing no piezoelectric property across thickness; (c) growing along [2$\bar{1}$ $\bar{1}$ 0] (a axis), top surfaces (0001), and side surfaces (01$\bar{1}$ 0). showing piezoelectric effect across thickness.*

The piezoelectric properties of ZnO nanobelt are known by using atomic force microscopy however, a challenge is there because of the sample's displacement due to the inverse piezoelectric effect by applying an electric field. Several techniques, including scanning probe microscopy (SPM), are employed to measure these small piezoelectric displacements. In particular, piezoresponse force microscope (PFM) is becoming a standard method to know the ferroelectric and piezoelectric phenomena. The PFM technique is based on the detection of local vibrations induced by an AC signal applied between the conductive tip of SFM and the bottom electrode of the sample. The local oscillations of the surface are transmitted to the tip and detected using a lock-in technique. The out-of-plane piezoresponse signal is extracted from the z-deflection signal given by the PSD (position sensitive detector) and represents the local oscillations perpendicular to the plane of the surface example; ferroelectric properties of individual barium titanate nanowires investigated by noncontact mode of SPM are made yet there are no piezoelectric measurement of one-dimensional nano-structures using the contact mode SPM.

The measurements of the piezoelectric properties of individual ZnO nanobelts using the PFM technique are described below. The result is compared with that of the (0001) bulk ZnO and *x*-cut quartz. It shows that the effective piezoelectric coefficient d_{33} for a ZnO nanobelt is larger than that of bulk ZnO which establishes. The basis for using ZnO nanobelts for nanoscale sensors and actuators.

Sample preparation: After coating (100) Si wafers with 100 nm Pd, ZnO nanobelts with typical dimension of tens of nanometers in thickness, hundreds of nanometers in width, and tens of micrometers in length are dispersed on the conductive surface. The whole surface is coated further, with another 5 nm Pd coating which serves is an electrode on the ZnO nanobelt to get a uniform electric field and avoid electrostatic effects. Ensure that the top and bottom surface of the nanobelt is not short-circuited after Pd deposition.

The ZnO nanobelt is located by AFM in tapping mode. If contact mode is used, the nanobelt is displaced by the tip during scanning, which causes a distorted image. Figure 19.26 is a three-dimensional image of an individual ZnO nanobelt lying on the surface as obtained in tapping mode, which shows a rectangular cross section. After locating the nanobelt, the tip is positioned to the center of the nanobelt.

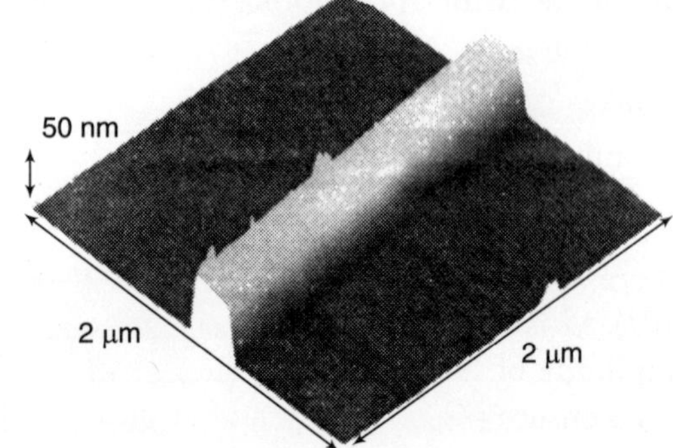

Figure 19.26 *AFM 3D (2μ m × 2 μ m) image of an individual nanobelt (360 nm in width and 65 nm in thickness) lying on the substrate surface, showing a rectangular cross section.*

Note that not every nanobelt lying on the surface is piezoelectric. The nanobelt showing piezoelectric characteristics grows along $[2\bar{1}\bar{1}0]$ with a (0001) top surface, as presented in Figure 19.26(c). The polar axis of the hexagonal wurtzite structured ZnO crystal is along [0001]. Piezoelectric measurements on the ZnO nanobelt are performed in contact mode at a single point with the addition of a function generator, a lock-in amplifier and a signal access module as shown in Figure 19.27.

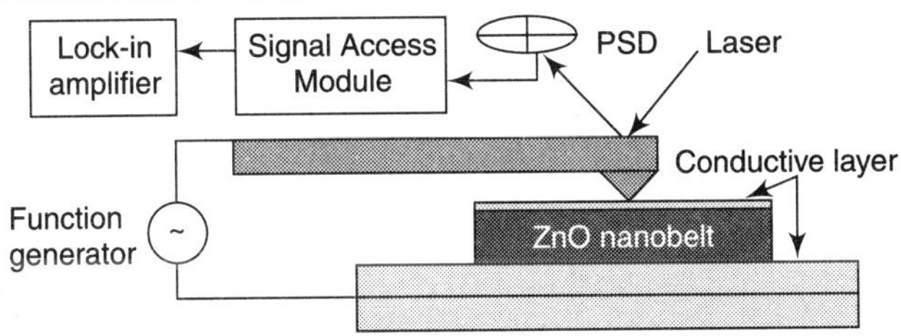

Figure 19.27 *Diagram of experimental setup.*

Here the conductive tip supplies current to the electrode and also measures the piezoelectric motion. The conductive tip is made by coating Pd (20 nm thick) on a etched silicon probe with nominal spring constant of 42 Nm^{-1} and tip radius of about 10 nm. A high stiffness cantilever is chosen to reduce the influence or electrostatic interaction with piezoelectric measurement.

The contact force between the AFM tip and nanobelt is ~1800 nN. This ensures that the measurement is in the strong-indentation regime as the piezoresponse in the strong-indentation regime is dominated by the d_{33} of the material. The typical resonance of the conductive tip is about 300 kHz. The frequency of the signal applied on the sample is from 30 kHz to 150 kHz; much higher than the low-pass cutoff frequency of the AFM topography feedback loop and lower than the cantilever resonance frequency. The frequency range chosen is from 1 to 16.7 kHz because higher the applied frequency the lower is the noise signal level, which is below 10^{-13} m in the chosen frequency range. The input signal is in the range of 1–4 V (RMS). The corresponding vertical deflection signal of the cantilever is recorded by lock-in amplifier. By multiplying the deflection signal with the calibration constant of the photodetector sensitivity,. the amplitude of the tip vibration is derived. The calibration constant is determined from the slope of the force-distance plot obtained after the scanner is calibrated using a standard grating. Since the scanner is independently calibrated with a known step height, the vertical deflection signal is calibrated to the known vertical displacement. The slope of the amplitude A_F versus the input signal U_f gives the effective piezoelectric coefficient d_e^{eff}.

$$A_f = V_1\ \delta = d_e^{\text{eff}} U_f \tag{1}$$

where A_f is vibration amplitude (in units of nm), V_f is vertical deflection signal of the cantilever (mV), δ is the calibration constant of the photodetector sensitivity (nm/V), and U_f is the amplitude of the testing ac voltage (V).

To ensure the reliability and accuracy of the measurement, effective piezoelectric coefficients of (0001) bulk ZnO and *x*-cut quartz (serving as a piezoelectric standard) are measured using the same PFM technique. As to (0001) bulk ZnO (5 × 5 × 0.5 mm) both side are coated with 100 mm Pd and the bottom side is attached to the conductive surface using conductive epoxy.

A similar process, is applied to *x*-cut quartz (ϕ 10 × 0.7 mm, gold coating of 150 nm on each side.

The results of piezoelectric measurements on the ZnO nanobelt with (0001) top surface, (0001) ZnO bulk, and x-cut quartz are shown in Figure 19.28(a). From the slopes of the curve, the effective piezoelectric coefficients are $d_{11} = 2.17$ pmV^{-1} for *x*-cut quartz and $d_{33} = 9.93$ pmV^{-1} for (0001) bulk ZnO, which are independent of frequency. Compared to the accepted values,

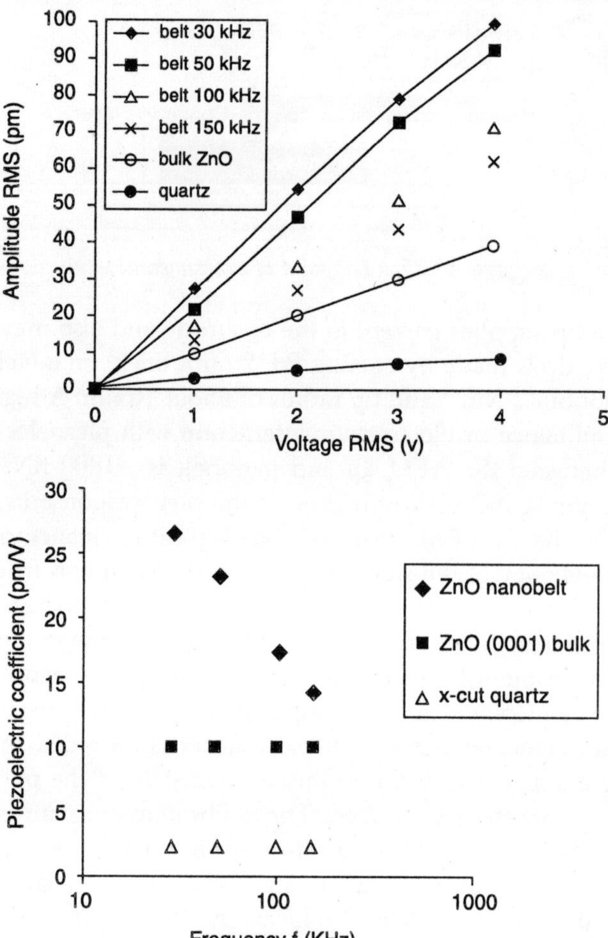

Figure 19.28 *(a) Piezoelectric measurements of ZnO nanobelt, bulk (0001) ZnO, and x-cut quartz. The linear relationship between amplitude and applied voltage is shown in every case, the slope of which gives the piezoelectric coefficient. (b) Frequency dependence ,of piezoelectric coefficient of ZnO nanobelt, bulk (0001) ZnO, and x-cut quartz. Only the piezoelectric coefficient of ZnO nanobelt is frequency dependent.*

$d_{11} = 2.3$ pmV^{-1} for quartz and $d_{33} = 12.4$ pmV^{-1} for ZnO,. the measured results for ZnO bulk and *x*-cut quartz, confirm the reliability and accuracy of the measuring technique. The effective piezoelectric coefficient of the ZnO nanobelt is, frequency dependent and varies from 14.3 to 26.7: pmV^{-1}, which is much larger than that of the bulk ZnO. One possible reason for the observed enhanced electromechanical response for the ZnO nanobelt is due to its perfect single crystallinity and freedom of dislocation, as inherent reduction in the piezoelectric coefficient due to internal defects is revealed in ultrathin PZT films.

An increase of piezoelectric response with decreasing feature size in epitaxial PZT thin film is achieved. The observed increase of piezoresponse amplitude is 300% with decreasing the film thickness from 200 to 100 nm, which is due to a change of domain configuration. But this mechanism is not applicable to the ZnO nanobelt of tens of nanometer in thickness, Further the different elastic boundary condition in piezoelectric measurement on the ZnO nanobelt and bulk sample contributes to the enhanced piezoelectric response. As to the ZnO nanobelt, there is no constraint at the interface between the bottom side and the conductive layer, while conductive epoxy is applied in the case of bulk sample. Hence, the effective piezoelectric coefficient of the ZnO nanobelt and bulk is given by equations 2 and 3 using lateral free and lateral full constraint boundary conditions, respectively.

$$d_{33\ \mathrm{belt}}^{\mathrm{eff}} \cong d_{33} \tag{2}$$

$$d_{33\ ,\mathrm{bulk}}^{\mathrm{eff}} \cong d_{33} - \frac{2S_{13}}{S_{11}+S_{12}} d_{31} \tag{3}$$

Assuming that d_{33} is the same in equations 2 and 3, it is interesting to estimate the magnitude of the second term in equation 3, which provides the effect of boundary condition. However, the components (S_{ij}) of compliance matrix (S) of ZnO are not directly available and are derived from the stiffness matrix C using the relationship $S = C^{-1}$.

$$S = \begin{bmatrix} 6.401 & -1.932 & -2.539 & 0 & 0 & 0 \\ -1.932 & 6.401 & -2.539 & 0 & 0 & 0 \\ -2.539 & -2.539 & 8.351 & 0 & 0 & 0 \\ 0 & 0 & 0 & 14 & 0 & 0 \\ 0 & 0 & 0 & 0 & 14 & 0 \\ 0 & 0 & 0 & 0 & 0 & 17 \end{bmatrix} \times 10^{-12} (m^2 N^{-1}) \tag{4}$$

Using the S_{ij} value from equation 4 and $d_{31} = -5.1$ pmV^{-1} the second term in equation 3 to be 5.795 pmV^{-1}. Taking $d_{33} = 12.4$ pmV^{-1} (the accepted piezoeletric coefficient of single-crystal ZnO), the conclusion that the boundary condition change the effective piezoelectric coefficient of ZnO up to 47%.

The frequency dependence of the piezoelectric coefficient is shown in Figure 19.28(b). Both ZnO bulk and *x*-cut quartz are frequency independent. As to the ZnO nanobelt, the higher the frequency (30–150 kHz), the lower the piezoelectric. coefficient d_{33}, which depends linearly on the logarithm of the applied field frequency. Such logarithmic frequency dependence is reported in ferroelectric materials such as PZT. Even though ZnO is not ferroelectric, the logarithmic frequency dependence is generally valid in random systems that have properties controlled by interface pinning, such as pinning of spontaneous polarization in ZnO, which is caused by surface charge due to the high surface-to-volume ratio of the ZnO nanobelt. Another possible reason for the unexpected frequency dependence originates from the imperfect electrical contact between' the bottom of the nanobelt and the conductive layer. With increasing the input frequency, the quality of electrical contact decreases, which induces the lower electromechanical response of the nanobelt.

(F) PIEZOELECTRIC AND INDUCED POTENTIAL DISTRIBUTION

To consider the potential distribution in the wire/belt based on piezoelectric induced potential distribution, the polarization is introduced in a belt as a result of elastic deformation. The relationship between strain (ε) and the local piezoelectric field (E) is given by $\varepsilon = dE$, where d is the piezoelectric coefficient. For a belt of thickness T and length L as shownin Figure 19.29(a). A under the displacement of an external force F from the AFM tip applied perpendicularly at the top of the belt ($z = L$), a strain field in the belt is formed. For any segment of the belt along its length, the local bending is described the following analytical analysis.

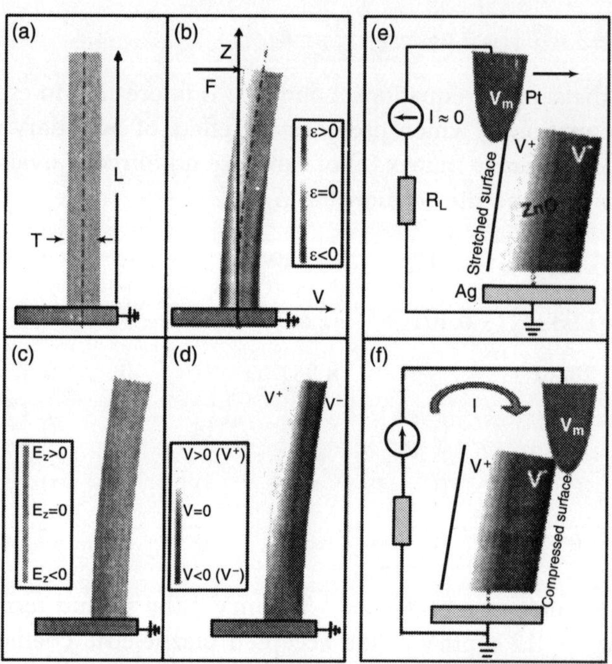

Figure 19.29 *(a) Definition of a belt. (b) Longitudinal strain ε_z distribution in the belt after being deflected by an AFM tip from the side. (a) The corresponding longitudinal piezoelectric induced electric field E_z distribution in the belt with. (d) Potential distribution in the belt as a result of piezoelectric effect, with the stretched and compressed side surfaces being positive and negative potentials, respectively. (e,f) Metal and semiconductor contacts between the AFM tip and the semiconductor ZnO belt at two reversed local contact potentials (positive and negative), showing reverse and forward biased Schottky rectifying behavior, respectively.*

Analytical Analysis of the Bending

For the following calculation, consider the in-plane bending of a nanowire by an external point force F perpendicular to the nanowire. For a short segment of nanowire, the local bending is approximated by a small arc characterized by a small angle θ and a local curvature of radius R (see Fig. 19.30), the local strain is

$$\varepsilon = \frac{H'G' - HG}{HG} = \frac{H'G' - HG}{EF} \approx \frac{(R+y)\theta}{R\theta} = \frac{y}{R}$$

Using the stress and strain relationship, we have:

$$\sigma = Y\varepsilon = Y\frac{y}{R} \qquad \text{(i)}$$

where Y is the elastic modulus, In the static state of the system the moments are balanced. The total moment of the bent nanowire is

$$M = \int y \cdot \sigma \cdot dA$$

By using equation (1)

$$M = \int \frac{Y}{R} \cdot y^2 \cdot dA = \frac{Y}{R} \cdot \int y^2 \cdot dA = \frac{Y}{R} \cdot I$$

we have $$\frac{M}{I} = \frac{Y}{R} = \frac{\sigma}{y} \qquad \text{(ii)}$$

From the geometry shown in Fig. 19.30(b), the following result is obtained under small angle approximation:

$$\frac{dy}{dz} = \tan\,\theta \approx \theta \quad \text{thus} \quad \frac{1}{R} = \frac{d\theta}{ds} \approx \frac{d\theta}{dz} = \frac{d}{dz}\left(\frac{dy}{dz}\right) = \frac{d^2y}{dz^2}$$

In combining with equation (2),

$$\frac{1}{R} = \frac{M}{YI} = \frac{d^2y}{dz^2} \qquad d\theta = \frac{1}{R} \cdot dz = \frac{M}{YI} \cdot dz \qquad \text{(iii)}$$

By the principle of balancing of moment: internal moment = external moment

$$M = YI\frac{d^2y}{dz^2} = F \cdot (L - z) \qquad \frac{d^2y}{dz^2} = \frac{M}{YI} = \frac{F \cdot (L-z)}{YI} \qquad \text{(iv)}$$

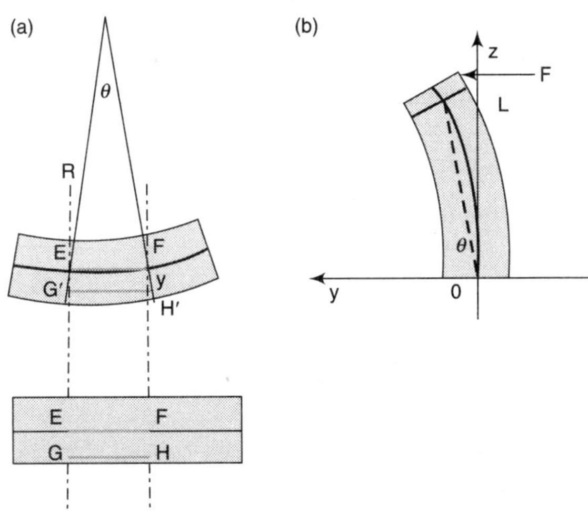

Figure 19.30 *(a) A segment of a nanowire for calculating the bending induced piezoelectric otential across the two side surfaces. (b) The definition of coordination system under an external displacement force F.*

where the external force is assumed to act approximately perpendicular to the nanowire at the top and L is the nanowire length, Equation (iv) determines the bending of the nanowire under the deflection of an external force at the front end.

Now, $$\frac{1}{R} = \frac{d^2y}{dz^2}$$

which is from the geometrical shape of the curved belt as described in calculus (see Figure 19.29(b)). The shape of the belt is described by the static deflection equation of the belt

and $$\frac{d^2y}{dz^2} = \frac{F(L-z)}{YI}$$

where Y is the elastic modulus of the belt and I is its momentum of inertia. The local strain in the belt is $\varepsilon = y/R$, and the corresponding electric field along z axis is (Figure 26(c).

Piezoelectric Induced Potential

From the definition of the piezoelectric coefficient (d): $d = \frac{\varepsilon}{E}$, the corresponding electric field along z-axis is

$$E_z = \frac{\varepsilon}{d} = \frac{y}{dR} \quad \text{(v)}$$

This is the electric field that dominates the potential distribution under small bending approximation and the ignorance of the electric field effect on local strain via the piezoelectric effect. Consider the potential at the side surface $y = a$ along the entire length of the nanowire:

$$E_z = \frac{a}{d \cdot R} \quad \text{(vi)}$$

For simplicity of analytical calculation to illustrate the physical principle, the potential at the two side surfaces is $y = \pm T/2$. By integrating the electric field along the entire length of the belt

$$V^{\pm} = \int E\, ds = \pm\int \frac{T}{2d}\frac{1}{R}\, ds = \pm\frac{T}{2d}\int d\theta = \pm\frac{a}{d}\theta_{max} \quad \text{(vii)}$$

where θ_{max} is the maximum deflection angle at the top of the wire.
Using equations (iii) and (iv); equations (vii) gives

$$V = \frac{a}{d}\int \frac{M}{YI} \cdot dz = \frac{a}{d}\int \frac{F \cdot (L-z)}{YI} \cdot dz = \frac{aF}{dYI}\int_0^L (L-z) \cdot dz = \frac{aFL^2}{2dYI} \quad \text{(viii)}$$

Together with the relationship between the maximum deflection y_m and the applied force: $F = \frac{3YIy_m}{L^3}$, the potential induced by piezoelectric effect at the stretched and compressed side surfaces, respectively is:

$$V = \frac{3ay_m}{2Ld} = \pm 3Ty_m/4Ld \quad \text{(ix)}$$

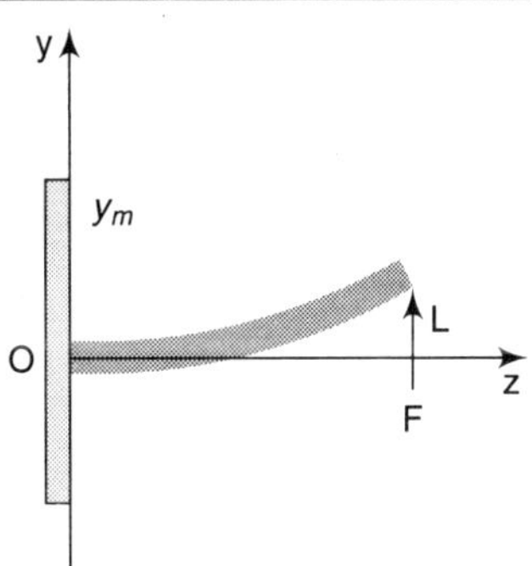

Figure 19.31

It must be pointed out that this is a qualitative estimation of the potential induced by PZ effect for understanding the physical process. Numerical calculation with considering the boundary conditions and dielectric screening is required to get quantitative results.

Deflection of the Nanowire Under External Force

Now consider the deflection of the nanowire under an externally applied force F perpendicular to the nanowire (see figure 19.31). From equation (iv) and boundary conditions:

$$y(z=0)=0;\ \frac{dy(z=0)}{dz}=0. \qquad \text{(x)}$$

Under small deflection and assuming that F is constant, integrating equation (iv) gives

$$\frac{dy}{dz}=\frac{F}{YI}\left(Lz-\frac{1}{2}z^2\right) \qquad \text{(xi)}$$

and

$$y=\frac{F}{YI}\left(\frac{1}{2}Lz^2-\frac{1}{6}z^3\right)z \qquad \text{(xii)}$$

For $z=L$, $y=y_m$

$$y_m=\frac{FL^3}{3YI} \quad F=\frac{3YIy_m{}^3}{L^3} \qquad \text{(xiii)}$$

The mechanical work done by the external force for bending of the nanowire:

$$W=\int\vec{F}\cdot\vec{dl}=\int_0^{y_m}F(y)dy=\int_0^{y_m}\frac{3YI}{L^3}\cdot y\cdot dy=\frac{3YI}{2L^3}y_m^2 \qquad \text{(xiv)}$$

Discharge of an RC Circuit

As shown in figure 19.32 on the right-hand side:

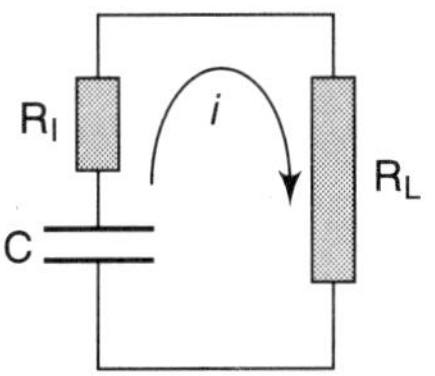

Figure 19.32

$$\frac{1}{C}\int i\cdot d_t+(R_L+R_I)\cdot i=0 \qquad \text{(xv)}$$

$$\frac{i}{C}+(R_L+R_I)\frac{di}{dt}=0 \qquad \text{(xvi)}$$

$$i=I_0\cdot e^{\frac{t}{(R_L+R_I)C}} \qquad \text{(xvii)}$$

For initial condition: $t=0$, $i=\dfrac{V_0}{R_L+R_I}$,

$$\therefore \qquad i=\frac{V_0}{R+r}\cdot e^{\frac{t}{(R_L+R_I)C}} \qquad \text{(xviii)}$$

The decay time constant $\tau = (R_L + R_I)C$

The electric power consumed by the resistor R (the output work):

$$\Delta W_{PZD} = \int i^2 R \cdot dt$$

$$= \frac{V_0^2 R_L}{(R_L + r_I)} \int_0^\infty e^{-\frac{2t}{(R_L + R_I)C}} \cdot dt$$

$$= \frac{V_0^2 R_L C}{2(R_L + r_I)}$$

For $R_I << R_L$

$$\Delta W_{PZD} \approx \frac{V_0^2 C}{2} \qquad \text{(xix)}$$

The elastic deformation energy of the nanowire dissipated in the first cycle of the resonance

To calculate the energy loss after a single cycle of vibration, it is assumed that the amplitude of the nanowire resonance decays with time as described by:

$$y_m = y_{mo} e^{-t/t_0} \qquad \text{(xx)}$$

where y_{mo} is the maximum deflection pushed by the AFM tip, and t_o is the decay constant of the vibration amplitude. Therefore, the elastic energy at a subsequent vibration from equation (xiv) is

$$W_{ELD} = \frac{3YI}{2L^3} y_m^2 = \frac{3YI}{2L^3} y_{m0}^2 \cdot e^{-2t/t_0} \qquad \text{(xxi)}$$

The energy dissipated after one cycle of vibration is

$$\Delta W_{ELD} = \frac{3YI}{t_0 L^3} y_{m0}^2 \cdot e^{-2t/t_0} \cdot \Delta t \qquad \text{(xxii)}$$

First the cycle: $t = 0$ and $\Delta t = \frac{1}{f}$:

$$\Delta W_{ELD} = \frac{3YI}{t_0 L^3} y_{m0}^2 \cdot \frac{1}{f} \qquad \text{(xxiii)}$$

Under the first order approximation, the life time of the resonance t_o is related to the quality factor Q and resonance frequency f by

$$t_0 \approx \frac{\sqrt{3}Q}{\pi f} \qquad \text{(xxiv)}$$

Therefore

$$\Delta W_{ELD} = \frac{3YI}{2L^3} y_{m0}^2 \cdot \frac{2\pi}{Q\sqrt{3}} = W_{\max} \cdot \frac{2\pi}{Q\sqrt{3}}$$

For ZnO nanobelts in a vacuum of 10^{-7} mm Hg, $Q = 600$.

(F) FIBER-BASED NANOGENERATORS AS FLEXIBLE AND FOLDABLE POWER SOURCES

The ceramic or semiconductor substrates used for growing ZnO NWs are hard, brittle, and cannot be used in areas that require a foldable or flexible power source, such as implantable biosensors in muscles or joints, and power generators built into walking shoes. It is necessary to use conductive polymers, which are likely to be biocompatible and safe, as substrates. Two advantages are offered by this approach. One is the cost-effective, large-scale solution to grow ZnO NW arrays at a temperature below 350 K. The other is the large degree of choice of flexible plastic substrates for growing aligned ZnO NW arrays, which plays an important role in flexible and portable electronics for harvesting low-frequency (10 Hz) energy from the environment, such as from body movement (e.g., gestures, respiration, or locomotion).

Recently fiber-based NGs are achieved. The design of the NG is based on the zigzag-electrode mechanism, but the zigzag electrode is replaced by an array of metal wires, as shown in Figure 19.33(a). By brushing the metal NW arrays across the ZnO NW arrays, the metal wires act like an array of AFM tips that induce the mechanical deformation to the ZnO nanowires and collect the charges. A unique advantage for ZnO NW arrays is that they can be grown at 350 K on substrates of any shape and any material. The metal NW arrays are made by metal coating of ZnO nanowire arrays grown on Kevlar fibers (Fig. 19.33(b). In practice, any

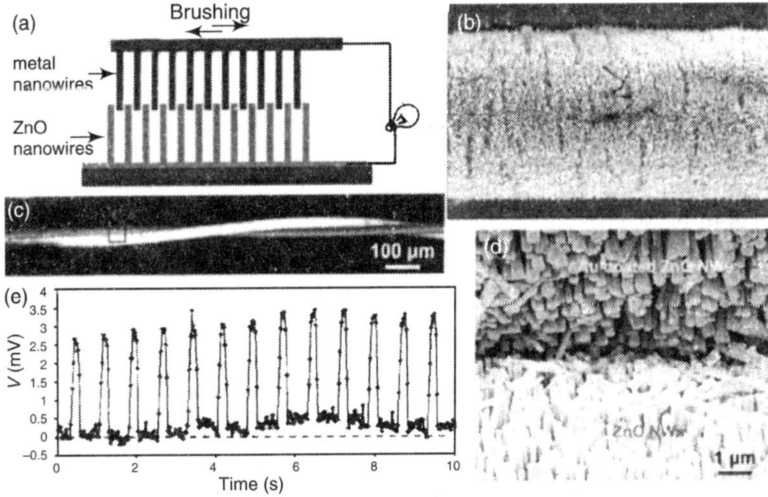

Figure 19.33 *The design and mechanism of the fiber-based NG as driven by a low-frequency, external vibration/friction/pulling force. (a) Idea of "two. brush" NG. One brush is made of ZnO nanowires, and the other brush is metal nanowires. (b) SEM image showing the distribution of NWs grown on a fiber surface. (c) An optical micrograph of a pair of entangled fibers, one of which is coated with Au (in darker contrast). (d) SEM image at the "teeth-to-teeth" interface of two fibers covered by NWs, with the top one coated with Au. The Au-coated NWs serve as the conductive "tips" that deflect the NWs at the bottom: a piezoelectric-semiconducting coupling - process, generates electric current. (e) The piezoelectric potential output by the two-fiber NG under the pulling and releasing of the top fiber by an external force.*

fiber works as long as it has good electrical conductivity. The metal to be coated is required to form a Schottky contact with ZnO. Entangling the two fibers, one coated with Au and one without, sets the principle of the fiber-based NG (Fig. 19.33). The NWs on the two fibers are "teeth-to-teeth" (Fig. 19.33(d)), a relative deflection by a distance as short as one NW size is sufficient to generate electricity. A cycled relative sliding between the two fibers produces output voltage and current owing to the deflection and bending of the ZnO nanowires (Fig. 19.33(e). This is the fiber-based NG, with potential for harvesting energy from body movement, muscle stretching, light wind, and vibration. It establishes the basis for building a "power-shirt" to wear.

(G) PLASTIC SUBSTRATES

The ZnO NWs are grown in solution as below. The plastic substrates used are 50 μm thick polyimide films, Zinc nitrate hydrate ($Zn(NO_3)_2{\cdot}6H_2O$) and hexamethylenetetramine.

Synthesis of Aligned ZnO Nanowires: The aligned ZnO NWs arrays are grown on a plastic film using a solution-based method. A thin, 100 nm, layer of Au is deposited on the plastic substrate by using thermal evaporation at 0.3-0.5 Å s^{-1} before the hydrothermal growth of ZnO NW arrays. The growth of ZnO NWs is conducted by suspending the Au-coated plastic substrate in a Pyrex glass bottle filled with an equal molar aqueous solution of $Zn(NO_3)_2 \cdot 6\ H_2O$ (0.01-0.04 M) and hexamethylenetetramine (0.01-0.04M) at temperatures between 330 and 350 K. The temperature, solution concentration, reaction time (1–72 h), and substrate surface quality are optimized for growing NW arrays with controlled dimensions and orientation. After reaction, the plastic substrates are removed from the solution rinsed with deionized water, and dried in air at 330–350 K overnight.

Morphology and Structure Characterization: The as-grown ZnO NWs are characterized with a scanning electron microscope at 5 and 10 kV and a transmission electron microscope 200 kV.

Electrical Measurement: Electrical measurements on the aligned NW array are conducted using an atomic force microscope.

Nanowire/Plastic Substrate Interconnection Reinforcement: To reinforce the interconnection between the. Au-coated plastic substrate and the ZnO microwire array, a matrix of spin-coated PMMA 200 nm thick is used to encapsulate the bottom portion of the NW array. The photoresist is spun onto the substrate at a speed of 3000 rpm for 35 s. The substrate is then baked at 390 K for 1 min to remove the solvent. U V light with a wavelength of 325 nm and an intensity of 3.1 $mWcm^{-2}$ is utilized to expose the as-coated sample for 10 min to crosslink the PMMA photoresist and also to expose the microwire tip portions for late circuit connection in the AFM measurement.

Figure 19.34 shows a series of scanning electron microscopy (SEM) and transmission electron microscopy (TEM) images of typical ZnO NW arrays grown on a conductive plastic substrate. Figure 19.34(a) shows a low-resolution, top-down view of the densely aligned NW arrays. The aligned NWs have a uniform diameter of 200-300 nm and a hexagonal cross section, as indicated by the magnified SEM image in Figure 19.34(b). The SEM image in Figure 19.34(c) displays ZnO NWs 2μm long aligned perpendicular to the plastic substrate. Figure 19.34(d) is a typical TEM image of a ZnO NW that has a diameter of 300 nm. The corresponding electron diffraction pattern indicates that the NW growth direction is [0001] and its side surfaces are $\{2\bar{1}\bar{1}0\}$.

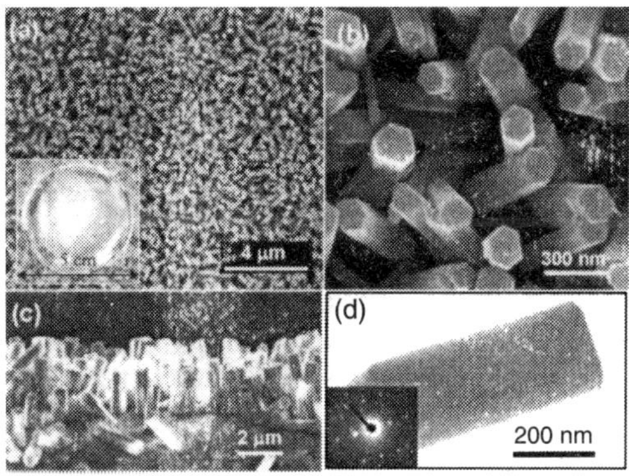

Figure 19.34 *A series of SEM and TEM image of typical ZnO NW arrays grown on a plastic substrate. (a) A low-magnification, top-down view of the densely aligned NW arrays and (b) a corresponding higher-resolution SEM image. The inset in (a) is an optical image of a sample of NWs on a polymer surface. (c) An SEM image of a side view of the sample displaying ZnO NWs aligned normal to the plastic substrate. (d) A typical TEM image of an individual ZnO NW; inset is the corresponding electron diffraction pattern of the selected area.*

By controlling reaction conditions such as temperature, concentration, pH value, reaction time, and plastic-substrate surface quality, NWs of different density distributions, dimensionality, and alignment are fabricated for electrical measurements. Figure 19.35(a) and (b) shows low-resolution and magnified SEM images, respectively, depicting sparsely grown ZnO NW arrays on a plastic substrate. The NWs have a density of 1 μm^{-2} on a Au-coated plastic substrate. The NWs are typically 100–350 nm wide and 1 μm long. The SEM image shown in the inset of Figure 19.35(b) is a side view of the NWs aligned perpendicularly to the plastic substrate.

Figure 19.35(c) is the experimental setup used for measuring the mechanically induced piezoelectric discharge from individual NWs. A conductive Si tip coated with a Pt film with a cone angle of 70° is used for AFM measurements. The rectangular cantilever has a calibrated normal spring constant of 1.857 N m^{-1}. In the AFM contact mode, a constant normal force of 5 nN is maintained between the tip and sample surface. When the tip is scanned over the top of the ZnO NWs, the tip height is adjusted according to the surface morphology and local contacting force. For the electric contact at the bottom of the NWs, Ag paste is applied to connect the Au film on the plastic substrate surface to the measurement circuit. Connecting Ag and ZnO produces an Ohmic contact. The output voltage across an external load of resistance R_L = 500 MΩ is continuously monitored as the tip is scanned over the NWs. In contact mode, as the tip is scanned over the vertically aligned NWs, the NWs are bent consecutively. The tip forces the elastic deflection of the oriented ZnO NWs and produces a charge separation and a voltage drop across the diameter of the NWs, with the stretched and compressed sides having positive and negative piezoelectric potentials, respectively. The center axis of the NW, as indicated by the dotted line, remains neutral.

As the conductive tip is scanned across the neutral axis, a discharge occurs when the tip touches the compressed side of the NW. A single NW that has a diameter of 300 nm produces an output voltage discharge of 45 m V (Fig. 19.35(d), which is the voltage drop across an

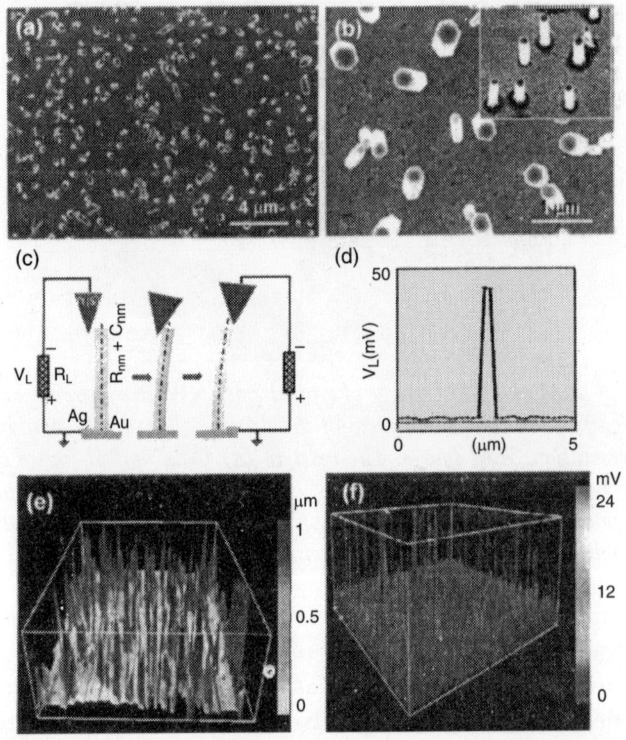

Figure 19.35 *(a) A low-magnification and (b) a higher-resolution SEM image showing a sparsely grown ZnO NW array on a plastic substrate. (c) Experimental setup for the conductive AFM measurement of the NW array. Note the polarity of the defined output voltage on the external load. (d) A typical voltage output profile for the AFM tip scanning over a single NW on the plastic substrate. (e) A 3D plot of the AFM topography image and (f) a voltage output profile obtained by scan. ning the AFM tip over a 40 μm × 40 μm area of aligned ZnO NWs on a plastic substrate.*

external resistor, converts the measured electrical current. In fact, the true output voltage is higher if the inner resistance of the NW is considered. Figure 19.35(d) is a typical voltage output for an AFM tip scanning over a single NW. Because of the limited scan speed of the AFM tip in comparison to the discharge time, only two data points are observed around the peak area, which shows that the true peak is twice as large as the measured 45 mV shown in this profile.

The scanning speed of the AFM tip is 53.51 μm s^{-1}, and the full width at half maximum (FWHM) of the peak in Figure 19.35(d) is 234 nm. Therefore, the lifetime decay constant of the circuit shown in Figure 19.35(c) is estimated to be $\tau_c = 4.4$ ms, based on the equation $\tau_c = (R_L + R_{nw}) C_{nw}$, where R_L and R_{nw} are the resistances for the external load (500 MΩ) and the ZnO NW, respectively, and C_{nw} is the capacitance of the NW and the measurement system. As R_{nw} is negligible in comparison to the external load; the capacitance of the NW can be calculated by $C_{nw} \approx \tau_c/R_L$. The output piezoelectric energy, W_o, for a single pulse is $W_o = 1/2 C_{nw} V_p^2 =$

$\tau_c V_p^2/2R_L = 8.9 \times 10^{-15}$ J, where V_p is the peak voltage. Here W_o only represents the harvested electrical energy from the first half cycle of the NW resonance that results from a single touch of the tip to the NW.

Power is calculated form the average energy generated within the resonance lifetime, and it is assumed that the energy generated for half of the resonance cycles is collected. The individual NWs are 300 nm in diameter and exhibit a hexagonal cross section, and are 1 μm in length. The first harmonic resonance frequency of the NW is calculated to be 70 MHz; considering the damping effect in air, the lifetime of a single NW is 50 μs. Therefore, the effective number of cycles required for energy harvesting is about 3500. By considering the decay of the vibration amplitude of a NW for the subsequent resonance, the output-power of a single NW is 10–20 pW.

Figure 19.35(e) is a 3D plot of a topographic AFM scan of a 40 μm × 40 μm area of the ZnO NW array. The scanning direction is from the front to the back, which is seen from the raised linear traces on the flexible plastic substrate. Because of the flexibility of the plastic substrate and in spite of the firm adhesion to the Ag paste, the substrate surface profile under tip contact is wavy. Relative to the plastic substrate, the NW array reveals to have a height distribution of 0.5–2.0 μm. The corresponding 3D plot of the voltage output image is shown in Figure 19.35(f). There are a number of sharp peaks ranging from 15 to 25 mV that represent voltage outputs. By counting the pulse numbers with respect to the number of NWs in the topography profile in Figure 19.35(e), the ratio of voltage peaks to the number of available NWs is 90:150 or about 60%, which suggests that the discharge events of the NWs captured by the AFM tip correspond to at least 50 % of the NWs. In fact, because of the limitation in the data-collecting speed of the atomic force microscope, which has a data step size of 156.2 nm in this measurement, the discharge peaks of some NWs are missed, possibly due to poor contact between the tip and the NWs and/or due to multiple contacts with neighboring NWs. Therefore the suitable bonding strength between the ZnO NWs and the polymer substrate and a uniform density distribution of NWs in the array improves the piezoelectric discharge efficiency.

The total number of NWs that effectively produced electrical energy output is calculated by assuming an average voltage peak height of 20 mV, and the density of ZnO NWs on the plastic substrate to be μm^{-2}. The power density per unit substrate area is 1–2 $pW\mu m^{-2}$, that is, 0.1–0.2 $mWcm^{-2}$, which is large enough to power a variety of devices that operate at lower power consumption, such as microelectromechanical systems (MEMS), nanoelectromechanical systems (NEMS), and other nanoscale devices.

Figure 19.36 displays simultaneously the recorded scanning height topography (blue line) and voltage-output profile for a line scan of the AFM tip across a ZnO NWarray. In the scanning range of 6.56 μm, at least four NWs are contacted based on the topography profile. The corresponding voltage-output profile also shows four peaks of heights 12, 15, 2, and 10 mV. In each case, the voltage-output peak begins to increase when the AFM tip touches the flat cross section of the NW, and the voltage peak reaches its maximum when the AFM tip reaches the side edge of the top flat cross section of the NW. When the AFM tip starts to cross the central axial line of the NW, the voltage discharge begins, which occurs between two data points corresponding to the center of the discharge peak. When the tip completely releases the NW tip, the discharge is determined by the characteristics of the external circuit.

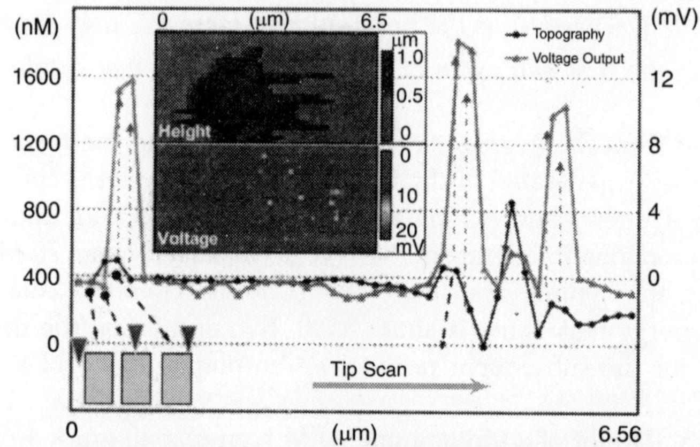

Figure 19.36 *The simultaneously recorded line scanning topography image (dotted line) voltage output profile of several aligned ZnO NWs. The inset in part of the figure shows a 2D height profile (top) and a simultaneously recorded voltage output image (bottom) obtained by scanning the tip across an area of 6.5 µm × 3.2 µm.*

The inset in Figure 19.36 displays the 2D AFM topography image and the corresponding voltage/current output image recorded simultaneously when the AFM tip scans over a 6.5 µm × 3.2 µm area. The height profile in the top image indicates that the local density of the NWs is so large that the AFM tip do not resolves them individually because of the tip size effect; thus, only the surrounding NW edges are resolved. The corresponding voltage-output profile reveals that the dots are distributed only at the extreme right hand side, indicating that the discharge occurs at the end of the tip scan over the NWs. The high density of the NW array prevents the deflection event from achieving completion without multiple contacts. Therefore, a reasonable NW density distribution that matches with the tip size and scanning speed is necessary for improving the voltage-output number.

Now NWs for 20–40 nm in diameter and 0.2–0.5 µm in length grown on a sapphire substrate, it is found that a single NW generates a 0.5 pW at 10mV of electrical power using AFM tip deflection at a 5 nN contact force. Here, NWs grown on a polymer substrate with diameters of 300 nm and lengths of 1 µm give rise to an output power of 5 pW at 45 mV. This value has improved by an order of magnitude.

To evaluate the power-generating capability of ZnO NW arrays grown on plastic substrates, NW arrays having different dimensions and orientation are tested to demonstrate the performance of the nanogenerator, and these arrays are shown in Figures 19.37 and 19.38. Figure 19.37(a) is a top-down view of an as-grown, densely packed, vertically aligned ZnO *microwire* array on a plastic substrate. The microwires are uniform with a width of 1–2µm and a height of 5–10 µm.

The SEM image in the inset is a side-on view of the aligned wire array, which reveals a good vertical alignment on the plastic substrate. The microwires are separated by 0.1–1 µm. The substrate surface area covered by the wires is 50–80% (the surface coverage in the case of Fig. 19.35 is only 10%). Under the conductive AFM test, using the same test parameters as those used for the system in Figure 19.35, it is found that the voltage discharge events occurr much less frequently than for the sparsely aligned NWS, hence indicate that a densely packed

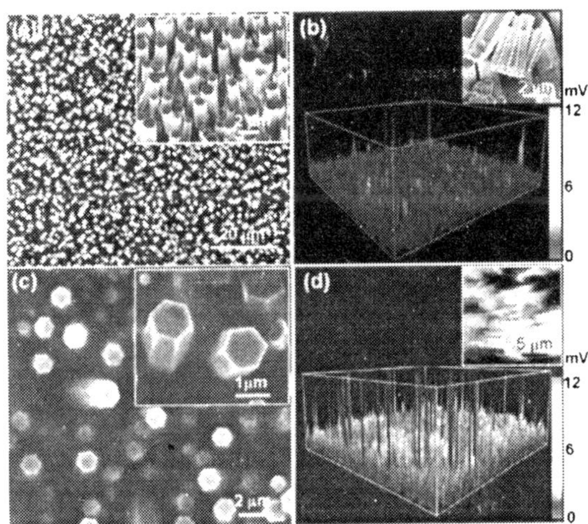

Figure 19.37 *(a) An SEM image showing a top-down view of the as-grown, densely packed, vertically aligned ZnO microwire array (inset: a side view showing the alignment of the microwires) on a plastic substrate and (b) the corresponding AFM voltage output profile (20 μm × 20 μm). The inset in (b) is an SEM image showing a portion of the microwires pushed down by the AFM tip. (c) An SEM image showing a top-down view of ZnO wires immersed in a PMMA coating on a plastic substrate (inset: an enlarged view of the PMMA-encapsulated microwires) and (d) the corresponding AFM voltage output profile from AFM measurements (inset: topography image of the 20 μm × 20 μm area).*

NW array is not the optimum choice for enhancing the power output of nanogenerators. Figure 19.37(b) is the corresponding 3D plot of the voltage image for an area of size 20 μm × 20 μm. Only a few (10) sharp voltage peaks emerge out of the noise level. So it suggests that the small normal tip force (5 nN) is sufficiently strong to deflect the ZnO microwires. To find out why only a few peaks are detected in the AFM scan in an area that has at least 50 wires, SEM is used. After tip scanning it found that only a few wires are left standing upright on the substrate; most of the wires in the array are pushed down by the AFM tip, as shown in the SEM image in the inset. Several possible reasons account for this. First, the adhesion between the Au-coated base and the ZnO microwires are so weak that it cannot bear the force applied by the AFM tip. The second reason is that the microwire is too strong to be elastically deformed; once it is pushed by the AFM tip, it tends to be displaced owing to a weak connection to the substrate rather than elastic deformation. Finally, the density of the wires is so high that there is not much room for the wires to be deflected by the AFM tip without touching neighboring wires; thus, the piezoelectric current, if any, slowly leaks out, resulting in a gradual discharge signal.

To improve the bonding between the wires and the substrate, a poly(methyl methacrylate) (PMMA) coating approximately 200 nm thick is introduced on the substrate after wire growth. Figure 19.37(c) shows the ZnO microwire array on a plastic substrate after the PMMA coating. It is seen that the bottom portion of the wire array is encapsulated by the PMMA coating (inset). The top portion of the wire array is clean after 10 min of UV irradiation following a spin-coating and soft-baking process. In Figure 19.37(c), the shorter wires are buried by the

PMMA coating, while most of the long microwires are standing on the surface. The result from AFM measurements is shown in Figure 19.37(d). The number of voltage peaks in an area of 20 μm × 20 μm increases to 50. The peak voltage output amplitude remains in the range of 6–12 mV. The inset is an AFM topography image of the corresponding microwire array affixed by PMMA. It is seen that the mierowires are firmly held onto the plastic substrate during the tip scan, suggesting that by using a reinforcing layer of polymer coating, the NW/substrate connection is significantly improved.

To determine the effect of NW shape and orientation on power generation, cone-shaped ZnO NWs and randomly oriented ZnO NWs on plastic substrates are grown and used for AFM measurements. Figure 19.38(a) shows as-grown, cone-shaped ZnO hexagonal NWs with 50 % vertical alignment on a plastic substrate. The nanocones typically have tip diameters of 100–200 nm, base diameters of 300–500 nm, and lengths of 3–5 μm (inset). Under the conductive tip deflection, only a few voltage output peaks in the range of 6–12 mV are observed (Fig. 19.38(b). Figure 19.38(c) is a typical SEM image of the as-grown ZnO NWs oriented randomly on the plastic substrate. The SEM image in the inset reveals the tilted hexagonal NWs rooted at the plastic substrate with dimensions of 200–600 nm in width and 1–2 μm in length. The AFM measurement result, shown in Figure 19.38(d), reveals an increased number of voltage output peaks of 6–12 mV thus suggests that as long as there is a way to bend the wires, a voltage/current signal is likely to be generated no matter what kind of orientation the wires adopt with respect to the plastic substrate. This result offers another advantage to energy harvesting utilizing large-scale NW arrays made by using solution-based chemistry.

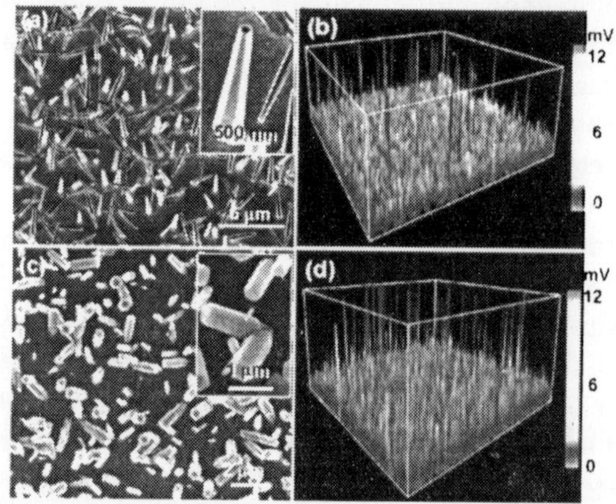

Figure 19.38 *(a) An SEM image of sparsely grown ZnO nanocones (inset: a magnified view of a single nanocone) and (b) the corresponding AFM voltage output profile (20 μm × 20 μm). (c) An SEM image of a typical top-down view of a randomly oriented ZnO NW array on a plastic substrate (inset: an enlarged view showing the randomly tilted NWs on a plastic substrate) and (d) the corresponding AFM voltage output profile (20 μ × 20 μm).*

The lead zirconium titania (PZT) ceramic is the most typical ceramic material for piezoelectric applications, and it is used for producing output voltage. In comparison to PZT, a NW array grown on a polymer substrate has a number of advantages for generating electricity. First, the NW-based nanogenerators are subjected to extremely large deformations, both on the NWs and the substrate, indicating that they can be used for flexible electronics as flexible/foldable power sources. Second, the large degree of deformation that the NWs bear affords a much larger volume density of power output. Third, in contrast to PZT, ZnO is a biocompatible, "biosafe" material, suggesting that it has a greater potential for use as an implantable power source in the human body. Fourth, the flexibility of the polymer substrate used for growing ZnO NWs makes it feasible to accommodate the flexibility of human muscles so that the mechanical energy (body movement, muscle stretching) in the human can be used to generate electricity. Finally, ZnO NW nanogenerators directly produce current because of their enhanced conductivity in the presence of oxygen vacancies.

(H) CdS NANOWIRES

CdS is a Piezoelectric semiconducting material with an energy band gap of about 2.5 eV. A wide range of applications have been demonstrated for one-dimensional CdS nanostructures, such as waveguide, photoconductor, logic gate, and field emitter. CdS NWs are also used for converting mechanical energy into electricity, therefore it can be used for nanoscale power devices.

Vertically CdS NW arrays are grown by two different approaches: hydrothermal and physical vapor deposition (PVD). The hydrothermal process followed the procedures of reported in literature with minor modifications. Cd foils and thiosemicarbazide are used as the Cd and S sources at a molar ratio of 3:2. An amount of 5 ml de-ionized water is first added into a Teflon-lined vessel of 25 ml capacity, and then ethylenediamine is added into the same vessel to reach 80% capacity of the vessel. The reaction is carried out at 450 K for 20 hours. As for the PVD approach, CdS powders and Au coated Si(111) wafer are used as the CdS source and collecting substrate, respectively. The temperature and pressure of the deposition chamber are set at 1220 K and 160 mmHg, respectively. Argon gas of 50 SCCM (SCCM denotes cubic centimeter per minute at STP) is used to carry the CdS vapors to the downstream of the growth chamber for deposition. The deposition is ran for 30 min at the peak temperature.

The morphology of the hydrothermally grown CdS NWs is characterized with scanning electron microscope (SEM). As shown in Fig. 19.39(a), these NWs are 150 nm in diameter and several micrometers in length. The x-ray diffraction (XRD) pattern of the as-grown CdS NWs is shown in Fig. 19.39(b). Also included in Fig. 19.39(b) are the XRD patterns of the reference cubic zinc blende (ZB) and hexagonal wurtzite (WZ) phases of CdS crystals for comparison purposes. All of the diffraction peaks, except the one located at 26.4° are indexed to the WZ phase. The diffraction peak at 2θ of 26.4° is a combined contribution from WZ (0002) and ZB (111) planes. The coexistence of the two phases is seen from the high-resolution transmission electron microscopy (HRTEM) image shown in Fig. 19.39(d), The NW is composed of alternating ZB and WZ phases, with alternating atomic layer stacking sequence from *ABC* to *AB* along the growth direction of the NWs. The growth direction is identified to be along ZB [111] and WZ <225>0001<225>. The streaks appearing in the corresponding selected area electron diffraction (SAED) pattern are a direct result of the stacking faults resulting from the phase transformation between ZB (ABC) and WZ (AB).

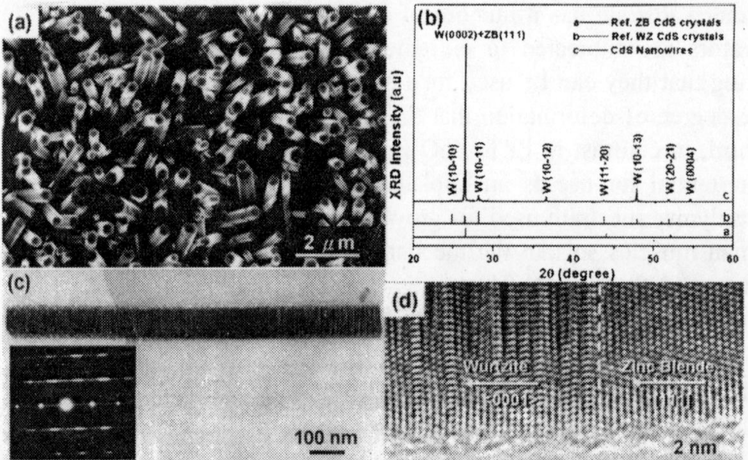

Figure 19.39 *(a and b) Top view SEM image and XRD pattern of the CdS nanowires prepared from the hydrothermal process, respectively. (c) TEM image of a single CdS nanowire, The inset is the corresponding SAED pattern. (d) HRTEM image of a single CdS nanowire.*

Piezoelectric measurements are performed in contact mode of an atomic force microscope (AFM) using a Pt coated Si tip with a cone angle of 70°. The cantilever has a spring constant of 1 Nm^{-1}. In AFM contact mode, a constant normal force of 5 nN and a scanning speed of 150.24 μms^{-1} are maintained between the tip and sample surface, By scanning the tip across the sample [Fig. 19.40(a)], output voltage was detected across an external load.

The process for generating the electric current is derived from the output voltage peak and the topological profile received by the tip when scanned across a NW. During the tip scans, no voltage output signals are observed; if the tip touches only the stretched side of the NW and does not lift up, to go beyond the central line of the NW to reach the compressed side, a negative voltage output signals detected when the tip goes beyond the NW to reach the compressed side of the NW, as shown in the topography (curve) and output voltage (curve) images of Fig. 19.40(b). This is clearly indicated by a delayed output in voltage signal in reference to the surface profile image of the NW The typical magnitude of the output voltage is

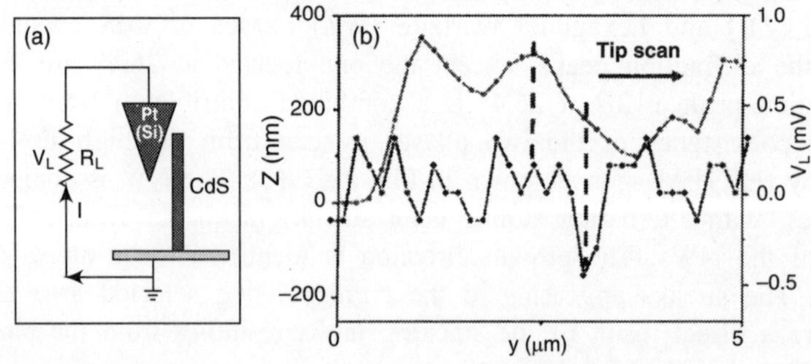

Figure 19.40 *(a) AFM measurement system. (b) Line profiles of topography and output voltage scanned across CdS nanowires.*

~ –(0.5–1) mV, where the negative sign means that the generated current flows from the tip to the NW The low voltage output is due to the phase transition between ZB and WZ phases along the growth direction of the NW, because WZ phase is piezoactive, while ZB phase is not. The presence of the ZB phase along the growth direction is disadvantageous to the piezoelectronic performance of the CdS NWs.

To improve the voltage output using solely the WZ phase, CdS NW arrays are prepared using a PVD process at a much higher temperature of 1220 K. The as-grown NWs are about 100 nm in diameter and over 1 μm in length, as shown in the TEM images of Figs. 19.41(a) and 19.41(b). The dot pattern of SAED shown in the inset of Fig. 19.41(c)

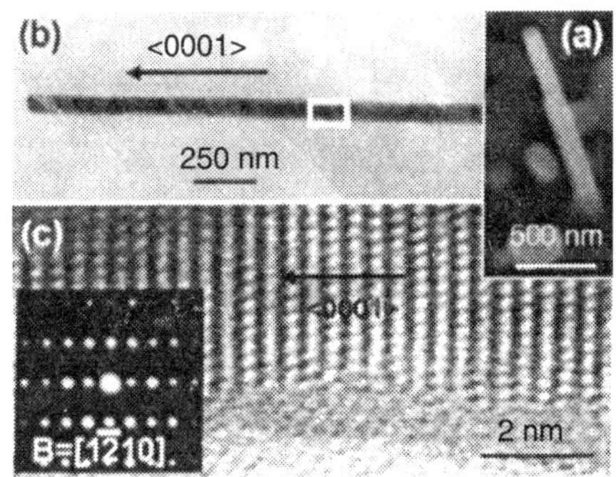

Figure 19.41 *(a and b) Side view SEM and TEM images of a single CdS nanowire produced from the PVD process. (c) HRTEM image of a single CdS nanowire at the marked region of (b). The inset is the corresponding SAED pattern of the nanowire recorded from the marked region in (b).*

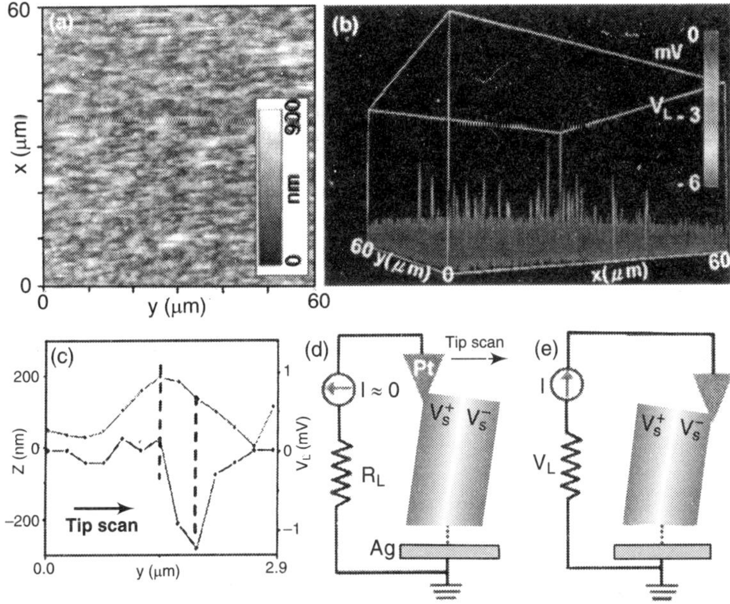

Figure 19.42 *(a and b) The topography and corresponding voltage output images of the nanowire arrays prepared from the PVD process, respectively. (c) Line profiles of topography and output voltage scanned across CdS nanowires. (d and e) Contact between the AFM tip and a semiconductor CdS nanowire at two reversed local contact potentials (positive and negative), showing reverse- and forward-biased Schottky rectifying behavior, respectively. The process in (d) is to generate and preserve the charges/potential, and the process in (e) is to discharge the potential through a flow of electrons from the circuit under the driving of the piezoelectric potential.*

reveals the single crystalline WZ phase of the NW. The HRTEM image of Fig. 19.41(c) reveals the NW grew along the ⟨0001⟩ direction. Further, there is no trace of the presence of the ZB phase in the NW.

The topography [Fig. 19.42(a)] and corres-ponding output voltage [Fig. 19.42(b) images across the load are recorded simultaneously when the AFM tip scans over the NW arrays. The magnitude of the voltage output is around –3 mV, much larger than that from the NW arrays produced with the hydrothermal process. Evidently, the pure WZ phase of the NWs from the PVD process gives higher voltage outputs.

The mechanism proposed previously for the ZnO NW based nanogenerators applies for the present case. The electron affinity of *n*-type CdS is 4.8 eV, while the work function of Pt is about 6.1 eV. There is a Schottky barrier formed at the Pt-CdS contact. In Fig. 19.42(d), as the AFM tip starts to deflect the NW, a positive potential V_s^+ produced at the stretched side of the NW, while a negative potential ($V_s^- < 0$) is induced at the compressed side. The potential of Pt metal tip is near zero, $V_m = 0$. No voltage signals are observed due to the presence of the reverse-biased Schottky barrier contact between the Pt tip and the stretched side of the *n*-type CdS NW ($\Delta V = V_m - V_s^+ < 0$). Figure 19.42(d) demonstrates the charge/potential accumulation process by bending a CdS NW. When the AFM tip goes beyond the central line of the NW and reached the compressed side, as shown in Fig. 19.42(e), negative voltage signals are produced because of the presence of the forward-biased Schottky barrier contact between the Pt tip and the compressed side of the CdS NW ($\Delta V = V_m - V_s^- > 0$). Figure 19.42(e) shows how the charges/potentials are released from the NW. The topography and voltage output images of Fig. 19.42(c) provide clear evidence for the above discussed charge accumulation and release processes when bending a CdS NW.

Nanopiezotronics

THE FUNDAMENTAL PRINCIPLES

Nanopoiezotronics utilizes the coupled piezoelectric and semiconducting property of nanowires and nanobelts for designing and fabricating electronic devices such as transistors and diodes. It has a wide range of applications in electro-mechanical coupled sensors and devices, nanoscale energy conversion for self-powered nanosystems, and harvesting/recycling of energy from environment.

Piezoelectricity (PZ) is a coupling between a material's mechanical and electrical behavior. When a piezoelectric material is squeezed, twisted, or bent, electric charges collect on its surfaces. Conversely, when a piezoelectric material is subjected to a voltage drop, it mechanically deforms. Many crystalline materials exhibit piezoelectric behavior, including quartz, wurtzite-structured crystals, rochelle salt, lead zirco-nate titanate ceramics, barium titanate, and polyvinylidene flouride (a polymer film). When such a crystal is mechanically deformed, the positive and negative-charge centers are displaced with respect to each other (Fig. 20.1). So

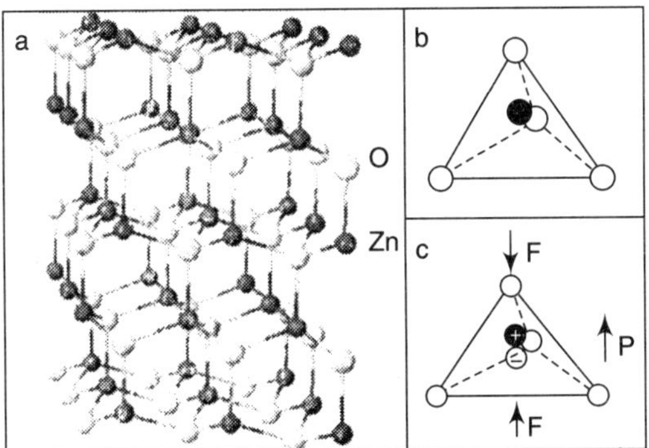

Figure 20.1 *(a) Structural model of wurtzite ZnO. b) Tetrahedral coordination between In and oxygen. c) Distortion of the tetrahedral unit tinder the compression of an external force, showing the displacement of the center of positive charge from that of the negative charge.*

while the overall crystal remains electrically neutral the difference in charge center displacements results in an electric polarization within the crystal. Electric polarization resulting from mechanical deformation is perceived as piezoelectricity.

For, nanopiezotronics, first the piezoelectric behavior of a single nanowire (NW) or nanobelt (NB) of a piezoelectric material such as ZnO or a vertical, straight ZnC NB (Fig. 20.2(a)), the deflection of the NB by an external applied force F results in (tensile) stretching of its outer surface (positive strain ε Fig. 20.2(b)) and compressing or its inner surface (negative strain ε Fig. 20.2(b)). An electric field E_z along the NB (z-direction) is then created inside the NB through the PZ effect, $E_z = \varepsilon_z/d$, where d is the PZ coefficient along the NB direction that is normally the positive c-axis of ZnO. The PZ-field direction is closely parallel to the z-axis (NB direction) at the tensile surface and antiparallel to the z-axis at the compressed surface (Fig. 20.2(c)). Under the first-order approximation, as a result of flipping of the PZ field across the width of the NB, the electric potential distribution from the compressed to the stretched side surface is approximately between V^-_p (negative) to V^+_p (positive). The electrode at the base of the NB is grounded. Note V^+_p and V^-_p; are the voltages produced by the PZ effect. The potential is created by the relative displacement of the Zn^{2+} cations with respect to the O^{2-} anions because of the PZ effect in the wurtzite crystal structure (see Fig. 20.1); thus, these ionic

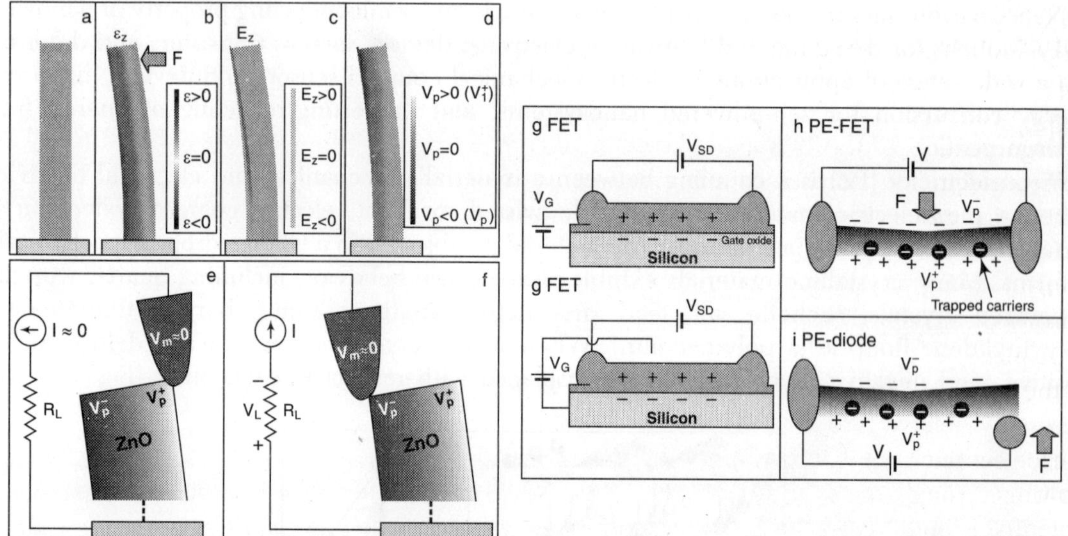

Figure 20.2 *The physical principle of nanopiezotronics, a-d) The principle of the piezoelectric nanogenerator a) schematic definition of a nanobelt (NB) and the coordination system; b) longitudinal strain ε_z distribution in the NB after being deflected by an atomic force microscope (AFM) tip from the side; c) the corresponding longitudinal piezoelectric induced electric field E_z distribution in the NB; d) potential distribution in the NB as a result of the piezoelectric effect; e) metal-semiconductor Schottky contact between the AFM tip and the semiconductor ZnO NB at reverse bias, f) metal-semiconductor Schottky contact between the AFM tip and the semiconductor ZnO N B at forward bias, g) field-effect transistors (FETs) using a single nanowire/nanobelt, with gate, source, and drain, h) The principle of the piezoelectric field-effect transistor (PE-FET). (i) The principle of the piezoelectric gated diode.*

charges do not freely move and cannot recombine without releasing the strain (Fig. 20.2(d)). The potential difference is maintained as long as the deformation is in place and no foreign free charges (such as from the metal, contacts) are injected. The process shown in Figure 20.2(a) to (d) is the fundamental principle of piezotronics for creating functional devices such as the nanogenerator, piezoelectric transistor, and piezoelectric diode.

Piezoelectric Field-Effect Transistor

Field-effect transistors (FETs) based on nanowires or nanotubes are one of the most important nanodevices. A typical NW FET is composed of a semiconducting NW that is is connected by two electrodes at its ends and is placed on a silicon substrate covered by a thin layer of gate oxide. A third electrode is built between the NW and the gate oxide (Fig. 20.3(a)). The gate voltage (V_G) traps and depletes the carriers in the NW. Thus by controlling the gate voltage, the flow of the current from the source to drain electrodes is achieved. An NW-based sensor is a source-drain structured NW FET without a gate, thus, a large portion of the NW is exposed to the environment. The mechanism of NW sensors for sensing gases, biomolecules, or eyen viruses relies on the creation of a charge depletion zone in the semiconductor NW by the surface-adsorbed sensing targets.

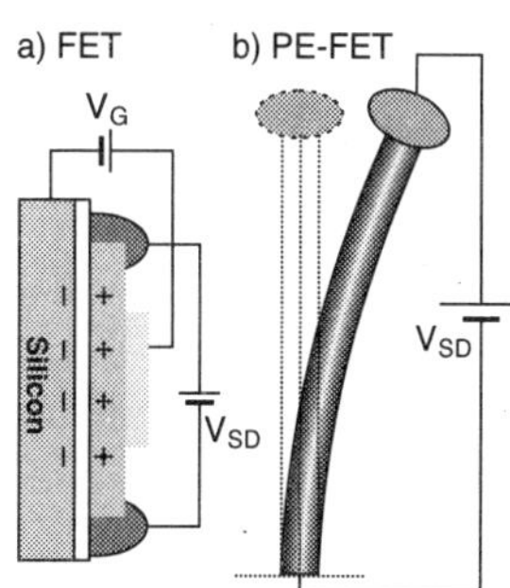

Figure 20.3 *The physical principle of the piezoelectric FET. a) Schematic of a conventional FET that uses a single nanowire/ nanobelt. with gate source, and drain (subscripts G, S, and D. respectively), b) The principle of the piezoelectric FET, in which the piezoelectric potential across the nanowire created by the bending force f_y replaces the gate of a conventional FET. The contacts 'at both ends are ohmic.*

A piezoelectric potential is created across the NW when it is deflected by an external force. If the potential is large enough, plays the role of V_G for the FET. By connecting a ZnO NW across two electrodes that can apply a bending force to the NW, the electric field created by piezoelectricity across the bent NW serves as the gate for controlling the electric current flowing through the NW (fig. 20.3(b). Once deformed by an external force, a piezoelectric potential is built across the bent NW, some free electrons in the n-type ZnO NW are trapped at the positive surface and become non-mobile charges, thus, lowering the effective carrier density in the NW. The trapped "free charges" do not deplete the charges produced by the piezoelectric effect, which are rigid, ionic, and affixed to the atoms. Even the positive potential side is partially screened by the trapped electrons while the negative potential side remains unchanged. The free electrons are repulsed away by the negative potential and leave a charge depletion zone around the compressed-surface side. Consequently, the width of the conducting channel in the ZnO NW becomes smaller and smaller while the depletion region becomes larger and larger with an increase of the NW bending. The resultant effect is a drastic decrease in conductivity of the NW. This is the proposed piezoelectric FET (PE-FET).

This piezoelectric field effect transistor (PE-FET) can be considered as a new type of FET, which can be turned on/off by applying mechanical force. As a result, it it is demonstrated as force sensor for measuring forces in the nanonewton range and even smaller.

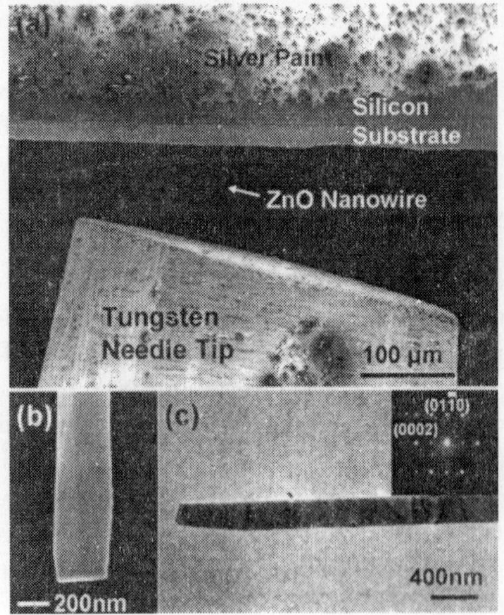

Figure 20.4 *(a) A low-magnification SEM image showing the setup for the in situ measurements. (b) A higher magnification SEM image showing the clean, tip of the ZnO nanowire. (c) A TEM image of a ZnO nanowire with single-crystal structure and growth direction [0001]; inset is the corresponding electron diffraction pattern.*

The experimental design is shown Figure 20.4(a) and the measurement is carried out in the chamber of a scanning electron microscope (SEM). Inside the SEM chamber, an x–y mechanical stage with a fine moving step of ~ 20 nm is located beside the SEM. The mechanical stage is controlled from the outside of the SEM. A tungsten needle with a machined and polished tip is attached on the stage and connected to the positive electrode of an external power source. ZnO NWs are synthesized by the well-established technique of thermal evaporation in a tube furnace. A single NW is prepared by aligning the NW on the edge of a silicon substrate using a probe station. The extended length of the NW, is ~ 100 μ m, while the other side of the NW is fixed onto the silicon substrate by conductive sliver paint, through which the NW, is connected to the negative electrode of the power source. The silicon substrate is placed on the stage with the NW pointing at the tungsten needle tip. Both the silver paint and NW have Ohmic contact with ZnO. When the required vacuum is achieved in the SEM chamber, the tungsten needle tip is moved to the center of the image screen. The ZnO NW is then controlled to approach the needle tip by controlling the SEM stage. Focusing the NW and the needle tip at the same time guaranteed that they are aligned at the same height level.

Before contact is made, the ZnO NW is examined under the SEM at a higher resolution. The length of the suspension part of the NW is 88.5 μ n. As shown in Figure 20.4(b), the NW's tip is flat and clean and the faceted side surfaces are observed. The width of this NW is measured to be 370 nm. TEM analysis reveal the single-crystal structure of the ZnO NW and the growth direction is along [0001] (Figure 20.4(c).

In order to achieve a good electrical contact between the ZnO NW and the tungsten tip, a field emission process is introduced at the first stage to clean the NW tip and welding the connection. During the field emission, the ZnO NW is kept ~ 10 μ m away from the tungsten surface, where 400 V is applied. Under this condition, the emission current reach as high as ~

1 μ A. After 1 min of emission, the high voltage is turned off and the ZnO NW is quickly moved toward the tungsten surface to make a contact. This process is very effective for making the Ohmic contact, due to high temperature generated at the tip by the emission process and the desorption of the surface contaminant.

The first contact of the NW with the W tip is shown in Figure 20.5(a) where the NW is already bent a little bit because a pushing force is necessary for a good electrical contact. The electron beam of the SEM is turned off and the *I–V* characteristic is measured by sweeping the voltage from –5 to 5 V. This eliminates the effect from the electron beam in SEM. After the measurement, the electron beam is turned back on again, and the NW is bent further by moving the SEM stage in-situ under direct imaging. The bending of the NW is recorded. Following such a procedure a sequential measurement is carried out. The five typical bending curvatures of the ZnO NW are shown in Figure 20.5(a) to (f), and their corresponding *I–V* curves are presented in Figure 20.5(f). The symmetric shape of the *I–V* curves indicate good Ohmic contacts at both ends of the NW. Among the five bending cases, the current drops significantly with the increase of bending (curves b–e in Figure 20.5(f)), indicating the decreased conductance with the increased strain.

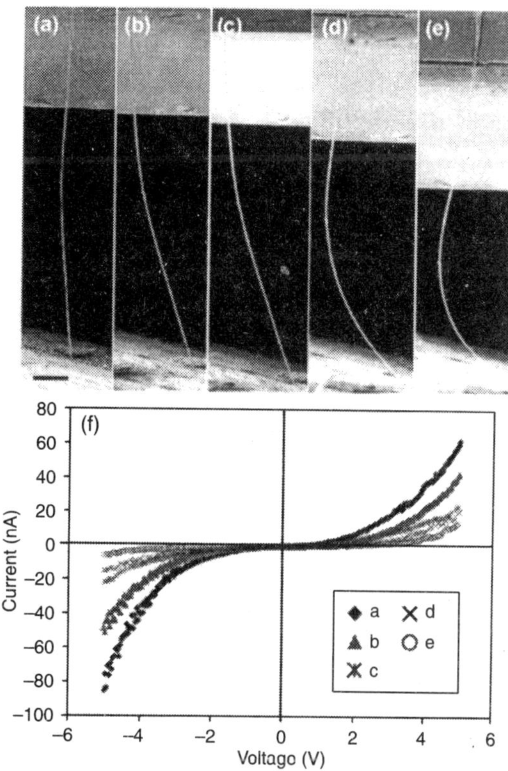

Figure 20.5 *(a–e) SEM images with the same magnification showing the five typical bending cases of the ZnO nanowire; the scale bar represents 10 μ m (f) Corresponding I–V characteristics of the ZnO nanowire for the five different bending cases. This is the I–V curve of the piezoelectric field effect transistor (PE-FET).*

This phenomenon is further confirmed by continuously changing the bending curvature under SEM observation. The applied voltage is fixed at +5 V and the current is continuously recorded while the ZnO NW is pulled back from large bending to almost straight and then pushed down again. The current variation is shown in Figure 20.6(a). When the NW is under significant compression, the current is only ~ 5 nA (stage I in Figure 20.6(a)). As the NW is slowly released and recover its straight shape, the current increases continuously with the decreasing of bending curvature, and the current is stabilized at ~ 40 nA (stage II in Figure 20.6(a)). The current drops Immediately when the NW is bent further (stage III in Figure 20.6(a)). This reveals a reversible sequence that the current passing through the ZnO NW at a fixed voltage is inversely proportional to its bending curvature. Higher magnification SEM images are taken at the contacting point when the NW is straight and highly bent (white rectangular boxes in the inset images of Figure 20.6(a), and the corresponding images are shown in Figure 20.6(b) and (c), respectively. It is observed that the NW is firmly attached to the tungsten needle surface without sliding, indicating the contacting is retained during bending process and does not cause any change in contact resistance.

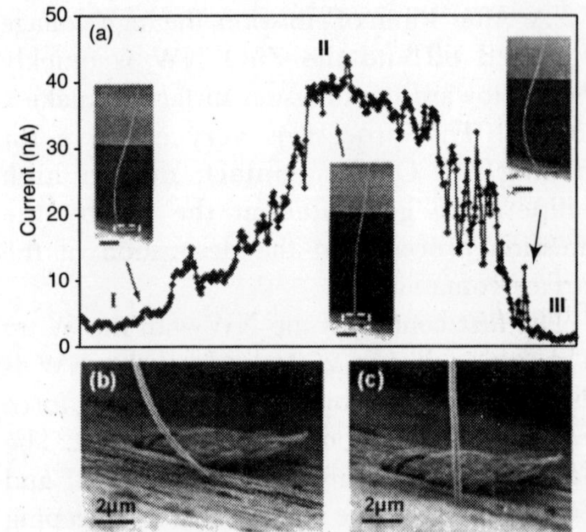

Figure 20.6 *(a) Current variation with a continuous releasing and bending process of the ZnO nanowire; insets are the three typical bending cases in stage I, II, and III, respectively; the scale bar represents 10 μ m. (b, c) SEM images of. the contacting point. between the ZnO nanowire tip and the tungsten needle surface when the nanowire was bent and released, showing no sliding and no losing.*

Now examine the mechanisms that are responsible for the change of conductance. When a semiconductor crystal is under strain, the change in electrical conductance is referred as the piezoresistance effect, which is caused by a change in band gap. width as a result of strained lattice. The change of resistance is given by

$$\delta\rho/\rho = \pi\rho l/l \qquad (1)$$

where ρ is the resistance, l is the original length, and π is the piezoresistance coefficient. This equation is for a crystal that is subjected to a homogeneous strain. A change in resistance is achieved by fixing a semiconductor strain sensor to the surface of the object, of which the strain is to be measured. The built up strain in the object is picked up by the semiconductor slab, which is either stretched or compressed homogeneously through its entire volume.

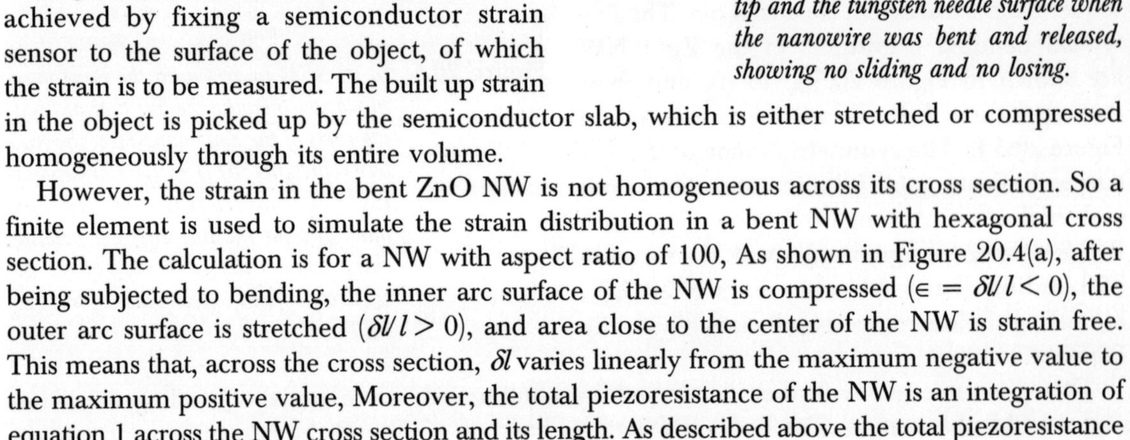

However, the strain in the bent ZnO NW is not homogeneous across its cross section. So a finite element is used to simulate the strain distribution in a bent NW with hexagonal cross section. The calculation is for a NW with aspect ratio of 100, As shown in Figure 20.4(a), after being subjected to bending, the inner arc surface of the NW is compressed ($\in = \delta l/l < 0$), the outer arc surface is stretched ($\delta l/l > 0$), and area close to the center of the NW is strain free. This means that, across the cross section, δl varies linearly from the maximum negative value to the maximum positive value, Moreover, the total piezoresistance of the NW is an integration of equation 1 across the NW cross section and its length. As described above the total piezoresistance of the bent NW is close to zero under the first-order approximation because of the nearly antisymmetric distribution of the strain across the width of the NW.

Therefore, the change in resistance of the NW as a result of bending is negligible. This result, cannot explain the observed result presented in Figures 20.5 and 20.6 that the resistance increased by a factor of ~ 7 after bending.

Now the possible mechanisms for explaining the observed phenomenon are: It is important to point out that ZnO is a material that simultaneously has semiconducting and piezoelectric properties. It is worth examination of the coupling between the two properties. A 75% decrease in in-plane conductance of a semiconducting Si slab is observed when it is sandwiched between two pieces of piezoelectric PZT crystals that are busted by an ac power across the thickness. The drop in the conductance of Si in the direction transverse to the propagation direction of the elastic wave in PZT is attributed to the trapping of free carriers at the surfaces of the silicon plate. This is a result of coupling between a semiconductor Si crystal and a PZT piezoelectric crystal. This coupling effect is also achieve in a single ZnO NW due to its semiconducting and piezoelectric dual properties. So a possible mechanism about the bending induced conductance change is shown in parts in Figure 20.7(b) and (c).

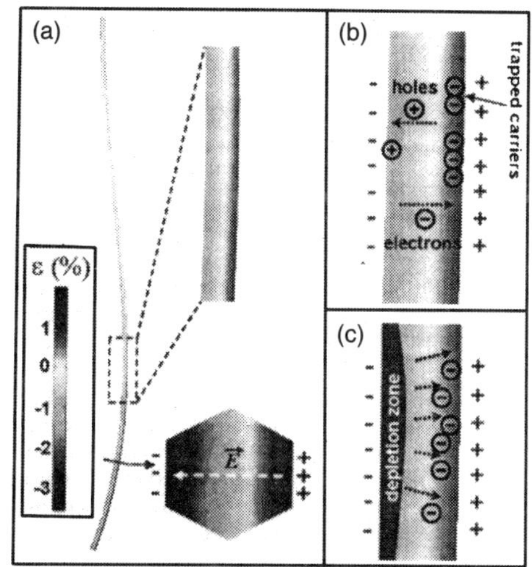

Figure 20.7 *Schematic diagrams showing the mechanisms responsible to the conductance change. (a) A finite element simulation of the strain distribution along the ZnO nanowire when it is bent. (b) The carrier trapping effect. (c) The creation of a charge depletion zone.*

A bent ZnO NW produces a positively charged and negatively charged surface at the outer and inner bending are surfaces of the NW due to the stretching and compression on the surfaces, respectively. These charges are induced by a piezoelectric effect, and the charges are static and nonmobile ionic charges. The local electric field is $E_p = \epsilon/d$, where d is the piezoelectric coefficient. Thus a small electric field is generated across the width of the ZnO NW, as shown in the cross-section image of the NW in Figure 20.7(a). Upon the build up of the electric field, two effects account for the reduction of the NW's conductance: carrier trapping effect and the creation of a charge depletion zone.

Similar to the Si-PZT system, when the piezopotential appears across the bent NW, some free electrons in the n-type ZnO NW are trapped at the positive side surface (outer arc surface) and become non-movable charges thus lowering the effective carrier density in the NW (Figure 20.7(b). If the positive potential side is partially neutralized by the trapped electrons, the negative potential remains unchanged. Hence, the piezo-induced electric field is retained across the width of the NW. This situation is of applying a gate voltage across the width of the ZnO NW as for a NW FET. The free electrons are repulsed away by the negative potential and leave a charge depletion zone around the compressed side, as shown in Figure 20.4(c). Consequently, the width of the conducting channel in the ZnO NW becomes smaller and smaller while the depletion region becomes larger and larger with the increase of the NW bending. The two effects presented in parts b and c of Figure 20.7 contribute to the dramatic drop of the conductance of the ZnO NW with the increase of bending. The maximum width that the depletion zone develops up to the strain-free plane (close to central axis of the NW) with consideration of the piezoelectric field, which naturally sets an upper limit to the effect contributed by the depletion charges.

Figure 20.7 is the working principle of a FET except that the gate voltage is produced by a piezoelectric effect. Therefore, the single ZnO across two Ohmic contacts is a piezoelectric field effect transistor (PE-FET), which is a unique coupling result of the semi-conducting and piezoelectric properties of ZnO.

Since the bending curvature of the NW is directly related to the force applied to it, a simple PE-FET force/pressure sensor is realized. The important step for calibrating the force/pressure sensor is quantitatively determine the force applied to the NW. As shown in the inset in Figure. 20.8 when the NW is pressed vertically, under small deflection angle approximation, the NW is deflected mainly due to the transverse force F_y, and the bending shape of the NW is given by.

$$\frac{d^2y}{dx^2} = \frac{F_y(L-x)}{YI} \quad (2)$$

where Y is the bending modulus, I is the momentum of inertia, and L is the total length of the NW. The shape of the bent NW is

$$y = \frac{F_y}{YI}\left(\frac{1}{2}Lx^2 - \frac{1}{6}x^3\right) \quad (3)$$

At the tip of the NW ($x = L$), the maximum deflection of the NW is

$$y_m = \frac{F_y L^3}{3YI} \quad \text{or} \quad F_y = \frac{3YIy_m}{L^3} \quad (4)$$

The bending modulus of ZnO NWs is 109 GPa and I is 6.62×10^{-28} m for this NW. Y_m is measured from the SEM images. Since the SEM image is a projection of the 3D structure, in order for an accurate measurement, the bending curve should be parallel to the projecting screen. Considering this effect, the tungsten needle surface is tilted 15° to the right-hand side and the initial bending is achieved by lateral movement of the NW instead of directly pushing downward. After parallel bending of the NW the stage is then pushed downward to achieve -further bending, Which is kept in the same plane. Since there are nonlinear *I–V* characteristics, the conductance is determined by the current measured at a constant 5 V potential. The total force versus conductance is plotted in Figure 20.8. At small bending, the decrease of conductance is almost linear to the bending force.

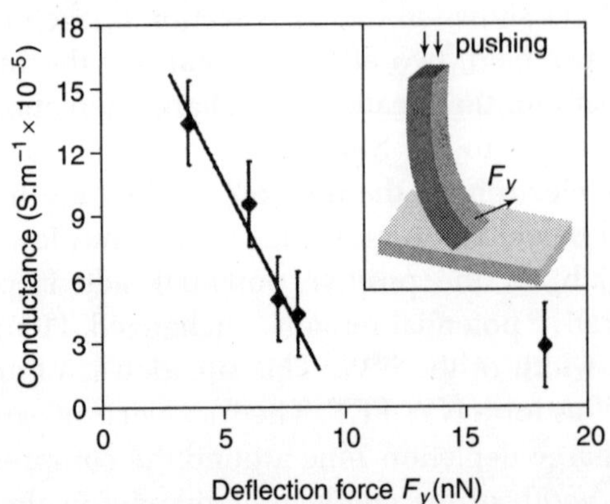

Figure 20.8 *A plot showing the relationship between the ZnO nanowire conductance as PE-FET and the deflection force derived from the experimental data, demonstrating a nanoforce or nanopressure sensor using a single nanowire. Inset is the schematic of the deflected NW*

With the increase of bending at higher transverse force (17 nN), the dropping tendency of the conductance is reduced. This is also shown in the two processes in Figure 20.8. At the beginning of bending, both contributions from the carrier trapping effect and the depletion zone increased with the bending curvature, thus producing a quick drop of the conductance. However, unlike the normal gate-controlled FETs, there is a positive potential on the outer side of the NW, which increases simultaneously with the negative potential at the inner side during the bending. The depletion region cannot extend beyond the neutral plane of the NW to the positive potential side. Thus, when the NW is bent to a significant degree, the piezoelectric field does not increase the size of the charge depletion zone and, hence, does not reduces the size of the conduction channel.

It is pointed out that the calculations in eqs 3 and 4 are for a NW that is subject to a small bending. The nonlinear effect has to be included if the degree of bending is large. This may cause significant change in the calibration of the measured force at larger applied forces. For practical application as force sensors, the measurements are recommended to be carried out under small deflections. For the data shown in Figure 20.8 the first four data points are most reliable for calibrating the sensor performance.

Piezoelectric Diode

A piezoelectric-field gated diode is demonstrated by using a two-probe technique. One probe holds one end of an NW that lay on an insulator substrate, the other probe pushes the NW from the other end by being in contact with the tensile surface, The W tips have ohmic contact with the NW. The *I–V* curve changes from a linear shape to rectifying behavior with an increase in the degree of NW bending (Fig. 20.11(a)), This is the piezoelectric diode (PE diode).

Rectifying/switching behavior of the PE diode: Consider the band diagram shown in Figure 20.9. Applying an external voltage of V_0 between the tip (T) and the root (R) of the NW, a difference in Fermi levels (E_F) between the tip and the root electrodes *I* created with the root being positive (Fig. 20.9(b)). When the lip pushes the NW from the tensile side, a positive V^+ and negative voltage, V^- are generated at the tensile and compressive surfaces, respectively, thus, the energy level of the conduction band (CB) of the NW next to the tip is lowered by qV^+, with q being the unit charge of the electron (Fig. 20.9(c)). In such a case, electrons easily flow from the tip side to the root side without experiencing any barrier, thus, the transport preserves ohmic behavior. This is the "forward" bias of the PE diode.

By switching the polarity of the applied voltage so that the tip has a higher voltage, the Fermi level of the tip is lowed by qV_0 in reference to that of the root electrode. Correspondingly, for_ $V_0^- < V^+$, a barrier of height $q(V^+-V_0)$ is created for the electrons to flow from the NW to the tip. The transition via thermal emission is $\exp[-q(V^+-V_0)/kT]$. Thus, for $V_0 < V^+$, the current decays quickly. This is the "reversed" bias of the PE diode. The threshold of the diode increases with increased NW bending.

Experiment

The ZnO NWs used for the experiment are grown using a vapor-liquid-solid process. A 2 nm thick Au thin film, which serves as tile catalyst for growth is deposited onto a (1120) Al_2O_3 substrate at room temperature in an electron-beam evaporation system 665 × 10^{-6} Pa). The experimental apparatus includes a horizontal tube furnace, a rotary pump system, and a gas supply system. A mixture of ZnO and graphite powders in a ratio of ZnO/C = 4:1 is placed in an alumina boat, which is heated to a peak temperature of 1370 K. The Al_2O_3 substrate is placed in the temperature zone of 1070 K for collecting the ZnO nanostructures. After the tube

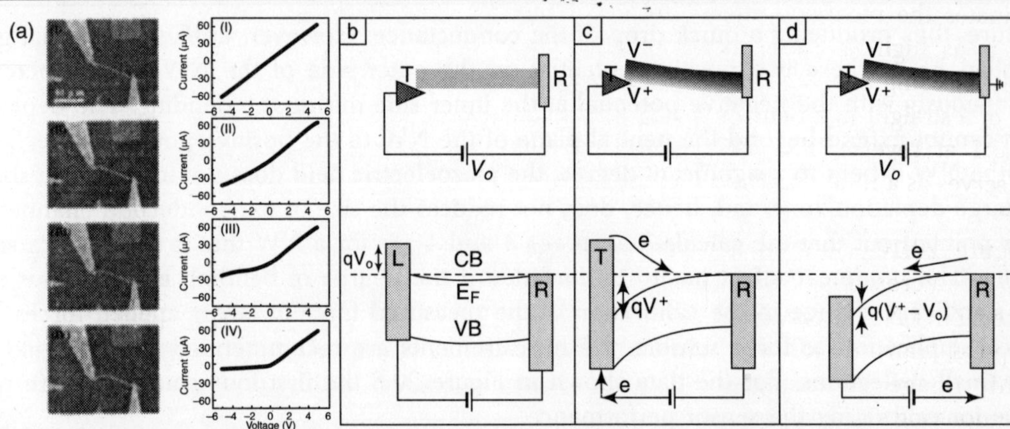

Figure 20.9 *The principle of the piezoelectric field diode, in which one end of an NW is fixed and enclosed by a metal electrode, and the other end is bent by a moving metal tip. Both ends have ohmic contact with ZnO. a) A sequence of SEM images of a ZnO wire at various bending angles and the corresponding I–V characteristics, with the fixed probe being positive and the moving tip negative b) Band diagram of an NW that has ohmic contact at the two ends when a constant voltage V_o is applied. with the tip (T) as negative and root (R) positive. The Fermi level of the lip is raised by qV_o (q is electronic charge). c) Forward bias case: Once the NW is pushed by the tip, a local positive piezoelectric potential at the tensile surface lowers the CB next to the tip, resulting in a forward flow of electrons from the tip through the NW to the root. d) Reverse biased case: When the applied voltage is switched in polarity, the Fermi level of the tip is lowered by qV_o A potential barrier of height $q(V^+-V_o)$ is created. Once the applied voltage $V_o < V^+$, the electrons have to transit over the barrier to reach the tip.*

is evacuated to a pressure of 133×10^{-3} Pa the samples are heated to 1370 K at a rate of 5 K min^{-1} and held at 1370 K for 60 min with a carrier gas of Ar and O_2 flowing through the tube. The substrate-bound ZnO NW arrays are mechanically scraped off and sonicated in ethanol, then deposited on a Si substrate covered with a Si_3N_4 layer as insulator.

The scraped NWs are placed on carbon-coated copper grids for TEM characterization. Morphology of grown ZnO nanostructures is performed with a TEM instrument operating at 200 kV and a field-emission scanning electron microscope. HRTEM images are obtained by using a field-emission TEM instrument, with a point-to-point resolution of 0.17 nm, operating at 300 kV.

The performance of *I–V* measurements and the manipulation of ZnO NWs are carried out in an MPNEM system. The two-terminal method is applied for electrical transport measurements. As all of the procedures are carried out at a pressure of 5×10^{-4} Pa. The MPNEM system is composed of several subsystems: (1) field-emission SEM gun 2) nanomanipulator system (four-needle experimental probing device, and 3) an *I–V* measurement system. Shown below is how an n-type ZnO NW can be used to produce a p-n junction that serves as a diode. The design is based on the mechanical pending of a ZnO NW. As a result, the potential energy barrier induced by piezoelectricity (ψpz) across the bent NW governs the electrical transport through the NW. To quantify ψpz, the current-voltage (*I–V*) characteristics received at different levels of deformation are included in theoretical calculations. The magnitude of the Piezoelectric barrier dominates the rectifying effect. The rectifying ratio is as high as 8.7:1 and is obtained by simply bending a NW. The operation current ratio of a straight to a bent ZnO NW is as high as 9.3:1 at reverse bias. This shows that the NW serves as a random access memory (RAM) unit.

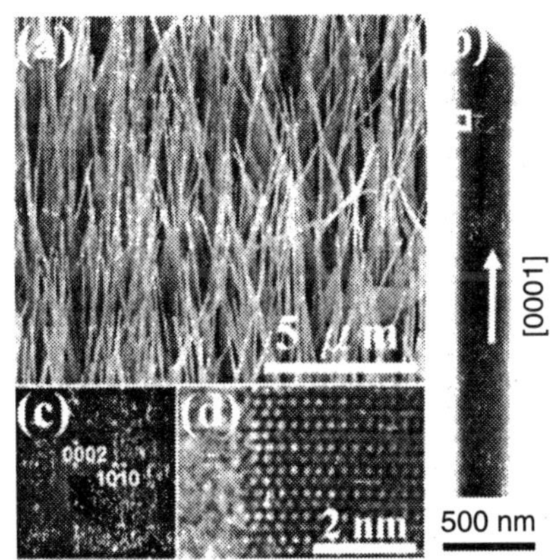

Figure 20.10 *An SEM image of a well-aligned ZnO NW array; b) TEM image of a single ZnO nanorod; c) Corresponding selected area electron diffraction pattern, and d) high-resolution TEM image of the ZnO nanowire at the boxed region marked in (b).*

Figure 20.10 a shows a typical scanning electron microscopy (SEM) image of a well-aligned ZnO NW array. Figure 20.10(b) shows a typical transmission electron microscopy (TEM) image of a single ZnO NW. The corresponding selected-area electron diffraction pattern (Fig. 20.10(c) confirms that the phase of the NWs is hexagonal wurtzite-structured ZnO. Figure 20.10(d) is a high-resolution TEM (HRTEM) image from the out region indicated in Figure 20.10(b), Figure 20.10(c) and (d). show that the ZnO NWs are single crystalline and free of dislocations. The growth direction of the ZnO NW is determined to be [0001].

I–V measurements and the manipulation of ZnO NWs are carried out in a multiprobe nanoelectronics measurement (MPNEM) system. The two-terminal method is applied for electrical transport measurements at high vacuum to minimize influences from the environment. The tungsten nanotip used for measuring the electrical transport of a nanowire is precoated with a Ti/Au (30 nm:30 nm) film by electron-beam evaporation to obtain Ohmic contact between Ti and ZnO. Creating Ohmic contact is a key step for the measurements. Focusing the NW and the W nanotip in the *MPNEM* system simultaneously in the same focal plane guarantees that they are contacted and placed at the same height on the surface of the Si_3N_4 layer.

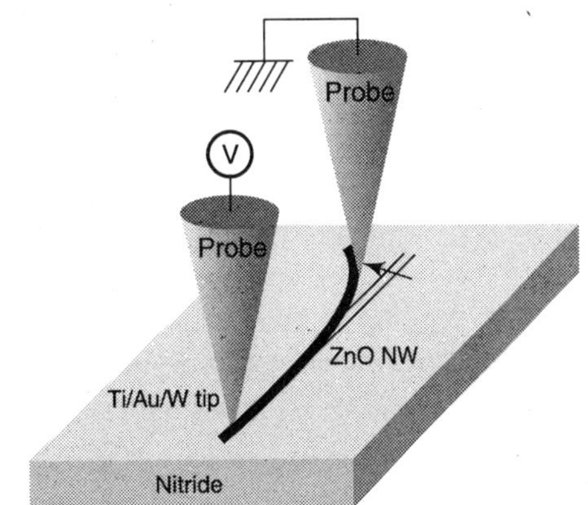

Figure 20.11 *Diagram of the nanomanipulation and in situ I–V measurement setup in the MPNEM system, in which SEM, which can provide higher resolution, is employed to locate the nanowires and navigate the nanotip to a certain region.*

Figure 20.11 illustrates a manipulator making a two-terminal connection to a single ZnO NW to measure its electrical transport properties. The lower nanotip is used to apply the voltage and measure the current through the NW. By controlling the lateral and longitudinal motions of the two nanotips with highly sensitive and sophisticated probing techniques, the manipulation of the NW and its I–V measurements are carried out in the scanning electron microscope. To eliminate the effect from the electron beam in the SEM instrument, the electron beam is turned off when taking I–V, measurements. The I–V characteristics are measured by sweeping the voltage from –5 to +5 V. After the measurement, the electron beam is turned on again to ensure the preservation of the shape and position of the NW.

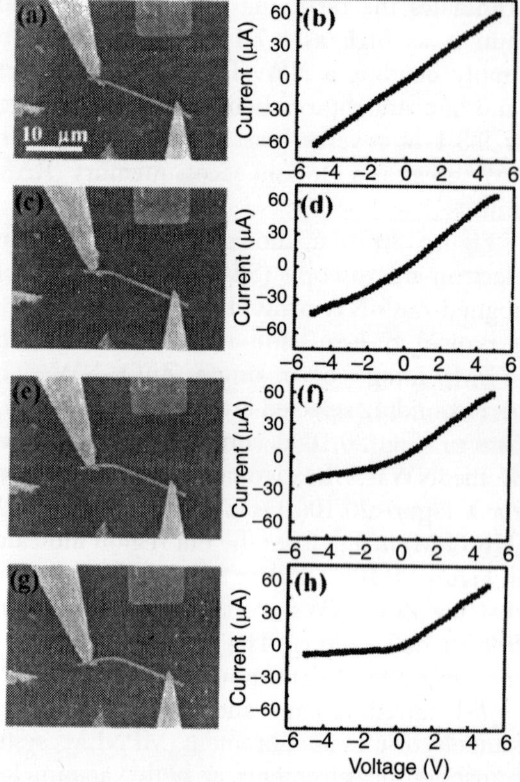

Figure 20.12 *The sequence of SEM images of the ZnO NW at various bending angles is shown on the left (a, c, e, g). The corresponding I–V characteristics are shown on the right (b, d, f, h).*

The NW is then bent further by moving the upper nanotip under the direct-imaging condition. Following the I–V measurements, the sequential images of the bent NW are captured. The typical sequence of the ZnO NW at various bending angles is shown in the left column of Figure 20.12. Their corresponding I–V characteristics are presented on the right hand side of Figure 20.12. At the first contact of the NW with the W tip (Fig. 20.12(a)), the NW is already bent a little because a pushing force is necessary for good electrical contact. The corresponding linear and symmetric I–V characteristic show that the Ti/Au-to-ZnO NW is an Ohmic contact rather than a double (back-to-back) Schottky contact (Fig. 20.12). As the bending proceeds (Fig. 20.12(c), (e) and (g), the electric current drops significantly with negative bias (Fig. 20.12(d), (f) anf (h) respectively) exhibits asymmetric I–V behaviors with the in creased strain. The nanotips are firmly attached to the NW without sliding, indicating that the contacts are well retained during the bending process and do not cause any change in contact resistance or contact area under bending. The reverse current drops severely at reverse bias voltage when the NW is bent further. When the NW is under significant bending, the reverse current at –5 V bias is as low as 6.6 μ A, and the rectifying ratio at $\pm$ 5 V is up to 8.7:1. Due to the large elasticity of the NW, the measurements of these devices are reversible. The details of electric current at reverse bias and the rectifying ratio at $\pm$ 5 V under a continuously changing bending curvature are listed in Table 20.1. The ZnO NW under bending shows certain rectifying I-V characteristics, similar to the result for a p-n junction. The substantial bending of the NW and the absence of a symmetric I-V show that electrical transport in the NW is governed by an internal field created by bending.

Table 20.1 *The electric current at reverse bias and the rectifying ratio at ± 5 V under a continuously changing bending curvature, together with an estimation of the piezoelectric potential energy barrier, ψ_{PZ} The measurement stages 1, 2, 3, and 4 correspond to Fig. 20.12(a), (c), (e) and (g) respectively.*

Measurement stage	*I at –5V [m A]*	*Rectifying ratio*	*Piezoelectric potential energy barrier [meV]*
1	–61.4	1:1	-
2	–42.0	1.6:1	9.5
3	–21.3	2.8:1	27.37
4	–6.6	8.7:1	57.66

A bent ZnO NW produces a piezoelectric electric field (E_{PZ}) along and across the NW due to the strain-induced piezoelectric effect. Consider the piezoelectric effect introduced in a single NW as a result of elastic deformation. For a simulated case in a NW of 490 nm thickness and 21 μ m length under the displacement of an external force F from the nanotip applied at the surface of the NW (Fig. 20.13(g)), the deflection of the ZnO NW creates a strain field. The outer surface is stretched (positive strain ε) and the inner surface is compressed (negative ε) in the area in contact with the nanotip, as shown in Figure 20.13(a). The magnitude of the deflection increases with the degree of bending, resulting in an increase in strain field. An electric field E_{PZ} along the NW is then created inside the NW through the piezoelectric effect, $E_{PZ} = \varepsilon/d$, where d is the piezoelectric coefficient. The piezoelectric field direction is closely parallel to the NW direction (z-axis) at the outer surface and antiparallel to the z-axis at the inner surface. With the increase in bending of the NW, the density of the piezoelectric charge on the surface also increase. The potential is created by the relative displacement of the Zn^{2+} cations with respect to the O^{2-} anions, a result of the piezoelectric effect in the wurtzite crystal structure; thus, these ionic charges cannot freely move and cannot recombine withour releasing the strain. The potential difference is maintained as long as the deformation is in place.

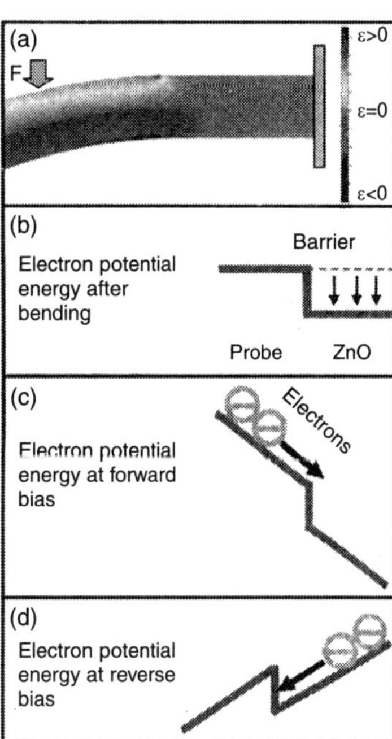

Figure 20.13 *(a) Longitudinal strain (e) distribution along a ZnO NW as a result of the piezoelectric effect when bent by a nanotip, with the stretched and compressed side surfaces being under positive and negative strain, respectively. Energy diagram of ZnO NW accounting for the applied bias and the piezoelectric field. b) Energy barrier built at the nanotip/nanowire interface due to the piezoelectric potential (V^+) at the stretched side. c) Current flow under forward bias. d) Current flow under reverse bias.*

Assume that, before bending, there is no energy barrier except for the resistance between Ti and ZnO. When the nanotip pushes a NW and bends it , as shown in Figure 20.11; a positive potential is produced at the stretched side of the NW due to the piezoelectric effect. As a result an energy barrier is produced at the interface between the tip and the NW, with the NW being at the higher potential as illustrated in Figure 20.13(b). Note that the actual potential distribution is not fixed due to the nonuniform distribution of strain. Considere, potential distrbution. The diagrams depicting the principle of operation are shown in Figure 20.13(c) and (d). For the bent ZnO NW under forward bias, the electrons are not blocked by this energy barrier, but under reverse bias, electrons overcome the energy barrier resulting from the piezoelectric electric field. This energy diagram corresponds electric transport measurements on bent ZnO. Such an energy barrier effectively serves as a p-n junction barrier at the interface, resisting the current flow from the tip to the NW but allowing current to flow from the NW to the tip. The magnitude of the barrier increase with the increase in the degree of bending, resulting in a drastic increase of rectifying effect due to the piezoelectric potential energy barrier. This is a simple piezoelectric gated diode. To quatify the barrier height produced by piezoelectricity, assume that the electric current also follows the typical *I–V* characteristics of a diode. The reverse current is given by Equation 1:

$$I_s = \mathrm{A}\, e^{-\psi_{PZ}/\kappa T}\,(\mathrm{e}^{\mathrm{eV/kT}}-1) \tag{1}$$

Where ψ_{PZ} is the potential energy barrier resulting from piezoelectricity, I_s is the reverse saturation current, V is the applied voltage, e is the electronic charge, k is the Biltzmann constant (8.617×10^{-5} $\mathrm{eVK^{-1}}$), T is the absolute temperature, and A is a constant. We consider the electric current at –5 V and room temperature with bending degree, thus the above equation is simplified as follows since the ($\mathrm{e^{eV/kT}}$–1) term is a constant:

$$\mathrm{I_s} = \mathrm{A}^* e^{-\psi_{PZ}/\kappa T} \tag{2}$$

where A* is a conxtant. It is assumed that the current is 61.4 μA as ψ_{PZ} is zero when there is no deflection observed (Table 12.1); thus A* is determined to be 61.4 μA. Therefore, by the given electric current from Table 12.1, this leads to an estimation of ψ_{PZ} to be9.82 meV for stage 2, 27.37 meV for stage 3, and 57.66 meV for stage 4, respectively. These are listed in Table 1. The calculations confirm that the magnitude of the piezoelectric barrier dominates the rectifying effect. The barrier height agrees well with the electrical energy output from a single ZnO NW for use as a NW based piezoelectric power nanogenerator.

Moreover, a mechanical force is applied to control the electric output to determine the “1” and “0” states (ON and OFF states, respectively; Fig. 12.14) Each NW array corresponds to a device element for memory application. With an applied –5 V bias, by applying a mechanical force, the operating

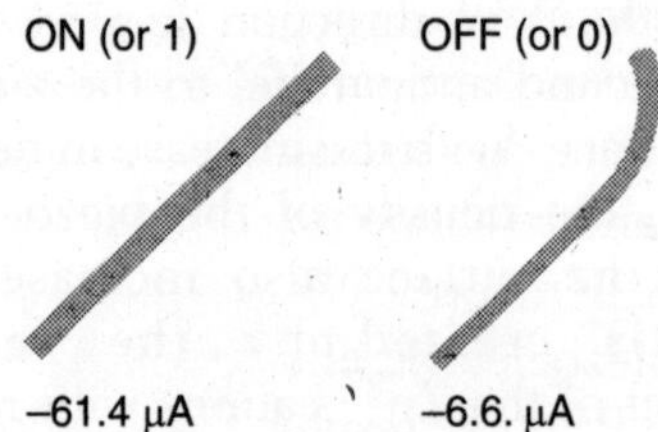

Figure 20.14 *Structures of the ZnO NW device element for RAM application in the “ON” (left) and “OFF” (right) states, showing the current under -5 V bias in each of the states, which gives an ON/OFF ratio of 9.3:1.*

current ratio between a straight and bent ZnO NW is as high as 9.3:1. This leads to the appearance of well-defined "1" and "0" states; that is, the shape will be highly sensitive to the electric current due to piezoelectricity. A device element, can be switched between these "1" and "0" states by mechanical force to receive different electric signals. On the basis of this switching mode, the elements as nanoscale electromechanical device are characterized.

Piezoelectric Nanosensors

As illustrated the PE-FET can be turned "on" or "off" by applying mechanical force/pressure to bend a NW/NB. A nanometer-sized force/pressure sensor is demonstrated for measuring forces in the nanonewton range and even smaller. Since the bending curvature of the NW is directly related to the force applied to it. the mechanical force can be retrieved from its bending shape, which is the principle of the force sensor, Based on the principle shown in Figure 20.15, a zero-power sensor can also be designed.

Piezoelectric Humidity Sensor

Piezoelectric humidity sensors are obtained by the following experiments. The devices are fabricated, using a single ZnO nanobelt (NB) bridging two Au electrodes. Poly(N-isopropylacrylamide) (PNIPAM) and poly(diallyldimethylammonium chloride) (PDADMAC) are functionalized alternatively layer-by-layer on the top surface of the NB (Fig. 20.15(a)). Measurements on such a conjugated device show a significant decrease in conductivity upon exposure to 12 and 85% relative humidity (RH) (Fig. 20.15(b)). To explore the origin of the reduced current upon vapor exposure series of experiments based on different functional materials on the devices are performed. First, the current response of an uncoated ZnO NB is examined. The result shows a gigantic increase in conductivity (Fig. 20.15(c)). Measurement of a conducting channel that contains only multilayers of polymers also show a drastic increase in conductivity (Fig. 20.15(d) which is in contrast to the response of the polymer-coated NB. as shown in Figure 20.15(b). The polymers used in the experiments, PNIPAM and PDADMAC, are very sensitive to enyironmental humidity changes. Upon exposure to water vapor, these polymers undergo a hydration process by absorbing a large amount of H_2O molecules. As a result, the volumes of these polymers increase by a significant amount. Because the polymers were coated only at one side of the ZnO NB. the swelling of the polymers then generated an asymmetric tensile stress at the contact surface with the ZnO NB. Consequently, the ZnO NB is bent into an are shape under this force, resulting in an asymmetric strain across the thickness of the NB. A bent ZnO NB produces piezoelectric potential across the NB because of the strain-induced piezoelectric effect, which traps and depletes the carriers in the NB. analogous to the case of the PE-FET, resulting in a decrease of conductivity. This is a different detection mechanism in comparison to the sensors based on FETs.

Piezoelectric Potential: How Long Does It Last?

For an insulating piezoelectric material, the piezoelectric potential is preserved as long as the strain is maintained. For a material like ZnO with the presence of oxygen vacancies and

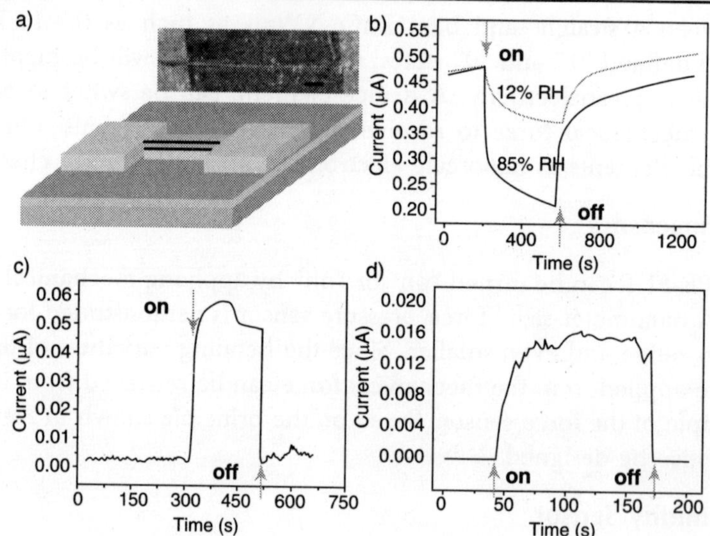

Figure 20.15 *(a) Illustration of multiplayer deposits f different polymers (onto the surface of a ZnO NB. surface by electrostatic attraction of the chargees on the polymers and an SEM image of such a polymer functionalized divice b) Responses of PNIPAM c) Current response of an uncoated ZnO NB on exposure to 85% RH. d) Current response of multi layered, polymers in RH vapour.*

surfaces, it has moderate conductivity. The question is: how, long will the piezoelectric potential be preserved after an NW is bent? An experiment is designed to measure the lifetime of the piezoelectric potential. The piezoelectric potential created in an NW across its width behaves like an electric field distribution across a parallel-plate capacitor (see Fig. 20.10(b)). A method to detect the presence and decay of the piezoelectric field as soon as the NW is bent is to continuously monitor the source-drain *I-V* transport properties of the PEFET as a function of time. This experiment is carried out by continuously applying a sweeping voltage through the NW and measuring its electric current. By plotting the current measured at a fixed bias of 2 V as a function of time (Fig. 20.16), a trend is seen about the recovery of conductance in the current-time (*I-t*) curve. The long lasting recovery of the current possibly indicates that the electronic effect caused by the piezoelectric characteristic of ZnO extends to as long as 100 s, before they reach a steady state. In addition, piezoresistance is a time-independent effect as long as the strain is fixed.

The slow-screening process is caused by the strong trapping effect of the electrons by vacancy/impurity/surface states in ZnO, which results in a slow release of the trapped charges, similar to the recovery of photoconductivity after UV-light excitation. It is possible that the piezoelectric potential created in the NW, if sufficiently large (> bandgap), excites the electrons in the valence bands to the conduction band or the vacancy/surface states, which hold the charges for an extended period of time, resulting in decreased conductivity. Alternatively, the effect caused by ,piezoelectric charges and potential hold for an equivalent length of time before being screened by external electrons to reach a steady state. This is relevant to the piezotronic devices proposed.

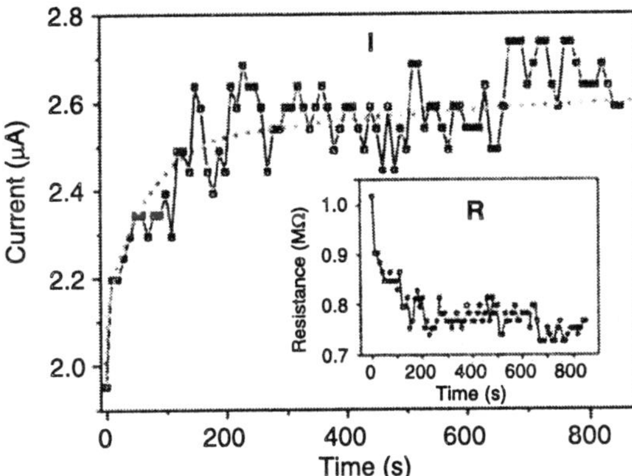

Figure 20.16 *Measuring the effective "lifetime" of the piezoelectric potential created in a ZnO wire after straining. For a fixed bias of 2 V, the current flowing through the bent wire was measured as a function of time. The current reached a steady state after approximately 150 s. The inset is an equivalent plot of the resistance-time (R-t) plot, corresponding to the I-t curve*

Furthermore, the trapped charges drop from the vacancy/surface states back to the valence band, possibly resulting in photon emission. This process is proposed as a piezoelectric-induced photon-emission effect.

Opto-piezotronic Devices

The response of a single CdD-NW-based NG to visible-light excitation is investigated. White light reduces the height of Schotty barrier existing at the tip-NW contact-mode change from schottky to quasi-ohmic. As a result, a positive voltage output is received when the tip contacts the stretched side of the NW. This is different from the result received when there is no light. This shows the effectiveness of using light to tune the out put of the NG. The triple coupling effect of semiconduting, piezoelectric, optoelectric properties, is useful for useful for designing new opto-piezotronic devices.

Perpectives

Based on the semiconducting and piezoelectric coupled properties, several important prototype devices, such as nanogenetors, PE-FETs, piezoelectric diodes, and piezoforce/pressure sensors, are demonstrated. These devices are the fundamental components of nanopiezotronics.

Piezotronics based on nanowires/nanobelts as the fundamental building blocks have the following unique advantages: First, the NW-based nanogenerators are subjected to extremely large deformation, so they can be used flexible electronics as a flexible/foldabel power source. Secondly, the large degree of deformation, can be withstood by the NWs and is likely to result in a larger volume density of power output. Third, ZnO is a biocompatible and biosafe material; it has great potential as an implantable power source within the human body. Fourth, the flexibility of the polymer substrate used for growing ZnO NWs/NBs makes it feasible to accommodate the flexibility of human muscles of that the mechanical energy (body movement, muscle stretching) in the human body to generate electricity can be used. Fifth, Zno NW/NB nanogenerators can directly produce current as a result of their enhanced conductivity with the presence of oxygen vacancies. Finally, ZnO is an environmentally "green" material. The phenomena demonstrated for Zno can also be applied to other wurtzite-structured materials, such as GaN and ZnS.

References

Z.L.Wang, Z.C Kang. Functional and Smart Materials—Structural Evolution and Structure Analysis, Plenum press, New York **1998**.

Z.W. Pan, Z.R. Dai, Z.L. Wang, *Science* **2001**, 291, 1947.

Z.R. Dai, J.L. Gole, J.D. Stout, Z.L. Wang, *J. Phys. Chem. B* **2001**, *106*, 1274.

M.H. Huang, S.Maim H.Q.Yan, Y.Y. Wu, H. Kind, E. Weber, R.Russo, P.D. Yang, *Science* **2001**, 292, 1897.

M.H. Huang, Y.Y. Wu, H. Feick, N. Tran, E. Weber, P.D. Yang *Adv.Mater.* **2001** *13*, 113.

Y.C.Kong. D.P. Yu, B. Zhang, W. Fang, S. Q. Feng, *Appl. Phys. Lett.* **2001**, 78, 407.

C.H. Liang, G.W. Meng, Y Lei. F.Phillipp, L.D. Zhang, *Adv. Mater.* **2001**, *13*, 1330.

Y.Q. Zhu, W.K. Hsu, M. Terrones, N.Grobert, H. Terrones, J.P. Hare,
H.W. Kroto, D. R. M. Walton, *J. Mater. Chem.* **1998**, *8*, 1859.

X.C. Wu. W.H.Song, W.D. Huange, M.H. Pu, B. Zhao, Y.P. Sun, J.J.
Du, *Chem. Phys. Lett.* **2000**, *328*,5.

H.Z. Zang Y.C. Kong, Y.Z. Wang, X. Du, Z.G. Bai, J.J. Wang D. P.
Yu, Y. Ding, Q.L. Hang, S. Q. Feng, *Solid State commun.* **1999**, *109*, 677.

C.H. Liang, G. W. Meng, G.Z. Wang, Y. W. Wang, L. D. Zhange, S. Y. Zhang, *Appl. Phys. Lett.* **2001**, *78*, 3202.

Z.G.Bai, D. P. Yu, H.Z. Zhang, Y.Ding, Y.P. Wang, X. Z. Gai, Q. L. Hang, G.C.Xiong, S.Q. Feng, *Chem Phys. Lett.* **1999**, *303*, 311.

Z.R. Dai, Z. W. Pan, Z.L. Wang, *Solid State commun.* **2001,** *118,* 351.

Z.W. Pan, Z. R. Dai, Z. L. Wang, *Appl. Phys. Lett.* **2002**, *80*, 309.

P.D Yang C.M. Lieber, *Science* **1996**, *273*, 1836.

R.S. Wangner, W. C. Ellis, *Appl. Phys. Lett.* **1964,** *4* 89.

Z.L. Wang, Z. W. Pan, *Adv. Mater.* **2002**, *14*, 1029.

Z.R. Dai, Z. W. Pan, Z.L. Wang, *J.Am.Chem. Soc.* **2002**, *124*, 8673.

Z.W, Pan, Z. R. Dai, C. Ma, Z. L. Wang, *J.Am.Chem.Soc.* **2002**, *124*, 1817.

J.P. Schaffer, A. Saxena, S. D. Antolovich, T.H. Sanders, Jr., S. B. Warner, The Science and Design of Engineering Materials, 2nd ed., McGraw-Hill, New York 1999, Ch.6.

M.S Arnold, P.Avouris, Z. W. Pan, Z. L. Wang, *J.Phys. Chem. B* 2003. *107*, 6.

E. Comini, G. Faglia, G.Sberveglieri, Z.W. Pan, Z.L. Wang, *Appl. Phys. Lett.* 2002, *81*, 1869.

X.D. Bai,E.G. Wang, P.X. Gao, Z.L. Wang, **2003**, unpublished.

S.Mao, M. Zhao, Z.L. Wang, **2002**. unpublished.

L.Shi, Z.L. Wang, unpublished.

W.Hughes, Z. L. Wang, **2003**, unpublished.

Nanowires and Nanobelts of Functional Materials (Ed: Z. L. Wang), Kluwer, Dordrecht, The Netherlands **2003**.

Metal and Semiconductor Nanowires (EdLZ. I,. Wang), Kluwer, Dordrecht, The Netherlands **2003**.

K.Kelm, W.Mader, *Z.Anorg. Allg. Chem.* **2005**, *631*. 2383.

R.Zboril, M. Mashlan, Barcova, M. Vujtek, *Hyperfine Interact.* **2002**
, *139/140*, 597.

E. Tronc. C. Chaneac, J. P. Jolivet, J.M. Greneche, *J. Appl. Phys.* **2005**. *98*. 053901.

J.Jin, K.Hashimoto, S. Ohkoshi, *J. Mater. Chem.* **2005**, *15*. 1067.

S.Ohkoshi, S.Sakuria, J.Jin, K. Hashimoto, *J. Appl. Phys.* **2005**, 97, 10K 312.

R.M. Cornell, U.Schwertmann, *The Iron Oxides: Structure, Properties, Reactions, Occurrence and Uses*, VCH, Weinheim, Germany **1996**.

A.lto, M. Shinkai, H. Honda, T. Kobayashi, *J, Biosci. Bioeng.* **2005**, *100*, 1.

H. Taguchi, *J.Phys. IV* **1997**, 7, 299.

V.Pillai, D. O.Shah, *J.Magn.Magn.Matter.* **1996**, *163*, 243.

M. Gich, A. Roig, C. Frontera, E. Molins, J. Sort, M. Popovici, G.Chouteau, D. Martiny Marero, J. Nogues, *J.Appl. phys.* **2005**, *98*, 044307.

M. Popvici, M. Gich, D.Niznansky, A. Roig, C. Savii, L. Casas, E. Molins, K.Zaveta, C. Enache, J.Sort, S. de Brion, G. Chouteau, *J. Nogues. Chem. Mater.* **2004**, *16*, 5542.

M. Gich, C. Frontera, A. Roig, J. Fonteuberta, E. Molins, N.Bellido, C.Simon, C.Fleta, *Nanotechnology* **2006**, *17*, 687.

J. R Morber, Y. Ding, M. Haluska, Y. Li, J. P. Liu, Z. L. Wang, R. L. Snyder, *J. Phys. Chem. B* **2006**. *110*, 21672.

D. J Pena, J.K.N. Mbindyo, A.J. Carado, T.E. Mollouk, C.D Keating, B. Razavi, and T.S. MayerL Template growth of photoconductive metal-CdSe-metal nanowires. *J.Phys. Chem.* **B106**, 7458 (2002).

J.K.N. Mbindyo, T.E. Mallouk, J.B. Mattzela, I.Kratochvilva, B. Razavi, T.N. Jackson, and T.S. Mayer: Template synthesis of metal nanowires containing nonlayer molecular junctions. *J.Am. Chem.Soc.* **124**, 4020 (2002).

A. Kolmakov, Y.Zhang, G.Cheng, and M. Moskovits: Detection of CO and O_2 using tin oxide nanowire sensor. *Adv.Mater.*15. 997 (2003).

T.B. Massalski: Binary Phase Diagrams, Vol. 1, (American Society for Metals, Metals Park, OH, 1986).

Y.Nuli, S.Zhao, and Q.Qin: Nanocrystalline tin oxide and nickel oxide film anodes for Li-ion battries. *J. Power Sources* **114**, 113. (2003).

J.Hu, T.W. Odom, and C.M. Lieber: Chemistry and physics in one dimension: Synthesis and properties of nanowire and nanotubes. *Acc. Chem. Res.* **32**, 435 (1999).

B.R. Martin, D.J. Dermody, B.D Reiss, M.Fang, L.A Lyon, M.J. Natan, and T.E, Mallouk: Orthogonal self-assembly on coloidal gold-platinum nanorods. *Adv.Mater.* **11**, 1021 (1999).

Z.W. Pan, Z.R. Dai, and Z.L. Wang: Nanobelts of semiconducting oxides *Science* **291**, 1947 (2001).

R.H. Baughman, A.A. Zakhidov, and W.A. de Neer: Carbon nanotube–The route toward application. *Science* **297**, 787 (2002).

Z.L. Wang: *Naniwires and Nanobelts—Materials. Properties and Devices* (Kluwer Academic Publishers, New York, 2003).

C.M. Lieber: Nanoscale science and technology Building big future from small things. *MRS Bull.* **28**, 486 (2003).

Y. Xia, P. Yang, Y. Sun, Y. Wu, B. Mayers, B. Gates, Y. Yin, F. Kim, and H. Yan: One-dimensiona nanostructures: Synthesis, characterization, and applications. *Adv.Mater.* **15**, 353 (2003).

A. Kolmakov and M. Moskovits: Chemical sensing and catalysis by one-demensional metal-oxide nanostructures. *Ann. Rev. Mater. Sci.* **34**, 151(2004).

D.Al-Mawlawi, C.Z. Liu, and M. Moskovits: Nanowires formed in anodic oxide templates. J.Mater. Res, **9**, 1014 (1994).

H. Masuda and M. Satoh: Fabrication of gold nanodot array using anodic porous alumina as an evaporation as an evaporation mask. *Jpn.J.App.Phys.* **35**, L126 (1996).

J.C. Hulteen and C.R. Martin: A general template-based method for the preparation of nanomaterials. *J. Mater. chem.* **7**, 1075. (1997).

S.R. Nicewarner-Pena, R.G. Freeman, B.D Reiss, L. He, D.J. Pena, I.A. Dalton, R. Cromer, C.D Keating, and M.J. Natan: Submicrometer metallic barcodes. *Science* **294**, 137 (2001).

N.I Kovtyukhova, B.R, Martin, J.K.N. Mbindyo, P.A. Smith.B. Razavi, T.S. Mayer, and T.E Mallouk: Layer-by-Layer assembly of rectifying junctions in and on metal nanowires. *J.Phys. Chem, B* **105**, 8762 (2001).

J.G. Wang, M.L. Tian, T.E. Mallouk, and M.H.W. Chan: Microstructure and interdiffusion of template-synthesized Au/Sn/Au junction nanowires. *Nano Letters* **4,** 1313 (2004).

H. Mehrer: Landolt-Bornstein numerical data and functional relationships in science and technology, Vol. 26, *Diffusion in Solid Metal and Alloys* (Springer-Verlag, Berlin, Germany, 1990).

N.Kanani: Electroplating: *Basic Principles. Process, and Practice.* (Elsevier Advanced Techmnology, Oxford, U.K., 2005).

S.Evoy: Dielectrophoretic assembly and integration of nanowire devices with functional CMOS operating circuitry. *Microelectron. Eng.* **75**, 31 (2004).

S. Hoeppener, R. Maoz, S.R. Cohen, L. Chi, H. Fuchs, and J.Sagiv: Metal nanoparticles, nanoparticles, nanowires, and contanct electrodes self-assembled on patterned monolayer templates–A bottom-up chemical approach. *Adv. Mater.* **14**, 1036(2002).

J.K.N. Mnindyo, B.D. Reiss, B.J. Martin, C.D. Keating, M.J. Natan, and T.E Molluk: DNA-directed assembly of gold nanowires on complementary surfaces. *Adv. Mater. 13*, 249 (2001).

J.J. Urban, J.E. Spanier, L. Ouyang, W.S. Yun, and H. Park: Single-crystal barium titanate nanowires. *Adv. Mater.* **15**, 423 (2003).

A.P. Li, F.Muller, A.Birner, N. Kielsch, and U. Gosele: Hexagonal pore arrays with a 50-420nm interpore distance formed by self-organization in anodic alumina. *J.Appl. Phys.***84,**6023 (1998).

Y.Wu, G.Cheng, K. Katsov, S.W. Sides, J. Wang, j. Tang, G.H. Fredickson, M. Moskovits, and G.D. Stucky: Composite mesostructures by nano-confinement. *Nature Mater.* **3**, 816 (2004).

V. Nagarajan, A.L. Roytburd, A. Stanishevsky, S. Prasertchoung, T.Zhou, L. Chen J. Melngailis, O. Auciello, and R. Ramesh: Dynamics of ferroelastic domains in ferroelectric thin films. *Nature Mater.* **2,** 43 (2003).

Z.L Wang (ed.), Nanowires and Nanobelts–materials, properties and devices, Vol, I: Metal and semiconductor Nanowires, Vol: II, Nanowires and Nanobelts of Functional Materials, Kluwer Academic Publisher, Boston (2003).

T.W. Odom, J. Huang, P.Kim, and C. M Lieber, *J.Phys. Chem. B* 104, 2794 (**2000**).

Wang Z.L 2004 *J.Phys.:Condens. Matter* **16** R829-58

Hsu Cl, Chang S J, Lin Y R, Li P C, Lin T S, Tsai S Y, Lu T H and Chen 1C 2005 *Chem. Phys. Lett.* **416** 75-8

Kar S, Pal B N, Chaudhuri S and Chakravorty D 2006 *J.Phys. Chem.* **B110** 4605-11

4] Wang Z L and Song J.H 2006 *Science* **312** 242-6

Gao P X, Song J, Liu J and Wang Z L 2006 *Adv. Mater.***19** 67-72.

Rubio-Sierra F J. Heckl W M and Stark R W 2005. *Adv. Eng. Mater.* 7-193-6

He J.H, Hsu J.H, Wang C W, Lin H N, Chen L J and Wang Z.L 2006 *J. Phy Chem.* **B110** 50-3.

Morber J R. Ding Y, Haluska M S, Li Y, Liu J P, Wang Z L and Snyder R L 2006 *J. Phys.Chem.* **B110** 21672-9

Cao, L. Garipcan B, Atchison J S, Ni C, Nabet B and Spanier J E 2006 *Nano Lett.* **6** 1852-7

Chang K W and Wu J J 2005 J. Mater. Res. 20 3397-403

Wagner R S and Ellis W C 1964 *Appl. phys.Lett.* **4** 89-90

Johansson J, Wacaser B A, Dick K A and Seifert W 2006 *Nanotechnology* **17** S355-61

Persson A I, Larsson M W. Sterstrom S, Ohlsson B J. Samyelson L and Wallenberg L R 2004 *Nat. Mater.* 3 677-81.

Dick K A, Deppert K, Karlsson L S, Larsson M W, Seifert W, Wallenberg L R and Samuelson L 2007 Mater. *Res. Bull.* **32** 127-33.

Dick K A, Deppert K, Karlsson L S, Wallenberg L R, Samuelson L and seifert W 2005 adv. *Funct. Mater.* **15** 1603-10.

Dick K A, Deppert K, Mårtensson T, Mandl B, Samuelson L and Seifert W 2005 *Nano Lett.* 5 761-4

Kamins T I, Williams, R S, Basile D P, Hesjedal T and Harris J S 2001 *J. Appl. Phys.* **89** 1008-16

Wang Y, Schmidt V, Senz S and Gösele U 2006 *Nat. Nanotechnol.* **1** 186-9

Colli A, Hofmann S,Ferrari A C, Ducati C, Martelli F, Rubini S, Cabrini S and Franciosi A 2005 *Appl. phys. Lett* **86** 153103.

Liu H S, Ishida K, Jin Z P and Du Y 2003. *Intermetallics* **11** 987-94

Wang X D, Song J H, Li P, Ryou J H, Dupuis R D, Summers C J and Wang Z L 2005 *J.Am.Chem. Soc.* **127** 7920-3

Cullity B D and Stock S R 2001 *Elements of X-Ray Diffraction* (Englewood Cliffs, NJ: Prentice-Hall)

Qi W H and Wang M P 2005 *J. Nanopart. Res*, **7** 51-7

Iwasaki H 1962 J. Phys. Soc. Japan **17** 1620-33

Buffat P and borel J P 1976 Phys. Rev. A 13 287-98 Makarow A V, Zbezhneva S G, Kovalenko V V and Rumyantseva M N 2003 Inorg. Mater. **39** 594-8

Wang H and Fischman G S 1994 J.Appl. Phys. 76 1557-62 Cheyssac P. Sacilotti M and Patriarche G 2006 *J. Appl. Phys.* **100** 044315.

K.P Huber, G.Herzberg, Constants of Diatomic Molecules, Van Nostrand Reinhold **(1979)**.

S.Zhou, M. Wu and W. Shu, "Finding Optimal Placements for Mobile Sensors: Wireless Sensor Network Topology Adjustment", Proceedings of the IEEE 6th Circuits and Systems Symposium of Emerging Technologies, Shanghai, China, May, 2004.

C. Montemago and G. Bachand, " Constructing Nanomechanical Devices Powered by Biomolecular Motors" Nanotechnology, 1999, pp 225-231.

Herwaarden A W V and Meijer G C M 1994 Thermal sensors Semiconductor Sensors ed S M Sze

C. Phoenix, *J. Evol. Tech.* 13 **(2003)**. Available at http// jet-press.org/volme13/Nanofactory.htm

Li, J. and Ng, H. T. (2004) Carbon nanotubeb sensors, in Encyclopedia of Nanoscience and Nanotechnology (Nalwa, H. S., ed.), American Scientific Publishers, Sandta Barbara, CA, Vol. 1, 591-601.

Narayanan A, Dan Y, Deshpande V, Lello N D, Evoy S and Raman S 2006 Dielectrophoretic integretion of nanodevices with CMOS VLSI circuitry IEEE Trans. Nanotechnol. **5** 101-9

Im H, Huang X-J, Gu B and Choi Y-K 2007 A dielectric-modulated field-effect tramsistor for biosensing Nat, Nanotechnol. **2** 430-4

Giersig M, Schäter A and pison U 2007 Diagnostic nanosensor and its use in medicine EP patent Specification 1811302

Albert K. J Lewis N.S, Schauer C.L, Sotzing G A, Stitzel S E, Vaid T P and Walt D R 2000 Cross-reactove cje,oca; Sensor arrays Chem. Rev.100 2595-626

Xu W, Vijayakrisnan N, Xie Y and Irwin M J 2004 Design of a nanosensor array architecture *ACM Proc. 14th ACM Great Lakes Symp. on VLSI (Boston, MA, April 2004)*

Fung C K M and Li W J 2004 Ultra-low-power polymer thin film encapsulated carbon nanotube thermal sensors *IEEE Conf. on Nanotechnology (Aug. 2004)* pp 158-60.

R.C. Merkle and R. A.Freitas, Jr., *J Nanosci. Nanotechnol.* 3, 319 **(2003)**.

D.J. Mann, J. Peng, R. A. Freitas, Jr., and R. C. Merkle, *J.Comput. Theor. Nanosci.* 1, 71 **(2004)**.

N. Oyabu, O. Custance, I. Yi, Y. Sugawara, and S. Morita, *Phys.Rev.Lett.* 90, 176102 **(2003)**.

S. W. Hal, L. Bartels, G. Meyer, and K.H. Rieder, *Phys. Rev. Lett*, 85, 277 **(2000)**.

D.V. Schroeder, *An Introduotion to Thermal Physics,* Addison-Wesley **(2000)**.

[Thundat *et al.* 2000] T. Thundat, E. Finto, Z. Hu, R. H. Ritchie, G. Wu, and A. Majumdar, "Chemical sensing Fourier space", Applied Physics Letters, Vol. 77, No. 24,pp. 4061-4063, December 11, 2000.

[Uchida *et al.* 1993] H. Uchida, D. H. Huang, J. Yoshinobu and M. Aono, "Single atom manipultion on the Si(111)7x7 surface by the scanning tunneling microscope (STM)", *Surface Science*, Vols. 287/288, pp 1056-1061, 1993.

Mitra S, Putnam C, Gauthier R, Halbach R, Halbach R, Seguin C and Salman A 2005 Evaluation of ESD characteristics for 65 nm SOI technology *IEEE Int. SOI Conf. (Honolulu HI, Oct. 2005)pp* 21-3

Shi J, Wang Z and Li HL 2006 Selfassembly of gold nanoparticles onto the surface multiwall carbon nanotubes functionalized with mercaptobenzen moieties *Springer J. Nanopart. Res.* 8 743-7

Zhang M, Chang P C H, Chai Y, Liang Q and Tang Z K 2006 Local silicon-gate carbon nanotube field effect transistors using silicon-on-insulator technology *Appl.Phys.Lett.*89 023116.

Ding B and Seeman N C 2006 Operation of a DNA robot arm inserted into a 2D DNA crystaline substrate *Science* **314** 1583-5

Jenkner M, Tartagni M, Hierlemann A and Thewes R 2004 Cell-based CMOS sensor and actuator arrays *IEEE J. Solid-State Circuits* **39** 2431-7

Zhen L S and Lu M S C 2005 A large-displacement CMOS-micromachined thermal actutor actuator with capacitive position sensing *Asian Solid-State Circuits Conf. (Nov. 2005)* pp 89-92

Liu Wm Wyk J D V and Odendall W G 2004 Design and evaluation of integrated electromagnetic power passives with vertical surface interconnections *IEEE Applied Power Electronics Conf. and Exposition (Blacksburg VA, Feb. 2004)* vol 2, pp 958-63

Roundy S, Wright P K and Rabaey J M 2006 *Energy Scavenging for Wireless Sensor Networks* (Berlin: Springer)

Takeuchi S and Shimoyama I 2002 Selective drive of electrostatic actuators using remote inductive powering, *Sensors Actuators* A **95** 269-73

Ghovanloo M and Najafi K 2004 Awide-band frequency-shift keying wireless link of inductively powered biomedical implants *IEEE Trans. Circuits Syst.* I **51** 2374-83

S.W. James and R.P. Tatam, *Optical Fiber Long Period Grating Sensors: Characteristics and Application* Meas.Sci.Technol. **14**, R49-R61, (2003).

S. khaliq, S W James and R P Tatam, *Enchanced sensitivity fiber optic ling period grating temperature sensor,* Meas. Sci. Technol. **13** 792-5 (2002)

V. Bhatia, *Application of long-period gratings to single and multi-parameter sensing, Opt.* Express, 4, 457-466 (1999).

R. Falciai, A.G. Mignani and A.Vannini, Long period gratings as solution cencentration sensors Sens. Actuators B-Chemical **74**, 74-77 (20001).

D.M Constantini, C.A.P. Muller, S.A. Vasiliev, H.G. Limberger, and R.P. Salathe *Tunable loss filter based on metal-coated ling-period fiber grating* IEEE Photonics Technol. Lett. **11**, 1458-1460 (1999)

J.N. Jang, S.Y. Kim, S.W. Kim, M.S. *Temperature insensitive long-period fibre gratings* Electron. Lett. **35**, 2134-2136 (1999)

Z.Zhang, M. Shiloach, S. Pilevar, C.C. Davis, J. Sirkis and Bentley, *Evanescent wave long period fiber Bragg grating as an immobilized antibody biosensor* Anal. Chem. **72**, 2895-2900 (2000).

N Rees, S. W James R P Tatam, and G J Ashwell., *Optical fiber long period grating with Langmuir-Blodgett thin film overlars,* Opt.Lett., **9**, 686-688 (2002).

K.Skjonnemand, PhD thesis, Cranfield University (2000).

I Del Villar, Machaerandio, I,R. Matias, F.J. Arreguie, *Deposition of overlays by electrostatic self-assembly in long-period fiber gratings*, Opt.Lett. **30** 720-722 (2005).

Z.Y.Wang, J.R. Heflin, R.H. Stolen, S.Ramachandran, *Highly sensitive optical response of optical fiber long period gratings to nanometer-thick ionic self-assembled multilayers* Appl.Phy.Lett. **86**, 223104 (2005)

A.K. Ghansk, K, Thyagaranjan and M.R. Shenoy, *Numerical-analysis of planar optical wave-guide using matrix approach* IEEE J.Lightwave Technol. **LT-5**, 660-667 (1987).

D. Gloge, *Weakly guiding fibers* Appl.Opt.**10**, 2252 (1971)

I.Del Villar, I. R. Matias, F.J Arregui, M. *Achaerandio Nanodeposition of Materials With Complex Refractive Index in Long-Period Fiber Gratings*, J. Lightwave Technol. **23**, 4192-4199 (2005)

A Cusano, A Iadicicco, P Pilla, L Contessa, S Campopiano, A Cutolo and M Giordano, *Cladding mode reorganization in high-refractive-index-coated long-period gratings: effects on the refractive-index sensitivity*. Opt.Lett. **30**, 2536-2538 (2005)

I. Del Villar, PjD Thesis, Universida Publica de Navarra, (2006)

Z. Wang, J.R. Helfin, R.H. Stoeln and S Ramachandran, *Analysis of the optical response of long period fiber gratings to nm-tick-film coatings,* Optics Express, **13**, 2808-2813, (2005).

I Ishaq, S W James, G J Ashwell, R P Tatam, *Modification of the refractive index response of long period gratings using thin film, overlays*, Sensors and Actuators B: Chemical, **107**, pp738-741 (2005)

I Del Villar, IR Matis, FJ Arregui, *Enhancement of sensitivity in long-period fiber gratings with deposition of low-refractive-index materials* Opt.Lett.**30,** 2363-2365 (2005)

A. Cusamno, P.Pilla, L.Contessa, A. Iadiccio, S. Campopiano, A. Cutolo, M. Giordano and G. Guerra, *High sensitivity optical chemosesnor based on coated long-period gratings for sub-ppm chmaical detection in water.* Appl.Phys.Lett. **87**, 234105 (2005).

B H Lee and J Nishii *Bending sensitivity of in-series long period fibre gratings* Opt.Lett. **23,** 1624-1626 (1998)

S W James, I Ishaq, G J Ashwell and R P Tatam, *Cascaded long period gratings with nano-structured coatings*, Opt.Lett., **30,** 2197-2199 (2005)

I Del Villar IR Matias, FJ Arregui, J Echeverria, RO Claus, *Strategies for fabrication of Hydrogen peroxide sensors based on electrostatic self-assembly (ESA) method,* Sens. Act, B-Chemical **108** 750-757 (2005)

I Del Villar IR Matias, FJ Arregui, R. O. Claus, *ESA-based in-fiber nanocative for hydrogen-peroxide detection* IEEE Trans. Nanotechnology **4** 187-193 2005.

FJ Arregui, IR Matia, KL Cooper, and R.O. Claus, *Fabrication of microgratings on the ends of standard optical fibers by the electrostatic self-assembly monolayer process,* Opt. Lett. **26** 131-133 2001.

X.D.Su, Y.J. Wu, W. Knoll, Biosens. Bioelectr. 21(2005) 719-726.

P.K. Dasgupta, H.L. Huang, G.F. Zhang, G.P. Cobb, Anal. Chim. Acta 58 (2002) 153-164.

T.M. Battaglia, et al., Anal. Chem. 77 (2005) 7016-7023.

X.Dai, O.Nekrassova, M.E.Hyde, R.G. Compton, Anal.Chem. 76 (2004) 5924-5929.

D.Melamed, Anal. Chim.Acta 532 (2005) 1-13.

D.Q Hung, O. Nekrassova, R.G. Compton, Talanta 64 (2004) 269-277.

H.J. Lee. A.W.Wark, Y.Li, R.M. Corn, Anal.Chem. 77 (2005) 7832-7837.

Z.Z Wang, T.Wilkop, Q. Cheng, Langumuir 21 (2005) 10292-10296.

Montemagno CD, Bachand GD. (1999). Constructing nanomechanical devices powered by biomolecular motors, Nanotechnology 10: 225-331.

Manning P, McNeil C. (2001) Microfabricated Multi-Analyte Ampermotric Sensors. *http:// nanocentre. ncl.ac.uk/*

F.J. Arregui, Y. Liub, I. R. Matiasa and R. O. *Claus Optical fibre humidity sensor using a nano Fabry-Perot cavity formed by the ionic self-assembly method*, Sensors and Actuators B: Chemical **59**, 54-59 (1999).

D.Flannery, *Fibre optic chemical sensing Lanmuir-Blodgett overlay waveguides, PhD Thesis, Cranfield University (1998).*

D. Flannery, S W James, R P Tatam and G J Ashwell, *pH Sensor using Langmuir-Blodgett overlays on polished optical fibre*, Opt.Lett., **22** , 567-569, (1997).

F.J. Grunthaner, A.J. Ricco, M.A Butler, A.L. Lane, C.P. Mckay, A.P. Zent, R.C. Quin. B.Murray, H.P. Klein, G.Y. Levin, R.W Terhune, M.L. Homer, A.Ksendzov, P.Niedermann, *Investigating the surface-chemistry of mars*, Anal.Chem.**67** A605-A610 (1995).

D.A. Jackson, *Recent Progress in monomode fibre-optic sensors* Meas.Sci.Technol. **5**, 621-638 (1994).

E.Udd Fiber Optic Smart Structures Wiley Interscience NY (1994).

N E Ress, S W James, S E Staines, G J Ashwell and R P Tatam, *Submicrometer fiber optic Fabry-Perot interferometer formed by the use of the Langmuir-Blodgett technique* Opt. Lett., **26**, 1840-1842, (2001).

F.J Arregui, I R Matias, Y J Liu, K M Lenahan, R O Claus *Optical fiber nanometer-scale Fabry-Perot interferometer formed by the ionic self-assembly monolayer process,* Opt. Lett. **24** 596-598 (1999).

Index